ANNUAL REVIEW OF PHYSICAL CHEMISTRY

ANNUAL REVIEW OF PHYSICAL CHEMISTRY

H. EYRING, *Editor*
University of Utah

C. J. CHRISTENSEN, *Associate Editor*
University of Utah

H. S. JOHNSTON, *Associate Editor*
University of California

VOLUME 16

1965

ANNUAL REVIEWS, INC.

PALO ALTO, CALIFORNIA, U.S.A.

ANNUAL REVIEWS, INC.
PALO ALTO, CALIFORNIA, U.S.A.

Library of Congress Catalog Card Number: A-51-1658

FOREIGN AGENCY

Maruzen Company, Limited
6, Tori-Nichome Nihonbashi
Tokyo

PRINTED AND BOUND IN THE UNITED STATES OF AMERICA BY
GEORGE BANTA COMPANY, INC.

PREFACE

This volume of the *Annual Review of Physical Chemistry* covers a wide field and includes a review of gas kinetics, for example, which has been missed in some of the recent volumes. The coverage is as complete as the ingenuity of the committee could provide in the allotted space. Subjects missed this time will be treated in future volumes.

The intention is to provide critical reviews even though this may prevent encyclopedic coverage in any one year. The authors have again reached the same high standard of performance that our readers have come to expect, and their generous unpaid services are deeply appreciated by students and colleagues throughout the world.

The committee thanks Carolyn Vanek for her effective support in editing this volume, and Dr. Stephen A. Greenberg for his preparation of the index. It is a continuing satisfaction to help in the preparation of this review.

THE COMMITTEE

ERRATUM

Volume 15

Page 78, line 5: *should read* . . . helium+Xenon, *not* helium+Argon.

CONTENTS

Linus Pauling

FIFTY YEARS OF PHYSICAL CHEMISTRY IN THE CALIFORNIA INSTITUTE OF TECHNOLOGY

By Linus Pauling

Research Professor, Center for the Study of Democratic Institutions, P.O. Box 4068, Santa Barbara, California

The California Institute of Technology assumed this name in 1920 (after its founding in 1891 it had used the earlier names Throop University, Throop Polytechnic Institute, and Throop College of Technology); and within a few years it began to be ranked among the world's leaders in research and instruction in physics and physical chemistry. In this article I shall discuss mainly the first two decades of its period of rapid development.

The idea of developing an outstanding institute of science and technology in Southern California was proposed by George Ellery Hale, Director of the Mt. Wilson Observatory, in 1907, and was immediately accepted by the Throop Board of Trustees, of which he was a member. Hale turned to his friend Arthur Amos Noyes, Director of the Research Laboratory of Physical Chemistry of the Massachusetts Institute of Technology, for help. Noyes spent a few months in Pasadena each year from 1913 to 1916, and in 1917 became a full-time member of the staff of Throop College and Director of the Gates Chemical Laboratory. The Gates Laboratory, the second building on the campus, was finished in 1917.

In 1918 Noyes and Hale brought Robert Andrews Millikan, Professor of Physics in the University of Chicago, to Pasadena for a visit, and in 1921 Millikan became Director of the Norman Bridge Laboratory of Physics. For a quarter of a century the Institute was administered by an Executive Council, of which Millikan was the chairman. The other members in the early 1920's were Noyes, Hale, and two local business men. By the late 1930's the Executive Council had increased from five to nine: Millikan (chairman), Hale, Thomas Hunt Morgan, Richard C. Tolman, Max Mason, William B. Munro (Professor of History and Government), and three business men (members of the Board of Trustees). Despite the large academic majority of the Executive Council, it retained the confidence of the Trustees over the whole period of extraordinary development of the Institute. It is my opinion that the adoption of this form of administration, which gave Hale, Noyes, and Millikan much freedom in putting their ideas about education and research into operation, was responsible for the great success of the Institute during the quarter-century beginning in 1920. The Executive Council was replaced by a president in 1946.

I first heard about Throop College from two students who began the academic year 1917–18 there, had academic trouble, and transferred to the School of Mines of the Oregon Agricultural College, where they became my classmates. Then in 1919–20, while I was working as full-time Assistant in

Quantitative Analysis during a one-year interim between my sophomore and junior years (the result of financial problems), a professor showed me an announcement of available fellowships in chemistry at Throop College and suggested that it would be a good place for me to study. I judge that these announcements, bearing the name of one of the country's leading physical chemists, A. A. Noyes, brought the College to the attention of many prospective graduate students. In early 1922, when I was finishing the work for the degree of B.S. in chemical engineering in the Oregon Agricultural College, I applied for a fellowship to C.I.T., Berkeley, and several other universities. I soon received offers from C.I.T. and Harvard. I accepted the appointment at C.I.T., and withdrew the other applications.

During the last forty years I have visited universities all over the world. I now have the opinion that I had the greatest good luck in having gone to Pasadena in 1922. I do not think that I could have found better conditions for preparation for a career in physical chemistry anywhere else in the world. Perhaps I would have got along just as well if I had gone to Berkeley. I did not have to decide between Pasadena and Berkeley—I simply accepted the first offer.

Two actions by Noyes support the foregoing thesis. First, he wrote at once, after I had accepted the fellowship, to point out that my course of study of physical chemistry in Corvallis, based on a simple textbook by Millard, was unsatisfactory. He sent me the proofsheets of the new edition of Noyes & Sherrill's *Chemical Principles*, and asked me to work all of the problems in the first nine chapters. This book was an excellent one for independent study. I worked the problems without any trouble during the evenings of the summer of 1922.

Noyes also suggested that my doctoral research be on the determination of the structure of crystals, under the supervision of Roscoe Gilkey Dickinson. I had read Irving Langmuir's 1919 papers on the electronic structure of molecules and had developed such an interest in the subject that I gave a seminar talk on molecular structure and the nature of the chemical bond in Corvallis in the fall of 1919. I borrowed the Bragg book on X-rays and crystal structure from the Oregon State Library, read it, and wrote Noyes that I accepted his suggestion.

I began my research on 1 October 1922. For years (since 1918) I had been speculating about the relation between the properties of substances and their molecular structure; the X-ray diffraction technique was beginning to provide the answers to many questions, and I was eager to get the answers to others. For three weeks I worked on the synthesis of LiH, with the idea of using X-ray intensities to test the presence of the hydrogen anion in the crystal; then I learned that this job had already been done, by Bijvoet and Karrsen in Holland. During the next month I made crystals of fifteen inorganic substances and subjected several of them to the first stages of X-ray investigation. None of them seemed to be suitable for a structure determination by the methods that were then in use. During the third month Dickinson carried me through the various steps in the complete structure determination of a crystal, the mineral molybdenite, MoS_2.

This achievement made a great impression on me. The process of structure determination involved a succession of logical arguments, which were presented to me by Dickinson in a meticulous way, with emphasis on rigor. Dickinson was an especially clear-headed and thoughtful scientist, strongly critical of carelessness and superficiality. I was pleased to learn that questions about the nature of the world could be answered by carefully planned and executed experiments. Also, I was pleased that the study of molybdenite should have led to a surprise: the molybdenum atom was found to be surrounded by six sulfur atoms at the corners of a trigonal prism, rather than of an octahedron, as might have been expected from the earlier structure determinations. This was the discovery of the trigonal prism as a coordination polyhedron around metal atoms.

Because of the small number of graduate students in Pasadena in 1922, it was possible for them to come into close contact with the members of the faculty. For example, together with two or three other graduate students and a faculty member, Ellis, I was taken by Noyes on some camping trips to the Palm Springs region, and all of the graduate students were the guests of Noyes in his oceanside house at Corona del Mar, 50 miles south of Pasadena, where a few years later the California Institute of Technology set up a marine laboratory. These occasions gave opportunity for the unhurried discussion of scientific and practical problems. Dickinson also took me with him on a number of trips to the desert during my first year as a graduate student. Moreover, the members of the staff were not overloaded; in 1922–23 I was the only graduate student being supervised by Dickinson.

In addition to inspiring them with the desire to carry on research, it is necessary in the training of young scientists to give them a good background of the knowledge that has already been obtained. The California Institute of Technology, despite its small size, was already carrying out this function in an admirable way in 1922. Although Noyes, as Director of the Gates Chemical Laboratory and a member of the Executive Council, was primarily responsible for the development of the Institute as a leading teaching and research school in physical chemistry, it was Richard Chace Tolman who, in my opinion, made the greatest contribution to the achievement of this goal. Tolman, who had come to Pasadena in 1921, had a thorough understanding of the new developments in physics and the conviction that chemical problems could be solved by the application of physical methods. In my first term, the fall of 1922, I found his course Introduction to Mathematical Physics, in which he laid emphasis on the basic principles and their quantitative application to moderately simple problems, to be especially valuable. His courses on advanced thermodynamics and on statistical mechanics applied to physical and chemical problems were outstanding for their clarity and thoroughness. But his greatest contribution to the training of the young physical chemists in Pasadena came, I believe, through the course called

Seminar in Physical Chemistry. During the three years when I was a graduate student this seminar involved working through three books: first, *The Origin of Spectra,* by Foote and Mohler; second, *Atomic Structure and Spectral Lines,* by Arnold Sommerfeld; and third, *Atombau und Spektrallinien,* by Sommerfeld (new edition, available then only in German). The other courses that I studied during my graduate years were the following: advanced algebra (Van Buskirk); higher dynamics (C. G. Darwin); chemical thermodynamics (Noyes; this course, in the fall of 1922, was the last one that he taught); advanced thermodynamics (Tolman, spring of 1923; a newly published book by Lewis and Randall was the text); kinetic theory (Robert A. Millikan); thermodynamics (Millikan); vector analysis (Harry Bateman); Newtonian potential theory (Bateman); thermodynamics (P. S. Epstein); quantum theory (P. Ehrenfest); physical optics and quantum theory (Epstein); functions of a complex variable (Bateman); and integral equations (Bateman). These courses helped me to overcome the handicap of lack of knowledge of physics and mathematics. The course work of other graduate students in physical chemistry was similar, but on the average not so heavy.

I learned a great deal also from the research conferences. The chemistry research conference, held once a week, was made lively by Tolman's efforts to find out the extent of understanding that had been achieved by the graduate students.

Only rarely did a visitor speak at a chemistry research conference. I remember that Fritz Haber was present at one, but did not speak. I do not remember any speaker from Berkeley, which is after all 420 miles away. The physics research conferences, on the other hand, often were held by visiting physicists, who sometimes also presented a series of special lectures. Among the visiting physicists in the early 1920's were H. A. Lorentz, C. G. Darwin, P. Ehrenfest, A. A. Michelson, R. C. Gibbs, S. Loria, A. Sommerfeld, M. Born, V. Bjerknes, C. V. Raman, J. Franck, and P. Langevin. There were two physics research conferences per week; often one was combined with the Physics and Astronomy Club, which met alternately at the Institute campus and the Pasadena laboratories of the Mt. Wilson Observatory.

The first three recipients of Ph.D. degrees from the Institute were chemists: Roscoe G. Dickinson (1920), Richard M. Bozorth (1922), and David F. Smith (1922). In 1924 six Ph.D. degrees were awarded in physics and three in chemistry (Richard M. Badger, R. Schumann, and Ernest H. Swift); in 1925 there were three in physics, one in mathematics, and four in chemistry (Paul H. Emmett, Linus Pauling, Albert L. Raymond, and Ernest C. White); and in 1926 nine in physics and six in chemistry (Gordon A. Alles, Sterling B. Hendricks, L. M. Kirkpatrick, C. H. Prescott, Oliver R. Wulf, and Don M. Yost).

By 1924 there were thirteen postdoctoral fellows in the Institute, of whom nine were National Research Fellows (four in chemistry—Arthur F. Benton, Philip S. Danner, George Glockler, and Maurice L. Huggins), and 43 graduate students, of whom sixteen were in chemistry, fourteen of them in physical chemistry. The number of undergraduate students at this time was about 500; it had reached 600 by 1937, and 700 by 1960. The number of graduate students was 220 (22 in chemistry) in 1937 and 678 (87 in chemistry) in 1963. In the early 1920's, research in physical chemistry and physics received its main support from a five-year grant of $30,000 per year that had been made by the Carnegie Institution of Washington for the support of research on the structure of matter and radiation.

Six papers were published as contributions from the Chemical Laboratory of Throop College of Technology. These papers, published between June 1915 and January 1920, included three on the properties of aqueous solutions of strong electrolytes, two on the determination of the structure of crystals by the X-ray diffraction method, and one on the analysis of metal-organic compounds. The first contribution from the Gates Chemical Laboratory of the California Institute of Technology, Number 7 in the series of papers in chemistry, was on chromoisomeric silver salts of pentabromophenol and a theory of chromoisomerism of solid compounds, by Howard J. Lucas and Archie R. Kemp.

These first seven papers constituted the beginning of three of the principal fields of research in physical chemistry in the Institute. The first two, written by Professor Stuart J. Bates, were on the osmotic pressure of ions and of undissociated molecules of salts in aqueous solution (published in the *Proceedings of the National Academy of Sciences* and in *the Journal of the American Chemical Society* in 1915). The field of ionic solutions was of great interest to Noyes, who had carried out an extensive series of experimental investigations on these solutions over a period of three decades. Noyes continued for several years in the California Institute of Technology the experimental work in this field that had been initiated by him in the Massachusetts Institute of Technology. In 1904 he had pointed out that some of the properties of solutions of strong electrolytes, such as the dependence of absorption spectrum on concentration, indicate that many of the substances are completely ionized, even in concentrated solutions. By 1922 he had reached the conclusion that the dependence of thermodynamic activity and electric conductance of these solutions on concentration is determined by the electrostatic interaction of the ions, and he embarked upon an experimental test of these ideas, based upon the statistical mechanical treatment of the problem that had been published by Milner in 1912 and 1913. While he was preparing this work for publication, the paper by Debye and Hückel on the interionic attraction theory of ionized solutions was published. Noyes then published a detailed discussion of this theory, including especially the comparison with experiment. He concluded that the properties of ionized solutions are satisfactorily accounted for by the theory, and he brought his long series of researches in this field to an end. Fifteen of the early publications from the Gates Chemical Laboratory were on this subject.

During later years Noyes continued his work in the fields of qualitative inorganic chemical analysis, chemical thermodynamics, and inorganic chemistry. He was especially interested in the prosecution of research by undergraduate students, and in 1936 and 1937 he published four papers on strong oxidizing agents in nitric acid solution, including the formation on bipositive silver in nitric acid by the action of ozone. The work described in these papers was carried out under his supervision by Clifford S. Garner, Thomas J. Deahl, Charles D. Coryell, Fred Stitt, and Alexander Kossiakoff. In 1925 he became interested in the interpretation of chemical properties of elements on the basis of the electronic structure of their atoms, and he prepared two papers in this field, with the assistance of a graduate student, Arnold O. Beckman. He also brought Professor John J. Abel and Dr. E. M. K. Geiling for one year to the California Institute of Technology from the Department of Pharmacology of Johns Hopkins University, to work on insulin. Abel reported the crystallization of insulin in a joint communication from the Gates Chemical Laboratory (Number 89) and the Department of Pharmacology of Johns Hopkins University. The isolation of insulin and study of its properties were carried out in Pasadena by two graduate students, Albert L. Raymond and Gordon Alles, both of whom later made important contributions in the field of pharmacology. In 1926 (Communication No. 99 from the Gates Chemical Laboratory), Alles published an account of his research on the comparative physiological action of some derivatives of guanidine. This work may well have been the inspiration for his later investigations that led to the discovery of benzedrine and other physiologically active amines.

C. Lalor Burdick has stated that the development of X-ray crystallography in the California Institute of Technology was due entirely to the imagination, vision, and conviction of Arthur A. Noyes. In 1913 Burdick was a graduate student in physical chemistry under Noyes at M.I.T., and was encouraged by him to go to Europe for doctoral work and postdoctoral study. After he had received his Ph.D degree in Basel he was advised by Noyes to work for six months in the X-ray laboratory of W. H. Bragg at University College in London. Burdick collaborated there with Dr. E. A. Owen in studying carborundum with use of the Bragg ionization chamber X-ray spectrometer. He then returned to M.I.T. and built an X-ray spectrometer. Noyes then invited him, in the latter part of 1916, to come to Pasadena. Burdick and James H. Ellis, a former M.I.T. student of Noyes who later became Associate Professor of Physical Chemical Research in the California Institute, built an improved spectrometer and used it to determine the crystal structure of chalcopyrite, $CuFeS_2$. The account of this work appeared in 1917 as Publication No. 3 of the Gates Chemical Laboratory in both the *Proceedings of the National Academy of Sciences* and the *Journal of the American Chemical Society.* This work and the simultaneous work of Albert W. Hull on the structure of metallic elements, done in the General Electric Research Laboratory, were the beginning of X-ray crystallography in the United States.

In 1917 Dickinson came to Pasadena from M.I.T., at Noyes' request, and continued the X-ray work. Within a few years he had carried out several important investigations. He showed, with E. A. Goodhue, that in the crystals $NaClO_3$ and $NaBrO_3$ the chlorate ion and the bromate ion are pyramidal in structure, and he determined the bond lengths to within 0.03 Å. He also verified the octahedral configuration of the hexachlorostannate ion, the square planar configuration of the tetrachloropalladite and tetrachloroplatinite ions, and the tetrahedral configuration of the zinc tetracyanide and cadmium tetracyanide ions. He and Albert L. Raymond, who was beginning his graduate work, made the first crystal structure determination of an organic compound, hexamethylenetetramine, in 1922 (published in 1923). In 1923 Richard M. Bozorth reported his structure determination of KHF_2, which provided the first X-ray evidence for the hydrogen bond. Bozorth also made structure determinations of some other crystals, including As_4O_6 and Sb_4O_6, which were the first inorganic crystals shown to contain discrete molecules.

For several years X-ray diffraction was the principal field of research in chemistry in the Institute. Of the 20 papers that had been published by the end of 1922, 15 were on the determination of the structure of crystals. As other fields of research began to be prosecuted the papers on the determination of the structure of crystals by X-ray diffraction dropped to about 20 per cent of the total, and then to about 10 per cent. During the last 50 years, about 400 papers on X-ray diffraction have been published from the Gates and Crellin Laboratories of Chemistry of the California Institute of Technology, representing the determination of the structure of about 400 crystals. Many American scientists received their training in X-ray crystallography in the California Institute of Technology; among the earlier ones were James B. Friauf, Sterling B. Hendricks, J. H. Sturdivant, L. Merle Kirkpatrick, Edwin McMillan, F. J. Ewing, Ralph Hultgren, Jack Sherman, J. L. Hoard, B. N. Dickinson, Harold P. Klug, Norman Elliott, David Harker, Franklin Offner, Kenneth S. Pitzer, Jurg Waser, Jerry Donohue, David Shoemaker, Kenneth Trueblood, and William N. Lipscomb.

Dickinson spent the academic year 1924–25 in the Cavendish Laboratory, learning the techniques of radiochemistry. On his return to Pasadena he gradually dropped the crystal structure field. He did some work with Tolman on the theory of the rate of chemical reactions, and, together with Dillon and Rasetti, he obtained the first Raman spectra of polyatomic gases. In 1935, with Don M. Yost, he studied the diffusion and absorption of neutrons in paraffin. The major part of his work during the rest of his life was in the field of photochemistry. Among his earlier students were A. L. Raymond, James B. Friauf, S. B. Hendricks, Allan C. G. Mitchell, Arnold O. Beckman, Robert T. Dillon, C. E. P. Jeffreys, S. Frederick Ravitz, John A. Leermakers, James L. Carrico, Nelson P. Nies, Ernest H. Lyons, H. Lotzkar, J. Norton Wilson, Edwin N. Lassettre, and Richard M. Noyes.

In his researches, carried out in part with graduate students, Tolman, who was Professor of Physical Chemistry and Mathematical Physics, attacked problems that ranged over a wide field: the general theory of relativity, the application of the old quantum theory to systems with motion that is only an approximation to conditionally periodic motion, the theory of the temperature coefficient of photochemical reaction rate, the entropy of diatomic gases, rotational heat capacity, the experimental determination of the mass of the electric carrier in metals, the entropy of supercooled liquids at the absolute zero, the principle of microscopic reversibility, experimental determination of the rate of decomposition of nitrogen pentoxide and of the rates of other chemical reactions, and especially the theoretical attack on the problem of the rate of unimolecular reactions.

The problem of the rate of chemical reactions and the mechanism of activation was one of the liveliest problems in physical chemistry during the decade beginning 40 years ago. The observed rates of some very rapid first-order reactions, such as the decomposition of nitrogen pentoxide, were so large that it seemed unlikely that the number of high-energy collisions between molecules was great enough to provide the equilibrium fraction of molecules with energy equal to the activation energy. Because of Tolman's interest in this problem, significant experimental contributions were made in this field by David F. Smith, Ernest C. White, Oliver R. Wulf, Martin E. Nordberg, Herman C. Ramsperger, William Ure, J. A. Leermakers, G. Waddington, Philip D. Brass, and William E. Vaughan. The attack on the theory of the rate of these reactions was made by Tolman in collaboration with R. G. Dickinson, Don M. Yost, and Herman C. Ramsperger, and the theoretical work was continued by Oscar K. Rice and Louis S. Kassel, as well as Ramsperger, during their stay in Pasadena as National Research Fellows. In the course of this work it was shown that the supply of activated molecules was not kept replenished in general through the absorption of infrared radiation, nor through individual bimolecular collisions of high energy, but rather, for molecules with several degrees of internal vibrational freedom, through a series of collisions of moderately high energy, such as to provide some molecules with an amount of internal energy equal to or greater than the activation energy.

Some branches of physical chemistry were taken up and then dropped. Arthur F. Benton came to Pasadena from Princeton in 1921, as a National Research Fellow. He worked on catalysis and surface chemistry, with Paul H. Emmett as a graduate student; but this field was dropped when they left in 1925. C. H. Prescott, Jr., made some measurements of chemical equilibrium at high temperatures; this field, too, was dropped at the end of the 1920's.

The work in physical chemistry became more and more the application of new physical techniques to the solution of chemical problems and the application also of the new physical theories, especially quantum theory and quantum mechanics. Richard M. Badger, who has been associated with the California Institute of Technology since 1918, presented an undergraduate thesis in 1921 with title *The Effect of Surface Conditions on the Intensity of*

X-ray Reflections From Crystal Planes, describing work that he had done with the Bragg ionization spectrometer that had been constructed by Burdick and Ellis. Then, after some work on the free energy of hydrogen cyanide and other substances, he moved into the field of molecular spectroscopy, to which he has made many contributions. Among his earlier research students and collaborators were C. Hawley Cartwright, J. W. Urmston, John L. Binder, Sho-Chow Woo, Lyman G. Bonner, John McMorris, Robert C. Barton, Charles M. Blair, Jr., Paul C. Cross, Simon H. Bauer, Austin L. Wahrhaftig, and Lloyd R. Zumwalt.

There was a general feeling of the excitement of discovery in those days. Quantum theory seemed to provide the answers to many puzzling questions, but it left others unanswered. For example, the moment of inertia and the vibrational frequency of the hydrogen molecule were thought to be known, and in fact were known, from the band spectrum of hydrogen, but the calculated values of the heat capacity of hydrogen gas at low temperatures could not be brought into agreement with experiment. Tolman and Badger compared the calculated heat capacity on the basis of many different assumptions about quantum weights of the rotational states with the experimental values, without finding a successful theory. The problem was not solved until 1927, when David M. Dennison, of the University of Michigan, showed that the assignment of spin one-half to the proton coupled with the assumption that transitions do not occur between even and odd rotational states of hydrogen during the time of an experimental measurement of the heat capacity led to a satisfactory theoretical curve.

I remember that, as a graduate student and a young postdoctoral fellow, I felt that the time had come when it would be possible to make a successful attack on many chemical problems by applying quantum theory and quantum mechanics, but I did not feel that I might make a successful attack on some problems in physics itself. As I think back on those years, I conclude that I was not able to distinguish clearly between the unsatisfactory nature of the existing physical theories and the unsatisfactory nature of my own understanding of them, and I tended to assume that my understanding was at fault. For example, in 1924 I listened with great interest to Professor Arnold Sommerfeld's account of his work on the fine structure of X-ray spectral lines. In his theory, which accounted in what seemed to be a thoroughly satisfactory way for the X-ray energy levels of atoms, he assigned to an atom values of two kinds of azimuthal quantum numbers of the electron orbits. These two kinds of quantum numbers, called the inner quantum number and the outer quantum number, determined the s,p,d, . . . splitting and also the fine structure of the levels. The s,p,d, . . . splitting was accounted for as the result of a difference in penetration of the inner shells for orbits with different eccentricity, and the fine structure as the result of a difference in the relativistic change in mass of the electron in orbits of different eccentricity. Why the second effect could occur without the first effect is hard to understand; but, as I remember the lectures and their reception, Sommerfeld himself failed to mention this difficulty and no one in the audience commented on it. Later on Millikan and Bowen wrote a paper entitled "A Great Puzzle in the Theory of Spectra," in which they emphasized this point, and the puzzle was resolved, of course, by the discovery of the spinning electron, which produces a fine-structure splitting of spectral lines identical with that calculated by Sommerfeld for the relativistic change in mass.

I held a National Research Fellowship for seven months, and resigned in February 1926 in order to work in Europe for 19 months as a Guggenheim Fellow. I wrote to Niels Bohr asking permission to come to work in his Institute in Copenhagen, and also to Sommerfeld, in Munich. Bohr did not answer the letter, but Sommerfeld did, and in this way the decision was made for me as to whether to go to Copenhagen or to Munich. Sommerfeld's lectures were outstanding, and I think it was good luck that led me to spend a year with him. I then spent a few weeks in Bohr's Institute and five months in Zurich, attending lectures by Schrödinger and Debye and working on the quantum mechanical theory of the chemical bond.

On my return to Pasadena in 1927 as Assistant Professor of Theoretical Chemistry, I presented a course on introduction to quantum mechanics, with chemical applications. I gave this course every year for many years, and I also gave it, as well as a course on the nature of the chemical bond, in the University of California in Berkeley, where I spent one or two months each spring for five years, beginning in 1929, as Visiting Lecturer in Chemistry and Physics. Other graduate courses that I taught in Pasadena were on the structure of crystals, the nature of the chemical bond, and the theory of the electric and magnetic properties of substances.

I had a surprise in 1930. The relations between the divisions of the California Institute of Technology were very good, and when, in 1930, my first book was published (*The Structure of Line Spectra*, with S. A. Goudsmit, now editor of the *Physical Review*, as co-author), I planned to give a course on this subject. However, my proposal was, according to Noyes, vetoed by Millikan, who, Noyes said, felt that this subject should be left to the physics department. Later on I thought that Noyes himself might have made the decision to keep me from straying out of chemistry. In fact, in early 1927, while I was in Munich, I received and accepted an offer of appointment as Assistant Professor of Theoretical Chemistry and Mathematical Physics, but when I arrived in Pasadena in October I found that Mathematical Physics had been dropped from my title.

In the spring and summer of 1930 I spent six months in Europe. I had been working on the structure of mica and other silicate minerals, and I first stayed for some time in the W. L. Bragg laboratory in Manchester, where I learned little except how to operate a Bragg X-ray spectrometer (the Pasadena spectrometer had been abandoned and replaced by photographic apparatus before my arrival in 1922). I then went to Munich, to continue my work on quantum mechanical problems. Fortunately I decided to visit Hermann Mark in his laboratory in Ludwigshafen. He showed me the electron-diffraction apparatus with which he and Wierl had determined the structure of gas molecules of carbon tetrachloride and benzene during the preceding year. I was overwhelmed by the possibilities of this new technique—for some time I had been looking for a diffraction method of determining the structure of molecules without having at the same time to determine the sometimes very complicated way in which the molecules are arranged relative to one another in a crystal. I asked Mark if he had any objection to my doing some electron-diffraction work in Pasadena. He said that he did not and that in fact he was not planning to continue the work much longer. On my return to Pasadena I asked a new graduate student, Lawrence O. Brockway, to build the apparatus, which he did, with the advice and help of Professor Badger.

Over the next 25 years the gas-molecule electron-diffraction technique was applied as an effective complement to the X-ray work with crystals. The first results obtained, after the check of the procedure with carbon tetrachloride, were the verification of the octahedral configuration and the determination of the bond lengths for the hexafluorides of sulfur, selenium, and tellurium. The linear arrangement of the three nitrogen atoms in methyl azide was then verified; the earlier crystal structure work on NaN_3 and KN_3 (Pauling and Hendricks, 1925) had shown the azide ion to be linear. Between 1932 and 1957 electron-diffraction determinations were made in Pasadena of the structure of the molecules of 225 substances. After Brockway went to the University of Michigan (1937) this work was in the charge of Verner Schomaker. Participants in the electron-diffraction work included J. Y. Beach, Paul C. Cross, Leslie E. Sutton, Simon H. Bauer, H. O. Jenkins, Fred T. Wall, Stanley M. Swingle, Dwight C. Carpenter, Kenneth J. Palmer, J. A. A. Ketelaar, Henri A. Levy, George C. Hampson, H. D. Springall, A. J. Stosick, Norman Elliott, H. D. Burnham, David P. Stevenson, James E. LuValle, Horace Russell, Jr., Robert A. Cooley, Alexander Kossiakoff, Max T. Rogers, Robert Spurr, Walter Gordy, John H. Saylor, Ralph Spitzer, W. J. Howell, Jr., Paul Giguere, Chia-Si Lu, Earl Malmberg, W. Shand, Jr., Jurg Waser, Jerry Donohue, William N. Lipscomb, Philip A. Shaffer, Jr., John M. O'Gorman, Stig Claesson, Kenneth W. Hedberg, J. D. Dunitz, Morton E. Jones, Edgar Heilbronner, J. N. Shoolery, R. S. Shulman, William F. Sheehan, Walter C. Hamilton, A. J. Stosick, William W. Wood, James A. Ibers, Jean A. Hoerni, and Darwin W. Smith. Sidgwick, at Oxford, had become greatly interested in our molecular-structure studies, and in the 1930's he sent four of his men (Jenkins, Sutton, Hampson, and Springall) to take part in it.

The first Pasadena experimental work in magnetochemistry was a study of nitroso compounds, carried out by E. Bright Wilson, Jr., in 1934, partially in order that his doctoral thesis should contain an experimental part. We borrowed a large water-cooled magnet for this purpose (and the later work on hemoglobin) from George Ellery Hale, who had it at his home and private laboratory in San Marino. The work on the nitroso compounds (which were found to be diamagnetic) had been suggested by the three-electron-bond theory of the triplet normal state of the oxygen molecule. This theory and the hybrid-orbital theory of the magnetic properties of complexes of the transition metals then suggested that a magnetic study be made of hemoglobin and its derivatives. I had thought that this work would tell whether the interaction between the oxygen molecules and the hemoglobin molecule was weak, leaving the oxygen molecule in the triplet state ("physical" force of attraction), or strong, causing all electrons to pair (formation of chemical bonds). The latter idea was verified; and in addition the surprising result was obtained that each iron atom has four unpaired electrons in hemoglobin and none in oxyhemoglobin. This work was done by Charles D. Coryell, as a postdoctoral fellow. The study of the difference in magnetic properties of different hemoglobin derivatives provided a new and useful way of getting information about the electronic structure of these substances and about the rates and equilibrium constants of their chemical reactions. A dozen papers were published in this field during the following years. Myoglobin, hemocyanin, and cytochrome *c* were also studied, the last by Hugo Theorell, who came to Pasadena for this purpose in the summer of 1939. In addition to Coryell and Theorell, Fred Stitt, Richard W. Dodson, Robert C. C. St. George, Charles D. Russell, Donald S. Taylor, and Allen Lein participated in this work.

The research on hemoglobin came to the attention of Karl Landsteiner in the Rockefeller Institute for Medical Research. He encouraged me to apply the new knowledge of molecular structure and the methods of physical chemistry to the problem of the structure of antibodies and the nature of serological reactions. Dan H. Campbell, David Pressman, and a number of students and postdoctoral fellows (Carol Ikeda, Miyoshi Ikawa, David H. Brown, John T. Maynard, Allan L. Grossberg, George G. Wright, Stanley M. Swingle, Arthur B. Pardee, John H. Bryden, Leland H. Pence) took part in an extensive series of experimental and theoretical investigations in this field during the 1940's. These studies showed that precipitating and agglutinating antibodies are bivalent (that is, that each one has two specific combining regions for the homologous antigens), and that blocking antibodies are univalent, and clarified the question of the nature of the forces between antibody and antigen.

The work on hemoglobin stimulated an interest in the question of the folding of polypeptide chains in proteins. I attacked the problem of the folding in alpha keratin with the use of molecular models in the summer of 1937, but without success, and I reached the conclusion (incorrect, in fact) that my assumptions (planar amide groups, bond lengths and angles as in simpler substances) were wrong. Robert B. Corey, who had come to Pasadena from the Rockefeller Institute on a postdoctoral fellowship in 1937, stayed on in order to settle this question by determining the crystal structure of amino acids and simple peptides. His 1938 paper on diketopiperazine reported the first detailed and correct structure determination of a peptide, and the 1939 paper by Gustav Albrecht and Corey on glycine reported the first one for an amino acid. Within a few years the work in Pasadena on many of these substances had provided precise and reliable knowledge of the dimensions of polypeptide chains, permitting the discovery of the principal ways of folding them. Participants in this work included E. W. Hughes, Verner Schomaker, J. Donohue, H. A. Levy, R. E. Marsh, David P. Shoemaker, Herman R. Branson, Joseph Kraut, K. N. Trueblood, A. B. Biswas, J. Norton Wilson, W. J. Moore, L. Katz, R. A. Pasternak, R. E. Barieau, C. S. Lu, and H. L. Yakel, Jr.

Don M. Yost carried on research in inorganic and physical chemistry in Pasadena over a period of forty years. His first work was on inorganic reactions in aqueous solutions, in collaboration with E. H. Swift, Herman F. Schott, R. J. White, Sidney Zabaro, Richard Pomeroy, W. R. Crowell, James M. Carter, and William H. Claussen. He then attacked the general problem of the thermodynamic properties of substances as determined by direct experiment and calculated from molecular constants. John McMorris, Charles E. Kircher, Jr., John B. Hatcher, Sho-Chow Woo, George O. Shull, Charles M. Blair, Jr., Thomas F. Anderson, Folke Skoog, William E. Stone, Norman Elliott, Robert C. Felt, John E. Sherborne, Carsten C. Steffens, Siegfried T. Gross, Walter D. Bonner, W. L. Gore, Clifford S. Garner, Emerson W. Green, Nelson P. Nies, Alan Beerbower, Edwin N. Lassettre, LeRoy H. Anderson, Don DeVault, Carrol M. Beeson, Russell N. Doescher, and Darrell W. Osborne participated in this work in the years before 1940. Yost also carried out work on the fine structure of X-ray absorption edges (with Ralph E. Winger), the properties of neutrons (with Dickinson), the use of radioisotopes (with Robert A. Cooley and Edwin McMillan), nuclear reactions caused by neutron or deuteron bombardment (with Louis N. Ridenour), the possible existence of fluorides of xenon (with Albert L. Kaye, 1933), and nuclear magnetic resonance.

Work in physical organic chemistry was begun in 1920 by Howard J. Lucas, and was continued for many years by him and his students, among whom were Archie R. Kemp, Archibald Y. Jameson, Hollis W. Moyse, Thomas P. Simpson, James M. Carter, William G. Young, Robert T. Dillon, W. F. Eberz, A. N. Prater, R. E. Morris, C. Ernst Redemann, Saul Winstein, and David Pressman. This work was at first hampered by Noyes' lack of interest in it. Although Lucas had published several papers, including the first one of an interesting series on electron displacement in the course of organic reactions, he was not included in the list of graduate faculty in the early years, including 1924–25. Noyes was interested in chemistry in relation to biology and medicine, but not in organic chemistry itself. After his death in 1936 the work in organic chemistry was rapidly expanded, with the help of a grant from the Rockefeller Foundation. Carl Niemann, Laszlo Zechmeister, and Edwin R. Buchman were appointed to the staff; later they were joined by John D. Roberts, George S. Hammond, John H. Richards, and Marjorie C. Caserio. Graduate work and research in organic chemistry, with emphasis on physical organic chemistry and natural-product chemistry, have now reached approximate equality with physical chemistry.

During recent years, work in physical chemistry in the California Institute of Technology has included electron-spin and nuclear magnetic resonance, molecular structure studies by optical and electron-impact spectroscopy, molecular-beam studies, photochemistry, molecular quantum mechanics, the coordination chemistry of metals, and X-ray investigations of metals and alloys and of inorganic and organic compounds, especially those related to nucleic acids and proteins. We may anticipate that the physical chemists of the California Institute of Technology will contribute as much to the understanding of the world during the next 50 years as during the last.

HOT ATOM CHEMISTRY[1]

By Richard Wolfgang
Yale University, New Haven, Connecticut

INTRODUCTION

Over the last five to ten years the complexion of chemical kinetics has undergone a very marked change. One aspect of this change is the more widespread use of beam techniques to study the dynamics of single collisions. Another is the exploration of chemical reactions occurring at high energies, in the electron-volt or bond-energy region. It is this topic of "hot" reactions with which this review is concerned.

Since well before the days of Bodenstein, the standard means of inducing chemical reaction has been to heat the reactants in an environment at thermal equilibrium. It is extremely difficult to produce, under properly controlled conditions, the very high temperatures that correspond to chemical energies. (Typical chemical bond strengths correspond to 10^4–10^{5}° K.) Furthermore, because of the exponential decline at the high energy end of the Maxwell-Boltzmann distribution, reactions near the threshold or activation energy mask those occurring at higher energies. As a result, the virtually universal reliance on thermochemical methods has, in effect, limited the scope of chemical kinetics to those reactions which occur at the lowest possible energy, that is, to so-called "threshold reactions." The idea of the energy dependence of a bimolecular reaction—its excitation function—has been quite foreign to classical chemical kinetics.

Thus, until about eight years ago, the only known general reaction of hydrogen atoms with hydrocarbons was abstraction, which occurs at and above activation energies of about 8 kcals (1/3 eV), e.g.,

$$T + CH_4 \rightarrow HT + CH_3$$

But other reactions in such systems are certainly conceivable, e.g.,

$$T + CH_4 \rightarrow CH_3T + H \quad \text{or} \quad CH_2T + H_2$$

Indeed, at higher kinetic energies of the hydrogen atom ($\sim$2–10 eV), these reactions are found to be comparable in importance to abstraction.

(The study of nuclear reactions has had a history almost the converse of that of chemical reactions. Interactions of accelerated, i.e., "hot" nuclei have been intensively studied for the last 30 years. On the other hand, thermonuclear processes have been experimentally difficult to realize, and as a result, knowledge of threshold processes has tended to lag.)

Recent progress in exploration of translationally hot species has been

[1] The survey of literature pertaining to this review was completed in November 1964.

linked to two techniques. Mass spectrometry has yielded some information on velocity dependence of reactions of gaseous ions (1). Such ionic collisions appear to be dominated by long-range ion-dipole forces and appear, at present, to have little in common with the hot reactions between neutral species which are discussed in this article. The principal technique for generating and studying hot atoms has been nuclear recoil. Certain nuclear transformations are capable of producing species of very high energy which, under properly controlled conditions, become available as hot atoms in the "chemical" energy region.[2] Although this method yields a very small number of hot atoms, reactions resulting in their chemical combination can readily be traced by use of their radioactivity.

The recoil technique results in an energy distribution of reactions which covers the entire chemical range. While this distribution is sufficiently well defined to be tractable by suitable kinetic expressions, it would obviously be preferable to study single collisions at defined energies using beam techniques. After a long delay (of the order of a century), such methods are finally becoming available for study of thermal reactions. Their introduction to the hot reaction field will be more rapid, but as yet no results are available.

In the meantime, recoil techniques have in the last few years enabled rapid progress in outlining the nature of hot atom reactions. The field is still in its infancy and only a few prototype systems have been studied with any degree of completeness. These are primarily hot hydrogen and hot carbon atoms. Some results on hot halogens are also available (although the bulk of the large literature on chemical effects of nuclear transformations in halogen systems is principally concerned with the nature of the reacting halogen and with radiation chemical effects rather than with hot atom chemistry per se.

This relatively short review will make frequent reference to a recently published long article on essentially the same topic (2). That article is an examination in detail which attempts to define, and where possible to resolve, the relevant questions and arguments in the field. The present review is much less ambitious and is intended to serve primarily as a brief introduction to the field and a critical summary of its present state. The reader is also referred to a recent article on chemical effects of nuclear transformations which, though not primarily concerned with hot atom chemistry, provides an idea of the early origins of the field (3).

[2] The terms "hot-atom chemistry" and "chemical effects of nuclear transformations" have, in the past, sometimes been considered synonymous. This is not correct; the latter is merely one technique which can be applied to the problem of hot atom reactions. Conversely the nuclear transformation method can be applied to fields other than hot atom chemistry. (For a recent review of chemical effect of nuclear transformations see Reference 3.) The practice, common in much of the early and some more recent work, of drawing conclusions on hot atom mechanisms from nuclear transformations occurring in poorly understood and controlled systems is to be avoided.

THE NUCLEAR RECOIL TECHNIQUE

With the exception of a few photochemical studies (4, 5, 6), all experiments on hot atoms have used the nuclear recoil technique. The basic method is extremely simple. A substance, which on nuclear transformation yields the desired hot atom, is mixed with the compound with which the hot atom is to react. Such a mixture must be homogeneous on the scale of the recoil range of the hot atom. The sample is exposed to the appropriate nuclear radiation for the period necessary to produce the required number of atoms, this irradiation being sufficiently short so that there is no appreciable macroscopic decomposition.

The transformed nuclei are generally produced as very high energy ions (usually in the kilovolt or million volt range). They lose energy in successive interactions with the medium and should become neutralized in the process. Eventually they drop into the "chemical" energy range (<100 eV), and in further collisions they may react and combine by a hot reaction. If they do not do so, they eventually reach thermal energies and will react as thermal atoms.

The products incorporating the hot atoms may be separated from each other and the reactant by any of several methods. Chromatographic methods, particularly gas chromatography, are usually preferred now (2). It has been recognized that the more traditional methods such as precipitation and distillation can give misleading results when applied to trace quantities. Much of the earlier work in the field of chemical effects of nuclear transformations is now suspect for that reason. The products, once separated, are invariably assayed by counting. Even if other methods had the sensitivity required to detect the trace quantities involved, none has the intrinsic discrimination possessed by counting techniques in responding only to nucleogenic species while ignoring the generally much higher background of similar non-nucleogenic impurities. (Thus, in working with hot C^{11}, it is typically necessary to assay 10^4 molecules of acetylene-C^{11} product in the presence of perhaps 10^{12} molecules of unlabeled acetylene impurity.) In the past few years, continuous flow counting methods, such as radio-gas chromatography, have become common (7).

Nuclear processes as sources of hot atoms.—The use of nuclear transformations as sources of hot atoms rests upon two facts which have become more fully appreciated only in recent years. One is that some processes produce species which react as hot or thermal ions rather than as hot atoms (8, 9). The other is that for most systems it can be predicted with a high degree of assurance whether or not there might be interference by ionic processes (2, 10).

The species initially produced by nuclear transformations are usually ionic. The question is whether they become neutralized by charge exchange processes before they lose enough energy to combine chemically. For species produced by processes having recoil energies of only hundreds of electron volts, such as radiative slow neutron capture (n, γ), it is usually difficult to be sure. But if a particle emission process, such as (γ, n), (n, $2n$), (n, p), etc. is used, very high recoil energies ($>10^4$ eV) result. The hot species will then undergo many charge-changing collisions before reaching the chemical energy range. It can then be shown, using the Mott-Massey adiabatic hypothesis, that when it does so, it will be as a hot atom in its ground or a low-lying excited state in nearly all cases (2).

The possibility of re-ionization due to delayed nuclear processes has been discussed. This may be important in systems involving medium and heavy nuclei (2).

This topic which has been more fully examined elsewhere (2) may, in part, be summarized by the statement that particle emission processes in relatively light nuclei are generally "safe" sources of hot atoms. There are no known exceptions to this rule.

Identification of hot processes (2).—In studying hot atom reactions, it is an essential and obvious requirement (but one which has not always been fulfilled) that interfering processes involving thermalized species be either suppressed or clearly defined. One criterion for identifying hot reactions is by their independence of scavenger action. Substances such as I_2 react efficiently with thermalized atoms and, even in small concentrations, provide an effective sink for them. On the other hand, a hot atom has only a few collisions in which it can react and is therefore unlikely to encounter and be affected by a scavenger molecule.

The addition of an inert moderator to a system reduces the number of collisions of hot atoms with reactant, but it has no effect on the thermal energy distribution and therefore on thermal reactions. Thus, addition of an excess of a noble gas to a system will suppress and hence identify hot processes (see Fig. 1). It is important to note that if a thermalized atom can yield the same product as the hot atom, a scavenger must be used in conjunction with a moderator (an example is the production of HT in reactions of T with hydrocarbons).

In cases where hot and thermal atoms yield similar products with comparable collision efficiencies, it is obviously difficult to establish a clear distinction between hot and thermal processes. This seems to be the situation in most carbon atom systems (see Fig. 1).

Radiation damage.—Excessive radiation damage can often obscure primary hot processes. Even in scavenged systems decomposition and transformation of product molecules can be serious. This problem is now well understood and appreciated (2).

THE KINETICS OF RECOIL-PRODUCED HOT ATOMS

When dealing with a single type of interaction, where the parameters describing the collision are specified within narrow limits, quantitative interpretation of data in terms of cross section or some related quantity is

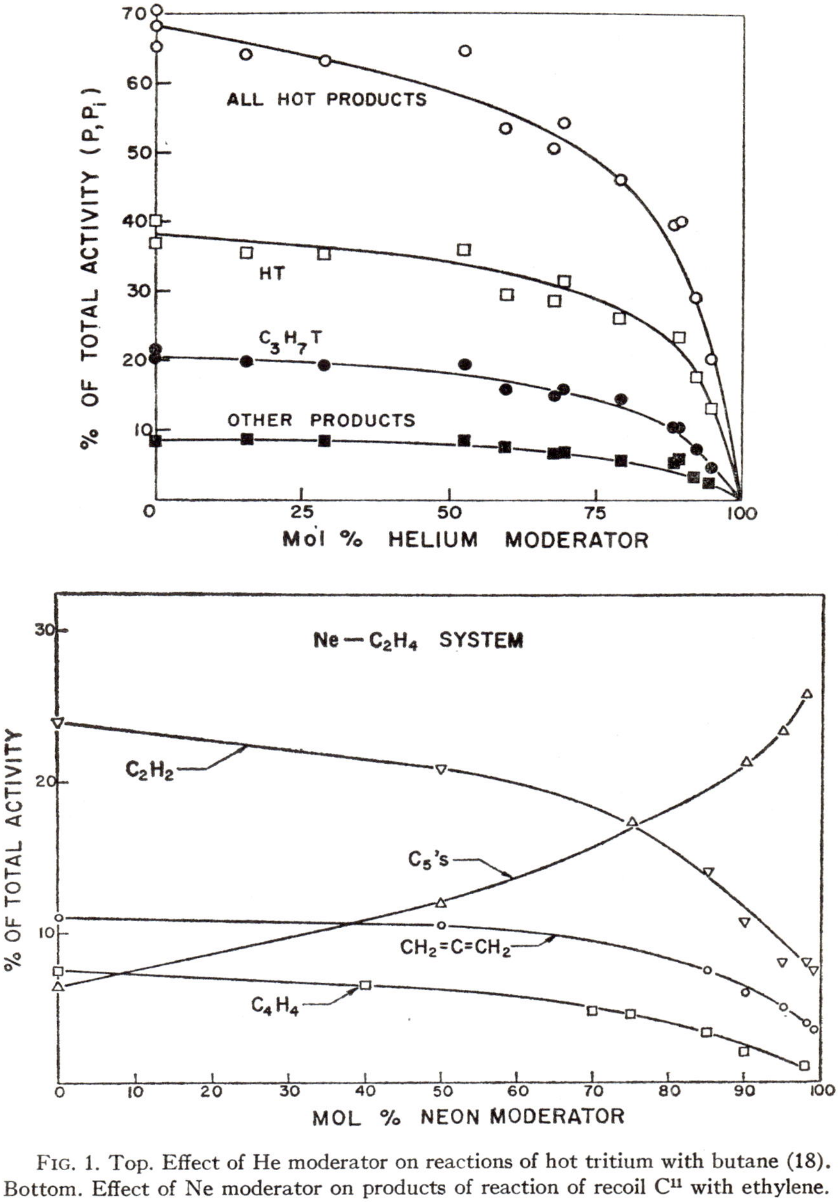

FIG. 1. Top. Effect of He moderator on reactions of hot tritium with butane (18). Bottom. Effect of Ne moderator on products of reaction of recoil C^{11} with ethylene. Note that in this case most products formed in the absence of moderator also appear in fully moderated systems. This indicates that they may be efficiently formed in thermal as well as hot reactions (2).

quite direct. This is the situation in the ideal (and as yet unrealized) beam experiment. But in the usual thermal experiment, the observations represent an average over a distribution in energies, this distribution being variable with the temperature. Similarly, hot atoms produced in a recoil experiment can collide and react over a range of energies, which in this case extends virtually to infinity on a chemical energy scale. The energy density of hot atom collisions is, of course, variable by dilution with moderators.

A kinetic treatment must serve to reduce data taken over a wide range of experimental conditions and hence energy distributions, to as few parameters as possible. These empirically determined parameters should preferably have a clear and fundamental significance, so that they can be calculated theoretically on the basis of any given model of the reaction mechanism. The results of a kinetic analysis can then serve to provide a test of the reaction model. The most familiar example of such a kinetic treatment is probably the collision theory of thermal reactions which results in expressions similar to the Arrhenius equation. These enable reduction of data taken over a range of temperatures to an "activation energy" and a "pre-exponential factor."

The total probability that a hot atom will react to combine while being slowed to thermal energies is

$$P = \sum_j \int_{E_2}^{E_1} f_j p_j(E) n(E) dE \qquad 1.$$

Here P is the fractional yield of all hot products, an experimentally determinable quantity; E_2 to E_1 is the energy range over which hot reaction can occur; f_j is the relative probability of collision with component j; $p_j(E)$ is the probability of reaction and combination in such a collision at energy E; and the collisional distribution function $n(E)dE$ is the number of collisions made by a hot atom between E and $E+dE$. An early theory for recoil reactions in liquids, developed by Miller, Gryder & Dodson (11), assumed a specific reaction mechanism—the cage model of Libby (12)—to evaluate this expression. The rather limited success of this and similar theories (13) based on the cage concept is probably due to an inadequacy in the underlying mechanistic model (14).

In the past few years a kinetic analysis of hot reactions has been developed by Estrup & Wolfgang (15), which does not assume any model for the reaction itself. Instead, the following simple assumption on the nature of the distribution function is made:

$$n(E)dE = \frac{1}{\alpha E} \left(1 - \sum_j \int_{E_2}^{E} f_j p_j(E) n(E) dE\right) \qquad 2.$$

The first term assumes that average energy loss is proportional to energy, the proportionality constant being α. The second term subtracts those atoms which have already reacted on reaching energy E. This approximation is of the nature of a limiting law—it will work best in highly moderated systems where the total probability of reaction is low. Using Equations 1 and 2, expressions can then be derived for any given system. For instance the total fractional probability, P, that a hot atom will react to combine in a system consisting of a reactant and an inert moderator is given by

$$-\frac{1}{\ln(1-P)} = \frac{\alpha_{\text{reactant}}}{I} + \frac{\alpha_{\text{moderator}}}{I}\frac{(1-f)}{f} \qquad 3.$$

Here the probability of collision with reactant is f and with moderator $(1-f)$.

$$I = \int_{E_1}^{E_2} \frac{p(E)}{E} dE \qquad 4.$$

is the so-called "reactivity integral." It is the area under an excitation function, $p(E)$ plotted against in E. Here, α is an energy loss parameter defined by

$$\alpha = -\left\langle \ln \frac{E(\text{after collision})}{E(\text{before collision})} \right\rangle \qquad 5.$$

P is measured at several concentrations of moderator corresponding to various values of f. On plotting the left side of Equation 3 against $(1-f)/f$, a straight line should result (see Fig. 2). From this, I and α_{reactant} can be determined in units of $\alpha_{\text{moderator}}$. Similar expressions can be derived for the yield of any individual product (see Fig. 2) and for hot atom interaction with reactive mixtures. Thus, data from recoil hot atom experiments can be expressed in terms of two types of parameters, one characterizing energy loss on collision and the other the reactivity integral.

The fundamental assumption of the kinetic theory concerns the distribution function (2). Frere (16) has shown this and derivative expressions to be rigorously correct under certain limiting conditions. (However, his treatment is restricted to hard-sphere collisions which Estrup has shown to be unrealistic assumption for hot atom collisions.) Gordus & Hsiung (17) have recently published a kinetic analysis of hot atom reactions which follows closely the formalism of neutron slowing-down theory. Again, with appropriate assumptions, results similar to those of Estrup & Wolfgang (15) were obtained. All these authors emphasize, though in varying degrees, that the kinetic theory has the greatest validity in the limiting case of very low fraction of reaction.

The kinetic theory has been applied to reactions of hot tritium atoms with a number of hydrocarbons (18, 19) and with mixtures of hydrogen and methane (15). It has also been used for analysis of the reaction of hot F^{18} with CF_4 (20) and, for certain reactions of recoil, bromine (21) and iodine (22). In all cases the fit obtained was good within the limits of experimental error. Thus, the kinetic theory seems to provide a useful framework not only in the limit, but over the full range of experimental conditions (18, 19). This is surprising; but then it is perhaps similarly surprising that the Arrhenius equation is as useful as it is, considering the assumptions required by the collision theory of thermal reactions.

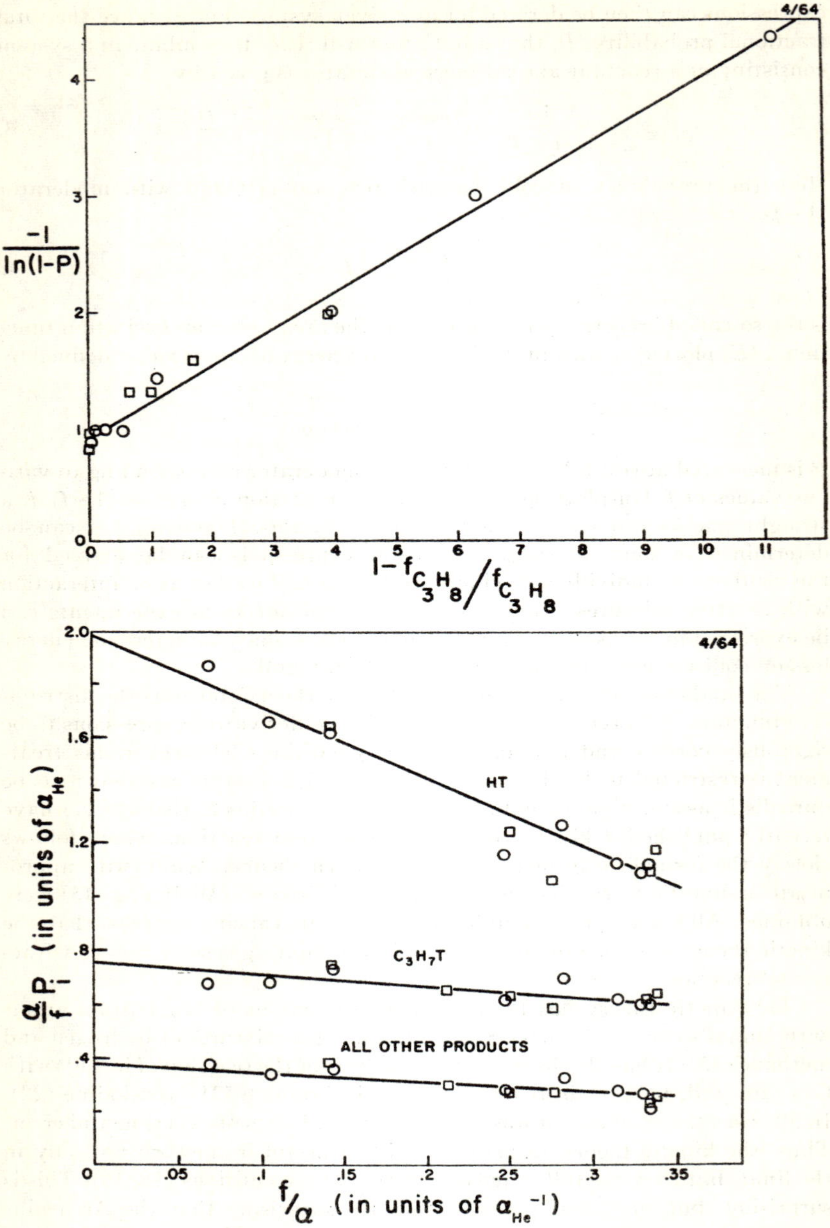

FIG. 2. Kinetic analysis of reactions of hot tritium atoms with butane. Top. Analysis of trend of total hot yield (P). Bottom. Analysis of individual hot yields (P_i) (18).

In summary, the kinetic theory is of value in the following respects:

(*a*) It can provide a rigorous test of experimental data. Because of the minimal number of parameters, essentially only two, bad results are difficult to accommodate. Thus, a small contribution to yields by thermal processes leads to significant deviations.

(*b*) It shows that the explicit form of excitation functions cannot be obtained using the recoil method. The excitation function $p(E)$ always appears in a form integrated over energy.

(*c*) It provides values of reactivity integrals I and energy loss parameters α. These are expressed in units of $\alpha_{\text{moderator}}$. To obtain absolute values of α and I, $\alpha_{\text{moderator}}$ must be determined independently. Estrup (23) has shown that this quantity is not well approximated by the usual hard-sphere expressions, but that it can be estimated using known interaction potentials.

(*d*) Ratios $\alpha_{\text{reactant}}/\alpha_{\text{moderator}}$ obtained by the kinetic analysis (18, 19, 20) indicate that collisions of hot atoms with molecules are very highly inelastic. Obviously at the energies involved there is efficient coupling between translational and internal degrees of freedom.

(*e*) The magnitudes of reactivity integrals show that, as expected, hot atom reactions can have very high collision efficiencies (18).

(*f*) From certain "energy-shadowing" effects the relative energies at which hot reactions occur can be determined (15, 18).

Extensions of the kinetic theory.—An extension of the original theory (15) to systems of two liquid reactants has been described by Milman (24). Using a different, but apparently equivalent, approach Kontis, Sanitrongs & Weston (25a) have treated certain ternary liquid mixtures. Filatov, Nesmeyanov & Chepyzhev (25b) have made kinetic analyses of certain recoil halogen reactions in liquid systems. It should be noted that this work incorporates some very drastic mechanistic assumptions largely based on elastic collision models which seem to have a rather limited range of validity.

Other approaches.—A method of numerical analysis of hot atom kinetics has been proposed by Rowland & Coulter (26). Monte Carlo methods are used to compute the paths of a large number of individual hot atoms. So far, this has only been done for the rather unrealistic case of hard-sphere collisions. However, the method is useful in providing a numerical test of the accuracy of the expressions obtained in the analytical kinetic theory.

Frere (27) has published a kinetic theory of hot atom reactions in a "multiplying" system where a hot atom may generate other hot atoms. No systems involving such cascades have yet been experimentally investigated.

Limitations of the kinetic theory.—It should be noted that the kinetic theory of hot reaction is only applicable to "primary" products (15). Thus, the stable products of hot carbon atom reaction cannot be directly treated by this formalism since they usually derive from secondary decomposition of a primary adduct.

While the present kinetic theory appears at this time to be quite adequate, future data of very high accuracy may require refinements which will take the form of additional parameters. Such elaboration will be interesting for its own sake but may not yield much new light on the hot reactions themselves. As the kinetic theory itself demonstrates, the necessity of having to average over a distribution limits the power of the recoil method to resolve fine detail.

THE HOT ATOM CHEMISTRY OF HYDROGEN

Introduction

Studies by El Sayed & Wolfgang (28) which recognized that tritium recoiling from nuclear processes, reacted by single-step atomic, rather than by ionic modes, initiated an intensive investigation of hot hydrogen atom chemistry. The system proved experimentally tractable in all respects and progress in the last few years has been rapid. Because of this, the reactions of hot hydrogen are now quite well understood and can serve as a useful prototype for hot atom chemistry in general (2).

Recoil tritium is most conveniently generated by slow neutron bombardment of Li^6 or He^3. The Li^6 $(n, \alpha)T$ reaction is the standard source for studies of liquid and solid systems. It generates long range 2.7 MeV tritium. This makes it possible to study the reactions of hot hydrogen with any compound by simply neutron-irradiating a relatively coarse mixture of the substance with a lithium salt. However, most of the studies on mechanism of reaction of hot hydrogen have been made on gases—both because of their inherent simplicity and because scavenger and moderator techniques are most readily used in gaseous systems. The He^3 $(n, p)T$ reaction, in a mixture of the compound to be studied with a little He^3, is then the most convenient source of tritium. The gas phase studies have also been expedited greatly by development of methods of analysis by gas chromatography coupled to internal gas flow counting (2, 7).

The reactions of recoil tritium in many systems have now been investigated. [A bibliography extending to early 1964 has been published in the recent detailed review article (2).] Unfortunately some of this work is not suitable as a basis for definitive conclusions regarding hot hydrogen atom reactions, because neither scavenger nor moderator techniques were used to distinguish those products, which were the result of interfering thermal, rather than hot, processes. (Thus, in the absence of scavenger, radiation-produced radicals will react with thermalized hot atoms to produce labeled molecules which may readily be mistaken for products of hot atom reactions.) This objection is almost unavoidable in solid phase systems in which scavenger cannot normally be introduced, and must, unhappily, also be made about many studies in fluid systems. Such work has often had as its primary purpose the practical one of preparing labeled compounds and is valuable in this respect. But conclusions on hot processes which are based on such studies must be treated with reserve, and in general are useful only as preliminary and tentative indications.

TABLE I

PRODUCTS OF REACTION OF RECOIL TRITIUM ATOMS WITH BUTANE AND BUTENE-1[a,b]

Reactant	Saturated products					Radical products		Unsaturated products				Other hot products
	HT	CH_3T	C_2H_5T	C_3H_7T	C_4H_9T	$CH_2T\cdot$	$C_2H_4T\cdot$	C_2H_3T	C_3H_5T	Butene-1	Butene-2	
Butane	40	1.8	0.9	0.5	22	1.5	0.8	1.3	0.3	0.0	—	<0.5
Butene-1	26	1.1	0.3	0	0	—	—	7.5	8.0	20	1.6	<1.0

[a] Yields as percentage of total T atoms available.
[b] Sources: Ref. 18 and prior work cited therein.

SYSTEMATICS OF HOT HYDROGEN ATOM REACTIONS

Reactions of hot hydrogen in all systems studied display very similar basic features. Systematics and mechanisms which are applicable with some modification to most compounds can therefore be conveniently discussed using the well-studied hydrocarbons as prototypes.

In most compounds, 40 to 90 per cent of hot hydrogen atoms react before thermalization. Their reactions closely follow the systematics originally found for methane (28). Table I, given as an example, shows the products obtained with butane and butene-1. A classification of reactions (29) follows.

Abstraction of atoms.—(*a*) Hydrogen atom abstraction, e.g.,

$$T + CH_3CH_2CH_2CH_3 \rightarrow HT + CH_3CH_2CH_2CH_2\cdot \text{ (or } CH_3CH_2\dot{C}HCH_3)$$

is normally the most efficient reaction of hot hydrogen. This is the threshold reaction and thus usually the only one observed for thermal hydrogen. As expected, it has a far higher collision efficiency when hot. Bond energy effects seem, in general, to be very small or negligible in hot atom reactions. However, Breckenridge, Root & Rowland (30) have shown that H abstraction efficiencies correlate roughly with the energy of the C–H bond. The effect appears to be of the order of 10 per cent/kilocalorie (still small by thermal reaction standards).

(*b*) Abstraction of other atoms has not yet been observed. There is no reason to believe this does not occur—it is only because the product exchanges hydrogen easily, particularly on a trace scale, that it has not been identified. However, the experimental problem is not insoluble, and future discovery and study of this type of reaction would be most welcome.

Substitution of a single entity.—These reactions are unknown (or very unusual) with thermal hydrogen but with hot atoms can be comparable in efficiency to abstraction. (*a*) Hydrogen atom displacement, e.g.,

$$T + CH_3CH_2CH_2CH_3 \rightarrow CH_3CH_2CH_2CH_2T + H$$

has been shown to occur predominantly with retention of configuration (i.e., non-Walden inversion) at the carbon atom substituted, at least in the relatively complex systems in which it has been studied. These include glucose (solid) (31), alanine (solid) (32), secondary butyl alcohol (gaseous) (33), and dichloroethylene (gaseous) (34). The gas phase experiments show that this retention is not due to any constraint imposed by a crystal or liquid cage.

(*b*) Halogen atom displacement, e.g.,

$$RCl + T \rightarrow RT + Cl$$

occurs readily in monohalomethanes. However, when the carbon atom bears a second substituent, the efficiency of this reaction is greatly diminished (see Table II) (35). Similar overall processes observed in aromatic compounds (36, 37) are probably due to the successive addition-dissociation mechanism (see *Addition to unsaturated systems*) (2).

TABLE II

PRODUCTS OF REACTION OF RECOIL TRITIUM ATOMS WITH FLUOROMETHANES (35)[a]

Reactant	Product				
	HT	CH_3T	CH_2TF	$CHTF_2$	CTF_3
CH_3F	24	4.4[b]	12	0	0
CH_2F_2	20	0	1.5[b]	5.5	0
CHF_3	8	0	0	0.7[b]	2.9
CF_4	0	0	0	0	0.2[b]

[a] Yields as percentage of total T atoms available.
[b] Denotes T-for-F substitution product.

(*c*) Radical displacement e.g.,

$$CH_3CH_2CH_2CH_3 + T \rightarrow CH_3T + CH_3CH_2CH_2\cdot; \quad \text{or} \quad CH_3CH_2T + CH_3CH_2\cdot$$

This reaction generally has a much lower efficiency than substitution of atoms (except in unsaturated systems where it may proceed by the addition-displacement reaction described below). Upon rupturing a C–C bond, the T atom inevitably has a strong preference to join the smaller of the two radicals. For instance, in neopentane, about ten times as much CH_3T as $(CH_3)_3CT$ is formed (38).

Substitution of two entities.—Two entities, either atoms or radicals, may be replaced, e.g.,

$$CH_3CH_2CH_2CH_3 + T \rightarrow CH_2T\cdot + H + CH_3CH_2CH_2\cdot; \text{ or } CH_3CH_2CH_2CHT\cdot + 2H$$

These reactions can proceed in a single step, or as a single displacement followed by unimolecular decomposition of the excited primary product (2, 35, 38, 39). Their yield is usually small except in certain systems where the single displacement may be inhibited (see below) or where the primary product is exceptionally unstable. The radical produced may dissociate further

$$CH_3CH_2CH_2CHT\cdot \rightarrow CH_2{=}CHT + CH_3CH_2\cdot$$

or may be detected by combination with scavenger

$$CH_2T\cdot + I_2 \rightarrow CH_2TI + I$$

Addition to unsaturated systems.—Addition to multiple bonds, as in the familiar thermal reaction, is also observed with hot hydrogen, e.g.,

$$T + CH_3CH_2CH{=}CH_2 \rightarrow CH_3CH_2CHTCH_2\cdot$$

Urch and Wolfgang (40) have shown that in this process a long-lived, well-defined intermediate is formed, which may be identified from the characteristic pattern of its decomposition, e.g.,

$$CH_3CH_2CHTCH_2\cdot \rightarrow CH\ CH_2\cdot + CHT{=}CH_2$$

(By contrast there is no convincing evidence for the formation of such an intermediate in processes detailed above.)

A kinetic analysis (18) has shown that while this sequence can occur thermally, with recoil tritium it occurs mainly as a hot atom reaction. The addition-decomposition reaction path can often form the same products as direct displacement, e.g.,

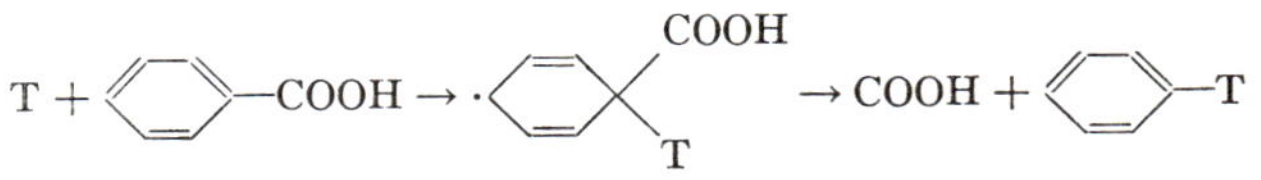

Automatically attributing such yields to the direct mechanism, as has occasionally been done, appears to be unjustified.

Essentially all reported reactions of hot hydrogen fall into these four categories. Only very low yields of other reactions such as "fragmentation" processes, in which many bonds are broken, have been reported. However, the yield ratios of these four reaction types vary sharply and characteristically from system to system.

We may summarize the features common to all these reactions as follows: (*a*) There are usually several reaction paths of comparable probability. (*b*) Only one or two bonds are broken in 98 per cent of the reactions in which the hot atom becomes bound (see Table I). (*c*) Products are generally left with residual internal excitation energies that are modest in terms of the energy available. This is evidenced by the absence of extensive isomerization or decomposition of product molecules. In a definitive experiment, Lee & Rowland (41) have shown that H-substitution in cyclobutane leaves the molecule with a median energy of the order of 5 eV. (*d*) Hot hydrogen atom reactions appear to be quite specific. Many conceivable products are absent or appear in small yield. The yield of apparently similar reactions can vary sharply from system to system. This specificity runs counter to the primitive notion that since, energetically, anything can happen in a hot atom process, everything does indeed occur in an indiscriminate manner.

The Present Model of Hot Hydrogem Atom Reactions

The unified reaction model discussed here is essentially that favored at Yale University (2, 10, 15b, 28, 29, 33, 35, 38). This theory, a fundamental aspect of which is the hypothesis that in activated systems primary reaction is a fast direct process, has been largely developed in discussions with F. S. Rowland and co-workers, who have explored the other possibility, that an intermediate may be formed that has a lifetime long enough to undergo 'reasonably normal' unimolecular decay (42, 43). The discussion on this critical question is summarized elsewhere (2) and it suffices to say that there now seems to be substantial agreement on the fundamentals of the model presented below (44).

Multiplicity of reaction paths.—Whereas thermal or threshold processes can normally proceed by only a few reaction paths and yield only one, or occasionally two, primary products, the reactions of hot hydrogen can directly lead to a variety of primary products. This situation could hardly be otherwise. If the interaction is represented by a multidimensional energy surface, there would be a number of passes and saddle points corresponding to the various possible products. In a threshold reaction, the system can, by definition, only cross the lowest saddle point. Only one product is thus generally possible, and the fraction of trajectories leading to reaction is small. This is equivalent to saying that, because of steric requirements on the collision complex, the collision efficiency is low. In a hot atom process, reaction trajectories are not limited to the lowest energy pass. This means that many configurations of the colliding system can lead to reaction, many products can be formed, and collision efficiencies are high. The expectation of high collision efficiencies has been confirmed by application of the kinetic theory to recoil hydrogen reactions (18).

Nature of the reactive hot collision.—There are two extreme models of collision of atoms with molecules. If the atom is relatively slow, the collision time is long enough for its effect to be transmitted through the valence bonds to all parts of the molecule. In effect, the atom thus collides with the whole molecule, even if that is quite large. However, thermal activation energies and preliminary results of photochemical hot atom experiments (6) indicate that characteristic hot atom reactions, such as displacement, only become efficient above 1 to 2 eV. When the velocity of a tritium atom corresponding to a few electron volts is divided into a distance of atomic dimensions, a characteristic collision time of the order of 10^{-14} sec is obtained. This is comparable to the relaxation time of a C–H bond. Thus, if a tritium atom is energetic enough to cause reaction, the time of its primary interaction is too short for much energy to be carried away from the collision site, and the collision will not usually have the extreme atom-molecule character.

At the other extreme, one may envisage that a highly energetic (>10–20 eV), very fast atom may interact with only a single atom of the molecule struck. The molecule then appears simply as a weakly coupled collection of atoms. One form of this atom-atom collision mechanism is the "billiard-ball" model proposed by Libby (12). Theoretical calculations of Cross & Wolfgang (45) show that gas-phase collision efficiencies for substitution reaction proceeding by this mechanism at the high energies where it would be valid are very low. This was confirmed by experimental measurements (45) of isotope effects in the system $T+CH_4$, CD_4.

It appears that collisions resulting in reaction and combination of the hot atom involve an intermediate mechanism. This is known as the Impact Model of hot atom reactions (38). Atoms with energies over 1 to 2 eV undergo a relatively fast (10^{-14} sec) localized interaction in which only a few atoms in the impact area are coupled to the collision. As a result only one or two bonds are broken—in accordance with the data. Since in a localized collision only a limited amount of energy can be transferred to the few bonds affected, the efficiency of reaction with combination may be expected to decline above 5 to 10 eV. For the same reason the products of such hot reaction will contain rather limited excitation energy (41) and will not undergo extensive isomerization or decomposition, again as observed.

Kinetic theory data (α values) show that collisions in the energy range where reaction occurs, but which do not result in combination, are of a similar character. Such collisions are inelastic, the hot atom losing a large fraction of its energy, although not all (15, 18, 46).

The relatively high energy (5 eV) of the hot atom precludes, in general, the possibility that after the initial fast interaction a long-lived complex is formed. Only if the molecule struck is unsaturated can it form a strong enough bond to hold the hot atom without simultaneous rupture of another bond. The radical formed by combination with an unsaturated compound may subsequently break up according to the laws governing unimolecular decomposition—as has been observed in hot hydrogen atom addition to alkenes (40). However, with a saturated system there is no force or bond sufficiently strong to keep the whole collision complex together for more than one vibrational period. The strong experimental and theoretical evidence indicating this absence of an internally equilibrated complex has been reviewed (2). The path taken by the hot reaction is therefore not determined by the usual minimum-energy considerations governing unimolecular decay (except following hot addition to unsaturates). Instead, the factors governing the course of a reaction occurring in a fast localized collision appear to be primarily steric, inertial, and inductive.

Steric factors: dependence of reaction mode or impact parameters.—Henchman, Urch & Wolfgang (29) have suggested that the path of the reaction is largely governed by the impact parameters of the collision. Such steric considerations suggest the four idealized modes of reaction (2) shown in Figure 3. They are: substitution with and without Walden inversion; abstraction by hot atom approach along the axis of the bond involved; and abstraction by a stripping mechanism.

Steric considerations also predict that attack at less accessible bonds should be less efficient. This is consistent with the fact that C–C bonds are

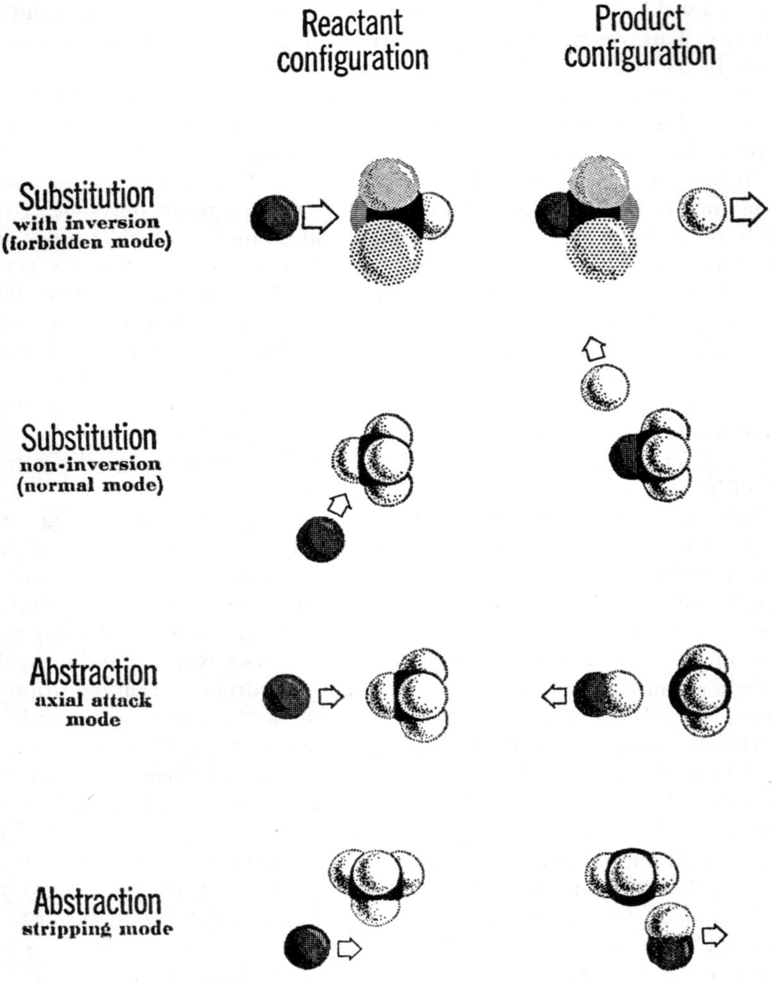

FIG. 3. Idealized modes of hot atom reaction (2).

less frequently ruptured than C–H bonds (38). The finding that substitution of the tertiary C–H bond in isobutane is less efficient than that at any one primary C–H bond has been ascribed to the same factor (47). Root & Rowland have shown, however, that such steric obstruction effects at C–H bonds are not generally of large magnitude (48).

At this point a note on grazing or stripping reactions is appropriate (2). In a glancing collision, little momentum is transferred and the energy of the hot atom is largely unavailable for driving the reaction. This has two consequences: (*a*) The energy range over which stripping can take place may be greater than usual; (*b*) unlike other hot processes, this one may well be quite appreciably dependent on the ΔE of this reaction. The bond-energy dependence found by Breckenridge, Root & Rowland (30) for H atom abstraction only might then be taken as evidence for this mechanism. [These authors feel, however, that the best interpretation of their data may instead involve competitive unimolecular-type decay of a longer-lived intermediate (51).] Apart from these results there are as yet no experimental data supporting a possible grazing mode of hot reaction. However, it is interesting to note that the occurence of such processes has very recently been established in thermal systems through beam studies of Datz, Herschbach et al. (49, 50) on alkali atoms.

Inertial factors.—Not all hot hydrogen atom reactions which seem sterically possible proceed efficiently. The reason appears to lie in inertial restrictions which follow naturally from the basic model of a fast, localized collision. These have been formulated as a "Golden Rule" of hot atom chemistry. As stated by Odum & Wolfgang, this is simply that a hot reaction will tend to be forbidden if it requires relaxation motions which are slow compared to the period of the collision (35).

Studies on optically active systems have shown that hot hydrogen substitution of hydrogen with Walden inversion is very improbable (see Fig. 1) in relatively complex molecules. This becomes reasonable if it is considered that the relaxation time for inversion of heavy ligands is relatively long (10^{-13} sec). Thus, no vacant orbital capable of binding the incident tritium atom can become available at the point of collision during the time of collision (10^{-14} sec). The hot atom is not captured and escapes. By contrast, hydrogen atom substitution without inversion does not involve motion of heavy ligands and is not forbidden (33).

Substitution of a heavy atom can be an efficient process, but is much less so if there is a second heavy substituent on the central carbon atom. An example is T for F substitution in CH_3F and CH_2F_2 (Table II). In both cases, the hot atom can break the C–F bond without materially displacing the heavy F atom in the time of collision. However, in CH_3F the H atoms can rapidly shift so that an orbital to capture the tritium becomes available. Such a quick relaxation is not possible if a second F atom is involved and the reaction tends to abort (35). Very similar reasoning has been used to account for the fact that when a T atom ruptures a C–C bond it perfers to combine with that carbon atom having fewer heavy substituents (35).

The peculiar fact that in some systems, such as CH_2F_2, substitution of two atoms still occurs or is even enhanced, even though single-atom substitution is forbidden, has also been attributed to the Golden Rule (35). In fact, this principle accounts for many of the specificities found in hot hydrogen atom reactions. It is even more remarkable that not a single case of a reaction violating this Golden Rule, which proceeds with normally high efficiency, appears to be known.

Inductive factors.—Rowland has pointed out (51) that inductive factors may be expected to have some effect on hot atom reaction probabilities. The nature and magnitudes of such inductive effects are not yet fully defined, but they are understood to denote electron availability or polarizability factors. Thus, a marked decrease in the probability of substituting any given H atom in the halomethane series CH_4, CH_3X, CH_2X_2, CHX_3 (see Table II) seems too large to attribute entirely to steric hindrance by the X atoms. Possibly the electron-withdrawing nature of the halogen (X) atoms makes the C–H bond less reactive. (However, inertial factors involving the halogen atoms may also be responsible here.) Conversely, electron-rich unsaturated bonds seem to be an especially attractive target for hot atom attack. Thus a recent analysis of ethylene and dichloroethylene data shows that the addition reaction with the double bond is several times as likely as all substitution and abstraction reactions occurring at the other bonds (2).

Isotope Effects

The effect of substituting deuterium for protium in molecules reacting with hot tritium atoms has been carefully studied in a number of systems. The observation that hydrogen substitution isotope effects are relatively small shows that the billiard-ball mechanism of reactions is not important (45). However, the use of isotope effects to study more subtle aspects of mechanism has been hampered by the fact that the possible origins of isotopic differences appear to be quite complex and are not yet well understood. Isotope effects in thermal reactions are principally caused by small, calculable changes in bond strengths resulting from zero-point energy differences. In view of the large amount of energy available, this type of isotope effect is less important in hot reactions, but other types of effects having other origins must be considered.

Jurgeleit & Wolfgang (52) have classified hot atom isotope effects as follows: (*a*) moderator isotope effects which modify energy loss in collisions in which the hot atom escapes (by changing the number of hot collisions, these will change the total probability of reaction); (*b*) reactive isotope effects, comprising all effects on collisions in which the hot atom actually combines.

Lee & Rowland (43) have shown that reactive effects are dominant in recoil tritium systems. Such effects are of the order of 15 to 35 per cent with protonated systems being the more reactive. The effect on hydrogen abstraction may, at least partly, be attributable to a zero-point energy isotope effect, since these reactions seem to show some sensitivity to bond energy. This would be reasonable if grazing reactions are involved.

Secondary isotope effects such as that in the reaction

$$T + CH_3F(CD_3F) \rightarrow CH_3T(CD_3T) + F$$

are now generally recognized (2, 52, 53) as being due to inertial effects on rotational-vibrational relaxation of the CH_3 (CD_3) group as required by the original formulation (35) of the Golden Rule (see above). Isotope effects on substitution by T of the H(D) atom itself are, oddly enough, less well understood. These could be due to the inertia of the atom being replaced. However, in the systems so far investigated, it could also be a secondary isotope effect, caused by isotopic differences in masses of other atoms in the molecule (53). Experiments on systems of the type CX_3H and CX_3D which would settle this question have not been done.

More complete reviews of data and speculation on hot hydrogen isotope effects will be found elsewhere (2, 53).

Condensed Systems

A number of factors may be expected to modify hot atom reactions in condensed systems: (*a*) The cage effect originally proposed by Libby for halogen hot atoms (12). Following collision and rupture of a bond, a hot atom which might otherwise escape is held in the vicinity of the radical and may combine with it. In effect, the time of collision is increased. This might be expected to enhance yields, particularly those of reactions forbidden by inertial factors (Golden Rule). (*b*) Steric effects will be modified by surrounding molecules. Furthermore certain configurational changes which can occur freely in isolated molecules may be hampered. (*c*) Improved intermolecular energy transfer in the condensed phase may stabilize certain products which might otherwise decompose.

Experimental data, however, indicate very little difference between recoil hydrogen reactions in gas, liquid, and solid. For instance, a study of the *trans*-hexene-2 system (54) showed significant differences only in products resulting from decomposition of butyl radical formed by addition reactions. This reflects the operation of factor (*c*) above. There is no evidence as yet that factors (*a*) and (*b*) are of importance for hot hydrogen. (Quite possibly the hydrogen atom is too small to be trapped in a solvent cage.)

An extensive series of studies on the fate of recoil tritium in a variety of condensed systems, particularly alcohols, aromatic and cyclic unsaturated compounds, has been made by Nesmeyanov, Dzantiev, Pozdeyev and coworkers (37, 55, 56, 57). Insofar as these results reflect hot atom reactions rather than radiation-induced and thermal processes, the same general pattern as for simpler gaseous systems was found. In addition, these workers studied the effect of energy migration on the survival of excited product molecules (56).

Summary: Theoretical Approaches to Hot Atom Reaction

The impact model, as described above, accounts for essentially all features of the hot atom reactions of hydrogen that have been found in experiments in which hot processes were well defined and controlled (i.e., in properly scavenged systems). In view of the specific nature of these reactions, this means that the evidence for the qualitative validity of the model is strong, and that there exists a sound basis for a quantitative theory of mechanism.

Little progress has been made as yet on such a quantitative theory of the reaction mechanism. However, there has been a start on two quite different approaches. In the first, classical calculations have been made for the very high energy region in which the interaction may be approximated as collisions between pairs of atoms only loosely coupled to the rest of the molecule. Reactions of the type

$$A + RB \rightarrow RA + B$$

proceeding by collision of atom A with B (R may be a radical), have been treated by Cross & Wolfgang (45) and subsequently in greater detail by Light & Horrocks (58). It was found that at the high energies where this model should hold (>20 eV), such reactions are only important if the masses of atoms A and B are very similar. Although both sets of calculation assumed hard-sphere (billiard-ball) collisions, which will tend to give overestimates, cross-sections were found to be quite low. While this type of mechanism may extend the reaction range to relatively high energies, its contribution to the integrated cross section will be small. A related mechanism involving a double collision within the collision complex has been treated by Bates (59).

A more common approach to chemical reaction mechanisms, and one which could be applicable to all energies, is to consider trajectories across a multidimensional potential energy surface. In hot atom reactions, however, the system is not limited to the threshold mode passing through the saddle point. All trajectories must be considered, requiring a staggering amount of computation. Calculations of this sort have recently been performed by Karplus, Porter & Sharma (60) for the reaction $D + H_2$ at relatively low energies. These studies indicated the absence of any kind of persistent reaction complex and showed an excitation function initially rising with energy. Future comparison with experimental hot atom data will be interesting. But it is clear that theoretical methods require much further development before they can catch up with experimental progress on systems of reasonable complexity.

THE HOT ATOM CHEMISTRY OF HALOGENS

Background

The study of chemical effects of nuclear transformations involving halogens has a long history, going back to the discovery of the Szilard-Chalmers effect. The realization that halogens produced in nuclear events could be used to study chemical reactions of high kinetic energy species led to the use of the term "hot-atom chemistry" by Libby.

The first theories on recoil halogen atom reactions did not emphasize actual direct hot atom substitution and combination reactions. Early work on alkyl halides by Suess (61), Libby (12), and others had shown that the probability of organic combination of the recoil halogen was much lower in gases than in liquids. This led to an emphasis on two-step processes involving the concept of the Franck-Rabinovitch solvent cage, rather than on direct reactions. Libby postulated that hot atom interactions at high energies with single atoms in a molecule (billiard-ball mechanism) (12), or at lower energies with the whole molecule (epithermal mechanism) (62), could create a radical and leave the hot atom with so little energy that it could not escape from the cage formed by the surrounding solvent molecules. After further collisional deactivation within the cage, it would then combine with the radical it had previously formed. Willard (63) favored the "random-fragmentation" theory in which the hot atom was postulated to produce a dense "brush-heap" of radicals near the end of its recoil track. The atom would then randomly combine with one (or more) of these radicals. In both these models, the primary hot reactions are those involving fragmentation of the molecule struck without simultaneous capture of the hot atom. Combination of the recoil atom would then be a thermal or near-thermal process occurring in the solvent cage.

Neither of these mechanisms would predict any combination of the hot atom in the gas phase since there would be no cage to keep the atom and radical together. The finding by Willard et al. that in some gas phase systems there could, after all, be appreciable organic combination of the recoil halogen was therefore significant. Willard suggested that this indicated the existence of one-step, hot atom substitution reactions (64), although subsequently he found evidence to favor the alternative concept that combination reaction of recoil species involved ion-molecule reactions (65). It now appears that, depending on the system, either, or sometimes both, of these hypotheses may apply.

These developments have been extensively reviewed by Willard (64–67), Campbell (3), and others (66). Their complexity indicates the difficulty we face in trying to decide which recoil halogen results, particularly those from liquid phase experiments, are relevant to this review, in that they provide firm information on direct hot atom reactions.

Recent Gas Phase Studies

Rack & Gordus have studied the reaction with methane of Br^{80} produced by the Br^{79} (n, γ) Br^{80} process (21). As discussed in an earlier section it is difficult to be sure on an a priori basis if species produced by radiative neutron capture (n, γ) will react as ions, excited atoms, or ground-state hot atoms. Rack & Gordus attacked this question using a number of moderators of different ionization potentials and analyzing their data according to the kinetic theory. They concluded that reactions leading to combination of the recoil bromine were entirely a consequence of the translational energy of the Br^{80} rather than of any ionic character. On the other hand, they found that I^{128} from the $I^{127}(n, \gamma)I^{128}$ process combined with methane partly as a hot atom and partly as a ground state or excited ion (22). In both cases the efficiencies of the hot atom reactions seemed to be quite low.

The interaction of I^{126} with methyl iodide formed by the $I^{127}(\gamma, n)I^{126}$ process has been studied by Cross & Wolfgang (68). Photoneutron emission results in high recoil energy, and the likelihood that the species produced will reach and react in the chemical energy region (<100 eV) as ions is therefore greatly reduced. Again low reaction efficiencies were found, in agreement with earlier work on reaction with methyl iodide of I^{128} formed by radiative neutron capture (2, 3, 67).

An exhaustive study of reactions with gaseous CF_4 of F^{18} produced by photoneutron emission (γ, n) has been made by Colebourne et al. (20, 69). A scavenger system consisting of ethylene and I_2 was found effective for removal of thermal fluorine. A complete kinetic analysis of data involving Ne, He, and Ar scavengers was made. Reactions entirely analogous to those in the $T+CH_4$ system were found.

$$F^{18} + CF_4 \rightarrow CF_3F^{18} + F$$
$$\rightarrow CF_2F^{18}_{\cdot} + 2F(\text{or } F_2)$$

However, reaction efficiencies are much lower than for hot hydrogen. Another striking difference is that the two atom displacement to give $CF_2F^{18}_{\cdot}$, despite its much higher energy requirement, has virtually the same yield as single atom substitution giving CF_3F^{18}.

Mechanism.—Cross, Colebourne & Todd (20, 68, 69) have offered two possible explanations for the generally low efficiency of gas phase hot halogen reactions studied so far. Both are based on the impact model of hot atom reactions as developed for hydrogen (2). (*a*) Because of the larger size and, for a given energy, lower velocity of the halogens, their collisions will not be as fast nor as localized as those of hydrogen atoms. There will therefore be more opportunity for the kinetic energy of the incident halogen to spread through the molecules. With the available energy more diffused it is less likely that any given bond will be broken, and the probability of reaction is thus decreased. At higher energies bonds will of course be broken. However, the product molecule may then contain so much energy as to be unstable. Formation of high energy products in which more than one bond has been broken will therefore be favored. It is precisely this that has been observed in the relatively high yield of $CF_2F^{18}_{\cdot}$ from $F^{18}+CF_4$ (69). (*b*) Steric hindrance would be more marked in the reaction of X at a C–X bond, and to a lesser extent of X at a C–H bond, than for T at a C–H bond. The probability of reaction with a single bond is thus decreased. However, the same steric considerations suggest that interaction with and rupture of "two" bonds might be relatively more likely (20).

Relevant data presently available are insufficient to decide the relative importance of these two factors. For the present these hypotheses must be considered as highly tentative; however, the evidence does suggest that the general impact model of direct hot reaction is valid in the gaseous hot halogen system.

Liquid Phase Studies

On surveying the enormous literature relating to interactions of recoil halogens in liquids (3, 67), it is apparent that the main concern in this field has been to try to sort out the various types of processes that affect the fate of the recoil species. The subject of this article—the actual bimolecular hot-atom collision—is only one of these processes. Because in the past it has usually not been possible to define clearly and to separate experimentally these events, surprisingly little definite information on the detailed mechanism of actual hot reactions has come from studies of chemical effects of nuclear transformations in liquids. (This is even more true for studies of chemical effects of nuclear transformations in solids. In most such experiments the concern has been with using the recoil atom to study the effect of imperfections in solids.)

Most studies on recoil halogens have involved their formation by (n, γ) reactions. As discussed in a previous section this often introduces some uncertainty as to what extent one is dealing with ionic rather than atomic reactions. An equally serious difficulty is that the usefulness of moderators and scavengers in experimentally defining hot reactions is much more restricted in liquids than in gases. There are as yet no inert substances known which are conveniently miscible with most organic liquids and could thus act as moderators. In liquid systems, scavengers can readily identify "diffusive" reactions undergone by thermalized hot atoms which have made many collisions before reaction. But it is much more difficult to distinguish between direct hot reactions and the two-step sequences of fragmentation and radical-recombination occurring within a solvent cage (70).

In past years there has been much spirited discussion on the relative merits of the billiard-ball and epithermal theories of Libby versus the random fragmentation hypothesis of Willard (63, 64, 66). The emphasis has changed recently. This is partly due to results on condensed phase reactions of recoil hydrogen and carbon which rule out random fragmentation as an important process for these hot atoms and indicate that the main function of the solvent cage is to de-excite the product into which the hot atom has been incorporated by direct reaction (2, 71).

The possibility that direct bimolecular substitution reactions may, after all, be the dominant process for certain recoil halogens in liquid media has received important support from comparative studies by Willard on gaseous and liquid alkyl chlorides (66). The similarity of products found suggests that mechanisms in the two phases are similar. Although no scavenger was used in this work, direct atomic substitution was taken as primarily responsible for the products formed in the vapor.

Milman (72) has recently reviewed the status of some of the dominant hypotheses in the recoil chemistry of alkyl halides. She first tentatively concludes that the results obtained using (n, γ) produced recoil iodine as well as bromine in alkyl halides are primarily due to atomic rather than ionic processes. This conclusion is largely based on work of Schuler (73) who found the reactions of (γ, n), (n, γ), (d, p), and $(n, 2n)$ produced recoil iodine in liquids to be similar. She then examines the question of whether combination of recoil halogens in liquids might be the result of a direct hot reaction rather than of a two-step process. Since, as has been mentioned, it is difficult for scavengers alone to make the distinction, this had to be done on the basis of the nature of the products. Milman found that with very few exceptions all products found in properly scavenged systems could be derived from direct reactions exactly analogous to those found for hot hydrogen. Thus, in the case of the very thoroughly studied Br^{80}-ethyl bromide system, all but a very small fraction of the scavenged yield can be accounted for on the basis of substitution by the Br^{80} for one or two atoms or groups in a single ethyl bromide molecule.

On the basis of present evidence, there is, however, no justification for assuming that direct hot reactions are the dominant process in "all" liquid alkyl halide systems. In particular, Geissler & Willard (74) have recently suggested that the fate of recoil iodine and possibly other species may be governed by localized radiolysis due to an Auger shower occurring at the end of the recoil track. Still another point of view has been advanced by Shaw (75) who finds evidence that pyrolysis-type reactions in a "hot-spike" at the end of the recoil track are important. Systems in which either of these two types of effects dominate are, of course, unsuitable as a source of information on hot atom reactions. For the present, therefore, hot atom interpretations of liquid phase recoil-halogen experiments must be regarded as speculative.

Kinetic theory.—If products in liquid systems are formed by direct hot reaction, some form of the kinetic theory ought to be applicable. As suitably inert moderators appear to be unavailable, Milman has studied hot reactions of Br^{80} with reactive binary mixtures over the entire concentration range (24). Such systems are more complex than single reactant-inert moderator mixtures, especially in the liquid phase, and application of the kinetic theory required several drastic, though not unreasonable, assumptions. The good fits and internal consistency obtained suggest that the treatment has validity. The α values obtained are relative rather than absolute (15a, 23), but their magnitude indicates a high energy loss per collision. This is most reasonable, since in the energy range of a few electron volts one would expect the large, relatively slow, Br^{80} to interact with a large part of the molecule. As Milman points out, such a collision is similar to that envisaged for reactive collisions in Libby's epithermal model (62).

Kontis et al. (25a) have recently made a kinetic analysis of recoil halogen reactions in certain ternary mixtures. The uncertainties in dealing with these complex systems make it difficult to evaluate the full significance of this work. Filatov, Nesmeyanov & Chepyzhev have published a kinetic analysis of mixed systems (25b). The diverse and drastic assumptions apparently made include facets of the hard sphere cage-recombination mechanism. Results presented do not provide a rigorous test of this treatment.

Mechanism.—If we tentatively accept the hypothesis that most products formed in liquid systems represent direct hot reaction, rather than complex other processes, the picture of hot halogen atom chemistry that emerges seems physically reasonable and well in accord with the more firmly based results in tritium and gaseous halogens. Milman (72) has pointed out that the systematics of liquid phase recoil halogens are quite similar to those of hot hydrogen, suggesting that the impact model applies.

However, in considering results of recoil halogen studies (3, 67, 76), it must be remembered that the collision period of a hot halogen atom is much longer than that of hot hydrogen. One consequence of this is to make inertial relaxation factors (Golden Rule of hot atom chemistry) much less restrictive. Thus, substitution with Walden inversion may be a more efficient process. Recent data of Rowland, Wai & Ting (77) on reaction of Cl^{38} with isomeric dichlorobutanes tend to support this expectation.

The probability of rupturing two bonds to form a radical incorporating the hot atom is much greater than in the case of hot hydrogen (72). This finding is similar to the conclusion on gas-phase halogens. The same hypothesis, that collision of the relatively large and slow hot halogen atoms favors formation of high energy products, should be applicable here.

This facet of the proposed mechanism of hot halogen reactions could also account for the basic finding, now a quarter-century old, that hot combination of recoil halogens is much more efficient in the condensed phase than in gases. If the product of a hot halogen reaction is indeed likely to be highly excited, its chances of survival will be greatly enhanced by the rapid collisional de-activation that can occur in a condensed medium. In this natural extension of the impact model to hot halogens in liquids, we emphasize the function of the solvent in de-exciting the primary products rather than in caging radical intermediates. However, it is interesting to note that Libby's epithermal mechanism becomes identical to the impact model if it is taken to the limit where the time interval between production of the radical by the hot atom, and its combination with that radical, is reduced to the magnitude of a bond-vibration period.

THE HOT CHEMISTRY OF POLYVALENT ATOMS

The chemistry of free carbon atoms, both hot and thermal, has been extensively studied. It has been reviewed in several articles (2, 71, 78), the latest of which, by this author, is of the same scope as the present review (78). This subject will therefore not be discussed here. It is worth pointing out here, however, that the hot and thermal properties of carbon atoms are qualitatively quite similar (see Fig. 1). This appears to be a consequence of the high intrinsic chemical potential of this species.

The chemistry of other polyvalent entities, notably N, P, As, and S, has also been extensively investigated by the recoil method (3). Results so far are not sufficient, in general, to determine which of the properties of these species are due to excess kinetic energy, i.e., to their being hot atoms.

ACKNOWLEDGMENTS

Critical discussions with C. MacKay, J. Dubrin, and L. Spicer were most helpful in the writing of this review. The work was supported by the U. S. Atomic Energy Commission.

LITERATURE CITED

1. Talrose, V. L., *Pure Appl. Chem.*, **5,** 455 (1962)
2. Wolfgang, R., *Progress in Reaction Kinetics*, **III** (G. Porter, Ed., Pergamon, Oxford, 1965)
3. Campbell, I. G., *Advan. Inorg. Chem. Radiochem.*, **5,** 135 (1963)
4. Schwarz, H. A., Hamill, W. H., and Williams, R. R., *J. Am. Chem. Soc.*, **74,** 6007 (1952)
5. Carter, R. J., Hamill, W. H., and Williams, R. R., *J. Am. Chem. Soc.*, **77,** 6457 (1955)
6. Martin, R. M., and Willard, J. E., *J. Chem. Phys.*, **40,** 2999, 3007 (1964)
7. Wolfgang, R., and Rowland, F. S., *Anal. Chem.*, **30,** 903 (1958)
8. Thompson, J. L., and Miller, W. W., *J. Chem. Phys.*, **38,** 2477 (1963)
9. Levey, G., and Willard, J. E., *J. Chem. Phys.*, **25,** 904 (1956)
10. El Sayed, M. A., Estrup, P. J., and Wolfgang, R., *J. Phys. Chem.*, **62,** 1356 (1958)
11. Miller, J. M., Gryder, J. W., and Dodson, R. W., *J. Chem. Phys.*, **18,** 579 (1950)
12. Libby, W. F., *J. Am. Chem. Soc.*, **62,** 1930 (1940); *ibid.*, **69,** 2523 (1947)
13. Capron, P. C., and Oshima, Y., *J. Chem. Phys.*, **20,** 1403 (1952)
14. Miller, J. M., and Dodson, R. W., *J. Chem. Phys.*, **18,** 865 (1950)

15a. Wolfgang, R., *J. Chem. Phys.*, **39,** 2983 (1963)

15b. An earlier, less complete, formulation: Estrup, P. J., and Wolfgang, R., *J. Am. Chem. Soc.*, **82,** 2661, 2665 (1960)

16. Frere, R., *Bull. Acad. Roy. Med. Belg.*, **XLVIII,** 216 (1962)
17. Hsiung, C., and Gordus, A. A., *J. Am. Chem. Soc.*, **86,** 2782 (1964); Hsiung, C., Verosub, K. L., and Gordus, A. A., *J. Chem. Phys.*, **41,** 1595 (1964)
18. Rosenberg, A. H., and Wolfgang, R., *J. Chem. Phys.*, **41,** 2159 (1964)
19. Urch, D. S., and Welch, M. J., *Symp. Chem. Effects Assoc. Nuclear Reactions, IAEA, Vienna, 1964* (In press)
20. Colebourne, N., Todd, J. F. J., and Wolfgang, R., *Symp. Chem. Effects Assoc. Nuclear Reactions, IAEA, Vienna, 1964* (In press)
21. Rack, E. P., and Gordus, A. A., *J. Phys. Chem.*, **65,** 944, 1855 (1961)
22. Rack, E. P., and Gordus, A. A., *J. Phys. Chem.*, **65,** 944 (1961)
23. Estrup, P., *J. Chem. Phys.*, **41,** 164 (1964)
24. Milman, M., *Radiochim. Acta*, **2,** 180 (1964)

25a. Kontis, S. S., Sanitrongs, P., and Weston, M., *Symp. Chem. Effects Assoc. Nuclear Reactions, IAEA, Vienna, 1964* (In press)

25b. Filatov, E. S., Nesmeyanov, An. N., and Chepyzhev, Yu. B., *Radiochemistry*, **6,** 595 (1964); *Moscow University Newsletter, Ser. II, Chemistry*, **4,** 13 (1964)

26. Rowland, F. S., and Coulter, P., *Radiochim. Acta*, **2,** 163 (1964)
27. Frere, R., *Bull. Acad. Roy. Med. Belg.*, **XLVIII,** 296 (1962)
28. El Sayed, M. F. A., and Wolfgang, R., *J. Am. Chem. Soc.*, **79,** 3286 (1957)
29. Henchman, M., Urch, D., and Wolfgang, R., *Can. J. Chem.*, **38,** 1722 (1960)
30. Breckenridge, W., Root, J. W., and Rowland, F. S., *J. Chem. Phys.*, **39,** 2374 (1963)
31. Keller, H., and Rowland, F. S., *J. Phys. Chem.*, **62,** 1373 (1958)
32. Kay, J. G., Malsan, R. P., and Rowland, F. S., *J. Am. Chem. Soc.*, **81,** 5050 (1959)
33. Henchman, M., and Wolfgang, R., *J. Am. Chem. Soc.*, **83,** 2991 (1961)
34. Urch, D., and Wolfgang, R., *J. Am. Chem. Soc.*, **83,** 2997 (1961)
35. Odum, R., and Wolfgang, R., *J. Am. Chem. Soc.*, **83,** 4668 (1961); *ibid.*, **85,** 1050 (1963)
36. White, R. M., and Rowland, F. S., *J. Am. Chem. Soc.*, **82,** 4713 (1960)
37. Pozdeyev, V. V., Nesmeyanov, A. N., and Dzantiev, B. G., *Radiochemistry*, **5,** 395 (1963)
38. Urch, D., and Wolfgang, R., *J. Am. Chem. Soc.*, **83,** 2982 (1961)
39. Tang, Y., Lee, E. K. C., and Rowland, F. S., *J. Am. Chem. Soc.*, **86,** 1280 (1964)
40. Urch, D., and Wolfgang, R., *J. Am. Chem. Soc.*, **81,** 2025 (1959); *Chem. Effects Nuclear Transformations, IAEA, Vienna, 1961*, **II,** *99*
41. Lee, E. K. C., and Rowland, F. S., *J. Am. Chem. Soc.*, **85,** 897 (1963)
42. Lee, J. K., Musgrave, P., and Rowland, F. S., *J. Am. Chem. Soc.*, **82,** 3545 (1960)

43. Lee, E. K. C., and Rowland, F. S., *J. Am. Chem. Soc.*, **85,** 2907 (1963)
44. Lee, E. K. C., Miller, G., and Rowland, F. S., *J. Am. Chem. Soc.*, **87,** 199 (1965)
45. Cross, R. J., and Wolfgang, R., *J. Chem. Phys.*, **35,** 2002 (1961)
46. Root, J. W., and Rowland, F. S., *J. Chem. Phys.*, **38,** 2030 (1963)
47. Odell, A., Rosenberg, A. H., Fink, R. D., and Wolfgang, R., *J. Chem. Phys.*, **40,** 3730 (1964)
48. Root, J. W., and Rowland, F. S., *J. Am. Chem. Soc.*, **84,** 3027 (1962)
49. Datz, S., and Minturn, R. E., *J. Chem. Phys.*, **41,** 1153 (1964)
50. Wilson, K. R., Kwei, G. H., Norris, J. A., Herm, R. R. Bireley, J. H., and Herschbach, D. R., *J. Chem. Phys.*, **41,** 1154 (1964)
51. Rowland, F. S. (Private communication)
52. Jurgeleit, H. C., and Wolfgang, R., *J. Am. Chem. Soc.*, **85,** 1057 (1963)
53. Lee, E. K. C., Root, J. W., and Rowland, F. S., *Symp. Chem. Effects Assoc. Nuclear Reactions, IAEA, Vienna, December 1964* (In press)
54. Lee, E. K. C., and Rowland, F. S., *J. Am. Chem. Soc.*, **84,** 3085 (1962)
55. Nesmeyanov, A. N., Tszyan, T., and Filatov, E. S., *Radiochemistry*, **5,** 515 (1963)
56. Pozdeyev, V. V., Nesmeyanov, A. N., and Dzantiev, B. G., *Kinetics Catalysis (USSR) (English transl.*, **4,** 318 (1963)
57. Pozdeyev, V. V., Nesmeyanov, A. N., and Dzantiev, B. G., *Radiochemistry*, **4,** 116, 215, 398, 404, 615 (1962); **6,** 314 (1964)
58. Light, J. C., and Horrocks, J., *Proc. Phys. Soc.*, **84,** 527 (1964)
59. Bates, D. R., Cook, C. J., and Smith, F. J., *Proc. Phys. Soc.*, **83,** 49 (1964)
60. Karplus, M., Porter, R. N., and Sharma, R. D., *J. Chem. Phys.*, **40,** 2033 (1964)
61. Suess, H., *Z. Physik. Chem. (Leipzig)*, **B45,** 297 (1939)
62. Fox, M., and Libby, W. F., *J. Chem. Phys.*, **20,** 487 (1952)
63. Willard, J. E., *Ann. Rev. Nucl. Sci.*, **3,** 193 (1953)
64. Willard, J. E., *Ann. Rev. Phys. Chem.*, **6,** 141 (1955)
65. Gordus, A. A., and Willard, J. E., *J. Am. Chem. Soc.*, **79,** 4609 (1957)
66. Willard, J. E., *Chem. Effects Nuclear Transformations, IAEA, Vienna, 1961*, **I,** 215–97.
67. Willard, J. E., *Symp. Chem. Effects Assoc. Nuclear Reactions, IAEA, Vienna, December 1964* (In press)
68. Cross, R. J., Jr., and Wolfgang, R., *Radiochim. Acta*, **2,** 112 (1964)
69. Colebourne, N., and Wolfgang, R., *J. Chem. Phys.*, **38,** 2782 (1963)
70. Harris, W. E., *Can. J. Chem.*, **39,** 121 (1961); also Ref. 66, p. 229
71. Wolf, A. P., *Advances in Physical Organic Chemistry*, **II** (1964)
72. Milman, M., *Radiochim. Acta.*, **1,** 15 (1962)
73. Schuler, R. H., *J. Chem. Phys.*, **22,** 2026 (1954)
74. Geissler, P. R., and Willard, J. E., *J. Phys. Chem.*, **67,** 1675 (1963)
75. Shaw, P. F. D., *Radiochim. Acta*, **1,** 177 (1963); Cole, A. J., Mia, M. D., Miller, G. E., and Shaw, P. F. D., *Symp. Chem. Effects Assoc. Nuclear Reactions, IAEA, Vienna, December 1964* (In press)
76. Nesmeyanov, A. N., and Filatov, E. S., *Radiochemistry*, **5,** 378 (1963); **3,** 610 (1961); **4,** 462, 613 (1962)
77. Wai, C. M., Ting, C. T., and Rowland, F. S., *J. Am. Chem. Soc.*, **86,** 2525 (1964)
78. Wolfgang, R., and MacKay, C., *Science* (In press)

DIAMAGNETIC ANISOTROPY OF ELECTRON GROUPS[1,2]

By A. A. Bothner-By
Mellon Institute, Pittsburgh, Pennsylvania

AND

J. A. Pople
Carnegie Institute of Technology, Pittsburgh, Pennsylvania

Introduction

The electrons of a molecule immersed in a uniform magnetic field, $\boldsymbol{H}_0$, move in a manner which corresponds to an intramolecular flow of current. If there are no unpaired electrons, this is the dominant effect of the field and gives rise to the phenomena of diamagnetism. The magnetic moment $\boldsymbol{\mu}$, generated by the induced currents is given by

$$\boldsymbol{\mu} = \chi \boldsymbol{H}_o \qquad 1.$$

while the field at the site of the *i*th nucleus in the molecule is given by

$$\boldsymbol{H}_i = (1 - \boldsymbol{\sigma}_i)\boldsymbol{H}_o \qquad 2.$$

where $-\boldsymbol{\sigma}_i \boldsymbol{H}_0$ is the secondary field at the nucleus due to the electron currents. Both $\boldsymbol{\chi}$, the susceptibility, and $\boldsymbol{\sigma}_i$, the shielding constant, defined by Equations 1 and 2, are tensor quantities; for neither the magnetic moment nor the secondary field at a nucleus need be parallel to the applied field for a particular molecular orientation. The average susceptibility χ and the average shielding constants σ, are just the average values of these tensors.

A number of direct measurements of the anisotropy of $\boldsymbol{\chi}$ were made some years ago by crystal and magnetic birefringence studies (Cotton-Mouton effect) (1, 2). Recently, new methods for measuring anisotropies of $\boldsymbol{\sigma}$ have become available. As the anisotropy of $\boldsymbol{\chi}$ of individual groups also plays an important part in predictive theories of $\boldsymbol{\sigma}$ (and, hence σ), there is now considerable general interest in this area. In this review we aim to survey the existing literature on these aspects of anisotropy.

Anisotropy of Diamagnetic Susceptibilities

If a molecule is placed in a uniform magnetic field with components $H_\beta(\beta = x, y, z)$, the induced diamagnetic moment is a vector μ_α (not necessarily parallel to the magnetic field) given by

[1] The survey of literature pertaining to this review was concluded in December 1964.

[2] This work was supported by grants from the Air Force Office of Scientific Research AF-AFOSR 19963 and the National Science Foundation. We are indebted to L. Gedansky for assistance in the preparation of Table I.

$$\mu_\alpha = \sum_\beta \chi_{\alpha\beta} H_\beta \qquad 3.$$

The set of quantities $\chi_{\alpha\beta}$ (six of which are independent, since it is easily shown that $\chi_{\alpha\beta} = \chi_{\beta\alpha}$) form the diamagnetic susceptibility tensor. If the axes are rotated so that χ_{xy}, χ_{yz}, χ_{zx} all vanish (principal axes), the three principal values χ_{xx}, χ_{yy}, and χ_{zz} may be different if the molecule is of low enough symmetry. The differences $\chi_{zz} - \chi_{xx}$ and $\chi_{zz} - \chi_{yy}$ are then anisotropies of the susceptibility tensor. For molecules with a threefold or higher principal axis, there is only one anisotropy

$$\Delta\chi = \chi_{\parallel} - \chi_{\perp} \qquad 4.$$

where $\chi_{||}$ is the susceptibility along the principal axis, and $\chi_\perp$ the (unique) value perpendicular to it. In general, a knowledge of the principal axes, the anisotropies, and the mean susceptibility

$$\chi = \tfrac{1}{3}(\chi_{xx} + \chi_{yy} + \chi_{zz}) \qquad 5.$$

is sufficient to derive the complete tensor.

Anisotropies from Crystal Data

Most direct experimental data on diamagnetic anisotropies have been derived from measurements on molecular crystals with lower than cubic symmetry. Such crystals have bulk diamagnetic anisotropy, and, if it is assumed that the susceptibility tensor for the crystal can be obtained by superposing those for the individual molecules (taking proper account of their orientation relative to crystal axes), then information about molecular components can be derived if the crystal structure is known. The amount of information that can be gained in this way depends on the type of crystal. In the monoclinic case, for example, symmetry requires that one crystal axis (*b*-axis) is a principal axis of the crystal susceptibility tensor, and there are only four observable quantities (the three principal susceptibilities and the inclination of the remaining principal axis to the *a*- or *c*-crystal axes). The number of unknowns in the molecular tensor, however, may be less than six. For planar molecules, for example (on which many measurements have been made), symmetry requires that one principal axis of the molecular tensor is perpendicular to the plane and only four components have to be determined. The connection between crystal and molecular tensors for the principal crystal systems has been discussed in detail by Lonsdale & Krishnan (3) considering the orientations of principal axes of the molecular susceptibility to crystal axes. A more direct procedure is the transformation of the complete tensors including off-diagonal terms (4). In certain cases, thermal motion of molecules leads to temperature-dependent crystal susceptibilities (5), but corrections can usually be applied.

A summary of experimental anisotropies derived from crystal measurements is given in Table I. Many of these measurements are on aromatic molecules, where there is a marked anisotropy with high diamagnetism perpendicular to the molecular plane. We shall return to this effect in later sections.

TABLE I

ANISOTROPIC SUSCEPTIBILITIES FROM CRYSTAL MEASUREMENTS

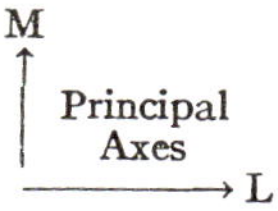	$10^6\chi_L$	$10^6\chi_M$	$10^6\chi_N$	Ref.
	−34.9	−34.9	−94.6	(5)
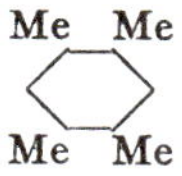	−77.3	−82.4	−143.9	(3)
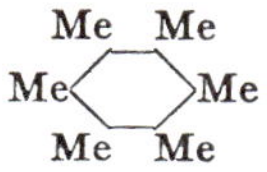	−101.1	−102.7	−163.8	(3)[a]
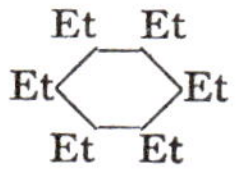	−165.0	−165.0	−231.0	(6)
	−78.3	−50.3	−120.2	(7)
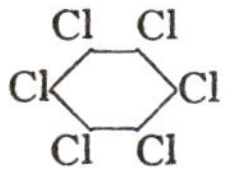	−128.0	−128.0	−182.0	(8)
	−97.1	−70.5	−136.7	(7)
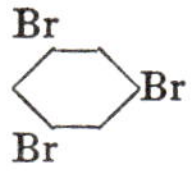	−122.5	−122.5	−170.0	(9)[b]
	−87.6	−59.9	−129.0	(7)
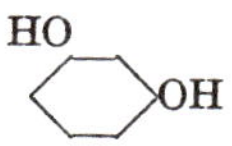	−49.2	−49.2	−103.2	(10)[c]
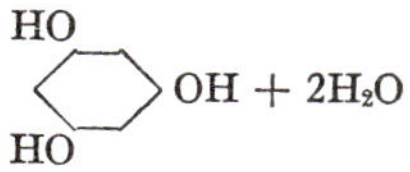	−83.1	−84.1	−120.2	(7)[d]

[a] L-direction displaced by about 10° from line joining opposite methyl groups.
[b] Slight in-plane anisotropy found but probably insignificant.
[c] In-plane components assumed equal.
[d] Contribution of water included.

TABLE I—(*Continued*)

Principal Axes (M ↑, → L)	$10^6\chi_L$	$10^6\chi_M$	$10^6\chi_N$	Ref.
	−64.0	−38.0	−106.0	(11)
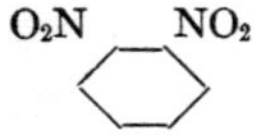	(−57.0)	(−45.0)	(−106.0)	(11)[e]
	−52.0	−43.0	−104.8	(7)
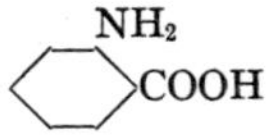	−57.7	−58.8	−120.5	(7)[f]
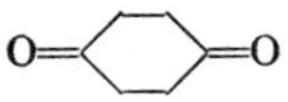	−23.0	−27.0	−65.2	(7)
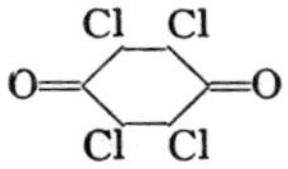	−84.4	−98.5	−138.5	(7)
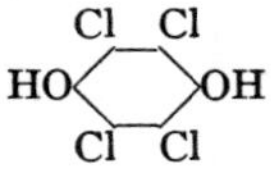	−103.0	−114.3	−144.9	(7)
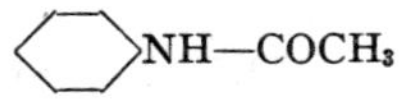	−55.8	−44.3	−116.9	(7)[f]
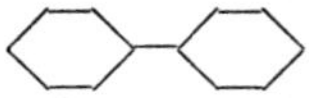	−67.7	−61.7	−183.8	(7)
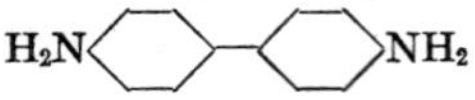	−76.0	−76.0	−184.0	(11)[e]
	−96.8	−88.1	−271.3	(3)
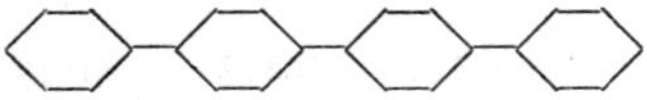	−122.0	−110.0	−372.0	(8)
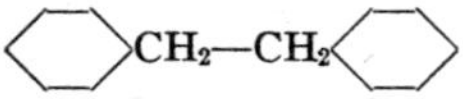	−90.8	−90.8	−202.3	(12)
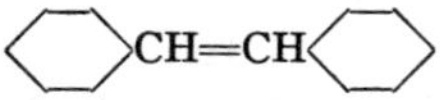	−85.8	−50.1	−209.6	(12)[f]
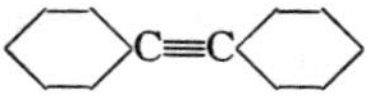	−81.5	−67.8	−198.3	(13)

[e] Data only partially analyzed.
[f] Principal axes approximate.

TABLE I—(*Continued*)

M ↑ Principal Axes → L	$10^6\chi_L$	$10^6\chi_M$	$10^6\chi_N$	Ref.
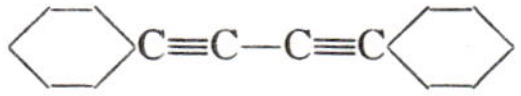	−109.4	−75.3	−206.7	(12)
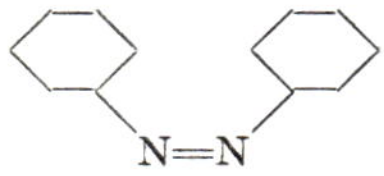	−99.4	−64.0	−177.4	(12)[f]
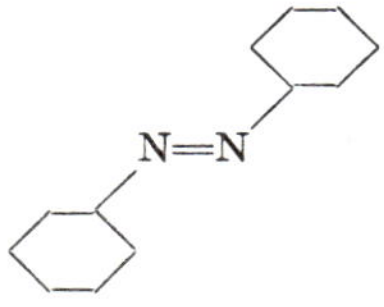	−78.4	−54.0	−193.4	(12)[f]
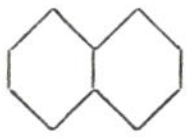	−54.7	−52.6	−173.5	(3)
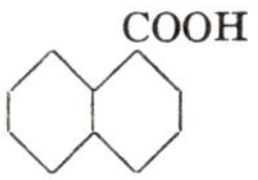	−70.5	−58.95	−192.5	(7)[f]
	−63.9	−51.9	−175.2	(7)[f]
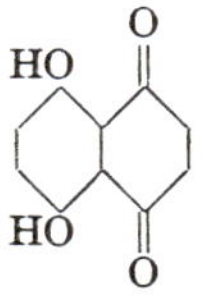	−57.3	−57.3	−140.1	(14)[c]
	−75.8	−62.6	−251.6	(3)
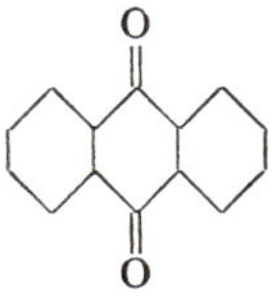	−76.1	−64.5	−217.6	(7)
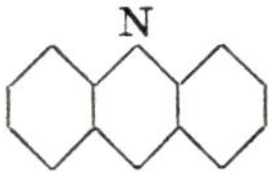	−61.4	−70.5	−238.0	(7)[g]

[g] Molecule not strictly planar.

TABLE I—(*Continued*)

M ↑ Principal Axes → L	$10^6\chi_L$	$10^6\chi_M$	$10^6\chi_N$	Ref.
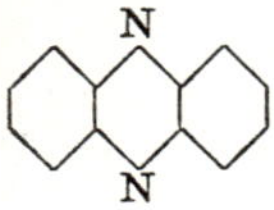	−46.2	−61.8	−243.0	(15)
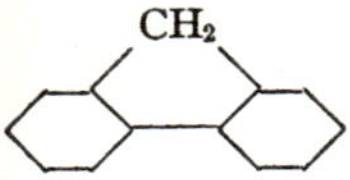	−72.6	−72.6	−193.6	(8)[c]
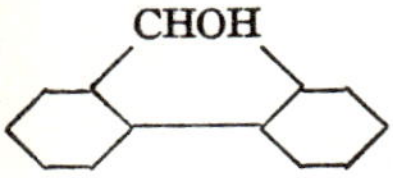	−68.0	−68.0	−189.0	(11)[e]
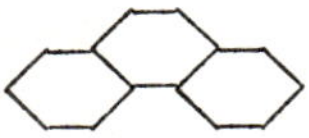	−74.0	−74.0	−240.0	(8)[c]
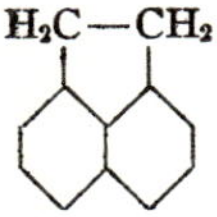	−72.0	−70.5	−185.5	(7)
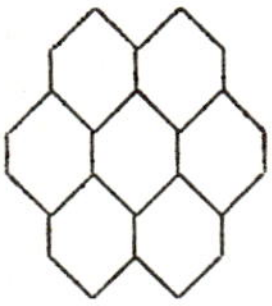	−90.0	−90.0	−480.0	(14)[c]
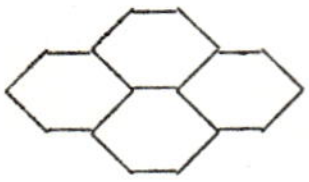	−80.6	−80.6	−303.0	(3)[e,e]
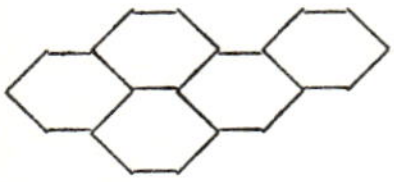	−100.0	−100.0	−350.0	(11)[c,e]
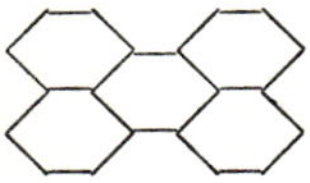	−80.0	−80.0	−320.0	(11)[c,e]
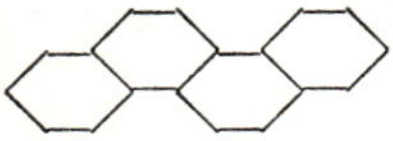	−88.0	−83.3	−310.8	(3)[f]

TABLE I—(*Continued*)

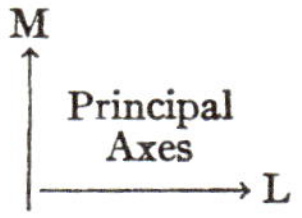	$10^6\chi_L$	$10^6\chi_M$	$10^6\chi_N$	Ref.
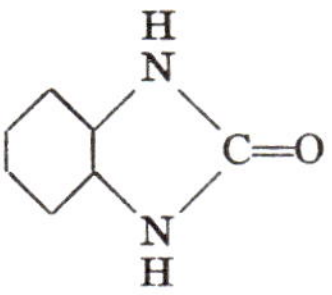	−62.0	−57.3	−124.2	(7)[f]
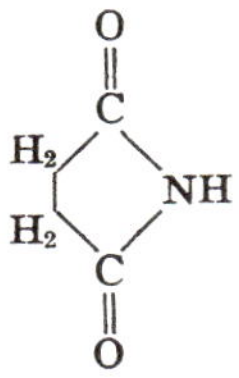	−45.1	−41.6	−55.1	(16)
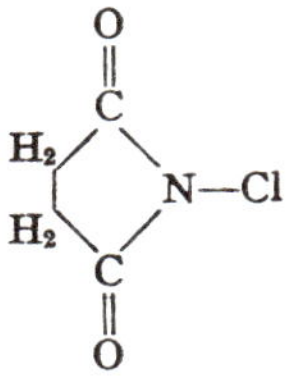	−64.5	−51.8	−78.0	(7)
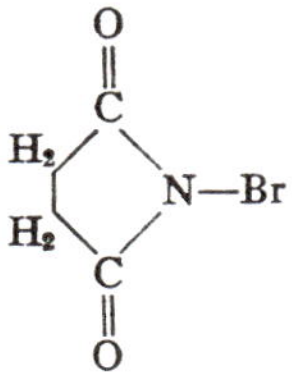	−76.0	−61.6	−87.3	(7)
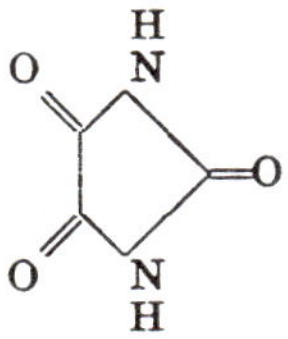	−28.6	−27.5	−49.9	(7)
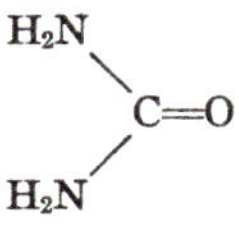	−31.8	−31.8	−37.0	(17)

TABLE I—(*Continued*)

M ↑ Principal Axes → L	$10^6\chi_L$	$10^6\chi_M$	$10^6\chi_N$	Ref.
HOOC—CH_2—CH_2—COOH	−53.1	−48.6	−60.7	(12)[f]
FERROCENE	−104.1	−108.6	−162.3	(20)[f]
HOOC—HC═CH—COOH	−43.3	−42.9	−62.3	(12)[f]
CH_3—CH═CH—CH═CH—COOH	−57.0	−49.0	−44.0	(18)[e]
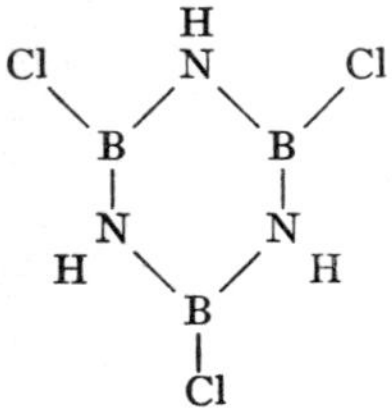	−97.6	−97.3	−115.3	(9)
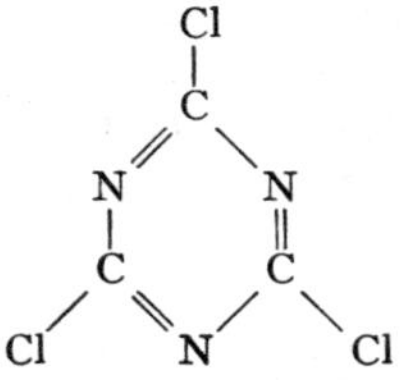	−70.9	−71.2	−101.3	(19)[e]

Anisotropies from magnetic birefringence.—The theory of the Cotton-Mouton effect (1) shows that the birefringence induced by a magnetic field perpendicular to a beam of light depends on $\boldsymbol{\chi}$ and $\boldsymbol{\alpha}$, the electric polarizability tensor. A limited amount of data exists (2), but deduction of accurate magnetic anisotropies is hindered by the unavailability of reliable values of $\boldsymbol{\alpha}$. A recent experiment on ethane (21) gives $\chi_{\|} - \chi_{\perp} = -4.9 \times 10^{-6}$ cm³/mole.

NMR Methods: Mean Shielding

There is a close connection between the magnetic susceptibility tensors $\boldsymbol{\chi}$, and the several nuclear shielding tensors $\boldsymbol{\sigma}_i$ in any molecule, which arises from the similarity of the quantum-mechanical operators used to calculate the expectation values χ_{vv} and σ_{vv} in each case. In the normal high-resolution NMR experiment, only a mean shielding ($\sigma_i = tr\,\boldsymbol{\sigma}/3$) is measured, and even this is not measured absolutely, but relative to some arbitrary reference: $\sigma_{obs} = \sigma_i - \sigma_{stand}$. McConnell (22) treated the problem, dissecting the total shielding into local and long-range contributions; the long-range contribution to the shielding by a given electron group, g, is

$$\sigma_g = (\chi_g \boldsymbol{R}\cdot\boldsymbol{R} - 3\chi_g\cdot \boldsymbol{RR})/NR^5 \qquad 6.$$

where N is Avogadro's number.

This derivation is based on the assumption that the electron group is distant enough, that the moment induced therein by the applied field may be represented with sufficient accuracy by a point dipole, and the secondary field produced at the shielded nucleus may be calculated on that basis.

Extensive efforts have been made to obtain information on χ for various electron groupings using this equation. The procedure is to compare two compounds, which ideally are of known geometry and identical except for the presence of the group in one and its absence or different orientation in the other. With the implicit assumption that changes in the shielding arising from other sources are negligible, or can be allowed for, application of Equation 6 then yields information about χ_g. If the electron grouping is axially symmetric, Equation 6 reduces to the simpler form,

$$\sigma_g = (1 - 3\cos^2\gamma)\Delta\chi_g/3NR^3 \qquad 7.$$

where σ_g is the mean shielding, γ is the angle between the symmetry axis of the electron group and the radius vector R between the group and the shielded nucleus, and

$$\Delta\chi_g = \chi_{\|} - \chi_{\perp} \qquad 8.$$

Two principal difficulties arise in the application of this method: (*a*) the contributions to the shieldings σ_g are small and significant essentially only with protons; (*b*) the dipolar approximation is really good only if the shielded nucleus is removed about six bond lengths from the electron group as has been shown by Didry & Guy (23). Under such conditions, however, the contributions to the shieldings are small, and other effects (electric field, etc.) give rise to shielding contributions of comparable but uncertain size. To illustrate, Wellman & Bordwell (24) have examined the contributions to the shift of the axial and equatorial protons in α-bromocyclohexanone. The reversal from the normal behavior in this case (the axial proton is observed at lower field) can not be explained as arising from magnetic anisotropy in the carbonyl group, and presumably arises from the effect of the internal electrostatic field in the molecule. Slomp & McGarvey (25) have examined a series of 6-methyl steroids (both β- and α-isomers) and found that the protons of the β-methyl groups were less shielded than those of the α-methyls. Explanations based on magnetic anisotropy failed, and Slomp & McGarvey suggest that the steric compression to which the β-methyls are subject results in second-order paramagnetism, which gives a downfield shift. A similar observation on a series of octahydrophenanthrenes is reported by Nagata, Teresawa & Tori (26). Cook & Danyluk (27) observe that acetylenic protons in *ortho*-substituted phenylacetylenes are deshielded relative to *meta* and *para* derivatives. This possibly arises from a paramagnetic contribution to the longitudinal susceptibility of the acetylenic group, brought about by interference of the *ortho*-substituent with the free π-electron precession about this axis.

Studies on steroids, which are available in great number, appear most promising: the geometry of the steroid nucleus is well established, and the effects of substituents have been shown to be nearly additive. The possibility of substitution in several sites which are distinct but bear the same spatial relation to the proton(s) whose chemical shift is being measured is a distinct advantage. Zürcher (28) has obtained and analyzed data on 265 steroids. He gives additive contributions for substituents and discusses the contributions of groups from "equivalent" (from the point of view of shielding) positions. Earlier assemblages of data on these compounds include those of Shoolery & Rogers (29), Cox, Bishop & Richards (30), Slomp & McGarvey (25), and Jacquesy, Lehr & Levisalles (31).

Aromatics.—These are characterized by a large diamagnetic susceptibility in the direction normal to the plane of the aromatic rings. The resultant difference in shielding between aromatic protons and olefinic protons was given a qualitative explanation by Pople (32) in 1956 using a dipolar model. The model was improved somewhat by Waugh & Fessenden (33) and by Bovey & Johnson (34), who substituted for the dipole a pair of current loops situated approximately at the points of maximum electron density in the π-orbitals of benzene. The current flowing in these loops may be calculated on the classical scheme given by Pauling (35), or by quantum mechanical schemes such as that originally presented by London (36), or modifications such as that of McWeeny (37). The latter are usually more successful. Jonathon, Gordon & Dailey (38) have analyzed completely the proton spectra of naphthalene, anthracene, pyrene, perylene, triphenylene, and coronene and compared the results with those expected using Pople's method of estimating ring currents (39). The trends in shifts are reproduced. House, Magin & Thompson (40) have studied the proton spectra of the diphenylnaphthalenes; in the case of 1,8-diphenylnaphthalene, in which the benzene rings are sterically constrained to lie face-to-face, the expected shift (upfield by 0.5 ppm) is observed. Curtin & Dayagi (41) observe the expected upfield shifts for the *ortho* protons in *cis*- vs. *trans*-1,2-diarylcyclopentanes, and similar shifts in *meso* and DL-3,4-diaryladipic esters which are of value in making configurational assignments. Overberger & Anselme (42) have examined the spectra of *cis*- and *trans*-1,2-diphenyl-cyclohexanes; the shifts parallel those in the cyclopentanes. Differential shielding of nonequivalent methylene protons in a variety of systems containing a benzene ring has been studied by Whitesides, Holtz & Roberts (43), who conclude that the shifts can all be explained with reasonable molecular geometries using the Johnson-Bovey calculations for the remote shielding contributions of the benzene ring. Longone & Chow (44) recorded the spectrum of a "multilayered paracyclophane" and find the protons on the inner benzene rings less shielded than those on the outer. Kurland & Wise (45) used the Bovey-Johnson tables to calculate the expected shift of the *ortho* protons in diphenyl as a function of the dihedral angle between the planes of the rings, then studied the observed shift as a function of temperature and accounted for the observed variations as an average over thermally accessible torsional states.

Two facets of the ring current problem are of special interest: (*a*) Most quantum mechanical calculations of the strength of the current relate it to degree of aromatic character. Thus the occurrence of large ring currents is expected in cyclooctadecanonaene (***a***, Fig. 1) and has been deduced from the NMR spectrum by Jackman et al. (46). The presence of a large ring current in compound ***b*** (Fig. 1), synthesized by Boekelheide & Phillips (47), is signaled by the pronounced upfield shift of the angular methyl groups. Similarly, compound ***c***, synthesized by Vogel (48), displays signals for the methylene protons at unusually high field and has a ring current. On the other hand, compound ***d***, synthesized by Staab & Binnig (49), would be predicted

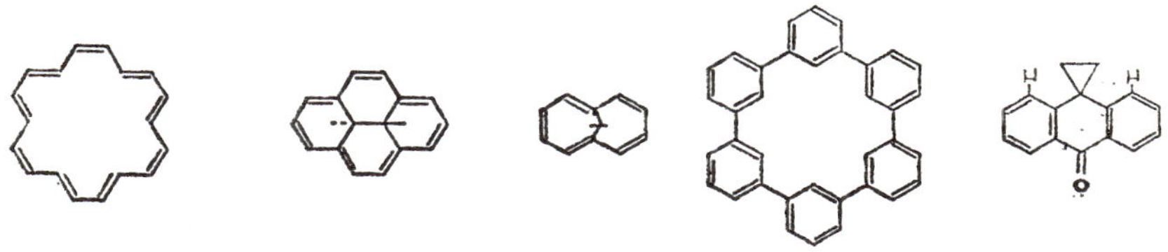

a *b* *c* *d* *e*

FIG. 1. Compounds investigated for ring current.

by the Pauling theory, but not by the quantum mechanical theory (37) to have a large ring current flowing in the macrocyclic ring. No evidence was obtained for such a current (49). (*b*) A second point of considerable interest is whether ring currents can flow in saturated systems, such as the cyclopropanes, -butanes, etc. There is some evidence to support the occurrence of a ring current in cyclopropanes. Thus, they have unusually large diamagnetic susceptibilities, as established by Lacher, Pollock & Park (50), and the NMR signals of protons located in proximity to them are often shifted in the direction predicted on the basis of such a model. Tadanier & Cole (51) have ascribed the lesser shieldings of 6-β(axial) protons as compared with 6-α protons in 3α, 5α-cyclosteroids to differential shielding by the cyclopropane ring. Sauers & Sonnet (52) have examined shifts in the isomeric adducts of carbethoxycarbene and bicycloheptadiene, in norbornyl and nortricyclyl derivatives, and come to a similar conclusion. Forsén & Norin (53) have noted upfield shifts of protons located near the threefold axis of cyclopropane rings in nortricyclyl derivatives and in the compound ***e*** (Fig. 1), and have calculated a value for the anisotropy (which should, however, be divided by 2). The magnitude of such a ring current has been discussed by Lacher, Pollock & Park (50), Patel, Howden & Roberts (54), and Burke & Lauterbur (55). The latter comment that a current of 3.5 electrons precessing with the Larmor frequency in a circular orbit of radius 1.1 Å accounts for the observed C^{13} and H^1 shieldings, as well as the excess isotropic susceptibility ($\sim -6\times10^{-6}$ cm^3/mole).

Acetylenes.—The shielding of protons falls off in the order aliphatic, acetylenic, olefinic; it was early suggested by Pople (56) that the acetylenic protons were unusually shielded by virtue of a contribution from the anisotropic susceptibility of the acetylenic group. A molecular orbital treatment (57) gave the value $\Delta\chi_{C\equiv C} = -19.4$ cm³/mole. There have been two attempts to ascertain the contribution to isotropic shielding by acetylenic groups taking into account other effects. Heel & Zeil (58), and later Zeil & Buchert (59), studied the proton shifts in a variety of ethyl- and iso-propylacetylenes, haloacetylenes, and nitriles. They assumed that the proton shifts were linearly dependent on the electronegativity of the substituent and, in addition, suffered shifts arising from the diamagnetic anisotropy of the substituents, which, however, were constant from one compound to another. By taking appropriate differences to eliminate the effect of electronegativity, they found anisotropies as follow: $-C\equiv C-Cl$, -37×10^{-6} cm³/mole; $-C\equiv C-Br$, -38×10^{-6} cm³/mole; $C\equiv C-I$, -35×10^{-6} cm³/mole; $-C\equiv C-H$, -34×10^{-6} cm³/mole; $-C\equiv N$, -34×10^{-6} cm³/mole.

Reddy & Goldstein (60) have used the observed C^{13}–H coupling constant as a measure of *s*-character, hence hybridization and local shielding in a C–H bond. To the extent that this is allowable, plots of observed chemical shift versus $J_{C^{13}-H}$ will give straight lines for families of compounds with similar structure. The vertical displacement of these lines should give an estimate of the shielding from remote sources, in or outside the molecule. In this way they estimate $\Delta\chi$ for the acetylenic triple bond and for the nitrile group as -16.5×10^{-6} cm³/mole.

Cross & Harrison (61) have studied a series of steroids with and without nitrile substituents in the 5α or 5β positions. In each case they observe a deshielding of the C-10 methyl protons, on substitution with $-C\equiv N$, although the dipolar formula predicts a shielding, whether the substituent is introduced α or β. The effect is presumably due to the electric field of the $-C\equiv N$ group; the shift *between* α and β derivatives is that expected.

Olefins.—Several suggestions have been made as to the probable form of the susceptibility tensor for a carbon-carbon double bond: Jackman (62) has suggested a large diamagnetism in the *x* direction (see Fig. 2 below). Conroy (63) has suggested a large diamagnetism in the *y* direction. Pople (57) suggests a paramagnetism in the *y* direction centered on the carbon

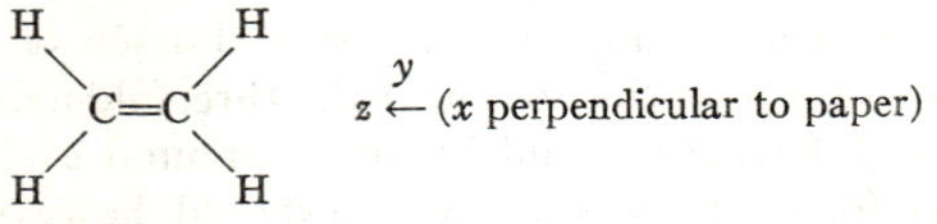

FIG. 2. Principal axes of ethylene.

atoms rather than in the middle of the bond. Most NMR measurements support the suggestions of Jackman and of Pople, but do not distinguish between these (both Jackman's and Pople's suggestions imply that protons near the *y* axis will be deshielded, those near the *x* axis shielded; Jackman's hypothesis implies a deshielding contribution for remote protons near the *z* axis, Pople's a shielding contribution for such protons. The latter predictions have not been tested). Thus, Wiberg & Nist (64) find the β-protons in cyclopentene deshielded with respect to cyclopentane by 0.4 ppm. Ayer, McDonald, & Stothers (65) have studied the spectrum of the adduct of maleic anhydride, maleopimaric acid, and several related compounds. In the adduct a double bond is rigidly held in such a position that an angular methyl group is close to the x axis; it is shielded in accord with the Pople and Jackman hypotheses. A similar conclusion was reached by Meyer & Huffman (66). Yamaguchi, Okuda & Nakagawa (67), in a study of morphine derivatives, find a large diamagnetism in the x direction, and using the dipolar formula, deduce $\Delta\chi = \chi_{xx} - \chi_{zz} = \chi_{xx} - \chi_{zz} = -8.2 \times 10^{-6}$ cm^3/mole. Snyder & Franzus (68), in a study of norbornadienesand norbornenes, note that bridge hydrogens located *syn* to a double bond are more shielded than those located *anti*, in accord with both Jackman and Pople. Tori et al. (69) have also studied norbornadienes and bicyclooctadienes, and have calculated an anisotropy centered on each trigonal carbon atom of $\sim -6 \times 10^{-6}$ cm^3/mole, χ_{yy} and χ_{zz} being assumed equal. Fraser (70) reached qualitatively similar conclusions on considering the shifts of 5-exo and 5-endo protons in bicyclo-(2, 2, 1) heptene-2 derivatives. Anet (71) notes a shift between the nonequivalent methylene protons of "frozen" cycloheptatriene, and proposes that the signal from the pseudoaxial H should occur at higher field than that of the pseudoequatorial H.

Carbonyl groups.—Jackman (62) has suggested a large diamagnetism in the direction normal to the nodal plane of the π-orbitals, as with ethylene. Pople has suggested a paramagnetism centered on the carbon atom, large in the y direction, and the largest diamagnetism on the O atom in the x direction. Again both hypotheses agree in predicting deshielding for remote protons close to the y-axis, shielding for remote protons close to the x-axis; they differ in predictions about the contribution to protons close to the z-axis in the region closer to the carbon. Jackman (62) has given several examples of substituted acrylic esters which are in accord with both hypotheses. Narasimhan & Rogers (72) have attempted to obtain information about $\chi_{C=O}$ from a study of the different chemical shifts in the substituted amides. They comment on the difficulties associated with the use of the dipolar formula, and suggest as possible values $\chi_{zz} \sim 0$, $\chi_{xx} \sim +2.5 \times 10^{-6}$, $\chi_{yy} = -6.6 \times 10^{-6}$ cm^3/mole, with the dipole located on the O atom. These models are qualitatively supported by the work of Williams, Bhacca & Djerassi (73) on some ketosteroids: The equatorial hydrogen in the 1-position of 11-ketosteroids is subject to a pronounced deshielding. In the same way the pseudo-equatorial proton in the 15-position of the steroid nucleus is deshielded by a keto group in the 7-position. Levenberg & Richards (74) have reported the proton shifts in a series of bridged substituted ferrocenes and have noted that in 2-acetyl-1,1′-trimethylene ferrocene, the 2′ proton, which is located essentially on the y-axis of the carbonyl group (see Fig. 3), is shielded by 0.5 ppm with respect to the average shift of the other protons, while the 3 proton is deshielded in harmony with Jackman's findings. Klose (75) has studied the chemical shift difference between the two methylene protons in a series of

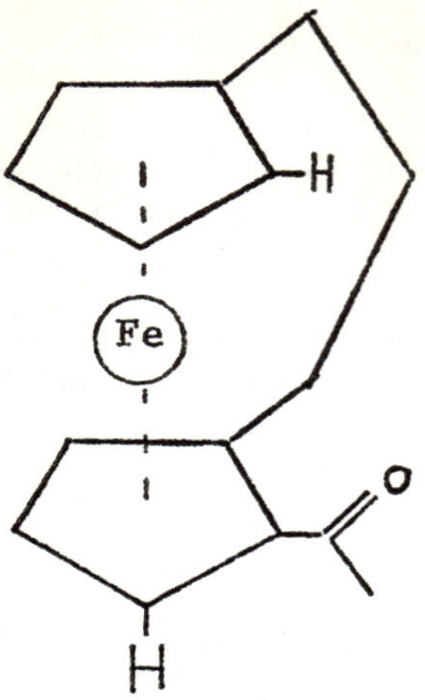

FIG. 3. 2-Acetyl-1,1′-trimethylene ferrocene.

α-methylene cyclanones of ring size 5 to 10 and 16. The shift between the protons may be calculated for any dihedral angle between the C=O and $C=CH_2$ groups assuming the anisotropy suggested by Narasimhan & Rogers (72) for the C=O group and that of Tillieu & Guy (76) for the C=C single bond; the experimentally observed values correspond to that calculated for the *s-cis* conformation for rings of 5 to 8 members, and to that calculated for the *s-trans* conformation for rings of 9 and 10 members. The latter conformations are supported by other evidence.

C–C and C–H single bonds.—The situation with respect to the evaluation of the effects of these groups on chemical shifts is far from satisfactory. It was suggested by Bothner-By & Naar-Colin (77) that the C–C bond might be assigned an anisotropy of about $+3.3\times10^{-6}$ cm³/mole, which would account for the chemical shift differences between cyclopentane, cyclohexane, and open-chain methylene compounds. This value was derived assuming an isotropic susceptibility for C–H bonds, however. The idea was extended and found to offer a qualitative explanation of observed proton shifts in a number of cases. For example, Jackman (62) noted that the shift between equatorial and axial protons in cyclohexane derivatives [0.478 ppm in cyclohexane itself: Bovey et al. (78)] could be accounted for on this basis; Musher (79) used essentially the same approach on data from 21 cyclic alcohols, and deduces the somewhat larger value $\Delta\chi_{C-C}=+5.0\times10^{-6}$ cm³/mole. Moritz & Sheppard (80), using the same approach and analyzing data on cyclohexane (axial *vs.* equatorial), methane, ethane, propane, and other simple open-chain hydrocarbons, axial *vs.* equatorial methyl groups, etc., suggest a value $\chi_{C-C}\sim+7.0\times10^{-6}$ cm³/mole. Hall (81) applies this model to carbohydrates, in order to evaluate both $\Delta\chi_{C-C}$ and $\Delta\chi_{C-O}$, and obtains $\Delta\chi_{C-C}=+3.9\times10^{-6}$ cm³/mole (and essentially the same value for $\Delta\chi_{C-O}$).

Musher (82) points out that if the same model is used in an attempt to rationalize the chemical shifts in ten bicycloheptanols, an absurd value of 15×10^{-6} cm³/mole is required for $\Delta\chi_{C-C}$. In fact, any value for $\Delta\chi_{C-C}$ larger than $+4.5 \times 15^{-6}$ cm³/mole seems absurd since the mean susceptibility is 3.0×10^{-6} cm³/mole, and larger values would require the bond to be paramagnetic in the longitudinal direction.

Some of the difficulty may be avoided by ascribing part of the chemical shift to electronegativity differences and allowing for an anisotropic susceptibility in C–H bonds, as suggested by Narasimhan & Rogers (72). Considering the case of propane, they suggest reasonable values of $\Delta\chi_{C-C}$ to be $+1.5$ to $+3.0 \times 10^{-6}$ cm³/mole and $\Delta\chi_{C-H}$ to be $+0.24 \times 10^{-6}$ to $+1.50 \times 10^{-6}$ cm³/mole [*i.e.*, those calculated by Tillieu (76)]. Zürcher (83) considers shifts of the angular methyls in 5α- and 5β-androstanes and suggests values of $\Delta\chi_{C-H}$ from 0 to $+1.5 \times 10^{-6}$ cm³/mole and $\Delta\chi_{C-C}$ from $+1.95$ to $+3.72 \times 10^{-6}$ cm³/mole. The possible contributions of remote C–H bond magnetic anisotropy to chemical shift has also been noted by Musher (84). Yamaguchi & Brownstein (85) consider the effect of both $\Delta\chi_{C-C}$ and $\Delta\chi_{C-H}$ on several methylcyclohexanes [data of Muller & Tosch (86)] and conclude that no reasonable values can account for the shifts so that other factors must be at work. Cavanaugh & Dailey (87) reached the same conclusion from an examination of data of alkyl derivatives. Davies (88) has discussed the evidence from chemical shift measurements, crystal susceptibilities, the Cotton-Mouton constant, and *ab initio* calculations, and suggests that $\Delta\chi_{C-C} \sim +4.0 \times 10^{-6}$, $\Delta\chi_{C-H} \sim +3.0 \times 10^{-6}$ cm³/mole. Zürcher (89) has also combined information from several sources: mean and crystal susceptibilities, electric polarizabilities as obtained from Raman intensities, and a relation given by Gans & Mrowka (90) relating α and χ. Gans & Mrowka's equation implies that the largest diamagnetic susceptibility is invariably at right angles to the largest electric polarizability: On the other hand, negative Cotton-Mouton constants, which have been experimentally observed, imply that the largest components of polarizability and susceptibility are parallel. There is thus some doubt about the general validity of the Gans-Mrowka equation. Zürcher obtains by this treatment $\Delta\chi_{C-C} = +1.04 \times 10^{-6}$ cm³/mole, $\Delta\chi_{C-H} \sim 0$. Reddy & Goldstein (91) apply their method of comparing shifts with $J_{C^{13}-H}$, and deduce $\Delta\chi_{C-C} \sim +10.0 \times 10^{-6}$.

The rather small shifts observed, the extreme likelihood that other factors are making contributions to these shifts of comparable magnitude, the approximate character of the dipolar equation, the inconsistency of the results obtained, and the absurdity of requiring χ_{ii} for C–C to be paramagnetic, lead one to view the values for $\Delta\chi_{C-C}$ and $\Delta\chi_{C-H}$ obtained by the NMR method with some reservation.

Other groups.—The presence of a 6-β-fluoro substituent in cholestenone derivatives causes a deshielding of the C_{-10} methyl protons by ~0.1 ppm; a 6-β-chloro substituent causes deshielding by ~0.25 ppm, according to Cross, Carpio & Ringold (92). Flautt & Erman (93) have compared proton chemical shifts in bornane, bornyl chloride, and isobornyl chloride, and conclude that they may be rationalized on the basis of a large positive $\Delta\chi$ for the C-Cl bond. Spiesecke & Schneider (94) have discussed the contribution of $\Delta\chi_{C-X}$ to proton shifts in alkyl and aryl halides, as have Schaefer, Reynolds & Yonemoto (95). C^{13} shifts in these compounds cannot be rationalized on this basis, as the contributions are much too small; at the same time, there is a certain parallelism between the proton and C^{13} chemical shifts. Schaefer & Yonemoto (96) consider $\Delta\chi$ for C-Cl and C-O bonds ($\sim+3\times10^{-6}$ cm^3/mole), the effect of electric field on shielding, and charge densities calculated by LCAO–MO with the empirical relation $\Delta\sigma=10^{-5}q$, where q is the charge in electrons/atom, and obtain a fit with chemical shift data on substituted ethylenes. Fort, Cheeseman, & Taylor (97) note a deshielding of a proton close in space to a chlorine substituent in 5 heterocyclic compounds, and suggest that $\Delta\chi_{C-Cl}$ and electron withdrawal may be contributing factors. In all of these cases the effect of electric field and steric compression are probably pronounced and obscure the effect of $\Delta\chi_{C-X}$ on the shift.

Effects of sulfur have been reported by Campaigne, Chamberlin, & Edwards (98), and by Tomoeda et al. (99); the former suggest that $\Delta\chi_{C-S}<0$. Arbouzov & Samitov (100) have investigated cyclic sulfites and suggest $\Delta\chi_{S\to O}=+13.5\times10^{-6}$ cm^3/mole. Pritchard & Lauterbur (101) had suggested $\Delta\chi_{S\to O}=+6.0\times10^{-6}$ cm^3/mole, correcting for the effect of the electric field.

Yamaguchi (102) has made an estimate of the three principal susceptibilities of the nitro group in *ortho*-substituted aromatics. The rotation of the nitro group about the C-N bond was estimated from ultraviolet absorption and related, using the dipolar formula to the shieldings of protons in *ortho* methyls. A correction for the effect of the electric field was made. Yamaguchi suggests $\chi_L\sim-11\times10^{-6}$, $\chi_T=-2\times10^{-6}$, and $\chi_\perp=-14\times10^{-6}$ cm^3/mole.

Anisotropic Shielding

In principle, there are two ways in which the measurement of the anisotropy of the shielding tensor may aid in evaluating $\boldsymbol{\chi}$, the magnetic susceptibility tensor for electron groups: (*a*) If the group is located remote from the nucleus in question, McConnell's dipolar approximation (Eq. 6) may be applied with less ambiguity than when the mean shielding is determined. (*b*) If local paramagnetic terms dominate the shielding, the paramagnetic terms may be related to the spin-rotation constant (103) and the rotational magnetic moment. The form of the rotational magnetic moment tensor for polyatomic molecules has been investigated by Eshbach & Strandberg (104).

Each of these approaches suffers from severe difficulties. In the first case, since the contribution to shielding from remote groups is small, the effect is significant and interpretable only for protons, where shieldings from other sources do not swamp out the contribution of interest. Unfortunately, it is just in this case that the measurement of shift anisotropies in crystals is most difficult, because dipolar broadening leads to line widths much larger than the chemical shifts and obscuring them.

The problem of determining shift anisotropies has been attacked in a most elegant and interesting manner by Englert & Saupe (105, 106), who have studied the proton NMR spectra of aromatic compounds dissolved in nematic melts (such as 4,4′-dialkoxyazoxybenzenes). The solutes are partially oriented in such phases, but retain sufficient freedom of motion so that intermolecular dipole-dipole broadening is reduced to a relatively small contribution. The degree of orientation may be deduced from the intramolecular dipole-dipole interaction, which now appears as fine structure. The shift between the mean shielding for benzene observed in isotropic solution and the weighted mean observed in the nematic melt is given by

$$\delta\sigma = S_{zz}(\sigma_{zz} - (\sigma_{xx} + \sigma_{yy})/2) \qquad 9.$$

where S_{zz} is the orientation factor

$$S_{zz} = \tfrac{1}{2}(3\langle\cos^2\theta_{zz}\rangle - 1) \qquad 10.$$

In this equation $\langle\cos^2\theta_{zz}\rangle$ is the mean square of the cosine of angle between the 6-fold axis and the applied magnetic field. From their experiments, Englert & Saupe deduce

$$\Delta\sigma = \sigma_{zz} - \sigma_{xx} = -(2.88 \pm 0.13) \times 10^{-6} \qquad 11.$$

for benzene, and $\Delta\sigma = -(4.91 \pm 0.25) \times 10^{-6}$ for 1,3,5-trichlorobenzene. The former figure agrees qualitatively with the estimates of Pople (32) and of Fessenden & Waugh (33); the latter appears large and may be related to the presence of the Cl substituents.

In the case of studies on shift tensors in F^{19}, C^{13}, etc., resonances, the difficulty lies in the theoretical linkage of the measured quantities with χ. Shifts have been observed both with partially oriented and completely crystalline samples. Thus Hunt & Meyer (107) have studied the fluorine shift tensor in methyl fluoride molecules trapped in a hydroquinone clathrate at 1.3°K. Orientation of the molecules was confirmed by observation of anisotropic dipolar splitting caused by the protons. The anisotropy in the shift was measured directly and was found to be $\Delta\sigma = \sigma_{\parallel} - \sigma_{\perp} = -66 \pm 8$ ppm. This interesting result is contrary to expectation, as there should be a paramagnetic contribution to the shielding perpendicular to the C-F bond. Wilson (108) has studied the anisotropy of the chemical shift of fluorine in stretched Teflon, and found $\Delta\sigma = |\sigma_{\parallel} - \sigma_{\perp}| = 216 \pm 6 \times 10^{-6}$.

Study of polycrystalline samples also yields information on the shift tensor in favorable cases. The observed signal may be asymmetric and the sense of the asymmetry is interpretable in terms of the sign of asymmetry of an axially symmetric shift tensor (109), or the line width may be studied as a function of applied field, and the anisotropy deduced from the magnitude of the field-dependent part of the width. Both methods have been applied by Blinc et al. (110) in a study of xenon tetrafluoride. The average value obtained for $\Delta\sigma = \sigma_{\parallel} - \sigma_{\perp} = 680 \times 10^{-6}$. Andrew & Tunstall (111) have analyzed the line shape from polycrystalline samples at 85°K to obtain $\Delta\sigma$ for fluorine resonances in $CFCl_2CFCl_2$ ($\Delta\sigma = (240 \pm 40) \times 10^{-6}$); CF_2BrCF_2Br ($\Delta\sigma = (260 \pm 40) \times 10^{-6}$); and 1,2,4-trifluorobenzene ($\Delta\sigma = (250 \pm 40) \times 10^{-6}$).

Lauterbur (112) has reported on the shift anisotropy of the C^{13} resonance in single crystals of calcite, where the magnetic nuclei are sufficiently dilute that dipolar broadening does not interfere with the direct measurement of the shift tensor. In this case, $\Delta\sigma = 75 \times 10^{-6}$, a figure which has been rationalized by Pople (113).

Ragle (114) reports $\Delta\sigma = -550 \times 10^{-6}$ for the V_2O_5 resonance.

Predictive Theories

Many authors have attempted to calculate diamagnetic properties of molecules, including anisotropic susceptibilities and chemical shifts, using the methods of quantum mechanics. The general theory of diamagnetic susceptibilities, treating the external magnetic field as a perturbation, was first developed by Van Vleck (115). He derived the formula (for the molar susceptibility component χ_{zz})

$$\chi_{zz} = -\frac{Ne^2}{4mc^2}\sum_i \overline{x_i^2 + y_i^2} + \frac{Ne^2}{2m^2c^2}\sum_{n(\neq 0)} \frac{|\langle 0|M_z|0\rangle|^2}{E_n - E_o} \qquad 12.$$

Here the first term (usually called the diamagnetic part) is proportional to the mean square distance $\overline{x^2+y^2}$ of electrons from the z-axis and corresponds to the uniform circulation of electron currents in spherical atoms or ions as dealt with by the classical theory of Langevin. The second part of Equation 12 (the temperature-independent paramagnetic term) is positive and gives the reduction of diamagnetism consequent on the mixing of the ground and excited states n by the magnetic field. M_z is the z-component of the total electronic angular momentum. A general formula for shielding constants comparable to Equation 12 was obtained by Ramsey (116).

The Van Vleck and Ramsey formulas have been comparatively little used in detailed calculations on susceptibilities, primarily because of the difficulty of including contributions of all excited states in the second part of Equation 12, particularly those in the continuum. An alternative method that avoids these difficulties is a variational technique of the type first proposed by Tillieu & Guy (76, 117). This uses a function

$$\psi = \psi_o(1 + \boldsymbol{H}\cdot\boldsymbol{g}) \qquad 13.$$

where ψ_o is the unperturbed wave function, and $\boldsymbol{g}$ is a vector function of electronic coordinates to be determined by variation methods. In practice, a function $\boldsymbol{g}$ is chosen containing adjustable parameters which can be optimized to give minimum energies. A simple form for $\boldsymbol{g}$ is a quadratic function of the electron coordinates, the origin being chosen at the center of charge of the electron distribution. When applied to a two-electron system, this variational technique leads (118) to the formula

$$\chi_{zz} = -\frac{Ne^2}{2mc^2}\left[\overline{x^2 + y^2} + \frac{\overline{(x^2 - y^2)}^2}{\overline{x^2 + y^2}}\right] \qquad 14.$$

the two parts corresponding to the diamagnetic and paramagnetic parts of the Van Vleck question. This result was first derived by Gans & Mrowka

(90) who further expressed the mean values $\overline{x^2}$, $\overline{y^2}$ and $\overline{z^2}$ in terms of electrical polarizabilities α using another approximate theory. Their result is

$$\chi_{zz} = -\frac{Ne^2}{8mc^2}(na_o)\left[\alpha_x^{1/2} + \alpha_y^{1/2} - \frac{(\alpha_x^{1/2} - \alpha_y^{1/2})^2}{\alpha_x^{1/2} + \alpha_y^{1/2}}\right] \qquad 15.$$

where a_0 is the Bohr radius, n the number of electrons and α_x, α_y, α_z components of the electrical polarizability.

Variational functions of the type shown in Equation 13 have the drawback that ψ is restricted to have nodal surfaces in the same positions as ψ_o, thereby limiting the flexibility of the variation (119). A more general approach to the variation method has been given by Karplus & Kolker (120); they calculate the paramagnetic parts of formulas such as Equation 12 using a single determinant of molecular orbitals as the unperturbed wave function. Other approaches to the susceptibility problem are those of Stevens, Pitzer & Lipscomb (121), who consider perturbation of the Hartree-Fock equations in the presence of a magnetic field, and that of Ishiguro & Koide (122) later refined by Hoyland (123).

Both the direct summation of terms in Equation 12 involving excited states and the variational techniques are attempts to evaluate the two parts of the susceptibility separately. For large molecules, these two parts become big and largely cancel (the signs being opposite) so that calculations become subject to considerable error. A method of avoiding difficulties of this sort is to use so-called "gauge-invariant atomic orbitals," where atomic orbitals are modified by multiplication by a complex factor depending on the gauge of the magnetic vector potential. This eliminates difficulties of the type encountered in direct calculations. An approach of this kind to problems of molecular diamagnetism was first made by London (36) in connection with ring currents in aromatic hydrocarbons, and it has since been applied to more general studies of magnetic properties by Hameka (124–130), Pople (56, 57, 113, 131), and others (132).

Some calculated values of suceptibilities and anisotropies for diatomic molecules are compared with experimental data and estimates in Table II. [A fuller discussion of calculations on H_2 is given by Das & Bersohn (133).] Both the variational and the gauge-invariant atomic orbital methods are reasonably satisfactory for the mean susceptibilities, but there are major variations in their predictions for anisotropies. The Hameka method is notably better in predicting the high negative anisotropy of the triple bond apparent in the nitrogen molecule. Unfortunately, there is no direct experimental data (except for H_2) on the diamagnetic anisotropies of these molecules. The estimated values quoted in the table are obtained (120) using theoretical values of χ_d and values of χ_p deduced from rotational magnetic moments.

For larger molecules, the total susceptibility has to be broken down either into bond or atomic contributions. Two methods for a priori calculations of these contributions have been developed. Tillieu & Guy (76, 134, 135), and later Baudet (136) have applied the variational technique (Eq. 13

TABLE II

SUSCEPTIBILITY CALCULATIONS ON DIATOMIC MOLECULES (UNITS 10^{-6} CM^3 MOLE^{-1})

Molecule	χ_{Calc}	χ_{exp}[a]	$(\chi\parallel - \chi\perp)_{calc}$	$(\chi\parallel - \chi\perp)_{est}$[a]	Reference and method
H_2	−4.02 −3.85 −3.99	−3.94	0.50 0.57 0.56	0.575	(120)—variational (118)—variational (126)—gauge invariant
Li_2	−32.97		3.25		(120)—variational
N_2	−13.8 −17.3 −12.81	−13.3±3	4.27 1.57 −5.09	−8.65	(150)—variational (120)—variational (130)—gauge-invariant
F_2	−18.6		7.44		(120)—variational
LiH	−9.13 and 9.15[b] −9.39		−0.45 −0.07	−2.14	(120)—variational (129)—gauge-invariant
HF	−8.45 and 8.65† −8.49	−8.6	+0.64 −0.13	+2.00	(120)—variational (128)—gauge-invariant
LiF	−13.69 and −15.34[b]		−4.90	−2.69	(120)—variational
CO	−15.65 and −18.11[b]		−0.74	−5.23	(120)—variational

[a] See Ref. 120. $(\chi\parallel - \chi\perp)_{est}$ are estimated values using part experimental and part theoretical data.
[b] Two values depending on choice of origin.

and 14) to pairs of electrons in localized bonds (separate calculations being made for the σ and π parts of multiple bonds and lone pairs). The other method is an approximate molecular orbital treatment based on gauge-invariant orbitals developed by Pople (57). This leads to atomic contributions, the anisotropies of which can be associated with the bonds formed by each atom. Table III summarizes the results obtained by both methods for hydrocarbon bonds, giving susceptibilities and anisotropies for the variation method and C–C anisotropies for the gauge-invariant method.

The figures quoted in Table III indicate that the theories make different predictions about anisotropies. This is particularly true for C=C where the gauge-invariant method predicts low diamagnetism in the in-plane direction perpendicular to the bond. The Tillieu-Guy treatment predicts low diamagnetism along the bond. There is some experimental evidence on this

TABLE III

SUSCEPTIBILITY AND ANISOTROPY CALCULATIONS FOR BONDS IN HYDROCARBONS (UNITS 10^{-6} EMU MOLE^{-1})

Bond	χ	$\chi_{xx}-\chi$	$\chi_{yy}-\chi$	$\chi_{zz}-\chi$	$\chi_{zz}-\frac{1}{2}(\chi_{xx}+\chi_{yy})$	Reference
$C_{IV}—C_{IV}$	−3.10	−0.40	−0.40	+0.81	+1.21	(117, 134, 135)
	—	0	0	0	0	(57)
$C_{III}=C_{III}$[a]	−5.68	−1.20	−0.35	+1.55	+2.32	(117, 134, 135)
	—	−2.9	+5.7	−2.9	−4.3	(57)
$C_{II}\equiv C_{II}$	−12.26	+0.54	+0.54	−1.07	−1.61	(117, 134, 135)
	—	+6.4	+6.4	−13.0	−19.4	(57)
$C_{IV}—H$	−4.02	−0.30	−0.30	+0.60	+0.90	(117, 134, 135)
$C_{III}—H$	−3.74	−0.16	−0.16	+0.33	+0.49	(117, 134, 135)
C_{II}–H	−3.33	+0.19	+0.19	−0.37	−0.52	(117, 134, 135)

[a] Axes in $C_{III}=C_{III}$ as in Figure 2; otherwise *z*-axis along the bond.

point from measured in-plane anisotropies (Table I) of naphthalene and anthracene (which may be regarded as made up of a number of partial double-bonds); the higher diamagnetism in the longitudinal (*L*) direction compared with the transverse (*M*) direction of these molecules corresponds to the Pople model (4). For the C≡C bond, both theories predict high diamagnetism along the axis, but the variational method gives a small numerical value. A relatively large negative value of $\Delta\chi$ is supported by the in-plane anisotropy of diphenylactylene (Table I). For single C–C and C–H bonds, the gauge-invariant theory gives zero anisotropy and only small values are found in the Tillieu-Guy results. There seems to be little theoretical justification for some of the large empirical values of $\Delta\chi_{CC}$ and $\Delta\chi_{CH}$ mentioned in the previous section.

In addition to diamagnetic anisotropies associated with localized bonds, aromatic molecules have large anisotropies due to "ring currents" in which mobile π-electrons move in interatomic cyclic paths if the magnetic field is perpendicular to the molecular plane. A quantitative semiclassical theory of this effect was given by Pauling (35), and this was later substantiated by the quantum-mechanical treatment of London (36, 137). The London method is a molecular orbital technique based on gauge-invariant atomic orbitals and is closely related to the well known Hückel theory of π-electrons. It has been applied with some success to a wide variety of hydrocarbons, notably by McWeeny (37, 138). These calculations give an anisotropy proportional to the molecular orbital resonance integral β, and when applied to a series of aromatic hydrocarbons, they give good agreement with experimental values of the ratio $\Delta\chi/\Delta\chi$ (benzene). Other extensions of the original London treatment include calculations on porphyrins and heterocyclic systems by Berthier et al. (139) and others (140, 141). Also Hall and colleagues (142–144) have extended the theory to the self-consistent approximation, in which partial account is taken of electron interaction. This is found to have relatively little effect on the results.

Although the London theory works well on relative values of diamagnetic anisotropies, direct calculations of $\Delta\chi$ (benzene) give too small a value, even if more refined methods of calculation are used (145, 146). Parr & Lykos (147) suggested that certain integrals in London's calculation were underestimated and that an "enhancement factor" had to be applied. However, recent developments (148, 149) indicate that only part of the benzene anisotropy is due to ring currents, about 30 per cent being due to local anisotropies comparable to those observed in other planar conjugated noncyclic systems (113).

LITERATURE CITED

1. Beams, J. W., *Rev. Mod. Phys.*, **4,** 131 (1932)
2. Chinchalkar, S., *Indian J. Phys.*, **6,** 1 (1931)
3. Lonsdale, K., and Krishnan, K. S., *Proc. Roy. Soc.* (*London*), *Ser. A*, **156,** 597 (1963)
4. Ferguson, A. F., and Pople, J. A., *J. Chem. Phys.* (To be published)
5. Hoarau, J., Lumbroso, N., and Pacault, A., *Compt. Rend.*, **242,** 1702 (1956)
6. Ganguli, N., *Z. Krist.*, **A93,** 42 (1936)
7. Lasheen, M. A., *Phil. Trans. Roy. Soc. London, Ser. A*, **256,** 357 (1964)
8. Krishnan, K. S., and Banerjee, S., *Phil. Trans. Roy. Soc. London, Ser. A*, **234,** 265 (1935)
9. Lonsdale, K., and Toor, E. W., *Acta Cryst.*, **12,** 1048 (1959)
10. Lonsdale, K., *Nature*, **137,** 826 (1936)
11. Banerjee, S., *Z. Krist.*, **A100,** 316 (1939)
12. Lonsdale, K., *Proc. Roy. Soc.* (*London*), *Ser. A*, **171,** 541 (1939)
13. Lonsdale, K., *J. Chem. Soc.*, 364 (1938)
14. Rogers, M. T., *J. Am. Chem. Soc.*, **69,** 1506 (1947)
15. Herbstein, F. H., and Schmidt, G. M. J., *Acta Cryst.*, **8,** 405 (1955)
16. Mason, R., *Acta Cryst.*, **14,** 720 (1961)
17. Lonsdale, K., *Proc. Roy. Soc.* (*London*), *Ser. A*, **177,** 272 (1941)
18. Lonsdale, K., Robertson, J. C., and Woodward, I., *Proc. Roy. Soc.* (*London*), *Ser. A*, **178,** 43 (1941)
19. Lonsdale, K., *Z. Krist.*, **A95,** 471 (1936)
20. Mathis, R., Sweeny, M., and Fox, M. E., *J. Chem. Phys.*, **41,** 3653 (1964)
21. Buckingham, A. D., Prichard, W. H., and Whiffen, D. H., *Chemical Communications*, 51 (1965)
22. McConnell, H., *J. Chem. Phys.*, **27,** 226 (1957)
23. Didry, J. R., and Guy, J., *Compt. Rend.*, **253,** 422 (1961)
24. Wellman, J. M., and Bordwell, F. G., *Tetrahedron Letters*, 1703 (1963)
25. Slomp, G., and McGarvey, B. R., *J. Am. Chem. Soc.*, **81,** 2200 (1959)
26. Nagata, W., Terasawa, T., and Tori, K., *J. Am. Chem. Soc.*, **86,** 3746 (1964)
27. Cook, C. D., and Danyluk, S. S., *Tetrahedron*, **19,** 177 (1963)
28. Zürcher, R. F., *Helv. Chim. Acta*, **44,** 1380 (1961); **46,** 2055 (1963)
29. Shoolery, J. N., and Rogers, M. T., *J. Am. Chem. Soc.*, **80,** 5121 (1958)
30. Cox, J. S. G., Bishop, E. O., and Richards, R. E., *J. Chem. Soc.*, 5118 (1960)
31. Jacquesy, J. C., Lehn, J. M., and Levisalles, J., *Bull. Soc. Chim. France*, 2444 (1961)
32. Pople, J. A., *J. Chem. Phys.*, **24,** 1111 (1956)
33. Waugh, J. S., and Fessenden, R. W., *J. Am. Chem. Soc.*, **79,** 846 (1957)
34. Johnson, C. E., and Bovey, F. A., *J. Chem. Phys.*, **29,** 1012 (1958)
35. Pauling, L., *J. Chem. Phys.*, **4,** 673 (1936)
36. London, F., *J. Chem. Phys.*, **5,** 837 (1937)
37. McWeeny, R., *Proc. Phys. Soc.* (*London*), **64A,** 261, 921 (1951); **65A,** 839 (1952); **66A,** 714 (1953)
38. Jonathon, N., Gordon, S., and Dailey, B. P., *J. Chem. Phys.*, **36,** 2443 (1962)
39. Pople, J. A., *Mol. Phys.*, **1,** 175 (1958)
40. House, H. O., Magin, R. W., and Thompson, H. W., *J. Org. Chem.*, **28,** 2403 (1963)
41. Curtin, D. Y., and Dayagi, S., *Can. J. Chem.*, **42,** 867 (1963)
42. Overberger, G. G., and Anselme, J. P., *Chem. Ind.* (*London*), 280 (1964)

43. Whitesides, G. M., Holtz, D., and Roberts, J. D., *J. Am. Chem. Soc.*, **86**, 2628 (1964)
44. Longone, D. T., and Chow, H. S., *J. Am. Chem. Soc.*, **86**, 3898 (1964)
45. Kurland, R. J., and Wise, W. B., *J. Am. Chem. Soc.*, **86**, 1877 (1964)
46. Jackman, L. M., Sondheimer, F., Amiel, Y., Ben-Efraim, D. A., Gaoni, Y., Wolovsky, R., and Bothner-By, A. A., *J. Am. Chem. Soc.*, **84**, 4307 (1962)
47. Boekelheide, V., and Phillips, J. B., *Proc. Natl. Acad. Sci.*, **51**, 550 (1964)
48. Vogel, E., and Roth, H. D., *Angew. Chem.*, **76**, 145 (1964)
49. Staab, H. A., and Binnig, F., *Tetrahedron Letters*, 319 (1964)
50. Lacher, J. R., Pollock, J. W., and Park, J. D., *J. Chem. Phys.*, **20**, 1047 (1952)
51. Tadanier, J., and Cole, W., *J. Org. Chem.*, **27**, 4610 (1962)
52. Sauers, R. R., and Sonnet, P. E., *Chem. Ind. (London)*, 786 (1963)
53. Forsén, S., and Norin, T., *Tetrahedron Letters*, 2845 (1964); *Acta Chem. Scand.*, **15**, 592 (1961)
54. Patel, D. J., Howden, M. E. H., and Roberts, J. D., *J. Am. Chem. Soc.*, **85**, 3218 (1963)
55. Burke, J. J., and Lauterbur, P. C., *J. Am. Chem. Soc.*, **86**, 1870 (1964)
56. Pople, J. A., *Proc. Roy. Soc. (London), Ser. A.*, **239**, 541, 550 (1957)
57. Pople, J. A., *J. Chem. Phys.*, **37**, 53, 60 (1962)
58. Heel, H., and Zeil, W., *Z. Elektrochem.*, **64**, 962 (1960)
59. Zeil, W., and Buchert, H., *Z. Physik. Chem. (Frankfurt)*, **38**, 47 (1963)
60. Reddy, G. S., and Goldstein, J. H., *J. Chem. Phys.*, **36**, 2644 (1962); **39**, 3509 (1963)
61. Cross, A. D., and Harrison, I. T., *J. Am. Chem. Soc.*, **85**, 3223 (1963)
62. Jackman, L. M., *Nuclear Magnetic Resonance Spectroscopy*, 112–30 (Pergamon, London, 134 pp., 1959)
63. Conroy, H., in *Advances in Organic Chemistry: Methods and Results*, 265–328 (Raphael, R. A., Taylor, E. C., and Wynberg, H., Eds., Interscience, New York, 504 pp., 1960)
64. Wiberg, K. B., and Nist, B. J., *J. Am. Chem. Soc.*, **83**, 1226 (1961)
65. Ayer, W. A., McDonald, C. E., and Stothers, J. B., *Can. J. Chem.*, **41**, 1113 (1963)
66. Meyer, W. L., and Huffman, R. W., *Tetrahedron Letters*, 691 (1962)
67. Yamaguchi, S., Okuda, S., and Nakagawa, N., *Chem. Pharm. Bull.*, **11**, 1465 (1963)
68. Snyder, E. I., and Franzus, B., *J. Am. Chem. Soc.*, **86**, 1166 (1964)
69. Tori, K., Hata, Y., Muneyuke, R., Takano, Y., Tsuji, T., and Tanida, H., *Can. J. Chem.*, **42**, 926 (1964)
70. Fraser, R. R., *Can. J. Chem.*, **40**, 78 (1962)
71. Anet, F. A. L., *J. Am. Chem. Soc.*, **86**, 458 (1964)
72. Narasimhan, P. T., and Rogers, M. T., *J. Chem. Phys.*, **31**, 1302 (1959)
73. Williams, D. H., Bhacca, N. S., and Djerassi, C., *J. Am. Chem. Soc.*, **85**, 2811 (1963)
74. Levenberg, M. I., and Richards, J. H., *J. Am. Chem. Soc.*, **86**, 2634 (1964)
75. Klose, G., *Mol. Phys.*, **6**, 585 (1963)
76. Guy, J., and Tillieu, J., *J. Chem. Phys.*, **24**, 1117 (1956)
77. Bothner-By, A. A., and Naar-Colin, C., *Ann. N. Y. Acad. Sci.*, **70**, 833 (1958)
78. Bovey, F. A., Hood, F. P., Anderson, E. W., and Kornegay, R. L.,*Proc. Chem. Soc.*, 146 (1964)
79. Musher, J. I., *J. Chem. Phys.*, **35**, 1159 (1961)
80. Moritz, A. G., and Sheppard, N., *Mol. Phys.*, **5**, 361 (1962)
81. Hall, L. D., *Tetrahedron Letters*, 1457 (1964)
82. Musher, J. I., *Mol. Phys.*, **6**, 93 (1963)
83. Zürcher, R. F., *Helv. Chim. Acta*, **44**, 1755 (1961)
84. Musher, J. I., *Discussions Faraday Soc.*, **34**, 65 (1962)
85. Yamaguchi, I., and Brownstein, S., *J. Phys. Chem.*, **68**, 1572 (1964)
86. Muller, N., and Tosch, W. C., *J. Chem. Phys.*, **37**, 1167 (1962)
87. Cavanaugh, J. R., and Dailey, B. P., *J. Chem. Phys.*, **34**, 1099 (1961)
88. Davies, D. W., *Mol. Phys.*, **6**, 489 (1963)
89. Zürcher, R. F., *Discussions Faraday Soc.*, **34**, 66 (1962); *J. Chem. Phys.*, **37**, 2421 (1962)
90. Gans, R., and Mrowka, B., *Königsberger Gelehrte Gesellschaft*, **12**, **1**, 30 (1935)
91. Reddy, G. S., and Goldstein, J. H., *J. Chem. Phys.*, **38**, 2736 (1963)
92. Cross, A. D., Carpio, H., and Ringold, H. J., *J. Med. Chem.*, **6**, 198 (1963)
93. Flautt, T. J., and Erman, W. F., *J. Am. Chem. Soc.*, **85**, 3213 (1963)
94. Spieseck, H., and Schneider, W. G., *J. Chem. Phys.*, **35**, 722, 731 (1961)
95. Schaefer, T., Reynolds, W. F., and

Yonemoto, Y., *Can. J. Chem.*, **41,** 2969 (1963)

96. Schaefer, T., and Yonemoto, T., *Can. J. Chem.*, **42,** 2318 (1964)
97. Fort, R. C., Jr., Cheeseman, G. W. H., and Taylor, E. C., *J. Org. Chem.* **29,** 2440 (1964)
98. Campaigne, E., Chamberlin, N. F., and Edwards, B. E., *J. Org. Chem.*, **27,** 135 (1962)
99. Tomoeda, M., Inuzuka, M., Furuta, T., and Takehashi, T., *Tetrahedron Letters*, 1233 (1964)
100. Arbouzov, B. A., and Samitov, Yu., *Tetrahedron Letters*, 473 (1963)
101. Pritchard, J. G., and Lauterbur, P. C., *J. Am. Chem. Soc.*, **83,** 2105 (1961)
102. Yamaguchi, I., *Mol. Phys.*, **6,** 106 (1963)
103. Chan, S. I., Baker, M. R., and Ramsey, N. F., *Phys. Rev.*, **136A,** 1224 (1964)
104. Eshbach, J. R., and Strandberg, M. W. P., *Phys. Rev.*, **85,** 24 (1952)
105. Saupe, A., and Englert, G., *Phys., Rev. Letters*, **11,** 462 (1963); *Z. Naturforsch.*, **19a,** 172 (1964)
106. Saupe, A., *Z. Naturforsch.*, **19a,** 161 (1964)
107. Hunt, E., and Meyer, H., *J. Chem. Phys.*, **41,** 353 (1964)
108. Wilson, C. W., III, *J. Polymer Sci.*, **61,** 403 (1962)
109. Jones, W. H., Jr., Graham, T. P., and Barnes, R. G., *Phys. Rev.*, **132,** 1898 (1963)
110. Blinc, R., Zuparicic, I., Maricic, S., and Veksli, Z., *J. Chem. Phys.*, **39,** 2109 (1963)
111. Andrew, E. R., and Tunstall, D. P., *Proc. Phys. Soc. (London)*, **81,** 986 (1963)
112. Lauterbur, P. C., *Phys. Rev. Letters*, **1,** 343 (1958)
113. Pople, J. A., *J. Chem. Phys.*, **38,** 1276 (1963)
114. Ragle, J. L., *J. Chem. Phys.*, **35,** 753 (1961)
115. Van Vleck, J. H., *Electric and Magnetic Susceptibilities* (Oxford University Press, New York, 384 pp., 1932)
116. Ramsey, N. F., *Phys. Rev.*, **79,** 699 (1950)
117. Tillieu, J., and Guy, J., *Compt. Rend.*, **239,** 1203 (1934)
118. Tillieu, J., *Ann. Phys. (Paris)*, **2,** 471, 631 (1957)
119. McLachlan, A. D., and Baker, M. R., *Mol. Phys.*, **4,** 255 (1961)
120. Karplus, M., and Kolker, H. J., *J. Chem. Phys.*, **38,** 1263 (1963)
121. Stevens, R. M., Pitzer, R. M., and Lipscomb, W. N., *J. Chem. Phys.*, **38,** 550 (1963)
122. Ishiguro, E., and Koide, S., *Phys. Rev.*, **94,** 350 (1954)
123. Hoyland, J. R., *J. Chem. Phys.*, **41,** 3135 (1964)
124. Hameka, H. F., *Z. Naturforsch.*, **14a,** 599 (1959)
125. Hameka, H. F., *J. Chem. Phys.*, **34,** 366 (1961)
126. Hameka, H. F., *Rev. Mod. Phys.*, **34,** 87 (1962)
127. Hameka, H. F., *Physica*, **28,** 908 (1962)
128. Hamano, H., Kim, H., and Hameka, H. F., *Physica*, **29,** 111 (1963)
129. Kim, H., Hamano, H., and Hameka, H. F., *Physica*, **29,** 117 (1963)
130. de la Vega, J. R., and Hameka, H. F., *J. Chem. Phys.*, **40,** 1929 (1964)
131. Pople, J. A., *Discussions Faraday Soc.*, **34,** 7, 68 (1962)
132. Raychaudhuri, A. K., and Sinha, S. K., *Mol. Phys.*, **7,** 473 (1964); Karplus, M., and Kolker, H. J., *J. Chem. Phys.*, **35,** 2235 (1961)
133. Das, T. P., and Bersohn, R., *Phys. Rev.*, **115,** 897 (1959)
134. Tillieu, J., and Guy, J., *Compt. Rend.*, **241,** 382 (1955)
135. Tillieu, J., and Guy, J., *Compt. Rend.*, **242,** 1279, 1436 (1956)
136. Baudet, J., *J. Chim. Phys.*, **58,** 228 (1961)
137. London, F., *J. Phys. Rad.*, **8,** 397 (1937)
138. Maddox, I. J., and McWeeny, R., *J. Chem. Phys.*, **36,** 2353 (1962)
139. Berthier, G., Mayot, M., Pullman, B., and Pullman, A., *J. Phys. Rad.*, **13,** 15 (1952)
140. Abraham, R. J., *Mol. Phys.*, **4,** 145 (1961)
141. Craig, D. P., Heffernan, M. L., Mason, R., and Paddock, N. L., *J. Chem. Soc. (London)*, 1376 (1961)
142. Hall, G. G., and Hardisson, A., *Proc. Roy. Soc. (London), Ser. A*, **268,** 328 (1962)
143. Hall, G. G., Hardisson, A., and Jackman, L. M., *Discussions Faraday Soc.*, **34,** 15 (1962)
144. Hall, G. G., Hardisson, A., and Jackman, L. M., *Tetrahedron*, **19,** 101 (1963)
145. Itoh, T., Ohno, K., and Yoshizumi, H., *J. Phys. Soc. (Japan)*, **10,** 103 (1955)
166. Caralp, L., and Hoarau, J., *J. Chim. Phys.*, **60,** 889 (1963)
147. Parr, R. G., and Lykos, P. G., *J. Chem. Phys.*, **28,** 361 (1958))
148. Dailey, B. P., *J. Chem. Phys.*, **41,** 2304 (1964)
149. Pople, J. A., *J. Chem. Phys.*, **41,** 2559 (1964)
150. Baudet, J., Tillieu, and Guy, J., *Compt. Rend.*, **244,** 1756 (1957)

TIME-CORRELATION FUNCTIONS AND TRANSPORT COEFFICIENTS IN STATISTICAL MECHANICS[1]

By Robert Zwanzig

National Bureau of Standards, Washington, D. C.

Introduction

In the last decade a remarkable and fundamental development has occurred in the statistical mechanical theory of transport processes, leading to time-correlation function expressions for transport coefficients. The purposes of this review are fourfold: (*a*) to explain what time-correlation functions are; (*b*) to describe various methods that have been used to relate transport coefficients to time-correlation functions; (*c*) to survey some applications of time-correlation functions, particularly with reference to problems arising in physical chemistry; and (*d*) to indicate some potentially fruitful areas for future investigation.

The general importance of time-correlation functions is demonstrated clearly by comparison with partition functions. In equilibrium statistical mechanics, as is well known, we can find all of the thermodynamic properties of any particular system if we can first calculate its partition function. Although the actual calculation of the partition function may be exceedingly difficult (as in the case of liquids), at least we know where to start. Time-correlation functions play the same role in the theory of transport processes. For example, there is an exact formula for the coefficient of shear viscosity in terms of a certain time-correlation function. The calculation may be hard, but we know where to start.

In one respect the analogy breaks down. The state of thermal equilibrium is unique, but there are many different kinds of nonequilibrium state. A single partition function gives all thermodynamic properties, but we need a different time-correlation function for each distinct transport process. The appropriate formulas have been obtained for all transport coefficients of current interest; later in this review we give examples of some and references to where the others can be found.

Description of Time-Correlation Functions

Before embarking on a detailed review of the historical background and applications, a brief description of the essential characteristics of time-correlation functions will be instructive.

Roughly speaking, we obtain a time-correlation function when we take some time dependent quantity $A(t)$, multiply it by another time dependent quantity $B(t')$ *evaluated at some different time* t', and then average the product over some equilibrium ensemble. A more precise definition follows.

Time-correlation functions occur in both classical and quantum statistical

[1] The survey of literature pertaining to this review was concluded in November 1964.

mechanics. The quantum mechanical theory is naturally more general and must include the corresponding classical formulas as limiting cases. Since the classical theory is somewhat easier to describe, we shall do this first, and then show what modifications have to be made in the quantum theory. (In the following part of this review, devoted to several methods used to relate transport coefficients to time-correlation functions, we begin with one derivation using quantum mechanics, and thereafter restrict the discussion to classical mechanics.)

In classical mechanics, the dynamical state of any system is specified by giving the coordinates $q_1, q_2, q_3, \cdots$ and the conjugate momenta $p_1, p_2, p_3, \cdots$ of all degrees of freedom at any particular time t. For convenience we often use the single symbol q or p to denote the set of all coordinates and momenta, and $\int dq$ or $\int dp$ to denote integration over all of coordinate or momentum space.

The time dependence of the coordinates $q(t)$ and momenta $p(t)$ is obtained by solving the equations of motion of the entire system. When Hamilton's equations of motion are used, there is one first-order differential equation for each coordinate or momentum; thus the solution contains one arbitrary constant for each coordinate or momentum. These constants are conveniently picked as the values $q(0)$ and $p(0)$ at the initial time $t=0$. (In the following, we often leave out the (0), with the understanding that when no specific time is mentioned, the initial time is meant.) Then the solution of the equations of motion takes the form of a function of t, q, and p; for example, $q(t; q, p)$ is the coordinate at time t, expressed as a function of the coordinates q and momenta p at the initial time $t=0$.

Now let us consider some function $A(p, q)$ of the state of the system. This function does not necessarily have to depend on all the variables. In one useful example, associated with the self-diffusion coefficient, the function A is just the momentum of the "tagged" molecule under observation. The numerical value of A at time t depends on the coordinates and momenta at t, and thus it depends on the time and on the initial values p and q, or

$$A(p(t), q(t)) = A(t; p, q) = A(t)$$

We frequently leave implicit the dependence on initial values and write just $A(t)$.

The time-correlation function of A is then defined as the ensemble average $\langle\ \rangle$ of the product of the initial A and the value of A at time t, $\langle A(0)A(t)\rangle$. On introducing the equilibrium phase space distribution function $f(p, q)$, the ensemble average is given more explicitly by

$$\langle A(0)A(t)\rangle = \int dp \int dq A(p, q) A(t; p, q) f(p, q)$$

In just the same way we may define time-correlation functions of different variables, viz. $\langle A(0)\ B(t)\rangle$

Because the choice of origin of the time scale is entirely arbitrary, and because the equilibrium ensemble distribution function is invariant to displacement of time, we have the obvious identity

$$\langle A(t)B(t')\rangle = \langle A(0)B(t'-t)\rangle$$

The formulas for transport coefficients to be discussed in this review all have the general structure of a one-sided Fourier transform (or a complex Laplace transform) of a time-correlation function,

$$\sigma(\omega) = \int_0^\infty dt \exp(-i\omega t)\langle \dot{A}(0)\dot{B}(t)\rangle$$

It should be observed that the variables in the time-correlation function are time derivatives, e.g., $\dot{A} = dA/dt$, and not the functions themselves. The variable ω in the transform may be regarded as a frequency. We suppose that it has a small negative imaginary part, if necessary, to provide convergence. (Usually one calculates transport coefficients in the limit of an infinitely large system—the "thermodynamic limit"; if this limit is taken first, there are no special convergence difficulties, and the small negative imaginary part is not needed.) In most applications, only the low frequency behavior

$$\lim_{\omega \to 0 - i\epsilon} \sigma(\omega)$$

is of interest. (We have indicated the small negative imaginary part explicitly in this limit.)

It may seem at first that in order to calculate a transport coefficient by means of time-correlation functions, one has to perform the almost impossible task of solving the general N-body problem. As we shall see, the situation is actually not so bad. For example, we need only the time dependence of certain special quantities $\dot{A}$ (p, q) and this information may be considerably easier to get than the complete solution of the equations of motion. (As a trivial but illustrative example, consider the function A = total energy. This is in fact a constant of the motion, so we know its time dependence without solving equations of motion at all.)

Furthermore, we need only the time dependence of an ensemble average. It turns out in many cases that the ensemble average eliminates a substantial amount of the complexity of the exact solution of the equations of motion. This is particularly clear in certain calculations on crystal lattices in the harmonic approximation (1).

A final simplification arises from the special emphasis put on low frequency components of time-correlation functions, which involve only rather gross features of the exact time dependence. For example, in calculating the viscosity of a gas at low density, we find that we do not need molecular collision trajectories in complete detail; only the asymptotes of these trajectories are relevant. The reason is that a collision has a very short duration, perhaps 10^{-12} sec, so that the complete history of a collision is associated with frequencies of the order of 10^{12} sec^{-1}.

The nature of the functions $A(p, q)$, the ensemble average, and the low frequency limit all serve to make calculations more practical. Later in this review we discuss some examples where calculations have been performed successfully.

The quantum mechanical expressions for transport coefficients are somewhat more complicated, although in a way that is not really relevant to most of the applications that we discuss here. We use units where $h = 2\pi$.

Five changes have to be made. (*a*) The functions A are replaced by their corresponding quantum mechanical operators. (*b*) The equilibrium phase space distribution function is replaced by the equilibrium density matrix ρ. (*c*) The ensemble average is calculated by taking the trace of the operator product,

$$\langle A \rangle = Trace\ (\rho A)$$

(*d*) The time derivative of the operator A is given by Heisenberg's equation of motion,

$$dA/dt = i[H, A]$$

where H is the Hamiltonian operator of the entire system. The solution of this equation of motion is written as the time dependent operator

$$A(t) = e^{itH} A e^{-itH}$$

(*e*) Finally, the definition of the time-correlation function is modified by introduction of the "Kubo transform." [This terminology is due to Jaynes (2).] The Kubo transform of an operator A is another operator $\tilde{A}$ defined by

$$\tilde{A} = \frac{1}{\beta} \int_0^\beta d\lambda (\exp \lambda H) A (\exp - \lambda H)$$

We use the standard abbreviation $\beta = 1/k_B T$, where k_B is Boltzmann's constant and T is absolute temperature.

With the preceding five changes, transport coefficients in quantum mechanics have the form

$$\sigma(\omega) = \int_0^\infty dt \exp(-i\omega t) \langle \tilde{\dot{A}}(0) \dot{B}(t) \rangle$$

From the standpoint of notation, the only evident difference between the classical and the quantum mechanical expressions lies in the presence of the Kubo transform; and at sufficiently high temperatures (as required for a classical limit) the Kubo transform of an operator is just the operator itself.

Derivation of Formulas

This part of the review deals principally with the various methods that have been used to establish connections between time-correlation functions and transport coefficients. Discussion of the resulting formulas and their use in physical chemistry is reserved for the following part.

Most of the discussion is about work published since 1950. Earlier work on correlation functions and transport coefficients was concerned primarily with the theory of Brownian motion and its relation to electrical noise (Nyquist noise). A chapter in the textbook *Statistical Mechanics* by Davidson (3) gives an elementary introduction to this material. A more thorough treatment may be found in *Elementary Statistical Physics* by Kittel (4); this book contains most of the important references to the earlier work. *Statistical Physics* by Landau & Lifshitz (5) is another valuable source.

Theoretical connections between time-correlation functions and transport coefficients have been investigated from many points of view. A convenient and very natural classification of methods distinguishes between transport processes driven by external forces and transport processes associated with internal (or thermodynamic) forces.

In the first category, the response of an equilibrium system to an externally imposed force, the theory is simple and straightforward. The standard reference here is a 1957 article by Kubo (6); a somewhat more pedagogic version is contained in Kubo's Boulder Lectures (7).

It should be noted, however, that the basic idea was in the air for some time before 1957. We mention in particular the work of Callen & Welton (8) and Callen & Greene (9) on the fluctuation-dissipation theorem. A review by Bernard & Callen (10) provides an excellent and comprehensive survey of this work. The textbooks of Landau & Lifshitz (5) and Kittel (4) discuss Callen's work.

Kubo's method for handling the external force problem has become so popular in the last few years that the results obtained not only by this method, but also by entirely unrelated methods, are often called "Kubo formulas." This terminology is unfortunate and should be avoided. The results referred to in this way were, for the most part, obtained by others and at earlier times. We mention in particular the work of Green (11, 12), to be described shortly.

Kubo's *method* is especially attractive because of its simplicity and its strongly operational character. Consider, for example, the application of this method to the theory of electrical conductivtiy.

In its most elementary form, a measurement of electrical conductivity consists of observing the electric current in response to an applied electric field. Because a single measurement is usually not enough to establish a law of nature—some accident, or even an extraordinarily large thermal fluctuation, may have occurred—the measurement is repeated many times, and the results are averaged. Then the conductivity is defined by the relation of average electric current $\langle \boldsymbol{J} \rangle$ to applied field $\boldsymbol{E}$,

$$\langle \boldsymbol{J} \rangle = \sigma \boldsymbol{E} + O(\boldsymbol{E}^2)$$

(The electric field is taken small enough that nonlinear effects are of no importance.)

More often, the frequency dependence of the electrical conductivity is of interest. To measure this quantity, one switches on an arbitrary time-dependent electric field $\boldsymbol{E}(t)$ at $t=0$, and one then measures the average current $\langle \boldsymbol{J}(t) \rangle$. The observed relation between $\langle \boldsymbol{J}(t) \rangle$ and $\boldsymbol{E}(t)$ is

$$\langle J(t)\rangle = \int_0^t dt' \Phi(t-t')E(t') + O(E^2)$$

where $\Phi(t-t')$ is an "after-effect function" showing the response at t to a disturbance at t'. Frequency components of the current and field are defined by Fourier analysis,

$$\langle J_\omega\rangle = \int_0^\infty dt \exp(-i\omega t)\langle J(t)\rangle$$

$$E_\omega = \int_0^\infty dt \exp(-i\omega t)E(t)$$

Note that the lower limit on these integrals is $t=0$ and not the more conventional $t=-\infty$; this is because the field and current are both zero before the field is switched on. By an easy rearrangement of the integrals, we obtain

$$\langle J_\omega\rangle = \hat{\sigma}(\omega)E_\omega$$

where $\hat{\sigma}(\omega)$ is the frequency dependent conductivity,

$$\hat{\sigma}(\omega) = \int_0^\infty dt \exp(-i\omega t)\Phi(t)$$

Kubo's statistical mechanical theory is just a general model of the above experiments. The method, while simple and direct, relies on certain abstract operator techniques that may not be familiar to most physical chemists. To show just how simple it is, we present here a version of the derivation that relies only on methods to be found in most elementary quantum mechanics textbooks. (The derivation, in this case, is no more difficult in quantum mechanics than it is in classical mechanics, so the quantum mechanical version is given for the sake of generality.)

We note first that there are three stages to an experiment on electrical conductivity: the initial preparation of the system, the switching on of the field, and the measurement of the current. Each of these stages has its analogue in the statistical mechanical theory.

In the first stage, the experimenter prepares the system in a "state of thermal equilibrium." It is important to know just what this means. The experimenter does not actually control the exact state of the system, in the sense of specifying in all detail the wave function of the system, or its position in phase space. All he can do is select a sample, for example a piece of wire, according to some prescription. To repeat the experiment he selects another sample, according to the same prescription, and so on. The most likely prescription for an experiment of this sort is that the samples are to be taken at random from a spool of wire, or perhaps from many spools of wire that are themselves chosen at random. That is, the experimenter takes samples from an ensemble. To say that the system is prepared in a state of thermal equilibrium is to say that repetitions of an experiment are performed on samples taken from a thermal equilibrium (canonical) ensemble.

Let H_0 be the total Hamiltonian operator of the system (in the absence of any applied field). The exact eigenfunctions ψ_n and eigenvalues E_n of H_0 are determined by

$$H_0\psi_n = E_n\psi_n$$

In the representation determined by these exact (unperturbed) eigenstates, the equilibrium density matrix is diagonal,

$$(\rho)_{mn} = \rho_n\delta_{mn}$$

and ρ_n is the probability of observing the system in the state ψ_n,

$$\rho_n = Q^{-1} \exp - \beta E_n$$

We use the standard abbreviation $\beta = 1/k_B T$. The partition function Q is

$$Q = \sum_n \exp - \beta E_n$$

Thus the average behavior observed in repeated experiments is found by calculating the actual behavior of a particular eigenstate ψ_n, followed by averaging over the canonical distribution of eigenstates.

In the second stage of an experiment, an external field $\boldsymbol{E}(t)$ is switched on. For convenience we suppose that the field is directed along a cartesian coordinate axis, labeled by $a = 1$, 2, or 3. The field interacts with the total electric dipole moment $\boldsymbol{M}$ of the system,

$$\boldsymbol{M} = \sum_j e_j \boldsymbol{R}_j$$

where e_j and $\boldsymbol{R}_j$ are the charge and position of the j-th particle. We assume that the field is spatially uniform over the system. Then the energy of interaction is described by the perturbation Hamiltonian

$$H_1(t) = - M_a E_a(t)$$

Because the field points in the a-th direction, only the a-th component of the moment appears in H_1.

The state $\psi(t)$ of the system at time t is determined by solving the time-dependent Schrodinger equation

$$i\frac{\partial\psi}{\partial t} = (H_0 + H_1(t))\psi(t)$$

subject to the initial condition that $\psi(0)$ equals ψ_n with probability ρ_n.

It is now an elementary calculation in time-dependent perturbation theory to find $\psi(t)$. To the first order in the applied field, the result is

$$\psi(t) = \psi_n e^{-iE_n t} + \sum_{m\neq n} \frac{1}{i}\int_0^t dt' e^{i(E_m - E_n)t'}[H_1(t')]_{mn}\psi_m e^{-iE_m t} + O(H_1^2)$$

In the third stage of the experiment, the current is measured. This corresponds, in the statistical mechanical theory, to calculating the expectation value of the current operator J, first with respect to the state $\psi(t)$ and then with respect to the canonical distribution of initial states.

The component J_b of electric current in the b-th direction is

$$J_b = \sum e_j(\dot{R}_j)_b = \dot{M}_b$$

Its average value at time t, for the given initial state ψ_n, is

$$\int dR\psi^*(t)J_b\psi(t) = (J_b)_{nn} + \sum_{m\neq n} \frac{1}{i}\int_0^t dt'(J_b)_{nm}e^{-i(E_m-E_n)(t-t')}[H_1(t')]_{mn}$$
$$- \sum_{m\neq n} \frac{1}{i}\int_0^t dt'(J_b)_{mn}e^{-i(E_n-E_m)(t-t')}[H_1(t')]_{nm} + O(H_1^2)$$

Next, we average this result over the canonical distribution of initial states; we put in the explicit formula for H_1 in terms of the electric moment and the electric field; and we rearrange the terms slightly. The result can be put in the form

$$\langle J_b(t)\rangle = \langle J_b\rangle_{eq} + \int_0^t dt'\phi_{ba}(t-t')E_a(t') + O(E^2)$$

where $\langle J_b\rangle_{eq}$ is the average current at equilibrium, and vanishes,

$$\langle J_b\rangle_{eq} = \sum_n \rho_n(J_b)_{nn} = 0$$

and ϕ_{ba} is the after-effect function giving the current in the b-th direction due to a field in the a-th direction,

$$\phi_{ba}(t) = \sum\sum_{m\neq n} \rho_n(i)^{-1}\{e^{i(E_m-E_n)t}(J_b)_{mn}(M_a)_{nm} - e^{i(E_n-E_m)t}(J_b)_{nm}(M_a)_{mn}\}$$

Matrix elements of the electric moment can be expressed in terms of matrix elements of the current by means of the Heisenberg equation of motion

$$(J)_{mn} = (\dot{M})_{mn} = -(i)^{-1}(E_m - E_n)(M)_{mn}$$

The time-dependent current operator $J(t)$ has matrix elements

$$[J(t)]_{mn} = e^{i(E_m-E_n)t}(J)_{mn}$$

With use of the preceding two equations, and by some further algebraic rearrangement, we obtain the following expression for ϕ_{ba}:

$$\phi_{ba}(t) = \sum_m\sum_n \rho_m(J_a)_{mn}[J_b(t)]_{nm}\frac{e^{-\beta(E_n-E_m)} - 1}{E_m - E_n}$$

Now we refer back to the formula given earlier for the Kubo transform, viz.

$$\tilde{J}_a = \frac{1}{\beta}\int_0^\beta d\lambda e^{\lambda H}J_a e^{-\lambda H}$$

The matrix elements of $\tilde{J}_a$ are

$$(\tilde{J}_a)_{mn} = \frac{1}{\beta}\int_0^\beta d\lambda e^{\lambda(E_m-E_n)}(J_a)_{mn}$$
$$= \frac{1}{\beta}\,\frac{e^{-\beta(E_n-E_m)}-1}{E_m-E_n}(J_a)_{mn}$$

Consequently we may write ϕ in the form

$$\phi_{ba}(t) = \beta\sum_m\sum_n \rho_m(\tilde{J}_a)_{mn}[J_b(t)]_{nm}$$

But by the rule of matrix multiplication and the definition of the canonical average, this is just

$$\phi_{ba}(t) = \beta\langle\tilde{J}_a J_b(t)\rangle$$

Thus $\phi(t)$ is a time-correlation function. The corresponding frequency dependent electric conductivity tensor is

$$\sigma_{ba}(\omega) = \beta\int_0^\infty dt\,\exp\,(-i\omega t)\langle\tilde{J}_a J_b(t)\rangle$$

Several comments about this remarkably simple theory are in order.

(*a*) Kubo's original derivation was considerably more concise than the one given here, largely as a result of his systematic use of abstract operator techniques.

(*b*) The corresponding formulas in classical statistical mechanics are obtained, as already pointed out, by omitting the Kubo transform, calculating the time dependence by means of classical mechanics, and performing the ensemble average by integration over phase space. In Kubo's article (6), the classical formula is also obtained by direct calculation using classical statistical mechanics, rather than by appeal to a correspondence argument.

(*c*) Often in discussions of the theory of irreversible processes, the question arises "Where was the irreversibility put into the theory?" In the present case the answer is simple. Irreversible behavior is surely observed in the real world. If statistical mechanics, as applied here, is a good model of phenomena in the real world, then irreversible behavior must be accounted for automatically, without any conscious effort to "put it in." (In this connection it should be noted that some of the famous irreversibility "paradoxes" are avoided in a very simple way by calculating time-correlation functions in the limit of an infinitely large system. This limit is mainly a mathematical convenience, serving to eliminate recurrences occurring at such distant times that they would never be observed anyway.)

(*d*) While the transport process under consideration was electric conductivity, the method obviously can be used for any process driven by an external field: for example, sedimentation in a centrifugal field. Kubo approached the general theory by way of its application to the theory of line width in nuclear magnetic resonance spectra (13).

(*e*) If we had calculated the response in the form of induced polarization rather than induced current, we would have obtained an expression for the dielectric response (macroscopic polarizability) instead of the conductivity. Inspection of the derivation will show that the principal difference in the results lies in the role of real and imaginary parts. That is, the imaginary part of the conductivity describes a dielectric response; and the imaginary part of the dielectric constant describes a conductive response. Thus, both kinds of response are accounted for at the same time. This topic, together with the further complications due to magnetic effects, has been discussed by Blatt & Matsubara (14). We return to the relation of time-correlation functions and dielectric theory in the following part of this review.

In connection with the last two remarks, we give here for the sake of completeness the most general result obtained by Kubo's method.

Let us consider situations where the perturbing Hamiltonian consists of a number of terms, each of which has the form of some property of the system A_j multiplied by some kind of field $E_j(t)$,

$$H_1 = -\sum_j A_j E_j(t)$$

(Because of the linearity of first order perturbation theory, the effects of the various fields are additive, so they can be treated independently.) The response of the system to the fields will be expressed in terms of certain arbitrary currents $\dot{B}_j = J_j$,

$$\langle J_j(t)\rangle = \sum_l \int_0^t dt' \phi_{jl}(t - t')E_l(t') + O(E^2)$$

or, in Fourier components (in analogy with the electrical conductivity theory)

$$\langle J_j(\omega)\rangle = \sum_l \sigma_{jl}(\omega)E_l(\omega)$$

The transport coefficients $\sigma_{jl}(\omega)$ are given by

$$\sigma_{jl}(\omega) = \beta \int_0^\infty dt \exp(-i\omega t)\langle \tilde{\dot{A}}_l(0)B_j(t)\rangle$$

which is the general form that we referred to in the first part of this review.

Thermal processes.—Transport coefficients associated with internal disturbances (or thermal forces) cannot be treated directly by Kubo's method. For example, there does not appear to be any convincing way of representing the effects of a temperature gradient by means of a perturbation Hamiltonian. However, many other methods have been investigated. We propose to describe here the essential features of some of them.

Among the variety of methods that have been devised for treating thermal processes, we may distinguish between six general approaches, grouped for convenience under the headings:

(*a*) indirect Kubo, (*b*) Fokker-Planck, (*c*) regression hypothesis, (*d*) local equilibrium, (*e*) external reservoir, and (*f*) prediction theory.

Before discussing these approaches, we observe that while rather different in character, they all lead to identical values for transport coefficients. For this reason, any expressed preference for one approach over the others must be regarded as a matter of individual taste in such questions as rigor, elegance, and simplicity.

[Certain apparent discrepancies between results obtained, by means of different methods, by Green (12), Mori (15, 16), and McLennan (17), have been resolved by Green (18) and McLennan (19). The point here was that time-correlation functions, being second moments of fluctuating quantities, must depend on the ensemble used. Thus, the time-correlation function of a particular quantity A, calculated in one ensemble, may correspond to the time-correlation function of a somewhat different quantity A' in another ensemble. On allowing for the effects of changing ensembles, the resulting formulas for transport coefficients give identical numerical values. Having once noted this potential source of confusion, we shall not refer to it again.]

The consensus of workers in this field seems to be that the indirect Kubo method is the most convincing approach presently available. Also, it is by far the easiest to explain, since it relies mainly on a clever application of Kubo's method for external forces. The principal references to the indirect Kubo method are by Luttinger (20), Kadanoff & Martin (21), Jackson & Mazur (22), Felderhof & Oppenheim (23), and Montroll (24).

The gist of the indirect Kubo method is best explained by an illustration drawn from hydrodynamics. At time $t=0$, an external body force $\boldsymbol{F}(\boldsymbol{R}, t)$ is applied to a fluid originally in thermal equilibrium. This force causes fluid motion, and in particular, it gives rise to a certain velocity field $\boldsymbol{v}(\boldsymbol{R}, t)$. The relation between the force and the response is most conveniently expressed in terms of Fourier components, viz.

$$\boldsymbol{F}(\omega, \boldsymbol{k}) = \int d^3R \int dt \boldsymbol{F}(\boldsymbol{R}, t) \exp(-i\boldsymbol{k}\cdot\boldsymbol{R} + i\omega t)$$

$$\boldsymbol{v}(\omega, \boldsymbol{k}) = \int d^3R \int dt \boldsymbol{v}(\boldsymbol{R}, t) \exp(-i\boldsymbol{k}\cdot\boldsymbol{R} + i\omega t)$$

On solving the hydrodynamic equations of motion (for sufficiently small forces that these equations may be linearized in deviations from equilibrium), one obtains a simple relation between $\boldsymbol{v}$ and $\boldsymbol{F}$,

$$\boldsymbol{v}(\omega, \boldsymbol{k}) = \sigma(\omega, \boldsymbol{k})\boldsymbol{F}(\omega, \boldsymbol{k})$$

When the force is transverse, i.e., when $\nabla\cdot\boldsymbol{F}=0$ and $\nabla\times\boldsymbol{F}\neq 0$, then the velocity field is also transverse. In this case, the coefficient $\sigma(\omega, \boldsymbol{k})$ is

$$\sigma(\omega, \boldsymbol{k}) = [\eta k^2 - i\omega\rho]^{-1}$$

The mass density of the fluid is ρ, and η is the coefficient of shear viscosity. We emphasize that the calculation leading to this equation is entirely macroscopic in character, being based solely on the known validity of the Navier-Stokes equation.

The coefficient of shear viscosity can be extracted from σ by first going to the limit of long wavelength (or small $\boldsymbol{k}$), and then going to the limit of low frequency, as follows:

$$\eta = \lim_{\omega \to 0} \lim_{k \to 0} \frac{\rho^2 \omega^2}{k^2} \, Re \, \sigma(\omega, \boldsymbol{k})$$

This can be seen, for example, by expanding σ in powers of k.

But $\sigma(\omega, \boldsymbol{k})$ can be obtained by Kubo's method! The force $\boldsymbol{F}(\boldsymbol{R}, t)$ is in fact externally imposed, and $\boldsymbol{v}(\boldsymbol{R}, t)$ is a dynamical response of the system to that force. Application of Kubo's method leads to a statistical mechanical formula for $\sigma(\omega, \boldsymbol{k})$. When the indicated limits are taken, a statistical mechanical formula for the coefficient of shear viscosity is obtained. The result, first obtained by Green (12) by an entirely different method, will be given in the part of this review devoted to applications.

An essential feature of the argument just outlined is that the force was taken to be transverse, with zero divergence and nonzero curl. This is somewhat unusual; most forces that one is accustomed to deal with are conservative, so that they are expressed as the gradient of a potential. But then they automatically have zero curl, and cannot be transverse. We may, however, imagine a fictitious transverse force of the following kind. Suppose that each molecule of the fluid is given an infinitesimal electric charge. Then the force exerted on these charges by the electric field of a plane polarized light wave is transverse.

The fact that this experiment can be performed only in thought and not in the laboratory is irrelevant. There is no reason to believe that the Navier-Stokes equation fails for fictitious transverse forces, any more than that Newton's equations of motion fail for such forces.

A similar thought experiment is required for thermal conductivity. In Luttinger's treatment (20), the force comes from the interaction of a varying gravitational field with the mass associated with the internal energy of a fluid ($E = mc^2$!).

While the indirect Kubo method is perhaps the most convincing one developed up to now, it serves only to re-derive results already obtained by other methods. The first theory to yield useful results is the one we categorize as the Fokker-Planck theory.

This approach is based on ideas that are familiar in the theory of Brownian motion. The movement of a Brownian particle in a fluid is governed by three different kinds of force: (*a*) a force due to external sources, for example, the force exerted on a charged particle by an electric field; (*b*) a force due to friction, generally given by Stokes' law; and (*c*) a force due to random weak interactions of the Brownian particle with the molecules of the surrounding fluid. Because of the random nature of part of the total force, the motion of the Brownian particle is irregular. This suggests the use of probability methods in describing Brownian motion.

Let $P(\boldsymbol{R}_0, t_0/\boldsymbol{R}, t)$ be the probability (per unit volume) of finding the

Brownian particle at position $\boldsymbol{R}$ at time t, given its position $\boldsymbol{R}_0$ at time t_0. This function is determined by the fundamental equation of the theory of Brownian motion, the Fokker-Planck equation

$$\frac{\partial P}{\partial t} = \nabla \cdot (\boldsymbol{A}P) + \nabla^2(DP)$$

Many texts give derivations of this equation: see Kittel (4).

The coefficients $\boldsymbol{A}$ and D appearing in the Fokker-Planck equation have a simple meaning. Denote by $\Delta \boldsymbol{R}$ the displacement of the Brownian particle during a time interval Δt. Then in the course of the derivation of the Fokker-Planck equation, one finds that $\boldsymbol{A}$ is expressed in terms of the ensemble average $\langle \Delta \boldsymbol{R} \rangle$ as follows:

$$\boldsymbol{A} = \lim_{\Delta t \to 0} \langle \Delta \boldsymbol{R} \rangle / \Delta t$$

That is, $\boldsymbol{A}$ is an average velocity. It must be noted that the limit $\Delta t \to 0$ has a special meaning in this context. The time interval Δt is to be small in a macroscopic sense, but still larger than all characteristic molecular times. That is, many individual interactions between the Brownian particle and the surrounding molecules will occur during Δt, but the particle will not move appreciably during Δt.

The coefficient D is determined by the second moment of the displacement, $\langle (\Delta \boldsymbol{R})^2 \rangle$, according to

$$D = \lim_{\Delta t \to 0} \langle (\Delta \boldsymbol{R})^2 \rangle / 6\Delta t$$

Higher moments of the displacement are treated as negligible in the customary derivation of the Fokker-Planck equation.

The formula just given for D is in fact a way of writing an important equation due to Einstein, $\langle R^2 \rangle_{av} = 6Dt$. Evidently D is just the coefficient of self-diffusion. This provides in principle a molecular theory of the diffusion coefficient which involves a time-correlation function. To see how this goes, let us replace the displacement by a time integral of the velocity of the Brownian particle,

$$\Delta \boldsymbol{R} = \int_0^{\Delta t} ds \boldsymbol{v}(s)$$

Then the definition of D leads to

$$D = \lim_{\Delta t \to 0} (6\Delta t)^{-1} \int_0^{\Delta t} ds \int_0^{\Delta t} ds' \langle \boldsymbol{v}(s) \cdot \boldsymbol{v}(s') \rangle$$

Note that the integrand contains the time-correlation function of the velocity of the Brownian particle. Because of invariance to any change in the origin of the time axis

$$\langle \boldsymbol{v}(s) \cdot \boldsymbol{v}(s') \rangle = \langle \boldsymbol{v}(0) \cdot \boldsymbol{v}(s' - s) \rangle$$

one of the time integrals can be performed immediately, leading to

$$D = \lim_{\Delta t \to 0} \frac{1}{3} \int_0^{\Delta t} ds (1 - s/\Delta t) \langle \boldsymbol{v}(0) \cdot \boldsymbol{v}(s) \rangle$$

Suppose now that the time-correlation function tends to zero over times of a molecular order of magnitude. Then, for time intervals Δt that are small but yet macroscopic, the term $s/\Delta t$ may be omitted—

$$D \cong \frac{1}{3} \int_0^{\Delta t} ds \langle \boldsymbol{v}(0) \cdot \boldsymbol{v}(s) \rangle$$

Parenthetically, we mention that the rigorous theory of self-diffusion leads to a very slightly different result,

$$D = \lim_{\epsilon \to 0} \frac{1}{3} \int_0^{\infty} dt \exp(-\epsilon s) \langle \boldsymbol{v}(0) \cdot \boldsymbol{v}(s) \rangle$$

The previous condition on the size of Δt is replaced by an extra limiting process.

In 1946, Kirkwood (25) presented a molecular theory of Brownian motion that was meant to apply to the motion of an individual molecule in a medium composed of similar molecules (e.g., self-diffusion in liquid argon). In comparison with the elementary theory just described, one essential difference is that Kirkwood's theory was concerned with Brownian motion in momentum space. He derived a Fokker-Planck equation in momentum space; and in the course of the derivation, he found a molecular formula for the friction constant.

Let $\langle \boldsymbol{p} \rangle$ be the average momentum of a molecule in a liquid, and let M be its mass. Then the friction constant is defined as the coefficient in the frictional retarding force on the molecule,

$$\frac{d\langle \boldsymbol{p} \rangle}{dt} = -\zeta \langle \boldsymbol{p} \rangle / M$$

Kirkwood's expression for ζ takes the form of a time-correlation function formula in the instantaneous force $\boldsymbol{F}(t)$ exerted on the molecule by its surroundings,

$$\zeta = \frac{1}{3} \beta \int_0^{\tau} dt \langle \boldsymbol{F}(0) \cdot \boldsymbol{F}(t) \rangle$$

The upper limit τ is supposed to be macroscopically short, but long compared with characteristic molecular times. For a review of Kirkwood's theory, see Rice & Frisch (26).

It is important to observe that the preceding result gives a "transport coefficient" associated with momentum, and that it involves a time-correlation function in the time derivative of the momentum. This is analogous to the formula for the diffusion coefficient D of a Brownian particle, associated with the position of the particle, which involves a time-correlation function in the time derivative of the position.

Kirkwood's theory is not exact, and there is still considerable question as to the limits of its validity. By a different method, however, Lebowitz & Rubin (27) have shown the correctness of Kirkwood's results in a limiting case, where the mass M of the Brownian particle is very much larger than the mass of any of the surrounding molecules. (This restriction actually excludes Kirkwood's application to monatomic fluids.) The exact formula for ζ in the heavy mass limit is

$$\zeta = \lim_{\epsilon \to 0} \frac{1}{3} \beta \int_0^\infty dt \exp(-\epsilon t) \langle \boldsymbol{F}(0) \cdot \boldsymbol{F}(t) \rangle$$

Because of the heavy mass limit, the motion is to be calculated with the Brownian particle held fixed in space.

In 1952 and 1954, Green (11, 12) presented an important generalization of Kirkwood's work. The new feature supplied by Green's theory was the introduction of Brownian motion ideas into the general theory of irreversible processes.

In experiments on transport processes, certain specific properties of a system are observed as functions of time. These properties may be, for example, the local mass density of a fluid, its velocity field, it local energy density, etc. Green refers to them as "gross variables." We shall denote these quantities by A_1, A_2, A_3, $\cdots$; and collectively, by A.

The gross variables, being dynamical quantities, are functions of time. On the average, their time dependence is described by well-known transport laws, e.g., Fick's law of diffusion. In any individual experiment, however, apparently random small fluctuations from the average behavior will be seen, as a result of the molecular many-body character of a macroscopic system. This situation is reminiscent of the one in Brownian motion, where the path of the Brownian particle is regular on the average, but subject to small random fluctuations.

Green introduced a probability distribution function $g(a; t)$ in the space of the gross variables: the probability of the event $a < A < a + da$ at time t is $g(a; t)da$. He then gave persuasive arguments that $g(a; t)$ should satisfy an equation having the Fokker-Planck structure. By arguments very much like those used by Kirkwood to derive a molecular expression for the friction constant that appears in the Fokker-Planck equation in momentum space, Green found molecular expressions for the various coefficients appearing in the Fokker-Planck equation in a-space. Starting from the Fokker-Planck equation, Green showed how the average behavior of the gross variables obeyed laws having exactly the same structure as the experimentally known transport laws. The transport laws obtained theoretically contain the coefficients that appear in the Fokker-Planck equation; by comparison, Green identified these coefficients with the various experimental transport coefficients.

By following the procedure just outlined, Green obtained expressions involving time-correlation functions for the coefficients of shear and bulk viscosity, thermal conductivity, diffusion, and thermal diffusion.

Other versions of this theory, extensions, and related material may be found in the book by DeGroot & Mazur (28), and in articles by Hashitsume (29), Lax (30), Onsager & Machlup (31), Uhlhorn (32), Van Kampen (33), and Yamamoto (34).

Green's theory is based on a number of assumptions, notably that the dynamical process of interest is a Markov process, and that the departure from thermal equilibrium is small. A rigorous justification of this theory, and an extension to non-Markoffian and nonlinear processes, has been given by Zwanzig (35). The quantum mechanical version of Zwanzig's theory was worked out recently by Sewell (36). Also, Mori (37) has obtained a "generalized Langevin" counterpart to the generalized Fokker-Planck theory.

At this point we must mention work bearing a strong family relation to the Fokker-Planck theory. Helfand (38) used a generalization of the Einstein formula $\langle R^2 \rangle = 6Dt$ to derive expressions for other transport coefficients, along the same lines as the argument we just presented for the coefficient of diffusion of a Brownian particle. In their textbooks, Landau & Lifshitz (39) presented a theory of fluctuations in electrodynamic and hydrodynamic systems that may be regarded as the "Langevin equation" counterpart to Green's Fokker-Planck theory.

The third method used to establish connections between transport coefficients and time-correlation functions is based on Onsager's hypothesis concerning the regression of fluctuations (40). For further discussion of this hypothesis and its use in other connections see Casimir (41), DeGroot & Mazur (28), and Wigner (42).

Onsager's hypothesis asserts that spontaneous fluctuations from thermal equilibrium decay, on the average, according to the transport laws governing the corresponding macroscopic variables. (Because deviations from the average are of the same order of magnitude as the average itself, this assertion says nothing at all about the time dependence of a spontaneous fluctuation in any individual case. Only the average is supposed to behave in a systematic way.)

Kubo, Yokota & Nakajima (43) based a derivation of formulas for thermal transport coefficients on this hypothesis. The possibility of doing so is suggested clearly (at least by hindsight!) in Onsager's work; he demonstrated the symmetry of the matrix of transport coefficients in coupled irreversible processes, i.e., the reciprocal relations. But in order to give a statistical mechanical treatment of some property of transport coefficients, one must have some kind of representation of the transport coefficients. Kubo, Yokota & Nakajima extracted the appropriate representation from Onsager's ideas.

To illustrate how the method works, we consider here the simplest example, where the macroscopic transport law concerns only the average value $\bar{a}(t)$ of a single dynamical quantity $a(t)$,

$$\frac{d\bar{a}(t)}{dt} = -M\bar{a}(t)$$

Thus M is a rate constant, or transport coefficient, for the decay of $\bar{a}(t)$. The quantity a is chosen so that its average value vanishes at thermal equilibrium.

The transport law describes an average behavior; but in any actual experiment, one follows the evolution of a particular system and a particular quantity $a(t)$. If experiments are repeated in a macroscopically reproducible way, the initial values $a(0)$ will be the same in each repetition. Later values $a(t)$ will differ slightly as a result of uncontrolled differences between initial states on the molecular level. The average $\bar{a}(t)$ in repeated experiments depends parametrically on the initial value $a(0)$ common to all repetitions.

The observation just made is put to use in this theory in the following way. Let us multiply both sides of the transport law by the initial value $a(0)$, and then average over an equilibrium ensemble of initial values, denoted by $\langle\rangle$:

$$\left\langle \frac{d}{dt}\,\bar{a}(t)a(0)\right\rangle = -M\langle \bar{a}(t)a(0)\rangle$$

The resulting equation now contains a double average: first a conditional average over repetitions for the given $a(0)$, and then an unconditional average over the equilibrium distribution of $a(0)$. Evidently the same average can be computed more directly. If $a(t)$ is the numerical value of a function $A(p(t), q(t))$ in phase space, then we may write the double average in the form of a time-correlation function,

$$\begin{aligned}\langle \bar{a}(t)a(0)\rangle &= \langle A(p(t), q(t))A(p(0), q(0))\rangle \\ &= \langle A(t)A(0)\rangle\end{aligned}$$

(It is perhaps easiest to verify this by working backwards from the right hand side.)

The regression hypothesis has just been used without being stated explicitly. The point is that when we average over an equilibrium distribution of initial $a(0)$, the major contribution comes from very small values of $a(0)$, of the order of spontaneous fluctuations. But then we want the transport law to be valid for such small initial values. This is precisely what the regression hypothesis is about.

As a result of the preceding manipulations, the transport law now has the form

$$\langle \dot{A}(t)A(0)\rangle = -M\langle A(t)A(0)\rangle$$

Let us integrate this equation over a small time interval t,

$$\int_0^t dt'\langle \dot{A}(t')A(0)\rangle = -M\int_0^t dt'\langle A(t')A(0)\rangle$$

If Mt is much smaller than unity, the right hand side can be simplified to

$$= -tM\langle A(0)A(0)\rangle + O(M^2t^2)$$

We assume that t is always small enough for this approximation to be good. On the left, we take advantage of time invariance to shift the time origin,

$$\langle \dot{A}(t')A(0)\rangle = \langle \dot{A}(0)A(-t')\rangle$$

and then we introduce another time derivative by

$$A(-t') = A(0) - \int_0^{t'} ds\dot{A}(-s)$$

This gives, for small enough t,

$$\int_0^t dt' \langle \dot{A}(0)A(0)\rangle - \int_0^t dt' \int_0^{t'} ds \langle \dot{A}(0)\dot{A}(-s)\rangle = -tM\langle A(0)A(0)\rangle$$

The first term on the left vanishes because

$$\langle \dot{A}A\rangle = \left\langle \frac{1}{2}\frac{d}{dt}A^2 \right\rangle$$

and because the equilibrium average of a rate of change must vanish. In the second term on the left, the time origin is shifted back by an interval s; this replaces $\langle \dot{A}(0)\dot{A}(-s)\rangle$ by $\langle \dot{A}(s)\dot{A}(0)\rangle$. Finally, the order of integration is reversed, the t' integration is performed immediately, and both sides of the equation are divided by t, yielding

$$\int_0^t ds(1 - s/t)\langle \dot{A}(s)\dot{A}(0)\rangle = M\langle A^2\rangle$$

This equation is very much like one already encountered in this review, and it is handled in the same way. Again, the decisive question is whether the time-correlation function decays fast enough. If it does, then we may drop the extra term s/t in the integrand, and replace the upper limit t by any time τ long enough for the integrand to have decayed to zero. In this case, we get

$$M = \int_0^\tau ds \langle \dot{A}(s)\dot{A}(0)\rangle / \langle A^2\rangle$$

The rate constant is now expressed by means of a time-correlation function in the time derivative of the quantity of interest.

The result of the derivation does not look quite like the earlier formulas for transport coefficients, because of the extra factor $\langle A^2\rangle$. The difference can be understood, however, by remembering that transport coefficients are usually defined as coefficients describing the response $J = da/dt$ to a thermodynamic force $\mathfrak{F}$,

$$J = L\mathfrak{F}$$

The force is the derivative of a nonequilibrium entropy $S(a)$ with respect to a deviation from equilibrium,

$$\mathfrak{F} = \frac{\partial S(a)}{\partial a}$$

The entropy, being a maximum at equilibrium, can be expanded in powers of a,

$$S(a) = S(0) - \frac{1}{2}ga^2 + \cdots$$

so that the transport law takes the form

$$J = -Lg\bar{a}$$

The quantity Lg corresponds to M. By means of equilibrium fluctuation theory (28), it is easy to show that

$$\langle A^2 \rangle = k_B g^{-1}$$

Then the transport coefficient L has the same structure as our earlier formulas,

$$L = \frac{1}{k_B}\int_0^\tau ds \langle \dot{A}(s)\dot{A}(0)\rangle$$

Along with the regression hypothesis, this theory involves another assumption. It is supposed to be possible to find a time τ that is long enough for the time-correlation function to decay to zero, but also short enough to fulfill the condition that $M\tau$ be much smaller than unity.

Kubo, Yokota & Nakajima (43), and also Kubo (7), have given plausible arguments that this assumption is safe. A more rigorous analysis by Zwanzig (44) led to the conclusion that the theory is valid whenever the transport process under consideration is "slow enough" in a mathematically well-defined sense. Thus the theory may be applied, e.g., to ordinary hydrodynamic processes, but perhaps not to extremely fast chemical reactions in shock waves. (This remark is conjectural—no numerical estimates have been made.)

The basic assumption of the theory, the regression hypothesis, can be justified only in the context of a more general theory. Van Kampen (33) has shown that it is correct in the context of the Fokker-Planck theory, at least for linear transport laws. In fact, the regression hypothesis is no more than a statement that transport laws linear for macroscopic departures from equilibrium are still linear, and with the same slopes, for microscopic departures from equilibrium. Any exception to this would be astonishing indeed.

The fourth method used to relate transport coefficients to time-correlation functions was invented in 1956 by Mori (15, 45); we call this the local equilibrium theory. Recently Longuet-Higgins (46) has presented a simple and pedagogically useful version of this theory.

The basis of this method is the observation that when a typical nonequilibrium system relaxes toward thermal equilibrium, it passes through a sequence of local equilibrium states. Consider, e.g., thermal conduction in a temperature gradient. Use of the notion of a temperature gradient implies that the nonequilibrium system can be described by a local temperature. But this in turn implies that the system can be treated as if it were locally canonical.

Andersen et al. (47) have discussed a class of systems for which relaxation proceeds through a sequence of Boltzmann distributions, as assumed in the local equilibrium theory. Their results, while exact, are based on treatment of dynamics as a Markoff process; and this is generally not correct except in certain special limiting cases.

As a very simple illustration of the local equilibrium theory, we consider the case where only a single macroscopic variable A (aside from the total energy H) is of interest. The ensemble mean values of these quantities are denoted by $\bar{A}$ and $\bar{H}$. In an equilibrium ensemble where $\bar{A}$ is held fixed by some kind of external constraint, the phase space distribution function $f(p, q)$ is

$$f(p, q) = Q^{-1} \exp(-\beta H - \gamma A)$$

The normalizing factor is the partition function $Q(\beta, \gamma)$,

$$Q(\beta, \gamma) = \iint dp dq \exp(-\beta H - \gamma A)$$

This distribution function gives maximum entropy under the constraints of fixed $\bar{H}$ and $\bar{A}$. The normal thermal equilibrium distribution function, appropriate when $\bar{A}$ is not constrained, is obtained by setting γ equal to 0. Thermal equilibrium mean values, with $\gamma = 0$, will be denoted as in the previous discussion by angular brackets $\langle\rangle$. The functional relation between the Lagrange multipliers β and γ, and the mean values $\bar{H}$ and $\bar{A}$, is given by the well known formulas

$$\bar{H} = -\frac{\partial \log Q}{\partial \beta}, \qquad \bar{A} = -\frac{\partial \log Q}{\partial \gamma}$$

The principal assumption of the local equilibrium theory is that the phase space distribution function maintains the above form for all times; the relaxation toward equilibrium due to removing the constraint fixing $\bar{A}$ is supposed to be described by a time-dependent coefficient $\gamma(t)$. The other coefficient β is independent of time because H is a constant of the motion.

This assumption is demonstrably not correct, except for the trivial case where A is also a constant of the motion. It has the further failing, that the mean value of the rate of change of A invariably vanishes. Very likely, however, the true distribution function of a relaxing system does not deviate appreciably from the locally canonical form, because the idea of local thermal equilibrium in a relaxing system appears to be experimentally justified.

In practice, the locally canonical distribution is used as follows. It is assumed to be correct at some initial time $t = 0$. Then the system is allowed to "age" for a certain time interval τ. This time is supposed to be long enough for many dynamical processes (e.g., collisions between molecules) to occur, but still short enough that the macroscopic state of the system has not changed appreciably. Finally, the mean value of the current $J = \dot{A}$ is measured at τ. This average is given explicitly by

$$\bar{J}(\tau) = Q^{-1} \iint dp dq \dot{A}(\tau) \exp(-\beta H - \gamma A)$$

Small deviations from equilibrium correspond to small γ. In fact, the thermodynamic force associated with the variable $\bar{A}$ can be shown to be just

$$\mathfrak{F} = k_B\gamma + O(\gamma^2)$$

This suggests an expansion of the average current in powers of γ, giving

$$\bar{J}(\tau) = \langle J(\tau)\rangle - \gamma\langle \dot{A}(\tau)A(0)\rangle + \gamma\langle J(\tau)\rangle\langle A(0)\rangle + O(\gamma^2)$$

It should be recalled that $\bar{J}$ is the current in the nonequilibrium ensemble, while $\langle J\rangle$ is the current at thermal equilibrium. The latter quantity vanishes, so that

$$\bar{J}(\tau) = -\gamma\langle \dot{A}(\tau)A(0)\rangle + O(\gamma^2)$$

By performing manipulations exactly like those already described in connection with the method based on the regression hypothesis, this may be transformed without further approximation into

$$\bar{J}(\tau) = L\mathfrak{F}(0)$$

The transport coefficient L is given by the same formula already obtained using the other approaches,

$$L = \frac{1}{k_B}\int_0^\tau ds\langle \dot{A}(s)\dot{A}(0)\rangle$$

By the assumption that τ is a macroscopically short time, the thermodynamic force at $t=0$ will hardly differ from its value at $t=\tau$, so that both sides of the transport law can be expressed as functions of the same time τ,

$$\bar{J}(\tau) = L\mathfrak{F}(\tau)$$

But if the phase space distribution function is locally canonical to a good approximation, then the result just obtained is independent of the choice of origin in time and is generally valid.

We shall not go into the various arguments that may be used to justify the preceding derivation. It may be of some interest, however, to notice the relation of the local equilibrium theory to the rigorous generalization of the Fokker-Planck theory set forth by Zwanzig (35). The idea behind this work was the separation of the phase space distribution function into a "relevant" part and an "irrelevant" part by means of a projection operator. The relevant part was chosen to be microcanonical in the gross variables; as is well known, there is usually no practical distinction between the microcanonical and canonical ensembles, so that the relevant part corresponds to Mori's locally canonical distribution. Then an exact kinetic equation was derived for the relevant part—this was put into the form of a generalized Fokker-Planck equation, but it could have been left just as well in the form of a kinetic equation in phase space. The effects of the "aging" process were accounted for in the time dependence of the irrelevant part of the distribution. This appeared at an intermediate step in the derivation of the kinetic equation for the relevant part.

The theoretical treatments of transport phenomena that we have just discussed all fall into two general categories, response to an externally imposed force, or free relaxation toward equilibrium. One important kind of irreversible behavior is not accounted for by these treatments: response to interaction with a heat bath. This gap is filled by the external reservoir theory.

The original intention of the external reservoir theory was to provide an obvious mechanism for the approach of a nonequilibrium system toward thermal equilibrium. This was accomplished by coupling the system to a heat bath or reservoir, in such a way that the interaction is a random process. The heat bath is supposed to be so large that it always stays in thermal equilibrium, no matter what the system is doing. In this connection, the principal references are a series of articles by Bergmann & Lebowitz (48–51).

It was then realized that external reservoirs could be used to maintain a system in a steady nonequilibrium state. For example, a steady temperature gradient could be set up by connecting the system to two reservoirs having different temperatures. By determining the various currents in this steady state, one then finds formulas for transport coefficients, e.g., thermal conductivity.

This program has been carried out particularly by McLennan (17, 52, 53). His review article provides an excellent summary of this work (54), so we shall not present any detail here. The results are the same as obtained by the other methods, so that one has learned at least that transport coefficients are independent of the method used to establish a departure from equilibrium.

Other articles covering much the same ground have been published by Kirkwood & Fitts (55), Lebowitz & Shimony (56), Bernard (57), and Lax (30).

The final method for treating irreversible processes to be discussed in this part of the review is the prediction theory due to Jaynes & Scalapino (58). This theory is an extension to time-dependent phenomena of Jaynes' earlier work (59) on the relation of information theory to statistical thermodynamics.

The gist of Jaynes' method is as follows. Suppose we know certain properties of a many-body system, e.g., its average energy and perhaps the averages of some other quantities A. How can we predict other properties of the same system? The answer, of course, has been known for a long time: we construct a canonical ensemble out of the specified mean values, and then we use the distribution function of this ensemble to calculate mean values of the other quantities. Jaynes pointed out that this familiar procedure may be regarded as an application of information theory, simply by using the Gibbs entropy of an ensemble as a measure of the amount of information one has about that ensemble. The "best" ensemble corresponding to given information about the system (in the form of mean values) is the one that describes this information without assuming any extra information—or it is the one having maximum Gibbs entropy consistent with the given information.

Jaynes proposed that nonequilibrium processes be treated in just the same way. Suppose that we know the history of certain mean values over some specified time interval. Then by maximizing the Gibbs entropy of the ensemble corresponding to this given information, we can construct a generalized canonical ensemble. This ensemble gives exactly the specified mean values, at least over the specified time interval. (The ensemble distribution function bears a very strong resemblance to Mori's local equilibrium distribution, although its source is different.)

But now we can use this generalized canonical ensemble to predict the future behavior of the system. For example, a heat current may have flowed steadily during the period of observation. The ensemble distribution function contains information about this current, in the form of a time-dependent Lagrange multiplier arising from the variational problem of maximizing entropy under the constraint of specified mean heat current. This Lagrange multiplier clearly involves the temperature gradient associated with the heat current. Now we calculate the "free," i.e. the unconstrained, future behavior of the heat current. In the first approximation, the unconstrained heat current turns out to be proportional to the temperature gradient; from the coefficient of proportionality, we get the thermal conductivity. The results of this and related calculations are in exact agreement with those obtained by other methods.

This concludes our review of six methods used to relate transport coefficients to time-correlation functions. Before turning to the next topic, however, we wish to express a few opinions as to the general inter-relations of these methods.

The indirect Kubo method appears to be the simplest and most direct. Its application is limited to those cases where we already know the structure of the relevant transport law, e.g., the Navier-Stokes equation. Further, we must also know how an externally imposed force (even though fictitious) will affect the motion of the system. As new and unfamiliar problems come up, this method may lose its attractiveness. For an example of a somewhat dubious application, consider the calculation of a reaction rate. With few exceptions, the structure of the transport laws of chemical kinetics is not known (because of the extremely complicated nature of most chemical reactions). Another somewhat similar example is in rheology, where there is still considerable argument as to how to take account of various kinds of non-Newtonian behavior.

The preceding remarks apply also to the method based on the regression hypothesis.

The Fokker-Planck theory (in the original form attributed to Green) and the local equilibrium theory of Mori are restricted to the treatment of only the low frequency parts of transport coefficients. While this does not exclude the usual hydrodynamic applications, it does exclude applications to certain interesting processes involving "memory effects," e.g., various rheological phenomena.

This restriction was removed in Zwanzig's generalized Fokker-Planck theory. This theory, however, is rather complicated and abstract; and no applications have been made so far.

As mentioned already, the results obtained by various authors differ in appearance also because of the use of different ensembles. In our opinion the most satisfactory explanation of the apparent difference, and the actual equivalence, is that of McLennan (19).

We have remarked several times that these methods all lead to the same results. This assertion must be qualified slightly. Some methods lead automatically to expressions for transport coefficients that involve integrations over a characteristic time interval τ, long compared with molecular times, and short compared with macroscopic times. Other methods (in my opinion more rigorous) lead to integrals over an infinite time interval, coupled usually with some limiting process (for example, to zero frequency). In a practical sense these two kinds of results should be regarded as identical, although they differ in formal appearance.

Applications of Time-Correlation Functions

Friction constant.—The frictional force on a Brownian motion particle (or B-particle) is $-\zeta \boldsymbol{v}$, where $\boldsymbol{v}$ is the velocity of the particle and ζ is the friction constant. Kirkwood (25) was the first to derive a statistical mechanical expression for ζ,

$$\zeta = \frac{1}{3}\beta \int_0^\tau dt \langle \boldsymbol{F}(0) \cdot \boldsymbol{F}(t) \rangle$$

The total force exerted on the B-particle by its surroundings at time t is $\boldsymbol{F}(t)$. The derivation of this formula lacks rigor; in distinction to the other transport coefficients to be discussed here, there is some question as to the general correctness of Kirkwood's result.

Many calculations of ζ, particularly for liquid argon, have been presented. An early one, typical of several others, was given by Kirkwood, Buff & Green (60) in connection with their theory of viscosity. Their calculation was based on the assumption that the time-correlation function decays exponentially,

$$\langle F(0) \cdot F(t) \rangle = \langle F(0) \cdot F(0) \rangle \exp(-\lambda t)$$

so that the time integral can be found immediately. They made the further assumption that $1/\lambda$, a natural time associated with Brownian motion, could be identified with the momentum relaxation time m/ζ. This leads directly to

$$\zeta^2 = \tfrac{1}{3}\beta m \langle F^2 \rangle = \tfrac{1}{3} m \langle \nabla^2 U \rangle$$

The second form is obtained by integration by parts; U is the total intermolecular potential felt by the B-particle.

The same formula was obtained, by a somewhat different argument, by Rice & Kirkwood (61). Their approach was based on the idea of diffusion in the space of a pair of particles. The pair force between molecules i and j at time t may be expressed as a function of the relative separation $\boldsymbol{R}_{ij}(t)$. These authors treated the change in $\boldsymbol{R}_{ij}$ during the time interval τ as small compared with typical intermolecular separations; they used a generalization of Einstein's Brownian motion equation for the mean squared separation during τ,

$$\langle \Delta \boldsymbol{R}_{ij}(\tau) \Delta \boldsymbol{R}_{ij}(\tau) \rangle = \boldsymbol{D}^{(2)}\tau$$

and they treated $\boldsymbol{D}^{(2)}$, the diffusion tensor in pair space, as a direct sum of diffusion tensors in the spaces of the individual molecules. Finally, they related $\boldsymbol{D}$ to ζ by the familiar expression $\boldsymbol{D} = k_B T/\zeta$.

Incidentally, it should be noted that the latter equation, which is valid for Brownian motion of a large particle in a bath of small particles, is probably not valid otherwise. For a discussion of the validity of formulas of this type, see Balescu (62), Swenson (63), and Zwanzig (64).

Several articles in Kirkwood's series on transport processes (65, 66) made use of a friction constant calculated in the linear trajectory approximation. Only the result of the approximation was given in these articles; the general theory, and some extensions, was discussed later by Helfand (67). In essence, the approximation consists of calculating the force $\boldsymbol{F}(t)$ on the B-particle as if the particle were moving in a straight line, unaffected by interactions with the environment. This approximation is in fact exact in the limit of weak interactions between particles, as shown first by Ross (68). It does not appear to be particularly sensible for dense gases or liquids, so modifications have been made to take account of hard core repulsions as well as weak attractions; see Rice & Allnatt (69) and Helfand (67).

We mention finally a curious acoustical model, in which the environment is treated as a hydrodynamic continuum. The mechanism of friction is radiation damping associated with the emission of sound waves. This approach was investigated by Fixman (70), Kotin (71), and Rice (72). Rice's article contains a good summary of the relations between various methods.

The friction constant has a special and annoying property not possessed by the other transport coefficients to be discussed here; the cutoff time τ cannot safely be replaced by infinity. If this is done, the whole integral vanishes. This peculiarity has been discussed by Suddaby & Gray (73), Helfand (67), and Zwanzig (44). Its main consequence is that physical intuition has to be used in the choice of τ. This is an important reason for doubting the general exactness of Kirkwood's theory.

There are, however, some exceptions. If the decay of the time-correlation function of momentum is governed by some parameter of slowness ϵ, then the corresponding friction constant is given exactly by

$$\zeta = \frac{1}{3}\beta \int_0^\infty dt \lim_{\epsilon \to 0} \langle F(0) \cdot F(t) \rangle$$

This formula applies in particular in the limit when the B-particle interacts weakly with its environment, so that ϵ is the interaction strength. In this case one obtains the linear trajectory result. It applies also when the B-particle is very much heavier than the surrounding molecules, so that ϵ is the mass ratio. These exact results were first obtained by Lebowitz & Rubin (27), and later by Zwanzig (44); they constitute the only rigorous formulas available for friction constants.

In the heavy mass limit, which might be termed "real" Brownian motion, ζ has been calculated explicitly in two extreme limits. Green (74), Mazo (75), and Andersen & Shuler (117) have treated the "Rayleigh" gas, where the mean free path of the molecules in the environment is much greater than the radius of the B-particle. In another extreme, the environment is a dense fluid treated as a hydrodynamic continuum undergoing fluctuations. The friction constant for this case, calculated by Zwanzig (76), is exactly as given by Stokes' law.

Kirkwood's theory of the friction constant for translational motion has its analogue for rotations. This topic has been treated by Steele (77).

Viscosity.—Formulas for the coefficients of shear viscosity η and bulk viscosity η_v were obtained first by Green (12) by means of the Fokker-Planck theory. Thereafter, the same results were found by most of the other procedures outlined in the previous section. We pass by a certain amount of controversy about the agreement between results obtained by different methods, which has now been settled, and present only the final results.

Time-correlation function expressions for shear and bulk viscosity, in classical mechanics, are given by

$$\eta = \frac{1}{V}\beta \int_0^\infty dt \langle J^{xy}(0) J^{xy}(t)\rangle$$

$$\eta_v = \frac{1}{V}\beta \int_0^\infty dt \frac{1}{9}\sum_a \sum_b \langle J^{aa}(0) J^{bb}(t)\rangle$$

Here V is the volume of the system. The tensor J^{ab}, (for a, $b=1, 2, 3$), is

$$J^{ab} = \sum_j \frac{p_{ja}p_{jb}}{m_j} + \sum_j R_{ja}F_{jb} - \langle J^{ab}\rangle$$

The a-th cartesian component of the position of the j-th molecule is R_{ja}; the corresponding component of momentum is p_{ja}; the a-th component of total force on the j-th molecule is F_{ja}; the mass of this molecule is m_j; and the subtracted part $\langle J^{ab}\rangle$ is a function of pressure $\bar{P}$, internal energy E, and particle number N,

$$\langle J^{ab}\rangle = V\delta_{ab}\left\{\bar{P} + \frac{\partial \bar{P}}{\partial N}(N - \langle N\rangle) + \frac{\partial \bar{P}}{\partial E}(E - \langle E\rangle)\right\}$$

This subtracted part, needed only for the bulk viscosity, takes account of the ensemble used for the averages. For example, if we use the microcanonical ensemble, then $N=\langle N\rangle$ and $E=\langle E\rangle$, so two terms drop out. With the canonical ensemble, the energy difference does not vanish, while with the grand canonical ensemble, both terms are present.

The first treatment of thermal transport coefficients by means of time-correlation functions appears in a virtually unnoticed article by Herzfeld (78) in 1958. His approach is based on the assumption that the time-correlation function decays exponentially. As in the example of the friction constant that we have already discussed, this assumption relates the transport coefficient to a relaxation time and to an equilibrium fluctuation. Herzfeld did not provide any theoretical discussion of the relaxation time. In calculating the equilibrium fluctuation, he made several approximations, e.g., neglect of certain correlations in space. For the shear viscosity, he obtained finally an expression very much like the one obtained by Eyring using the transition state theory. While Herzfeld's theory is not conclusive, it is very attractive and deserves further investigation.

Several authors have discussed the equivalence of the time-correlation function method and the more familiar Boltzmann equation method for calculating the viscosity of gases at low densities: Mori (79), Fujita (80), and McLennan & Swenson (81). It turns out that in evaluating the time-correlation function, one finds at an intermediate stage that one has to solve a certain integral equation. This equation is identical with an equation appearing in the first Chapman-Enskog solution of the Boltzmann equation; it is the equation that has been used many times to calculate the viscosity of low density gases, and the results are known to be in excellent agreement with experiment.

To go to the next order in density, one must take proper account of triple collisions. Approximate calculations based on time-correlation functions have been reported by Ono & Shizume (82) and Wainwright (83). An exact treatment was given by Kawasaki & Oppenheim (84). Their calculation was based on a generalization of a procedure used by Zwanzig (85) to discuss self-diffusion in gases. The main result obtained by Kawasaki & Oppenheim, as in the two-body case, was that a certain integral equation appeared at an intermediate stage. This equation is a generalization of the Chapman-Enskog equation to the next order in density. The same equation is obtained from solution of the generalized Boltzmann equation derived by Choh & Uhlenbeck (86). The equation contains a kernel, analogous to the collision cross section for binary interactions, but containing also three body interactions. As yet no one has calculated this kernel, so that no numerical results have been obtained.

Ernst (118) has recently presented a somewhat different demonstration of the equivalence of the time-correlation approach and the generalized Boltzmann equation approach.

Thermal conductivity.—The coefficient of thermal conductivity, in classical mechanics, is

$$K = \frac{1}{3Vk_BT^2}\int_0^\infty dt\langle S(0)\cdot S(t)\rangle$$

The "heat current" S appearing here is given most conveniently in the form

$$S = \frac{d}{dt} \sum_j (H_j - h_j) R_j$$

where H_j is the contribution of the j-th molecule to the total Hamiltonian of the system, and h_j is the enthalpy per molecule. (The potential energy of interaction of pairs of molecules is divided equally between partners.)

References given in the discussion of viscosity, concerned with the question of density expansions, are equally relevant to the theory of thermal conductivity of gases. For example, it is known that the above formula gives results identical with the conventional Boltzmann equation approach, in the binary collision regime.

Considerable attention has been devoted to the use of time-correlation function methods in calculating thermal conductivity of crystal lattices (due to anharmonic phonon-phonon scattering or impurity scattering). We shall not discuss this work here, but merely mention some names: Hardy (87), Schieve & Peterson (88), and Maradudin (89).

A characteristic property of metals is expressed in the Wiedemann-Franz law, which relates the thermal conductivity of a metal to its electric conductivity. In a beautiful application of time-correlation function methods, Chester & Thellung (90) have shown that this law is an exact consequence of the one-electron theory of metals.

Chemical reaction rates.—Consider a chemical reaction of the form

$$\nu_1 A_1 + \nu_2 A_2 + \cdots \rightleftarrows \nu_1' B_1 + \nu_2' B_2 + \cdots$$

The stoichiometric coefficients ν_i of the reactants are taken to be positive, and the ν_j' (with primes) of the products are taken to be negative. Let $N_1, N_2, \cdots$ and $M_1, M_2, \cdots$ be the number of molecules of reactants and products present in the mixture; and use ΔN_j and ΔM_j for deviations from thermal equilibrium values. Because of the structure of the rate equation, the various changes in composition are all related by

$$\frac{\Delta N_1}{\nu_1} = \frac{\Delta N_2}{\nu_2} = \cdots = \frac{\Delta M_1}{\nu_1'} = \frac{\Delta M_2}{\nu_2'} = \cdots = \Delta N$$

which defines ΔN. Thus we may use the single variable ΔN to describe the extent to which the reaction has taken place at any time.

The thermodynamic driving force for the reaction is the chemical affinity $\mathcal{A}$,

$$\mathcal{A} = -\sum \nu_i \mu_i - \sum \nu_j' \mu_j'$$

where μ and μ' are the chemical potentials of the reactants and products. For small departures from equilibrium, the rate of the reaction is expressed (in the characteristic Onsager form of irreversible thermodynamics) as

$$\frac{d\langle \Delta N \rangle}{dt} = G\mathcal{A}$$

The rate is a constant times the driving force. Then G is the rate constant of the reaction. (The forward and reverse rates can be obtained from knowledge of G and of the equilibrium constant of the reaction.)

Having written the rate equation in the standard form, it is easy to see that G is given by

$$G = \beta \int_0^\infty dt \langle \tilde{\dot{N}}(0) \dot{N}(t) \rangle$$

The rate constant is determined by the time-correlation function of the instantaneous time derivative dN/dt.

This result was derived first by Yamamoto (91). A later review by Aroeste (92) contains a similar treatment.

Up to now, the only progress made along these lines has been the demonstration that the time-correlation function can be calculated in terms of reaction cross sections at low density, leading to results known from more intuitive arguments (91, 92).

Nuclear magnetic resonance.—The earliest, and most fruitful, application of time-correlation function techniques to problems of irreversible behavior were made in connection with NMR line shapes. The basic formula is due to Kubo & Tomita (13). Let $\chi_{\mu\nu}(\omega)$ be the frequency dependent magnetic susceptibility tensor of a material, let $\boldsymbol{M}$ be the total magnetic moment of the system, and let V be its volume. Then χ is expressed by means of the time-correlation function of the rates of change of magnetic moments by

$$\chi_{\mu\nu}(\omega) = \frac{\beta}{i\omega V} \int_0^\infty dt e^{-i\omega t} \langle \tilde{\dot{M}}_\nu(0) \dot{M}_\mu(t) \rangle$$

An alternative form, obtained by integrating by parts, is

$$\chi_{\mu\nu}(\omega) = \frac{\beta}{V} \langle \tilde{M}_\nu M_\mu \rangle - \frac{\beta i\omega}{V} \int_0^\infty dt e^{-i\omega t} \langle \tilde{M}_\nu(0) M_\mu(t) \rangle$$

This contains $\boldsymbol{M}$ instead of $d\boldsymbol{M}/dt$, and shows clearly the relation of $\chi(\omega)$ to the familiar static result.

The substantial literature dealing with applications of this formula is reviewed regularly, in these volumes and elsewhere: so we just mention here a few articles that appear to have special interest.

Argyres & Kelley (93) have recently given a thorough and general analysis of the theory of resonance and relaxation in spin systems, using modern quantum statistical mechanical methods.

Oppenheim, Bloom & Torrey (94–96) developed a theory of nuclear spin-lattice relaxation times in gases and liquids. An interesting feature of this work is the use of a "constant acceleration approximation" in the calculation of time-correlation functions. (From their point of view, the linear trajectory approximation that we referred to earlier may be termed a "constant velocity approximation.")

Electric conductivity.—The time-correlation function formula for the electric conductivity tensor $\sigma_{\mu\nu}(\omega)$ is

$$\sigma_{\mu\nu}(\omega) = \frac{\beta}{V}\int_0^\infty dt e^{-i\omega t}\langle \bar{J}_\nu(0) J_\mu(t)\rangle$$

where J is the total electric current density in the system. This formula was given by Kubo in 1957.

Many calculations of σ have been based on Kubo's formula. Most of this effort has been directed at the electric conductivity of metals; see, in particular, articles by Montroll & Ward (97), Berger, Van Leuwen & Verboven (98), and Janner (99). We do not comment on this work, except to note that as usual, the results agree with those obtained by more elementary methods in the limiting cases where elementary methods apply (e.g., low density of scattering impurities).

Of considerable interest to physical chemists is a treatment of electric conductivity of ionic solutions, due to Friedman (100). He obtained a cluster expansion of the above formula for σ, modeled on an expansion found earlier by Montroll & Ward (97). This approach, while not yet thoroughly explored, is likely to lead to a much more detailed molecular picture of ionic conductivity than any other theory presently available.

Dielectric susceptibility.—In applying Kubo's method to the calculation of dielectric susceptibility, a special problem comes up, associated with the long range character of dipolar forces. This problem is familiar in the theory of the static dielectric constant, and is discussed fully, for example, in Fröhlich's book (101).

The difficulty can be put as follows. The most natural way to apply Kubo's method is to impose an electric field on a macroscopic body, and to then determine its total induced electric moment. The ratio of these quantities is called the net polarizability of the body. But, as is well known, the net polarizability does not depend only on the dielectric susceptibility of the material, but also on the shape of the body. For this reason, if one wants a time-correlation function theory of dielectric susceptibility, one must specify the shape of the body for which the time-correlation function is to be calculated.

Consider a spherical specimen of the material of interest, and, for simplicity, consider only an isotropic material, so that the dielectric susceptibility $\epsilon(\omega)$ is a scalar quantity. The net polarizability $\alpha_s(\omega)$ of the sphere is related to ϵ by

$$\frac{\epsilon(\omega) - 1}{\epsilon(\omega) + 2} = \frac{4\pi}{3}\alpha_s(\omega)$$

Kubo's method gives for α_s the formula

$$\alpha_s(\omega) = \frac{\beta}{3V}\langle M^2\rangle - \frac{\beta i\omega}{3V}\int_0^\infty dt e^{-i\omega t}\langle M(0)\cdot M(t)\rangle$$

where M is the total instantaneous electric moment of the sphere. (The resemblance of this formula to one given previously for magnetic susceptibility will be evident.) The preceding equation appears to have been given first by Glarum (102).

Glarum's article contains also an interesting discussion of the relation between macroscopic decay times (i.e., those associated with the decay of macroscopic polarization) and microscopic relaxation times (i.e., those associated with such molecular processes as re-orientation of individual molecules).

Starting with the above formula for $\epsilon(\omega)$, Zwanzig (103) investigated the effects of dipolar interactions on the relaxation process. He considered, as a model, a rigid lattice of permanent point dipoles, undergoing isotropic Brownian motion (as in Debye's theory of dielectric relaxation). The calculation took account of the torques exerted on dipoles as a result of interactions with electric fields due to other dipoles. It was found that as soon as one observed deviations from ideal Debye-Langevin behavior of the static dielectric constant, one observed also deviations from ideal Debye relaxation.

In 1956, Fano (104) presented a quantum mechanical theory of dielectric behavior in dense media. He obtained some very interesting and significant results concerning the relation of ϵ to the time-correlation function of the charge density. In particular, he observed that one gets either ϵ or its reciprocal, depending on whether or not one includes certain long range electrostatic interactions in the Hamiltonian governing the time dependence of the time-correlation function. The same observation has been made, in a somewhat different context, by Ambegaokar & Kohn (105, 106).

The theory of pressure broadening, i.e., the effects of gas collisions on spectral line shapes, has been investigated by means of time-correlation functions by Fano (107). A technical feat of this work was the derivation of a relation between the scattering operator in the quantum mechanical Liouville representation and the corresponding scattering operator in the Schrodinger representation.

Inelastic neutron scattering.—The cross section for inelastic scattering of neutrons by macroscopic systems may be expressed in terms of time-correlation functions, as shown by Van Hove (108). We do not give all the details; in essence, the coherent part of the scattering, associated with momentum transfer $\hbar q$ and energy transfer $\hbar\omega$, is proportional to

$$S_c(q, \omega) = \int \left\langle \sum_l e^{-iq\cdot\mathbf{R}_l(0)} \sum_j e^{iq\cdot\mathbf{R}_j(t)} \right\rangle e^{-i\omega t} dt$$

This is evidently the frequency spectrum of the q-th Fourier component of the local density of the material. The incoherent part of the scattering is proportional to

$$S_i(q, \omega) = \int \sum_l \langle e^{-iq\cdot R_l(0)} e^{iq\cdot R_l(t)} \rangle e^{-i\omega t} dt$$

Note that the cross terms ($j \neq l$) present in S_c are omitted from S_i.

The extensive literature on these quantities has been reviewed many times, particularly in the proceedings of symposia organized by the International Atomic Energy Agency (109). For some recent advances in the detailed theoretical analysis of the above time-correlation functions, see articles by Nelkin & Ghatak (110), Yip & Nelkin (111), Nossal (112), Van Leeuwen & Yip (113), and Singwi (114).

The incoherent part of the scattering can be related to the time-correlation function of the current density of a particular molecule in the medium; by using this relation, Zwanzig (115) showed how $S_i(q, \omega)$ is related to a generalized self-diffusion coefficient $D(\boldsymbol{q}, \omega)$.

Rahman (116) solved the equations of motion of 864 argon atoms in a box, by means of high speed digital computers, under conditions appropriate to liquid argon. Among many interesting results, he calculated the time-correlation function of velocity and its frequency spectrum. These quantities differ substantially from previous expectations. Rahman's calculations provide what is probably the most detailed "experimental" information currently available about dynamical processes in liquids.

Future Developments

In the final part of this review, I make the dangerous attempt to predict some future developments in this field. To a large extent, the following predictions are described better as wishful thinking; these are some of the things that I would like to see worked out.

With respect to practical calculations of transport coefficients, little progress can be expected in the next few years. The easy problems were done long ago by other methods (e.g., viscosity of a gas at low density). The time-correlation function method provides a more rigorous basis for the older theory, but it adds little of practical value. Attempts are being made to evaluate the triple collision contribution to transport coefficients of gases, but the theory looks extremely hard. Even if it is successful, such calculations probably will not replace direct measurement as the most economical method of getting numbers.

There seems to be little question as to the correctness of time-correlation expressions for the standard transport coefficients. I expect that a fair amount of attention will be given to the rigorous treatment of transport processes involving nonlocal effects of the sort described by frequency and wave vector dispersion. Investigations of this type may help to bridge the gap between macroscopic hydrodynamics and molecular dynamics. That is, hydrodynamic processes at sufficiently high frequency and short wavelength involve the dynamics of individual molecules to a much greater extent than at the familiar low frequencies and long wavelengths.

Another direction in which the present theory will be extended is to include nonlinear irreversible processes. It is known already that nonlinear effects are important in some rheological situations, for example, the flow of polymer solutions; but so far there is very little worked out statistical mechanical theory of the appropriate hydrodynamic equations. A modest beginning would be to estimate the magnitude of rate of shear needed to get appreciable nonlinear effects in, say, a gas at low density.

Nonlinear effects are of the greatest importance in chemical kinetics; the theory due to Yamamoto that was described earlier was limited to small deviations from equilibrium. One would like to be able to estimate how large a deviation from equilibrium must be to invalidate familiar methods. Also, in the kinetics of very fast reactions, it is probably important to take into account frequency dispersion of the rate constant. This has never been done.

Perhaps the most valuable use of time-correlation functions will be in relating different transport coefficients and thermodynamic properties. An example is the discussion of the Wiedemann-Franz law by Chester & Thellung. This kind of investigation is particularly valuable when it does not require explicit calculation of transport coefficients, but only knowledge of the structure of the transport coefficient formulas.

One example which will prove of great value is in the theory of ionic solutions. Electric conductivity is given by a well defined time-correlation function expression. Experiments and previous theoretical work suggest a strong relation between ionic conductivity and viscosity. The viscosity coefficient is also given by a time-correlation function expression. It is natural to hope that some direct connection will be found, independent of special models or detailed calculations.

The success of Eyring's theory of absolute reaction rates is well known. In some circles, however, this theory is in disrepute because it does not stand as the first step in a well defined sequence of approximations. I expect that in the next few years, the relation of Eyring's methods to time-correlation function methods will be clarified. Probably, the theory of viscosity in dense media will be developed by finding some kind of "reaction path" in phase space which will dominate the time dependence of the appropriate time-correlation function.

In conclusion, I expect that most of the progress in this field during the next few years will come from analysis of the structure of time-correlation function formulas, rather than from actual calculation.

LITERATURE CITED

1. Rubin, R. J., *Phys. Rev.*, **131,** 96 (1963)
2. Jaynes, E. T. (Private communication)
3. Davidson, N., *Statistical Mechanics* (McGraw-Hill, New York, 1962)
4. Kittel, C., *Elementary Statistical Physics* (Wiley, New York, 1958)
5. Landau, L. D., and Lifshitz, E. M., *Statistical Physics* (Addison-Wesley Publ. Co., Inc., Reading, Mass., 1958)
6. Kubo, R., *J. Phys. Soc. Japan*, **12,** 570 (1957)
7. Kubo, R., *Lectures Theoret. Phys, (Boulder)*, **1,** 120 (1958)
8. Callen, H. B., and Welton, T. A., *Phys. Rev.*, **83,** 34 (1951)
9. Callen, H. B., and Greene, R. F., *Phys. Rev.*, **86,** 702 (1952); **88,** 1387 (1952)
10. Bernard, W., and Callen, H. B., *Rev. Mod. Phys.*, **31,** 1017 (1959)
11. Green, M. S., *J. Chem. Phys.*, **20,** 1281 (1952)
12. Green, M. S., *J. Chem. Phys.*, **22,** 398 (1954)
13. Kubo, R., and Tomita, K., *J. Phys. Soc. Japan*, **9,** 888 (1954)
14. Blatt, J. M., and Matsubara, T., *Progr. Theoret. Phys. (Kyoto)*, **21,** 696 (1959)
15. Mori, H., *Phys. Rev.*, **112,** 1829 (1958)
16. Mori, H., *Progr. Theoret. Phys. (Kyoto)*, **28,** 763 (1962)
17. McLennan, J. A., Jr., *Phys. Fluids*, **3,** 493 (1960)
18. Green, M. S., *Phys. Rev.*, **119,** 829 (1960)
19. McLennan, J. A., Jr., *Progr. Theoret. Phys. (Kyoto)*, **30,** 408 (1963)
20. Luttinger, J. M., *Phys. Rev.*, **135A,** 1505 (1964)
21. Kadanoff, L. P., and Martin, P. C., *Ann. Phys.*, **24,** 419 (1963)
22. Jackson, J. L., and Mazur, P., *Physica*, **30,** 2295 (1964)
23. Felderhof, B. U., and Oppenheim, I. (Private communication)
24. Montroll, E. W., *Rendiconti Della Scuola Internazionale Di Fisica Enrico Fermi*, **X,** 217 (Societa Italiana Di Fisica, Bologna, 1960)
25. Kirkwood, J. G., *J. Chem. Phys.*, **14,** 180 (1946)
26. Rice, S. A., and Frisch, H. L., *Ann. Rev. Phys. Chem.*, **11,** 187 (1960)
27. Lebowitz, J. L., and Rubin, E., *Phys. Rev.*, **131,** 2381 (1963)
28. deGroot, S. R., and Mazur, P., *Non-Equilibrium Thermodynamics* (North-Holland, Amsterdam, 1962)
29. Hashitsume, N., *Progr. Theoret. Phys.*, **8,** 461 (1952); **15,** 369 (1956)
30. Lax, M., *Rev. Mod. Phys.*, **32,** 25 (1960)
31. Onsager, L., and Machlup, S., *Phys. Rev.*, **91,** 1505 (1953); *ibid.* 1512
32. Uhlhorn, U., *Arkiv Fysik*, **17,** 193, 233, 257, 273, 343, 361 (1960)
33. Van Kampen, N. G., *Physica*, **23,** 707, 816 (1957)
34. Yamamoto, T., *Progr. Theoret. Phys. (Kyoto)*, **10,** 11 (1953)
35. Zwanzig, R., *Phys. Rev.*, **124,** 983 (1961)
36. Sewell, G. L. (Preprint)
37. Mori, H. (Preprint)
38. Helfand, E., *Phys. Rev.*, **119,** 1 (1960)
39. Landau, L. D., and Lifshitz, E. M., *Electrodynamics of Continuous Media;* and *Fluid Mechanics* (Addison-Wesley Publ. Co., Inc., Reading, Mass., 1960)
40. Onsager, L., *Phys. Rev.*, **37,** 405 (1931); **38,** 2265 (1931)
41. Casimir, H. B. G., *Rev. Mod. Phys.*, **17,** 343 (1945)
42. Wigner, E. P., *J. Chem. Phys.*, **22,** 1912 (1954)
43. Kubo, R., Yokota, M., and Nakajima, S., *J. Phys. Soc. Japan*, **12,** 1203 (1957)
44. Zwanzig, R., *J. Chem. Phys.*, **40,** 2527 (1964)
45. Mori, H., *J. Phys. Soc. Japan*, **11,** 1029 (1956)
46. Longuet-Higgins, H. C., *Mol. Phys.*, **6,** 65 (1963)
47. Andersen, H. C., Oppenheim, I., Shuler, K. E., and Weiss, G. H., *J. Math. Phys.*, **5,** 522 (1964)
48. Lebowitz, J. L., and Bergmann, P. G., *Ann. Phys.*, **1,** 1 (1957)
49. Bergmann, P. G., and Lebowitz, J. L., *Phys. Rev.*, **99,** 578 (1955)
50. Lebowitz, J. L., *Phys. Rev.*, **114,** 1192 (1959)
51. Willis, C. R., and Bergmann, P. G., *Phys. Rev.*, **128,** 391 (1962)
52. McLennan, J. A., Jr., *Phys. Rev.*, **115,** 1405 (1959)
53. McLennan, J. A., Jr., *Phys. Fluids*, **4,** 1319 (1961)
54. McLennan, J. A., Jr., in *Advan. Chem. Phys.*, **5,** 261 (1963)
55. Kirkwood, J. G., and Fitts, D. D.,

J. Chem. Phys., **33,** 1317 (1960)
56. Lebowitz, J. L., and Shimony, A., *Phys. Rev.*, **128,** 1945 (1962)
57. Bernard, W., *Phys. Rev.*, **128,** 421 (1962)
58. Jaynes, E. T., and Scalapino, D. (Private communication)
59. Jaynes, E. T., *Phys. Rev.*, **106,** 620 (1957); **108,** 171 (1957)
60. Kirkwood, J. G., Buff, F. P., and Green, M. S., *J. Chem. Phys.*, **17,** 988 (1949)
61. Rice, S. A., and Kirkwood, J. G., *J. Chem. Phys.*, **31,** 901 (1959)
62. Balescu, R., *Physica*, **27,** 693 (1961)
63. Swenson, R. J., *Physica*, **29,** 1174 (1963)
64. Zwanzig, R., *Lectures Theoret. Phys.* (*Boulder*), **3,** 106 (1960)
65. Zwanzig, R. W., Kirkwood, J. G., Stripp, K. F., and Oppenheim, I., *J. Chem. Phys.*, **21,** 2050 (1953)
66. Zwanzig, R. W., Kirkwood, J. G., Oppenheim, I., and Alder, B., *J. Chem. Phys.*, **22,** 783 (1954)
67. Helfand, E., *Phys. Fluids*, **4,** 681 (1961)
68. Ross, J., *J. Chem. Phys.*, **24,** 375 (1956)
69. Rice, S. A., and Allnatt, A. R., *J. Chem. Phys.*, **34,** 2144 (1961)
70. Fixman, M., *J. Chem. Phys.*, **28,** 397 (1958)
71. Kotin, L. (Doctoral thesis, Harvard Univ., Cambridge, Mass., 1960; Referred to in ref. 72)
72. Rice, S. A., *Mol. Phys.*, **4,** 305 (1961)
73. Suddaby, A., and Gray, P., *Proc. Phys. Soc.* (*London*), **A75,** 109 (1960)
74. Green, M. S., *J. Chem. Phys.*, **19,** 1036 (1951)
75. Mazo, R. A., *J. Chem. Phys.*, **35,** 831 (1961)
76. Zwanzig, R., *J. Res. Natl. Bur. Std.*, **68**(B), 143 (1964)
77. Steele, W. A., *J. Chem. Phys.*, **38,** 2404 (1963)
78. Herzfeld, K. F., *J. Chem. Phys.*, **28,** 595 (1958)
79. Mori, H., *Phys. Rev.*, **111,** 694 (1958)
80. Fujita, S., *J. Math. Phys.*, **3,** 359 (1962)
81. McLennan, J. A., Jr., and Swenson, R. J., *J. Math. Phys.*, **4,** 1527 (1963)
82. Ono, S., and Shizume, T., *J. Phys. Soc. Japan*, **18,** 29 (1963)
83. Wainwright, T., *J. Chem. Phys.*, **40,** 2932 (1964)
84. Kawasaki, K., and Oppenheim, I., *Phys. Rev.*, **136**(A), 1519 (1964)
85. Zwanzig, R., *Phys. Rev.*, **129,** 486 (1963)
86. Choh, S. T., and Uhlenbeck, G. E., *The Kinetic Theory of Phenomena in Dense Gases* (Doctoral thesis, Univ. Michigan, Ann Arbor, Mich., 1958)
87. Hardy, R. J., Doctoral thesis, Lehigh University (1962); *Phys. Rev.*, **132,** 168 (1963)
88. Schieve, W. C., and Peterson, R. L., *Phys. Rev.*, **126,** 1458 (1962)
89. Maradudin, A. A., *J. Am. Chem. Soc.*, **86,** 3405 (1964)
90. Chester, G. V., and Thellung, A., *Proc. Phys. Soc.*, **77,** 1005 (1961)
91. Yamamoto, T., *J. Chem. Phys.*, **33,** 281 (1960)
92. Aroeste, H., in *Advan. Chem. Phys.*, **6,** 1 (1964)
93. Argyres, P. N., and Kelley, P. L., *Phys. Rev.*, **134A,** 98 (1964)
94. Oppenheim, I., and Bloom, M., *Can. J. Phys.*, **39,** 845 (1961)
95. Bloom, M., and Oppenheim, I., *Can. J. Phys.*, **41,** 1580 (1963)
96. Oppenheim, I., Bloom, M., and Torrey, H. C., *Can. J. Phys.*, **42,** 70 (1964)
97. Montroll, E. W., and Ward, J. C., *Physica*, **25,** 423 (1949)
98. Berger, P., Van Leeuwen, J. M. J., and Verboven, E., *Physica*, **29,** 1409 (1963)
99. Janner, A., *Helv. Phys. Acta*, **36,** 857 (1963)
100. Friedman, H. L., *Physica*, **30,** 509, 537 (1964)
101. Fröhlich, H., *Theory of Dielectrics* (Oxford Univ. Press, London, 1949)
102. Glarum, S., *J. Chem. Phys.*, **33,** 1371 (1960)
103. Zwanzig, R., *J. Chem. Phys.*, **38,** 2766 (1963)
104. Fano, U., *Phys. Rev.*, **103,** 1202 (1956)
105. Ambegaokar, V., and Kohn, W., *Phys. Rev.*, **117,** 423 (1960)
106. Ambegaokar, V., *Phys. Rev.*, **121,** 91 (1961)
107. Fano, U., *Phys. Rev.*, **131,** 259 (1963)
108. Van Hove, L., *Phys. Rev.*, **95,** 249 (1954)
109. *Proc. Symp. Inelastic Scattering Neutrons Solids Liquids, 1960 and 1962* (Intern. At. Energy Agency, Vienna, 1961 and 1963)

110. Nelkin, M., and Ghatak, A., *Phys. Rev.*, **135A,** 4 (1964)
111. Yip, S., and Nelkin, M., *Phys. Rev.*, **135A,** 1241 (1964)
112. Nossal, R., *Phys. Rev.*, **135A,** 1579 (1964)
113. Van Leeuwen, J. M. J., and Yip, S., *Phys. Rev.* (In press)
114. Singwi, K. S., *Phys. Rev.*, **136A,** 969 (1964)
115. Zwanzig, R., *Phys. Rev.*, **133A,** 50 (1964)
116. Rahman, A., *Phys. Rev.*, **136A,** 405 (1964)
117. Andersen, K., and Shuler, K. E., *J. Chem. Phys.*, **40,** 633 (1964)
118. Ernst, M. H. J. J., *Transport Coefficients from Time Correlation Functions* (Doctoral thesis, Univ. Amsterdam, 1964)

CATALYSIS BY COORDINATION COMPOUNDS[1,2]

By Jack Halpern

Department of Chemistry, University of Chicago, Chicago, Illinois

Introduction

Among the most important developments in the field of catalysis in recent years are those relating to the discovery and study of a variety of new catalytic reactions of transition metal ions and coordination compounds. This general field encompasses both homogeneous and heterogeneous systems and includes a variety of reactions, notably those involving the catalytic activation of inert saturated molecules (e.g., H_2), as well as a variety of addition reactions—the hydrogenation of olefins and related substrates, the catalytic oxidation of olefins, polymerization by catalysts of the Ziegler-Natta type, hydroformylation, hydration of acetylenes, etc.

The present review is concerned with this general field. The scope of the subject precludes a comprehensive review of the whole field and makes it necessary to restrict the coverage to certain classes of reactions. Those selected include: homogeneous catalytic hydrogenation; oxidation of olefins; hydration of acetylenic compounds; and certain aspects of catalytic polymerization. The first of these is among the simplest and most extensively studied classes of reactions in this general field and is reviewed in some detail; the review of the remaining topics is much more selective and is confined to recent developments of a rather fundamental nature. The selection of topics and of emphasis reflects both the reviewer's interests and his attempt to present, insofar as is possible at this stage, a systematic picture of a field in which the connection between various phenomena is only partly understood.

Of necessity, certain other classes of reactions, although connected with the theme of this review, have had to be omitted from consideration. Among these are the great variety of polymerization and cyclopolymerization reactions of acetylenes catalyzed by transition metal complexes, metal ion-catalyzed hydrolysis and related reactions, metal ion-catalyzed autoxidations, etc.

Insofar as the catalytic phenomena which comprise the subject of this review touch on many aspects of the bonding and structure of coordination compounds of the transition metals, they draw heavily on recent developments in inorganic chemistry, notably those connected with the ligand field theory. A recent review by Nyholm (1) on the structure and reactivity of transition metal compounds is particularly pertinent to this theme.

[1] The survey of literature pertaining to this review was concluded in December 1964.

[2] Support of this work by the National Institutes of Health (USPHS Grant No. GM-10662) and by the Petroleum Research Fund, administered by the American Chemical Society, is gratefully acknowledged.

Homogeneous Catalytic Hydrogenation

Until fairly recently the known examples of catalysts for reactions of molecular hydrogen were solids, with few exceptions, and their catalytic activity was widely believed to be related to properties, such as electronic band structure, or to a particular geometric arrangement of atoms characteristic of the solid state. The first clear-cut demonstration of a homogeneously catalyzed hydrogenation reaction in solution is due to Calvin (2) who showed in 1938 that copper (I) acetate, in quinoline solution, catalyzes the homogeneous reduction by molecular hydrogen of substrates such as copper (II) or benzoquinone. This remained almost the only example of a homogeneous hydrogenation catalyst for some time, and it is only during the past decade that the scope of this subject has expanded considerably with the discovery and study of a large number of such catalysts, most of them ions or complexes of transition metals.

More than a hundred publications have appeared during this period related to the study of homogeneously catalyzed hydrogenation reactions and directed at the elucidation in considerable detail of the mechanisms of such reactions, the factors governing catalytic activity, and the connection between catalytic activity in these systems and those involving heterogeneous or enzymic hydrogenation catalysts, or catalysts for the activation of other inert molecules, e.g., hydrocarbons. The majority of these publications have appeared since this subject was last reviewed in 1959 [Halpern (3, 4)], and it is with these recent developments that this review will be for the most part concerned. Among the most significant of these developments are the discovery and study of a number of homogeneous catalysts for the hydrogenation of olefins and related substrates.

Copper(II).—Halpern, Macgregor & Peters (5, 6) have examined the catalysis by copper(II) of the oxidation of H_2 by substrates such as chromium(VI). In aqueous perchloric acid solution, the reaction follows the rate law, $-d[H_2]/dt = k_1k_2[H_2][Cu^{2+}]^2/(k_{-1}[H^+]+k_2[Cu^{2+}])$ where $k_1 = 1.0 \times 10^{13}$ exp $[-26{,}600/RT]M^{-1}sec^{-1}$ and $(k_{-1}/k_2) = 0.25$ (at 110°). This is in accord with the mechanism,

$$Cu^{2+} + H_2 \underset{k_{-1}}{\overset{k_1}{\rightleftarrows}} CuH^+ + H^+ \quad 1.$$

$$CuH^+ + Cu^{2+} \xrightarrow{k_2} 2Cu^+ + H^+ \quad 2.$$

$$3Cu^+ + Cr(VI) \xrightarrow{\text{fast}} 3Cu^{2+} + Cr(III) \quad 3.$$

At sufficiently high ratios of Cr(VI) to Cu^{2+}, the direct oxidation of CuH^+ by Cr(VI) was also observed [Hahn & Peters (7)]. A similar mechanism was invoked by McDuffie et al. (8) to account for the catalysis by copper(II) of the reaction between H_2 and O_2 in aqueous solution. Evidence for the reversible heterolytic splitting of H_2 by Cu^{2+}, according to step (1), has also been provided by studies of the copper(II)-catalyzed isotopic exchange reaction, $D_2+H_2O \rightarrow HD+HDO$ [Hahn (9a)]. The dependence of the effects of complexing of Cu^{2+}, on its catalytic activity for the oxidation of H_2 by Cr(VI) has been examined by Peters & Halpern (10). The observed sequence of catalytic activities, butyrate $\approx$ propionate $>$ acetate $> SO_4^{=} > Cl^- > H_2O >$ glycinate, ethylenediamine, was interpreted in terms of the opposing effects of ligand basicity and complex stability. A decrease in the reactivity of copper(II) alkanoates toward H_2 in going from water to solvents of lower polarity such as heptanoic acid or octadecane was observed by Chalk & Halpern (11) and attributed to the effect of charge separation in the transition state.

Reaction (1) has also been invoked by Halpern et al. (12) to account for the oxidation of Cu^+ by H atoms, i.e., $Cu^+ + H \rightarrow CuH^+$; $CuH^+ + H^+ \rightarrow Cu^{2+} + H_2$. A similar mechanism was suggested for the oxidation of Fe^{2+} by H atoms and has received some experimental support [Czapski, Jortner & Stein (13)].

Silver(I).—The kinetics of the silver(I)-catalyzed oxidation of H_2 by Cr(VI) were examined by Webster & Halpern (14). In aqueous perchloric acid solution the rate law was found to be $-d[H_2]/dt = k_1[H_2][Ag^+]^2/(k_{-1}[H^+] + k_2[Ag^+]) + k_3[H_2][Ag^+]^2$ where $k_1 = 1.2 \times 10^{12} \exp[-24{,}000/RT]$ M^{-1} sec^{-1}, $(k_{-1}/k_2) = 2.3 \times 10^7 \exp[-14{,}000/RT]$ and $k_3 = 6.8 \times 10^7 \exp[-14{,}700/RT]$ M^{-2} sec^{-1}. The first term in the rate law was attributed to a path, analogous to that for Cu^{2+}, in which the rate-determining step involves the heterolytic splitting of H_2, i.e.,

$$Ag^+ + H_2 \rightleftarrows AgH + H^+ \qquad 4.$$

while the second term was identified with a path involving the homolytic splitting of H_2, i.e.,

$$2Ag^+ + H_2 \rightarrow 2AgH^+ \qquad 5.$$

Alternative steps not involving the formation of hydride intermediates, such as $2Ag^+ + H_2 \rightarrow 2Ag$ (or Ag_2) $+ 2H^+$, are ruled out on energetic grounds [Halpern (15)]. AgH^+ has also been invoked as an intermediate in the reduction of Ag^+ by H atoms [Halpern et al. (12)] and in the Ag^+-catalyzed oxidation of H_2 by MnO_4^- which has the rate law $k[H_2][Ag^+][MnO_4^-]$ suggesting the rate-determining step $H_2 + Ag^+ + MnO_4^- \rightarrow AgH^+ + MnO_4H^-$ [Webster & Halpern (16)]. The reversibility of the step described by equation (4) has been confirmed by H_2-D_2O exchange experiments [Webster & Halpern (14)].

In solvents more basic than water (e.g., pyridine or dodecylamine) or in aqueous solutions containing basic ligands which complex with Ag^+ (e.g., acetate, fluoride, or amines), the path involving heterolytic splitting of H_2 is apparently greatly enhanced compared with that involving homolytic splitting, and the reverse of reaction 4 is inhibited so that the rate law for the oxidation of H_2 by Ag(I) becomes $k[H_2][Ag(I)]$ [Webster & Halpern

(17), Wright, Weller & Mills (18), Wilmarth & Kapauan (19)]. The effect of complexing Ag^+ with amines on its reactivity toward H_2 has been examined in some detail by Halpern & Milne (20). The order of reactivity was found to be tertiary amines > secondary amines > primary amines > ammonia, corresponding to an inverse dependence of the reactivity on the stability of the silver-amine complex and a direct dependence on the basicity of the amine [Nagy & Simandi (21)]. This suggests that the reaction involves displacement of an amine ligand (2), which in turn stabilizes the proton generated by the heterolytic splitting of H_2, i.e.,

$$(L{-}Ag{-}L)^+ + H_2 \rightarrow \left[\begin{array}{ccc} L{-}Ag^+ & \cdots & L \\ \vdots & & \vdots \\ H^- & \cdots & H^+ \end{array}\right]^+ \rightarrow (L{-}Ag{-}H) + (HL)^+ \qquad 6.$$

The very large enhancement of the reactivity of Ag^+ toward H_2, by F^-, observed by Beck, Gimesi & Farkas (22), has been similarly explained. The possibility of proton transfer through a water molecule has also been considered in these interpretations.

Copper(I).—The rate law for the copper(I) acetate-catalyzed *para-ortho*-H_2 conversion, or for the oxidation of H_2 by substrates such as copper(II) or quinone, in quinoline solution, has been found by several workers (23–26), in line with Calvin's original observations, to be $k[H_2][Cu(I)]^2$. This suggests that H_2 is split homolytically according to $2Cu(I) + H_2 \rightleftarrows 2(Cu^I - H)$. The possibility that a dimeric Cu(I) species is the active catalyst has been considered, but there appears to be no reliable evidence for this. Several workers including Calvin & Wilmarth (25), Wright & Weller (27), Wright, Weller & Mills (18), and Parris & Williams (28) have examined the effects of anion and solvent variation on this system, with discrepant results (18, 28) in some cases.

Evidence has also been reported for the catalytic activation of H_2 by copper(I) in aqueous solution [Dunning & Potter (29); Hahn & Peters (9b)], and in heptanoic acid, diphenyl, and octadecane [Chalk & Halpern (11)]. The reduction by H_2 of copper(II) sulfate in aqueous solution and of copper(II) heptanoate in heptanoic acid solution according to the rate law (for the catalytic path) $k[H_2][Cu(I)]$, suggesting heterolytic splitting of H_2, i.e.,

$$Cu(I) + H_2 \rightleftarrows CuH + H^+ \qquad 7.$$

In all of these cases the reactivity of Cu(I) toward H_2 was higher than that of Cu(II) by factors ranging from 20 to 40. On the basis of kinetic evidence it was concluded [Chalk & Halpern (11)] that copper(I) heptanoate is partially dimerized in heptanoic acid solution and that only the monomer is catalytically active. The reversibility of reaction **7** in the absence of a reducible substrate such as copper(II) has been confirmed by isotopic exchange experiments (11).

At sufficiently high temperatures, CuH generated by reaction **7 is** capable apparently of reducing other substrates including carboxylic acids. Thus, Stouthamer (30) has found that copper salts catalyze the hydrogenation of carboxylic acids to alcohols at temperatures above 250° according to the rate law $k[Cu][H_2]$. The reaction was attributed to reduction of the carboxyl groups by CuH generated by reaction 7. The enhancement of the catalytic activity of copper salts by salts of certain other metals, notably cadmium, was observed but not explained.

Mercury(II) and Mercury(I).—Both mercury(II) and mercury(I) were found by Korinek & Halpern (31) to oxidize H_2 homogeneously according to the rate laws $k_1[H_2][Hg^{2+}]$ and $k_2[H_2][Hg_2^{2+}]$, respectively, where $k_1 = 4.2 \times 10^{10} \exp[-18{,}100/RT]$ and $k_2 = 1.2 \times 10^{11} \exp[-20{,}400/RT]$ M^{-1} sec^{-1}. The participation of hydrido-metal intermediates, analogous to CuH^+, AgH, etc., is doubtful in these cases and receives no support on either kinetic or thermodynamic grounds. Failure to observe a significant increase in the solubility of mercury in water with increasing H^+ concentration, even at very high acidities [Halpern & Harkness (32)], suggests that HgH^+ is unstable with respect to Hg_{aq} and H^+, and hence that the primary reactions of Hg^{2+} and Hg_2^{2+} with H_2 may well be simply two-electron reductions of the metal ion coupled with proton transfer to the solvent, i.e., $Hg^{2+} + H_2 \rightarrow Hg_{aq} + 2H^+$; $Hg_2^{2+} + H_2 \rightarrow 2Hg_{aq}$ (or Hg_2) $+ 2H^+$. The corresponding processes in the cases of Cu^{2+}, Cu^+, and Ag^+ are ruled out on energetic grounds [Halpern (15)]. The reactivity of Hg^{2+} toward H_2 is increased by complexing with sulfate but, in contrast to Cu^{2+}, is reduced by complexing with acetate, chloride, or bromide [Korinek & Halpern (33)], presumably because of the much higher stabilities of these complexes.

Ruthenium(III).—Harrod, Ciccone & Halpern (34) found that in aqueous HCl solution, the oxidation of H_2 by substrates such as Ru(IV) and Fe(III) is catalyzed by Ru(III) according to the rate law $-d[H_2]/dt = k[H_2][Ru(III)]$ where $k = 4.0 \times 10^{14} \exp[-23{,}800/RT]$ M^{-1} sec^{-1}. In the absence of an oxidizing substrate, Halpern & James (35) and Schindewolf (36) observed that Ru(III) also catalyzes the isotopic exchange between D_2 and H_2O according to the same rate law. The ratio of HD to H_2 in the exchange product was found to increase with the H^+ and Cl^- concentrations in approximate accord with the heterolytic splitting mechanism,

$$D_2 + RuCl_6^{3-} \xrightarrow{k_1} D^+ + Cl^- + DRuCl_5^{3-}$$

$$DRuCl_5^{3-} \xrightarrow[H^+ + Cl^-]{k_{-1}} HD$$

$$DRuCl_5^{3-} \xrightarrow[H_2O]{k_2} HRuCl_5^{3-} \rightarrow H_2$$

which predicts $[HD]/[H_2] = (k_{-1}/k_2)[H^+][Cl^-]$. The study of isotopic exchange patterns is thus capable, in favorable cases, of providing fairly direct evidence of a heterolytic splitting mechanism and of identifying the displaced ligand, even in the absence of an explicit knowledge of the nature of the catalytic species or of the kinetics. In this connection, an early suggestion of Rittenberg & Krasna (37) that the hydrogen-activating enzyme hydrogenase splits H_2 heterolytically was based on an essentially similar observation, namely that the initial product of the enzyme-catalyzed H_2-D_2O exchange is predominantly HD rather than D_2.

Rhodium(III).—Harrod & Halpern (38) have reported that in aqueous HCl solution, Rh(III) catalyzes the oxidation of H_2 by Fe(III) according to the rate law $-d[H_2]/dt = k[H_2][Rh(III)]$ where, in 3M HCl, $k = 2.3 \times 10^{15}$ exp $[-25{,}200/RT]$ M^{-1} sec^{-1}. It seems likely that the mechanism is analogous to that described above for Ru(III), involving the heterolytic splitting of H_2 in the rate-determining step $RhCl_6^{3-} + H_2 \rightarrow HRhCl_5^{3-} + H^+ + Cl^-$. The proposed intermediate $HRhCl_5^{3-}$ finds a close analogy in the stable hydrido complex $[HRh(trien)Cl]^+$ (where trien = triethylenetetraamine) prepared by Gillard & Wilkinson (39) by the reduction of $[Rh(trien)Cl_2]^+$ with BH_4^-. Gillard et al. (40) have also suggested that the catalysis by H_2 of the substitution reaction $[Rh(py)_3Cl_3] + py \rightarrow [Rh(py)_4Cl_2]Cl$ (py = pyridine), is due to an intermediate hydrido complex $[HRh(py)_3Cl_2]$ formed by the reaction $[Rh(py)_3Cl_3] + H_2 \rightarrow [HRh(py)_3Cl_2] + HCl$, and have reported that $RhCl_3$ in methanol solution catalyzes the homogeneous hydrogenation of hexene. Of related interest is the catalysis by $RhCl_3$ and other group-8 metal complexes, of double bond migration reactions of olefins, recently reported by Harrod & Chalk (41) and attributed to reversible addition of a hydrido-metal complex to the olefin.

Ruthenium(II).—Halpern, Harrod & James (42) have reported that ruthenium(II) chloride in aqueous HCl is a homogeneous catalyst for the hydrogenation of certain olefinic compounds such as maleic and fumaric acids. In the case of fumaric acid, a ruthenium(II)-olefin complex is formed which reacts with H_2 to yield succinic acid according to the rate law $k[H_2][Ru(II)\text{-olefin}]$, with $\Delta H^{\ddagger} = 18$ kcal/mole and $\Delta S^{\ddagger} = -8$ eu. Reduction of fumaric acid with D_2 in H_2O solution yields undeuterated succinic acid, whereas reduction with H_2 (or D_2) in D_2O solution yields predominantly DL-2,3-dideuteriosuccinic acid. The hydrogen which adds to the double bond thus originates predominantly from the solvent, and the addition is predominantly *cis*. Olefins containing isolated (i.e., nonactivated) double bonds are not hydrogenated under these conditions, but their complexes with Ru(II) catalyze D_2-H_2O exchange at about the same rate as the fumaric acid complex undergoes hydrogenation. The mechanism depicted in Figure 1 was proposed to explain these observations. It resembles, in a number of respects, that proposed by Burwell et al. (43) for the heterogeneous catalytic hydrogenation of olefins on chromium oxide gel. Of related interest are the Ru(II)-catalyzed hydrogen transfer reactions of allyl alcohol recently reported by Nicholson & Shaw (44).

Iridium(I).—Vaska & DiLuzio (45) have reported that the d^8 square

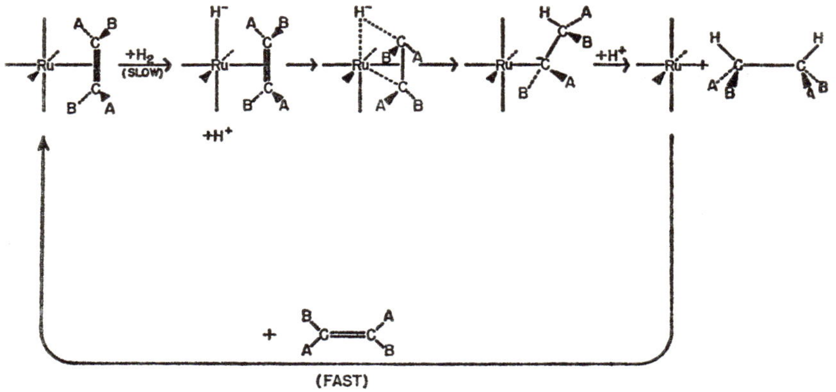

FIG. 1. Mechanism of the ruthenium(II) chloride-catalyzed hydrogenation of fumaric acid.

planar complex $[Ir^{I}Cl(CO)(PPh_3)_2]$, in benzene solution, reacts with H_2 at room temperature to yield the dihydrido complex $[Ir^{III}H_2Cl(CO)(PPh_3)_2]$. This is apparently an example of a general class of reactions, $[Ir^{I}Cl(CO)(PPh_3)_2]+XY \rightarrow [Ir^{III}XYCl(CO)(PPh_3)_2]$, which have also been observed for XY = HCl, Cl_2, and O_2, and represents still another mechanism of activation of H_2 which might be expected to operate with other d^8 square planar complexes but of which this is the only clear example as yet. Some isotopic exchange studies relating to this system have also been reported by Vaska (46).

Palladium(II).—Halpern, Harrod & Potter (47) reported that $PdCl_4^{=}$ catalyzes the oxidation of H_2 by Fe(III) in aqueous solution according to the rate law $-d[H_2]/dt = k[H_2][PdCl_4^{=}]$ where $k = 6.6 \times 10^{11}$ exp $[-20{,}000/RT]$ $M^{-1}sec^{-1}$. This was interpreted in terms of a mechanism involving the heterolytic splitting of H_2, i.e., $PdCl_4^{=} + H_2 \rightarrow HPdCl_3^{=} + HCl$, to yield a reactive intermediate analogous to the stable hydrido complex $HPdCl(PEt_3)_2$ reported by Chatt, Duncanson & Shaw (48). It is possible, however, that in this case also the formation of a dihydrido complex by a reaction analogous to that described above for iridium(I) is involved as an intermediate step, i.e., $Pd^{II}Cl_4^{2-} + H_2 \rightarrow [Pd^{IV}H_2Cl_4^{=}] \rightarrow HPd^{II}Cl_3^{=} + HCl$. Maxted & Ismail (49) reported that various salts, including the acetates of copper(II), nickel(II), silver(I), mercury(II), zinc, cadmium sodium, and calcium, activate palladium(II) chloride as a catalyst for the hydrogenation of ethyl crotonate in aqueous solution. Although metallic palladium is formed under the conditions of hydrogenation, it was claimed that the reaction is not heterogeneously catalyzed.

Platinum(II).—Cramer et al. (50) have reported the homogeneous hydrogenation of ethylene and acetylene in methanol solutions using a complex platinum(II)-tin(II) chloride catalyst; it was suggested that the active platinum-tin complex may contain $(SnCl_3)^-$ ligands attached through platinum-tin bonds. $SnCl_2$ was also found to labilize $PtCl_4^=$ and to catalyze the reaction with ethylene to form Zeise's salt. Bailar & Itatani (51) have reported related studies using mixtures of $(Ph_3P)_2PtCl_2$ and $SnCl_2$ as homogeneous catalysts for the hydrogenation of methyl oleate, methyl linoleate, and soybean oil methyl ester. An early observation by Flynn & Hulburt (52) suggesting homogeneous hydrogenation of ethyleneplatinous chloride is also of interest in this connection.

Pentacyanocobaltate(II).—The reversible uptake of H_2 by aqueous solutions of $Co(CN)_5^{3-}$, observed by several workers (53, 54), has been attributed by King & Winfield (55) to the reaction

$$2[Co^{II}(CN)_5]^{3-} + H_2 \rightleftarrows 2[HCo^{III}(CN)_5]^{3-} \qquad 8.$$

The kinetics and thermodynamics of this reaction have been examined by De Vries (56), whose data yield the rate law $-d[H_2]/dt = k[H_2][Co(CN)_5^{3-}]^2$ (where $k = 0.8\ M^{-2}sec^{-1}$ between 0° and 35°), and the following thermodynamic parameters, $K_{25°} = 1.6 \times 10^5\ M^{-1}$, $\Delta H° = -11.2$ kcal/mole, $\Delta S° = -14$ eu. The H–Co bond dissociation energy in $HCo(CN)_5^{3-}$ is thus close to 57 kcal/mole.

Aqueous solutions of $Co(CN)_5^{3-}$ spontaneously liberate H_2 as a result of the reaction sequence

$$2[Co(CN_5)]^{3-} + H_2O \rightarrow [HCo(CN)_5]^{3-} + [Co(CN)_5OH]^{3-} \qquad 9.$$

$$2[HCo(CN)_5]^{3-} \rightarrow 2[Co(CN)_5]^{3-} + H_2 \qquad 10.$$

The kinetics of reaction (9) have been examined by De Vries (56) who reported the rate law $-d[HCo(CN)_5^{3-}]/dt = k[HCo(CN)_5^{3-}]^2$ where $k = 13 \exp[-4{,}000/RT]\ M^{-1}sec^{-1}$. Griffith & Wilkinson (57) have confirmed the presence of a hydrido complex, presumably $[HCo(CN)_5]^{3-}$, in aqueous solutions of $Co(CN)_5^{3-}$ by NMR measurements.

$Co(CN)_5^{3-}$ is also a homogeneous catalyst for the isotopic exchange of D_2 with H_2O [Mills, Weller & Wheeler (54)] and for the oxidation of H_2 by substrates such as H_2O_2 and $Fe(CN)_6^{3-}$ [Kwiatek, Mador & Seyler (58)]. All these reactions apparently proceed through step 8.

Kwiatek, Mador & Seyler (58) have reported that $Co(CN)_5^{3-}$ is also a homogeneous catalyst for the hydrogenation of certain conjugated olefins such as butadiene, styrene, cinnamic acid, etc.; monoolefins and nonconjugated dienes were not hydrogenated. The following mechanism was proposed for the hydrogenation of butadiene,

$$2[Co(CN)_5]^{3-} + H_2 \rightarrow 2[HCo(CN)_5]^{3-} \qquad 8.$$

$$[HCo(CN)_5]^{3-} + C_4H_6 \rightarrow [C_4H_7Co(CN)_5]^{3-} \qquad 11.$$

$$[C_4H_7Co(CN)_5]^{3-} + [HCo(CN)_5]^{3-} \rightarrow C_4H_8 + 2[Co(CN)_5]^{3-} \qquad 12.$$

Kwiatek & Seyler (59) have also prepared the intermediate $[C_4H_7Co(CN)_4]^{3-}$

by reaction of $Co(CN)_5^{3-}$ with crotyl bromide and characterized it by NMR as a σ-bonded butenyl complex which undergoes rearrangement with loss of CN^- to form a tetracyano-π-allylcobaltate(III) complex, i.e.,

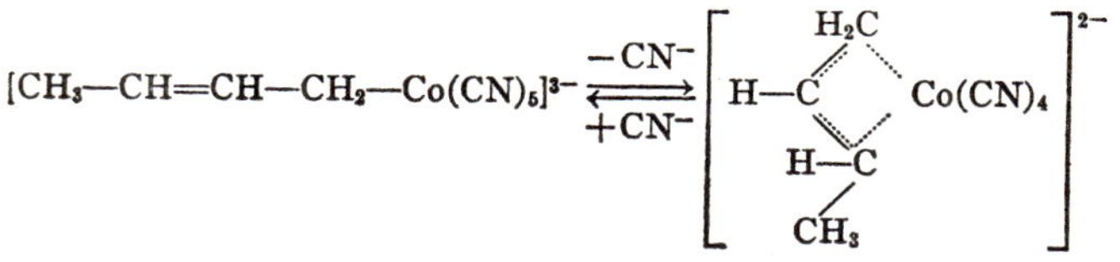

The observation that the product of hydrogenation of butadiene at low CN^- concentrations ($CN^-/Co^{2+} < 5.5$) is predominantly *trans*-2-butene, whereas that at higher CN^- ($CN^-/Co^{2+} > 6$) is predominantly 1-butene, was explained in terms of this equilibrium.

Somewhat different behavior was reported by Simandi & Nagy (60) for the $Co(CN)_5^{2-}$-catalyzed hydrogenation of cinnamic acid. On the basis of kinetic observations the following mechanism involving a radical-anioninter-mediate ($H\dot{S}$) was proposed ($S = PhCH_2{=}CHCOO^-$; $H\dot{S} = PhCH_2\dot{C}HCOO^-$; $H_2S = PhCH_2CH_2COO^-$)

$$2[Co(CN)_5]^{3-} + H_2 \rightleftarrows 2[HCo(CN)_5]^{3-} \qquad 14.$$

$$[HCo(CN)_5]^{3-} + S \rightarrow [Co(CN)_5]^{3-} + H\dot{S} \qquad 15.$$

$$[HCo(CN)_5]^{3-} + H\dot{S} \rightarrow [Co(CN)_5]^{3-} + H_2S \qquad 16.$$

Other recent studies on $Co(CN)_5^{3-}$-catalyzed hydrogenation reactions include those of Mabrouk, Dutton & Cowan (61) on the hydrogenation of sorbic acid, and those of Murakami et al. (62, 63) on a variety of organic substrates.

Metal carbonyls.—Marko (64) has examined the hydrogenation of propionaldehyde to propyl alcohol in toluene solution using $Co_2(CO)_8$ as catalyst. At 150° the rate law was found to be $d[CH_3CH_2CH_2OH[/dt = k[CH_3CH_2CHO][Co][H_2][CO]^{-2}$ (where [Co] = total cobalt concentration). A mechanism related to that originally suggested by Heck & Breslow (65) for the hydroformylation of olefins was proposed, which includes the essential steps

$$HCo(CO)_4 \rightleftarrows HCo(CO)_3 + CO$$

$$HCo(CO)_3 + RCH{=}O \rightarrow RCH_2{-}OCo(CO)_3 \xrightarrow{H_2} RCH_2OH + HCo(CO)_3$$

$$RCH_2OCo(CO)_3 \underset{-CO}{\overset{+CO}{\rightleftarrows}} RCH_2OCo(CO)_4$$

Marko (66) has also proposed a closely related mechanism to account for the hydrogenation of olefins which accompanies hydroformylation. According to this interpretation, the two processes result from competing reactions of a common intermediate, $RCo(CO)_4$, in accord with the following scheme.

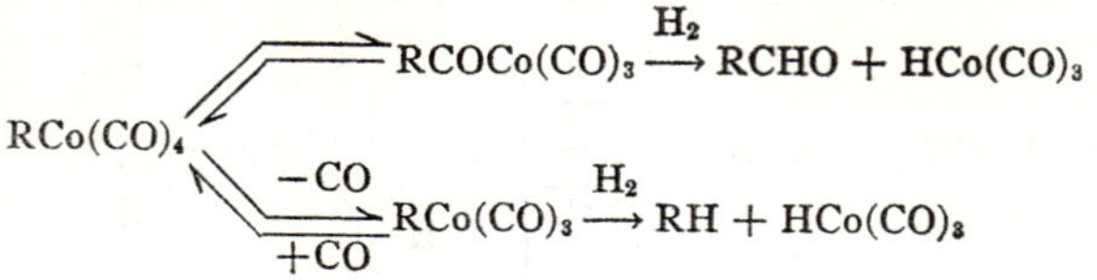

The conclusion that the reduction of aldehydes to alcohols under hydroformylation conditions is a homogeneous reaction has been questioned by Aldridge & Jonassen (67), who have suggested instead that the reduction of the aldehyde is effected not by H_2 but by $HCo(CO)_4$, which in turn is produced by a heterogeneous reaction between $Co_2(CO)_8$ and H_2, catalyzed by metallic cobalt. Kinetic evidence, as well as evidence based on the susceptibility of the reaction to poisons, were cited in favor of this view.

Frankel et al. (68) have examined the hydrogenation of metal linoleate in cyclohexane solution using iron pentacarbonyl as catalyst. Hydrogenation apparently proceeds homogeneously to yield monoenoic fatty esters, while at the same time some isomerization of the reactant occurs to yield a conjugated diene which in turn reacts with the $Fe(CO)_5$ to form a $Fe(\pi\text{-diene})(CO)_3$ complex. It was suggested that the latter is a catalytic intermediate in the hydrogenation reaction. In line with this, it was found that the $Fe(\pi\text{-diene})(CO)_3$ complex was a more active hydrogenation catalyst than $Fe(CO)_5$ itself, a result also reported by Ogata & Misono (69).

Ziegler-type catalysts.—Sloan, Matlack & Breslow (70) have reported the homogeneous hydrogenation of olefins using a variety of Ziegler-type catalysts, e. g., $Cr(\text{acetylacetonate})_3{-}Al(i{-}C_4H_9)_3$, and have suggested a mechanism involving addition of a metal hydride complex to the olefin followed by hydrogenolysis of the metal-carbon bond. Ziegler-type catalysts, e. g., $Ni(\text{2-ethylhexanoate}_2{-}Al(C_2H_5)_3$ have also been reported by Lapporte & Schuett (71) to act as homogeneous catalysts for the hydrogenation of aromatic compounds, e.g., benzene to cyclohexane.

Other transition metal salts.—Tulupov (72) has reported the homogeneous hydrogenation of cyclohexene and other olefins using salts of Ni(II), Cr(III), Mn(II), Co(II), and Fe(III) as catalysts. Hahn (9), on the other hand, reported that the sulfate or perchlorate salts of Ni(II), Cr(III), Co(II), and Fe(III) were catalytically inactive for the oxidation of H_2 by chromium(VI) in aqueous solutions at temperatures up to 268° C.

Boranes.—DeWitt, Ramp & Trapasso (73, 74) have reported the hydrogenation of olefins under homogeneous conditions using tri-*n*-butyl-borane as catalyst. The following mechanism was proposed:

$$BR_3 + H_2 \rightarrow R_2BH + RH;\ R_2BH + R'CH{=}CH_2$$
$$\rightarrow R'CH_2CH_2BR_2;\ R'CH_2CH_2BR_2 + H_2 \rightarrow R'CH_2CH_3 + R_2BH$$

The last step apparently is rate-determining. Hydrogenolysis of boron-alkyl compounds and catalytic hydrogenation of olefins has also been discussed by Köster, Bruno & Binger (75). Other recently reported examples of hydrogenolysis of metal-carbon bonds include the reaction $Li_3Cr(C_6H_5)_6+H_2 \rightarrow Li_3Cr(C_6H_5)_5H+C_6H_6$ [Hein & Weiss (76)] and hydrogenolysis reactions of LiC_6H_5, $Mn(C_6H_5)_2$ and $(C_6H_5)Ti(CH_3)_2$ [Clauss & Bestian (77)]. Podell, Petree & Zietz (78) have discussed the mechanisms and rates of hydrogenolysis of organometallic compounds and suggested that the ease of hydrogenolysis is related to the polarity or ionic character of the metal-carbon bond, hence the electronegativity of the metal. Of related interest is the reaction $(Et_3P)_2Pt(GePh_3)_2+H_2 \rightarrow (Et_3P)_2Pt(H)GePh_3+Ph_3GeH$ recently reported by Cross & Glocking (79), which involves hydrogenolysis of a metal-germanium bond.

Base-catalyzed hydrogenation reactions.—Base catalysis of the *para-ortho*-H_2 conversion in aqueous solution, and of the D_2-H_2O isotopic exchange reaction has been reported by several workers (80–84). The rate law $k[H_2][OH^-]$ (where $k=4.7\times10^{13}$ exp $[-23,800/RT]$ $M^{-1}sec^{-1}$) has been interpreted in terms of a reversible heterolytic splitting of H_2, i.e.,

$$H_2 + OH^- \rightleftarrows H^- + H_2O \qquad 17.$$

Although the postulation of a free H^- ion as an intermediate in this reaction has been questioned (82), recent studies by Flournoy & Wilmarth (84) in concentrated alkali solutions, which reveal a close thermodynamic similarity between the transition state and metal hydroxides, suggest that heterolytic cleavage of the H_2 bond is nearly complete in the transition state, and that the latter closely resembles a solvated H^- ion.

Analogous catalysis by NH_2^- of p-H_2 conversion and D_2-NH_3 exchange has been observed in liquid ammonia at $-50°$ [Wilmarth & Dayton (85); Dirian et al. (86)]. Walling & Bollyky (87) have also reported catalysis by t-butoxide of the hydrogenation of benzophenone to benzhydrol in several organic solvents, presumably through a mechanism involving a hydride ion intermediate, i. e.,

$$RO^-+H_2 \rightleftarrows ROH+H^-;\ H^-+R_2C{=}O \rightleftarrows R_2CHO^-;\ R_2CHO^-+ROH \rightleftarrows R_2CHOH+RO^-.$$

These authors have suggested that the fairly easy hydrogenolysis of organometallic compounds already discussed, may involve a similar attack on H_2 by the strongly basic carbanion or carbanionoid reagent.

Several attempts (83, 87) to find analogous acid-catalyzed reactions of the type $H^++D_2 \rightarrow HD+D^+$; or alternatively, $ROH+H^+ \rightarrow R^++H_2O$ followed by $R^++H_2 \rightarrow RH+H^+$, have yielded negative results.

Kinetic isotope effects.—Relative rates of reaction of H_2 and D_2 with several of the species cited above have been reported, including OH^- (36), NH_2^- (36), $Co(CN)_5^{3-}$ (36), Ru(III) (36), Cu(II) (11), and Cu(I) (11, 26). Despite the variety of mechanisms represented by these reactions the value of

k_{H_2}/k_{D_2} was found in every case to be close to 1.5. Harrod & Halpern (88) have also measured the relative rates of reaction of H_2 with a number of species in H_2O and D_2O. Values of k_{H_2O}/k_{D_2O} for the reaction of H_2 with $Cu(OAc)_2$, $PdCl_4^{2-}$, $RhCl_6^{3-}$, and MnO_4^- were found to range from 0.90 to 1.0, whereas for the reaction of H_2 with Cu^{2+}, Hg^{2+}, Hg_2^{2+}, and Ag^+, the values of k_{H_2O}/k_{D_2O} were all in the range 1.25 ± 0.1. The latter isotope effect apparently is general for aquo-cations and is independent of whether homolytic (Ag^+) or heterolytic (e.g., Cu^{2+}) splitting of H_2 is involved, despite the fact that the latter mechanism (i.e., $Cu^{2+} + H_2O + H_2 \rightarrow CuH^+ + H_3O^+$), in contrast to the former, assigns to a water molecule the specific role of proton acceptor.

Interpretation of homogeneous hydrogenation catalysis.—Two themes contribute to an understanding of the catalytic reactions described above: (*a*) the elucidation of the catalytic mechanism, i.e., recognition of the sequence of elementary steps which comprise the overall reaction; and (*b*) the understanding of the factors which influence the rate of reaction, i. e., the dependence of catalytic activity, for a given catalytic mechanism, on the structure and electronic configuration of the catalyst.

The studies reviewed above relate principally to the first of these themes and have led to the resolution of many of the reactions cited into their component steps so that the elementary step involving the reaction of H_2 is recognizable. More limited progress has been made, however, in respect of the second of the above themes, namely that of understanding how the reactivity toward H_2 depends on the electron configuration of the catalyst. Early attempts [Halpern (3, 4)] to relate catalytic activity of metal ions to their electron configuration, or to generalized electron acceptor [Halpern (4); Nagy & Simandi (89] or electron donor [Bikhovsky & Jemkin (90)] properties have met with only limited success. More clearly, catalytic activity in many of the systems cited is linked to the stabilities and labilities of the hydrido-metallic complexes which participate as intermediates in the catalytic reaction and whose formation provides the driving force for the splitting of H_2. Our understanding of the properties of such hydrido complexes has been greatly extended in recent years by the discovery and study of many stable analogues which are susceptible to direct examination [Chatt (91); Green (92)]. One of the conclusions which has emerged from these studies is that the hydrogen in such complexes commonly acts effectively as an anionic ligand (although its bonding to the metal undoubtedly involves a high degree of covalency) occupying a normal coordination position. Thus $HPtCl(PEt_3)_2$, like $PtCl_2(PEt_3)_2$ (from which it is derived by reaction with a suitable hydride donor), is to be regarded as a square planar complex of the d^8 ion, Pt^{2+} [Chatt (91)].

An understanding of catalytic activity in these systems depends upon an appreciation of the factors affecting the stabilities and labilities of such hydrido complexes of transition metals. These are still far from well understood. The considerations invoked by Jaffé & Doak (93) and by Chatt & Shaw (94), in respect of the stabilities of organo complexes of transition metals, would also appear to be relevant to the corresponding hydrido complexes and, indeed, have led to the successful preparation of closely related parallel series of stable hydrido and organo complexes, e. g., $HPtCl(PEt_3)_2$ and $CH_3PtCl(PEt_3)_2$ [Chatt (91, 94)]. The "stability" of these complexes appears to be both thermodynamic and kinetic in origin, the latter factor being of particular importance according to Chatt & Shaw (91, 94) in the stabilization of hydrido complexes by π-bonding ligands, notably tertiary phosphines and, to a lesser degree, CN^-, CO, and unsaturated hydrocarbons.

Catalytic activity in these systems, as in others, requires that the catalytic intermediate be labile and that its thermodynamic stability lie within rather narrowly defined limits. Unless it is sufficiently stable, it will not form readily, while if its stability is too great, its subsequent transformation to the products will be slow. Among the transition metals, the most effective catalysts for the activation of H_2 are likely to be those forming fairly stable but highly labile hydrido complexes. It is thus not surprising that hydrido complexes such as $HPtCl(PEt_3)_2$, which are stabilized by tertiary phosphine ligands which confer a high degree of substitution-inertness on the complex, are generally not effective catalytic intermediates. On the one hand they are not formed readily by direct reaction with H_2, e.g., $PtCl_2(PEt_3)_2 + H_2 \rightarrow HPtCl(PEt_3)_2 + HCl$; at the same time once formed they are relatively unreactive. With metal ions of the platinum groups, the halides appear to be much more effective than are tertiary phosphines in achieving the compromise between stability and lability necessary for effective catalytic activity. CO, CN^- and, in one instance recently reported [Cramer et al. (50)], $SnCl_3^-$ are other ligands which appear to be particularly effective in "activating" transition metal ions.

The mechanism of formation of the hydrido intermediates, as well as their stability and lability, enters into a consideration of the reactivity of metal complexes toward H_2, since such reactions may give rise to hydrides by several routes. In the studies cited above, at least three such routes have been recognized: (*a*) heterolytic splitting of H_2, e. g., $Ru^{III}Cl_6^{3-} + H_2 \rightarrow HRuCl_5^{3-} + HCl$; (*b*) homolytic splitting of H_2, e. g., $2Co^{II}(CN)_5^{3-} + H_2 \rightarrow 2HCo^{III}(CN)_5^{3-}$; and (*c*) addition of H_2 to form a dihydride, e. g., $Ir^{I}Cl(CO)(PPh_3)_2 + H_2 \rightarrow Ir^{III}H_2Cl(CO)(PPh_3)_2$.

The first of the above mechanisms is essentially a substitutional process in which there is no change in the oxidation state of the metal ion. This type of mechanism appears to be fairly widespread, particularly in media of moderately high basicity where an additional driving force is provided by stabilization, in some cases apparently by the displaced ligand, of the released H^+ ion. This factor, along with the lability of the complex and stability of the corresponding hydrido derivative, appears to govern reactivity toward H_2 in such cases. The other two mechanisms correspond formally to oxidation processes in which the oxidation number of the metal increases. The fact that the catalytic complexes in these cases are coordinately unsaturated so that entry of a hydride ligand into the coordination shell does not involve displacement of another ligand undoubtedly contributes to the catalytic activity of these complexes. Although only one instance of the last type of mechanism has been found so far, it might be expected to operate more generally with other square planar d^8 complexes. The expected order of reactivity corresponding to the decreasing tendency of such complexes to undergo oxidation and expand their coordination number is (subject to modification by the ligands present) Ir(I), Rh(I), Co(I) > Pt(II) > Pd(II) > Ni(II) > Au(III).

Catalytic activation of other inert saturated molecules.—The close parallel between the stability (and lability) patterns [Chatt (91)] of the hydrido and organo complexes of transition metals suggests that it should be possible to effect the activation of single bonds involving carbon through catalytic mechanisms analogous to those which operate in the activation of H_2, but involving the formation of organo rather than hydrido complexes. To date, this appears to have been achieved only in a few cases, most of which involve the splitting of carbon-halogen bonds by coordinately unsaturated complexes. Examples of this are ($R = CH_3$, etc.):

$2Co(CN)_5^{3-} + RI \rightarrow RCo(CN)_5^{3-} + Co(CN)_5I^{3-}$ [Halpern & Maher 95)];

$(PEt_3)_2PtRI + RI \rightarrow (PEt_3)_2PtR_2I_2$ [Chatt, Duncanson & Shaw (96)];

$IrCl(CO)(PPH_3)_2 + RI \rightarrow Ir(CH_3)ClI(CO)(PPh_3)_2$ [Heck 97)].

Catalytic hydrogenation of olefins and related substrates.—In the case of reactions involving the homogeneous catalytic hydrogenation of organic substrates such as olefins, the mechanisms have been established only for a few such reactions, comparison of which reveals significant differences. One type of mechanism which appears to have fairly widespread applicability involves addition of a metal hydride to the olefin, i. e.,

$$M{-}H + \rangle C{=}C\langle \rightarrow M{-}\overset{|}{\underset{|}{C}}{-}\overset{|}{\underset{|}{C}}{-}H \qquad 18.$$

followed by one of several possibilities e.g.,

$$M{-}\overset{|}{\underset{|}{C}}{-}\overset{|}{\underset{|}{C}}{-}H + H_2 \rightarrow M{-}H + H{-}\overset{|}{\underset{|}{C}}{-}\overset{|}{\underset{|}{C}}{-}H;$$

$$M{-}\overset{|}{\underset{|}{C}}{-}\overset{|}{\underset{|}{C}}{-}H + H^+ \rightarrow M^+ + H{-}\overset{|}{\underset{|}{C}}{-}\overset{|}{\underset{|}{C}}{-}H;$$

or

$$\mathrm{M{-}\overset{\diagdown\ \diagup}{\underset{\diagup\ \diagdown}{C{-}C}}{-}H + M{-}H \rightarrow 2M + H{-}\overset{\diagdown\ \diagup}{\underset{\diagup\ \diagdown}{C{-}C}}{-}H}$$

A number of hydrometallation reactions analogous to 18, involving the addition of metal hydrides, such as $HPtCl(PEt_3)_2$ [Chatt & Shaw (98)], $(CH_3)_2SnH_2$ [Clark & Kwon (99)], $HMn(CO)_5$ [Treichel, Pitcher & Stone (100)], and $HCo(CN)_5^{3-}$ [May & Wilkinson (101)], to olefins, dienes, acetylenes, etc., have been observed, but certain aspects of the mechanisms of such reactions and of the factors affecting reactivity are not yet understood. Among the questions still to be resolved is whether coordination of the olefin to the metal as a π-bonded ligand (as in the case of the Ru(II)-catalyzed hydrogenation reaction cited) is a general feature of such "insertion" reactions.

The addition of metal hydrides to olefins is but one example of a much wider class of "insertion" reactions, other known examples of which involve the insertion of molecules such as CO, acetylene, dienes, SO_2 into metal-oxygen, metal-carbon, and metal-nitrogen bond. This subject has been recently reviewed by Heck (102). The catalytic processes discussed below, in addition to the examples of homogeneous catalytic hydrogenation already cited, serve to illustrate the widespread role of such insertion reactions in catalysis.

Catalytic Hydration of Acetylene

This general subject has recnetly been reviewed by Miocque, Hung & Yen (103). Mercury(II) salts have been widely used as catalysts for the hydrogenation of acetylenic compounds for many years, but many features of the mechanism of this reaction remain to be resolved. On the basis of isotopic tracer studies, which revealed that all the hydrogen derived from the solvent in the mercury(II)-catalyzed hydration of acetylene is incorporated into the methyl group of the acetaldehyde product and that this incorporation is subject to a considerable kinetic isotope effect, Rekasheva & Samchenko (104) have proposed a mechanism in which the slow step is the addition of water to a complexed acetylene molecule, i. e., $(CH{\equiv}CH)HgX_2 + H_2O \rightarrow (CH_2{=}CHOH)HgX_2$. These observations rule out several earlier suggestions concerning the mechanism of this reaction. Budde & Dessy (105) have reported evidence for a bis-(diphenylacetylene)mercury(II) complex as a critical intermediate in the hydration of phenyl acetylene catalyzed by mercury(II) perchlorate. A new catalyst for the hydration of acetylenic compounds, ruthenium(III) chloride, has been reported by Halpern, James & Kemp (106). The rate was found to be first order each in ruthenium(III) and in acetylene, and the dependence of the rate on the Cl^- concentration suggested that the principal catalytic species were $Ru(H_2O)_2Cl_4^-$ and

$Ru(H_2O)Cl_5^{2-}$ while $RuCl_6^{3-}$ was inactive, thus implying a requirement for a ligand water molecule. The following mechanism was proposed for this reaction,

$$RuCl_5OH_2^{2-} \xrightarrow[-H^+]{RC{\equiv}CR'} [Cl_5Ru{-}CR{=}C(OH)R']^{3-} \xrightarrow{H_3O^+} RuCl_5OH_2^{2-} + RCH_2COR' \qquad 19.$$

Oxidation of Olefins by Palladium(II) Chloride

The principal recent studies on this system include those of Smidt (107); Matveev et al. (108); Vargaftik, Moiseev & Syrkin (109–111); and Henry (112). Several kinetic studies (110–112) reveal a rate law for the oxidation of ethylene to acetaldehyde, of the form $k[C_2H_4][PdCl_4^{=}][H^+]^{-1}[Cl^-]^{-2}$ with a reduction in rate by a factor of 4 in going from H_2O to D_2O (111). These observations have been reconciled by Henry (112) in terms of the following mechanism in which the rate-determining step is the rearrangement of a π-olefin complex (I) to a σ-bonded, β-hydroxy-ethyl palladium complex (II).

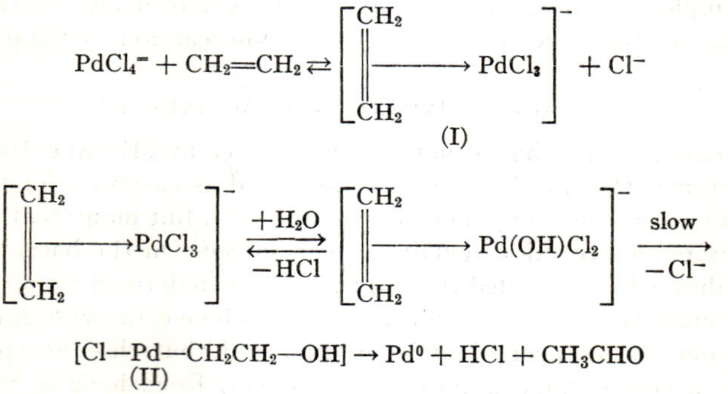

20.

21.

In the presence of a suitable oxidant, such as copper(II), Pd° is reoxidized to Pd(II) so that a catalytic oxidation cycle, involving Pd(II) as a catalyst, is established (107, 108). The essential features of the above mechanism can also be applied to the formation of vinyl esters by the palladium(II)-oxidation of olefins in acetic acid [Moiseev, Vargaftik & Syrkin (113); Stern & Spector (114)].

Of related interest are studies by Halpern with Harkness and Nakamura (115, 116) on the oxidation of CO by mercury(II) and silver(I) which have been interpreted in terms of a mechanism involving "insertion" of CO into a metal-oxygen bond, i. e., M–OH+CO→M–COOH.

CATALYSIS OF POLYMERIZATION BY COORDINATION COMPOUNDS

The polymerization of olefins by catalysts of the Ziegler-Natta type continues to be an active field of investigation while many questions relating to the constitution of these catalysts, the nature of the active sites, and the mechanism of their action, remain to be resolved [Dawans & Teyssie (117); Ketley & Werber (118)]. Much of this work is beyond the scope of this review, and attention will be confined here to a few restricted themes which are most closely related to the subjects already considered.

A recent paper by Cossee (119) further elaborates the mechanism proposed earlier by him (120) for the polymerization of olefins by Ziegler-Natta catalysts. In this mechanism, the essential features of which are depicted below, the active site is assumed to be a transition metal ion in the surface layer of the solid catalyst (e.g., $TiCl_3$) which is essentially octahedrally coordinated, one coordination position being vacant and one occupied by the growing polymer chain (R). The monomeric olefin is adsorbed in the catalyst by coordination, in the vacant position, as a π-bonded ligand, followed by a rearrangement resulting in "insertion" of the monomer between the metal and the σ-coordinated polymer.

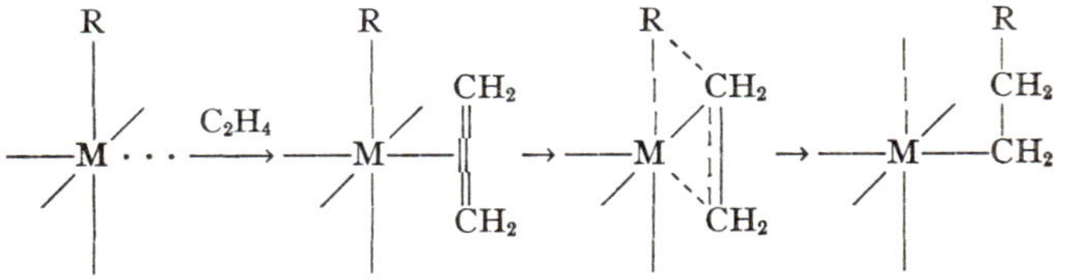

22.

It is suggested that the latter step is rate-determining and the factors affecting the rate of this step, along with other features of the mechanism, are examined in some detail from the standpoint of the ligand field theory. According to this mechanism the aluminum alkyl constituent of the catalyst is not involved directly in the propagation steps, its role presumably being in alkylating the transition metal, in chain transfer, and as a scavenger. In studies related to this theme, Arlman (121) has examined the interaction of solid $ScCl_3$ with isooctane solutions of $Zn(C_2H_5)_2$ labelled with Zn^{65} and C^{14}. Alkylation of the transition metal halide was demonstrated by the finding that the ethyl/zinc ratio at the surface exceeded two.

Arlman (122) has examined the geometry of the active site in catalysts derived from $TiCl_2$ and $TiCl_3$ in the context of the Cossee mechanism, and Arlman & Cossee (123) have shown how the mechanism, making use of the structural features of the catalyst surface, accounts for the stereospecificity of the polymerization of α-olefins to yield isotactic polymers. In line with the conclusion that a solid catalyst is required for stereospecific polymerization of olefins, soluble catalysts, such as those derived from VCl_3-$Sn(C_6H_5)_4$-$AlBr_3$ [Phillips & Carrick (124)] or (π-$C_5H_5)_2TiCl_2$-$(CH_3)_2AlCl$ [Breslow & Newburg (125)] which are chemically related to those of the Ziegler-Natta type and effective for the polymerization of ethylene, have not proved effective thus far for the stereospecific polymerization of α-olefins.

Bestian et al. (126) have recently described a catalyst of the composition CH_3TiCl_3-CH_3AlCl_2 for the polymerization of ethylene at low temperatures (-50 to $-100°$). This catalyst is effective in solvents containing chlorinated hydrocarbons, and although chemically similar to typical Ziegler-Natta catalysts, differs from the latter in at least two respects, (*a*) the titanium remains in the $+4$ oxidation state, and (*b*) polymers of low molceular weight and branched structures are produced. It was suggested that the results are incompatible with either a cationic or anionic mechanism, and a mechanism involving a π-olefin complex, possibly $[TiCl_2(CH_3)(C_2H_4)]^+$, as the catalytic intermediate was proposed.

Other recent developments in the field of polymerization catalysis by coordination compounds, which are of interest in the present context, relate to the stereospecific polymerization of butadiene in polar solvents (water, ethanol, etc.) or emulsions by salts of group 8 metals. The first example of such a catalyst was described by Rinehart et al. (127), who reported that rhodium(III) chloride or nitrate catalyzes the polymerization of butadiene in water or alcohol to crystalline *trans*-1,4-polybutadiene. According to Canale et al. (128), salts of other group-8 metals, including Pd(II), Ir(III), Ru(III), Co(II), also catalyze the polymerization of butadiene, the steric course of polymerization being dependent on the nature of the metal ion and on the ligands. Teyssie & Dauby (129) have reported that 1,3-cyclohexadiene accelerates the $RhCl_3$-catalyzed polymerization of butadiene, suggesting the participation of a lower oxidation state of rhodium, possibly rhodium(I) (or alternatively a hydrido complex of rhodium). Natta, Dall'Asta & Motroni (130) have reported the polymerization of cyclobutene in aqueous emulsion to a crystalline polycyclobutylenamer-2, also using rhodium (III) chloride as catalyst. These catalyst systems differ from earlier organometallic polymerization catalysts including those of the Ziegler-Natta type in a number of respects, among which are their stability in media such as water or alcohols containing active hydrogen. Studies so far reported shed little light on their mechanism, but it seems likely, as several workers [Canale et al. (128); Teyssie (131)] have suggested, that π-diene or π-allyl complexes, which are characteristic of these metal ions, are involved as intermediates.

Concluding Remarks

Much work remains to be done to clarify the mechanisms of many of the catalytic reactions which have been discussed in this review, and the study of these and related systems currently comprises one of the most active and rapidly expanding areas in the field of catalysis. Even at this stage certain general principles underlying the great catalytic versatility of coordination compounds of the transition metals are already apparent. Among these are the following: (*a*) the ability of transition metals to stabilize molecular fragments and reaction intermediates, including such species as H, alkyl groups, allyl groups, etc. by coordination; (*b*) their ability to coordinate ligands through either σ- or π-bonding resulting in the "insertion" rearrangements which are believed to play a role in many catalytic addition reactions; and (*c*) the great lability, toward both insertion and displacement, of the bonds formed by transition metals, particularly with carbon and hydrogen.

LITERATURE CITED

1. Nyholm, R. S., *Proc. Intern. Congr. Catalysis, 3rd, Amsterdam, 1964* (North-Holland Publishing Co., Amsterdam, in press)
2. Calvin, M., *Trans. Faraday Soc.*, **34**, 1181 (1938); *J. Am. Chem. Soc.*, **61**, 2230–34 (1939)
3. Halpern, J., *J. Phys. Chem.*, **63**, 398–403 (1959)
4. Halpern, J., *Advan. Catalysis*, **9**, 301–70 (1959)
5. Peters, E., and Halpern, J., *Can. J. Chem.*, **33**, 356–64 (1955); *J. Phys. Chem.*, **59**, 793–96 (1955)
6. Halpern, J., Macgregor, E. R., and Peters, E., *J. Phys. Chem.*, **60**, 1455–56 (1956)
7. Hahn, E. A., and Peters, E., *Can. J. Chem.*, **39**, 162–70 (1961)
8. McDuffie, H. F., Compere, E. L., Stone, H. H., Woo, L. F., and Secoy, C. H., *J. Phys. Chem.*, **62**, 1030–36 (1958)

9a. Hahn, E. A., *Kinetics of Copper Reduction by Hydrogen from Aqueous Solution* (Doctoral thesis, Univ. British Columbia, Vancouver, B.C., 1963)

9b. Hahn, E. A., and Peters, E., *J. Phys. Chem.*, **69**, 547–52 (1965)

9. Hahn, E. A., *Kinetics of Copper Reduction by Hydrogen from Aqueous Solution* (Doctoral thesis, Univ. British Columbia, Vancouver, B.C., 1963)
10. Peters, E., and Halpern, J., *Can. J. Chem.*, **34**, 554–62 (1956)
11. Chalk, A. J., and Halpern, J., *J. Am. Chem. Soc.*, **81**, 5846–52 (1959); *ibid.*, 5852–54 (1959)
12. Halpern, J., Czapski, G., Jortner, J., and Stein, G., *Nature*, **186**, 629–31 (1960)
13. Czapski, G., Jortner, J., and Stein, G., *J. Phys. Chem.*, **65**, 960–64 (1961)
14. Webster, A. H., and Halpern, J., *J. Phys. Chem.*, **61**, 1239–45 (1957)
15. Halpern, J., *Advan Catalysis*, **9**, 302–11 (1957)
16. Webster, A. H., and Halpern, J., *Trans Faraday Soc.*, **53**, 51–60 (1957)
17. Webster, A H., and Halpern, J., *J. Phys. Chem.*, **61**, 1245–48 (1957)
18. Wright, L., Weller, S., and Mills, G. A., *J. Phys. Chem.*, **59**, 1060–64 (1955)
19. Wilmarth, W. K., and Kapauan, A. F., *J. Am. Chem. Soc.*, **78**, 1308–11 (1956)
20. Halpern, J., and Milne, J. B., *Proc. Intern. Congr. Catalysis, 2nd, Paris, 1960*, **1**, 445–57 (Editions Technip, Paris, 1961)
21. Nagy, F., and Simandi, L., *Acta Chim. Acad. Sci. Hung.*, **38**, 373–83 (1963)
22. Beck, M. T., Gimesi, I., and Farkas, J., *Nature*, **197**, 73 (1963); Beck, M. T., and Gimesi, I., *Magy. Kem. Folyoivat*, **69**, 552–55 (1963)
23. Weller, S., and Mills, G. A., *J. Am. Chem. Soc.*, **75**, 769–77 (1953)
24. Wilmarth, W. K., and Barsh, M. K., *J. Am. Chem. Soc.*, **75**, 2237–42 (1953)
25. Calvin, M., and Wilmarth, W. K., *J. Am. Chem. Soc.*, **78**, 1301–5 (1956)
26. Wilmarth, W. K., and Barsh, M. K., *J. Am. Chem. Soc.*, **78**, 1305–7 (1956)
27. Wright, L. W., and Weller, S., *J. Am. Chem. Soc.*, **76**, 3345–46 (1954)
28. Parris, M., and Williams, R. J. P., *Discussions Faraday Soc.*, **29**, 240–47 (1960)
29. Dunning, W. J., and Potter, P. E., *Proc. Chem. Soc.*, 244–45 (1960)
30. Stouthamer, B., *Preparation of Unsaturated Fatty Alcohols Through the Homogeneous Catalytic Hydrogenation of Unsaturated Fatty Acids* (Doctoral thesis, Delft, Netherlands, 1964)
31. Korinek, G. J., and Halpern, J., *J. Phys. Chem.*, **60**, 285–89 (1956)
32. Halpern, J., and Harkness, A. C. (Unpublished results)
33. Korinek, G. J., and Halpern, J., *Can. J. Chem.*, **34**, 1372–81 (1956)
34. Harrod, J. F., Ciccone, S., and Halpern, J., *Can. J. Chem.*, **39**, 1372–76 (1961)
35. Halpern, J., and James, B. R., *Abstr. Am. Chem. Soc., 142nd Meeting, Atlantic City, N.J., September 1962* (Unpublished results)
36. Schindewolf, U., *Ber. Bunsenges.*, **67**, 219–28 (1963)
37. Rittenberg, D., and Krasna, A., *Discussions Faraday Soc.*, **20**, 185 (1955)
38. Harrod, J. D., and Halpern, J., *Can. J. Chem.*, **37**, 1933–35 (1959)
39. Gillard, R. D., and Wilkinson, G., *J. Chem. Soc.*, 3594–99 (1963)
40. Gillard, R. D., Osborn, J. A., Stockwell, P. B., and Wilkinson, G., *Proc. Chem. Soc.*, 284–85 (1964)
41. Harrod, J. F., and Chalk, A. J., *J. Am. Chem. Soc.*, **86**, 1776–79 (1964)
42. Halpern, J., Harrod, J. F., and James,

B. R., *J. Am. Chem. Soc.*, **83**, 753 (1961); and unpublished results; Halpern, J., *Am. Chem. Soc., 141st Meeting, Washington, March 1962*
43. Burwell, R. L., Littlewood, A. B., Cardew, M., Pass, G., and Stoddart, C. T. H., *J. Am. Chem. Soc.*, **82**, 6272–80 (1960)
44. Nicholson, J. K., and Shaw, B. L., *Proc. Chem. Soc.*, 282–83 (1963)
45. Vaska, L., and DiLuzio, J. W., *J. Am. Chem. Soc.*, **84**, 679–80 (1962)
46. Vaska, L., *Proc. Intern. Conf. Coord. Chem., 8th, Vienna, 1964*
47. Halpern, J., Harrod, J. F., and Potter, P. E., *Can. J. Chem.*, **37**, 1446–50 (1959)
48. Chatt, J., Duncanson, L. A., and Shaw, B. L., *Chem. Ind. (London)*, 859 (1958)
49. Maxted, E. B., and Ismail, S. M., *J. Chem. Soc.*, 1750–52 (1964)
50. Cramer, R. D., Jenner, E. L., Lindsey, R. V., Jr., and Stolberg, U. G., *J. Am. Chem. Soc.*, **85**, 1691–92 (1963)
51. Bailar, J. C., Jr., and Itatani, H., *Proc. Symp. Coord. Chem., Tihany, Hungary, 1964*, 373–96
52. Flynn, J. H., and Hulburt, H. M., *J. Am. Chem. Soc.*, **76**, 3393–96 (1954)
53. King, N. K., and Winfield, M. E., *J. Am. Chem. Soc.*, **80**, 2060–65 (1958)
54. Mills, G. A., Weller, S., and Wheeler, A., *J. Phys. Chem.*, **63**, 403–10 (1959)
55. King, N. K., and Winfield, M. E., *J. Am. Chem. Soc.*, **83**, 3366–73 (1961)
56. De Vries, B., *J. Catalysis*, **1**, 489–97 (1962)
57. Griffith, W. P., and Wilkinson, G., *J. Chem. Soc.*, **553**, 2757–62 (1959)
58. Kwiatek, J., Mador, I. L., and Seyler, J. K., *Advan. Chem. Ser.*, **37**, 201–15 (1963)
59. Kwiatek, J., and Seyler, J. K., *Proc. Intern. Conf. Coord. Chem., 8th, Vienna, 1964*, 308–12
60. Simandi, L., and Nagy, F., *Proc. Symp. Coord. Chem., Tihany, Hungary, 1964*, 329–49
61. Mabrouk, A. F., Dutton, H. J., and Cowan, J. C., *J. Am. Oil Chemists' Soc.*, **41**, 153–58 (1964)
62. Murakami, M., *Proc. Conf. Coord. Chem., 7th, Sweden ,1962*, 268–70
63. Murakami, M., Kawai, R., and Suzuki, K., *Nippon Kagaku Zasshi*, **84**, 669–73 (1963)
64. Marko, L., *Proc. Chem. Soc.*, 67–8 (1962)
65. Heck, R. F., and Breslow, D. S., *Proc. Intern. Congr. Catalysis, 2nd, Paris, 1960*, 671–85 (Editions Technip, Paris, 1961); Breslow, D. S., and Heck, R. F., *Chem. Ind. (London)*, 467 (1960)
66. Marko, L., *Chem. Ind. (London)*, 260–61 (1962)
67. Aldridge, C. L., and Jonassen, H. B., *J. Am. Chem. Soc.*, **85**, 886–91 (1963)
68. Frankel, E. N., Emken, E. A., Peters, H. M., Davison, V. L., and Butterfield, R. O., *J. Org. Chem.*, **29**, 3292–97 (1964)
69. Ogata, I., and Misono, A., *Bull. Chem. Soc. Japan*, **37**, 439–40 (1964); **37**, 900–2 (1964)
70. Sloan, M. F., Matlack, A. S., and Breslow, D. S., *J. Am. Chem. Soc.*, **85**, 4014–18 (1963)
71. Lapporte, S. J., and Schuett, W. R., *J. Org. Chem.*, **28**, 1947–48 (1963)
72. Tulupov, V. A., *Zh. Fiz. Khim.*, **31**, 519–21 (1957); *ibid.*, **32**, 727–29 (1958); *ibid.*, **36**, 1617 (1962)
73. DeWitt, E. J., Ramp, F. L., and Trapasso, L. E., *J. Am. Chem. Soc.*, **83**, 4672 (1961)
74. Ramp, F. L., DeWitt, E. J., and Trapasso, L. E., *J. Org. Chem.*, **27**, 4368–72 (1962)
75. Köster, R., Bruno, G., and Binger, P., *Ann. Chem.*, **644**, 1–22 (1961)
76. Hein, F., and Weiss, R., *Z. Allgem. Anorg. Chem.*, **295**, 145–52 (1958); Hein, F., *Proc. Intern. Congr. Pure Appl. Chem., Munich, 1959*
77. Clauss, K., and Bestian, H., *Ann. Chem.*, **654**, 8–19 (1962)
78. Podall, H. E., Petree, H. E., and Zietz, J. R., *J. Org. Chem.*, **24**, 1222–26 (1959)
79. Cross, R. J., and Glocking, F., *Proc. Chem. Soc.*, 143 (1964)
80. Wirtz, K., and Bonhoeffer, K. F., *Z. Physik. Chem. (Leipzig)*, **177**(A), 1 (1936)
81. Abe, S., *Sci. Papers Inst. Phys. Chem. Res. (Tokyo)*, **38**, 287–97 (1941)
82. Wilmarth, W. K., Dayton, J. C., and Flournoy, J. M., *J. Am. Chem. Soc.*, **75**, 4549–53 (1953)
83. Miller, S. L., and Rittenberg, D., *J. Am. Chem. Soc.*, **80**, 64–65 (1958)
84. Flournoy, J. M., and Wilmarth, W. K., *J. Am. Chem. Soc.*, **83**, 2257–62 (1961)
85. Wilmarth, W. K., and Dayton, J. C.,

J. Am. Chem. Soc., **75**, 4553–56 (1953)

86. Dirian, G., Botter, F., Ravoire, J., and Grandcollot, P., *J. Chem. Phys.*, **60**, 139–47 (1963)
87. Walling, C., and Bollyky, L., *J. Am. Chem. Soc.*, **86**, 3750–52 (1964)
88. Harrod, J. F., and Halpern, J., *J. Phys. Chem.*, **65**, 563 (1961)
89. Nagy, F., and Simandi, L., *Acta Chim. Acad. Sci. Hung*, **38**, 213–24 (1963)
90. Bikhovsky, V. K., and Jemkin, O. N., *Kinetika i Kataliz*, **1**, 374–78 (1960)
91. Chatt, J., *Proc. Chem. Soc.*, 318–26 (1962); *J. Chem. Soc.*, 5075–84 (1962)
92. Green, M. L. H., *Angew. Chem.*, **72**, 719–25 (1960)
93. Jaffé, H. H., and Doak, G. O., *J. Chem. Phys.*, **21**, 196–200 (1953)
94. Chatt, J., and Shaw, B. L., *J. Chem. Soc.*, 705 (1959)
95. Halpern, J., and Maher, J. P., *J. Am. Chem. Soc.*, **86**, 2311 (1964)
96. Chatt, J., Duncanson, L. A., and Shaw, B. L., *Chem. Ind.* (*London*), 859 (1958)
97. Heck, R. F., *J. Am. Chem. Soc.*, **86**, 2796–99 (1964)
98. Chatt, J., and Shaw, B. L., *J. Chem. Soc.*, 5075–84 (1962)
99. Clark, H. C., and Kwon, J. T., *Can. J. Chem.*, **42**, 1288–93 (1964)
100. Treichel, P. M., Pitcher, E., and Stone, F. G. A., *Inorg. Chem.*, **1**, 511–17 (1962)
101. May, M. J., and Wilkinson, G., *Nature*, **1167** (1964)
102. Heck, R. F., *Symp. Inorg. Reaction Mechanisms, Kansas, 1964* (Advan. Chem. Ser., in press)
103. Miocque, M., Hung, N. M., and Yen, V. Q., *Ann. Chim.* (*Paris*), **8**, 157–74 (1963)
104. Rekasheva, A. F., and Samchenko, I. P., *Dokl. Akad. Nauk SSSR*, **133**, 1340–43 (1960)
105. Budde, W. L., and Dessy, R. E., *J. Am. Chem. Soc.*, **85**, 3964–70 (1963)
106. Halpern, J., James, B. R., and Kemp, A. L. W., *J. Am. Chem. Soc.*, **83**, 4097–98 (1961)
107. Smidt, J., *Chem. Ind.* (*London*), 54–61 (1962)
108. Matveev, K. J., Odyakov, V. F., Suzdalnitskaya, Yu. V., Bukhtoyarov, I. F., and Emelyanova, O. A., *Kinetika i Kataliz*, **3**, 661–73 (1962)
109. Vargaftik, M. N., Moiseev, I. I., and Syrkin, Ya. K., *Dokl. Akad. Nauk SSSR*, **139**, 1396–99 (1961)
110. Vargaftik, M. N., Moiseev, I. I., and Syrkin, Ya. K., *Dokl. Akad. Nauk SSSR*, **147**, 399–402 (1962)
111. Moiseev, I. I., Vargaftik, M. N., and Syrkin, Ya. K., *Izv. Akad. Nauk SSSR Otd. Khim. Nauk*, **6**, 1144–45 (1963)
112. Henry, P. M., *J. Am. Chem. Soc.*, **86**, 3246–50 (1964)
113. Moiseev, I. I., Vargaftik, M. N., and Syrkin, Ya. K., *Dokl. Akad. Nauk SSSR*, **133**, 377 (1960)
114. Stern, E. W., and Spector, M. L., *Proc. Chem. Soc.*, 370 (1961)
115. Harkness, A. C., and Halpern, J., *J. Am. Chem. Soc.*, **83**, 1258 (1961)
116. Nakamura, S., and Halpern, J., *J. Am. Chem. Soc.*, **83**, 4102 (1961)
117. Dawans, F., and Teyssie, P., *Bull. Soc. Chim. France*, **10**, 2376–92 (1963)
118. Ketley, A. D., and Werber, F. X., *Science*, **145**, 667–73 (1964)
119. Cossee, P., *J. Catalysis*, **3**, 80–88 (1964)
120. Cossee, P., *Tetrahedron Letters* (17), 12–16 (1960); in *Advances in the Chemistry of the Coordination Compounds*, 241 (Kirschner, S., Ed., Macmillan, New York, 1961)
121. Arlman, E. J., *Proc. Intern. Congr. Catalysis, 3rd, Amsterdam 1964* (North-Holland Publishing Co., Amsterdam, in press)
122. Arlman, E. J., *J. Catalysis*, **3**, 89–98 (1964)
123. Arlman, E. J., and Cossee, P., *J. Catalysis*, **3**, 99–104 (1964)
124. Phillips, G. W., and Carrick, W. L., *J. Am. Chem. Soc.*, **84**, 920–25 (1962)
125. Breslow, D. S., and Newburg, N. R., *J. Am. Chem. Soc.*, **79**, 5072 (1957); *ibid.*, **81**, 81–86 (1959)
126. Bestian, H., Clauss, K., Jensen, H., and Prinz, E., *Angew. Chem.*, **2**, 32–41 (1963)
127. Rinehart, R. E., Smith, H. P., Witt, H. S., and Romeyn, H., Jr., *J. Am. Chem. Soc.*, **83**, 4864–65 (1961); *ibid.*, **84**, 4145–47 (1962)
128. Canale, A. J., Hewett, W. A., Shryne, T. M., and Youngman, E. A., *Chem. Ind.* (*London*), 1054–55 (1962); Canale, A. J., and Hewett, W. A., *J. Polymer Sci., Pt. B*, **2**, 1041–45 (1964)
129. Teyssie, P., and Dauby, R., *J. Polymer Sci., Pt. B*, **2**, 413–16 (1964)
130. Natta, G., Dall'Asta, G., and Motroni, G., *J. Polymer Sci., Pt. B*, **2**, 349–51 (1964)
131. Teyssie, P., *Compt. Rend.*, 2846–49 (1963)

GAS CHROMATOGRAPHY[1,2]

By T. Gäumann

Institut de chimie-physique, EPUL, Lausanne, Switzerland

Since the last review on this subject by Habgood (1), the number of publications on gas chromatography has continued to increase at a rate of at least 30 per cent per annum. For 1964, about 1500 papers on the subject were anticipated. At the same time, the field of application is widening from physical chemical subjects such as adsorption characteristics (2), polymerization processes (3), and reaction kinetics on inhomogeneous surfaces (4) to detection of leaks in vacuum lines (5) and life on Mars (6). As there are a number of careful and rather complete reviews on the subject being regularly published, for example in *Analytical Chemistry* (7, 8), no attempt has been made to cover the entire field within this review. We will lay stress on the use of gas chromatography as a qualitative analytical tool.

In order to be able to separate substances, it is important to achieve a high resolving power of the equipment. Apart from experimental skill and some luck, knowledge of the theoretical background, which is being reviewed, is necessary. After the substance has left the separating column, it must be indicated by a detecting system. The different kinds used are, in most cases, badly understood in terms of the processes that give rise to a signal. Beside the possibility of using several detectors that respond to different classes of substances, one has the possibility of performing chemical reactions during some part of the analysis. The combination of the gas chromatographic apparatus with other equipment to achieve a qualitative analysis has become an important facility. The ever increasing amount of retention data published demands some universal system for their presentation. Some special separation applications will conclude this review.

A number of books on gas chromatography appeared in the last few years. Knox (9) gives a short introduction, whereas Littlewood (10) and Purnell (11) cover the subject in a rather complete fashion. The main difference between the latter two books may be that the former prefers to use the theoretical background to explain the experimental results, whereas, in the latter, the theory is illustrated by experimental facts. The well-known book by Bayer (12) appeared in a second edition; Kaiser (13) finished an inexpensive three-volume paperbound edition, the second volume being solely concerned with capillary gas chromatography, and the third volume containing

[1] The survey of literature pertinent to this review was concluded in the autumn of 1964 and mainly covered the two-year period 1962 to 1964. Earlier work is included wherever necessary.

[2] Work supported by the Schweizerische Nationalfond zur Förderung der wissenschaftlichen Forschung.

a wealth of gas chromatographic tables and formulae. Zhukhovitski & Turkel'taub (14) wrote a volume that summarizes Russian work in the field. Other books are written by several authors (15) or include only a chapter on the subject (16). Jeffery & Kipping (17) concentrate on gas analysis by gas chromatography, and Miller (18) on the experimental stage of the art. Compilation of the available data (19) as well as a guide to gas chromatography literature (20) form an important task. Several international meetings on the subject took place and the records have been published (21–27). The second volume of a journal on gas chromatography, which regularly contains notes and records of the meetings, will appear in 1964 (28).

Theory.—Fundamental aspects of the theory of gas chromatography have been summarized in a series of articles by Khan (29). The plate-height dependence H of the flow velocity u is a composite function of several terms, as has been shown by van Deemter, Zuiderweg & Klinkenberg (30):

$$H = A + B/u + C \cdot u$$

Exact calculations of the constants A, B, and C were made for different models in a series of publications. Dubský & Krejči (31) used a statistical approach. $A = 2 \cdot \lambda \cdot d_p$ is related to the eddy diffusion in the column, d_p being the particle diameter, while λ is of the order of unity. This term sometimes turns out to be negative, which means that the "classical" eddy-diffusion concept cannot be correct. This is shown by Giddings & Robison (32) by using inert glass-bead columns in which the plate height is sometimes less than the particle diameter. The authors propose a coupling theory of the eddy diffusion that emphasizes the interaction between the geometrical parameter λ and various nongeometrical quantities such as flow velocity and diffusion rate. They compare the theoretical results of earlier papers with experiments. This led DeFord, Loyd & Ayers (33) to propose a simpler equation that contains no A-term. Their B-term accounts for the axial diffusion, a C_g-term for nonequilibrum in the gas phase and a C_l-term for nonequilibrium conditions in the liquid phase. This probably represents the simplest possible form of equation.

Knox & McLaren (34) determined the B-term experimentally with a standard deviation of 2 per cent. A narrow band of unabsorbed solute is injected into the column and eluted part way down. The flow is arrested, and the band is allowed to spread by diffusion for a fixed time. Afterwards, the concentration profile is measured in the usual way. Giddings & Schettler (35) propose to use either a least-square computer program or an incremental method by changing the pressure conditions to evaluate the different terms.

It is generally acknowledged by now that the C-term should be split in two parts C_l and C_g, depending on the mass transfer in the liquid and gaseous phase. If the respective volume fraction of one theoretical plate of the gaseous and liquid phase is given by F_g and F_l, the capacity ratio is given by $k = K \cdot F_l/F_g$, where $K = c_l/c_g$ is the linear partition coefficient. Perret &

Purnell (36) studied the dependence of C_l and C_g upon the capacity ratio. According to their findings, the maximum possible resolution in packed columns has not yet been reached experimentally. Since the filling of packed columns is still more an art than a science, this leads one to question if the comparison between theoretical and experimental results is always meaningful. The connection between C_l and the capacity ratio depends also on the type of carrier gas, as has been shown by Barr & Sawyer (37). For helium used as a carrier gas, the expected dependence $k/(1+k)^2$ of C_l can be found, but there is practically no dependence of C_l on k for argon and nitrogen.

With a series of very nice experiments, Littlewood (38) demonstrates that one major source of dispersion across packed columns is uneven flow of carrier gas. To develop a theory, one has to make simplifications that are not always justified. One of these concepts concerns the liquid distribution in the column. A paper by Giddings (39) deals with the uniformity of the liquid film in capillary columns and the structure of the solid support in ordinary packed and glass-bead columns. The ordinary solid support contains a higher percentage of capillary that adsorbs liquid, the former being responsible for the plate height. The liquid around the contact points of the glass beads normally controls the plate height. By cutting sawtooth-shaped wells of calculated dimensions, an improvement could be made. The structure of the solid support material causes interactions between different flow channels. This should be calculable by a generalized nonequilibrium theory (40). The same author gives a review of different methods of approach to calculate the C-term (41). The above-mentioned idea of liquids in capillary pores of the solid support led Gordon, Krige & Pretorius (42) to develop a theory for the C_l-term in terms of easily measurable parameters. Giddings (43) compared gas chromatography and liquid chromatography and found that the latter should show an advantage of several orders of magnitude in the maximum number of plates attainable. This has not yet been proved by experiment.

Preparative gas chromatography using large scale columns has become increasingly popular in the last few years. Giddings used the nonequilibrium theory to obtain equations for the plate height of such columns (44). The plate height can be correlated with the usual flow characteristics as well as with lateral diffusivity, column porosity, and cross-sectional variations in liquid loading. The latter depends on the particle size uniformity across the column. Experiments to show this nonuniformity have been conducted by Giddings & Fuller (45). They filled various vertical columns of 5 cm inside diameter with celite or glass beads of relatively narrow-mesh range (80/100 mesh, 120 to 170 mesh) by different processes, fixed them with gelatin and measured the variation of mean particle size with distance from the center. The mean deviation of the particle diameter is naturally quite large, but the results definitely indicate an accumulation of the larger particles on the outside of the column cross section. The effect of column length on the separation ability of a large scale column has been the subject of two papers by the same author (46, 47). The plate height increases in proportion to column length and should finally level off to a limiting value.

The liquid amounts per injection in preparative columns are usually in the order of milliliters. Sawyer & Purnell (48) discuss sample-size and feed-volume effects in the light of known theories, and show that the commonly used sample volumes are mostly too large for maximum efficiency of the columns. When the substance is being injected, it will occupy the volume, not of one, but of several theoretical plates. Calculations for the symmetrical enlargements of peaks for this case are done by Glückauf (49). Kalmanovskii & Zhukhovitzkii (50) discussed the practical problems of separating two substances with Gaussian distributions. Said (51) pointed out an error in Glückauf's (52) well-known formula for separating two substances. In the case of overlapping peaks, quantitative analysis can be made by estimation of the half-height by means of the retention time (53). The problem of peak asymmetry is discussed by Ettre (54). Halász (55) suggests the use of a performance parameter to define a column. Higgins & Smith (56) correctly remark that in most preparative problems—contrary to analytical problems—the peak form is better represented by a square rather than a Gaussian elution curve. This is especially true because one will normally purposely overload the column in order to save expensive column material. The authors discuss different kinds of overloading. They distinguish between condensation overloading, feed-volume overloading, nonlinear isotherm overloading, enthalpy overloading, and radial profile overloading. It is hoped that their definition of column-packing procedures (bulk packing, snow packing, snow-storm packing, mountain packing) will be generally accepted. The shape of the elution peak was calculated as a function of the nonideality of the solution by Cruickshank & Everett (57) and as a function of sample input profile by Reilley, Hildebrand & Ashley (58).

Gordon, Krige & Pretorius have started a series wherein they calculate the influence of the different parameters on the theory of preparative linear gas chromatography. In the first paper (59), they consider the role of flow velocity. The preparative efficiency increases with the flow velocity and ultimately reaches a maximum which is larger than the corresponding theoretical value for the preparative column. The column length should be as short as possible when all other variables are kept constant (60). When the column is operated at the maximum-flow velocity, the preparative efficiency increases almost linearly with column length. Column diameter is the subject of the third paper (61). The problem of a continuous gas chromatography apparatus has been discussed by Giddings (62) and Golay, Hill & Norem (63).

Kováts (64) suggested the use of capillary columns of wider than normal diameter in order to increase the throughput capacity. Jentzsch & Hövermann (65) considered the theory and practice of such columns of 1 mm internal diameter. The influence of the carrier gas flow rate is demonstrated.

Thijssen (66) suggests that the height equivalent to a transfer unit would be a better measure of the separation properties of an open-hole column than height equivalent to a theoretical plate. Its dependence on different physical parameters is calculated.

An eluted peak is distorted even in an ideal case because of the pressure change during the passage of the solute through a theoretical plate. This effect is taken into consideration by Scott (67). Lengthening of the column means at the same time increasing the column pressure which is usually considered as detrimental. Giddings (68) has shown that this is not necessarily true and his findings are substantiated by Sternberg (69). Henly, Rose & Sweeny (70) calculated a general expression relating the partition coefficients in a multicomponent mixture to concentration under the assumption of the perfect gas law.

The theoretical background of using two different liquid phases, the three different methods of making such columns, and their advantages and disadvantages are calculated and discussed by Keller & Stewart (71) and Waksmundzki & Suprynowicz (72).

Several authors consider the problem of minimizing time for gas chromatographic analysis of complex mixtures (73, 74). The role of pressure gradients (75) and the general theory of minimum time operation (76) are discussed by Giddings. The optimum column length for a fixed time of analysis is calculated by Karger & Cooke (77). For short analysis, the time of injection plays an important role in governing attainable resolution, as is shown by Guiochon (78). Rooney & Aznavourian use a digital computer to calculate the optimum conditions (79).

Detectors.—A comprehensive survey of different detectors has been given by Littlewood (10), Gaziev et al. (80), Schoedler (81), Johnson (82), and Halász (83). Zlatkis (84) and Lovelock (85, 86) concentrate on ionization detectors. The problem of calibration and determination of linearity of very sensitive detectors has been treated by several authors. Burnett & Swoboda (87) saturate a diluent gas with a vapor of a very dilute solution. Henry's law is obeyed at these concentrations and the concentration in the gas phase is calculable. Fowlis & Scott (88) describe a vapor dilution apparatus which provides known relative concentrations of vapor in a gas.

The use of the differences in thermal conductivities as a function of concentration (katharometer) or flow (gas-density balance) is still one of the most simple and widely used systems, a review of which has been prepared by Purnell (11). The drawback is the limited sensitivity, which might be somewhat improved by using alternating bridge current (89). Several things affect the sensitivity. Ackman et al. (90) observed carbonaceous fouling of the filaments in spite of regular cleaning. The black deposit could not be burnt in an oxygen stream. Scott & Han (91) modified a cell to get a better independence of change in flow rate. Introduction of a shunt and a series resistance to the Wheatstone bridge, formed by the filaments, reduces the baseline drift. Optimum values for the resistors have been given by Littlewood

(92). Manka (93) uses both sides of a thermistor conductivity cell for two columns in parallel (silica gel and molecular sieve).

Response factors for thermal conductivity cells were calculated by Mecke & Zirker (94) for helium or hydrogen as carrier gas. They used the Chapman-Enskog theory and find that the limiting value for heavy gases is proportional to the means value of the collision cross section. Reasonable agreement with experimental values is obtained. Hoffmann (95) and Luy (96) correlate sensitivity factors with kinetic theory of gases and use the relative sensitivities experimentally obtained to get new kinetic gas data. Smith & Bowden (97) divide the response function into a product of three factors. They give the calculation of the two instrumental ones and use existing theory (98) for the thermal conductivity of a binary mixture. Jamieson (99) gives relative detector responses for a series of ethers and acetals.

The gas-density balance has the advantage of easily calculable response. Nerheim (100) gives, in a good way, the interrelationship between density, flow, and electrical response that form the basis for the construction of such a detector. Guillemin & Auricourt (101) have measured the experimental performance of this detector very extensively.

A hydrogen flame has a very low concentration of gaseous ions. If an organic substance is added, the ion current increases sharply and linearly over a wide range of concentrations. This makes the detector a very popular one. Ongkiehong (102) was among the first to test the performance of this detector very thoroughly. Middlehurst & Kennett (103) give a description of different detectors and amplifiers associated with them, including the use of alternating current to modulate the flame signal and to avoid the problem of dc-amplifier stability. With the present-day type of detectors, they figure out a theoretical limit of detection of about 10^{-14} g/s of ethylene. With their ac-triode detector, they arrive at a practical value of sensitivity fifty times larger than this, together with a drift corresponding to about 10^{-15} g/s ethylene per hour. Construction of a detector is described by Leipnitz & Könnecke (104). When silicon oil is used as liquid phase, a silicon deposit may develop on the detector electrodes giving rise to additional noise. According to Winger (105), this can be removed by passing 2 ml samples of Freon 12 through the column and the detector. The question, whether air or oxygen should be used as fuel gas, has been discussed by Halász and coworkers (106). They show that the signal increases about tenfold by using oxygen instead of air, but the signal-to-noise ratio stays the same. This means that oxygen does not give a higher sensitivity, but because the ion current is bigger, one can use smaller input resistances in the dc-amplifier. Besides being less troublesome, one is able to decrease the associated time constant of the amplifier.

Flame-ionization detectors have the disadvantage—in common with all ionization detectors—that the specific response to a substance cannot be calculated in advance. Ettre (107) compared the relative molar responses of hydrocarbons obtained by several authors and found some disagreement.

In a subsequent paper (108), he measured the relative molar response to a wide variety of compounds. The homologous series follow a linear relationship with the carbon number. The same result is found by Perkins et al. (109, 110), but the response to different functional groups varies widely. Dewar (111) made a theoretical approach to this problem. The thesis is put forward that the detector response would be easier to correlate when there is an excess of oxygen in the supply. Oxygen accessibility is discussed with respect to burner geometry and to molecular structure. Sternberg, Gallaway & Jones (112) studied the influence of several parameters on the response at some length. They favor chemi-ionization to thermal ionization by solid carbon aggregates. The proposal is made that the specific response is mainly defined by the number of carbon atoms in the molecule; the effect of specific groups can be taken into account by using "effective carbon numbers" given in the text. The complicated reaction schemes in the flame do not allow one to make a straightforward calculation. The already very wide concentration range of linear response is further widened by the observation of Cohen, Parzen & Bailey (113) that the response above the linear range is logarithmic for another two decades. A mathematical derivation is given for this peculiar kind of response for an overloaded detector.

The hydrogen flame detector can also be used to monitor high molecular weight effluents from liquid chromatographic columns (114, 115). The principle is to extract part of the effluent, e.g., by a moving wire, to evaporate the solvent, and to bring the substance in a hydrogen flame. Karmen (116) puts a mesh, coated with NaOH, in the flame. Any effluent from the gas chromatography column containing phosphorus or halogen will increase the rate of volatilization of sodium detected in a second flame. The detector gives the same kind of specific response as the electron capture detector (see later).

Another possible type of sensitive ionization detector contains a radioactive foil. If a noble gas, e.g., argon, is used as carrier gas, the high energy radiation will produce metastable states of the carrier gas. These metastable states are able to ionize any impurity present with an ionization potential lower than the energy of the metastable state (e.g., 11.6 eV for argon, 19.8 eV for helium). The so-called argon-ionization detector has become popular because of its simplicity. Wahlroos (117) discusses the characteristics of a tetrode detector. Teranishi, Nimmo & Corse (118) and Mahadeva, Iyer & Mittal (119) use strontium-90 as a radioactive emitter in order to be able to use the detector at high temperatures (240 to 500°). In another paper, the latter authors (120) use thallium-204 as source of β-rays. Promethium-147 has been preferred by Russian workers (121, 122). Schmidt-Küster & Wiesner (123) discuss the different possibilities of arranging the radioactive foil.

The problem of response to different classes of chemical compounds is possibly even worse for the argon detector than for the hydrogen-flame detector. For lower fatty acid series, Böttcher, Clemens & van Gent (124) find a response which is logarithmically related to the number of carbon atoms in the molecule, in contrast to Lovelock's equation (125). Rouayheb, Folmer & Hamilton (126) find that in the six to twelve carbon atoms range the area under the peak is proportional to the weight percentage of the substance. In this connection it should be mentioned that the peak height response for the argon detector is relatively independent of the carrier gas flow. This is in contrast to the other popular detectors which demand a very careful control of the gas flow and makes it a very handy detector for routine analysis. Sharpless (127) gets a specific response to methyl esters of dicarboxylic acids that alternates with carbon number. Nelson & Milun (128) operate their detector somewhere in between an argon detector and an electron capture detector to get specific response to steroids. According to Kipping & Jeffrey (129), the argon detector can detect nitric oxide with a sensitivity of about 0.01 μl. It is probably fair to say that the unpredictable response of the argon detector forms its major drawback and no way has yet been found to get around it.

Some time ago Haahti, Nikkari & Kulonen (130) published a paper in which they showed that this type of detector can work without a radioactive source. With all probability they used field emission effects to get ionization. The performance of such a detector has been described by Wahlroos (131).

Overloading of an argon detector can lead to several difficulties. If the concentration gets too high, a discharge may diminish the sensitivity for quite a while and perhaps the cell requires cleaning. After a large peak, the sensitivity of the detector may be diminished for several minutes. Upham, Lindgren & Nichols (132) get S-shaped calibration curves for their detector. Sims (133) shows that with appropriate graphic treatment the linear range of the detector can be enlarged.

Detailed study of the ionization mechanism has received much attention. Since it is mainly a radiation chemical problem, it can only be mentioned briefly here. Shahin & Lipsky (134) constructed a coaxial micro-ionization chamber for this kind of study. The dependence on applied potential and solute concentration form the basis of a paper by Knapp & Meyer (135). Strickler & Arakawa (136) show that optical emission forms an important part of the detection process. The emission spectrum shows three maxima at 1300, 1800, and 2000 Å with different quenching properties. They suspect that an excited state with an energy of 14.5 eV may also play an important role. They ascribe the well-known band at 3093 Å to the OH-radical from water as an impurity. A detector that operates in different modes and that can be used for the detection of permanent gases at low field strengths and of organic vapors at high ones is the subject of a publication by Lipsky & Shahin (137, 138).

The argon detector has a much smaller sensitivity for permanent gases because their ionization potential is higher than 11.6 eV. According to Galwey (139), the sensitivity for these gases can be increased by using very critically controlled conditions for the gas flow and the detector temperature. This might possibly be attributed to field emission effects. Another possibility that gives a rather high sensitivity has been described by Lesser (140). He feeds the detector constantly with a small amount of organic vapor (e.g., from a solid piece of 1,2,4,5-tetrachlorobenzene between the column and the detector) to produce a fixed amount of ionization current. The presence of a permanent gas will result in a decrease of ion current caused by the scavenging or quenching action of these gases. Lipsky & Shahin (141) work with xenon and krypton as carrier gases and find, surprisingly, an increase of sensitivity in the order Ar < Kr < Xe. The simplest solution to the detection of permanent gases would be the use of helium as carrier gas, but its purification is apparently difficult. Karmen, Giuffrida & Bowman (142) compare helium and argon as carrier gas. Serpinet (143), Parish & Parsons (144), and Wiseman (145) claim a very high sensitivity with helium. Váradi (146) uses it not only for the detection of permanent gases, but also for general use. A special multigrid detector which operates in vacuum can perform several functions at the same time.

Taylor (147) describes radiation hazards which may arise with the use of radioactive detectors. This paper should be read by anyone using this kind of equipment.

One is not necessarily forced to use a radioactive source in order to operate argon or helium detectors. Beside the field emission effect mentioned above (130, 131), it is possible to use a self-sustained discharge. Cells that operate from a dc-discharge have been described by Karmen & Bowman (148, 149), Bresler & Popev (150), Yamane (151), and Evrard, Thevelin & Joossens (152). Construction of detectors which use an electrodeless, high-frequency spark are given by Sternberg & Poulsen (153) and Andrew, Phillips & Semlyen (154). Roesler (155) gave preliminary results with a photoionization detector.

A detector with a radioactive source can be operated at low voltages with argon or another permanent gas as carrier. When a substance capable of trapping electrons passes through the detector, the chance of recombination of an ion pair increases with a corresponding decrease of the ionization current. The signal is thus reversed in polarity compared with the normal ionization detector. This so-called electron capture detector, developed by Lovelock & Lipsky (156), has the advantage of extremely high sensitivity and specific response. Its drawback is its very limited linear range. It seems that pulsing the detector improves its performance. Goodwin, Goulden & Reynolds (157) use a commercial argon ionization detector operated at less than 300 V as the electron capture detector. This type of detector and its operational performance has been described by several authors (158–161). Smith & Merritt (162) work with a three-electrode device and superimposed dc- and ac-voltages.

Using a pulsed detector and argon as the carrier gas, Smith & Fidiam (163) get a detection limit for nitrogen of 10^{-10} g/s. This detector finds a very important application in the measurement of lead alkyls as well as in pesticides. Because one does not get a signal for the pure hydrocarbons, the detector responds specifically for lead compounds. A parallel combination with another detector, e.g., a photoionization detector, can be used to measure the major components (164). Barrall & Ballinger (165) compare the coaxial- and the parallel-plate construction and do not find a major difference except for a somewhat lower selectivity of the former. The selective response is complicated, the accuracy is of the order of a few per cent; often the signal-to-concentration relationship is nonlinear. The sensitivity is of the order of 10 μl tetramethyl lead per gallon gasoline in a half-hour analysis. These figures are substantiated by Dawson (166). A specific measurement of iodides in the radiation chemistry of hydrocarbon-iodine mixtures has been made by Adloff & Gurgeniat (167). Aromatic hydrocarbons may also give rise to a signal (168, 169), therefore there is a possibility of measuring their electronegativity. Lovelock, Simmonds & van den Heuvel (170) measured the affinity of steroids for electrons with thermal energies [compare (128)]. The specific response of the electron capture detector, which also varies with the applied field strength, lends itself to qualitative analysis. Much work along this line has been done by Landowne & Lipsky. They show that for the detection of aliphatic halides the temperature of the detector is, beside other factors, of prime importance (171). By using a specific derivative, they were able to make group analyses. Sterols can be determined as their monochloroacetates with a sensitivity of the order of 10^{-11} to 10^{-13} g/s (172). The different halogeno acetates do not show a regularity in specific response. Pulsing the detector increases the linear range with some loss in sensitivity. The 2,4-dinitrophenyl group may be used as the active site for the analysis of amino acids (173). Linearity (1:4000) is comparable to the steroid derivatives and sensitivity is somewhat higher. Extremely high sensitivity poses some problems in regard to the irreversible adsorption in the column. The substance seen by the detector may not always be the one injected. Trace quantities of metal chelates are determined by Ross (174). Albert (175) compares the electron capture detector with the flame ionization detector for this problem. The limit of detection of the former is from 8 to 8000 times lower, but the range of linear response was only 1:100.

By operating a radioactive detector with hydrogen or helium as the carrier gas together with the appropriate voltages and geometrical arrangements, it can be used as a cross-section detector. It has the advantage of linear operation throughout the normal per cent range of concentrations. This detector has been developed mainly by Lovelock and his co-workers (176–179) and Abel (180, 181). Its limited sensitivity need not necessarily be a drawback, and it can be used from the separation of permanent gases and low boiling hydrocarbons up to the analysis of steroids.

Many physical and chemical effects can be used to measure the effluent of a column quantitatively or qualitatively. Novak & Janák (182) constructed the pneumatic analogue of the Wheatstone bridge. The substance passes through one capillary arm of the bridge and the corresponding change in pressure is recorded. In common with the gas-density balance, the effective measuring part is not exposed to the substance. The absorption of ultraviolet light (183), ultrasonic waves (184, 185), and the change in dielectric constant as measured in the microwave range (186) can be used as detecting effects. Microcoulometry (187, 188) competes with the electron capture detector in the analysis of pesticides. Sensitivity is claimed to be greater than with the argon detector (189). With oxygen as the carrier gas, one can burn a substance selectively on platinum wire (190). The sensitivity is several orders of magnitude higher than that of a conventional thermal conductivity detector (191). The carbon dioxide produced may be measured by the resulting increase in the electrical conductance when it is dissolved in water (192). Carbon dioxide as the carrier gas can be quantitatively absorbed, allowing a volumetric detection of permanent gases. The preparation of very pure carbon dioxide is described by Sirat (193), and the operation of a chromatograph of this type by Kateman & Rijks (194). The use of thermoionic ionization gauges has been described by Váradi & Ettre (195) and Hudson, King & Brandt (196). A very sensitive and selective detector for fluorine compounds, Br_2, NO_2, etc., has been developed by Gudzinowicz & Smith (197). The above-mentioned gases release krypton-85, adsorbed on a quinol clathrate, the liberated radioactive krypton is subsequently counted. The Hersch cell provides a specific detector for the measurement of oxygen (198). Eluted substances can be adsorbed on a plate, e.g., of a balance (199), a semiconductor film that changes its resistance (200) or a piezoelectric crystal, where the change in resonance frequency can be measured very accurately (201).

Qualitative analysis.—The gas chromatographic column will eventually separate two compounds, but it will not give any information as to the structure of these compounds. Retention data (see later) provide some information; but, even with a widely experienced operator, this will always be preliminary. The detector acts as a sensor for the eluted substance; it will give some information on the concentration, but, with rare exceptions, no information can be gained about the structure of the eluted compound.

One possibility for getting such information is to perform a chemical reaction with the mixture, either before, during, or after the analysis. Such reactions often take place inadvertently. If they happen during the passage through the column, the detector signal is characteristically changed. This might produce a peak-broadening, as calculated by Klinkenberg (202), or two more or less separated peaks, the secondary peak being unusually broad (203).

One of the simplest chemical reactions that can be carried out in front of the column is a catalytic hydrogenation or dehydrogenation (204–206). It is not limited to hydrocarbons, but can be used to determine the carbon skeleton of various compounds, such as amines, ketones, alcohols, etc. (207, 208). Catalytic conversion of carbon monoxide to methane allows one to use the sensitive flame ionization detector (209).

A very nice procedure is to use two detectors with a short reactive column in between them. The missing or displaced peak in the second chromatogram will belong to the substance that has reacted in the column. One does not lose gas chromatographic resolution and can realize immediately what has occurred on the column. Different columns are easily interchanged, the only condition being that the first detector must be nondestructive. Schenck & Eisma (210) and Adlard & Whitham (211) use molecular sieves to extract the *n*-alkanes; Ferrin, Chase & Hurn (212) use two detectors in parallel, one of them preceded by the reactive column in order to discriminate against unsaturates. Rowan (213) designed several classes of reactions to characterize a hydrocarbon mixture. Oxidation (214) and pyrolysis (215) are used to determine the structure of hydrocarbons. Ikeda, Simmons & Grossman (216) remove alcohols with specially treated boric acid.

A reactive column in front of the analytical column allows one to measure the products evolved, such as hydrogen from active boron-hydrogen bonds with wet molecular sieves (217), or from traces of water on a calcium carbide bed (218). The reaction can already be performed in the syringe. The advantage of this very promising method is that one can use the same column under the same conditions with and without reaction. The shifted peak belongs to the substance which has reacted. Anders & Mannering (219) used an injection with anhydrides to produce esters. Haken (220) compares acetoanhydride with acetylchloride for this purpose. Langer & Pantages (221) prepared the trimethylsilyl ether derivatives of alcohols. Hoff & Feit (222) developed a whole scheme of reactions and used it to identify the low-boiling compounds of wood smoke (223).

It is easy to carry out reactions with an eluted substance which has passed through a nondestructive detector. Halogen derivatives can be characterized by their blue-green flame when burned in the presence of copper (224). Mikkelsen & Spencer (225) describe special techniques. Casu & Cavallotti (226) use a series of test papers. Janák (227, 228) has developed a very fine method for condensing the effluent on paper or a thin layer in order to get multidimensional chromatography. Gas chromatography is used to separate the mixture according to the boiling points of its substituents; thin layer- or paper-chromatography will separate them according to polarity. Kaiser (229) automatized the method.

The ideal form of qualitative analysis would be to separate the mixture with the help of the gas chromatographic column, to trap the different fractions and analyse them by spectroscopic methods. Unfortunately with complex mixtures which will not be completely separated, and with amounts less than 0.1 mg, this procedure generally does not work very well. It is desirable to use a direct combination of the gas-liquid chromatography apparatus for some kind of spectroscopy. The difficulties with this combination are that the sample is always diluted in the carrier gas and the width of the eluted peak will be of the order of a fraction of a minute. Packed columns allow higher substance concentrations in the eluted gas stream; capillary columns have the advantage of higher resolution. The ideal but costly instrument in this connection is the mass spectrometer. A detailed review of this very highly developed technique is outside the scope of this review and the reader is referred to the newer literature (230–235). Infrared spectrometers are almost universally operated with beam modulation by means of mechaniical choppers. Because of noise problems, the chopping frequency is limited to the order of 10 cps, resulting in a time constant of a second or more. Several authors tried to make the best of this situation (236–239). Wilks & Brown (240) operate at a scan speed of 15 sec from 2.5 to 7 μ at some sacrifice in resolution. Bartz & Ruhl (241) get the whole spectrum in the same time by using two single-beam monochromators in parallel. A Beckman DU spectrophotometer has been modified to allow scanning from 1600 to 2100 Å in 6 sec with a resolution of 0.15 Å and a high sensitivity (242).

Retention data.—Within a short time, a large amount of data on retention times has been published, but the presentation of these data leaves much to be desired. Several systems have been developed using different kinds of reference substances. It seems to the author that the use of nonreactive paraffins should be by far the best choice. Evans & Smith (243, 244) proposed *n*-nonane as a standard reference substance and tested it at some length. Kováts (245) used the whole homologous series of *n*-alkanes as reference standards. His system of retention indices is now widely acknowledged (246–248), and a review, covering the results, rules, and tables so far published, has been written (249). Churchwell & Zlatkis (250) gave some newer results on hydrocarbons. Van den Dool & Kratz (251), Guiochon (252), and Habgood (253) expanded the range of application to temperature-programmed chromatograms. Strickler & Kováts (254) calculated the necessary column resolution for a given difference in the retention indices of a pair of substances.

Several points are of interest in connection with the determination of retention data or indices. The problem of determining the column hold-up has been treated by Gold (255). He recommends measuring a series of low boiling hydrocarbons and extrapolating to the virtual zero. Smith (256) showed that the retention time of air does not coincide with the column hold-up because of adsorption or solution processes. According to Amaya & Sasaki (257), a single component has a shorter retention time than the same substance injected in a mixture. The solid support may interact with the solute (258, 259). The retention index on polar-liquid phases may depend on the thermal treatment of the phase (260) and its percentage on the solid carrier (261). Even the carrier gas has been shown to have some effect on the degree of separation (262), possibly because of gas imperfections. Since all these difficulties affect the different retention systems in the same way, the use of Kováts' retention-index system is highly recommended. The next step in standardization would be the choice of liquid phases in order to make the tabulation of data easier, but this still seems to be far away.

Isotope separation.—Gas-liquid chromatography has been a strong competitor to mass spectrometry for the analysis of complex mixtures ever since its beginning. Isotope analysis has up to the present remained within the domain of the latter, but gas-liquid chromatography is now taking over part of this field. Separation of the hydrogen isotopes by this technique has become a straightforward matter. Glückauf & Kitt (263) and Phillips, Owens & Hemlin (264) used heat-displacement techniques to achieve this result. A column of iron-coated (265) or activated (266, 267) alumina or one filled with a molecular sieve and operated at $-78°$ (268) is used to separate H_2, HD, and D_2. It is important to have inactive column materials if the ortho-para-ratio has also to be determined. Etching of a glass capillary with an aqueous ammonia solution at elevated temperatures gives a column that elutes the isotopes in the order *p*-H_2, *o*-H_2, HD, *o*-D_2, *p*-D_2 at $-196°$ (269). The method of deuterium determination has been used by Arnett & Duggleby (270) to measure the deuterium content of water that has been decomposed over CaH_2. The separation of H_2, HT, and T_2 can be achieved on palladium on flint-quartz, silicagel, etc. (271). The same technique allowed the enrichment of low-level tritium samples for counting purposes (272). Carter & Smith (273) separate all hydrogen isotopes by using ferric oxide on an alumina column to obtain a rapid *o*-*p*-equilibrium. The order of elution is $H_2:HD:HT:D_2:T_2=1:1.22:1.42:1.49:1.85:2.08$.

On a column with $AgNO_3$-ethylene glycol as the stationary phase (274), the separation of ethylene-d_0, -d_2, and -d_4 has been performed by Cvetanović, Duncan & Falconer (275) and Dubrin, MacKay & Wolfgang (276). Lee & Rowland (277) were able to distinguish between the different position isomers of tritiated propanes. Different deuterated hydrocarbons in the C_4 to C_8 range have been separated from their protonated counterparts on squalene as liquid phase (278). Root, Lee & Rowland (279) managed to separate CD_4 from CH_4 and C_4D_{10} from C_4H_{10} using recycling gas chromatography. Gäumann was able to separate dodecanes differing by two deuterium atoms and found regularities in the retention index that enables the determination of the deuterium content of a hydrocarbon (280). Bruner et al. (281–284) described the preparation of capillary columns to separate deuterated benzenes and cyclohexanes in order to measure their thermodynamic properties. The isotope separations that have been achieved up to the present show this field to be a very promising one.

Isomers.—In general, there is always a liquid phase that allows the separation of a pair of isomers. If there are many isomers, it may be necessary to use several liquid phases. The separation of olefins that differ in the position of the double bond and in the *cis-trans* configuration has been achieved for butenes (285), hexenes (286), octenes (287, 288), and some nonenes (289). Hively (290) showed that the ratio of retention times of *cis-* and *trans-* compounds becomes larger with the increasing polarity of the liquid phase.

This means that one is likely to use liquid phases as polar as possible if a complex mixture of isomeric olefins is to be separated. There is not always a relation between the boiling point and the retention time on polar columns such as $AgNO_3$-glycol, as has been shown for menthenes by Hückel & Jennewein (291). They also studied the series of octanes. The same rules apply to stereoisomers, as demonstrated by Komers & Kochloefl (292) for alkylcyclohexanols, and by Feltkamp & Kraus (293) for decalene-diols-1,4.

The resolution of diastereomers, i.e., *dl*- and meso-forms, is also possible. Tannenbaum & Wick (294) studied several phases in order to separate methyl-2,4-dimethylheptanoates, diethylene glycol being the only usable phase. Stern, Atkinson & Jennings (295) separated 2,3-dichlorobutane on a preparative scale on different liquid phases. Schmid et al. (296) got a separation of meso- and rac-2,3-diethylsuccinic anhydride and other compounds on Apiezon-L. Gil-Av & Nurok (297) concentrated on several lactates, using different liquid phases. It is not generally acknowledged in hydrocarbon analysis that paraffins, substituted at adjacent carbon atoms, also give two separate peaks. For dodecanes, this has been shown to be the case (298); the identity of the two peaks has been proved by their mass spectrum. On nonpolar columns, two methyl groups give a separation of the order of 3.5 units in the retention index for the meso- and the rac-form; on polar columns this figure is about five units (299). The figures for two ethyl groups are one and two, respectively.

The separation of optically active isomers is apparently much more difficult. Goldberg & Ross (300) tried optically active liquid phases without success. Sievers, Moshier & Morris (301) got a partial resolution of *dl*-tris-(1,1,1,5,5,5-hexafluoro-2,4-pentane-dianato-) chromium(III).

LITERATURE CITED

1. Habgood, H. W., *Ann. Rev. Phys. Chem.*, **13**, 259–80 (1962)
2. Petrova, R. S., Khrapova, E. V., and Shcherbakova, K. D., in *Gas Chromatography*, 18–25 (See ref. 24)
3. Guyot, A., Blanc, C., Daniel, J.-C., and Trambouze, Y., *Compt. Rend.*, **253**, 1795–97 (1961)
4. Roginskii, S. Z., and Rosenthal, A. L., *Kinetika i Kataliz*, **5**, 104–12 (1964)
5. Ortega, I., and Cyrus, J. D., *Anal. Chem.*, **35**, 1649–51 (1963)
6. Oyama, V. I., *Nature*, **200**, 1058–59 (1963)
7. Dal Nogare, S., and Juvet, R. S., Jr., *Anal. Chem.*, **34**, 35R–47R (1962)
8. Juvet, R. S., Jr., and Dal Nogare, S., *Anal. Chem.*, **36**, 36R–51R (1964)
9. Knox, J. H., *Gas Chromatography* (Methuen, London, 126 pp., 1962)
10. Littlewood, A. B., *Gas Chromatography* (Academic Press, New York, 514 pp., 1962)
11. Purnell, H., *Gas Chromatography* (Wiley, London, 441 pp., 1962)
12. Bayer, E., *Gas-Chromatographie* (Springer, Heidelberg, 324 pp., 1962)
13. Kaiser, R., *Chromatographie in der Gasphase* (Bibliographisches Institut, Mannheim, 3 vol., 577 pp., 1960–62., English translation by P. H. Scott, Butterworths, London, 1963)
14. Zhukovitskii, A. A., and Turkel'taub, N. M., *Gazovaya Khromatografiya* (Gostoptekhizdat, Moskwa, 1962)
15. Szymanski, H. A., Ed., *Lectures in Gas Chromatography* (Plenum Press, New York, 282 pp., 1963)
16. Heftmann, C., Ed., *Chromatography* (Reinhold, New York, 1961)
17. Jeffery, P. G., and Kipping, P. J., *Gas Analysis by Gas Chromatography* (Pergamon, Oxford, 228 pp., 1964)
18. Miller, J. M., *Experimental Gas Chromatography* (Gow-Mac Instrument Co., Madison, Wis., 103 pp., 1963)
19. Prepared by J. S. Lewis, *Compilation of Gas Chromatographic Data*, Spec. Tech. Publ. No. 343 (ASTM, Philadelphia, Pa., 1963)
20. Signeur, A. V., *Guide to Gas Chromatography Literature* (Plenum Press, New York, 250 pp., 1963)
21. Zhukovitskii, A. A., Ed., *Gazovaya Khromatografiya* (Izd. A. N. SSSR, Moskwa, 1960); 1st Conf. USSR on Gas-Liquid Chromatography
22. Brenner, N., Callen, J. E., and Weiss, M. D., Eds., *Gas Chromatography* (Academic Press, New York, 719 pp., 1962); 3rd Intern. Symp., 1961
23. Szymanski, H. A., Ed., *Progress in Industrial Gas Chromatography* (Plenum Press, New York, 1, 236 pp., 1962); Proc., 3rd, Ann Gas Chromatography Inst., Buffalo, N.Y., 1961
24. van Swaay, M., Ed., *Gas Chromatography* (Butterworths, London, 409 pp., 1962); Proc. 4th Symp.
25. Angele, P., and Struppe, H. G., Ed., *Gas-Chromatographie 1963* (Akademie Verlag, Berlin, 1963)
26. Fowler, L., Ed., *Gas Chromatography* (Academic Press, New York, 270 pp., 1963); 4th Intern. Symp.
27. Goldup, A., Ed., *Gas Chromatography* (The Institute of Petroleum, London, 1964); Preprints, 5th Intern. Symp. on Gas-Liquid Chromatography
28. *Journal of Gas Chromatography*, 2 (1964)
29. Khan, M. A., *Laboratory Practice*, **10** (Aug. & Oct., 1961, Jan., Feb. & March 1962)
30. van Deemter, J. J., Zuiderweg, F. J., and Klinkenberg, A., *Chem. Eng. Sci.*, **5**, 271 (1956)
31. Dubský, H., and Krejči, M., *Collection Czech. Chem. Commun.*, **1964**, 1706–10
32. Giddings, J. C., and Robison, R. A., *Anal. Chem.*, **34**, 885–90 (1962)
33. DeFord, D. D., Loyd, R. J., and Ayers, B. O., *Anal. Chem.*, **35**, 426–29 (1963)
34. Knox, J. H., and McLaren, L., *Anal. Chem.*, **36**, 1477–82 (1964)
35. Giddings, J. C., and Schettler, P. D., *Anal. Chem.*, **36**, 1483–89 (1964)
36. Perrett, R. H., and Purnell, J. H., *Anal. Chem.*, **35**, 430–39 (1963)
37. Barr, J. K., and Sawyer, D. T., *Anal. Chem.*, **36**, 1753–59 (1964)
38. Littlewood, A. B., in *Gas Chromatography* (See ref. 27)
39. Giddings, J. C., *Anal. Chem.*, **34**, 458–65 (1962)
40. Giddings, J. C., *ibid.*, 1186–92
41. Giddings, J. C., *ibid.*, **35**, 439–49 (1963)
42. Gordon, S. M., Krige, G. J., and Pretorius, V., *Anal. Chem.*, **36**, 750–53 (1964)

43. Giddings, J. C., *Anal. Chem.*, **36,** 1890–92 (1964)
44. Giddings, J. C., *J. Gas Chromatog.*, **1,** 38–42 (April 1963)
45. Giddings, J. C., and Fuller, E. N., *J. Chromatog.*, **7,** 255–58 (1962)
46. Giddings, J. C., *J. Gas Chromatog.*, **1,** 12–21 (Jan. 1963)
47. Giddings, J. C., and Jensen, G. E., *J. Gas Chromatog.*, **2,** 290–95 (1964)
48. Sawyer, D. T., and Purnell, J. H., *Anal. Chem.*, **36,** 457–60 (1964)
49. Glückauf, E., *Trans. Faraday Soc.*, **60,** 729 (1964)
50. Kalmanovskii, V. I., and Zhukovitskii, A. A., in *Gas Chromatography* (See ref. 27)
51. Said, A. S., *J. Gas Chromatog.*, **1,** 20–24 (June 1963)
52. Glückauf, E., *Trans. Faraday Soc.*, **51,** 34 (1955)
53. Blaustein, B. D., and Feldman, G. M., *Anal. Chem.*, **36,** 65–70 (1964)
54. Ettre, L. S., *J. Chromatog.*, **11,** 267–70 (1963)
55. Halász, I., Hartmann, K., and Heine, E., in *Gas Chromatography* (See ref. (27)
56. Higgins, G. M. C., and Smith, J. F., in *Gas Chromatography* (See ref. 27)
57. Cruickshank, A. J. B., and Everett, D. H., *J. Chromatog.*, **11,** 289–94 (1963)
58. Reilley, C. N., Hildebrand, G. P., and Ashley, J. W., Jr., *Anal. Chem.*, **34,** 1198–213 (1962)
59. Gordon, S. M., Krige, G. J., and Pretorius, V., *J. Gas Chromatog.*, **2,** 241–45 (1964)
60. Gordon, S. M., Krige, G. J., and Pretorius, V., *ibid.*, 246–51
61. Gordon, S. M., Krige, G. J., and Pretorius, V., *ibid.*, 285–89
62. Giddings, J. C., *Anal. Chem.*, **34,** 37–39 (1962)
63. Golay, M. J. E., Hill, H. I., and Norem, S. D., *Anal. Chem.*, **35,** 488–91 (1963)
64. Kováts, E.sz., *Informal Symp., Inst. Hydrocarbons, Birmingham, 1961* [Quoted by Ettre et al., *J. Gas Chromatog.*, **1,** 7 (Feb. 1963)]
65. Jentzsch, D., and Hövermann, W., *J. Chromatog.*, **11,** 440–51 (1963)
66. Thijssen, H. A. C., *J. Chromatog.*, **11,** 141–50 (1963)
67. Scott, R. P. W., *Anal. Chem.*, **36,** 1455–61 (1964)
68. Giddings, J. C., *Anal. Chem.*, **36,** 741–44 (1964)
69. Sternberg, J. C., *Anal. Chem.*, **36,** 921–22 (1964)
70. Henly, R. S., Rose, A., and Sweeny, R. F., *Anal. Chem.*, **36,** 744–50 (1964)
71. Keller, R. A., and Stewart, G. H., *Anal. Chem.*, **36,** 1186–91 (1964)
72. Waksmundzki, A., and Suprynowicz, Z., in *Gas Chromatography* (See ref. 27)
73. Ayers, B. O., Loyd, R. J., and DeFord, D. D., *Anal. Chem.*, **33,** 986–91 (1961)
74. Boeke, J., and Parke, N. G., III, in *Gas Chromatography*, 391–422 (See ref. 22)
75. Giddings, J. C., *Nature*, **191,** 1291–92 (1961)
76. Giddings, J. C., *Anal. Chem.*, **34,** 314–19 (1962)
77. Karger, B. L., and Cooke, W. D., *Anal. Chem.*, **36,** 985–91 (1964)
78. Guiochon, G., *Anal. Chem.*, **35,** 399–400 (1963)
79. Rooney, T. B., and Aznavourian, W., *Anal. Chem.*, **36,** 2112 (1964)
80. Gaziev, G. A., Porshneva, N. V., Lu, P.-C., Yanovskii, M. I., Turkel' taub, N. M., and Kornyakov, V. S., in *Gazovaya Khromatografiya*, 307 (See ref. 21)
81. Schoedler, C., *Bull. Soc. Chim. France*, **1962,** 2315
82. Johnson, R. E., in *Lectures in Gas Chromatography*, 65–86 (See ref. 15)
83. Halász, I., *Anal. Chem.*, **36,** 1428–30 (1964)
84. Zlatkis, A., in *Lectures in Gas Chromatography*, 87–104 (See ref. 15)
85. Lovelock, J. E., *Research and Development*, Oct. 1961
86. Lovelock, J. E., *Anal. Chem.*, **33,** 162–78 (1961)
87. Burnett, M. G., and Swoboda, P. A. T., *Anal. Chem.*, **34,** 1162–63 (1962)
88. Fowlis, I. A., and Scott, R. P. W., *J. Chromatog.*, **11,** 1–10 (1963)
89. Purcell, J. R., and Keeler, R. N., *Rev. Sci. Instr.*, **31,** 304–6 (1960)
90. Ackman, R. G., Burgher, R. D., Sipos, J. C., and Odense, P. H., *J. Chromatog.*, **9,** 531–34 (1962)
91. Scott, D. S., and Han, A., *Anal. Chem.*, **33,** 160 (1961)
92. Littlewood, A. B., *J. Sci. Instr.*, **37,** 185–88 (1960)
93. Manka, D. P., *Anal. Chem.*, **36,** 480–82, 2218 (1964)
94. Mecke, R., and Zirker, K., *J. Chromatog.*, **7,** 1–12 (1962)
95. Hoffmann, E. G., *Anal. Chem.*, **34,** 1216–22 (1962)
96. Luy, H., *Z. Anal. Chem.*, **194,** 241 (1963)

97. Smith, B. D., and Bowden, W. W., *Anal. Chem.*, **36,** 82–86 (1964)
98. Reid, R. C., and Sherwood, T. K., *The Properties of Gases and Liquids* (McGraw, New York, 1958)
99. Jamieson, G. R., *J. Chromatog.*, **8,** 544–46 (1962)
100. Nerheim, A. G., *Anal. Chem.*, **35,** 1640–44 (1963)
101. Guillemin, C. L., and Auricourt, F., *J. Gas Chromatog.*, **1,** 24–29 (Oct. 1963)
102. Ongkiehong, L., *Gas Chromatog., Proc. Symp., 3rd, Edinburgh, 1960,* 7 (Butterworths, London, 1960)
103. Middlehurst, J., and Kennett, B., *J. Chromatog.*, **10,** 294–302 (1963)
104. Leipnitz, W., and Könnecke, H.-G., *J. Prakt. Chem.*, **18,** 110–12 (1962)
105. Winger, R., *Gas Chromatog. Newsletter*, **4,** 3 (Appl. Physics Laboratories, Calif., 1963)
106. Bruderreck, H., Schneider, W., and Halász, I., *Anal. Chem.*, **36,** 461–73 (1964)
107. Ettre, L. S., in *Gas Chromatography,* 307–27 (See ref. 22)
108. Ettre, L. S., *J. Chromatog.*, **8,** 525–30 (1962)
109. Perkins, G., Jr., Rouayheb, G. M., Lively, L. D., and Hamilton, W. C., in *Gas Chromatography,* 269–85 (See ref. 22)
110. Perkins, G., Jr., Laramy, R. E., and Lively, L. D., *Anal. Chem.*, **35,** 360–62 (1963)
111. Dewar, R. A., *J. Chromatog.*, **6,** 312–23 (1961)
112. Sternberg, J. C., Gallaway, W. S., and Jones, D. T., in *Gas Chromatography,* 231–67 (See ref. 22)
113. Cohen, E. N., Parzen, E., and Bailey, D. M., *J. Gas Chromatog.*, **1,** 14–18 (Aug. 1963)
114. Haahti, E., Nikkari, T., and Kärkkäinen, J., in *Gas Chromatography* (See ref. 27)
115. James, A. T., Ravenhill, J. R., and Scott, R. P. W., in *Gas Chromatography* (See ref. 27)
116. Karmen, A., *Anal. Chem.*, **36,** 1416–21 (1964)
117. Wahlroos, O., *Acta Chem. Scand.*, **16,** 1039–40 (1962)
118. Teranishi, R., Nimmo, C. C., and Corse, J., *Anal. Chem.*, **32,** 896 (1960)
119. Mahadeva, R., Iyer, R. M., and Mittal, J. P., *J. Chromatog.*, **11,** 404–5 (1963)
120. Iyer, R. M., and Mittal, J. P., *J. Sci. Ind. Res. (India)*, **21D,** 19–22 (1962)
121. Frank, Y. A., and Yanovskii, M. I., *Kinetics Catalysis (USSR)*, **2,** 274 (1961)
122. Gaziev, G. A., Oziraner, S. N., Yanovskii, M. I., and Kornyakov, V. S., *Zh. Fiz. Khim.*, **5,** 1150 (1961)
123. Schmidt-Küster, W. J., and Wiesner, L., *Erdöl Kohle,* **15,** 193–96 (1962)
124. Böttcher, C. J. F., Clemens, G. F. G., and van Gent, C. M., *J. Chromatog.*, **3,** 582–84 (1960)
125. Lovelock, J. E., *J. Chromatog.*, **1,** 35 (1958)
126. Rouayheb, G. M., Folmer, O. F., Jr., and Hamilton, W. C., *Anal. Chim. Acta,* **26,** 378–90 (1962)
127. Sharpless, N. E., *J. Chromatog.*, **12,** 401–3 (1963)
128. Nelson, J. P., and Milun, A. J., *Chem. Ind. (London)*, **1962,** 1722–23.
129. Kipping, P. J., and Jeffery, P. G., *Nature,* **200,** 1314 (1963)
130. Haahti, E., Nikkari, T., and Kulonen, E., *J. Chromatog.*, **3,** 372–73 (1960)
131. Wahlroos, O., *Acta Chem. Scand.*, **15,** 708–9 (1961)
132. Upham, F. T., Lindgren, F. T., and Nichols, A. V., *Anal. Chem.*, **33,** 845–49 (1961).
133. Sims, R. P. A., *J. Chromatog.*, **8,** 538–40 (1962)
134. Shahin, M. M., and Lipsky, S. R., *Anal. Chem.*, **35,** 1562–66 (1963)
135. Knapp, J. Z., and Meyer, A. S., *Anal. Chem.*, **36,** 1430–41 (1964)
136. Strickler, T. D., and Arakawa, E. T., *J. Chem. Phys.*, **41,** 1783–89 (1964)
137. Lipsky, S. R., and Shahin, M. M., *Nature,* **197,** 625–26 (1963)
138. Shahin, M. M., and Lipsky, S. R., *Anal. Chem.*, **35,** 467–74 (1963)
139. Galwey, A. K., *Talanta,* **9,** 1043–52 (1962)
140. Lesser, R., *Angew. Chem.*, **72,** 775–77 (1960)
141. Lipsky, S. R., and Shahin, M. M., *Nature,* **200,** 566–67 (1963)
142. Karmen, A., Giuffrida, L., and Bowman, R. L., *J. Chromatog.*, **9,** 13–20 (1962)
143. Serpinet, J., *Anal. Chim. Acta,* **25,** 505–6 (1961)
144. Parish, R. V., and Parsons, W. H., *Chem. Ind. (London)*, **1961,** 1951.
145. Wiseman, W. A., *Nature,* **192,** 964–65 (1961)
146. Váradi, P. F., in *Gas Chromatography,* 195–205 (See ref. 22)

147. Taylor, M. P., *J. Chromatog.*, **9,** 28–33 (1962)
148. Karmen, A., and Bowman, R. L., *Nature*, **196,** 62–63 (1962)
149. Karmen, A., and Bowman, R. L., in *Gas Chromatography*, 189–93 (See ref. 22)
150. Bresler, S. E., and Popov, A. G., *Zh. Fiz. Khim.*, **37,** 1178–82 (1963)
151. Yamane, M., *J. Chromatog.*, **9,** 158, 162–72 (1962)
152. Evrard, E., Thevelin, M., and Joossens, J. V., *Nature*, **193,** 59–60 (1962)
153. Sternberg, J. C., and Poulson, R. E., *J. Chromatog.*, **3,** 406–10 (1960)
154. Andrew, T. D., Phillips, C. S. G., and Semlyen, J. A., *J. Gas. Chromatog.*, **1,** 27 (Jan. 1963)
155. Roesler, J. F., *Anal. Chem.*, **36,** 1900–3 (1964)
156. Lovelock, J. E., and Lipsky, S. R., *J. Am. Chem. Soc.*, **82,** 431–33 (1960)
157. Goodwin, E. S., Goulden, R., and Reynolds, J. G., *Analyst*, **86,** 697–709.
158. Lovelock, J. E., and Gregory, N. L., in *Gas Chromatography*, 219–29 (See ref. 22)
159. Washbrooke, P. F., *Chemiker Ztg.*, **86,** 377 (1962)
160. Lovelock, J. E., *Anal. Chem.*, **35,** 474–81 (1963)
161. Ashworth, R. J., *J. Gas Chromatog.*, **2,** 252–55 (1964)
162. Smith, V. N., and Merritt, E. J., *Anal. Chem.*, **34,** 1476–82 (1962)
163. Smith, V. N., and Fidiam, J. F., *Anal. Chem.*, **36,** 1739–44 (1964)
164. Lovelock, J. E., and Zlatkis, A., *Anal. Chem.*, **33,** 1958–59 (1961)
165. Barrall, E. M., and Ballinger, P. R., *J. Gas Chromatog.*, **1,** 7–13 (Aug. 1963)
166. Dawson, H. J., Jr., *Anal. Chem.*, **35,** 542–45 (1963)
167. Adloff, J. P., and Gurgeniat, P. *J. Chromatog.*, **12,** 96–98 (1963)
168. Lovelock, J. E., Zlatkis, A., and Becker, R. S., *Nature*, **193,** 540 (1962)
169. Wentworth, W. E., and Becker, R. S., *J. Am. Chem. Soc.*, **84,** 4263–66 (1962)
170. Lovelock, J. E., Simmonds, P. G., and van den Heuvel, W. J. A., *Nature*, **197,** 249 (1963)
171. Landowne, R. A., and Lipsky, S. R., *Anal. Chem.*, **34,** 726–30 (1962)
172. Landowne, R. A., and Lipsky, S. R., *ibid.*, **35,** 532–35 (1963)
173. Landowne, R. A., and Lipsky, S. R., *Nature*, **199,** 141–43 (1963)
174. Ross, W. D., *Anal. Chem.*, **35,** 1596–98 (1963)
175. Albert, D. K., *Anal. Chem.*, **36,** 2034–35 (1964)
176. Lovelock, J. E., *Nature*, **187,** 49-50 (1960)
177. Lovelock, J. E., Shoemake, G. R., and Zlatkis, A., *Anal. Chem.*, **35,** 460–65 (1963)
178. Simmonds, P. G., and Lovelock, J. E., *Anal. Chem.*, **35,** 1345–48 (1963)
179. Lovelock, J. E., Shoemake, G. R., and Zlatkis, A., *Anal. Chem.*, **36,** 1410–15 (1964)
180. Abel, K., *Anal. Chem.*, **36,** 954–55 (1964)
181. Abel, K., and deSchmertzing, H., *Anal. Chem.*, **35,** 1754–56 (1963)
182. Novak, J., and Janák, J., *Symp. Gas-Chromatog., IV, Leuna/DDR*, **1963,** 100
183. Merritt, J., Comendant, F., Abrams, S. T., and Smith V. N., *Anal. Chem.* **35,** 1461–64 (1963)
184. Testerman, M. K., and McLeod, P. C., in *Gas Chromatography*, 183–88 (See ref. 22)
185. Noble, F. W., Abel, K., and Cooke, P. W., *Anal. Chem.*, **36,** 1421–27.
186. Johansson, G., *Anal. Chem.*, **34,** 914–16 (1962)
187. Coulson, D. M., Cavanagh, L. A., DeVries, J. E., and Walther, B., *Instrumental Methods for the Analysis of Food Additives*, 183 (Butz, W. H., and Noebels, H. J., Eds., Interscience, 1962)
188. Cavanagh, L. A., Coulson, D. M., and DeVries, J. E., *ibid.*, 195–99
189. DeVries, J. E., Coulson, D. M., and Cavanagh, L. A., *Symp., Submicrogram Experimentation, 1960 Arlington*, 219 (Cheronis, N. D., Ed.)
190. Folmer, O. F., Jr., Yang, K., and Perkins, G., Jr., *Anal. Chem.*, **35,** 454–59 (1963)
191. Schay, G., Székely, G., and Traply, G., in *Gas Chromatography*, 179–83 (See ref. 24)
192. Piringer, O., and Pascalau, M., *J. Chromatog.*, **8,** 410–12 (1962)
193. Sirat, N., *Bull. Soc. Chim. France*, **1962,** 1388-89.
194. Kateman, G., and Rijks, J. A., *J. Chromatog.*, **14,** 13–21 (1964)
195. Váradi, P. F., and Ettre, K., *Anal. Chem.*, **34,** 1417–22 (1962)

196. Hudson, B. E., King, W. H., and Brandt, W. W., in *Gas Chromatography*, 207–18 (See ref. 22)
197. Gudzinowicz, B. J., and Smith, W. R., *Anal. Chem.*, **35**, 465–67 (1963)
198. Phillips, T. R., Johnson, E. G., and Woodward, H., *Anal. Chem.*, **36**, 450–52 (1964)
199. Bevan, S. C., and Thorburn, S., *J. Chromatog.*, **11**, 301–6 (1963)
200. Seiyama, T., Kato, A., Fujiishi, K., and Nagatani, M., *Anal. Chem.*, **34**, 1502–3 (1962)
201. King, W. H., *Anal. Chem.*, **36**, 1735–39 (1964)
202. Klinkenberg, A., *Chem. Eng. Sci.*, **15**, 255–59 (1961)
203. Kallen, J., and Heilbronner, E., *Helv. Chim. Acta*, **43**, 489–90 (1960)
204. Smith, B., Larson, G., and Rydén, J., *Acta Chem. Scand.*, **17**, 313–18 (1963)
205. Klesment, I. R., Rang, S. A., and Eisen, O. G., *Neftekhimiya*, **3**, 864–70 (1963)
206. Okamoto, T., and Onaka, T., *Chem. Pharm. Bull. (Japan)*, **11**, 1087–88 (1963)
207. Okamoto, T., and Onaka, T., *ibid.*, 1086
208. Beroza, M., and Sarmento, R., *Nature*, **196**, 768–69 (1962); *Anal. Chem.*, **34**, 1801–11 (1962); **35**, 1353–57 (1963)
209. Porter, K., and Volman, D. H., *Anal. Chem.*, **34**, 748–51 (1962)
210. Schenck, P. A., and Eisma, E., *Nature*, **199**, 170 (1963)
211. Adlard, E. R., and Whitham, B. T., in *Gas Chromatography*, 371–90 (See ref. 22)
212. Ferrin, C. R., Chase, J. O., and Hurn, R. W., in *Gas Chromatography*, 423–29 (See ref. 22)
213. Rowan, R., Jr., *Anal. Chem.*, **33**, 658–65 (1961)
214. Smith, B., Ohlson, R., and Olson, A.-M., *Acta Chem. Scand.*, **16**, 1463–67 (1962)
215. Dhont, J. H., *Nature*, **192**, 747–48 (1961)
216. Ikeda, R. M., Simmons, D. E., and Grossman, J. D., *Anal. Chem.*, **36**, 2188 (1964)
217. Putnam, R. F., and Myers, H. W., *Anal. Chem.*, **34**, 486–88 (1962)
218. Knight, H. S., and Weiss, F. T., *Anal. Chem.*, **34**, 749–51 (1962)
219. Anders, M. W., and Mannering, G. J., *Anal. Chem.*, **34**, 730–33 (1962)
220. Haken, J. K., *J. Gas Chromatog.*, **1**, 30 (Oct. 1963)
221. Langer, S. H., and Pantages, P., *Nature*, **191**, 141–42 (1961)
222. Hoff, J. E., and Feit, E. D., *Anal. Chem.*, **35**, 1298–99 (1963)
223. Hoff, J. E., and Kapsalopoulou, A. J., *J. Gas Chromatog.*, **2**, 296–99 (1964)
224. Chovin, P., Lebbe, J., and Moureu, H., *J. Chromatog.*, **6**, 365–67 (1961)
225. Mikkelsen, L., and Spencer, S. F., in *Lectures in Gas Chromatography*, 177–98 (See ref. 15)
226. Casu, B., and Cavallotti, L., *Anal. Chem.*, **34**, 1514–16 (1962)
227. Janák, J., *J. Gas Chromatog.*, **1**, 20–23 (Oct. 1963)
228. Janák, J., *J. Chromatog.*, **15**, 15–28 (1964)
229. Kaiser, R., *Z. Anal. Chem.*, **205**, 284–98 (1964)
230. Stenhagen, E., *Z. Anal. Chem.*, **205**, 109–24 (1964)
231. Ryhage, R., *Anal. Chem.*, **36**, 759–64 (1964)
232. Watson, J. T., and Biemann, K., *Anal. Chem.*, **36**, 1135–37 (1964)
233. Teranishi, R., Buttery, R. G., McFadden, W. H., Mon, T. R., and Wasserman, J., *Anal. Chem.*, **36**, 1509–12 (1964)
234. McFadden, W. H., and Day, E. A., *Anal. Chem.*, **36**, 2362–63 (1964)
235. Banner, A. E., Elliott, R. M., and Kelly, W., in *Gas Chromatography* (See ref. 22)
236. Chang, S. S., Ireland, C. E., Tai, H., and Staley, A. E., *Anal. Chem.*, **33**, 479 (1961)
237. Haslam, J. J., and Wilks, H. A., *Analyst*, **86**, 44 (1961)
238. Flett, M. S. C., and Hughes, J., *J. Chromatog.*, **11**, 434–39 (1963)
239. Behrendt, S., *Nature*, **201**, 70 (1964)
240. Wilks, P. A., Jr., and Brown, R. A., *Anal. Chem.*, **36**, 1896–99 (1964)
241. Bartz, A. M., and Ruhl, H. D., *Anal. Chem.*, **36**, 1892–96 (1964)
242. Kaye, W., and Waska, F., *Anal. Chem.*, **36**, 2380–81 (1964)
243. Evans, M. B., and Smith, J. F., *J. Chromatog.*, **5**, 300–7 (1961); **6**, 293–311 (1961); **8**, 303–7, 541–44 (1962); **9**, 147–53 (1962)
244. Evans, M. B., *J. Chromatog.*, **11**, 401–3 (1963); **12**, 2–7 (1963)
245. Kováts, E. sz., *Helv. Chim. Acta*, **41**, 1915 (1958)
246. Struppe, H. G., in *Gas-Chromatographie 1963*, 378 (See ref. 25)

247. Inst. of Petroleum, GLC Discussion Group, Data Subcommittee (See ref. 27)
248. Dhont, J. H., *Nature*, **198,** 990–91 (1963)
249. Kováts, E. sz., in *Advances in Chromatography* (Giddings, J. C., and Keller, R. A., Eds., M. Dekker Inc., New York, 1965)
250. Churchwell, R. L., and Zlatkis, A., *J. Gas Chromatog.*, **2,** 275–76 (1964)
251. van den Dool, H., and Kratz, P. D., *J. Chromatog.*, **11,** 463–71 (1963)
252. Guiochon, G., *Anal. Chem.*, **36,** 661–63 (1964)
253. Habgood, H. W., *Anal. Chem.*, **36,** 663–65 (1964)
254. Strickler, H., and Kováts, E. sz., *J. Chromatog.*, **8,** 289–92 (1962)
255. Gold, H. J., *Anal. Chem.*, **34,** 174–75 (1962)
256. Smith, J. F., *Nature*, **193,** 679 (1962)
257. Amaya, K., and Sasaki, K., *Tokyo Kogyo Shikensho Hokoku*, **57,** 503 (1962)
258. Craig, B. M., in *Gas Chromatography*, 37–56 (See ref. 22)
259. Scholz, R. G., and Brandt, W. W., in *Gas Chromatography*, 7–26 (See ref. 22)
260. Chen, C., and Gacke, D., *Anal. Chem.*, **36,** 72–76 (1964)
261. Pecsok, R. L., de Yllana, A., and Abdul-Karim, A., *Anal. Chem.*, **36,** 452–57 (1964)
262. Goldup, A., Luckhurst, G. R., and Swanton, W. T., *Nature*, **193,** 333–34 (1962)
263. Glückauf, E., and Kitt, G. P., *Proc. Intern. Symp. Isotope Separation, Amsterdam, 1957*, 210 (1958)
264. Phillips, T. R., Owens, D. R., and Hemlin, A. G., *Nature*, **192,** 1067–68 (1961)
265. Shipman, G. F., *Anal. Chem.*, **34,** 877–78 (1962)
266. Venugopalan, M., and Kutschke, K. O., *Can. J. Chem.*, **41,** 548–49 (1963)
267. van Hock, W. A., and Emmett, P. H., *J. Phys. Chem.*, **64,** 673–75 (1960)
268. Botter, F., de la Perrière, G., and Tishchenko, S., *Comm. Energie At. (France)*, *Rappt.* **1962** (1961); *Chem. Abstr.*, **55,** 25591
269. Mohnke, M., and Saffert, W., in *Gas Chromatography*, 216–21 (See ref. 22)
270. Arnett, E. M., and Duggleby, P. McM., *Anal. Chem.*, **35,** 1420–24 (1963)
271. Smith, H. A., and Carter, E. H., *Tritium in the Physical and Biological Sciences*, **1,** 121–33 (IAEA, Wien, 1962)
272. Crespi, M. B. A., and Perschke, H., *Intern. J. Appl. Radiation Isotopes*, **15,** 569 (1964)
273. Carter, E. H., and Smith, H. A., *J. Phys. Chem.*, **67,** 1512–16 (1963)
274. Muhs, M. A., and Weiss, F. T., *J. Am. Chem. Soc.*, **84,** 4697 (1962)
275. Cvetanović, R. J., Duncan, F. J., and Falconer, W. E., *Can. J. Chem.*, **41,** 2095 (1963)
276. Dubrin, J., MacKay, C., and Wolfgang, R., *J. Am. Chem. Soc.*, **86,** 959–60 (1964)
277. Lee, E. K. C., and Rowland, F. S., *Anal. Chem.*, **36,** 2181 (1964)
278. Falconer, W. E., and Cvetanović, R. J., *Anal. Chem.*, **34,** 1064–66 (1962)
279. Root, J. W., Lee, E. K. C., and Rowland, F. S., *Science*, **143** 676 (1964)
280. Gäumann, T., *Helv. Chim. Acta* (In press)
281. Bruner, F., and Cartoni, G. P., *Anal. Chem.*, **36,** 1522–26 (1964)
282. Bruner, F., and Cartoni, G. P., *J. Chromatog.*, **10,** 396 (1963)
283. Liberti, A., Cartoni, G. P., and Bruner, F., in *Gas Chromatography* (See ref. 27)
284. Liberti, A., Cartoni, G. P., and Bruner, F., *J. Chromatog.*, 8–14 (1963)
285. Higgins, C. E., and Baldwin, W. H., *Anal. Chem.*, **36,** 473–75 (1964)
286. Mikhailova, E. A., and Diachenko, A. I., *Dokl. Akad. Nauk SSSR*, **144,** 1056 (1962)
287. Vigdergauz, M. S., and Afanasiev, M. I., *Neftekhimiya*, **3,** 425–29 (1963)
288. Bendel, E., Kern, M., Janssen, R., and Steffan, G., *Angew. Chem.*, **74,** 905 (1962)
289. Závada, J., and Sicher, J., *Proc. Chem. Soc.*, **1963,** 96–97
290. Hively, R. A., *Anal. Chem.*, **35,** 1921–24 (1963)
291. Hückel, W., and Jennewein, C.-M., *Z. Anal. Chem.*, **205,** 202–16 (1964)
292. Komers, R., and Kochloefl, K., *Collection Czech. Chem. Commun.*, **28,** 46–54 (1963)
293. Feltkamp, H., and Kraus, W., *Ann. Chem.*, **651,** 11–17 (1962)
294. Tannenbaum, S. R., and Wick, E. L., *J. Org. Chem.*, **27,** 2650–51 (1962)

295. Stern, R., Atkinson, E. R., and Jennings, F. C., *Chem. Ind.* (*London*), **1962**, 1758–59.
296. Nagarajan, K., Weissmann, C., Schmid, H., and Karrer, P., *Helv. Chim. Acta*, **46**, 1212–31 (1963)
297. Gil-Av, E., and Nurok, D., *Proc. Chem. Soc.*, **1962**, 146–47.
298. Widmer, H., and Gäumann, T., *Helv. Chim. Acta*, **46**, 944–54 (1963)
299. Gäumann, T., *Helv. Chim. Acta* (In press)
300. Goldberg, G., and Ross, W. A., *Chem. Ind.* (*London*), **1962**, 657–58
301. Sievers, R. E., Moshier, R. W., and Morris, M. L., *Inorg. Chem.*, **1**, 966–67 (1962)

SOLUTIONS OF NONELECTROLYTES[1]

By Ian Brown

Division of Physical Chemistry, C.S.I.R.O., Melbourne, Australia

In the last few years, considerable attention has been devoted to studies of dilute solutions by gas chromatography, to the properties of mixtures of chain molecules, and to solutions containing associated components. In this review, some of the recent experimental work in these three fields is described and an attempt is made to assess their present status.

There have also been some symposia and publications of general interest. The plenary lectures and a list of original papers have now been published (1) from the symposium on Thermodynamics and Thermochemistry held at Lund, Sweden, last year. These include a summary by McGlashan (2) of the session on calorimetry of nonreacting systems. The papers and discussion from the symposium on Molecular Interactions in the Liquid Phase held in Paris last year have also been reported (3), and an interesting collection of papers has been presented (4) in honour of the 80th birthday of Peter Debye. Reviews have been written by McGlashan (5) on the equilibrium properties of liquid mixtures, by McLaughlin (6) on the thermal conductivity of liquids and condensed gases, by Smith (7) on the hydrogen bond, and by Himmelblau (8) on diffusion of dissolved gases in liquids. Books have been written by Rowlinson (9), by Van Ness (10), and by Swietoslawski (11).

Thermodynamic Properties of Solutions by Gas Chromatography

In 1955, Martin (12) suggested that gas chromatography provides perhaps the easiest of all means for studying the thermodynamics of the interaction of a volatile solute with a nonvolatile solvent, and he predicted that its potential value for providing this type of data should be very great. This prediction, with certain reservations, has been confirmed.

This technique is ideally suited for obtaining values of solute activity coefficients at infinite dilution, γ_1°, and it is quicker than conventional static methods. Values of γ_1° obtained by this method have been found to be independent of the apparatus and most of the operational factors used. These values have been compared with values obtained by extrapolation to infinite dilution of data obtained by static methods. Reasonable agreement was found (13–16) for hydrocarbon solutes in hydrocarbon or slightly polar stationary-phase solvents. Reasonable agreement was also found (13, 17) for polar solutes in hydrocarbon solvents provided that adsorption of the solute on the support used for the solvent was negligible.

The simple expression used to calculate γ_{p1}°, when deviations from ideality of the gas plus vapour phase can be ignored, is given by Equation 1.

[1] This review is based on material received at the Library of the Chemical Research Laboratories to the end of December 1964.

$$\gamma_{p1}^{\circ} = (N_2RT)/(p_1^{\circ}V_{N1}) \quad 1.$$

where N_2 is the number of moles of solvent on the column, p_1° is the vapour pressure of the pure solute at column temperature $T^{\circ}K$; V_{N1} is the net corrected retention volume of the solute given by Equation 2, and R is the gas constant.

$$V_{N1} = V_{R1}^{\circ} - V_M = KV_L \quad 2.$$

where V_{R1}° is the retention volume of the solute at column temperature corrected to mean volume pressure, V_M is the corrected dead volume of the column proper; K is the partition coefficient of the solute between the liquid and gas phases, and V_L is the volume of the solvent on the column at column temperature.

The assumptions underlying the derivation of Equation 1 have been discussed in detail by various authors (13, 16, 18); Anderson & Napier (15) have discussed the difficulties of measuring accurate values of V_M and Everett & Stoddart (19) have studied the effects of sample size, pressure drop, and flow rate. Allowance for the deviations of the solute vapour from ideal gas behaviour was made by Adlard, Kahn & Whitham (20). Everett & Stoddart (19) used Equation 3 for obtaining the corrected activity coefficient γ_{f1}° where allowance is made for the deviations from ideal behaviour of the solute in the gas plus vapour mixture.

$$\ln \gamma_{f1}^{\circ} = \ln \gamma_{p1}^{\circ} - p_1^{\circ}/RT(B_{11} - v_1^{\circ}) + \bar{p}/RT(2B_{13} - B_{33} - v_1) \quad 3.$$

where B_{11}, B_{33}, and B_{13} are, respectively, the second virial coefficients of the solute, gas, and cross term in the expression for the virial coefficient of the mixed vapour. The mean column pressure is $\bar{p}$, v_1° is the molar volume of the liquid solute, and v_1 is its partial molar volume at infinite dilution in the stationary phase. Where measured values of B_{13} are not available, these authors recommend using the principle of corresponding states (21) rather than the arithmetic mean approximation.

Values of γ_{f1}° employing these refinements were calculated by Everett & Stoddart for solutions of hydrocarbons in dinonylphthalate, and they were found to agree to better than ± 1 per cent with values obtained using static methods by Ashworth & Everett (22). Recent careful measurements by Cruickshank, Everett & Westaway (23), in a study of solutions of six hydrocarbons in squalene at 30°, show that γ_{f1}° values do not vary significantly with changes in particle size of liquid phase support or with the ratio of phase to support. They give an improved method for obtaining the retention volume corresponding to zero pressure gradient along the column and conclude that, for all systems studied, values of γ_{f1}° from initial retention volumes as defined by Keulemans (24) agree with the values extrapolated from static measurements, but those from peak retention volumes do not agree.

The importance of the factor B_{13} was further investigated by Desty et al. (25), who studied the variation of solute retention with the nature of the carrier gas and mean column pressure for a variety of hydrocarbon solutes in squalene. They showed the advantage of using helium as carrier gas rather than a gas which gives an appreciable value of B_{13} with the solute. They showed that extrapolation of retention volumes to zero $\bar{p}$ gave the same value for different gases. These authors confirm the statement of Everett & Stoddart (19) that gas chromatographic methods could be used for the determination of B_{13} for gas plus vapour mixtures if sufficiently precise retention volumes could be measured.

Measurements of $\gamma_{f_1}{}^\circ$ using this technique are not limited to nonvolatile stationary phases. By saturating the carrier gas with solvent, Kwantes et al. (13, 17) measured values of $\gamma_{p1}{}^\circ$ for methyl ethyl ketone and *n*-butanol in both *n*-heptane and water, and Burnett (26) measured values for alcohols and ketones in water. For accurate measurements of $\gamma_{f_1}{}^\circ$ for such systems, it will be necessary to employ more complicated corrections for vapour deviations which allow for the volatile solvent.

In summing up the reliability of $\gamma_{f_1}{}^\circ$ values determined by gas chromatography, three main types of solution have to be considered. For nonpolar or slightly polar solutes in nonpolar or slightly polar solvents where $\gamma_{f_1}{}^\circ$ is of the order of from 0.1 to 10, it can be concluded that values of $\gamma_{f_1}{}^\circ$, which agree accurately with carefully measured static values, can be obtained if all the above-mentioned precautions are taken. For polar solutes in nonpolar or slightly polar phases, accurate values of $\gamma_{f_1}{}^\circ$ can also be obtained if truly inert supports are used for the solvent in packed columns (17), or if well-coated, open tubular columns are used (27). With packed columns and active supports, peak distortion and changes in retention of polar solutes are found (28) on varying the phase to support ratio. These effects are largely because of adsorption of solutes on the support. The relative merits of various inert supports have been studied by Kirkland (29, 30).

The situation is not yet completely clear for measurements made on strongly polar stationary phases where $\gamma_{f_1}{}^\circ$ may be considerably above unity for nonpolar solutes or considerably below unity for polar solutes. Retention volumes and the order of elution of solutes have been found to vary with the ratio of polar phase to support. It has been suggested by Martin (31) that this effect is caused by appreciable contributions to the retention volume because of adsorption of the solute at the liquid surface, and he proposed Equation 4 to relate the contributions from solution and adsorption.

$$V_N = KV_L + k_a A_L \qquad 4.$$

where k_a is an adsorption coefficient and A_L is the area of liquid surface. He showed Equation 4 to hold for hydrocarbons on two liquid phases. He also studied (32) the excess surface concentration of solutes and found a linear relation between values found by gas chromatographic and surface tension measurements. Martin's explanation has not been completely accepted (33, 34).

Pecsok, de Yllana & Abdul-Karim (35) have recently extended the measurements of Martin and confirm his conclusions, but they find values of k_a

about 25 per cent lower. Matire, Pecsok & Purnell (36) have made careful measurements of activity coefficients and surface tensions as a function of concentration for benzene and cyclohexane in β,β'-thiodipropionitrile at 30°, and their work substantiates well the hypothesis that excess surface concentrations may play an important part in determining the value of $\gamma_{f_1}{}^\circ$ measured chromatographically. It would be interesting to investigate this problem for solutions which show strong negative deviations on electron-donor and electron-acceptor types of phases. If this phenomenon is of general significance, it complicates the study of polar solutions by this method, but it may provide a new tool for experimental study of the Gibbs adsorption equation.

A systematic study of the thermodynamic properties of dilute solutions by gas chromatography and by other means would serve a threefold purpose. It would provide data to test existing theories of dilute solutions and thus stimulate the development of new theories. Secondly, it would lead to the development of a scientific approach to the selection of suitable stationary phase liquids for difficult separations. Thirdly, it would be of considerable use in the selection of solvents for commercial separation processes, as is illustrated by the work of Deal & Derr (37). A number of such systematic studies has been carried out in which attempts have been made to relate the retention volumes of solutes to the properties of the solute and solvent.

Matire (38) used an extension of the Hildebrand-Scatchard treatment of regular solutions to predict γ° and, thus, retention volumes. He divided the solubility parameter into nonpolar and polar parts and added a factor to allow for negative deviations. Ashworth & Everett (22) showed that γ° for solutions of seven low molecular weight hydrocarbons in dinonylphthalate and in squalene could be interpreted by means of equations of the Miller-Guggenheim or Flory-Huggins type. They expressed the deviations in the form of Equation 5.

$$\ln \gamma = \ln \gamma^{\text{ath}} + \ln \gamma^{\text{th}} \qquad 5.$$

where γ^{ath} accounts for effects due to differences in the size of solvent and solute molecules and is independent of temperature, and γ^{th}, which allows for the balance of interaction energies between the molecules, is temperature dependent. They found that their experimental results were best represented by Equations 6 and 7.

$$\gamma_1{}^{\text{th}} = \chi\phi_2{}^2 \qquad 6.$$

$$\gamma_1{}^{\text{ath}} = \ln\left[(1 - \phi_2)/x_1\right] - z/2 \ln\left[1 - (1 - 1/r)2\phi_2/z\right] \qquad 7.$$

where ϕ_2 is the volume fraction of solvent, x_1 is the mole fraction of solute, z is the number of nearest neighbours (taken as 6), and r was made equal to the ratio of molar volumes of the two liquids. The factor χ is related to the so-called interchange energy.

Matire (18) also employed a relation equivalent to Equation 5. He first assumed the solution to be regular in order to evaluate the thermal contribution and then assumed it to be athermal in order to evaluate the other contribution. For the thermal contribution he used Equation 8.

$$\ln \gamma_1^{th} = v_1{}^*/RT(\delta_1 - \delta_2)^2 \qquad 8.$$

where $v_1{}^*$ is the "apparent" molar volume of the solute in the solution. The symbols δ_1 and δ_2 are the solubility parameters of the solute and solvent, respectively. For the athermal term he used the Miller-Guggenheim formulation in the form of Equation 9.

$$\ln \gamma_1^{ath} = \ln 1/r - z/2 \ln (1 - 2/z + 2/zr) \qquad 9.$$

He made z equal 12 and r equal $v_2{}^\circ/v_1{}^*$ where $v_2{}^\circ$ is the molar volume of the pure liquid solvent. Matire tested this approach by evaluating $v_1{}^*$ and δ_2, as adjustable parameters, by fitting Equation 8 to the activity coefficient data of Desty & Swanton (39) for 12 nonpolar solutes in five aromatic solvents, for which values of r were close to unity. Using the values found for $v_1{}^*$, and δ_2 for dinonylphthalate from other data (19), he compared his own measured $\gamma_1{}^\circ$ values for ten of the same solutes in this ester with the calculated values and found agreement to within 3 per cent. Further work on the prediction of γ° values is given by Matire (40).

In a study of stationary phase solvents and their influence on the retention volumes of solutes of various types, Brown (41) paid particular attention to solvents containing a high proportion of polar groups, and he pointed out that ΔG of Equation 10 is a measure of the interaction of the functional group X of the solute $R \cdot X$ with the solvent, if no specific orientational entropy effects occur.

$$\Delta G = -RT \ln (V_{NRX}/V_{NRH}) \qquad 10.$$

where V_{NRX} and V_{NRH} are the net retention volumes of the polar solute $R \cdot X$ and its homomorphic hydrocarbon $R \cdot H$ in the polar solvent.

A similar approach was used for solvents by Littlewood (42) in an interesting study of the factors determining the retention volumes of monofunctional solutes in monofunctional hexadecyl compounds. In these solvents the dipoles are relatively dilute in comparison with those containing a higher proportion of polar groups. This fact together with the use of hexadecane as the "homomorphic" solvent enabled a study to be made of the effect of dipole interaction between the functional group of the solute and that of the solvent on retention volumes. Littlewood showed that the retention volume of a polar solute in a polar hexadecyl derivative is governed by two factors a and b, and he defined two quantities r_a and r_b such that r_a is the ratio of the retention of an alkane on hexadecane to that on a polar hexadecyl derivative, and r_b/r_a is the ratio of the retention of a polar solute on the polar hexadecyl derivative to that on hexadecane. Using reasoning analogous to that of Brown (41), it can be shown, when the retentions are determined per mole of solvent, that log r_b is a measure of the free energy of interaction between the polar group on the solute and the polar group on the solvent. Littlewood has shown empirically that log r_b equals $0.04\, \mu_1\, \mu_2{}^{1.2}$ where μ_1 and μ_2 are the dipole moments of the solvent and solute, respectively (provided the polarizability and hydrogen bonding effects are small). He also discussed the effects of solute and solvent polarizability and hydrogen bonding on the retention volumes, and the factors governing the relation found between the free energy and entropy of solution.

A very thorough study has been made by Ducros (43) of the retention volumes and derived thermodynamic properties of six types of homologous series of solutes on the homologous series of solvents: di-*n*-butyl-, di-*n*-amyl-, di-*n*-hexyl-, and di-*n*-heptyl *d*-tartrates at eight temperatures from 70° to 140°. The thermodynamic properties of solutions of a number of nonpolar and polar solutes in a series of hydrocarbon solvents have been determined by Bergeron (44), who correlated values of $\gamma_1°$ with the molar volumes of the components.

A session of the symposium on gas chromatography held at Brighton, England, this year was devoted to the exploitation of specific molecular interactions in gas chromatography. However the papers and discussion on this subject have not yet appeared.

In addition to the above studies of the thermodynamic properties of solutions by this technique, gas chromatography has been used to a considerable extent in the determination of other physical and thermodynamic data, and many studies have been made of the adsorption of volatile compounds on the surface of a variety of solids. These investigations are, however, beyond the scope of this review.

Mixtures of Chain Molecules

A considerable amount of new experimental data has been published, particularly for mixtures of liquid normal alkanes. These data have been used to test the validity of the principle of congruence and the principle of corresponding states for mixtures of chain molecules.

The principle of congruence.—This principle, which was first formulated by Brønsted & Koefoed (45) from a study of the excess free energy G^E of mixtures of *n*-alkanes, has been shown to hold reasonably well for heats of mixing H^M (46), for volume changes on mixing V^M (47) of liquids, for the second virial coefficients of gas mixtures (48), and for liquid-vapour distribution ratios (49).

This principle requires that any thermodynamic property of a mixture of chain molecules from a given homologous series, at a given temperature and pressure, depends only on the average chain length $\bar{n}$. Mixtures with the same value of $\bar{n}$ are said to be congruent and, if $\bar{n}$ is integral, are also congruent with the appropriate pure homologue.

For the volume change on mixing of paraffins, the principle appears to be obeyed very well for mixtures containing different combinations of chain length at a given temperature, but a discrepancy has been found between measured values of V^M and those calculated from the properties of the pure components. Desmyter & van der Waals (50) attributed this to errors in the reported densities of the pure alkanes and to values of V^M being measured in the absence of air while the molar volumes of the pure components were measured with air-saturated materials. This discrepancy was also found by Gomez-Ibanez & Liu (51) who used the graphical method of Hijmans (52). They concluded that the explanation must lie elsewhere. It is not always possible to make this type of comparison. For example, Diaz Pena & Martin (53) have measured heats of mixing for methanol+n-alcohols up to decanol and have found the data for these mixtures to be congruent. The enthalpies of the pure liquid alcohols are not known with sufficient accuracy to enable the experimental values of H^M to be compared with those calculated from the enthalpies of the pure alcohols.

Values of H^M for methanol+n-dodecanol, which were presented at the Lund symposium, have now been published by Diaz Pena & Martin (54), who have used the results to compare the various graphical and analytical relations which have been proposed to represent data for congruent systems. Brown, Fock & Smith (55) have assumed congruence to be valid for the enthalpies of the lower alkanes and their mixtures to estimate H^M for ethane+hexane and propane+hexane at 25° and at a pressure of 41.7 atm where ethane is liquid.

Recent measurements by Holleman (56) of V^M for 13 mixtures of n-alkanes, covering values of carbon number n from 6 to 62 at temperatures from 51° to 126° and his measurements of H^M (57, 58) of eight mixtures with n from 6 to 36 and over a temperature range from 40° to 106°, have enabled a thorough test to be made of the principle of congruence. He concludes that the principle is to be regarded as an approximate rule, which, for V^M is satisfied almost quantitatively, with only minor deviations occurring between the experimental values of V^M and those calculated from the molar volumes of the pure substances.

Holleman stated in his paper at the Lund symposium that the principle of congruence did not hold for H^M of some alkane mixtures at 76°. From the results presented in his thesis, Holleman concludes that this principle does hold for heats of mixing. It is to be regarded as a first approximation, with reasonably good agreement for measured values between different mixtures at the same temperature, particularly at low temperatures and for mixtures having the same volatile component. Appreciable differences were estimated to occur between experimental values of H^M of alkane mixtures and those calculated from the enthalpies of the pure components. Holleman showed that an essential condition for the validity of the principle for excess properties of n-alkane mixtures is that the parameters q, r, and c (see below) are all linear functions of the chain length n, and he suggests that the differences mentioned above may be because of small deviations from linearity of one, or more, of these parameters.

The principle of corresponding states for chain molecules.—Prigogine and his co-workers (59) formulated a principle of corresponding states for liquid chain molecules which was reformulated in a phenomenological way by Hijmans (60). He employed three parameters q, r, and c to characterize the liquid alkanes. These represent the number of reference segments contributing to the energy, volume, and entropy, respectively, the same amount as one chain molecule in the corresponding state. He found that for n-alkanes, they were very close to a linear function of n, and he concluded that the principle of corresponding states, as formulated for chain molecules, applies to the pure n-alkanes. A similar treatment has been applied to oligomer and polymer liquids (61).

Holleman & Hijmans (62) combined the principle of congruence with the principle of corresponding states to give a principle of corresponding states for the excess properties of mixtures of chain molecules. This principle enables the calculation, from the excess properties of a given mixture of chain molecules at one temperature, of the excess properties of a corresponding mixture at a different temperature. The recent measurements of V^M for a large number of alkane mixtures by Holleman (56) (see above) have been used to show that the principle of corresponding states for the excess function V^M holds quantitatively for mixtures of n-alkanes.

The testing of this principle for heats of mixing of alkanes has been controversial. Two years ago, Holleman & Hijmans (62) tested this principle using the limited data then available (46, 63), and they concluded that, as a consequence of the principle, H^M must have the same sign throughout the liquid region. This was in contrast to the prediction of McGlashan et al. (64) that H^M for alkane mixtures will become negative at temperatures above 65°, and the finding of Davenport & Rowlinson (65) that such mixtures have lower consolute temperatures near the critical temperature of the more volatile component. Brown, Fock & Smith (55) considered that the treatment of Holleman & Hijmans only implies that the reduced H^M values for a system at one temperature will apply, at another temperature, to some other specific system, irrespective of the sign of H^M involved.

The measurements of McGlashan and co-workers (66) and the recent extensive measurements of Holleman (58) show that H^M for alkane mixtures do indeed change sign with an increase in temperature, and the minimum values shift to a higher mole fraction of volatile component as the temperature is raised. Holleman concluded (58) that the principle of corresponding states for excess functions, which depends on the validity of the principle of congruence, is at least approximately valid for heats of mixing. It has been pointed out (55) that the available data of H^M of n-alkane mixtures, covering a wide range of n and T, indicate a point of inflexion in the curve of liquid enthalpy versus n, which occurs at about n equals 6 at 25° and shifts to a higher value of n as the temperature is raised to about n equals 15 at 100°.

Bhattacharyya, Patterson & Somcynsky (67) have tested the principle of corresponding states as formulated by Hijmans (60) for pure alkanes and have used it to predict the temperature and composition dependence of V^M, H^M, and G^E for alkane mixtures. They showed that the curve of reduced configurational energy versus reduced temperature, although close to a straight line, is S-shaped, and they discuss this in relation to the corresponding curve for the reduced configurational heat capacity C_P*. They point out that, if $C_P{}^*$ were to follow the cell theory, H^M would be negative at all temperatures. They conclude that the principle gives qualitative agreement between experimental and predicted values of the above excess properties and that lack of exact agreement is attributable to errors in the reducing procedure or is a reflection of deviations from the principle of corresponding states.

A different, less empirical approach was taken by Flory, Orwoll & Vrij (68) in a study of chain molecule liquids. They formulated a partition function based on a model consisting of a linear sequence of segments and expressed the intermolecular energy in a form which avoids the restrictions of the cell model. The reduced equation of state which follows has reduced variables defined by three parameters related to the volume per segment, the interaction per segment, and the external degrees of freedom per segment. These are evaluated from experimental values of the volume, coefficient of thermal expansion, and compressibility of liquid *n*-alkanes. The reduced equation of state was successful in correlating p, V, T data for liquid alkanes with n from 6 to about 40.

In a second paper (69), Flory and his co-workers have extended this theory to mixtures of homologous chain molecules and have calculated, for the (6+16) mixture of alkanes, values of V^M which were in good agreement with measured values at 20° and in reasonable agreement with those at 50°. Calculated values of the excess chemical potentials were found to be in excellent agreement with experimental data. Their theory correctly predicts the decrease in H^M with temperature, leading to negative values at sufficiently high temperature, and accounts for the appearance of lower consolute temperatures for alkane mixtures with large differences in size. This study establishes the significance of characteristics of the liquid state in determining the thermodynamic properties of liquid mixtures. This treatment has been recast by Flory & Abe (70) in a form useful in determining the excess properties of mixtures of small, approximately spherical molecules.

These studies show the necessity for new, very accurate measurements of the molar volumes, enthalpies, and other properties of the highly purified members of various homologous series of liquids over a range of temperature and possibly over a range of pressure. This would enable a study to be made of the differences between measured values of the excess properties and those calculated from the pure components. These new data would facilitate the suggestion of Holleman (58) that the formulation of a new principle is required to replace the principle of congruence, and thus allow a reassessment to be made of the principle of corresponding states for the excess properties of mixtures of chain molecules.

The lattice model.—The lattice model has been used by Everett & Munn (71) in a recent study of the properties of alkane mixtures. They have compared the very precise values of the activity coefficients and heats of mixing of McGlashan, Williamson & Morcom (72, 73) for the system hexane+hexadecane, with values calculated using the lattice model with the parameters

z, r, and w independent of concentration, but allowed to vary with temperature. (z is number of nearest neighbours, r is ratio of sizes of component molecules, and w is related to the energies of interaction between like and unlike molecules.)

The activity coefficient γ was separated into two parts according to an equation similar to Equation 5; the configurational term was calculated from the Miller-Guggenheim equation and was subtracted from the experimental value of $\ln \gamma$ to obtain $\ln \gamma^{th}$. Values of χ were then calculated using Equation 11 for values of $z=9$, 12, and ∞ and for a series of values of r. The value of r was found which gave the best fit of the activity coefficient data at each temperature.

$$\ln \gamma_1^{th} = \chi(\phi_2')^2 \tag{11}$$

where $\phi_2' = qx_2/(x_1+qx_2)$, x is mole fraction and q is given by $qz=rz-2r+2$.

The values of the parameters found were then used to calculate values of H^M which were in excellent agreement with experimental values and had the correct dependence on concentration, in contrast to the results obtained for the first mixing process of McGlashan, Morcom & Williamson (64) who used temperature independent values of z and r. Values of H^M were also calculated for the system n-octane+n-tetracosane at temperatures from 51° to 106° and were found to be in excellent agreement with measured values of Holleman (57). It was found that the Flory-Huggins equation, with r and χ as adjustable parameters, represented the data with an accuracy at least as high as that found using more complicated theories.

Bhattacharyya et al. (67) consider that the mixing process, where components and the solution are at equal reduced p, V, and T, is the most appropriate process for comparison with the lattice model and that the interchange energy may be negligible. They suggest that if this is so, the process of mixing at "constant volume per segment" of McGlashan et al. (64) must introduce, as an artifact, a large positive interchange free energy in order to obtain the required value of G^E.

In an extension of earlier work (74), an interesting comparison has been made by Diaz Pena (75) of the thermodynamic properties of the system cyclohexane+carbon tetrachloride for four mixing processes with those implied by the lattice model. The mixing processes were similar to those considered by McGlashan et al. (64). He used data for the isothermal compressibility of the components measured by Diaz Pena & Cavero (76) and found excellent agreement with the lattice model for the mixing process "at constant volume per element," at four temperatures from 10° to 55°, for U^E, F^E, and TS^E. The values of the interchange energy u and the interchange free energy w gradually increased and decreased, respectively, as the temperature increased.

A useful, semiempirical approach has been proposed by Blanks & Prausnitz (77) for estimating the solubility of polymers in polar and nonpolar solvents. An important, systematic study of 27 mixtures of 12 hydrocarbon components was made by Lundberg (78) who measured values of H^M at 25° and at 50°.

Associated Solutions

In the last few years, considerable effort has been devoted to the study of solutions which contain associated components, and those in which association can occur between unlike components because of dipole interaction, hydrogen bonding, or electron donor-acceptor interactions.

The excess thermodynamic properties of these mixtures may lie between two extremes. On the one hand G^E, H^M, S^E, and V^M may all have large positive values because of the breakdown of one strongly associated component on dilution with an "inert" solvent. On the other hand, they may all have large negative values because of strong association between the two componets which are not (appreciably) associated themselves. These two extremes are probably never encountered alone in practice, experimental values being a resultant of both processes taking place simultaneously. Where these are approximately equal, we find S-shaped concentration dependence of some of the excess properties, and we may even have an approximately "ideal" solution of associated components. For example, Brown, Fock & Smith (79, 80, 81) found that the positive extreme in behaviour predominates in the system $CH_3NO_2+CCl_4$, the negative one in CH_3NO_2 CH_3COCH_3, and for the system $CH_3NO_2+CH_3CN$, although V^M is negative, G^E, H^M, and S^E are all substantially zero over the whole concentration range. McGlashan (2) has emphasized the value of H^M and its temperature dependence in studying complex formation in these solutions.

Ideal associated solutions.—In a study of the system dioxan (A)+chloroform (B), McGlashan & Rastogi (82) explained their results on the assumption that the four species A, AB, AB_2, and B form an ideal mixture. In an attempt to assess the magnitude of the nonspecific interactions in this system, McKinnon & Williamson (83) have studied the system dioxan$+CCl_4$ and found that H^M is also negative in agreement with the results of Ott, Goates & Mangelson (84). Thus, the nonspecific interactions in the former system are small and the assumption of McGlashan & Rastogi is reasonably valid. Scott et al. (85) separated the positive physical contribution to H^M from the negative, chemical contribution for the system $C_7F_{15}H+CH_3COCH_3$. This work prompted McGlashan (5) to point out the folly of attempting to treat solutions as "ideal associated" by ascribing the whole of the excess functions to complex formation.

The theoretical conditions for which binary mixtures of associated liquids form ideal or athermal mixtures has been studied by Gross & Wilson (86). Kehiaian with Sosnkowska-Kehiaian (87–91) discuss the behaviour of excess properties of ideal and regular associated mixtures of various species.

Polar+nonpolar mixtures.—There have been a number of studies of solutions of polar, associated compounds in hydrocarbon solvents (see below, alcohol solutions). At the Lund Symposium (2), Schäfer discussed heats of mixing and their variation with temperature for the systems of methyl chloride and bromide with propene and of acetone+hexane and+heptane in terms of dipole association of the polar component. Edwards & Schäfer (92) have determined G^E for acetone+isooctane at four temperatures from 30° to 50°. The liquid-vapour equilibria of acetone+n-hexane at 35° to 55° have been determined by Kudryavtseva & Susarev (93). Two studies, unfortunately at constant pressure, have been made of azeotropes of nitromethane plus a series of n-alkanes (C_7 to C_{12}) (94) and the liquid-vapour equilibria of pyridine+n-alkanes (C_7 to C_{10}) (95).

The association of methyl, ethyl, and n-propylamine in solution in n-paraffins has been thoroughly studied by Wolf, Höpfner & Höpfner (96), who determined activity coefficients at a number of temperatures from −55° to +20° and, from these, the relative concentrations of the associated species present and their heats of association. The solubility parameters and activity coefficients of long chain alkylamines in organic solvents have been studied by Kertes (97). A direct means of studying complexing forces of associated compounds with high dipole moments has been suggested by Jakobsen & Brasch (98), who found a number of such compounds to have broad absorption bands in the far infrared between 50 and 88 cm^{-1}.

Alcohol solutions.—The most outstanding experimental work on alcohol solutions is that presented by Savini & Van Ness at the Lund Symposium (1, see also 2). They describe a remarkable continuous dilution calorimeter for rapid measurement of H^M over a range of temperature and concentration down to a mole fraction of 0.001 with an accuracy of ±0.3 per cent. Both integral and partial molar heats of mixing may be measured directly in the same experiments. They gave a large number of values of H^M for ethanol +benzene at 45°, which show a reversal in curvature of the plot for H^M/x_1x_2 against alcohol mole fraction x_1 below 0.02. Extensive data were also given for ethanol+toluene and ethanol+n-heptane at 25°, 45°, and 60°. Their calorimeter has also been shown (99) to be suitable for measurements of partly miscible systems. We look forward to the early publication of further details of this work.

A comparison of H^M data for benzene+n-alcohol mixtures with those for benzene+n-alkanes shows that the alkyl group to solvent interactions make a major contribution to H^M values of the former mixtures. This led Brown, Fock & Smith (55) to determine H^M of a number of n-alcohols in n-hexane at 25, 35, and 45°. They found that the contribution made by the hydroxyl groups (per mole of alcohol) to the enthalpy of any liquid mixture of an n-alcohol with an n-alkane, at a fixed temperature, was given by a single function of the concentration of hydroxyl groups in the mixture. They also showed that the theory of Barker & Smith (100) was inadequate to explain quantitatively the excess properties of mixtures of alcohols with benzene or with n-hexane, even when a mixing process at constant volume was considered and the free energy of interchange was allowed to be temperature dependent. Scatchard & Satkiewicz (101) have measured total pressures of the system ethanol+cyclohexane at five temperatures from 5° to 65°, using the precision static apparatus of Scatchard, Wilson & Satkiewicz (102) and using a new expression given by Wilson (103) to represent G^E. They have calculated from this data, at four temperatures and at even mole fractions, values of G^E/RT, H^M/RT, and S^E/RT. The interpolated values of H^M/RT at 25° are from 5 to 15 per cent higher than the calorimetric values of Schnaible, Van Ness & Smith (104) and even higher than those of Goates, Snow & James (105).

The excess thermodynamic properties of *n*-propanol+cyclohexane have been determined by Brzostowski, Hardman & Warycha (106–108) who measured liquid-vapour equilibria, V^M, and isothermal compressibilities at a number of temperatures. They compared values of the excess Planck function $Y^E/R = -G^E/RT$ with those calculated from a simple form of the lattice theory. Liquid-vapour equilibria of ethanol+*n*-hexane at 35, 45, and 55° have been measured by Kudryavtseva & Susarev (93).

Aqueous solutions of alcohols have also received attention. The system butan-2-ol+water has been studied by Altsybeeva et al. (109) who determined values of G^E, H^M, and V^M at a number of temperatures, and the system propan-1-ol+water by Belousov & Morachevskii (110), who determined G^E, H^M, and S^E at four temperatures and V^M at three temperatures.

The association of alcohols in various solvents has also been investigated by other methods. For example: infrared methods have been used by Fruwert, Petzold & Geiseler (111) to determine the effect of the solvents CCl_4, $CHCl_3$, C_6H_6, and C_6H_5Cl on the distribution of *r*-mers in *n*-propanol and by Walkup et al. (112) who studied I.R. shifts of a number of alcohols (C_1 to C_5) in *p*-dioxan. Magnetic resonance shifts of the OH proton of 3-ethylpentan-3-ol in CCl_4 was studied by Martin (113). Dipole moment measurements at four temperatures of a number of alcohols in CCl_4 have been made by Huyskens, Gillerot & Zeegers-Huyskens (114) in an attempt to estimate the polymers of alcohols present in solution. Dielectric constants of *n*-decyl alcohol in six nonpolar liquids were measured by Klages & Obermayer (115), who found the dipole moments followed a curve which was determined by the concentration of OH groups. The dielectric properties of pure alcohols have also been studied by Dannhauser & Bake (116) for eight lower alcohols from 25° to their critical temperatures. They determined $\Delta G, \Delta H$, and ΔS for hydrogen bond formation at 25°.

Polar+polar mixtures.—A new and unusual kind of specific intermolecular interaction has been reported by Kreglewski et al. (117–121) which he has named the "oxygen bridge." He found that trifluoroacetic anhydride was, contrary to expectation, miscible with acetone and showed that this system has large negative values of G^E, H^M, S^E, and V^M. This behaviour is apparently due to a specific interaction between the acceptor keto oxygens of the fluoro compound and the donor oxygen of acetone. He also found that systems trifluoroacetic anhydride with acetic anhydride and with acetonitrile have a negative value of H^M.

The excess thermodynamic properties of acetone+methyl iodide have been measured by Low & Moelwyn-Hughes (122) to supplement earlier measurements (123) of the heat capacity and compressibilities of this system. They calculated excess functions for mixing at both constant p and constant V and found that TS^E varied in an unusual way with composition. This may be due, in part, to errors in equilibrium data at low mole fractions of methyl iodide, because liquid samples from the still were taken after the liquid stream had mixed with the condensed vapour stream [compare Brown & Ewald (124)]. Moelwyn-Hughes & Thorpe (125, 126) have made a systematic study of intermolecular interactions between piperidine, tetrahydropyran, and cyclohexane, whose molecules are similar in size and shape, but only differ in their functional groups. They determined H^M, V^M, heat capacities, dielectric constants, and adiabatic and isothermal compressibilities of the components and their mixtures at temperatures from 20° to 60°. For the two mixtures of polar compounds with cyclohexane, they have calculated Kirkwood's g factor, and for tetrahydropyran+cyclohexane and chloroform+CCl_4, they have determined the dipolar contributions to the excess properties. For mixtures containing piperidine, the presence of hydrogen bonds was shown by NMR, and they suggest that the N–H – – – O bonds are weaker than the N–H – – – N bonds.

Diaz Pena & Haya (127) have measured V^M at 30° for a similar trio; cyclohexanol, cyclohexanone, cyclohexane. They found positive values of V^M for the two systems with cyclohexanone and a positive value for cyclohexanol+cyclohexane at low and negative values at high alcohol concentration.

Solutions of p-dioxan in polar liquids have also received attention. Values of G^E, H^M, and S^E for dioxan+nitromethane have been estimated by Stadnicki (128) from liquid-vapour equilibria at four temperatures from 30° to 98°. Heats of formation of complexes of dioxan with halogens were measured by Guy & Pritchard (129). Goates, Ott & Mangelson (130–132) have studied solid phase equilibria of dioxan+anisole and various tetrahalides with aromatic hydrocarbons and ethers. The technique of polarography has been used by Peover & Davies (133, 134) in the study of aromatic complexes.

A series of interesting experimental studies has been made by Kay and co-workers (135–137) of liquid-vapour phase relations in the critical region of binary systems that form azeotropes. Three groups of systems, each composed of members of an homologous series with a common azeotrope-forming compound, were found to be characterised by a pattern of phase behaviour distinctly different from that for nonazeotropic systems. The groups studied were: perfluoroheptane+n-alkanes, acetone+n-alkanes, and benzene+n-alcohols.

Liquid-liquid equilibria.—Interesting investigations have been made on the critical solution behaviour of systems with large deviations from ideality. The conditions governing the upper and lower critical solution temperatures

(UCST and LCST), previously discussed by Kehiaian (138), have been extended (139) to ideal associated mixtures of the type A+B+AB. A theoretical treatment of binary solutions of sulphur in nonelectrolyte solvents has been given by Scott (140), based on the Flory-Huggins equation and a consideration of two species of liquid sulphur S_λ and S_μ which are the "monomer" and "polymer," respectively, of S_8 ring units. Various values of an interchange energy parameter lead to calculated UCST and LCST in good qualitative agreement with experimental measurements.

The influence of pressure on the miscibility of binary systems has been studied by Schneider (141, 142), who found an interesting pattern of behaviour for the unmixing temperature of mixtures of methyl-pyridines and methyl-piperidines with water and D_2O at pressures up to 5000 bars. He found butyl glycol+water has a closed miscibility gap which diminishes and eventually disappears with an increase in p. In three systems which show a closed miscibility gap, a new phenomenon was found in which the LCST increases to a maximum and then decreases as the pressure is raised. He also studied (143) the system acetonitrile+water.

Mixtures containing perfluorinated compounds have continued to provide new and useful information. The CST of C_5, C_6, and C_7 paraffins in perfluorotributylamine have been determined by Munson (144) and have been compared with the corresponding data for other perfluorinated solvents. The solubilities and entropy of solution of a number of gases in this amine, in a trichloro-trifluoroethane, and in isooctane have been studied by Hiraoka & Hildebrand (145). Croll & Scott (146) have continued earlier work on solutions of small fluorocarbon molecules. They have determined G^E close to the CST and have evaluated the constants in Equation 12 for these and earlier measurements.

$$G^E/RT = x_1x_2[a_0 + a_1(x_1 - x_2) + a_2(x_1 - x_2)^2 \cdots] \qquad 12.$$

The miscibility diagram was determined for $C_2H_6+CF_4$, and this system, like many other fluorocarbon+alkane mixtures, showed values of G^E much greater than those calculated from the regular solution equation. They conclude that this anomaly arises from the poor approximation of a geometric mean interchange energy. The system $C_2H_6+CClF_3$, however, does not show this anomaly. The unmixing temperatures were measured for $CH_2F_2+CClF_3$, CHF_3+CClF_3, and CHF_3+CF_4, which all show skewing of the G^E versus x curve in a direction contrary to prediction, and it is suggested that this may be due to hydrogen bonding. They discuss the flat shape of the G^E curves and show that this is not consistent with proximity to the CST. This casts doubt on the validity of the hypothesis that this flattening is due to clustering. Shinoda & Hildebrand (147) determined the mutual solubility at the standard temperature of 25° for mixtures of perfluoroheptane with CCl_4 and CS_2.

A large amount of experimental evidence has shown that, for binary liquid mixtures close to the critical solution region, the coexistence curves have the form of Equation 13

$$X_1 - X_2 = K(T_c - T)^{1/n} \qquad 13.$$

where $(X_1 - X_2)$ is a measure of the difference between coexisting phases and the value of n is close to 3. Work on some systems (148), for example, suggests that this relation does not hold at temperatures very close to the CST, and in this region, the coexistence curve has a flat top. For the systems of n-octane with perfluoro-n-octane, trifluoroacetic acid and its anhydride, Kreglewski (149) found n to vary appreciably from the value 3 as the temperature approached the CST. He suggests that the coexistence curve will always appear to have a flat top, although its real existence is doubtful.

An outstanding experimental study has been made by Thompson & Rice (150) of the system perfluoromethylcyclohexane+CCl_4, using very carefully purified components and with temperature measurements to $\pm 0.00005°$. Measurements of the density of the upper phase and of the position of the meniscus showed, when allowance was made for density gradients due to gravity effects, that the coexistence curve follows a cubic relation to within 0.0001° of the CST with no evidence of a flat region. Ultrasonic absorption measurements on this and other systems have been made by Kruss (151), who found that intermolecular correlation lengths agreed with those determined from light scattering measurements. His results indicate that, near the CST, the system may already be in two phases, not distinctly separated, but regions in a type of emulsion.

Critical opalescence of n-C_7F_{16}+n-C_7H_{16} has been studied by Chu (152) by means of visible light scattering. He has summarized recent results for other systems, and these are discussed in terms of the Debye theory of critical opalescence (153) and the solubility parameter theory of Hildebrand & Scatchard. The findings substantiate the proposal of Scott (154) on the breakdown of the geometric mean law for this type of system. The results of X-ray scattering measurements at temperatures near the CST by Brady (155) indicate long-range inhomogeneities for the system C_7F_{16}+isooctane. Light-scattering measurements on 2,6-dimethyl pyridine+water, a typical system with a LCST, have been made by Pancirov & Brumberger (156), whose findings are not in agreement with the suggestion of Münster & Schneeweiss (157), that deviations occur from the Ornstein-Zernike-Debye theory.

An interesting solubility phenomenon was discovered by Cuseva & Parnov (158). They found that the solubility of certain hydrocarbons in D_2O increases with rise in temperature, while their corresponding solubilities in water, which are about 10 per cent higher than those in D_2O, decrease with rise in temperature.

The work reviewed shows the value of making accurate measurements of a number of different quantities, over wide ranges of temperature, pressure, and composition, on a limited number of carefully chosen systems. It also shows the advantages of studying mixtures of a given component with a number of members of an homologous series. Three factors have helped in the evaluation of the individual contributions to nonideal behaviour of a given liquid mixture due to various causes. These are: the use of thermodynamic data together with spectroscopic and other data; the use of measurements on homomorphic systems; and the realization that many simple molecules, previously thought to be inert, can take part in specific intermolecular interactions. The measurement of the excess properties of vapour mixtures would be a useful complement to the study of mixtures of the liquids. New experimental data have enabled an assessment to be made of the limitations of some hypotheses and theories of solution; these results should be useful in the formulation of better theories of solution.

LITERATURE CITED

1. Symposium on Thermodynamics & Thermochemistry, Lund, Sweden, July 1963, *Pure Appl. Chem.*, **8**, 89–214 (1964)
2. McGlashan, M. L., *Pure Appl. Chem.*, **8**, 157 (1964)
3. Interactions Moléculaires en Phase Liquide, 13e Réunion de la Societe de Chimie Physique, Paris, June 1963, *J. Chim. Phys.*, **61**, 15–269 (1964)
4. Debye, Peter J. W., *J. Am. Chem. Soc.*, **86**, 3405–562 (1964)
5. McGlashan, M. L., *Ann. Rept. Chem. Soc. (London)*, **59**, 73 (1963)
6. McLaughlin, E., *Chem. Rev.*, **64**, 389 (1964)
7. Smith, J. W., *Sci. Progr. (London)*, **52**, 97 (1964)
8. Himmelblau, D. M., *Chem. Rev.*, **64**, 527 (1964)
9. Rowlinson, J. S., *The Perfect Gas*, **5**, Topic 10 (Intern. Encyclopaedia Phys. Chem., Chem. Phys., Pergamon, Oxford, 1963)
10. Van Ness, H. C., *Classical Thermodynamics of Nonelectrolyte Solutions* (Pergamon, Oxford, 1964)
11. Swietoslawski, W., *Azeotropy & Polyazeotropy* (Macmillan, New York, 1964)
12. Martin, A. J. P., *Analyst*, **81**, 52 (1956)
13. Kwantes, A., and Rijnders, C. W. A., in *Gas Chromatography 1958*, 125 (Desty, D. H., Ed., Butterworths, London, 1958)
14. Adlard, E. R., and Khan, M. A., in *Gas Chromatography 1958*, 135 (See ref. 13)
15. Anderson, J. R., and Napier, K. H., *Australian J. Chem.*, **10**, 250 (1957)
16. Porter, P. E., Deal, C. H., and Stross, F. H., *J. Am. Chem. Soc.*, **78**, 2999 (1956)
17. Hofstee, T., Kwantes, A., and Rijnders, C. W. A., *Intern. Symp. Distillation*, 105 (Inst. Chem. Engrs., London, 1960)
18. Matire, D. E., *Gas Chromatography, Proc. Symp., 4th*, **33** (Fowler, L., Ed., Academic Press, London, 1963)
19. Everett, D. H., and Stoddart, C. T. H., *Trans. Faraday Soc.*, **57**, 746 (1961)
20. Adlard, E. R., Kahn, M. A., and Whitham, B. T., *Gas Chromatography 1960*, 251 (Scott, R. P. W., Ed., Butterworths, London, 1960)
21. Guggenheim, E. A., and McGlashan, M. L., *Proc. Roy. Soc. (London), Ser. A*, **206**, 448 (1951)
22. Ashworth, A. J., and Everett, D. H., *Trans. Faraday Soc.*, **56**, 1609 (1960)
23. Cruickshank, A. J. B., Everett, D. H., and Westaway, M. T., *Trans. Faraday Soc.*, **61**, 235 (1965)
24. Keulemans, A. I. M., *Gas Chromatography*, 16 (Verver, C. G., Ed., Reinhold, New York, 1959)
25. Desty, D. H., Goldup, A., Luckhurst, G. R., and Swanton, W. T., *Gas Chromatography 1962*, 67 (van Swaay, M., Ed., Butterworths, London, 1962)
26. Burnett, M. R., *Analyt. Chem.*, **35**, 1567 (1963)
27. Adlard, E. R., Kahn, M. A., and Whitham, B. T., *Gas Chromatography 1962*, 84 (van Swaay, M., Ed., Butterworths, London, 1962)
28. Eggersten, F. T., and Knight, H. S., *Analyt. Chem.*, **30**, 15 (1958)
29. Kirkland, J. J., *Gas Chromatography, Proc. Symp., 4th*, 77 (Fowler, L., Ed., Academic Press, London, 1963)

30. Kirkland, J. J., *Gas Chromatography 1964*, 146 (In preprints) (Goldup, A., Ed., Butterworths, London, 1964)
31. Martin, R. L., *Analyt. Chem.*, **33**, 347 (1961)
32. Martin, R. L., *ibid.*, **35**, 116 (1963)
33. Ottenstein, D. M., *J. Gas Chromatog.*, **1**(4), 11 (1963)
34. Purnell, J. H., *Gas Chromatography*, 419 (Wiley, New York, 1962)
35. Pecsok, R. L., de Yllana, A., and Abdul-Karim, Aziz, *Analyt. Chem.*, **36**, 452 (1964)
36. Matire, D. E., Pecsok, R. L., and Purnell, J. H., *Nature*, **203**, 1279 (1964)
37. Deal, C. H., and Derr, E. L., *Ind. Eng. Chem., Process Design Develop.*, **3**, 394 (1964)
38. Matire, D. E., *Analyt. Chem.*, **33**, 1143 (1961)
39. Desty, D. H., and Swanton, W. T., *J. Phys. Chem.*, **65**, 766 (1961)
40. Matire, D. E., *An Investigation of the Thermodynamics of Dilute Non-electrolyte Solutions by Gas Chromatography* (Doctoral thesis, Stevens Inst. Tech., Hoboken, N. J., 1963)
41. Brown, I., *J. Chromatog.*, **10**, 284 (1963)
42. Littlewood, A. B., *Analyt. Chem.*, **36**, 1441 (1964)
43. Ducros, M., *Publ. Sci. Tech. Min. Air. (France), Notes Tech.*, **126** (1963)
44. Bergeron, E. P., Jr., *Thermodynamic Properties of Solutions by Gas Chromatographic Methods* (Doctoral thesis, Louisiana State Univ., New Orleans, La., 1963)
45. Brønsted, J. N., and Koefoed, J., *Kgl. Danske Videnskab. Selskab, Mat.-Phys. Medd.*, **22**(17), 1 (1946)
46. McGlashan, M. L., and Morcom, K. W., *Trans. Faraday Soc.*, **57**, 907 (1961)
47. McGlashan, M. L., *Mol. Phys.*, **4**, 87 (1961)
48. Barker, J. A., and Linton, M., *J. Chem. Phys.*, **38**, 1853 (1963)
49. Nederbragt, G. W., and de Jong, J. J., *Appl. Sci. Res., Sect. A*, **3**, 125 (1953)
50. Desmyter, A., and van der Waals, J. H., *Rec. Trav. Chim.*, **77**, 53 (1958)
51. Gomez-Ibanez, J. D., and Liu, Chia-Tsun, *J. Phys. Chem.*, **67**, 1388 (1963)
52. Hijmans, J., *Mol. Phys.*, **1**, 307 (1958)
53. Diaz Pena, M., and Martin, F. F., *Anales Real Soc. Espan. Fis. Quim. (Madrid)*, **59**, 323, 331 (1963)
54. Diaz Pena, M., and Martin, F. F., *ibid.*, **60**, 9 (1964)
55. Brown, I., Fock, W., and Smith, F., *Australian J. Chem.*, **17**, 1106 (1964)
56. Holleman, Th., *Physica*, **29**, 585 (1963)
57. Holleman, Th., *ibid.*, 274
58. Holleman, Th., *A Principle of Corresponding States for the Excess Functions of Mixtures of Chain Molecules* (Doctoral thesis, Univ. Amsterdam, 1964)
59. Prigogine, I., Bellemans, A., and Naar-Colin, C. J., *J. Chem. Phys.*, **26**, 751 (1957)
60. Hijmans, J., *Physica*, **27**, 433 (1961)
61. Simha, R., and Havlik, A. J., *J. Am. Chem. Soc.*, **86**, 197 (1964)
62. Holleman, Th., and Hijmans, J., *Physica*, **28**, 604 (1962)
63. van der Waals, J. H., *Rec. Trav. Chim.*, **69**, 949 (1950)
64. McGlashan, M. L., Morcom, K. W., and Williamson, A. G., *Trans. Faraday Soc.*, **57**, 601 (1961)
65. Davenport, A. T., and Rowlinson, J. S., *Trans. Faraday Soc.*, **59**, 78 (1963)
66. Friend, J. A., Larkin, J. A., Maroundas, A., and McGlashan, M. L., *Nature*, **198**, 683 (1963)
67. Bhattacharyya, S. N., Patterson, D., and Somcynsky, T., *Physica*, **30**, 1276 (1964)
68. Flory, P. J., Orwoll, R. A., and Vrij, A., *J. Am. Chem. Soc.*, **86**, 3507 (1964)
69. Flory, P. J., Orwoll, R. A., and Vrij, A., *ibid.*, 3515
70. Flory, P. J., and Abe, A., *J. Am. Chem. Soc.*, **86**, 3563 (1964)
71. Everett, D. H., and Munn, R. J., *Trans. Faraday Soc.*, **60**, 1951 (1964)
72. McGlashan, M. L., and Williamson, A. G., *Trans. Faraday Soc.*, **57**, 588 (1961)
73. McGlashan, M. L., and Morcom, K. W., *Trans. Faraday Soc.*, **57**, 581 (1961)
74. Diaz Pena, M., and McGlashan, M. L., *Trans. Faraday Soc.*, **57**, 1511 (1961)
75. Diaz Pena, M., *Anales Real Soc. Espan. Fis. Quim. (Madrid)*, **59**, 273 (1963)
76. Diaz Pena, M., and Cavero, B., *Anales Real Soc. Espan. Fis. Quim. (Madrid)*, **60**, 357 (1964)
77. Blanks, R. F., and Prausnitz, J. M., *Ind. Engr. Chem., Fundamentals*, **3**, 1 (1964)

78. Lundberg, G. W., *J. Chem. Eng. Data*, **9,** 193 (1964)
79. Brown, I., and Fock, W., *Australian J. Chem.*, **9,** 180 (1956)
80. Brown, I., and Smith, F., *Australian J. Chem.*, **13,** 30 (1960)
81. Brown, I., and Smith, F., *ibid.*, **15,** 9 (1962)
82. McGlashan, M. L., and Rastogi, R. P., *Trans. Faraday Soc.*, **54,** 496 (1958)
83. McKinnon, I. R., and Williamson, A. G., *Australian J. Chem.*, **17,** 1374 (1964)
84. Ott, J. B., Goates, J. R., and Mangelson, N. F., *J. Chem. Eng. Data*, **9,** 203 (1964)
85. Anderson, D. L., Smith, R. A., Myers, D. B., Alley, S. A., Williamson, A. G., and Scott, R. L., *J. Phys. Chem.*, **66,** 621 (1962)
86. Gross, P., and Wilson, G. L., *Z. Physik. Chem. (Frankfurt)*, **41,** 304 (1964)
87. Kehiaian, H., *Bull. Acad. Polon. Sci., Ser. Sci. Chim.*, **11,** 479, 487, 493 (1963)
88. Kehiaian, H., and Sosnkowska-Kehiaian, K., *Bull. Acad. Polon. Sci., Ser. Sci. Chim.*, **11,** 549, 583, 591 (1963)
89. Kehiaian, H., *Bull. Acad. Polon. Sci., Ser. Sci. Chim.*, **12,** 77, 255, 323 (1964)
90. Kehiaian, H., and Sosnkowska-Kehiaian, K., *Bull. Acad. Polon. Sci., Ser. Sci. Chim.*, **12,** 425, 433 (1964)
91. Kehiaian, H., *Bull. Acad. Polon, Sci., Ser. Sci. Chim.*, **12,** 497 (1964)
92. Edwards, J. L., and Schäfer, K., *Z. Naturforsch.*, **19,** 136 (1964)
93. Kudryavtseva, L. S., and Susarev, M. P., *J. Appl. Chem. USSR* (English Transl.), **36,** 1419 (1963)
94. Malesinska, B., and Malesinski, W., *Bull. Acad. Polon. Sci., Ser. Sci. Chim.*, **11,** 469 (1963)
95. Maczynska, Z., and Zieborak, K., *Bull. Acad. Polon. Sci., Ser. Sci. Chim.*, **11,** 219 (1963)
96. Wolf, von H., Höpfner, A., and Höpfner, H. M., *Ber. Bunsenges. Physik. Chem.*, **68,** 410 (1964)
97. Kertes, A. S., *J. Inorg. Nucl. Chem.*, **26,** 1764 (1964)
98. Jakobsen, R. T., and Brasch, J. W., *J. Am. Chem. Soc.*, **86,** 3571 (1964)
99. Savini, Ch. G., and Van Ness, H. C. (Personal communication from Prof. Van Ness)
100. Barker, J. A., and Smith, F., *J. Chem. Phys.*, **22,** 375 (1954)
101. Scatchard, G., and Satkiewicz, F. G., *J. Am. Chem. Soc.*, **86,** 130 (1964)
102. Scatchard, G., Wilson, G. M., and Satkiewicz, F. G., *J. Am. Chem. Soc.*, **86,** 125 (1964)
103. Wilson, G. M., *J. Am. Chem. Soc.*, **86,** 127 (1964)
104. Schnaible, H. W., Van Ness, H. C., and Smith, J. M., *Am. Inst. Chem. Engrs. J.*, **3,** 147 (1957)
105. Goates, J. R., Snow, R. L., and James, M. R., *J. Phys. Chem.*, **65,** 335 (1961)
106. Brzostowski, W., and Hardman, T. M., *Bull. Acad. Polon. Sci., Ser. Sci. Chim.*, **11,** 447 (1963)
107. Brzostowski, W., and Warycha, S., *Bull. Acad. Polon. Sci., Ser. Sci. Chim.*, **11,** 539 (1963)
108. Brzostowski, W., *Bull. Acad. Polon. Sci., Ser. Sci. Chim.*, **11,** 545 (1963)
109. Altsybeeva, A. I., Belousov, V. P., Ovtrakht, N. V., and Morachevskii, A. G., *Zh. Fiz. Khim.*, **38,** 1242 (1964)
110. Belousov, V. P., and Morachevskii, A. G., *Zh. Fiz. Khim.*, **38,** 184 (1964)
111. Fruwert, J., Petzold, D., and Geiseler, G., *Z. Phyik. Chem. (Frankfurt)*, **38,** 184 (1964)
112. Walkup, J. H., Lyford, J., Margardt, G., and Robinson, G. W., *Trans. Kentucky Acad. Sci.*, **24,** 101 (1964)
113. Martin, J., *Compt. Rend.*, **256,** 3651 (1964)
114. Huyskens, P., Gillerot, G., and Zeegers-Huyskens, Th., *Bull. Soc. Chim. Belges*, **72,** 666 (1963)
115. Klages, G., and Obermayer, K., *Z. Naturforsch.*, **18**(a), 1125 (1963)
116. Dannhauser, W., and Bake, L. W., *J. Phys. Chem.*, **40,** 3058 (1964)
117. Kreglewski, A., *Bull. Acad. Polon. Sci., Ser. Sci. Chim.*, **11,** 301 (1963)
118. Wyrzykowska-Stankiewicz, D., and Kreglewski, A., *Bull. Acad. Polon. Sci., Ser. Sci. Chim.*, **11,** 465 (1963)
119. Kreglewski, A., and Woycicki, W., *Bull. Acad. Polon. Sci., Ser. Sci. Chim.*, **11,** 645 (1963)
120. Woycicki, W., and Kreglewski, A., *ibid.*, **12,** 263 (1964)
121. Woycicki, W., and Trebicka, B., *Bull. Acad. Polon. Sci., Ser. Sci. Chim.*, **12,** 319 (1964)
122. Low, D. I. R., and Moelwyn-Hughes, E. A., *Proc. Roy. Soc. (London), Ser. A*, **281,** 366 (1964)
123. Low, D. I. R., and Moelwyn-Hughes, E. A., *Proc. Roy. Soc. (London), Ser. A*, **267,** 384 (1962)

124. Brown, I., and Ewald, A. H., *Australian J. Sci. Res., Ser. A*, **3**, 306 (1950)
125. Moelwyn-Hughes, E. A., and Thorpe, P. L., *Proc. Roy. Soc. (London), Ser. A*, **277**, 423 (1964)
126. Moelwyn-Hughes, E. A., and Thorpe, P. L., *ibid.*, **278**, 574 (1964)
127. Diaz Pena, M., and Haya, M., *Anales Real Soc. Espan. Fis. Quim. (Madrid)*, **60**(B), 423 (1964)
128. Stadnicki, J. S., *Roczniki Chem.*, **38**, 827 (1964)
129. Guy, C. A., and Prichard, H. O., *J. Phys. Chem.*, **68**, 1250 (1964)
130. Goates, J. R., Ott, J. B., and Mangelson, N. F., *J. Chem. Eng. Data*, **9**, 330 (1964)
131. Mangelson, N. F., Goates, J. R., and Ott, J. B., *J. Phys. Chem.*, **68**, 1260 (1964)
132. Goates, J. R., Ott, J. B., and Mangelson, N. F., *ibid.*, 2617
133. Peover, M. E., *Trans. Faraday Soc.*, **60**, 417, 479 (1964)
134. Peover, M. E., and Davies, J. D., *Trans. Faraday Soc.*, **60**, 476 (1964)
135. Jordan, L. W., Jr., and Kay, W. B., *Chem. Eng. Progr., Symp. Ser.*, **59**, 44, 46 (1963)
136. Kay, W. B., *J. Phys. Chem.*, **68**, 827 (1964)
137. Skaates, J. M., and Kay, W. B., *Chem. Eng. Sci.*, **19**, 431 (1964)
138. Kehiaian, H., *Bull. Acad. Polon. Sci., Ser. Sci. Chim.*, **10**, 569, 579, 585 (1962)
139. Kehiaian, H., *ibid.*, **11**, 153, 279, 285 (1963)
140. Scott, R. L., *J. Phys. Chem.*, **69**, 261 (1965)
141. Schneider, G., *Z. Physik. Chem. (Frankfurt)*, **37**, 333 (1963)
142. Schneider, G., *ibid.*, **39**, 187 (1963)
143. Schneider, G., *ibid.*, **41**, 327 (1964)
144. Munson, M. S. B., *J. Phys. Chem.*, **68**, 796 (1964)
145. Hikaoka, H., and Hildebrand, J. H., *J. Phys. Chem.*, **68**, 213 (1964)
146. Croll, I. M., and Scott, R. L., *J. Phys. Chem.*, **68**, 3853 (1964)
147. Shinoda, K., and Hildebrand, J. H., *J. Phys. Chem.*, **68**, 3904 (1964)
148. Atack, D., and Rice, O. K., *J. Chem. Phys.*, **22**, 382 (1954)
149. Kreglewski, A., *Bull. Acad. Polon. Sci., Ser. Sci. Chim.*, **11**, 91 (1963)
150. Thompson, D. R., and Rice, O. K., *J. Am. Chem. Soc.*, **86**, 3547 (1964)
151. Kruss, P., *Can. J. Chem.*, **42**, 1712 (1964)
152. Chu, B., *J. Am. Chem. Soc.*, **86**, 3557 (1964)
153. Debye, P. J. W., *J. Chem. Phys.*, **31**, 690 (1959)
154. Scott, R. L., *J. Phys. Chem.*, **62**, 136 (1958)
155. Brady, G. W., *J. Chem. Phys.*, **40**, 2474 (1964)
156. Pancirov, R., and Brumberger, H., *J. Am. Chem. Soc.*, **86**, 3562 (1964)
157. Münster, A., and Schneeweiss, Ch., *Z. Physik. Chem.*, **37**, 369 (1963)
158. Cuseva, A. N., and Parnov, E. I., *Radiokhimiya*, **5**, 507 (1963)

FAST REACTIONS IN SOLUTION[1,2]

By Hans Strehlow

Max-Planck-Institut für physikalische Chemie, Göttingen, Germany

During the past year, numerous publications again have appeared dealing with rapid reactions in solution. In contrast with Hammes' recent review (1), this chapter is concerned with the development and improvement of experimental techniques as well as with their application in different types of chemical reactions. However, because of lack of space, enzymatic reactions will not be considered. Radiolysis induced reactions are discussed elsewhere in this volume. Understandably, the borderline between "fast" and "slow" reactions is somewhat diffuse.

The whole field has been surveyed in an excellent monograph by Caldin (2), which contains, in very readable form, an introduction to the experimental techniques and results obtained up to 1963. Reviews on more restricted topics have been given by Noyes (3) on the effect of diffusion rates in chemical kinetics, by Weller (4) on fast reactions of excited molecules, and by Eigen et al. (5) on rate constants of protolytic reactions. In addition, Eigen has contributed surveys on fast elementary steps in chemical reaction mechanisms (6), on the kinetic behavior of hydrogen-bonded systems (7), and on ionic- and charge-transfer reactions in solutions (8). A short review on relaxation spectrometry has been published by Eigen & Kustin (9).

Experimental techniques.—In this first section, experimental techniques and theoretical considerations for the evaluation of rate constants from observed physical quantities are reviewed.

In the field of flow technique, Thompson & Gordon (10) have proposed the use of a simple, inexpensive stopped-flow apparatus with a time constant of about 30 msec. The concentration change is followed spectrophotometrically with an adjusted Beckmann DU spectrophotometer. The extinction is recorded by an oscilloscope. Kresze, Winkler & Meiners (11) describe a flow apparatus designed with spectrophotometric detection for the measurement of fast organic reactions in the time range between a few milliseconds and minutes.

In the field of relaxation spectrometry, new developments can also be reported. Swinehart & Castellan (12) demonstrate the usefulness of relaxation methods to relatively slow reactions. The linearization of rate equations greatly simplifies the mathematical treatment of experimental results. The authors investigated the chromate-dichromate reaction by a simple concentration jump technique. The chromium concentration was changed in about

[1] The survey of literature pertaining to this review was completed in November 1964.

[2] The following abbreviations will be used: bipy (bipyridine); EDTA (ethylenediamine tetraacetic acid); phen (1,10-phenanthroline).

5 sec by a simple mixing technique, and the change of pH with time was followed spectrophotometrically. Relaxation times were of the order of 10 sec. Data obtained were consistent with the proposed mechanisms and with measurements by Schwarzenbach & Meier (13). A new relaxation technique for reactions involving a change of electric polarization has been developed by Bergmann, Eigen & de Maeyer (14). The dielectric loss of a low conducting solution is measured (15) in the presence of a high electrostatic field (2×10^5 V/cm) in the frequency range 0.3 to 50 Mc sec^{-1}. A dimerization reaction $2X \rightleftarrows X_2$ involving a large change of dipole moment leads to a chemical contribution to the dielectric constant which is proportional to $(\mu_{x_2}{}^2-\mu_x{}^2)^2$, where μ_{x_2} and μ_x are the dipole moments of the dimer and the monomer respectively. The very small increment to the loss angle which results has a maximum at a frequency $\nu=1/2\pi\tau$ where τ is the relaxation time of the reaction. This technique requires a low conductivity of the solution and a large difference of the dipole moments of the reacting species. It has been applied to the dimerization of ϵ-caprolactam in benzene and in carbon tetrachloride. The formation of the dimer was found to be a diffusion controlled reaction, whereas the rate of monomerization was of the order of 10^8 sec^{-1}. A technique like this furnishes useful information on the kinetics of hydrogen bond formation in proteins and nucleic acids.

An ultrasonic absorption apparatus for the frequency range 1 to 60 Mc sec^{-1} has been described by Atkinson, Kor & Jones (16). Preliminary measurements on aqueous $MnSO_4$ solutions are reported with two absorption maxima at 3 and 35 Mc sec^{-1}.

To increase the often poor signal-to-noise ratio (S/N) in relaxation experiments, Rüppel, Bültemann & Witt (17) propose the use of a repetitive technique. With n impulses, a gain of $\sqrt{n}$ is obtained for S/N. An experimental setup is described for the excitation of photochemically active systems with a series of small flashes ($\sim$2 J). The effect of sampling on the signal-to-noise ratio is demonstrated. Another suggested technique, advanced by Bauer & Berg (18), is a combined concentration-temperature jump method. Solution and solvent at different temperatures are mixed in a stopped-flow apparatus. The relaxation is followed polarographically. For some selected reaction schemes especially of interest in enzyme kinetics, expressions for the relaxation time are given by Alberty et al. (19). More extensive compilations are available in mimeographed form from the authors (Dept. of Chemistry, Univ. Wisconsin, Madison 6, Wis.).

An increase of the reaction time range for kinetic measurements by line broadening has been achieved in both NMR and ESR spectroscopy. Thus, average lifetimes for protons in an exchange reaction $AH+B \rightleftarrows BH+A$ can be determined by an NMR double resonance technique as devised by Forsén & Hoffman (20). The resonance line AH (or BH) is suddenly saturated by a strong radio frequency field, and the approach to the new equilibrium value of the resonance BH (or AH) can be observed. From these time constants the average lifetime τ_B (or τ_A) can be obtained if T_{1B} (T_{1A}) is measured separately. This technique is applicable for reactions with half lifetimes of the same order of magnitude as T_1, which is normally of the order of seconds but can be considerably shortened by the addition of a trace of paramagnetic substances. Then the lower limit of lifetimes measurable by this technique is determined by the chemical shift between AH and BH protons. This method has been applied in preliminary experiments to the proton transfer reaction between salicylaldehyde and 2-hydroxyacetophenone (21) and to the exchange kinetics of protons in acetylacetone from the hydroxylic via the ketonic methylene to the olifinic protons (22).

Allerhand & Gutowsky (23), in a very interesting paper, discuss the suitability of the Carr-Purcell spin echo technique for the investigation of fast chemical reactions. There are some advantages of this method as compared to the conventional NMR technique. The useful time range is much extended since there is no effect of the magnet inhomogeneity in the spin echo technique. Furthermore, from measurements of the effective T_2 as a function of the pulse frequency, values of $T_2°$ (relaxation time T_2 in absence of exchange) and of the chemical shift between the exchanging species may be obtained. In the conventional stationary method, this information is sometimes lacking, and rate data have to be calculated from estimated chemical shifts and $T_2°$ values. Therefore, only approximate results may be obtained. Allerhand & Gutowsky applied their technique to the measurement of hindered rotation of *N,N'*-dimethyltrichloracetamide and of dimethylcarbamylchloride. For the measurement of proton exchange rates with the spin echo technique, Powles & Strange (24) provide an analysis for systems with more than one NMR line. Measurements on methanol are reported. In the evaluation of rate constants from NMR exchange line narrowing, a correction (25) of a previous paper (26) by Alexander should also be considered.

In ESR technique, two methods are presented by Fessenden (27) for the measurement of radical lifetimes which are of the order of milliseconds and, thus, much shorter than the necessary time constants of an ESR spectrometer. One technique uses a rotating sector; the other is a sampling technique. The two approaches are applied to the measurement of the rate of recombination of ethyl radicals generated by pulsed electron-beam radiolysis in liquid ethane. The rate constant is $(1.7 \pm 0.4) \times 10^8$ M^1 sec^{-1}. Values obtained by the two methods differ by 7 per cent.

Two new techniques which seem to be quite promising have been added to the electrochemical methods for the measurement of fast reaction rates (28–32). One of these techniques, called High Level Faradaic Rectification, has been introduced by Barker & Nürnberg (28–31). Rates of dissociation for 12 weak acids in 1*M*LiCl at 20° C are reported. The dissociation field effect in the electric double layer has been corrected. Except for salicylic and lactic acid, the rate of recombination has been found to be diffusion controlled. The suggested reason for the exceptions is intramolecular hydrogen bonding. These results are in general agreement with presently accepted data (e.g., 5), but contrast with many electrochemical measurements on proton transfer reactions published earlier. The reasons for such accidental failures of the electrochemical procedures are not of a basic nature and can be avoided if they are critically applied (33). It is especially the electrochemical proton transfer rate measurement which is liable to lead to erroneous conclusions. Published data on complex formation and electron transfer reactions measured electrochemically are generally more reliable. Interesting improvements of the chronopotentiometric method for the determination of moderately fast reaction rates have been developed by Hoffmann & Jaenicke (34). In the case of different, reducible, kinetically important species, inconveniently different transition times can be made more similar by using a linearly increasing current density instead of a constant one. Thus, the rate of dissociation of the complex boric acid hydroxyhydroquinone was measured. The same authors propose another electrochemical technique (32) which allows for the measurement of a monomolecular (or pseudomonomolecular) rate constant k in the limits $2.5\times10^9c^2 > k > 10^{-1}$ sec^{-1}, where c is the concentration of the reducible species ($c \leq 10^{-2}M$). A substance is reduced in a short, rectangular cathodic pulse, and the product is oxidized in an anodic pulse which immediately follows. This double-chronopotentiometric technique has been applied to the evaluation of high rate constants for some coupling reactions of naphthosulphonates and aromatic amines, which are important in color photography development.

Further experimental potentialities (for which practical application in fast reaction kinetics has not yet been realized) have been proposed (35, 36, 37). Cover & Meites (35) theoretically discuss automatic and coulometric titrations for studying the kinetics of the complex formation. The electrode potential can depart from the equilibrium value when the rate of reaction is not fast enough as compared to the rate of titration. For different reactions coupled to the titration reaction proper, the electrometric titration curve has been evaluated for different rates of titration. Reaction rate constants higher than those determined by conventional methods can thus be obtained. In the case of reaction time not much shorter than the polarographic drop time, Reinert (36) has solved the involved diffusion equation. Thus, the potentiostatic measurement of currents on a dropping mercury electrode can be used to determine the rate of moderately fast reactions, provided that the sudden start of the reaction has been accomplished (e.g., by a stopped-flow technique). With a new technique developed by Becker & Köhler (37), the experimental boundary conditions better approximate those assumed in the mathematical calculation of kinetically determined polarographic currents.

Quite a few papers have appeared dealing with the intricate mathematical treatment of the various electrochemical techniques for measuring fast chemical reactions. The mathematical theory of constant-voltage methods has been summarized by Reinert (38), while that of polarography has been dealt with concisely by Brdička, Hanuš & Koutecký (39). Savéant & Vianello (40) calculate the current-time relationship for linear-voltage sweep chronoamperometry for the case of the reducible species formed by a monomolecular (or pseudomonomolecular) reaction. The ranges of the equilibrium and rate constant for which this technique is applicable are supplied in diagrams by the authors. The theory of stationary electrode polarography with preceding, following, or catalytic chemical reactions using single scan and periodic scan polarography is presented by Nicholson & Shain (41). Both reversible and irreversible charge transfer on the electrode are considered. Tables useful for the evaluation of rate constants for experimental data are included. Multiple-step first-order reactions preceding the electrochemical reduction, two-step preceding or following reactions, and catalytic reactions combined with preceding or following reaction on the ac polarographic current are investigated by Hung, Delmastro & Smith (42). Polarographic phase-angle studies in ac polarography have been shown by the authors to be of diagnostic value in the determination of reaction mechanisms. The rate constant of dimerization reactions, k, can be obtained from controlled potential electrolysis for the reaction system

$$\mathrm{O} + \mathrm{n_1e} \rightarrow \mathrm{I}, \qquad 2\mathrm{I} \xrightarrow{k} \mathrm{D}, \qquad \mathrm{I} + \mathrm{n_2e} \rightarrow \mathrm{R}$$

according to a theoretical analysis by Meites (43). For a plane electrode, Groden, Aylward & Hayes (44) calculate kinetically determined limiting currents for the case

$$\mathrm{A} + \mathrm{ne} \rightleftarrows \mathrm{B} \quad \text{and} \quad \mathrm{B} \xrightarrow{\text{irreversible}} \mathrm{C}$$

Tables for evaluation are presented. Mecke (45) has demonstrated mathematical equivalency for the approaches of Koutecký (46) and of Henke & Hans (47) to the problem of kinetically determined polarographic currents for a homogeneous first-order reaction preceding the electrochemical reduction. A very general treatment of the electrochemical behavior of systems

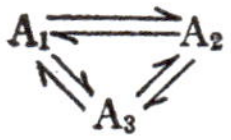

involving only first-order reactions and electron transfer has been contributed by Ashley & Reilley (48). The results can be applied to reactions preceding or following the proper electrode reaction, and to the so-called catalytic currents (parallel kinetics). Different experimental techniques and evaluation procedures are discussed. If in alternating current electrolysis a product will react irreversibly to form electroinactive products from the efficiency as a function of current frequency, the kinetic order of the irreversible reaction may be obtained as indicated by Pinkerton (49). A theoretical treatment for obtaining the rate of complex formation has been contributed by Buck (50). According to this author, from the dependence of the polarographic half wave potential on the log (*ligand concentration*), rates of moderately fast reactions emerge. The effect of the curvature of a hanging mercury drop electrode on kinetically determined currents is evaluated by Levich, Grabowski & Filinovski (51). A reinvestigation of the effect of the diffuse double layer on electrochemically determined rate constants of fast reactions has been conducted by Rangarajan (52).

Protolytic and hydrolytic reactions.—Rate constants of protolytic reactions in aqueous solution have been reviewed by Eigen et al. (5). Their work calls attention to an impressive list of different reactions studied mostly by relaxation methods during the last decade. In most cases, the rate of reactions involving the recombination of a proton with a base or a hydroxyl ion with an acid proves to be diffusion controlled and to be of the order of 10^{10} to 10^{11} M^{-1} sec^{-1}. However, some noticeable exceptions to this rule do occur. As shown by Eigen & Kruse (53) in an important paper, the rate of recombination of protons (hydroxyl ions) with bases (acids) is dependent only to a slight degree on steric effects and on the charge type of the base (acid) involved. Furthermore, this latter dependence can be deduced from Debye's formula for diffusion controlled reactions between ions. More important than the electrostatic effect is the influence of internal hydrogen bond chelation. Thus, rate constants which are some orders of magnitude lower than 10^{10} M^{-1} sec^{-1} are observed in the recombination of OH^- ions with acids, the protons of which are involved in strong internal hydrogen bonds. Other non-diffusion controlled protolytic reaction rates are found with CH-pseudoacids and with proton transfer reactions when the pK difference for the two acids involved energetically favors the back reaction.

As mentioned previously, the rate of recombination of protons with twelve different anions has been reported by Barker & Nürnberg (28). Polarographic and chronopotentiometric measurements on *o*-phthalic acid are interpreted by Buck (54) as kinetic currents limited by the rate of the reaction

$$C_6H_4(COOH)_2 + H^+ \overset{k}{\rightleftarrows} C_6H_4(COOH)_2H^+$$

The rate of formation of the superacid (dissociation constant $K = 24.4$ M) is found to be about 10^6 M^{-1} sec^{-1} if corrected for the influence of the electric double layer potential at the electrode. However, because of this extrapolation and since the reaction layer thickness is only about 7 Å, the results can be considered as approximate only. This is an interesting case of proton exchange between acid-base pairs of comparable strength. For the rate of neutralization of bimaleinate anion with OH^- in unbuffered solution, 3.6×10^{10} M^{-1} sec^{-1} was obtained polarographically by Mairanovski (55). Nishiyama, Maruyama & Hamaguchi (56) polarographically determined the rate of dissociation and recombination of terephthalaldehydic acid in 0.2 M phosphate buffer between pH 5.5 to 7.5 to be 5×10^3 sec^{-1} and 1.5×10^8 M^{-1} sec^{-1}, respectively, which seems to be more than one order of magnitude lower than expected from the theory of diffusion controlled reaction rate. Nürnberg, van Riesenbeck & von Stackelberg (57) have developed an electrochemical method for the estimation of recombination rates of protons with anions of weak acids. The reduction of hydrogen ions at the electrode is catalyzed by 2,2′-bipyridine or pyridine so that dissociation of the acid is the rate determining step. It was found that the recombination of protons with doubly ionized citric acid, with $H_2PO_4^-$, and with HSO_4^- is a diffusion controlled reaction.

Increasingly, advantage is being taken of the NMR technique in elucidating proton transfer processes. The rate of protonation and deprotonation of the weak base sulfamide has been obtained from NMR line broadening by Garrett, Tao & Jolly (58). The protonation rate was 3.2×10^7 M^{-1} sec^{-1}, whereas the deprotonation with OH^- proved to be diffusion controlled. Analogous data for the sulfamate ion were found to be about 10^{10} and 10^8 M^{-1} sec^{-1}. The proton transfer of a series of protonated aromatic substances in HF has been investigated by MacLean & Mackor (59) by NMR line-width broadening. Inter- and intramolecular proton transfer was observed. Proton transfer from H_3O^+ to HF_2^- in HF was found to be a diffusion controlled reaction. The rate of proton exchange between ammonium and carboxyl sites in a solution of methylamine in acetic acid is 230 sec^{-1} at 25° C, according to NMR line-width measurements by Grunwald & Price (60). Further investigation by the same authors (61) supplied analogous rate constants for NH_3, $(CH_3)_3N$, and $(HOCH_2)_3CNH_2$ in CH_3COOH, which are 6080 sec^{-1}, 945 sec^{-1}, and 4.1×10^4 sec^{-1}, respectively. A detailed discussion of the reaction mechanism is given by the authors. With NMR line broadening, Krakower & Reeves (62) measured the rate of proton transfer between *o*-chlorophenol and methanol. The constant for the second order rate process $PH^* + MH \rightleftarrows PH + MH^*$ was found to be 900 sec^{-1} at 25° C (concentrations given as mole fractions). The activation energy was found to be 4.58 kcal/mole. Addition of water or methanol, or both, increased the rate. The formation of an ion pair from the hydrogen bonded species

$$PH \cdots \overset{H}{O}\text{—}CH_3$$

is assumed to be the slow step. With stronger intramolecular hydrogen bonding (methylsalicylate, salicylaldehyde, *o*-hydroxyacetophenone), the rate of proton transfer is greately decreased. The same authors (63), from a broad NMR line, averaged over OH and NH^+ in a mixture of quinoline and trifluoracetic acid, reveal information for the proton transfer in this system. The exchange of the Lewis acid BF_3 between different ethers as bases has been studied with NMR line broadening by Rutenberg, Palko & Drury (64). Becker (65), from NMR line broadening, calculated the dehydration rate constant of hydrated pyruvic acid in 3*M* HCl solution. The rate of the protolytic nitro-aci tautomeric reaction of nitrocyclohexane was investigated polarographically by Zaitsev, Turyan & Zaitseva (66). The Brönsted relation is valid for the recombination and dissociation rates with different buffer systems. The highest rate, 480 M^{-1} sec^{-1}, is found for the recombination of the aci-anion with H_3O^+. Carrington & Smith (67) observed with ESR techniques that the proton exchange on neutral pyrogallol radical (free electron on oxygen in position 2) is slow compared with the hyperfine splitting 2.38×10^6 sec^{-1}. This is explained by hydrogen bonding. At lower pH, the proton transfer leading to the pyrogallol cation (pK$\sim$2) accelerates, and broadened lines appear. With other polyoxybenzenes, hyperfine splitting from hydroxyl protons could not be noted.

Three further papers are concerned with the protolytic behavior of molecules in an excited state. Excited perylene is a much stronger acid than the same molecule in the ground state. Therefore, the excited molecule transfers protons to bases during its short lifetime (6×10^{-9} sec), and the resultant quenching competes with fluorescence. From measuring the relative intensity of fluorescence as a function of base concentration, reaction rates for the proton transfer are obtained, as reported by Leonhardt & Weller (68). This proton transfer proves to be diffusion controlled in the polar solvents acetonitrile, dimethylformamide, formamide, and methanol with an effective distance of the reacting species of about 7 Å. The reaction is fast since the resulting ions are stabilized by solvation. In nonpolar solvents (benzene, methylcyclohexane), the rate is smaller by orders of magnitude, and the effective distance is much smaller. The dissociation of extremely strong acids (excited pyrene-ammonium ions pK$=-2$ to -6) in different solvents was investigated as a function of temperature by Urban & Weller (69) with the fluorescence method. As expected, there is no influence of the back reaction. The observed dependence of rate and activation energy on the solvents is discussed on the basis of the assumption of a partial change of solvation during the fast formation of the lyonium ion-base pair. According to Mataga & Kaifu (70), the rate of proton transfer from excited β-naphthol to triethylamine is a diffusion controlled reaction in cyclohexane and in benzene ($k\sim8\times10^9$ M^{-1} sec^{-1} and $\sim3\times10^9$ M^{-1} sec^{-1}, respectively).

Theoretical considerations relevant to proton transfer reactions have been published by Weiss (71) and by Allen & Patrick (72). Weiss proposes that the proton transfer in acid-base reactions proceeds predominantly by a quantum-mechanical effect. An improvement of the Smoluchowski equation for a diffusion controlled encounter has been presented by Allen & Patrick. This theory is better suited for large molecules.

Electron transfer reactions.—Considerable attention has been focused on electron transfer reactions during the last year because of existing theoretical considerations which permit the prediction of rate constants or at least the correlation of different rate constants. We are indebted especially to Marcus for developing this theory (73–76) and also for reviewing the field so competently (77). Endicott & Taube (78) have conducted a kinetic investigation of the electron transfer from $Ru(NH_3)_6^{2+}$ to various Co(III) complexes in order to test the Marcus theory for outer sphere electron transfer. Since $Ru(NH_3)_6^{2+}$ is a very stable complex in both oxidation states, it is more likely to be a better test system than the more labile $Cr(bipy)_3^{2+}$ which was used previously. The kinetics of the fast reactions were followed spectrophotometrically at low concentrations. In addition, the reduction of $Ru(NH_3)_6^{3+}$ by certain reducing agents was measured. The results obtained confirm that $Ru(NH_3)_6^{2+}$ is oxidized in an outer sphere mechanism and are consistent (with few exceptions) with the Marcus theory. Campion, Purdie & Sutin (79), using a flow technique, investigated the electron transfer for the following redox systems: Ce(IV)—$W(CN)_8^{4-}$; Ce(IV)—$Fe(CN)_6^{4-}$; Ce(IV)—$Mo(CN)_8^{4-}$; $IrCl_6^{2-}$—$W(CN)_8^{4-}$; $IrCl_6^{2-}$—$Fe(CN)_6^{4-}$; $IrCl_6^{2-}$—$Mo(CN)_8^{4-}$; $Mo(CN)_8^{3-}$—$W(CN)_8^{4-}$; $Mo(CN)_8^{3-}$—$Fe(CN)_6^{4-}$; $Fe(CN)_6^{3-}$—$W(CN)_8^{4-}$; Ce(IV)—$Fe(phen)_3^{2+}$; Ce(IV)—$Fe(phen)_2(CN)_2$; Ce(IV)—$Fe(phen)(CN)_4^{2-}$; Ce(IV)—$Fe(bipy)_3^{2+}$; Ce(IV)—$Fe(bipy)_2(CN)_2$; and Ce(IV)—$Fe(bipy)(CN)_4^{2-}$. The rates were found to be first order with respect to both components, with their rate constants ranging between $>10^8$ to 4.3×10^4 M^{-1} sec^{-1}.

With a flow technique, Diebler & Sutin (80) investigated some oxidation reduction reactions involving manganese(III) ion. The rate constants obtained by Sutin and co-workers (79, 80) were correlated with equilibrium constants and with known rate constants from other redox reactions and found to be in reasonable agreement with values calculated from Marcus' theory. Hurwitz & Kustin (81) investigated the kinetics of the reaction $Ox_1^{2-}+Red_2^{2+}\rightleftarrows Red_1^{3-}+Ox_2^{3+}$ with the temperature jump technique. Ox_1^{2-} is $IrCl_6^{2-}$, $IrBr_6^{2-}$, or $RhCl_6^{2-}$; Red^{2+} are substituted phenanthroline or bipyridine complexes of Fe(II) or Ru(II). The reaction rate is diffusion controlled (2×10^9 M^{-1} sec^{-1}) when the equilibrium constant is >1. This is a situation similar to the relationship between proton transfer rates and the ΔpK of the two acids, which has been examined by Eigen et al. (5). The electron transfer between the two anions $[Co(II)O_4W_{12}O_{36}]^{-6}$ and $[Co(III)O_4W_{12}O_{36}]^{-5}$ has been studied by Rasmussen & Brubaker (82) with a tracer method (^{60}Co) at 0° C and ionic strength 0.6 M. The rate constant is (0.63 ± 0.016) M^{-1} sec^{-1} and about five to six orders of magnitude lower than the rate estimated with Marcus' theory (73–77). Bächmann & Lieser (83) investigated the kinetics of electron transfer from ferric sulphato complexes to ferrous ion with Fe^{59} tracer in dilute solutions. These reactions proceed according to an inner sphere mechanism. The rate constants k of these and similar reactions depend on the difference of the stability constants K of the ferric and ferrous complexes: $\log_{10}k=1.1+0.5|\Delta\log K|$ [see Marcus (77)]. From this equation, it is possible to estimate stability constants from kinetic measurements which otherwise are not easily obtainable. From flow technique measurements, Dulz & Sutin (84) conclude that the chloride catalyzed oxidation of Cr^{2+} by Fe^{3+} proceeds along two paths, one being $FeCl^{2+}+Cr^{2+}\rightarrow[Fe\cdots Cl\cdots Cr]^{4+}\rightarrow Fe^{2+}+CrCl^{2+}$ with an inner sphere activated complex. The second path probably advances via an outer sphere complex. Using a flow technique, Nancollas & Sutin (85) studied the electron transfer reaction between Co(III) and Fe(II) in the presence of chloride ions. Without the chloride ions, the rate was found to be 300 M^{-1} sec^{-1} in 3.0 $HClO_4$ at 25° C. The chloride catalyzed reaction proceeds via an inner sphere activated complex with a chloride ion bridge between the cobalt and iron. The reduction of vanadium(V) by vanadium(III) has been investigated by Daugherty & Newton (86). The dependence on pH of the reaction rate (order of magnitude 10^2 to 10^5 M^{-1} sec^{-1}) indicates four parallel paths via the four activated complexes $(VO)_2O^{2+}$, $(VO)_2OH^{3+}$, $(VO)_2^{4+}$, and $(VO)_2H^{5+}$ (formulas, not structures). These results are consistent with former, indirectly obtained data. Newton & Baker (87, 88) found an intermediate hydrolytic dimer VOV^{4+} in the oxidation of V(II) with V(IV). About 65 per cent of the reaction goes through this moderately fast step, the rest probably proceeding by way of an outer sphere activated complex directly to V(III). Similar results have been obtained by Olver & Ross (89) in H_2SO_4 solutions by polarographic catalytic currents. However, to explain the basic catalysis observed, these authors propose a more reactive species, $V(OH)HSO_4^{2+}$, the formation constant of which is estimated to be of the order 1 M^{-1}. The reduction of a series of carboxylatopentammine-cobalt(III) complexes by Cr(II) was studied by Gould & Taube (90). Normally, the rate of such reactions is slow (0.08 to 0.3 M^{-1} sec^{-1}). The rates are strongly increased if the carboxylate ligand contains a carbonyl-, another carboxyl-, or a phenolic OH-group. The influence of pH, and of chelation on the rate constants of electron transfer through such groups, is discussed. With dicarboxylic acids and with phenolic and pyridine acids, the specific rate of electron transfer is much higher on the basic path than on the acidic path. Sykes (91) reports that the reduction of μ-peroxo complexes of cobalt(III) $[(NH_3)_5CoO_2(NH_3)_5]^{5+}$ ($\equiv M$) with ferrous ion to yield the analogous Co(II) complexes is slow, with a rate constant of the order 10^{-2} M^{-1} sec^{-1}. The reaction is strongly catalyzed by complex-forming ions X with catalytic effectiveness $F^- > Cl^- > SO_4^{2-} > Br^- > NO_3^- > OH^-$. A parallel reaction path with FeX^+ and MX^{4+} as the catalytic active intermediates is proposed by the author. Sheppard (92) reports that the reduction of neptunium(VI) by vanadium(III) in perchloric acid at 25° C can be expressed with the rate law

$$k = k_0 + \frac{k_1}{[H^+]}$$

where $k_0 = (6.3 \pm 1.2)$ M^{-1} sec^{-1} and $k_1 = (20.3 \pm 0.5)$ sec^{-1} at 25° C. Candlin, Halpern & Trimm (93) investigated the reduction kinetics of different pentamine cobalt(III) complexes by Cr^{2+}, $Cr(bipy)_3^{2+}$, V^{2+}, and Eu^{2+} with a stopped-flow technique. From the dependence of the reaction rate on the sixth ligand in the cobalt pentamine complex, the authors conclude that with Cr^{2+} and Eu^{2+}, the reduction occurs with an inner sphere mechanism, whereas with $Cr(bipy)_3^{2+}$ and V^{2+}, an outer sphere mechanism is probable. Lower limits for the rate of electron transfer have been obtained by Dietrich & Wahl (94) for some redox systems by NMR line broadening. For the systems $Os(bipy)_3^{2+}/Os(bipy)_3^{3+}$, ferrocene/ferrocinium, and ferroin/ferrin, rate constants $>5\times10^4$, $>1\times10^5$, and $>1\times10^5$ M^{-1} sec^{-1} were found, respectively (95). Espenson (96) determined the rate constant for the reaction

$$5CrI^{2+} + IO_3^- + 6H^+ \rightarrow 5Cr^{3+} + 3I_2 + 3H_2O$$

which proved to be first-order with respect to $[CrI^{2+}]$ and $[I_2]$; that is, the reaction is autocatalytic. The rate constant k_{ox} as defined by $-d\,[CrI^{2+}]/dt = k_{ox}[CrI^{2+}][I_2]$, was found to be $(7.8 \pm 1)M^{-1}$ sec^{-1} at 20° C and ionic strength 1.00 M.

Some investigators have made use of ESR line broadening for the determination of electron transfer rates. Hirota & Weissman (97) measured electron transfer rates between some ketyls and their ketones. The rate constants obtained for monomers ranged between 10^8 and 5×10^8 M^{-1} sec^{-1} in different solvents. The authors found that the alkali metal nucleus and the spin are transferred together. The rate of exchange with ketyl dimers is $<10^5$ M^{-1} sec^{-1}, too slow for the measurement with ESR. The electron spin exchange in potassium peroxylamine disulfonate ($K_2(SO_3)_2NO$) has been studied by Jones (98). A reaction rate of about 1.3×10^9 M^{-1} sec^{-1} was obtained. This investigation was undertaken primarily to demonstrate the appropriateness of the simplified Bloch treatment in such cases. From ESR line broadening, Reynolds (99) estimated the rate of electron transfer between 2,2′-bipyridine and its negative ion in 1,2-dimethoxyethane. Activation energy and entropy were 10 kcal mole^{-1} and 5 cal deg^{-1} mole^{-1}, respectively. An activated complex [bipy · · · K^+ · · · bipy$^-$] is proposed. Electron transfer reactions in "living polymers" (that is, polymers which are not terminated and which are able to grow) have been reviewed by Szwarc (100).

Metal complexes.—Also in the field of the kinetics of metal complex formation, many papers have appeared during 1964. In many cases again, the exchange of a solvent molecule from the inner solvation sphere was found to be the rate determining step. With NMR techniques, this exchange can be followed directly. The average lifetime of a water molecule in the inner hydration sphere of an Mn(II) ion has been determined by Mazitov (101) from NMR relaxation times to be about 2×10^{-8} sec. The whole water molecule exchanges with the bulk water. The rate of exchange of CH_3OH between the bulk of solution and the solvation sheaths of Ni^{2+} (102) and Co^{2+} (103) ions was investigated by Luz & Meiboom. Results obtained at low temperatures from line broadening agreed quite well for Co^{2+} and sufficiently for Ni^{2+} with data obtained from T_1 and T_2 relaxation time measurements (104). It is concluded that the proton exchange by protolytic reactions is negligible compared to the exchange of whole molecules. The measurements were carried out between -80 and $+80$° C. The rate varies in that interval from slow (<1 sec^{-1}) to very fast (nearly total averaging of the two lines). The rate of exchange at 25° C was found to be 1.8×10^4 sec^{-1} and 1.0×10^3 sec^{-1} for Co^{2+} and Ni^{2+}, respectively. Arrhenius parameters are also reported. The rate of CH_3OH exchange in the $Co(CH_3OH)_5Cl^+$ complex at varying temperatures has been studied by Luz (105) with different NMR techniques. At low temperatures, the NMR spectrum shows separate lines for the bulk CH_3OH and for the *trans*- and *cis*-ligand molecules (*trans* and *cis* refer to the position of the Cl ligand). The rate of exchange of *cis*-CH_3OH at -63° C is 510 sec^{-1} with an activation energy of 14.5 kcal mole^{-1}. The *trans*-CH_3OH exchanges more slowly.

With non-NMR methods, too, further information about solvent molecule exchange has been accumulated. With the temperature jump technique, Swift (106) found relaxation times in Co^{2+} and Zn^{2+} solutions of the order 0.1 sec. From spectral considerations, from entropy of activation, and from concentration independence, this relaxation was attributed to the reaction $Me(H_2O)_4^{2+} \rightleftarrows Me(H_2O)_6^{2+}$. Hammes & Morrell (107) have analyzed the complex binding and rate constants for the formation of nickel(II) and cobalt(II) phosphate complexes. The slow step of the reaction was interpreted as being the dissociation of the water molecule from the inner hydration sphere. Higher charged ligands (polyphosphate anions) were found to loosen these water molecules, thereby increasing the rate of dissociation. The relaxation times measured by the temperature jump method ranged from 50 μsec to 3 msec. With the pressure jump relaxation technique, Hoffmann, Stuehr & Yeager (108) investigated the rate of formation of a series of metal complexes. The rate of formation of $BeSO_4$ was found to be somewhat slower than that indicated by Strehlow & Becker (109). (This result has meanwhile been confirmed by the latter authors.) The complex formation of Be^{2+}, Ni^{2+}, Co^{2+}, Cu^{2+}, and Zn^{2+} with ligands such as *d*-tartrate, H-citrate, acetate, glycinate, oxalate, ascorbate, H_2EDTA, glycolate, formate, chloroacetate, and malonate has also been studied by Hoffmann, Stuehr & Yeager (108). The results are interpreted on the basis of the association mechanism proposed by Eigen (110). The former authors have given a rather complete analysis of the formation kinetics of the inner sphere complexes of nickel malonate and of $BeSO_4$. From flow experiments, Nancollas & Sutin (111) have also concluded that for the formation of the nickel monooxalato complex, the loss of a water molecule from the inner coordination shell of the cation is the rate determining step. The rate constants for the reactions

$$Ni^{2+} + A^{2-} \underset{k_{1d}}{\overset{k_{1f}}{\rightleftarrows}} NiA \quad \text{and} \quad Ni^{2+} + HA^{-} \underset{k_{2d}}{\overset{k_{2f}}{\rightleftarrows}} NiA + H^{+}$$

(A^{2-} = oxalate ion) at ionic strength 0.10 and at 25° C were found to be $k_{1f} = 7.5 \times 10^4\ M^{-1}\ \text{sec}^{-1}$, $k_{2f} = 5 \times 10^3\ M^{-1}\ \text{sec}^{-1}$, $k_{1d} = 3.6\ \text{sec}^{-1}$, and $k_{2d} = 1.5 \times 10^3\ M^{-1}\ \text{sec}^{-1}$. The activation energies were $\Delta H_{1f} = \Delta H_{2f} = 14$ kcal mole^{-1}. Eigen, Geier & Kruse (112), from temperature jump measurements, found that the replacement of OH in CH_3HgOH with different anions proceeds in an SN_2 mechanism. This finding contrasts with the SN_1 mechanism often observed in the complex formation of hydrated cations with various ligands.

The rate of water exchange of Ni(II) at a mercury-aqueous solution interface has been measured polarographically by Hush & Scarrot (113). A review on the kinetics of ligand dissociation from metal complexes, as investigated by electrochemical techniques, has been contributed by Gerischer (114). The rate of dissociation of Zn(II) nitrilotriacetate in NH_3-NH_4Cl buffer has been determined polarographically by Kodama & Hashimoto (115). From NMR widths, Pearson & Lanier (116) derived rate constants or lower limits of rate constants for ligand exchange with complexes of Mn(II), Fe(II), Co(II), Ni(II), and Cu(II). The ligands used were glycol, ammonia, ethylenediamine, glycine, sarcosine, and dimethylglycine. The order of exchange is second-order in the case of Cu(II), pointing to an SN_2 mechanism, which is common for planar structures. The other metals form octahedral complexes with an SN_1 exchange mechanism. The order of reaction rates is as found in similar cases: Mn > Fe > Co > Ni < < Cu. Substituent effects on the dissociation rate of Ni(II)-pyridine complexes have been investigated by Moore & Wilkins (117), using a stopped-flow apparatus. With different substituted 3- or 4-pyridines, a linear correlation exists between log k_{diss} and pK_{HL}, where K_{HL} is the acidity constant of the pyridinium ion and a linear dependence of log k_{diss} on the Hammett σ value of the bases is observed. The stability of the complexes—as far as it is known—is apparently determined mostly by the various rates of dissociation. Rorabacher & Margerum (118) investigated ligand exchange reactions between Ni-triethylenetetramine or Ni-tetraethylenepentamine and EDTA. The reaction was faster than the dissociation of the polyamine complexes. From the experimental results obtained, strong evidence emerges in favor of a multistep mechanism. In alternate steps there occur the occupying of an aquated coordination site by EDTA (starting with a carboxylic group) and the replacement of a polyamine occupied site by water in such a way that in the whole series of reaction steps, four or five coordination sites are always occupied by an organic ligand. The formation of the first EDTA-N-Ni bond is the rate determining step. This mechanism is also applicable to similar ligand exchange reactions. The reaction $HgCl_2 \rightleftharpoons HgCl^+ + Cl^-$ has been investigated by Eigen & Eyring (119) using the dispersion of the second Wien effect. The recombination rate constant $k_r \sim 7 \times 10^9$ M^{-1} sec^{-1} probably is only negligibly influenced by hydrolytic processes. It is comparable to similar rates with Ba^{2+} and Pb^{2+} and higher than corresponding values for smaller ions. According to Godfrey, Heffernan & Kerr (120), the average lifetime of the species $HgCl_2$, $HgCl_3^-$, and $HgCl_4^{2-}$ in $\sim 1M$ LiCl should be shorter than 10^{-3} sec since only one NMR line of ^{199}Hg is observed with a chemical shift of 210 ppm at 11.07 Mc sec^{-1} between $HgCl_2$ and $HgCl_4^{2-}$. The complex formation of the bivalent transition metal ions with the dye pyridine-2-azo-*p*-dimethylaniline was studied with the temperature jump method by Wilkins (121). Formation rate constants between 3×10^4 and 1.3×10^7 M^{-1} sec^{-1} and dissociation rate constants between 0.2 and 3×10^4 sec^{-1} were obtained. Again, the elimination of a water molecule is important for the formation rate. The reaction of Ni^{2+} with 1,10-phenanthroline has been reinvestigated at different temperatures by Barrett & Smith (122), using a simple stopped-flow technique. Essential agreement with the results of Margerum, Bystroff & Banks (123) was obtained. The exchange of Co^{2+} and Cd^{2+} between their EDTA complexes and their chlorides has been investigated by Cendales et al. (124) with a tracer technique. Valleau (125) has developed the theory of ultrasonic absorption relaxation of two interacting reactions by introducing a "relaxation viscosity." This author with Turner (126) measured the ultrasonic absorption in an aqueous solution of $CdCl_2$ at different frequencies. From the relaxation times thus observed, the rate constants for the reactions $CdCl_2 + Cl^- \rightleftarrows CdCl_3^-$ and $CdCl_3^- + Cl^- \rightleftarrows CdCl_4^{2-}$ were estimated at different temperatures.

Besides the work of Luz & Meiboom (103–105) on the CH_3OH exchange from inner solvation sheaths of cations, two other papers report fast reactions in nonaqueous media. Rate constants for the reaction

$$2AlX_3 \underset{k_{21}}{\overset{k_{12}}{\rightleftarrows}} AlX_4^- + AlX_2^+$$

in nitrobenzene at 25° C have been measured by Wendt (127) with the pressure jump technique. (X is Cl or Br, respectively.) The rate constants are $k_{21} = (2.85 \pm 0.4)10^4$ M^{-1} sec^{-1} and $k_{12} \approx 130$ M^{-1} sec^{-1} for the chloride, and $k_{21} = (380 \pm 60)$ M^{-1} sec^{-1} and $k_{12} \approx 8$ M^{-1} sec^{-1} for the bromide. The complex formation between Cd^{2+} and Br^- ions in molten mixtures of $NaNO_3$-KNO_3 at 263° C has been studied by Bockris et al. (128) by chronopotentiometry. For example, the average lifetime for $CdBr_2$ has been found to be 0.28 sec.

Radical reactions.—Many radical reactions prove to be diffusion controlled. Although radiolysis experiments are not the topic of this review, four papers of general interest should be mentioned. Diffusion controlled rates for the quenching process are theoretically treated by Burton and coworkers (129, 130) for the case of instantaneous excitation, rectangular pulse excitation, and the stationary state for the excited molecule. Experiments for the quenching of irradiated radiolysis (γ, UV) in terphenyl by CCl_4 lead to a diffusion controlled reaction in benzene. Rates in cyclohexane ($>3.10^{11}$ M^{-1} sec^{-1}) were found to be higher than those predicted by current theory. The high rate may not pertain to bimolecular reactions in the conventional sense. In an investigation by Gordon et al. (131), the decay of the concentration of hydrated electrons obtained by electron pulse radiolysis was followed by measuring optical absorption as a function of time. The rate of combination with 24 different ions or molecules in aqueous solution was found to be diffusion controlled or nearly so. Exceptions are the reactions with formate, acetate, formic and acetic acid, and formaldehyde.

The status of diffusion kinetics as studied with radiolysis experiments has been reviewed by Schwarz (132). Gol'-danskii, Firsov & Shantarovich (133) find that the reaction of positrons with different inorganic ions occurs with a diffusion controlled rate. Some papers concerned with the kinetic behavior of radicals not produced by radiolysis are of interest in this review. Radical transfer reactions were studied by Jackson (134) with yield determinations. Radical A produced by a thermal first-order decomposition reacted either to form A_2 or with a solvent molecule to produce B·, which again reacted to AB or B_2. From the yields of A_2, B_2, and AB, the rate of the reaction

$$A\cdot + HB \rightarrow AH + B\cdot$$

can be evaluated if the other rates involved are known. This method has been employed to obtain kinetic data on the formation of *p*-chlorbenzyl radical from *p*-chlorotoluene and benzyl radicals which were formed by thermal decomposition of dibenzylmercury (135). The radical anion of nitrobenzene $C_6H_5NO_2^-$ was developed in aqueous solution by galvanostatic electrolysis by Kastening (136). By dismutation, this anion forms nitrobenzene and phenylhydroxylamine. Rate constants for two parallel reactions were obtained: $k_1 = 5\times10^8$ [H$^+$] sec^{-1} and $k_2 = (1+4\times10^{12}[H^+])M^{-1}$ sec^{-1}, and a corresponding mechanism was proposed. Polarographically, the rate of oxidation of the radical anion with different oxidants was investigated (136). From ESR line broadening of the alkali radical ion complex of pyracene in tetrahydrofuran at $-30°$ C, the rate of the intramolecular reaction $M^-\cdot K^+ \rightleftarrows K^+\cdot M^-$ was estimated to be of the order of 10^6 sec^{-1} by de Boer & Mackor (137). The rate of exchange $M^-\cdot Me^{*+} + Me^+ \rightleftarrows Me^+\cdot M^- + Me^{*+}$ was roughly 10^9–10^{10} M^{-1} sec^{-1} in 50 per cent tetrahydrofuran and dimethoxyethane at 20° C. The effect of the motion of methyl groups on the ESR line widths of π-electron free radicals has been theoretically calculated by Freed & Fraenkel (138). Investigating the rate of recombination of ketyl anions with a combination of flashlight technique and polarography, Berg (139) found that the rate constants vary from <1.4 M^{-1} sec^{-1} for the perinaphthenone ketyl to $>1.3\times10^5$ M^{-1} sec^{-1} for benzophenone ketyl (0.1*N* NaOH, 40 per cent *i*-propanol, 25° C). Martin, Rummel & Gross (140) studied the kinetic behavior of NO_3 radicals produced by flash photolysis of Ce(IV)-nitrate (flash duration: 8 μsec, flash energy: 100 J). The oxidation of Ce(III) by NO_3 proceeds with $k = 1.7\times10^6$ M^{-1} sec^{-1} at 23° C with a very small energy of activation. In highly concentrated acetic acid, the process $CH_3COOH + NO_3 \rightarrow$ Prod. is faster with $k = 2.3\times10^2$ M^{-1} sec^{-1}. Using a flash photolysis technique, Kato, Morita & Koizumi (141) studied the photobleaching of methylene blue. An excited triplet state reacts with the ground state (rate constant k_3) to form an intermediate X (a loosely combined pair of semiquinone and a half oxidized state), the spectrum of which was obtained. In this reaction, X decomposes into leuco dye or ground state dye with rate constant k_4. The authors found $k_3 \sim 10^{10}$ M^{-1} sec^{-1} and $k_4 \sim 10^3$ sec^{-1}. Further flash photolysis work on methylene blue was undertaken by Matsumoto (142).

Miscellaneous reactions.—Additional fast chemical reactions which have been reported in 1964 and do not fit into the classes of reactions dealt with above are considered in this section. Some investigations on fast bromination reactions with low bromine concentrations have been performed. The rate of bromination of 2,3-dimethyl-1,4-diazopinium cation was measured at 0° C by Bell & Marshall (143). The bromine concentration at 10^{-6} to 10^{-9} was followed with a platinum redox electrode. The rate constant, independent of pH and Br$^-$ concentration, was found to be $(4.4\pm0.2)\times10^6$ M^{-1} sec^{-1}. The further bromination of the monobromo compound is slower and depends on pH and [Br$^-$]. The monobromination process has an energy of activation of 9.1 kcal $mole^{-1}$. For the second step, a parallel bromination occurs on the cation and free base. The latter reveals a reactivity ten times that of the cation. Similarly, by following low bromine concentration in an aqueous acetone solution by redox potential measurements, Bell & Davis (144) determined the kinetics of bromination of the acetone enol. From these and earlier measurements, the authors conclude that the more reactive the organic species, the less the selectivity in rates between different reagents (in this case, $k = 1.03 \times 10^7\ M^{-1}\ sec^{-1}$ for Br_2 and $2.8 \times 10^6\ M^{-1}\ sec^{-1}$ for Br_3^- as brominating agents at 25° C). The bromination of 8-quinolinol and 2-methyl-8-quinolinol was studied by Kozak, Fernando & Freiser (145) by an electrochemical method. Bromine was produced by galvanostatic pulses, and the bromination after the pulses was followed amperometrically. Rates were different for the protonated form and the free bases. Whereas the formation of the monobromoderivatives depended only on the bromine concentration for the further bromination, the tribromide ion Br_3^- also was found to be active. The rate constants for the first bromination in the protonated form were especially high ($\sim 5 \times 10^8\ M^{-1}\ sec^{-1}$). With the same technique the kinetics of the bromination of phenetole has been investigated by Kozak & Fernando (146). In this study, k was found to be $(4.8 \pm 0.2) \times 10^4\ M^{-1}\ sec^{-1}$ and $(1.1 \pm 0.2) \times 10^4\ M^{-1}\ sec^{-1}$ at 25° and 0° C, respectively. A linear correlation of rate constants of bromination for various substituted N,N-dimethylaniline and the corresponding Hammett σ values has been proposed by Dubois & Uzan (147). The rate constants were obtained with the electrochemical low-concentration-technique and ranged between 3×10^7 to $10^{11}\ M^{-1}\ sec^{-1}$.

Rate constants for the polymerization reactions

$$2B(OH)_3 + B(OH)_4^- \underset{k_{32}}{\overset{k_{23}}{\rightleftharpoons}} B_3O_3(OH)_4^- + 3H_2O$$

and

$$2B(OH)_3 + B_3O_3(OH)_4^- \underset{k_{43}}{\overset{k_{34}}{\rightleftharpoons}} B_5O_6(OH)_4^-$$

have been obtained at 25° C with the temperature jump technique by Anderson, Eyring & Whittaker (148). In this investigation k_{23} turns out to be about $3 \times 10^3\ M^{-2}\ sec^{-1}$ (at pH >7), and k_{34} is 4.8×10^2 and $2.0 \times 10^2\ M^{-2}\ sec^{-1}$ (at pH 6 and 7.4). At lower pH, k_{23} increases, suggesting acid catalysis. The rates found were overall rates. For the separation into elementary steps, sufficient experimental data are not available. Ultrasonic absorption measurements in liquid K-Na alloys by Jarzynski & Litovitz (149) are explained with the relaxation of the reaction $2Na + K \rightleftharpoons Na_2K$ near the melting point of the alloy. The frequency range covered was 65 to 135 Mc sec^{-1}. By a combination of ESR and polarographic data, Lynn, Rinker & Corcoran (150) obtained the rate constant 40 sec^{-1} for the monomerization of dithionite ion in aqueous solution. From NMR line widths, Nakagawa &

Tori (151) estimate the average lifetime of a benzene molecule in a sodium dodecyl micelle to be $\leq 10^{-4}$ sec. Yamamoto (152) reports a broadening of NMR proton lines of triethylaluminium at low temperatures. The reaction is a dimerization with two CH_2 groups bridging two Al atoms. At $-61°$ C, the average lifetime is of the order 10^{-2} sec. Because of second-order splitting, a precise evaluation could not be performed. The rate of intermolecular exchange of methyl groups in trimethylthallium in different solvents has been measured by Maher & Evans (153) with NMR. The bimolecular rate constant in benzene, 1.0×10^4 M^{-1} sec^{-1}, is higher by a factor of approximately 30 in mixed vinyl-methyl thallium compounds. Unexpectedly, the exchange of the vinyl groups is still faster. Oki, Iwamura & Hayakawa (154) measured the rate of inversion of some O,O′-bridged biphenyls with NMR line broadening. Evidence is given for rapid interconversion of boat and chair conformation of the monopositive and the mononegative ion of 1,2,3,6,7,8-hexahydropyrene from ESR line broadening by de Boer & Praat (155). Loewenstein et al. (156) found from NMR line broadening that the activation energy E_A for the hindered internal rotation in some thioamides (25 to 35 kcal $mole^{-1}$) is much greater than for the corresponding rotation in amides (6 to 18 kcal $mole^{-1}$). A dependence of E_A on the solvent is probably caused by an association of amide and solvent. Brown & Hollis (157) observe in NMR spectra a rapid exchange of the NO group in alkyl nitrites with the OH proton of alcohols. They doubt that the *cis-trans*-interconversion occurs with hindered internal rotation as previously assumed and propose the following dissociation-recombination mechanism: $CH_3–O–NO \rightleftarrows CH_3–O^- + NO^+$.

Increasingly, fast reaction techniques are being applied to the investigation of elementary steps in inorganic, organic, and biochemical reactions, and complicated mechanisms are being elucidated. In many cases today, theoretical predictions are possible. The arsenal of experimental techniques has been so enlarged that now for most reactions at least one well-suited method is available.

LITERATURE CITED

1. Hammes, G. G., *Ann. Rev. Phys. Chem.*, **15,** 13 (1964)
2. Caldin, E. F., *Fast Reactions in Solution* (Blackwell, Oxford, 1964)
3. Noyes, R. M., *Progr. Reaction Kinetics*, **1,** 129 (1963)
4. Weller, A., *Progr. Reaction Kinetics*, **1,** 187 (1963)
5. Eigen, M., Kruse, W., Maass, G., and de Maeyer, L., *Progr. Reaction Kinetics*, **2,** 287 (1964)
6. Eigen, M., *Pure Appl. Chem.*, **6,** 97 (1963)
7. Eigen, M., *Naturwissenschaften*, **50,** 426 (1963)
8. Eigen, M., *Ber. Bunsenges. Physik. Chem.*, **67,** 753 (1963)
9. Eigen, M., and Kustin, K., *ICSU Rev. World Sci.*, **5,** 97 (1963)
10. Thompson, R., and Gordon, G., *J. Sci. Instr.*, **41,** 480 (1964)
11. Kresze, G., Winkler, J., and Meiners, J., *Z. Anal. Chem.*, **200,** 351 (1964)
12. Swinehart, J. H., and Castellan, G. W., *Inorg. Chem.* **3,** 278 (1964)
13. Schwarzenbach, G., and Meier, J., *J. Inorg. Nucl. Chem.*, **8,** 302 (1958)
14. Bergmann, K., Eigen, M., and de Maeyer, L., *Ber. Bunsenges. Physik. Chem.*, **67,** 819 (1963)
15. Bergmann, K., *Ber. Bunsenges. Physik. Chem.*, **67,** 826 (1963)
16. Atkinson, G., Kor, S. K., and Jones,

R. L., *Rev. Sci. Instr.*, **35**, 1271 (1964)
17. Rüppel, H., Bültemann, V., and Witt, H. T., *Ber. Bunsenges. Physik. Chem.*, **68**, 340 (1964)
18. Bauer, E., and Berg, H., *Naturwissenschaften*, **51**, 460 (1964)
19. Alberty, R. A., Yagil, G., Diven, W. F., and Takahashi, M., *Acta Chem. Scand.*, **17**, Suppl. 1, 34 (1963)
20. Forsén, S., and Hoffman, R. A., *J. Chem. Phys.*, **39**, 2892 (1963)
21. Forsén, S., and Hoffman, R. A., *ibid.*, **40**, 1189 (1964)
22. Forsén, S., and Hoffman, R. A., *Acta Chem. Scand.*, **17**, 1787 (1963)
23. Allerhand, A., and Gutowsky, H. S., *J. Chem. Phys.*, **41**, 2115 (1964)
24. Powles, J. G., and Strange, J. H., *Mol. Phys.*, **8**, 169 (1964)
25. Alexander, S., *J. Chem. Phys.*, **40**, 2741 (1964)
26. Alexander, S., *ibid.*, **38**, 1787 (1963)
27. Fessenden, R. W., *J. Phys. Chem.*, **68**, 1508 (1964)
28. Barker, G. C., and Nürnberg, H. W., *Naturwissenschaften*, **51**, 191 (1964)
29. Nürnberg, H. W., and Barker, G. C., *Naturwissenschaften*, **51**, 191–92 (1964)
30. Nürnberg, H. W., and Dürbeck, H. W., *Z. Anal. Chem.*, **205**, 217 (1964)
31. Barker, G. C., Nürnberg, H. W., and Bolzan, J. A., *Ber. Kernforsch. Jülich*, **137**, CA (1963)
32. Jaenicke, W., and Hoffmann, H., *Z. Elektrochem.*, **66**, 803, 814 (1962)
33. Strehlow, H., *Z. Elektrochem.*, **64**, 45 (1960)
34. Hoffmann, H., and Jaenicke, W., *Z. Elektrochem.*, **66**, 7 (1962)
35. Cover, R. E., and Meites, L., *J. Phys. Chem.*, **67**, 1528 (1963)
36. Reinert, K.E., *Ber. Bunsenges. Physik. Chem.*, **66**, 379 (1962)
37. Becker, M., and Köhler, G., *Ber. Bunsenges. Physik. Chem.*, **67**, 690 (1963)
38. Reinert, K. E., *Z. Elektrochem.*, **66**, 379 (1962)
39. Brdička, R., Hanuš, V., and Koutecký, J., *Advances in Polarography* (Zuman and Kolthoff, Eds., Interscience, New York, 1962)
40. Savéant, J. M., and Vianello, E., *Electrochim. Acta*, **8**, 905 (1963)
41. Nicholson, R. S., and Shain, I., *Anal. Chem.*, **36**, 706 (1964)
42. Hung, H. L., Delmastro, J. R., and Smith, D. E., *J. Electroanal. Chem.*, **7**, 1 (1964)
43. Meites, L., *J. Electroanal. Chem.*, **5**, 270 (1963)
44. Groden, C. M., Aylward, G. H., and Hayes, J. W., *Australian J. Chem.*, **17**, 16 (1964)
45. Mecke, J., *Monatsber. Deut. Akad. Wiss. Berlin*, **4**, 296 (1962)
46. Koutecký, J., *Collection Czech. Chem. Commun.*, **18**, 311 (1953)
47. Henke, K.-H., and Hans, W., *Z. Elektrochem.*, **59**, 6761 (1955)
48. Ashley, J. W., Jr., and Reilley, C. N., *J. Electroanal. Chem.*, **7**, 253 (1964)
49. Pinkerton, R. C., *J. Electrochem. Soc.*, **111**, 596 (1964)
50. Buck, R. P., *J. Electroanal. Chem.*, **5**, 295 (1963)
51. Levich, V. G., Grabowski, Z. Ch., and Filinovski, V. Yu., *Dokl. Akad. Nauk SSSR*, **151**, 1379 (1963)
52. Rangarajan, S. K., *Can. J. Chem.*, **41**, 1469 (1963)
53. Eigen, M., and Kruse, W., *Z. Naturforsch.*, **18b**, 857 (1963)
54. Buck, R. P., *Anal. Chem.*, **35**, 1853 (1963)
55. Mairanovski, S. G., *Dokl. Akad. Nauk SSSR*, **149**, 1373 (1963)
56. Nishiyama, M., Maruyama, M., and Hamaguchi, H., *Bull. Chem. Soc. Japan*, **37**, 616 (1964)
57. Nürnberg, H. W., van Riesenbeck, G., and von Stackelberg, M., *Collection Czech. Chem. Commun.*, **26**, 126 (1961)
58. Garrett, M., Tao, T., and Jolly, W. L., *J. Phys. Chem.*, **68**, 824 (1964)
59. MacLean, C., and Mackor, E. L., *Discussions Faraday Soc.*, **34**, 165 (1962)
60. Grunwald, E., and Price, E., *J. Am. Chem. Soc.*, **86**, 2965 (1964)
61. Grunwald, E., and Price, E., *ibid.*, 2970
62. Krakower, E., and Reeves, L. W., *Trans. Faraday Soc.*, **59**, 2528 (1963)
63. Krakower, E., and Reeves, L. W., *Spectrochim. Acta*, **20**, 71 (1964)
64. Rutenberg, A. C., Palko, A. A., and Drury, J. S., *J. Phys. Chem.* **68**, 976 (1964)
65. Becker, M., *Ber. Bunsenges. Physik. Chem.*, **68**, 669 (1964)
66. Zaitsev, P. M., Turyan, Ya. I., and Zaitseva, Z. V., *Kinetika i Kataliz*, **4**, 564 (1963)
67. Carrington, A., and Smith, I. C. P., *Mol. Phys.*, **8**, 101 (1964)
68. Leonhardt, H., and Weller, A., *Ber.*

Bunsenges. Physik. Chem., **67**, 791 (1963)
69. Urban, W., and Weller, A., *Ber. Bunsenges. Physik. Chem.*, **67**, 787 (1963)
70. Mataga, N., and Kaifu, Y., *Mol. Phys.*, **7**, 137 (1963)
71. Weiss, J. J., *J. Chem. Phys.*, **41**, 1120 (1964)
72. Allen, P. E., and Patrick, C. R., *Makromol. Chem.*, **72**, 106 (1964)
73. Marcus, R. A., *Discussions Faraday Soc.*, **29**, 21 (1960)
74. Marcus, R. A., *J. Chem. Phys.*, **24**, 966 (1956)
75. Marcus, R. A., *ibid.*, **26**, 872 (1957)
76. Marcus, R. A., *J. Phys. Chem.*, **67**, 853 (1963)
77. Marcus, R. A., *Ann. Rev. Phys. Chem.*, **15**, 155 (1964)
78. Endicott, J. F., and Taube, H., *J. Am. Chem. Soc.*, **86**, 1686 (1964)
79. Campion, R. J., Purdie, N., and Sutin, N., *Inorg. Chem.*, **3**, 1091 (1964)
80. Diebler, H., and Sutin, N., *J. Phys. Chem.*, **68**, 174 (1964)
81. Hurwitz, P., and Kustin, K., *Inorg. Chem.*, **3**, 823 (1964)
82. Rasmussen, P. G., and Brubaker, C. H., Jr., *Inorg. Chem.*, **3**, 977 (1964)
83. Bächmann, K., and Lieser, K. H., *Ber. Bunsenges. Physik. Chem.*, **67**, 802 (1963)
84. Dulz, G., and Sutin, N., *J. Am. Chem. Soc.*, **86**, 289 (1964)
85. Nancollas, G. H., and Sutin, N., *J. Am. Chem. Soc.*, **86**, 1453 (1964)
86. Daugherty, N. A., and Newton, T. W., *J. Phys. Chem.*, **68**, 612 (1964)
87. Newton, T. W., and Baker, F. B., *Inorg. Chem.*, **3**, 569 (1964)
88. Newton, T. W., and Baker, F. B., *J. Phys. Chem.*, **68**, 228 (1964)
89. Olver, J. W., and Ross, J. W., *J. Phys. Chem.*, **66**, 1699 (1962)
90. Gould, E. S., and Taube, H., *J. Am. Chem. Soc.*, **86**, 1318 (1964)
91. Sykes, A. G., *Trans. Faraday Soc.*, **59**, 1334 (1963)
92. Sheppard, J. C., *J. Phys. Chem.*, **68**, 1190 (1964)
93. Candlin, J. P., Halpern, H., and Trimm, D. L., *J. Am. Chem. Soc.*, **86**, 1019 (1964)
94. Dietrich, M. W., and Wahl, A. C., *J. Chem. Phys.*, **38**, 1591 (1963)
95. Dietrich, M. W., and Wahl, A. C., *J. Chem. Phys.*, **39**, 2376 (1963)
96. Espenson, J. H., *Inorg. Chem.*, **3**, 968 (1964)
97. Hirota, N., and Weissman, S. I., *J. Am. Chem. Soc.*, **86**, 2537 (1964)
98. Jones, M. T., *J. Chem. Phys.*, **38**, 2892 (1963)
99. Reynolds, W. L., *J. Phys. Chem.*, **67**, 2866 (1963)
100. Szwarc, M., *Proc. Roy. Soc. (London), Ser. A*, **279**, 260 (1964)
101. Mazitov, R. K., *Compt. Rend. Acad. Sci. URSS*, **152**, 375 (1963)
102. Luz, Z., and Meiboom, S., *J. Chem. Phys.*, **40**, 1066 (1964)
103. Luz, Z., and Meiboom, S., *ibid.*, 1058
104. Luz, Z., and Meiboom, S., *ibid.*, 2686
105. Luz, Z., *J. Chem. Phys.*, **41**, 1748, 1756 (1964)
106. Swift, T. J., *Inorg. Chem.*, **3**, 526 (1964)
107. Hammes, G. G., and Morrell, M. L., *J. Am. Chem. Soc.*, **86**, 1497 (1964)
108. Hoffmann, H., Stuehr, J., and Yeager, E., *Naval Res. Rev.*, **27** (1964)
109. Strehlow, H., and Becker, M., *Z. Elektrochem.*, **63**, 457 (1959)
110. Eigen, M., *Z. Elektrochem.*, **64**, 115 (1960)
111. Nancollas, G. H., and Sutin, N., *Inorg. Chem.*, **3**, 360 (1964)
112. Eigen, M., Geier, G., and Kruse, W., in *Essays in Coordination Chemistry*, Exptl. Suppl. IX, 164 (Schneider, W., Anderegg, G., and Gut, R., Eds., Birkhäuser, Basel, Stuttgart, 1964)
113. Hush, N. S., and Scarrot, J. W., *J. Electroanal. Chem.*, **7**, 26 (1964)
114. Gerischer, H., *Chem. Ingr.-Tech.*, **36**, 666 (1964)
115. Kodama, M., and Hashimoto, M., *Bull. Chem. Soc. Japan.*, **35**, 1802 (1962)
116. Pearson, R. G., and Lanier, R. D., *J. Am. Chem. Soc.*, **86**, 765 (1964)
117. Moore, P., and Wilkins, R. G., *J. Chem. Soc.*, **1964**, 3454
118. Rorabacher, D. B., and Margerum, D. W., *Inorg. Chem.*, **3**, 382 (1964)
119. Eigen, M., and Eyring, E. M., *Inorg. Chem.*, **2**, 636 (1963)
120. Godfrey, P. D., Heffernan, M. L., and Kerr, D. F., *Australian J. Chem.*, **17**, 701 (1964)
121. Wilkins, R. G., *Inorg. Chem.*, **3**, 520 (1964)
122. Barrett, P. F., and Smith, W. MacF., *Can. J. Chem.*, **42**, 934 (1964)
123. Margerum, D. W., Bystroff, R. I., and Banks, C.V., *J. Am. Chem. Soc.*, **78**, 4211 (1956)
124. Cendales, M., Casabianca, M., Arroyo, A., and Nassiff, S. J., *Intern. J.*

Appl. Radiation Isotopes, **14**, 295 (1963)
125. Valleau, J. P., *Can. J. Chem.*, **42**, 1174 (1964)
126. Valleau, J. P., and Turner, S. J., *Can. J. Chem.*, **42**, 1186 (1964)
127. Wendt, H., *Ber. Bunsenges. Physik. Chem.*, **68**, 29 (1964)
128. Bockris, J. O'M., Inman, D., Reddy, A. K. N., and Srinivasan, S., *J. Electroanal. Chem.*, **5**, 476 (1963)
129. Yguerabide, J., Dillon, M. A., and Burton, M., *J. Chem. Phys.*, **40**, 3040 (1964)
130. Mullin, C. R., Dillon, M. A., and Burton, M., *J. Chem. Phys.*, **40**, 3053 (1964)
131. Gordon, S., Hart, E. J., Matheson, M. S., Rabani, J., and Thomas, J. K., *Discussions Faraday Soc.*, **36**, 193 (1964)
132. Schwarz, H. A., *Radiation Res.*, Suppl. 4, 89 (1964)
133. Gol'-danskii, V. I., Firsov, V. G., and Shantarovich, V. P., *Dokl. Akad. Nauk SSSR*, **155**, 636 (1964)
134. Jackson, R. A., *J. Chem. Soc.*, **1963**, 5281
135. Jackson, R. A., *ibid.*, 5284
136. Kastening, B., *Electrochim. Acta*, **9**, 241 (1964)
137. de Boer, E., and Mackor, E. L., *J. Am. Chem. Soc.*, **86**, 1513 (1964)
138. Freed, J. H., and Fraenkel, G. K., *J. Am. Chem. Soc.*, **86**, 3477 (1964)
139. Berg, H., *Z. Chem.*, **2**, 237 (1962)
140. Martin, T. W., Rummel, R. E., and Gross, R. C., *J. Am. Chem. Soc.*, **86**, 2595 (1964)
141. Kato, S., Morita, M., and Koizumi, M., *Bull. Chem. Soc., Japan*, **37**, 117 (1964)
142. Matsumoto, S., *Bull. Chem. Soc. Japan*, **37**, 491, 499 (1964)
143. Bell, R. P., and Marshall, D. R., *J. Chem. Soc.*, **1964**, 2195
144. Bell, R. P., and Davis, G. C., *J. Chem. Soc.*, **1964**, 902
45. Kozak, G. S., Fernando, Q., and Freiser, H., *Anal. Chem.*, **36**, 298 (1964)
146. Kozak, G. S., and Fernando, Q., *J. Phys. Chem.*, **67**, 811 (1963)
147. Dubois, J. E., and Uzan, R., *Tetrahedron Letters*, **2397** (1964)
148. Anderson, J. L., Eyring, E. M., and Whittaker, M. P., *J. Phys. Chem.*, **68**, 1128 (1964)
149. Jarzynski, J., and Litovitz, T. A., *J. Chem. Phys.*, **41**, 1290 (1964)
150. Lynn, S., Rinker, R. G., and Corcoran, W. H., *J. Phys. Chem.*, **68**, 2363 (1964)
151. Nakagawa, T., and Tori, K., *Kolloid-Z.*, **194**, 143 (1964)
152. Yamamoto, O., *Bull. Chem. Soc. Japan*, **37**, 1125 (1964)
153. Maher, J. P., and Evans, D. F., *J. Chem. Soc.*, **1963**, 5534
154. Oki, M., Iwamura, H., and Hayakawa, N., *Bull. Chem. Soc. Japan*, **36**, 1542 (1963)
155. de Boer, E., and Praat, A. P., *Mol. Phys.*, **8**, 291 (1964)
156. Loewenstein, A., Melera, A., Rigny, P., and Walter, W., *J. Phys. Chem.*, **68**, 1597 (1964)
157. Brown, H. W., and Hollis, D. P., *J. Mol. Spectr.*, **13**, 305 (1964)

THE SOLUTION CHEMISTRY OF SIMPLE FUSED SALTS[1]

By O. J. Kleppa

Department of Chemistry and Institute for the Study of Metals, The University of Chicago, Chicago, Illinois

INTRODUCTION

Several years have passed since the *Annual Review of Physical Chemistry* gave special recognition to the field of fused-salt chemistry [Blomgren & Van Artsdalen (1)]. The intervening period has seen very significant developments in this field. Some of these developments have been the subject of review articles devoted to selected topics [Flood (2); Gruen & McBeth (3); Bloom (4); Mackenzie (5); Wait & Janz (6); Corbett & Duke (7)]; other developments have been recorded in the proceedings of conferences such as the 1960 conference organized by the New York Academy (8) and the 1961 Discussion of the Faraday Society (9). The proceedings of a conference which emphasized the recent work of Roumanian (Murgulescu et al.) and Russian investigators (Delimarskii, Beleaev, Markov, et al.) was published in Bucuresti in 1963 (10). Unfortunately, this volume is printed in the Roumanian language. Thus it is somewhat inaccessible to most investigators.

Another notable publishing event is the appearance in 1961 of an English translation of the book *Electrochemistry of Fused Salts* by Delimarskii & Markov (11). This work gives detailed accounts of many aspects of the electrochemistry of fused salts with particular reference to the work of Russian investigators up to about 1958.

Of special significance has been the publication during 1964 of two monographs which give general coverage to most aspects of fused-salt chemistry (12, 13). The editors of the monographs have had the collaboration of many of the leaders in the field, and in many respects the two books are complementary rather than overlapping.

These various books, review articles, and conferences indicate that the field of fused-salt chemistry has come of age. They also demonstrate that this is no longer a single field, but rather a series of loosely interrelated subfields. As a consequence, no single investigator can speak with authority in more than a modest fraction of the whole field. In view of this, we shall entirely refrain from any attempt at encyclopedic coverage.

Instead we have selected for attention the subfield of the equilibrium solution chemistry of simple fused salts. Even in this restricted area we shall be selective rather than encyclopedic in our references to the literature. For more extensive discussions, particularly of the earlier literature and of more

[1] The survey of literature pertaining to this review was concluded in December 1964.

complex systems, the reader is referred to the recent monographs, and especially to the chapters by Blander (14) and by Førland (15).

THE PROCESS OF FUSION

It has been recognized for some time that the liquid ionic state, although still rather inaccessible by rigorous theoretical methods, shows important similarities to the corresponding ionic solids. For this reason we shall review briefly the changes that accompany the process of fusion.

From a thermodynamic point of view, the process of fusion in most alkali halides is associated with an entropy change of the order of 6 cal/degree-mole. This is roughly twice the value observed in the rare gases, and is consistent with the fact that these ionic salts contain two distinctly different particles. During the past few years, the heats of fusion of many simple halides have been measured or remeasured, notably by Dworkin & Bredig (16–19), while a redetermination for all of the monovalent nitrates has been reported by Kleppa & McCarty (20). Other measurements of the enthalpies of fusion, with particular reference to possible "premelting" phenomena, have been made by Janz and co-workers (21–25). When compared to the total lattice energy of the ionic salt, the heat of fusion is a very modest fraction, usually of the order of 1 to 3 per cent.

On the other hand, the volume changes on fusion, as known, e.g., from the works of Schinke & Sauerwald (26, 27), show a large range of variation. For the simple alkali halides, they are quite large, characteristically of the order of 10 to 30 per cent of the molar volume. For the alkaline earth halides, on the other hand, they tend to be significantly smaller; in fact the volume change is about 1 per cent for $CaCl_2$.

For the monovalent nitrates as a group, the variation in the volume change on fusion is particularly great—from slightly negative in the case of rubidium nitrate to about 21 per cent for lithium nitrate (27). For silver nitrate and potassium nitrate, the dependence of the volume change on pressure has been studied by Owens and co-workers (28, 29). For KNO_3, they find that the volume change becomes zero at about six kilobars.

In view of the Coulombic repulsion between ions of the same sign and the attraction between ions of opposite sign, it can be predicted that at moderate temperatures each ion in a fused salt will surround itself with nearest-neighbor ions of the opposite sign. Thus the salt is expected to exhibit a "short-range order" related to that of the ionic solids. This intuitive picture has been confirmed and refined in recent years through careful X-ray and neutron diffraction work [e.g., Levy & Danford (30)]. For simple alkali halides, one finds that, in spite of the large volume expansion on fusion, the average distance between oppositely charged nearest neighbors is slightly reduced in the liquid state, compared to the same distance in the solid. Similarly, the average distance between equally charged next-nearest-neighbor ions is slightly increased.

It is generally recognized that in diffraction work on liquids, reliable coordination numbers are more difficult to obtain than average distances. This is inherent in the evaluation of both quantities from the experimentally determined radial distribution function. Nevertheless, it appears to be well established that for the liquid alkali halides, the number of nearest neighbors in the melt at temperatures not too far above the melting point is about 4 to 5. This should be compared with the coordination number characteristic of the solid sodium chloride structure. The number of next nearest neighbors in the melt appears to be about 10, as compared to 12 in solid sodium chloride.

Further insights into changes associated with the process of fusion have been developed through the application of a variety of different spectroscopic methods, such as ultraviolet and visible spectroscopy, infrared and Raman work, and, most recently, from nuclear magnetic resonance and from magneto-optical studies.

The present status of ultraviolet and visible spectroscopy in fused salts has been reviewed by Smith (31) and by Gruen (3, 32). The major research effort in this area has been aimed at the spectra of transition metal ion solutes in various fused-salt media, while somewhat less effort seems to have been devoted to the pure salts. The pure alkali halides exhibit strong absorption bands in the ultraviolet, which extend into the vacuum ultraviolet region and which are very difficult to measure. These bands are attributed to a transfer of charge from the halide ion to the metal ion. The absorption edges of the charge-transfer band in pure alkali halides characteristically show "red shifts" on fusion of the order of 0.5 eV compared to the room temperature values.

For salts of the non-octet ions (silver, thallium, lead, etc.), the tail of the charge-transfer band may extend into the visible region. In these cases the process of fusion characteristically yields a liquid which is more deeply colored than the corresponding solid.

The fields of infrared and Raman spectroscopy in fused salts have been reviewed by Wait & Janz (6), and by James (33). On the whole, investigations in this area have been aimed at two problems:

(*a*) How does the characteristic vibrational spectrum of stable and well-characterized complex species (NO_3^-, CNS^-) depend on the medium in which it is located (melt, solid, aqueous solution)? An example is the study by Baddiel & Janz (34) of the thiocyanates. As expected, it is found that the changes tend to be quite small; i.e., the nitrate ion in the melt differs little from the nitrate ion in the parent solid or in water.

(*b*) Are there new complex species in the melt formed by association among the ionic species present? An example is the Raman work by Balasubrahmanyan & Nanis (35) on liquid $PbCl_2$–KCl.

The significance of spectral information relating to the second problem has been discussed by Wilmshurst (36). Wilmshurst notes that since the relaxation time of the "lattice" in a typical simple fused salt is long ($\sim 10^{-10}$ sec) compared to the time of a molecular vibration, vibrational spectroscopy is of doubtful value in proving the existence of complex associated species. Similarly, ultraviolet, visible, and electron spin resonance spectroscopy will not be capable of distinguishing between an ion "sitting" on a "lattice" site and a complex kinetic entity. Among the common spectroscopic techniques, nuclear magnetic resonance alone involves periods of observation which are sufficiently long to offer the possibility of positive identification of such entities. As far as is known, this technique has not been applied for this purpose in fused-salts media. However, it has proved a valuable tool in the study of fusion, and of chemical bonding in predominantly ionic melts.

The first successful application of nuclear magnetic resonance to fused salts was by Rowland & Bromberg (37). During the past year further developments have occurred. For ^{205}Tl in a variety of different thallium salts, both in the solid and in the liquid state, Hafner & Nachtrieb (38) have studied the paramagnetic chemical shifts, $\Delta\omega/\omega$, relative to the hypothetical resonance line of the same ion in free space. These shifts primarily represent a measure of the magnetic interaction of the applied field with excited paramagnetic states which are mixed into the ground state of the free ion through its interaction with the neighboring ions. Quantitatively, Hafner & Nachtrieb interpret the chemical shifts in terms of Ramsey's theory (39), which leads to the following expression:

$$\Delta\omega/\omega = (16/3)\mu_B{}^2\langle 1/r^3\rangle(B/\Delta E) \qquad 1.$$

Here μ_B is the Bohr magneton, $\langle 1/r^3\rangle$ is the average over the radial part of the wave function of the excited paramagnetic state, B is a measure of the "covalency" through the sum of the overlaps between the wave function of the thallium ion with the neighboring anions, while ΔE is the energy difference between the ionic ground state and some excited state.

Both for crystalline and molten thallium salts, Hafner & Nachtrieb find that the chemical shifts increase significantly in the sequence $NO_3^- \langle Cl^- \langle Br^- \langle I^-$, i.e., in the sequence of increasing covalency. They also find an increase in the chemical shift with increasing temperature, and in almost all salts of Tl^+, a substantial paramagnetic shift at the melting point. As a consequence, Hafner & Nachtrieb conclude that fusion in the thallium salts is associated with an increase in the covalent contribution to the chemical bonding between nearest neighbors. Similar considerations may be expected to apply in salts of other non-octet ions. However, no detailed information is as yet available for simple ionic salts such as the alkali and alkaline earth halides.

Zarzycki & Naudin (40) have reported a series of measurements of the molar refractivity and the molar magnetic rotation (Faraday effect) in ionic salts in the solid and the liquid state. From these data they derive information on the red shifts of the ultraviolet absorption bands and on the details of the nearest-neighbor coordination in the liquid state. Related structural information may be inferred from the nuclear magnetic resonance chemical shifts [Hafner & Nachtrieb (41)].

THEORIES OF FUSED SALTS

Through the multipole expansion of quantum mechanics, we are led to recognize in simple ionic liquids several contributions to the pair potential acting between two ions, i and j, of charge z_i and z_j and of polarizability α_i and α_j, as a function of the interionic distance r [e.g., Stillinger (42)]

$$\phi_{ij}(r) = A_{ij}e^{-B_{ij}r} + z_i z_j e^2/r - \left[(z_i e)^2\alpha_j + (z_j e)^2\alpha_i\right]/2r^4 - C_{ij}/r^6 + \text{higher order terms} \qquad 2.$$

The terms on the right-hand side represent:

(*a*) Short-range repulsive potential. The constants A and B may be expected to be the same as for the solid state.

(*b*) Charge-charge interaction (point-charge) term.

(*c*) Charge-dipole term.

(*d*) Dipole-dipole term. The coefficient C_{ij} may be approximated by the London formula for the dispersion energy

$$C_{ij} = (3/2)\alpha_i\alpha_j I_i I_j/(I_i + I_j)$$

where I_i is the ionization potential of the ion.

(*e*) Higher-order terms, such as dipole-quadrupole terms.

In order to obtain the total cohesive energy of the ionic melt, a summation must be carried out over all the pair interactions in the system. The summation of terms (*a*), (*b*), and (*d*) is expected to be essentially additive. However, the charge-dipole term (*c*) involves the field created by the charges and induced dipoles surrounding a given ion. Thus this quantity has the character of a many-body term. This introduces an element of great complexity in any rigorous theoretical treatment of even a very simple fused salt. In approximate treatments, this difficulty is usually side stepped through the introduction of a dielectric constant, κ.

Additional complexity arises if the salt in question is not a simple ionic medium. This is presumably the case for many salts which have proved particularly attractive from an experimental point of view, such as the salts of the highly polarizable ions silver, thallium, lead, and cadmium. Evidence for this is provided, e.g., by the lattice energies of these salts, which cannot be fully accounted for in terms of ionic and dispersion interactions alone [Ladd & Lee (43)], and by the recent nuclear magnetic resonance work of Hafner & Nachtrieb (38).

So far, the most ambitious theoretical attack on the problem of the equilibrium properties of fused salts is the work of Stillinger et al. (44) who used the rigorous but formal method of distribution functions. A general review of the present status of fused-salt equilibrium theory is given by Stillinger (42).

The rigorous theories are not yet developed to a point where they provide a basis for the prediction and correlation of equilibrium data. For this purpose and for related purposes in the area of transport phenomena, a variety of more approximate theories have been proposed. A survey of the success or lack of success of the approximate theories in predicting and correlating data has been made by Bloom & Bockris (45).

In the area of fused-salt solution chemistry, the most successful theoretical approach has been based on the methods of corresponding-states theory [Pitzer (46)]. For fused salts this work goes back to the work of Reiss & Mayer (47) on the surface tension of fused salts which, in turn, is based on the earlier theory of Reiss, Frisch & Lebowitz (48) on the molecular theory of hard-sphere fluids. A more general theory of corresponding states for fused salts was developed by Reiss, Mayer & Katz (49).

Reiss, Mayer & Katz assume that the equilibrium properties of a fused salt can be related to those of a model ionic melt. This melt is defined through the following assumptions:

(*a*) The role of the long-range Coulomb forces is to generate a locally ordered structure, in which, on the average, a negative ion is surrounded by positive ions and vice versa.

(*b*) Because of the local structure and the Coulomb repulsion between ions of the same sign, short-range interactions between ions of like sign are neglected.

(*c*) The pair potential between ions i and j is taken to be of the form

$$\phi_{ij} = \infty; \qquad r \leq d \tag{3a.}$$

$$\phi_{ij} = (z_i e)(z_j e)/\kappa r; \qquad r > d \tag{3b.}$$

where κ is a dielectric constant. It will be noted that with this choice of potentials each salt is fully characterized by the specification of a single parameter of length, d. A convenient choice for this parameter is the sum of the ionic radii $r_+ + r_-$.

Reiss, Mayer & Katz find that with this choice of interionic potential the following expressions are obtained for the reduced pressure π, the reduced temperature τ, and reduced volume θ, for a charge-symmetrical salt ($z_i = z_j$):

$$\pi = \kappa d^4 p / z^2$$

$$\tau = d\kappa T / z^2$$

$$\theta = V / d^3$$

One can write a reduced equation of state for any salt for which the stated assumptions are valid.

On the basis of the provisional assumption of $\kappa = 1$, Reiss et al. find their theory to be confirmed in several ways. For example, they show that the reduced melting points for the simple alkali halides, $\tau_m = dT_m/z^2$, all fall near the value 3×10^{-5} deg cm. Among these salts the only significant deviations found are for the lithium halides, which have values near 2.2×10^{-5} deg cm. This is interpreted to reflect that the size relations betweeen cations and anions in these cases are such that assumption (*b*) above may be violated. Other confirmations of the theory are found for the vapor pressures and surface tensions, thermal expansion, etc. [McQuarrie (50); Mayer (51)]. A recent application of the rigid-sphere theory of Reiss and co-workers for the calculation of various equilibrium properties of pure fused salts is reported by Yosim & Owens (52). As expected, the theory of Reiss et al. fails to account for the behavior of salts of the highly polarizable ions silver, thallium, etc., for which polarization, dispersion, and small covalent contributions to interionic bonding must be included.

MIXTURES OF FUSED SALTS

As a consequence of the electrostatic forces between the ions there will, as pointed out above, be no mutual randomization of cations and anions. This led Temkin (53) to postulate that in an "ideal fused-salt mixture" the different types of anions will be randomly distributed among the anions, while similarly the different types of cations will be randomly distributed among the cations.

In a mixture containing n_A moles of cation A, n_B moles of cation B, n_X moles of anion X, and n_Y moles of anion Y, the integral entropy of mixing will, according to Temkin, be

$$\Delta S = -R(n_A \ln N_A + n_B \ln N_B + n_X \ln N_X + n_Y \ln N_Y) \qquad 4.$$

In this expression, N represents the "ionic fractions," defined through

$$N_A = n_A/(n_A + n_B), \text{ etc.}$$

For the ideal fused-salt mixture, we also have $\Delta H = \Delta U = \Delta V = 0$, while, of course

$$\Delta G = -T\Delta S$$

In real mixtures, departures from these relations generally will be found.

From a phenomenological point of view, simple mixtures of fused salts may be classified according to the following scheme:

(*a*) Systems with symmetrical charge structure containing a common ion. This ion may be either the anion, as in (Na–K)Cl or the cation as in Na(Cl–Br).

(*b*) Systems with unsymmetrical charge structure containing a common ion. Again this ion may be the anion as in (Ca–Na)Cl or the cation as in Na(SO_4–Cl).

(*c*) Systems containing two cations and two anions. These systems may be further described as charge symmetrical (Na^+, K^+, Cl^-, Br^-), or charge unsymmetrical (e.g., Na^+, Ca^{++}, Cl^-, Br^-).

This list suggests, in the order of increasing complexity, what is the basic problem in fused-salt solution chemistry: how does the behavior of each class of mixed system depend on the characteristic properties of the components, i.e., on such parameters as the size, charge, and structure of the participating ions?

A variety of different equilibrium and nonequilibrium methods have been applied in order to provide answers to this question. These are all high-temperature analogues of room-temperature methods. For a review, see, e.g., Førland (15).

CHARGE-SYMMETRICAL MIXTURES CONTAINING A COMMON ION

Until recently, the most comprehensive information on charge-symmetrical fused-salt mixtures has been the data of Kleppa et al., who determined the enthalpies of mixing of the binary monovalent nitrates (54–57) by means of direct high-temperature reaction calorimetry. For a few of these systems, data have been reported also on the volume change on mixing as obtained by direct volume increment measurements [Powers, Katz & Kleppa (58)].

In cases where the enthalpy of mixing was studied as a function of composition, the results were represented by empirical expressions of the type

$$\Delta H^M = N_1N_2(a + bN_1 + cN_1N_2) \qquad 5.$$

Here ΔH^M is the molar enthalpy of mixing, N_1 and N_2 are the mole fraction of the two components (N_1 refers to the salt with the smaller cation), while, for a given system, a, b, and c are constants.

In the binary alkali nitrates, the following characteristics of the constants a, b, and c were noted:

The parameter a is always negative and becomes more negative, the larger the difference in the size of the two cations. All systems studied in detail showed negative values of b. This implies that the limiting enthalpy of solution of the nitrate with the larger cation in the nitrate with the smaller cation is always more negative than the opposite process. The constant c in Equation 5 is always negative and increases sharply with increasing difference in size between the two cations. This was attributed by Kleppa & Hersh (55) to departures from random mixing within the cation "sublattice."

It was found that there is an approximate empirical relation between the excess enthalpy of the molten alkali nitrate mixtures and the quantity $\delta = (d_1 - d_2)/(d_1 + d_2)$ where d_1 and d_2 are the interionic distances characteristic of the two pure salts:

$$\Delta H^M = -N_1N_2 140\delta^2 \text{ kcal/mole} \qquad 6.$$

The distances d were obtained from the approximation

$$d_{MeNO_3} = r_{Me^+} + r_{NO_3^-}$$

The behavior of the alkali nitrate—silver nitrate and alkali nitrate—thallium nitrate systems is more complex, since they exhibit both positive and negative enthalpies of mixing. However, it was noted that it is possible to represent the results for these nitrate systems by approximate empirical expressions of the type

$$\Delta H^M = N_1N_2(+U_1\delta - 140\delta^2) \qquad 7.$$

The second term is the same one as for the binary alkali nitrates, while U_1 was different for the thallium and silver systems. This representation was based on a particular choice of the ionic radii for the silver (1.21 Å) and thallium ions (1.59 Å). However, these values probably are too large (58a). This suggests that the relation between ΔH^M and δ for these systems actually should have a form different from equation 7.

For the binary mixtures of sodium nitrate with the other alkali nitrates, Powers et al. (58) found positive excess volumes, which could be represented empirically by the approximate relation

$$\Delta V^M = N_1 N_2 V' \delta^4$$

where $V' = 2.2 \times 10^3$ cm^3/mole. It is significant that these excess volumes are positive, while the corresponding excess enthalpies are all negative. In the language of conformal solution theory [Longuet-Higgins (59)], this indicates that these mixtures cannot be accounted for by any first-order theory, which predicts the same sign for all excess properties.

Theories of charge-symmetrical fused-salt mixtures.—The data on the binary alkali nitrates were rationalized on the basis of very crude models, originally proposed by Førland (60) and by Blander (61). In Førland's model, one considers only a linear assembly of three hard-sphere ions, two cations, and a common anion. Blander extended this model by considering a single solute cation in an infinite linear chain of the solvent salt. Both models, while grossly oversimplying the physical picture, indicate a decrease in potential energy on mixing. This is attributed in large measure to the reduction in the Coulomb repulsion between next-nearest-neighbor cations.

Later, Lumsden (62) assumed that the principal source of the energy of mixing is the interaction of the anion with the polarization field in which it is placed. This field was considered to be zero in the pure molten salt and to have a calculable value in the mixture. While Lumsden's method of calculating the polarization energy may be criticized, his work was significant in drawing attention to the polarization interactions.

A modified, conformal solution theory for the heats of mixing of fused salts was developed by Reiss, Katz & Kleppa (63), based on the model ionic melt of Reiss, Mayer & Katz (49).

Application of the methods of conformal solution theory [Longuet-Higgins (59); Brown (64)] to charge-symmetrical fused-salt mixtures gives the following result. The first-order term cancels. The second-order term gives a functionally simple expression for the enthalpy of mixing

$$\Delta H^M = N_1 N_2 \Omega(T, P)[(d_1 - d_2)/d_1 d_2]^2 = N_1 N_2 \Omega \delta_{12}^2 \qquad 8.$$

Here $\Omega(T, P)$ is a so-far-unevaluated, complicated integral which is a negative function of temperature and pressure, while d_1 and d_2 are the interionic distances which characterize the two salts. This expression is very similar to the empirical relation found for the alkali nitrates (Eq. 6).

The authors emphasize that this expression represents only the second-order term of their theory. The higher-order terms were considered too complicated to evaluate. However, later on, Blander (65) was able to simplify the calculation of the third- and fourth-order terms.

Very recently, an extension of the conformal solution theory has been given by Davis & Rice (66). They assume the same properties of the model ionic melt as the theory of Reiss et al. However, their model includes also short-range dispersion interactions as a perturbation to the pair potential of the original theory. In the theory of Davis & Rice the general form of the pair potential between the ions is written as

$$\phi = (1/d)f(r/d) + \xi V(r) \qquad 9.$$

The first term is a generalized form of the interionic potential [Blander (65)], while ξ represents the coupling parameter for the particular van der Waals type interaction considered. $V(r)$ is defined as $V(r) = -r^{-n}$. Examples of ξ and n pertinent to the ion-ion interactions are:

(*a*) Ion-induced dipole-ion interaction: $n=4$; $\xi = e^2(\alpha_A + \alpha_B)/2$, where α_A and α_B are the polarizabilities of ions A and B, respectively, and e is the ionic charge.
(*b*) Van der Waals-London interaction: $n=6$, $\xi = C_{AB}$ where C_{AB} may be determined by various approximate expressions such as that given by London (67).

According to the theory of Davis & Rice, the thermodynamic properties of the mixture depend both on the parameter of length d, and on the coupling parameter ξ. The latter takes into account those properties of the cations and anions, such as polarizability and charge, which are important for the dispersion interactions.

In the Davis & Rice theory, the following expression is obtained for the enthalpy of mixing

$$\Delta H^M = N_1N_2(U_0 + U_1\delta_{12} + U_2\delta_{12}^2) \qquad 10.$$

The unevaluated coefficients U_0, U_1, and U_2 are complicated functions of ξ. However, they may be described qualitatively in terms of the types of interactions on which they depend, as follows:

(*a*) U_0: this term is positive and determined largely by the change in van der Waals interactions between next-nearest-neighbor cations. Terms of this type were previously calculated for fused-salt mixtures by Blander (68) and by Lumsden (62).
(*b*) U_1: the leading term in the expression for this coefficient depends on the difference between the nearest-neighbor anion-cation dispersion interactions in each of the two components of the mixture.
(*c*) U_2: this coefficient, while primarily represented by the coefficient $\Omega(T,P)$ of the theory of Reiss et al., includes also a term which depends on the sum of the anion-cation dispersion interactions in the two components of the mixutre.

For systems in which the dispersion interactions described by $\xi V(r)$ are negligibly small compared to the Coulomb potential energy, the theory of Davis & Rice reduces to that of Reiss et al. This appears to be the situation in the binary alkali nitrates. However, in systems which contain highly polarizable cations such as Ag^+ and Tl^+, the dispersion interactions are of greater importance. This will be reflected in increased importance of the terms that contain U_0 and U_1 in Equation 10.

Binary alkali halides.—For many reasons, it has proved extremely difficult to study the binary alkali halides by electrochemical methods. Therefore, until recently, the only thermodynamic information on these liquid systems consisted of phase diagram data and scattered cryoscopic measurements. The data were inadequate to provide the basis for a deeper understanding of these important solution systems. However, since 1960, calorimetric data have become available through the works of Aukrust et al. (69) and of Gilbert (70), and in particular through a very recent comprehensive study of the binary alkali chlorides, bromides, and iodides by Hersh & Kleppa (71).

Hersh & Kleppa find that the enthalpies of mixing for all the binary alkali halides, as well as for the previously studied binary alkali nitrates, can be represented to a first approximation by the following semiempirical relation:

$$\Delta H^M = N_1 N_2 (U_0^{++} - 340\delta_{12}^2) \text{ kcal/mole} \qquad 11.$$

In this expression the second term on the right-hand side, which is Coulombic in origin, corresponds to the term $U_2\delta_{12}^2$ in Equation 10. The first term on the right-hand side (U_0^{++}) differs from system to system. It represents a theoretical estimate of the contribution to the mixing enthalpy arising from the dispersion energy between next-nearest-neighbor cations [Lumsden (62); Blander (68)].

For large differences in size between the two cations, the Coulomb term predominates, and the enthalpy of mixing is negative. However, for small values of δ_{12} the term U_0^{++} may predominate, in which case the sign of the enthalpy of mixing may be positive. This behavior is actually observed, e.g., for the rubidium–cesium and potassium–cesium halides.

We give in Figure 1 a plot of $4\Delta H_{0.5}^M - U_0^{++}$ plotted against δ_{12}^2, taken from (71). Although the uncertainties associated with the calculation of U_0^{++} are quite large, it is indicated that all the studied systems, with the possible exception of (Li–K)F [Gilbert (70)], are reasonably well represented by Equation 11.

Hersh & Kleppa found no evidence in their data on the binary alkali halides for a term linear in δ_{12}. This suggests that in these solutions the term $U_1\delta_{12}$ in Equation 10 is relatively small. However, this cannot be considered proved, since it is not known how to make a realistic numerical estimate of the contributions to the mixing enthalpy which arise from nearest-neighbor dispersion interactions.

Mixtures involving silver and thallium halides.—Information on the solution behavior of the liquid silver bromide–alkali bromide systems has been available since the pioneering investigation of electromotive force by Hildebrand & Salstrom (72). Unfortunately the lack of corresponding data for the alkali halide systems has prevented a full interpretation of these early results. This situation is somewhat improved through the publication of

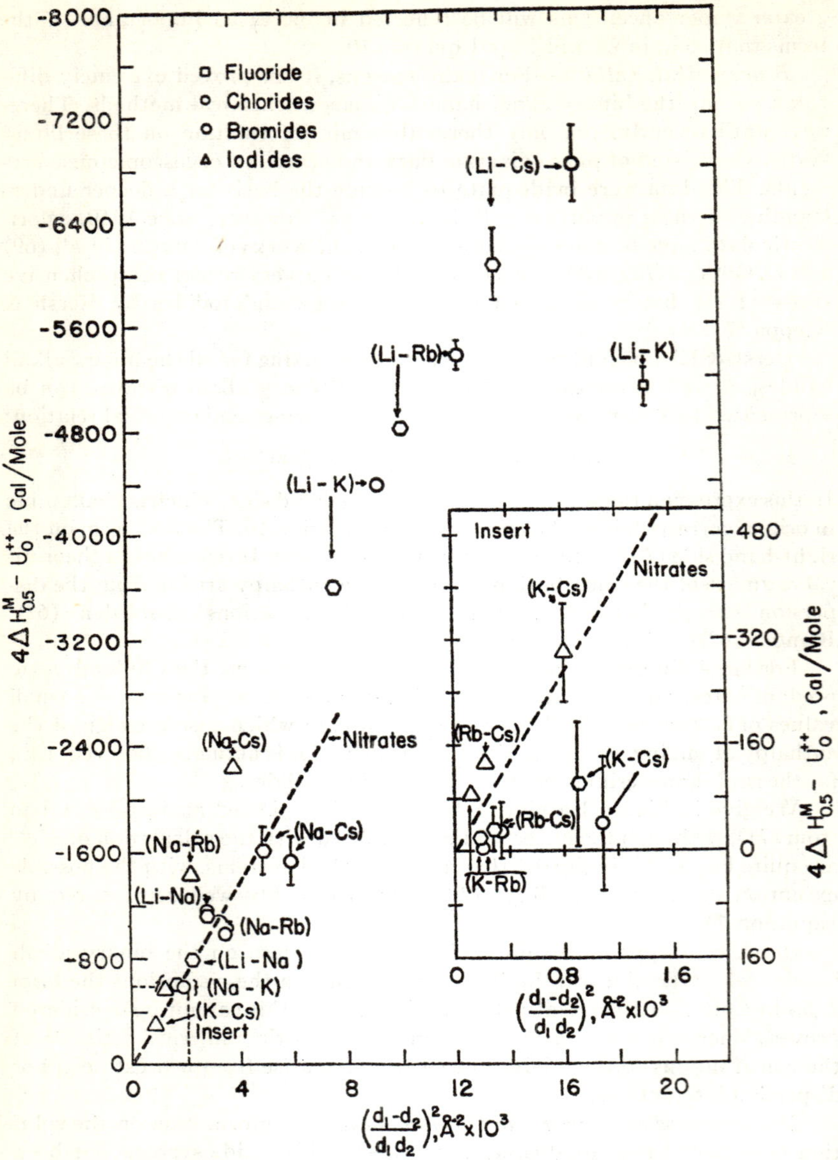

FIG. 1. Dependence of the "corrected" enthalpy of mixing, $4\Delta H_{0.5}{}^{M} - U_0{}^{++}$ for the binary alkali nitrates and halides on the parameter δ_{12} [from Hersh & Kleppa (71)]. The indicated limits of error principally reflect the uncertainty in the calculation of $U_0{}^{++}$.

calorimetric data by Hersh & Kleppa on the liquid silver chloride–alkali chloride systems (71).

It is found that it is not possible to rationalize the enthalpies of mixing for the silver halide–alkali halide systems in terms of the Coulombic and next-nearest-neighbor van der Waals interactions. However, the enthalpies of mixing for the silver chloride–alkali chloride mixtures may be represented semi-empirically through an expression of the type

$$\Delta H^M = N_1 N_2 (U_0^{++} + 40\delta_{12} - 340\delta_{12}^2) \text{ kcal/mole} \qquad 12.$$

Here the first and third terms on the right-hand side are the same as in Equation 11, while the second term, which is linear in δ_{12}, is new. Through this term, the lithium chloride–silver chloride system becomes more endothermic than predicted from Equation 11 alone, since lithium is a smaller ion than silver. Similarly the potassium–silver, rubidium–silver, and cesium–silvers systems are more exothermic than given by Equation 11, the exothermic contributions increasing in the order indicated. A re-analysis of the enthalpy data for the silver nitrate–alkali nitrate systems (56) yielded the expression

$$\Delta H^M = N_1 N_2 (U_0^{++} + 12\delta_{12} - 340\delta_{12}^2) \text{ kcal/mole} \qquad 13.$$

This expression differs from Equation 12 only through the smaller value of the term linear in δ_{12}. Equations 12 and 13 are of the form predicted by the theory of Davis & Rice. However, the relatively small difference in polarizability between the chloride and nitrate ions [Tessman, Kahn & Shockley (73)], and the large difference between the linear terms in Equations 12 and 13, make it difficult to explain the behavior of these systems in terms of ionic and dispersion interactions alone. Therefore the authors attribute the linear terms in Equations 12 and 13 to the partially covalent bonding between nearest-neighbor ions in the silver salts. This covalent contribution is expected to increase in the order $NO_3^- < Cl^- < Br^- < I^-$. If this interpretation is correct, one may predict that the linear term in Equations 12 and 13 should be somewhat larger for the silver bromide–alkali bromides than for the corresponding chlorides. Unpublished enthalpy data for the alkali bromide–silver bromide systems indicate that this may indeed be the case [Hersh, Navrotsky & Kleppa (74)].

Actually, an explanation along these lines of the thermodynamic behavior of the silver bromide–alkali bromide systems had already been proposed by Hildebrand & Salstrom (72) and elaborated further by Thurmond (75). Very recent nuclear magnetic resonance data on liquid thallium halide–alkali halide systems reported by Hafner & Nachtrieb (76) seem to point in the same direction. These authors find that the resonance frequency of the ^{205}Tl nucleus in molten mixtures of the thallous halides with the corresponding alkali halides varies linearly with the mole fraction of added halide. However, the direction of the chemical shift compared to the pure salt depends upon the radius of the alkali cation, being "upfield" for the halides of the smaller ions, lithium and sodium, and "downfield" for the larger ions, potassium, rubidium, and cesium. For a given mole fraction of alkali halide, the chemical shift is a linear function of the alkali cation radius. These results are interpreted in terms of the influence of the alkali ions on the overlap between the thallium ion and the neighboring halide ion. This overlap (i.e., the covalency) is increased, compared to the pure salt, for potassium, rubidium, and cesium, but reduced for sodium and lithium. This behavior is completely analogous to that indicated by the enthalpy data for the liquid silver halide–alkali halide mixtures.

Mixed anion–common cation systems.—It has been known for some time that simple mixed anion–common cation systems tend to show relatively small departures from ideality. For example, on the basis of a pioneering application of the oxygen electrode in a fused salt medium, Flood et al. (77) found that the liquid system Na_2CO_3–Na_2SO_4 is essentially ideal. For many other systems, near-ideal behavior can be inferred from phase diagram information. However, not until the past two years have systematic data become available which permit a somewhat more detailed discussion of this type of solution system. The new information consists in part of calorimetric measurements, in part of exchange equilibrium data.

The calorimetric investigations of Kleppa and co-workers (78, 79, 80) have covered two categories of mixtures. On the one hand, this work has consisted of solution and dilution experiments on higher melting chlorides, bromides, and other salts dissolved in the corresponding nitrates at temperatures below 500° C. On the other, it has involved preliminary work on alkali halide–halide systems at temperatures up to about 800° C. The results may be summarized as follows.

All the investigated alkali nitrate–chloride and nitrate–bromide mixtures have small positive enthalpies of mixing, with the largest values observed for the sodium systems. The positive enthalpies tend to decrease as the size of the alkali ion increases in the sequence $Na^+ > K^+ > Rb^+ > Cs^+$. The data for the liquid alkali chloride–bromide systems show even smaller positive enthalpies of mixing than the corresponding nitrate–halide systems.

The exchange investigations by Toguri and co-workers (81, 82) are based on exchange equilibria between HCl(*g*), HBr(*g*), and Me(Cl,Br) (*l*), where Me is Li, Na, K, Rb, Cs, Ca, Mg. From the dependence of the apparent equilibrium constant for the exchange equilibrium on the composition of the melt, one can derive information on the excess free energy of mixing in the liquid mixture. It is found that Mg(Cl,Br) and Ca(Cl,Br) have small negative deviations from ideality, while all the alkali chloride–bromide mixtures show small positive deviations. In particular, it is found that these deviations increase somewhat in the sequence Li(Cl, Br) < Na(Cl, Br) < K(Cl, Br). The data for the rubidium and cesium systems are less accurate, but show smaller positive deviations than the potassium system.

For these mixed anion–common cation systems which have been investigated in some detail, the two anions differ little in size. Therefore, the negative Coulomb term, which often tends to dominate the solution behavior of the mixed cation—common anion systems, undoubtedly will be quite small. For combinations of anions of more widely different size, say fluoride–chloride, fluoride–bromide, or fluoride–iodide mixtures, this situation may be different. In such cases, negative deviations from ideality may be expected, even in mixed anion systems, as may be inferred, e.g., from the phase diagram data of Haendler, Sennett & Wheeler (83) on the Li(F,Cl) system. Another example in this category is the $KClO_4$–KNO_3 liquid system, for which Kleppa & Meschel (79) report small negative enthalpies of mixing.

Otherwise, small positive deviations from ideality are expected to be a common occurrence in mixed anion systems. In some cases the magnitude of the positive deviations may perhaps be estimated from the van der Waals–London interaction between next-nearest-neighbor anions, as outlined above for the mixed cation case. In addition, it has been suggested [Flood (84)] that, because of the size relations between the (usually) large anions and the (usually) small cations, a significant positive contribution may often arise from the packing of the two anions on the anion "sublattice." Actually, in view of the positive excess volumes found by Powers et al. (58) for the binary alkali nitrates, it is possible that a small packing effect may be present even in mixed cation–common anion systems.

Finally, it should be noted that covalency effects related to those indicated for the silver halide–alkali halide mixtures undoubtedly may occur also in mixed anion systems. This may possibly account for the small negative enthalpies of mixing in the silver nitrate–silver halide systems (78).

Entropies of mixing.—It is often assumed that simple fused-salt mixtures have ideal entropies of mixing. Frequently these assumptions are based on early observations relating to the silver bromide–alkali bromides, for which systems Hildebrand & Salstrom found that the entropies of mixing are ideal within their experimental error. In fact, it is believed that these mixtures were the first known examples of "regular solutions" in the Hildebrand sense.

In view of the availability of reasonably reliable excess free energy and enthalpy data for some fused-salt mixtures, it is now possible, in a few cases, to give more precise estimates of the excess entropies of mixing.

For Li(Cl,Br), Na(Cl,Br), and K(Cl,Br), Toguri, Flood & Førland (82) compare the excess free energies from their own exchange equilibrium measurements with the corresponding enthalpy data reported by Kleppa, Hersh & Toguri (80). In this manner they obtain the values −0.02, −0.06, and −0.11 cal/deg mole for the excess entropies of mixing at the 50–50 composition.

Hersh & Kleppa (71) similarly compare the enthalpy data for (Li,Ag)Cl with corresponding excess free energies derived from the electromotive force studies of Salstrom et al. (85) and Panish et al. (86). This comparison yields the value −0.04 cal/deg mole for the excess entropy of mixing in the equimolar mixture. Ketelaar (87) has measured the excess free energies of mixing in liquid silver nitrate–alkali nitrate mixtures by a galvanic cell without transference which is reversible in the formation of NO_3^- (87a). A comparison of Ketelaar's excess free energies for lithium–silver and sodium–silver nitrate with the available enthalpy data indicates an excess entropy of mixing in these systems of about -0.08 cal/degree mole at the 50–50 composition.

From these comparisons it appears that small negative excess entropies may be a common occurrence both in simple mixed anion, and in mixed cation, systems. However, there clearly is a great need for additional data in this area before a more general discussion can be given.

Charge-Unsymmetrical Mixtures Containing a Common Ion

This is a large, and as yet not well understood, area of fused-salt solution chemistry. Formally, it extends from relatively simple mixtures (e.g., alkali plus alkaline earth salts, halides plus sulfates) to quite complex systems, such as those formed by the borates, silicates, and phosphates. While these two categories may be considered to represent two extremes in a fairly continuous spectrum, the present discussion will be restricted to the simpler types of mixtures.

During the past decade, a great many equilibrium and electromotive force investigations of simple charge-unsymmetrical mixtures have been reported. A number of these have been done by Russian workers, who have made extensive use of the electromotive force formation and decomposition cell methods pioneered by Hildebrand & Salstrom (72). References to the Russian work may be found in the book by Delimarskii & Markov (11) and in the review by Blander (14). Other data on charge-unsymmetrical systems have been obtained through vapor pressure measurements on systems containing volatile species such as $PbCl_2$, $CdCl_2$ [e.g., Bloom (4)] and $FeCl_2$ [Beusman (88)], and also by the study of decomposition pressures, as in the work by Førland on systems containing calcium carbonate (60). Finally, significant contributions in this area have been made by phase diagram analysis, notably by members of the Norwegian school [e.g., Førland (15)] and in the recent cryoscopic studies of solutions in liquid sodium fluoride by Cantor et al. (89–92). As a result of these investigations, certain general statements can be made regarding the behavior of simple charge-unsymmetrical systems containing a common anion.

(*a*) When a salt of a simple divalent ion, e.g., Ca^{++}, is mixed with the corresponding alkali salts, the systems will show increasing negative departures from ideality with increasing size of the alkali metal ion.

(*b*) For a given alkali metal cation, the negative deviation from ideality tends to increase as the size of the divalent ion is reduced.

(*c*) If the alkali ion is replaced by another singly charged ion of comparable size (i.e., Na^+ by Ag^+, or Rb^+ by Tl^+), the negative deviation from ideality is reduced (or the positive deviation increased).

(*d*) If the divalent alkaline earth ion is replaced by a more polarizable ion of comparable size (i.e., Sr^{++} by Pb^{++}), the negative deviation from ideality is increased.

In a phenomenological sense these observations are well established. Qualitatively they may be rationalized in terms of the competition between the two cationic species for the common anion. This competition is aided by a small size, a high charge, and a high polarizability of the cation. For alkali–alkaline earth nitrates, Kleppa noted empirically (93) that the ionic potential of the cation Z/r is a rough quantitative measure of the ability of the cation to bind its anion neighbors.

Very recently, a theory for simple charge-unsymmetrical fused-salt mixtures has been given by Davis (94). In this theory, mixtures of the type $(P^+ - Q^{Z+})X^-$ are considered, and the analysis represents an extension of the conformal solution theory of Reiss et al. (63) to this more complicated case. In an unpublished note (December 1964), Davis (95) has extended this theory to the more general case of mixtures of the type $(P^{+Z_1} - Q^{+Z_2})X^{-Z_A}$. According to the Davis theory, there should, for charge-unsymmetrical mixtures with the same values of τ and θ, be a first-order contribution to the excess thermodynamic properties which depends roughly linearly on the previously defined parameter $\delta_{12} = (d_1 - d_2)/d_1 d_2$. As shown by McCarty, Hersh & Kleppa (96), this is in agreement with available calorimetric data on the limiting heats of solution of the alkaline earth nitrates in the alkali nitrates.

Davis concludes that the mixing properties of simple charge-unsymmetrical nitrate mixtures are adequately described in terms of the Coulomb-plus-hard-core interaction potential. This is consistent with the success of Reiss et al. (63) in accounting for the heats of mixing of the binary alkali nitrates, and amounts to a neglect of polarization and dispersion forces. For systems containing more polarizable cations, or smaller anions than the large nitrate ion, one clearly should consider also these additional terms. To a first approximation, this may perhaps be achieved by taking into account only the changes in van der Waals interaction between next-nearest-neighbor cations as proposed for the charge-symmetrical case by Lumsden (62) and by Blander (68). However, as we noted above in our discussion of the silver halide systems, we may also have to consider contributions to the thermodynamic properties which are essentially covalent in character.

At the present time, data are not available which permit a serious test of the Davis theory for systems other than the nitrates. For example, detailed thermodynamic information has as yet not been obtained for the important alkaline earth–alkali halide systems. However, some recent data reported by McCarty & Kleppa (97) on the lead chloride–alkali chloride systems may be relevant. These show that even for these systems, which contain the highly polarizable ion Pb^{++}, the limiting heats of solution vary in a roughly linear manner with the parameter δ_{12}.

The entropy of mixing.—The problem of the entropy of mixing in charge-unsymmetrical ionic systems has been the subject of considerable interest. Specifically, it has been questioned whether the Temkin model is applicable to these systems. Some of the arguments have been based on analogies with ionic solid solutions. It is well known that in the solid state, impurity "molecules" of a charge structure different from that of the host ionic solid will give rise to local departures from electroneutrality. Electroneutrality can be restored, e.g., through the creation of lattice vacancies of appropriate type.

Førland (60) has carried the notion of distinct "vacancies" into the liquid state and has proposed several quasi-lattice models which assume their existence. However, when he attempts to compare the entropies of mixing calculated from his models with experimental entropy data derived from equilibrium information alone, he is unable to reach very firm conclusions. The recently reported calorimetric enthalpy values for the alkali chloride–lead chloride mixtures, combined with earlier electromotive force free-energy data for the same systems [Lantratov & Alabyshev (98); Markov, Delimarskii & Panchenko (99)], have permitted a fresh look at this problem. This comparison indicates that the lithium chloride–lead chloride and sodium chloride–lead chloride systems, which both show quite small departures from ideality, also seem to have relatively small (positive) excess entropies; i.e., for these systems the entropies of mixing appear to be fairly well represented by the Temkin model. This conclusion parallels that previously reached by Førland for the near-ideal liquid mixture $AgBr$–$PbBr_2$ and for $CaCO_3$–Na_2CO_3 (60). For solutions of Ag_2S in liquid AgCl, i.e., for a mixed anion–common cation system, similar behavior has been reported by Bell & Flengas (100).

On the other hand, for potassium chloride–lead chloride and rubidium chloride–lead chloride, which systems both show significant negative deviations from ideality, fairly large positive excess entropies are found. It is of interest that these positive excess entropies correlate with the positive excess volumes reported by Bloom for these systems (4). However, unless additional assumptions are made, they cannot be reconciled with the quasi-lattice models proposed by Førland.

Systems Containing Two Cations and Two Anions

Thermodynamics and theory.—As an illustration of the special problems presented by these reciprocal fused-salt systems, we will take the mixed system which contains the four ions Na^+, K^+, Cl^-, and Br^-. This is a three-component system in the sense of thermodynamics; i.e., a mixture of any composition may be prepared by bringing together suitable amounts of three of the four salts NaCl, NaBr, KCl, and KBr. Let us now consider the formation (at constant P, T) of one mole of the ternary mixture $Na_xK_{1-x}Cl_yBr_{1-y}$ (l) from the three components NaCl (l), NaBr (l), and KBr (l):

$$y\text{NaCl} + (x - y)\text{NaBr} + (1 - x)\text{KBr} = \text{Na}_x\text{K}_{1-x}\text{C}_{1y}\text{Br}_{1-y} \qquad 14.$$

This process may be broken down into a series of consecutive steps, which when added will yield the overall mixing reaction. We will demonstrate this by reversing the mixing procedure. Schematically, the various steps are illustrated in Figure 2.

$$\text{Na}_x\text{K}_{1-x}\text{Cl}_y\text{Br}_{1-y} = x\text{Na}(\text{Cl}_y\text{Br}_{1-y}) + (1 - x)\text{K}(\text{Cl}_y\text{Br}_{1-y}) \qquad 15.$$

$$x\text{Na}(\text{Cl}_y\text{Br}_{1-y}) = xy\text{NaCl} + x(1 - y)\text{NaBr} \qquad 16.$$

$$(1 - x)\text{K}(\text{Cl}_y\text{Br}_{1-y}) = (1 - x)y\text{KCl} + (1 - x)(1 - y)\text{KBr} \qquad 17.$$

$$(1 - x)y\text{KCl} + (1 - x)y\text{NaBr} = (1 - x)y\text{NaCl} + (1 - x)y\text{KBr} \qquad 18.$$

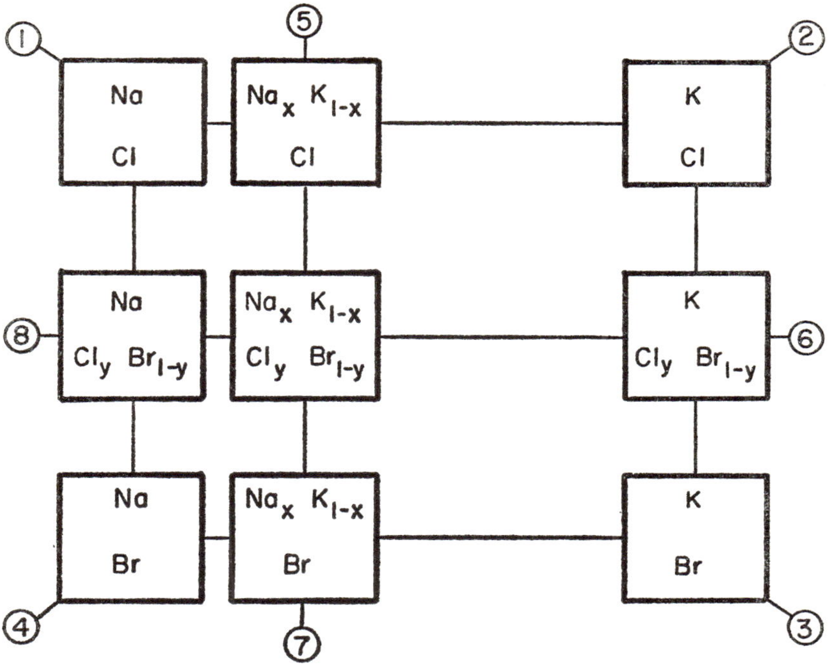

Ionic Fractions

$$x = N_{Na^+} = \frac{n_{Na^+}}{n_{Na^+} + n_{K^+}} \quad ; \quad y = N_{Cl^-} = \frac{n_{Cl^-}}{n_{Cl^-} + n_{Br^-}}$$

$$(1-x) = N_{K^+} = \frac{n_{K^+}}{n_{Na^+} + n_{K^+}} \quad ; \quad (1-y) = N_{Br^-} = \frac{n_{Br^-}}{n_{Cl^-} + n_{Br^-}}$$

FIG. 2. Schematic diagram showing the formation of the ternary fused-salt mixture $Na_xK_{1-x}Cl_yBr_{1-y}$ from the pure salts.

Alternatively, we may unmix along a different path:

$$Na_xK_{1-x}Cl_yBr_{1-y} = yCl(Na_xK_{1-x}) + (1-y)Br(Na_xK_{1-x}) \qquad 19.$$

$$yCl(Na_xK_{1-x}) = xyNaCl + y(1-x)KCl \qquad 20.$$

$$(1-y)Br(Na_xK_{1-x}) = (1-y)xNaBr + (1-y)(1-x)KBr \qquad 21.$$

Clearly

$$(14) + (15) + (16) + (17) + (18) = (14) + (19) + (20) + (21) + (18) = 0 \qquad 22.$$

Let X be any extensive thermodynamic quantity, let us denote the molar change in X according to Equation 14 with $\Delta X_{1234}{}^M$, and let the symbol $\Delta X_{ij}{}^M$ represent the corresponding molar change for the binary or quasi-binary mixing processes starting from i and j in Figure 2. Finally, let $\Delta X_{13}{}^0$ be the standard change in X associated with the metathetical reaction

$$NaCl(l) + KBr(l) = NaBr(l) + KCl(l) \qquad 23.$$

In view of Equation 22, we have

$$\Delta X_{1234}{}^M = (1-x)y\Delta X_{13}{}^0 + x\Delta X_{14}{}^M + (1-x)\Delta X_{23}{}^M + \Delta X_{68}{}^M \qquad 24a.$$

$$= (1-x)y\Delta X_{13}{}^0 + y\Delta X_{12}{}^M + (1-y)\Delta X_{34}{}^M + \Delta X_{57}{}^M \qquad 24b.$$

We now define a molar ternary excess quantity, $\Delta X_{1234}{}^E$, through the following relations

$$\begin{aligned}\Delta X_{1234}{}^E &= \Delta X_{68}{}^M - y\Delta X_{12}{}^M - (1-y)\Delta X_{34}{}^M \\ &= \Delta X_{57}{}^M - x\Delta X_{14}{}^M - (1-x)\Delta X_{23}{}^M\end{aligned} \qquad 25.$$

Introducing this quantity and rearranging terms in Equation 24, we may write

$$\begin{aligned}\Delta X_{1234}{}^M = (1-x)y\Delta X_{13}{}^0 &+ x\Delta X_{14}{}^M + (1-x)\Delta X_{23}{}^M + y\Delta X_{12}{}^M \\ &+ (1-y)\Delta X_{34}{}^M + \Delta X_{1234}{}^E\end{aligned} \qquad 26.$$

Thus, we have reduced the thermodynamic problem associated with simple reciprocal mixing processes to the sum of a metathetical term, four binary terms plus a ternary excess term $\Delta X_{1234}{}^E$. Only this last term is a novel feature. It should be the principal objective of work on reciprocal mixtures to provide information on this quantity, and to relate it to the characteristic properties of the component species.

This purely thermodynamic analysis provides a convenient starting point for a review of recent theoretical and experimental work on simple reciprocal fused-salt solutions.

The appropriate thermodynamic equations for an ideal reciprocal mixture were given already by Temkin (see above). From Equation 4 above, we get for the ideal case in our new notation:

$$\begin{aligned}\Delta G_{1234}{}^M &= -T\Delta S_{1234}{}^M \\ &= RT[x \ln x + (1-x)\ln(1-x) + y\ln y + (1-y)\ln(1-y)]\end{aligned} \qquad 27.$$

where the ionic fractions are defined in Figure 2. For this case we clearly have $\Delta X_{1234}{}^E = 0$ for all extensive properties X.

The first detailed theory of nonideal reciprocal fused-salt mixtures was given by Flood, Førland & Grjotheim (101, 102), following a study of the thermal decomposition of mixed dichromates by Flood & Muan (103). This theory is a quasi-thermodynamic one, based on a cycle related to the one given above. However, it involves a mass-action approach and emphasizes activities and activity coefficients, rather than the integral thermodynamic quantities. As a consequence the derived expressions appear in a somewhat different form. For convenience in comparing their theory with experiments,

Flood, Førland & Grjotheim introduce simplifying assumptions. These are equivalent to neglecting the binary terms and the ternary excess terms in Equation 26. This inadequacy in the original theory was rectified by Førland (60), who re-introduced the binary terms on an *ad hoc* basis.

During the past year, a general theory of reciprocal fused-salt mixtures has been developed by Blander & Yosim (104), who have applied the methods of conformal solution theory to the special case of reciprocal mixtures. Blander & Yosim show that the first five terms on the right-hand side of Equation 26 follow directly from a straightforward application of this theory. The term for the metathetical step is first order, while the four binary terms in Equation 26 are those already derived by Reiss et al. (63). In addition, conformal solution theory provides a new second-order ternary excess quantity. The leading term in this quantity depends on composition through the product of the four ionic fractions, $x(1-x)y(1-y)$, as might also be expected from its definition through Equations 24a and 24b above. In the theory of Blander & Yosim the proportionality constants for $\Delta X_{1234}{}^E$ represent the difference between two complicated and unevaluated integrals.

As in the case of simple binary systems, the nonideal reciprocal fused-salt mixtures are expected to exhibit deviations from random distribution of the ions. This problem was recognized by Flood et al., who suggested a treatment along the lines of the quasi-chemical theory of Guggenheim (105). More explicit calculations of the influence of departures from random mixing on the solution properties have been made by Blander & Braunstein on the basis of a quasi-lattice model (106, 107, 107a). These calculations have been discussed in considerable detail in the recent review article by Blander (14). Here we shall confine ourselves to stating the more significant conclusions.

Blander's "symmetric approximation" is based on the usual assumption of additivity of pair interactions between nearest neighbors, and is essentially equivalent to the quasi-chemical theory. In this approximation the ternary excess terms are predicted to be simply related to departures from random mixing, and one obtains for the ternary excess free energy, $\Delta G_{1234}{}^E = -x(1-x)y(1-y)(\Delta G^0)^2/2ZRT$. ΔG^0 is the standard free-energy charge associated with the appropriate metathetical step, while Z is an effective coordination number.

In Blander's "asymmetric approximation" the assumption of the additivity of pair interactions between nearest neighbors is relaxed. As a result the simple symmetrical dependence of $\Delta X_{1234}{}^E$ on the product $x(1-x)y(1-y)$ is lost. This approximation allows for the case in which the energy change associated with the formation of the combination AX from A^+ and X^- is different from that associated with the formation of AX_2^- from $AX+X$ or A_2X^+ from AX and A^+. This situation is expected to arise when A^+ and X^- have a tendency toward formation of specific complex entities with clearly defined geometry. Thus the asymmetric approximation is applicable to situations analogous to those encountered in the field of aqueous complex chemistry.

Comparison with experiment.—The original theory of Flood et al. has proved useful in approximate treatments of a variety of different fused-salt solution problems. However, since this theory neglects all terms other than that associated with the metathetical step, it cannot be expected to yield truly quantitative agreement with experiment.

A strong point in the Flood theory is that it can readily be generalized so that it can be applied to mixtures which contain more than three components. This requires only the introduction of one additional metathetical term for each additional component. As a consequence, this theory offers a basis for the treatment of very complicated multicomponent systems, such as those encountered in the chemistry of silicates and slags [Flood & Grjotheim (108); Grjotheim (109)]. A recent example is the work by Flood & Toguri (110) on the application of the law of mass action to metal–slag equilibria.

The significance of the binary terms in Equation 26 is well illustrated in recent studies by Blander et al. (111, 112). These investigations involve electromotive force measurements of solubility products and heats of solution for silver halides in liquid alkali nitrates according to processes of the following type:

$$AgX(s) + MeNO_3(l) = \text{dilute solutions in } MeNO_3 \qquad \text{a}$$

Blander analyses this overall process in terms of a metathetical step

$$AgX(s) + MeNO_3(l) = MeX(l) + AgNO_3(l) \qquad \text{b}$$

plus two binary steps

$$MeX(l) + MeNO_3(l) = \text{infinite dilute solution in } MeNO_3 \qquad \text{c}$$

$$AgNO_3(l) + MeNO_3(l) = \text{infinite dilute solution in } MeNO_3 \qquad \text{d}$$

He finds that his experimentally determined heats of solution for the overall process, ΔH_a, are well represented by the sums $\Delta H_b + \Delta H_c + \Delta H_d$, calculated from data reported in the literature. The dominant term in these sums is ΔH_b, which in the considered cases represents about 95 per cent of ΔH_a. It follows that for these examples the treatment of Flood et al. is a very reasonable first approximation. A related study is that by Seward & Field of the solubility of silver chloride and thallium bromide in fused-nitrate solvents (113, 114). Another illustration of the applicability of this approximate treatment, refined by the inclusion of the binary terms, is provided by Toguri, Flood & Førland (115) in a study of exchange equilibria involving systems of the type [A, B (Br, Cl), HCl, HBr], where A and B are drawn from the set Li, Na, K.

Calorimetric work which relates to the studies of Blander et al. and of Seward & Field has been reported by Hu, Ko & Hepler (116) and, more extensively, by Meschel & Kleppa (117). The latter authors measured the heats of solution for a series of alkali and alkaline earth halides in alkali, silver, and thallium nitrates. By correcting the measured enthalpies of solution by means of the appropriate binary terms, the enthalpy changes were evaluated for the metathetical steps at high temperature. For simple reciprocal alkali halide–alkali nitrate systems, this term varies linearly with $(1/d_1+1/d_2-1/d_3-1/d_4)$, where d_i is the interionic distance characteristic of salt i. This dependence on interionic distance illustrates the applicability of the simple corresponding-states theory of Reiss et al. On the other hand, deviations are found for systems involving salts of silver and thallium. This reflects the extra-Coulombic contributions to cohesion in these salts.

So far, there have been no comprehensive tests of the conformal solution theory for reciprocal mixtures of Blander & Yosim (104). However, a first step has been taken by Blander & Topol (118), who have examined available equilibrium and electromotive force data for the LiF–KCl and $NaNO_3$–AgCl systems. The authors conclude that the experimental data can be explained only if the ternary excess terms are taken into account. For concentrated solutions in the simple ionic system LiF–KCl, the ternary excess terms are of significant magnitude and relatively insensitive to temperature. Similarly, even for dilute solutions of AgCl in $NaNO_3$, the ternary excess terms are found to be quite large, but also strongly temperature dependent. The authors relate these observations, on the one hand, to departures from random mixing, and on the other to the known tendency of the ions Ag^+ and Cl^- to form well-characterized complex species in nitrate melt solvents. The latter example is one taken from a series of electromotive force investigations by Blander, Braunstein and their co-workers. These have been aimed specifically at the study of complex association formation between nonoctet ions (Ag^+, Cd^{++}, Pb^{++}) and halide ions in various fused-nitrate solvents. Earlier work in this area was reviewed by Blander (14). More recent studies are reported in papers by Braunstein et al. (119–123) and by Hill and co-workers (124, 125).

LITERATURE CITED

1. Blomgren, G E., and Van Artsdalen, E. R., *Ann. Rev. Phys. Chem.*, **11**, 273–306 (1960)
2. Flood, H., *Pure Appl. Chem.*, **5**, 529–41 (1962)
3. Gruen, D. M., and McBeth, R. L., *Pure Appl. Chem.*, **6**, 23–47 (1963)
4. Bloom, H., *Pure Appl. Chem.*, **7**, 389–406 (1963)
5. Mackenzie, J. D., *Advan. Inorg. Chem. Radiochem.*, **4**, 293–318 (1962)
6. Wait, S. C., and Janz, G. J., *Quart. Rev.*, **17**, 225–42 (1963)
7. Corbett, J. D., and Duke, F. R., *Technique of Inorganic Chemistry*, **1**, 103–56 (Jonassen, H. B., and Weissberger, A., Eds., Interscience, New York-London, 1963)
8. Molten Salts, *Ann. N. Y. Acad. Sci.*, **79**, 759–1098 (1960)
9. *Discussions Faraday Soc.*, **32**, 1–268 (1961)
10. *Chimia Fizica a Electroliţilor Topiţi* (Acad. Republicii Populare Romine, Bucuresti, 1963)
11. Delimarskii, Iu. K., and Markov, B. F., *Electrochemistry of Fused Salts* (Transl. from Russian by Peiperl, A., Sigma Press, Washington, D. C., 1961)
12. Blander, M., Ed., *Molten Salt Chemistry* (Interscience, New York-London, 1964)
13. Sundheim, B. R., Ed., *Fused Salts* (McGraw, New York, 1964)
14. Blander, M., Ed., in *Molten Salt Chemistry*, 127–237 (Interscience, New York-London, 1964)
15. Førland, T., 63–164 (see ref. 13)
16. Dworkin, A. S., and Bredig, M. A., *J. Phys. Chem.*, **64**, 269–72 (1960)

17. Dworkin, A. S., and Bredig, M. A., *J. Phys. Chem.*, **67**, 697–98 (1963)
18. Dworkin, A. S., and Bredig, M. A., *J. Phys. Chem.*, **67**, 2499 (1963)
19. Dworkin, A. S., and Bredig, M. A., *J. Chem. Eng. Data*, **8**, 416–17 (1963)
20. Kleppa, O. J., and McCarty, F. G., *J. Chem. Eng. Data*, **8**, 331–32 (1963)
21. Janz, G. J., Kelly, F. J., and Perano, J. L., *Trans. Faraday Soc.*, **59**, 2718–22 (1963)
22. Janz, G. J., and Kelly, F. J., *J. Phys. Chem.*, **67**, 2848–50 (1963)
23. Janz, G. J., Kelly, F. J., and Perano, J. L., *J. Chem. Eng. Data*, **9**, 113 (1964)
24. Janz, G. J., Neuenschwander, E., and Kelly, F. J., *Trans. Faraday Soc.*, **59**, 841–45 (1963)
25. Janz, G. J., and Perano, J. L., *Trans. Faraday Soc.*, **60**, 1742–44 (1964)
26. Schinke, H., and Sauerwald, F. D., *Z. Anorg. Allgem. Chem.*, **287**, 313–24 (1956)
27. Schinke, H., and Sauerwald, F. D., *ibid.*, **304**, 25–36 (1960)
28. Owens, B. B., *J. Chem. Phys.*, **39**, 1053–54 (1963)
29. Babb, S. E., Chaney, P. E., and Owens, B. B., *J. Chem. Phys.*, **41**, 2210–11 (1964)
30. Levy, H. A., and Danford, M. D., 109–25 (see ref. 12)
31. Smith, G. Pedro, 427–505 (see ref. 12)
32. Gruen, D. M., 301–39 (see ref. 13)
33. James, D. W., 507–33 (see ref. 12)
34. Baddiel, C. B., and Janz, G. J., *Trans. Faraday Soc.*, **60**, 2009–12 (1964)
35. Balasubrahmanyan, K., and Nanis, L., *J. Chem. Phys.*, **40**, 2657–62 (1964)
36. Wilmshurst, J. K., *J. Chem. Phys.*, **39**, 1779–88 (1963)
37. Rowland, T. J., and Bromberg, J. P., *J. Chem. Phys.*, **29**, 626–30 (1958)
38. Hafner, S., and Nachtrieb, N. H., *J. Chem. Phys.*, **40**, 2891–98 (1964)
39. Ramsey, N. F., *Phys. Rev.*, **78**, 699–703 (1950)
40. Zarzycki, J., and Naudin, F., *Rev. Hautes Temp. Refract.*, **1**, 121–37 (1964)
41. Hafner, S., and Nachtrieb, N. H. (Private communication)
42. Stillinger, F. H., 1–108 (see ref. 12)
43. Ladd, M. F. C., and Lee, W. H., in *Progress in Solid State Chemistry*, **1**, 37–82 (Reiss, H., Ed., Macmillan, New York, 1964)
44. Stillinger, F. H., Kirkwood, J. G., and Wojtowicz, P. J., *J. Chem. Phys.*, **32**, 1837–45 (1960)
45. Bloom, H., and Bockris, J. O'M., 1–62 (see ref. 13)
46. Pitzer, K. S., *J. Chem. Phys.*, **7**, 583–90 (1939)
47. Reiss, H., and Mayer, S. W., *J. Chem. Phys.*, **34**, 2001–3 (1961)
48. Reiss, H., Frisch, H. L., and Lebowitz, J. L., *J. Chem. Phys.*, **31**, 369–80 (1960)
49. Reiss, H., Mayer, S. W., and Katz, J. L., *J. Chem. Phys.*, **35**, 820–26 (1961)
50. McQuarrie, D. A., *J. Phys. Chem.*, **67**, 518–19 (1963)
51. Mayer, S. W., *J. Chem. Phys.*, **40**, 2429–30 (1964)
52. Yosim, S. J., and Owens, B. B., *J. Chem. Phys.*, **41**, 2032–36 (1964)
53. Temkin, M., *Acta Physicochem. URSS*, **20**, 411–20 (1945)
54. Kleppa, O. J., *J. Phys. Chem.*, **67**, 1937–40 (1960)
55. Kleppa, O. J., and Hersh, L. S., *J. Chem. Phys.*, **34**, 351–58 (1961)
56. Kleppa, O. J., Clarke, R. B., and Hersh, L. S., *J. Chem. Phys.*, **35**, 175–80 (1961)
57. Kleppa, O. J., and Hersh, L. S., *J. Chem. Phys.*, **36**, 544–47 (1962)
58. Powers, B. F., Katz, J. L., and Kleppa, O. J., *J. Phys. Chem.*, **66**, 103–5 (1962)
58a. Blander, M., *J. Chem. Phys.*, **41** (1964) (In press)
59. Longuet-Higgins, H. C., *Proc. Roy. Soc. (London)*, *A*, **205**, 247–69 (1951)
60. Førland, T., On the Properties of Some Mixtures of Fused Salts, *Norg. Tekn. Vitenskapsakad.*, *Ser. 2*, No. 4 (1957)
61. Blander, M., *J. Chem. Phys.*, **34**, 697–98 (1961)
62. Lumsden, J., *Discussions Faraday Soc.*, **32**, 138–46 (1961)
63. Reiss, H., Katz, J. L., and Kleppa, O. J., *J. Chem. Phys.*, **36**, 144–48 (1962)
64. Brown, W. B., *Proc. Roy. Soc. (London)*, *A*, **240**, 561–80 (1957)
65. Blander, M., *J. Chem. Phys.*, **37**, 172–73 (1962)
66. Davis, H. T., and Rice, S. A., *J. Chem. Phys.*, **41**, 14–24 (1964)
67. London, F., *Z. Physik. Chem.*, **B11**, 222–51 (1930)
68. Blander, M., *J. Chem. Phys.*, **36**, 1092–93 (1962)
69. Aukrust, E., Bjørge, B., Flood, H.,

and Førland, T., *Ann. N. Y. Acad. Sci.*, **79**, 830–37 (1960)
70. Gilbert, R. A., *J. Phys. Chem.*, **67**, 1143–45 (1963)
71. Hersh, L. S., and Kleppa, O. J., *J. Chem. Phys.*, **42**, 1309–22 (1965)
72. Hildebrand, J. H., and Salstrom, E. J., *J. Am. Chem. Soc.*, **54**, 4257–61 (1932)
73. Tessman, J. R., Kahn, A. H., and Shockley, W., *Phys. Rev.*, **92**, 890–95 (1953)
74. Hersh, L. S., Navrotsky, A., and Kleppa, O. J. (Unpublished results)
75. Thurmond, C. D., *J. Am. Chem. Soc.*, **75**, 3928–32 (1953)
76. Hafner, S., and Nachtrieb, N. H., *J. Chem. Phys.*, **42**, 631–36 (1965)
77. Flood, H., Førland, T., and Motzfeldt, K., *Acta Chem. Scand.*, **6**, 257–69 (1952)
78. Kleppa, O. J., and Meschel, S. V., *J. Phys. Chem.*, **67**, 668–71 (1963)
79. Kleppa, O. J., and Meschel, S. V., *J. Phys. Chem.*, **67**, 2750–53 (1963)
80. Kleppa, O. J., Hersh, L. S., and Toguri, J. M., *Acta Chem. Scand.*, **17**, 2681–87 (1963)
81. Førland, T., Saether, B., Fykse, O., Block-Bolten, A., Toguri, J. M., and Flood, H., *Acta Chem. Scand.*, **16**, 2429–33 (1962)
82. Toguri, J. M., Flood, H., and Førland, T., *Acta Chem. Scand.*, **17**, 1502–8 (1963)
83. Haendler, H. M., Sennett, P. S., and Wheeler, C. M., *J. Electrochem. Soc.*, **106**, 264–68 (1959)
84. Flood, H., *Discussions Faraday Soc.*, **32**, 168 (1961)
85. Salstrom, E. J., Kew, T. J., and Powell, T. M., *J. Am. Chem. Soc.*, **58**, 1848–50 (1936)
86. Panish, M. B., Blankenship, F. F., Grimes, W. R., and Newton, R. F., *J. Phys. Chem.*, **63**, 668–71 (1959)
87. Ketelaar, J. A. A., *J. Chim. Phys.*, **61**, 44–49 (1964)
87a. Ketelaar, J. A. A., and Dammers-De Klerk, A., *Rec. Trav. Chim.*, **83**, 322–28 (1964)
88. Beusman, C., *Oak Ridge Natl. Lab. 2323* (Office of Tech. Services, U. S. Dept. Commerce, Washington, D. C., 1957)
89. Cantor, S., *J. Phys. Chem.*, **65**, 2208–10 (1961)
90. Cantor, S., and Carlson, T. S., *J. Phys. Chem.*, **66**, 2711–13 (1962)
91. Cantor, S., and Ward, W. T., *J. Phys. Chem.*, **67**, 1868–70 (1963)
92. Cantor, S., and Ward, W. T., *J. Phys. Chem.*, **67**, 2766–69 (1963)
93. Kleppa, O. J., *J. Phys. Chem.*, **66**, 1668–71 (1962)
94. Davis, H. T., *J. Chem. Phys.*, **41**, 2761–66 (1964)
95. Davis, H. T. (Unpublished work, Dec., 1964)
96. McCarty, F. G., Hersh, L. S., and Kleppa, O. J., *J. Chem. Phys.*, **41**, 1522–23 (1964)
97. McCarty, F. G., and Kleppa, O. J., *J. Phys. Chem.*, **68**, 3846–52 (1964)
98. Lantratov, M. F., and Alabyshev, A. F., *Zh. Prikl. Khim.*, **26**, 263–76 (1953)
99. Markov, B. F., Delimarskii, Iu. K., and Panchenko, I. D., *Zh. Fiz. Khim.*, **28**, 1987–98 (1954)
100. Bell, M. C., and Flengas, S. N., *J. Electrochem. Soc.*, **111**, 569–75 (1964)
101. Flood, H., Førland, T., and Grjotheim, K., *The Physical Chemistry of Melts*, 46–62 (Inst. Mining and Metallurgy, London, 1953)
102. Flood, H., Førland, T., and Grjotheim, K., *Z. Anorg. Allgem. Chem.*, **276**, 289–315 (1954)
103. Flood, H., and Muan, A., *Acta Chem. Scand.*, **4**, 364–73 (1950)
104. Blander, M., and Yosim, S. J., *J. Chem. Phys.*, **39**, 2610–16 (1963)
105. Guggenheim, E. A., *Mixtures* (Oxford Univ. Press, 1952)
106. Blander, M., *J. Phys. Chem.*, **63**, 1262–65 (1959)
107. Blander, M., and Braunstein, J., *Ann. N. Y. Acad. Sci.*, **79**, 838–52 (1960)
107a. Blander, M., *J. Chem. Phys.*, **34**, 432–38 (1961)
108. Flood, H., and Grjotheim, K., *J. Iron Steel Inst.*, **171**, 64–70 (1952)
109. Grjotheim, K., *J. Iron Steel Inst.*, **178**, 354–56 (1954)
110. Flood, H., and Toguri, J. M., *Trans. AIME*, **227**, 525–29 (1963)
111. Blander, M., Braunstein, J., and Silverman, M. D., *J. Am. Chem. Soc.*, **85**, 895–97 (1963)
112. Blander, M., and Luchsinger, E. B., *J. Am. Chem. Soc.*, **86**, 319–22 (1964)
113. Seward, R. P., and Field, P. E., *J. Phys. Chem.*, **68**, 210–12 (1964)
114. Seward, R. P., and Field, P. E., *J. Phys. Chem.*, **68**, 1611–13 (1964)
115. Toguri, J. M., Flood, H., and Førland, T., *The Law of Mass Action*, A Centenary Volume, 131–49 (Oslo, 1964)

116. Hu, T., Ko, H. C., and Hepler, L. G., *J. Phys. Chem.*, **68**, 387–90 (1964)
117. Meschel, S. V., and Kleppa, O. J., *J. Phys. Chem.*, **68**, 3840–45 (1964)
118. Blander, M., and Topol, L. (To be published)
119. Braunstein, J., and Hagman, R. E., *J. Phys. Chem.*, **67**, 2881–82 (1963)
120. Manning, D. L., Blander, M., and Braunstein, J., *Inorg. Chem.*, **2**, 345–47 (1963)
121. Braunstein, J., and Minano, A. S., *Inorg. Chem.*, **3**, 218–21 (1964)
122. Thomas, C., and Braunstein, J., *J. Phys. Chem.*, **68**, 957–59 (1964)
123. Hess, J. M. C., Braunstein, J., and Braunstein, H., *J. Inorg. Nucl. Chem.*, **26**, 811–20 (1964)
124. Woolner, E. S., Jr., and Hill, D. G., *J. Phys. Chem.*, **67**, 1571–73 (1963)
125. Alvares-Funes, A. R., and Hill, D. G., *J. Phys. Chem.*, **67**, 1573–76 (1963)

THERMODYNAMICS[1,2]

By G. R. Somayajulu,[3] A. P. Kudchadker,[4] and B. J. Zwolinski
Department of Chemistry, Texas A&M University, College Station, Texas

INTRODUCTION

One of the most striking characteristics of modern thermochemistry and thermodynamics is its current strength and vitality in all areas of science. Chemists, engineers, physicists, and biologists are boldly and vigorously applying thermodynamic techniques and procedures primarily developed for normal ranges of the common laboratory variables of pressure and temperature into the most extreme ranges of these variables from absolute zero to millions of degrees in plasma work, and from low pressures to undreamed of pressures of several hundred thousand bars and in some cases a million kilobars. Furthermore, at one-time scientists cautiously discussed the application of thermodynamic notions to states out of equilibrium, whereas currently, in the words of Jaynes (1), information theory developments are defining accessible and nonaccessible nonequilibrium states that can be treated confidently with established thermodynamic techniques and its extensions.

If we accept that chemistry is the empirical discipline dealing with the structural forms of matter and their transformations, then according to Tisza (2), thermodynamics, in the broadest sense of the word, is a conceptualization of chemistry in much the same way that classical mechanics is the conceptualization of astronomy. In this article Tisza further extends his axiomatic approach to Gibbs' theory of macroscopic thermodynamics in equilibrium which is not unlike Carathéodory's axiomatic investigation of the classical Clausius-Kelvin engineering approach to thermodynamics. He also develops the conceptual structure of a generalized thermodynamics that extends in successive approximations from a purely macroscopic theory through statistics to quantum mechanics.

In chemistry proper, one of the most gratifying developments in recent years has been in the availability of more and better precise measurements on compounds of higher and higher purity. Constantly one hears of new thermal laboratories springing into being in government, academic and industrial locales. Even more gratifying to a thermodynamist is the full realization that statistical thermodynamic procedures wedded to quantum mechanical results are providing us with deeper insight into the details of molecular struc-

[1] The survey of literature pertaining to this review was concluded in December 1964.

[2] Partially supported by the Robert A. Welch Foundation.

[3] Postdoctoral research associate, MCA Research Project, Chemical Thermodynamic Properties Center.

[4] Graduate research associate, API Research Project 44, Chemical Thermodynamic Properties Center.

ture and at the same time providing us with reliable numerical data for practical calculations in the laboratory and in industrial processes.

This review will concern itself primarily with the following main topics: (*a*) status of fundamental and derived constants in thermodynamics; (*b*) critical data compilations; (*c*) mathematical theory of isomers; (*d*) enthalpies of formation and bond energies; (*e*) boiling point correlations; and (*f*) recent thermodynamic data.

STATUS OF FUNDAMENTAL AND DERIVED CONSTANTS IN THERMODYNAMICS

If we can envisage an "ideal" measurement of a quantity in a laboratory free of all errors, it will still be a function of the defined fundamental units of measurement, such as length, mass, and time. In many instances the precisely measured quantity is further combined with certain fundamental constants such as Avogadro's number, the gas constant *R*, and others, before it is reported in the literature. Thus, it is imperative that an internally consistent and widely acceptable set of "best" values of these fundamental constants be available and used by every scientist throughout the world. Only then will we be in a position to make relevant comparisons and judgments of related observations by several different investigators. A less satisfactory but acceptable practice is to have each investigator precisely spell out the constants and units of measurement employed in his study, whatever these may be.

It is truly remarkable that only 13 years ago, the Committee on Physical Chemistry of the National Research Council (U. S.) recommended such a uniform set of values of fundamental constants to be used in physical chemistry (3). Prior to that time, the 1926 constants of the International Critical Tables were more or less generally accepted in this country. Recent changes in the absolute temperature of the ice point ($273.1500 \pm 0.0001°$ K$\equiv$0° C), the atomic mass scale (carbon-12 exactly), and in certain of the experimentally measured fundamental constants led to a reexamination of this 1952 set of fundamental constants by the NRC Committee on Fundamental Constants who recommended a revised set of values in 1963 (4). The history and development of these studies and deliberations have been presented recently by Rossini (5, 6).

For the purposes of physical chemistry in general, and thermochemistry and thermodynamics in particular, it is convenient to group the set of fundamental constants into three categories:

(*a*) defined constants (ten exact values by definition)
(*b*) basic constants (five experimentally determined constants)
(*c*) derived constants (six constants coupled to basic constants through known physical relations)

The 1963 values taken from Rossini's report (7) are compared with the 1952 values in Tables I, II, and III. It should be noted that the ice point and the liter are now defined constants. Also, the value of the gas constant in

TABLE I

VALUES OF "DEFINED" CONSTANTS

Constant	Symbol	1963 Value (exact, by definition)	1952 Value
Unified atomic mass unit	u	1/12 times the mass of an atom of ^{12}C	
Mole	mol	The amount of a substance of specified chemical formula containing the same number of formula units (atoms, molecules, ions, electrons, or other entities) as there are atoms in 12 grams (exactly) of the pure nuclide ^{12}C	
Standard acceleration of gravity, in free fall	g	980.665 cm s^{-2}	Same
Normal atmosphere, pressure	atm	1,013,250 dyne cm^{-2}	Same
Absolute temperature of the triple point of water[a]	T_{tp}	273.16° K	273.17
Thermochemical calorie	cal	4.184 J	4.1840
International steam calorie	cal_{IT}	4.1868 J	4.18674
Inch	in	2.54 cm	2.54000508
Pound, avoirdupois	lb	453.59237 g	453.5924277
Liter	l	1000 cm^3	1000.028 ±0.004

[a] The difference between the temperature of the triple point of water and the so-called "ice point" (temperature of equilibrium of solid and liquid water saturated with air at one atmosphere) is accurately known (6): T (triple point)−T (ice point) $=0.0100 \pm 0.0001°$ K.

Table II is given to one more significant figure. It is hoped that the 1963 set of fundamental constants will be found acceptable to IUPAC, IUPAP, and other appropriate bodies.

CRITICAL DATA COMPILATIONS

Critical tables of data, whether in chemical thermodynamics or in the broad area of physical sciences, provide ready access to numerical results of scientific work and are thus basic to progress in fundamental and applied re-

TABLE II

VALUES OF "BASIC" CONSTANTS

Constant	Symbol	1963 Value, with estimated uncertainty	1952 Value, with estimated uncertainty
Velocity of light in vacuo	c	2.997925×10^{10} cm s^{-1} ± 0.000003	2.997902×10^{10} ± 0.000013
Avogadro number	N	6.02252×10^{23} molecules mol^{-1} ± 0.00028	6.02380×10^{23} ± 0.00016
Faraday constant	$\mathfrak{F}$	96,487.0 coulomb equiv^{-1} ± 1.6 23,060.9 cal volt^{-1} equiv^{-1} ± 0.4	96,493.1 ± 1.0 23,062.4 ± 0.2
Planck constant	h	6.6256×10^{-27} ergs ± 0.0005	6.62377×10^{-27} ± 0.00027
Pressure-volume product for one mole of gas at 0° C and zero pressure	$(PV)^{P=0}_{T_{0°C}}$	2271.06 J mol^{-1} ± 0.12 22413.6 cm^3 atm mol^{-1} ± 1.2	2271.16 ± 0.04 22,414.6 ± 0.4

TABLE III

VALUES OF "DERIVED CONSTANTS"

Constant	Symbol	1963 Value, with estimated uncertainty		1952 Value, with estimated uncertainty
Elementary charge	$e = \mathfrak{F}/N$	4.80298×10^{-10} cm$^{3/2}$g$^{1/2}$s^{-1} ± 0.00020	(esu)	4.802232×10^{-10} ± 0.000110
Gas constant	$R = \frac{(PV)^{P=0}_{T_{0°C}}}{T_{0°C}}$	8.31433 ± 0.00044 1.98717 ± 0.00011	J deg^{-1} mol^{-1} cal deg^{-1} mol^{-1}	8.31439 ± 0.00034 1.98719 ± 0.00008
Boltzmann constant	$k = R/N$	1.38054×10^{-16} ± 0.00009	erg deg^{-1} molecule^{-1}	1.380257×10^{-16} ± 0.000067
Second radiation constant	$c_2 = hc/k$	1.43879 ± 0.00009	cm deg	1.438676 ± 0.000091
Einstein constant relating mass and energy	$Y = c^2$	8.987554×10^{13} J g^{-1} ± 0.000018 2.148076×10^{13} cal g^{-1} ± 0.000004		8.987416×10^{13} ± 0.000081 2.148044×10^{13} ± 0.000019
Constant relating wave number and energy	$Z = nhc$	11.96255 ± 0.00038 2.85912 ± 0.00009	J cm mol^{-1} cal cm mol^{-1}	11.96171 ± 0.00026 2.858917 ± 0.000062

search. By critical tables it is commonly understood to mean tables of critically selected "best" and authoritative values for measured or derived properties of substances, maintained substantially up to date by revision at appropriate intervals. Several years ago it became evident that the revision of the seven 1926–1930 volumes of the International Critical Tables was not possible, but would be instead replaced by an array of loosely coordinated smaller compilations, such as the API Research Project 44, the NBS Circular 500, and others coordinated through a central office, the NRC Office of Critical Tables. The new ICT tables thus represent a collection of separate compilations covering small areas of science which are periodically revised and published under the auspices of separate private and government organizations. The first volume of a central index that permits access to the contents of certain of these publications was published in 1962 entitled the *Consolidated Index of Selected Property Values—Physical Chemistry and Thermodynamics* through the NRC Office of Critical Tables (8). Chemical scientists whether engaged in experimental or theoretical work should have easy access to this volume in the office and laboratory.

An important step to accelerate work on critical data for science was taken in 1963, when the Director of the Office of Science and Technology announced the establishment of a National Standard Reference Data System to be administered by the National Bureau of Standards (9). This system integrates to a single point of responsibility the data-compiling activities on the national scene while seeking cooperation and advice from the NRC Office of Critical Tables and other private and scientific groups. The Office of the NBS National Standard Reference Data Program was established late in 1963 and is now operative. A broad international program in this area needs to be organized. The activity in this area is reflected in three recent conferences dealing with critical data studies. Through the co-sponsorship of the National Research Council Subcommittee on Chemical Thermodynamics and of the NRC Office of Critical Tables, two conferences entitled the "Conference on Critical Tables of Thermodynamic Data" and "Conference on Standard Reference Data for Thermodynamics" were held at the National Academy of Sciences, Washington, D. C., in March 1963 and May 1964, respectively. A second Gordon Conference on critical data entitled "Scientific Information Problems in Research: Critical Tables and Progress in Science" was held in New Hampton, N. H., in July 1964. At this conference Professor Koichiro Nakanishi from Shinshu University, Nagano, Japan, announced the publication of the first volume of an annual publication entitled *Bussei-Teisu*, Vol. I, 1963 (in Japanese) on physical and thermodynamic properties of substances and mixtures by the Committee of Physical Constants of the Society of Chemical Engineers of Japan. The chairman is Professor Mitsuho Hirata, Tokyo Metropolitan University. The purpose of this publication is to prepare an annual compilation of new physicochemical data and, whenever possible, to critically evaluate and prepare standard reference data. In Volume I the committee has recommended values on simple physical properties (normal freezing and boiling points, refractive index, and density data) of certain primarily lower members of the following aliphatic homologous series of compounds: alcohols, acids, esters, ethers, aldehydes, ketones, and amines. Volume II is scheduled for publication in October 1964.

Skinner has presented a masterful discussion (5) of the nature and importance of "key" heat of formation data. In Table IV, we have selected enthalpies of formation at 298.15° K for 24 "key" substances based on the 1961 atomic masses and the 1963 fundamental constants. These values are basic in thermochemical calibrations and calculations leading to the determination of enthalpies of formation of organic compounds.

TABLE IV

ENTHALPIES OF FORMATION OF "KEY" COMPOUNDS

Substance	ΔH°_f, 298.15, kcal/mole			References
	Crystal	Liquid	Gas	
C, diamond	0.4533			(10)
CO			−26.416	(10)
CO_2			−94.051	(10)
H_2O		−68.315	−57.796	(10)
HF		− 71.65	−64.8	(10)
HCl			−22.062	(10)
HBr			− 8.70	(10)
HI			+ 6.33	(10)
H_2SO_4		−194.548		(10)
$H_2SO_4 \cdot 115\ H_2O$		−212.192		(10)
H_3PO_4	−305.7	−302.8		(10)
HNO_3		− 41.61	−32.28	(10)
CS_2		+21.65	+28.26	(11)
CF_4			−220.0[a]	(12, 13)
NH_3			−11.07	(10)
P_4H_{10}, hex	−713.2			(10)
SiO_2, quartz	−217.8[a]			(14, 15, 16)
SiF_4			−386.0	(14, 15)
Benzoic acid	− 92.036			NBS sample 39h
Succinic acid	−224.75			(17, 18)
Hippuric acid	−145.47			(19)
M-Trifluorotoluic acid	−253.68			(20)
p-Fluorobenzoic acid	−139.57			(21)
Thianthrene	+ 43.12			(22)

[a] Average value.

THE MATHEMATICAL THEORY OF ISOMERS

Ever since chemists in the early nineteenth century showed that formation of chemical compounds is not a capricious and fortuitous process, but that it proceeds in an orderly fashion according to the laws of multiple and reciprocal proportions, chemists and their mathematical colleagues have been fascinated with the problem of calculating the number of structural isomers. Naturally if we simply define molecules in terms of valence rules, the empirical formulae can then be treated as formal mathematical models which predict an infinite number of isomers. Since chemists investigate but do not philosophize, we need to define the mathematical problem in terms of appropriate conservation laws and boundary conditions as laid down by chemistry and physics. Only then can we answer such questions as, "what are the total kinds of compounds and their isomers that in principle existed in the past, exist now, and will exist in the future?" The solution of the problem of isomers is currently important for the following reasons:

(*a*) it provides a logical internally consistent framework for classification of compounds into families, classes, and series of related compounds important for purposes of nomenclature, storage, retrieval, and processing of chemical data;
(*b*) it identifies "key" compounds and homologous series that may be important for minimizing experimental work and correlating measured properties;
(*c*) it readily differentiates geometrical parameters from physicochemical assumptions in molecular models used in correlation schemes;
(*d*) it is important in analyses of complex equilibria.

Recently the authors in analyzing data for halogenated methanes and ethanes calculated the following numbers of mathematically possible isomers: halogenated methanes, $C_nF_\alpha Cl_\beta Br_\gamma I_\delta H_\epsilon$, 69 structural isomers when $n=1$; halogenated ethanes, 629 structural isomers when $n=2$. If the optical isomers (*dextro*, *levo*, and *meso*) are considered, then the total number of isomers is 74 and 1034, respectively.

In this section a brief review of the various mathematical approaches are summarized.

As long ago as 1875, Cayley (23) a mathematician, derived formulas using graph theory (24, 25) for the calculation of the number of isomeric hydrocarbons of the methane series and the number of isomeric alcohols of the methanol series. Unfortunately, however, some of the numbers calculated by him, i.e., the numbers of isomeric dodecanes, tridecanes, and tridecylalcohols, were not numerically correct. Since then various attempts (26) have been made by both mathematicians and chemists to correct Cayley's errors, but in 1957 Iri (27) proved the correctness of Cayley's method and ascribed the errors to certain slips in numerical calculations made by Cayley. Iri recalculated Cayley's numerical tables and found that Cayley's method can be used to calculate the numbers of structural isomers of various kinds of organic compounds.

Henze & Blair (26) developed recursion type mathematical formulas for the calculation of the number of structural isomers from the carbon content, both among the paraffinic hydrocarbons and their monosubstituted products. They also developed (28) a series of recursion relations permitting the calculation of the number of stereoisomeric and nonstereoisomeric monosubstitution products from their carbon content. By combining the formulas previously derived they then derived (29) formulas to calculate the number of stereoisomeric paraffinic hydrocarbons. Their results for the methane series and methanol series are given in Table V. The numbers calculated for the hydrocarbons agree exactly with the numbers obtained by actual writing of

TABLE V

NUMBER OF STRUCTURAL ISOMERS OF THE METHANE AND METHANOL SERIES

Carbon content	Number of isomeric alkanes	Number of isomeric alcohols			
		Primary	Secondary	Tertiary	Total
1	1	1	0	0	1
2	1	1	0	0	1
3	1	1	1	0	2
4	2	2	1	1	4
5	3	4	3	1	8
6	5	8	6	3	17
7	9	17	15	7	39
8	18	39	33	17	89
9	35	89	82	40	211
10	75	211	194	102	507
11	159	507	482	249	1,238
12	355	1,238	1,188	631	3,057
13	802	3,057	2,988	1,594	7,639
14	1,858	7,639	7,528	4,074	19,241
15	4,347	19,241	19,181	10,443	48,865
16	10,359	48,865	49,060	26,981	124,906
17	24,894	124,906	126,369	69,923	321,198
18	60,523	321,198	326,863	182,158	830,219
19	147,284	830,219	849,650	476,141	2,156,010
20	366,319	2,156,010	2,216,862	1,249,237	5,622,109

C_nH_{2n+2}	
C_{25}	36,797,588
C_{30}	4,111,846,763
C_{40}	62,491,178,805,831

their structural formulas through the tetradecanes. The totals for the alcohols agree exactly through the dodecyl alcohols.

Pólya (30), using group theory, solved a general problem of symmetry in organic compounds that can be used to calculate the number of structural isomers in various kinds of organic compounds (31). With this theorem, Hill (32) showed that it is possible to include various kinds of isomerism in a single discussion. The number of isomers calculated by this procedure is the total number of all types. Pólya (31) applied his method to find the number of isomers for a given ring, such as, benzene, naphthalene, anthracene, phenanthrene, thiophene, cyclic hydrocarbons, etc., where the degree of substitution is generalized. The alternate method of generalization, that is, to find the number of isomers for a given substitution or type of substitution where the ring is generalized, was investigated by Hill (33). Hill (34) also discussed the problem of finding the number of structural isomers in any simple symmetrical ring of n members for any value of n and any kind of substitution; i.e., where both the ring and the type of substitution are generalized.

Kimball (35) listed the possible spatial arrangements for compounds in which coordination numbers from one to eight are exhibited. Pólya's theorem may be used to calculate the number of isomers in coordination compounds. Using models and the method of Lunn & Senior (36), Marchi, Fernelius & McReynolds (37) developed isomer tables for compounds of elements exhibiting coordination number 8.

Senior (38) has investigated the conditions that the molecular formula of a compound must necessarily fulfill in order to be a unimer, i.e., one without any structural isomers. These conditions are set forth in nine theorems (39), and a complete list of unimeric types is given. It is important to note that unimers play a crucial role in the development of structural organic chemistry.

Graph theory has been recently applied to the treatment of chemical equilibria by Van Wazer and his associates (40). Dayhoff, Lippincott & Eck (41) have computed the concentrations of a large number of compounds of biological interest which would be present in the atmosphere under many combinations of temperature, pressure, and elemental composition. These studies throw light on the "rotational theory" of the evolution of chemical reactions and the origin of life.

ENTHALPIES OF FORMATION AND BOND ENERGIES

Several attempts were made in recent years to calculate the energies of the molecules using quantum mechanical, semi-empirical, and empirical methods. For most of the molecules studied, experimental values for the electronic energies are not available. Several authors preferred to use for comparison the enthalpies of atomization referred to 25° C. Such values include the thermal energies of the atoms and molecules and the zero-point energies of the fundamental modes of nuclear vibration. The general consensus (42, 43, 44) now is that, as far as the bond additivity schemes are concerned, it does not make much difference whether we use the electronic energies or the enthalpies of atomization appropriate to 0° K or 298.15° K. Although a final word on this point remains to be said, in this review we choose the enthalpies of atomization at 298.15° K.

The following values in kcal/g atom (10) for the enthalpies of formation of atoms at 298.15°K were used: C, 171.29; H, 52.10; F, 18.88; Cl, 29.08; Br, 26.74; I, 25.54; O, 59.56; S, 66.64; N, 112.98.

The data on enthalpies of formation are from API Research Project 44 tables (45), unless stated otherwise.

Proposed Models and Their Approximations

The earliest of all the models, viz., the Fajan's model (46), introduced in 1920, associates the energy of a molecule with the constituent bonds. Approximations based on the Fajan's model may be described as the zeroth-order approximations. Posed with the problem of nontransferability of the bond energy terms from one molecule to another, several authors sought to classify the bonds. Laidler (47) sought to explain the energies of the paraffins on the basis of three types of C–H bond and one type of C–C bond. Tatevskii (48) used ten types of C–C bond and three types of C–H bond. Bond length-bond energy relationships were used by several authors (49–54) to obtain the bond energy values for bonds in olefins and acetylenes. Somayajulu (55), following Fox & Martin (56), used Sutherland's (57) relationship between bond length, bond energy, and force constant to obtain bond energy values applicable for bonds in olefins and acetylenes. Several semi-empirical methods were used by others (58–61) to calculate bond energies. Some of the above methods are discussed in a later section.

Zahn (62), as long ago as 1934, recognized the inadequacy of the assumption of bond additivity and suggested association of part of the energy of the molecule with the bonds and part with pairs of bonds to a single atom. Dewar & Petit's (63) work lends support to Zahn's model. Zahn's model, which may serve as the first-order approximation, was the basis for Allen's scheme (43). As a further modification, Allen incorporated in his scheme the mutual or co-operative interactions between trios of bonds attached to a single atom (referred to hereafter as the trigonal interactions). Allen's scheme, with the inclusion of the trigonal interactions, follows from a model (LCBO approximation) suggested by Brown (64). The next step in the estimation of the energies of the paraffinic hydrocarbons involves corrections for the steric repulsion between the hydrogen atoms in close contact such as 1, 4 and 1, 5 C–H bonds in *gauche* configurations (43, 65). Tatevskii & Papulov (66) made attempts to show the equivalence of the pair bond interaction scheme to the bond additivity scheme. The chemical bond in molecular quantum mechanics was discussed by Allen & Shull (67).

In what follows next we shall examine the formulas based on the above mentioned models and the "magic" formula proposed by Mulliken (68).

Other formulas not examined in this review may be found in a recent review by Skinner & Pilcher (44).

Zahn's formula.—Zahn proposed the following expression for the total binding energy of a molecule:

$$E = \sum_{ij}^{\text{bonds}} R_{ij} + \sum_{i}^{\text{atoms}} \sum_{j,j'}^{\text{pairs}} R_{jij'}$$

where R_{ij} is the energy of the bond between the atoms X_i and X_j, and $R_{jij'}$ is the energy to be associated with the pair of bonds X_j–X_i–$X_{j'}$. These two types of energies are indeterminate as far as the thermochemical data are concerned without a priori calculations. However, under the simple condition,

$$R_{jij'} - 1/2(R_{jij} + R_{j'ij'}) = 0 \qquad 1.$$

the atoms X_i and X_j with valences v_i and v_j respectively, may be considered to contribute additional energies $\frac{1}{2}(v_i-1)R_{jij}$ and $\frac{1}{2}(v_j-1)R_{iji}$ respectively, independent of the numbers of bonds of the various types present. The energy of the molecule may then be expressed as the sum of apparent bond energies E_{ij}, given by the expression:

$$E_{ij} = R_{ij} + 1/2(v_i - 1)R_{jij} + 1/2(v_j - 1)R_{iji}$$

The R_{ij} values may be called the realistic bond energies.

If Condition 1 is not valid, we may write,

$$R_{jij'} - 1/2(R_{jij} + R_{j'ij'}) = \delta_{jij'}$$

so that

$$E = \sum_{ij}^{\text{bond types}} N_{ij}R_{ij} + \sum_{i}^{\text{atoms}} \sum_{j,j'}^{\text{pairs}} [1/2(R_{jij} + R_{j'ij'}) + \delta_{jij'}]$$

and in turn

$$E = \sum_{ij}^{\text{bond types}} N_{ij}E_{ij} + \sum_{i}^{\text{atoms}} \sum_{j,j'}^{\text{pairs}} \delta_{jij'}$$

where $\delta_{jij'}$ is the correction term representing the deviation from additivity of pair bond interaction energies and differs from zero only when $j \neq j'$.

On the basis of Zahn's model, we may write the following expressions for the paraffins, C_nH_{2n+2}

$$E_{C-H} = R_{C-H} + 3/2R_{H-C-H}$$
$$E_{C-C} = R_{C-C} + 3R_{C-C-C}$$
$$\delta_{H-C-C} \equiv R_{H-C-C} - 1/2(R_{H-C-H} + R_{C-C-C})$$
$$E(CH_4) = 4E_{C-H}$$
$$E(C_2H_6) = 6E_{C-H} + E_{C-C} + 6\delta_{H-C-C}$$
$$E(C_3H_8) = 8E_{C-H} + 2E_{C-C} + 10\delta_{H-C-C}$$

For the general case of a paraffinic hydrocarbon, C_nH_{2n+2}, we may write

$$E(C_nH_{2n+2}) = (2n + 2)E_{C-H} + (n - 1)E_{C-C} + (3n_p + 4n_s + 3n_t)\delta_{H-C-C} \qquad 2.$$

where n_p, n_s and n_t represent respectively the number of the primary, secondary, and tertiary C atoms in the molecule.

Zahn's formula with the inclusion of trigonal interactions.—Before we discuss the merits and demerits of the Zahn's formula, we may derive the formula with the inclusion of the trigonal interactions, defined previously.

There are four kinds of trios in paraffins, HHH, HHC, HCC, and CCC, leading to trigonal interaction terms, such as R_{HHH}, etc., so that by defining new realistic bond energy terms R'_{C-H} and R'_{C-C}, we have the following new expressions for apparent bond energies,

$$E_{C-H} = R'_{C-H} + 3/2R_{H-C-H} + R_{HHH}$$
$$E_{C-C} = R'_{C-C} + 3R_{C-C-C} + 2R_{CCC}$$

where,

$$R_{HHH} \equiv R_{C-H} - R'_{C-H}; \qquad R_{CCC} \equiv 1/2(R_{C-C} - R'_{C-C})$$

Now we introduce two correction terms δ_{HHC} and δ_{HCC}, defined by,

$$\delta_{HHC} \equiv R_{HHC} - 2/3R_{HHH} - 1/3R_{CCC}$$
$$\delta_{HCC} \equiv R_{HCC} - 1/3R_{HHH} - 2/3R_{CCC}$$

and obtain the expressions for the energies of the alkanes in the form,

$$E(CH_4) = 4E_{C-H}$$
$$E(C_2H_6) = 6E_{C-H} + E_{C-C} + 6\delta_{H-C-C} + 6\delta_{HHC}$$
$$E(C_3H_8) = 8E_{C-H} + 2E_{C-C} + 10\delta_{H-C-C} + 8\delta_{HHC} + 2\delta_{HCC}$$
$$E(C_nH_{2n+2}) = (2n + 2)E_{C-H} + (n - 1)E_{C-C} + (3n_p + 4n_s + 3n_t)\delta_{H-C-C} + (3n_p + 2n_s)\delta_{HHC} + (2n_s + 3n_t)\delta_{HCC}$$

By combining the correction terms for pair bond interaction and trigonal interactions into three new interaction contants, δ_p, δ_s, and δ_t, defined by:

$$\delta_p \equiv \delta_{H-C-C} + \delta_{HHC}$$
$$\delta_s \equiv \delta_{H-C-C} + 1/2\delta_{HHC} + 1/2\delta_{HCC}$$
$$\delta_t \equiv \delta_{H-C-C} + \delta_{HCC}$$

we can simplify the expression for the energy of a paraffin to read

$$E(C_nH_{2n+2}) = (2n + 2)E_{C-H} + (n - 1)E_{C-C} + 3n_p\delta_p + 4n_s\delta_s + 3n_t\delta_t \qquad 3.$$

Thus the inclusion of trigonal interactions in Zahn's model affords corrections for the primary, secondary, and tertiary character of the carbon atom. This may be interpreted to mean that in the Zahn's model with the inclusion of trigonal interactions (to be called hereafter the "new Zahn's model") the value of δ_{H-C-C} depends upon whether the C–H bond is primary, secondary, or tertiary and the constant value for δ_{H-C-C}, as envisaged in the simple theory of Zahn, cannot be used.

Substituting the value of E_{C-H} obtained from the enthalpy of atomization of methane, and the value of E_{C-C} obtained from the enthalpy of atomization of diamond in the expression for the energy of ethane, we may calculate the value for δ_p. From the constant increment in the enthalpy of atomization per methylene group, the value of $\delta_p + \delta_t$, and in turn the values of δ_s and δ_t may be determined. Following are the values (in kcal/mole) for the various parameters: $E_{C-H} = 99.395$; $E_{C-C} = 85.42$; $\delta_p = -1.06$; $\delta_s = -0.95$; $\delta_t = -0.84$.

The new Zahn's formula (Equation 3) can be transformed into Allen's formula as shown below.

Allen's formula.—Allen's formula for the energy of a paraffin may be derived if we write

$$E_{C-H} = R'_{C-H} + 3/2R_{H-C-H} + R_{HHH} \text{ (as before)}$$
$$E'_{C-C} = R'_{C-C} + 6R_{H-C-C} - 3R_{H-C-H} + 6R_{HHC} - 4R_{HHH}$$

and introduce two constants, Γ and Δ, defined by

$$\Gamma \equiv R_{H-C-H} + R_{C-C-C} - 2R_{H-C-C} + 2R_{HHH} + 4R_{HHC} + 2R_{HCC}$$
$$\Delta \equiv R_{CCC} - 3R_{HCC} + 3R_{HHC} - R_{HHH}$$

We obtain

$$E(C_nH_{2n+2}) = (2n+2)E_{C-H} + (n-1)E'_{C-C} + (3n_p + n_s - 6)\Gamma + (n_t + 4n_q)\Delta \quad 4.$$

The term $(3n_p + n_s - 6)$ represents the number of C–C–C pairs, and the term $(n_t + 4n_q)$ represents the number of nonbonded CCC trios in the molecule. The value of $E'_{C-C} = 79.05$ kcal/mole. The parameters Γ and Δ have the values (in kcal/mole) $\Gamma = 2.57$ and $\Delta = -0.67$. Equations 3 and 4, which are equivalent, have the same accuracy in the prediction of the enthalpies of atomization of the alkanes. It is important to note that this is the first justification of Allen's formula through the general Zahn's formula involving trigonal interactions.

Laidler's formula.—The new Zahn's formula can be readily transformed into Laidler's Formula 5 by rewriting Equation 3 as follows:

$$E(C_nH_{2n+2}) = (n-1)E_{C-C} + 3n_p(E_{C-H} + \delta_p) + 2n_s(E_{C-H} + 2\delta_s) + n_t(E_{C-H} + 3\delta_t) \quad 5.$$

where the terms $(E_{C-H} + \delta_p)$, $(E_{C-H} + 2\delta_s)$ and $(E_{C-H} + 3\delta_t)$ represent respectively the strengths of the primary, secondary, and tertiary C–H bonds. These bonds have the following values,

$$E_{C-H}\text{ (primary)} = 98.335 \text{ kcal/mole}$$
$$E_{C-H}\text{ (secondary)} = 97.495 \text{ kcal/mole}$$
$$E_{C-H}\text{ (tertiary)} = 96.875 \text{ kcal/mole}$$

Skinner & Pilcher (44) have shown that Allen's formula 4 can be transformed into Laidler's formula by evaluating Laidler's parameters as follows:

$$E_{C-H}\text{ (primary)} = E_{C-H} - 1/2\Gamma - 1/3\Delta$$
$$E_{C-H}\text{ (secondary)} = E_{C-H} - \Gamma - \Delta$$
$$E_{C-H}\text{ (tertiary)} = E_{C-H} - 3/2\Gamma - 2\Delta$$
$$E_{C-C} = E_{C-C} + 3\Gamma + 2\Delta$$

Thus Equations 3, 4 and 5 are mathematically equivalent and yield identical results for the alkanes. The calculated enthalpies of atomization are in good agreement with experiment for the *n*-alkanes but deviate for the branched hydrocarbons. These deviations, of course, can be accounted for in terms of repulsion between 1, 4 C–H bonds, 1, 5 C–H bonds, etc., in *gauche* conformations. Allen (43) deducts 0.5 kcal/mole for each pair of 1,4 C–H bonds *gauche* to one another. Skinner (65) refined the method of estimation of steric repulsion energy on the basis of an "angle release mechanism" suggested by him. In Table VI we have compared the errors in the prediction of the enthalpies of atomization on the basis of Equations 3, 4, and 5 with the steric repulsion terms calculated by Skinner (65). Although there seems to be general agreement between the above two quantities, we consider that the theory of steric repulsion is in its infancy, awaiting further development.

Other correlation procedures (48, 69, 70, 71) which explicitly consider steric effects have recently been compared for branched alkanes by Skinner & Pilcher (44).

Allen's formula was applied by McCullough & Good (72) to alkanethiols

TABLE VI

ERRORS IN THE PREDICTED VALUES OF ENTHALPIES OF ATOMIZATION OF ALKANES[a]

Compound	ΔH°_a (obs.) at 298.15° K kcal/mole	Error (calc.-obs.) kcal/mole	[*S*] kcal/mole
Propane	955.49	0.35	
n-Butane	1236.31	−0.06	
2-Methylpropane	1238.31	−0.16	
n-Pentane	1516.65	0.01	
2-Methylbutane	1518.57	−0.01	0.33
2,2-Dimethylpropane	1521.32	0.38	
n-Hexane	1797.10	−0.03	
2-Methylpentane	1798.80	0.17	0.33
2,3-Dimethylbutane	1799.63	1.24	1.50
n-Heptane	2077.52	−0.04	
2-Methylhexane	2079.23	0.15	0.33
2,2,3-Trimethylbutane	2081.59	2.83	3.90
n-Octane	2357.94	−0.05	
2-Methylheptane	2359.62	0.17	0.33
2,2,3,3-Tetramethylbutane	2362.11	5.86	7.50
n-Nonane	2638.35	−0.05	
3,3-Diethylpentane	2639.05	4.29	4.96
2,2,3,3-Tetramethylpentane	2640.31	8.07	8.74
2,2,3,4-Tetramethylpentane	2640.25	6.89	
2,2,4,4-Tetramethylpentane	2641.44	6.94	
2,3,3,4-Tetramethylpentane	2640.07	7.07	8.30

[a] This table uses Equations 3, 4, and 5 and their comparison with the steric repulsion terms estimated by Skinner (65). For each C_n alkane only, the least-branched and the most highly branched isomers were chosen.

and other sulfur compounds. Skinner (65) applied Allen's scheme to many other types of paraffins and also to substituted paraffins. Pilcher, Pell & Coleman (73) applied Allen's scheme to ethers. Since the Equations 3 and 5 are mathematically equivalent to Allen's equation (Equation 4), both of these equations could be applied to the substituted compounds. Among these, Equation 5 is the one commonly used by those who wish to treat the energy of a molecule as the sum of contributions from the bonds. An analogous equation based on the pair bond interaction theory is the new Zahn's equation. We consider the new Zahn's equation to be the simplest of the lot and conceptually the best. Also, we have investigated the role of the trigonal interactions by analysing the data for the substituted methanes with and without the trigonal interactions as shown in Table VII. The expressions used for the energies of the substituted methanes with the inclusion of the trigonal interactions are as follows:

$$E(CH_3X) = 3E_{C-H} + E_{C-X} + 3\delta_{H-C-X} + 3\delta_{HHX}$$
$$E(CH_2X_2) = 2E_{C-H} + 2E_{C-X} + 4\delta_{H-C-X} + 2\delta_{HHX} + 2\delta_{HXX}$$
$$E(CHX_3) = E_{C-H} + 3E_{C-X} + 3\delta_{H-C-X} + 3\delta_{HXX}$$
$$E(CX_4) = 4E_{C-X}$$

From reference to Table VII we see that little advantage is gained through the inclusion of trigonal interactions. The role of the trigonal interactions thus appears to be overemphasized.

TABLE VII

ENTHALPIES OF ATOMIZATION OF SUBSTITUTED METHANES (298.15° K)

Compound	ΔH°_a (in kcal/mole)		
	Observed (74)	Calculated without trigonal interactions	Calculated with trigonal interactions
CH_3F	404.97	405.82	
CH_2F_2	418.75	420.11	420.78
CHF_3	442.63	440.44	
CH_3Cl	376.27	376.74	
CH_2Cl_2	356.05	355.71	355.47
$CHCl_3$	334.63	334.51	
CH_3Br	362.84	363.60	
CH_2Br_2	329.97	330.39	330.67
$CHBr_3$	299.12	297.94	
CCl_3F	343.81	344.70	
CCl_2F_2	379.21	380.84	381.45
$CClF_3$	424.01	421.54	

The bond energies and correction terms used for the calculation of the enthalpies of atomization of the substituted methanes, paraffins, and other compounds are presented in Table VIII.

TABLE VIII

BOND ENERGIES AND PAIR BOND INTERACTION CONSTANTS BASED ON NEW ZAHN'S MODEL (298.15° K)

Bond	E (kcal/mole)	Bond pair	$-\delta$ (kcal/mole)
C–C	85.42	H–C_p–C	1.06
C–H	99.395	H–C_s–C	0.95
		H–C_t–C	0.84
C–F	116.70	H–C–F	3.02
C–Cl	78.28	H–C–Cl	−0.09
		F–C–Cl	2.28
C–Br	66.56	H–C–Br	0.38
C–I	53±1	H–C–I	1.
S–S	63.40		
S–H	87.83	H–S–S	2.85

Application of Zahn's model to olefins and aromatic hydrocarbons.—Zahn's model yields the following expressions for ethene and 1,3-butadiene:

$$E(C_2H_4) = 4R_{=C-H} + R_{C=C} + 2R_{H-\underset{\|}{C}-H} + 4R_{H-C=C}$$

$$E(C_4H_6) = 6R_{=C-H} + 2R_{C=C} + R_{=C-C=} + 2R_{H-\underset{\|}{C}-H} + 6R_{H-C=C} + 2R_{H-\underset{\|}{C}-\underset{\|}{C}} + 2R_{C-C=C}$$

We may now introduce a constant $\delta_{=C-H}$, defined by,

$$\delta_{=C-H} \equiv R_{=C-H} - 1/2(R_{=C-C=} + R_{H-H}) + R_{H-\underset{\|}{C}-H} + R_{H-C=C} - R_{H-\underset{\|}{C}-\underset{\|}{C}} - R_{C-C=C}$$

and write

$$2E(C_2H_4) = E(C_4H_6) + E(H_2) + 2\delta_{=C-H}$$

By using the above expression and the enthalpies of atomization of ethene, 1,3-butadiene, and hydrogen, the value of $\delta_{=C-H}$ was found to be -0.67 kcal/mole.

For the energy of the Kekulé form of benzene we may write the following expression:

$$3E(C_2H_4) = E(C_6H_6;\ \text{Kekulé form}) + 3E(H_2) + 6\delta_{=C-H}$$

The resonance energy of benzene may be now obtained as follows:

$$\begin{aligned}\Delta H^\circ_a(C_6H_6) - \Delta H^\circ_a\ (\text{Kekulé form}) &= 1320.52 - 1298.83 \\ &= 21.69\ \text{kcal/mole}\end{aligned}$$

Nearly the same value was obtained for the resonance energy of benzene by Dewar & Schmeising (51), but their method was criticized by Skinner & Pilcher (44).

Using E_{ij} values we may write the following expressions for the energies of ethylene, 1,3-butadiene, 1,3,5-hexatriene, and benzene:

$$E(C_2H_4) = 4E_{=C-H} + E_{C=C}$$
$$E(C_4H_6) = 6E_{=C-H} + 2E_{C=C} + E_{=C-C=} + 2\delta_{=C-H}$$
$$E(C_6H_8) = 8E_{=C-H} + 3E_{C=C} + 2E_{=C-C=} + 4\delta_{=C-H}$$
$$E(C_6H_6) = 6E_{=C-H} + 3E_{C=C} + 3E_{=C-C=} + 6\delta_{=C-H} + RE$$

Linnett (75) has made efforts to eliminate phrases such as "attainment of octet," " stabilization by resonance," "exchange forces," "hybridization," etc. On the basis of Zahn's model the resonance energy of benzene may be expressed by means of two new correction terms.

The E_{ij} values in olefins cannot be determined from the thermochemical data alone, however, approximate values which conform to some assumption may be used. Some methods for obtaining the E_{ij} values for olefins are discussed in a later section.

Mulliken's formula.—Mulliken (68) has shown, using both the LCAO-MO and VB methods, that the enthalpy of atomization of a molecule with all electrons paired but without the correction for the zero-point vibrational energy may be represented, to a first approximation, by the equation:

$$D^\circ_0 = \Sigma X_{ij} - 1/2\Sigma Y_{kl} + 1/2\Sigma K_{mn} - \Sigma VSE + RE \qquad 7.$$

where RE is the resonance energy, if present; $\Sigma\, VSE$ is the energy of excitation of the atoms in the molecule to their respective valence states; X_{ij} represents the strength of the electron-pair bond involving the atomic orbitals i and j; Y_{kl} represents the interaction terms between homogeneous nonbonded pairs of electrons (σ, σ; π, π); and K_{mn} represent the interaction terms between heterogeneous nonbonded pairs of electrons (σ, π). Both X and Y values may be expressed as functions of the overlap integrals S and are given by the following equations:

$$X_{ij} = A_j S_{ij} \bar{I}_{ij}/(1 + S_{ij}) \qquad 8.$$
$$Y_{kl} = A_k S_{kl}^2 \bar{I}_{kl} \qquad 9.$$

where A values are coefficients to be adjusted empirically and the $\bar{I}$ values are the mean ionization potentials.

The S_{ij} quantities are to be calculated theoretically. These may be taken from the theoretically computed tables (76). The Y_{kl} quantities are of three kinds: (*a*) lone-pair: lone-pair, Y_{kl} quantities; (*b*) lone-pair: bond-pair, Y_{kl} quantities; and (*c*) bond-pair: bond-pair, Y_{kl} quantities. The Y_{kl} quantities may be calculated using Equation 9.

The K values may be calculated theoretically or may also be estimated. Valence state energies can be calculated with the help of spectroscopic data on atoms (77). The mean ionization potential can be obtained in two steps by (*a*) averaging intraatomically, for each AO, over valence state I values for all electrons in the same valence shell as the given AO, and (*b*) then averaging interatomically. The mean ionization potentials may be taken from Hinze & Jaffe's (78) article.

Following Mulliken, Pedley (79) expresses the enthalpy of atomization of a paraffin by means of the equation

$$\Delta H^\circ_a = \Sigma X_{C-H} + \Sigma X_{C-C} - \Sigma R_{H_1H_1} - \Sigma R_{H_1H_2} - \Sigma R_{C_1H_2} - \Sigma R_{C_1C_3}$$

where the repulsive terms are as follows: $R_{H_1H_1}$, between hydrogen atoms attached to the same carbon atom; $R_{H_1H_2\,(cis)}$ and $R_{H_1H_2\,(trans)}$, for hydrogen atoms on adjacent carbon atoms; $R_{C_1H_2}$, the repulsion between a carbon atom and a hydrogen atom attached to an adjacent carbon atom; and $R_{C_1C_3}$, the repulsion between two methyl or methylene groups attached to the same carbon atom.

Pedley found that the HH, CH, and CC repulsion terms could be represented by the equations:

$$R_{HH} = 78.4/r_{HH}^4 \text{ (kcal/mole)}$$
$$R_{CH} = 113.0/r_{CH}^4 \text{ (kcal/mole)}$$
$$R_{CC} = 376.1/r_{CC}^4 \text{ (kcal/mole)}$$

so that with appropriate interatomic distances he obtained the following values for the repulsive terms in alkanes (kcal/mole):

$$R_{H_1H_1} = 7.70$$
$$R_{H_1H_2(cis)} = 2.05$$
$$R_{H_1H_2(trans)} = 0.90$$
$$R_{C_1H_2} = 5.26$$
$$R_{C_1C_3} = 10.59$$

With these values Pedley calculated the strengths of C–H and C–C bonds to be 111.04 kcal/mole and 102.70 kcal/mole, respectively. The enthalpies of atomization of paraffins on the basis of the above constants were calculated by Pedley but they were found to differ appreciably from the experimental values. Pedley associated the above discrepancy with the deviation of the H–C–H and C–C–C bond angles from the tetrahedral. Lide, Jr. (80), finds the H–C–H angle in propane to be 3.4° smaller than the tetrahedral value, while the C–C–C angle is 2.9° larger than the tetrahedral. The C–C–C angles in the higher normal paraffins fall in the range 111.9°–112.9°. These deviations may be explained on the basis of changes in the hybridization of the carbon orbitals (81). Zahn's model ignores such minor changes in bond angles and bond lengths from the normally assumed values and absorbs the effects of such deviations into the pair bond interaction terms. This is justified by Mizushima & Shimanouchi's (82) estimated values for the energy necessary to change the C–C–C bond angle by 10° as 0.7 kcal/mole and for the energy necessary to change the C–C bond distance by 0.1 Å as 2.9 kcal/mole.

An electrostatic model for covalent molecules proposed by Searcy (83) considers the effects of dissimilar atoms and the lone-pair electrons on the bond angles. Searcy estimated the electrostatic repulsion numbers (ERN) for various substituents. His model enables the prediction of bond angles with a considerable degree of accuracy. It appears that Searcy's procedure might be used for estimating the pair-bond interaction terms for the Zahn model.

Internal Rotation Barriers

Considerable progress has been made in recent years on the evaluation of the internal rotation barriers in several molecules. The Hückel theory, with an extended set consisting of 2*s* and 2*p* carbon and 1*s* hydrogen orbitals and with inclusion of overlap and all interactions, was shown by Hoffmann (84) to yield a good qualitative solution of most hydrocarbon conformational problems. For C–C, 1.54 Å, C–H, 1.10 Å and tetrahedral angles, the difference in energy between an eclipsed and staggered form of ethane was computed to be 4.0 kcal/mole. The experimental value is 2.875 ± 0.125 kcal/mole (85). Pitzer and Lipscomb (86), using SCF approximation and Slater orbitals, have calculated the energies for both the eclipsed and staggered forms of ethane. The calculated value of the barrier to internal rotation is 3.3 kcal/mole in good agreement with experiment.

Mulliken's formula 9 was used by Cignitti & Allen (87) for the evaluation of the nonbonded HH repulsion energies of the eclipsed and staggered forms of ethane. They found the contribution of all HH repulsions to internal rotation barrier to be 1.34 kcal/mole.

Simmons & Williams (88) used Mulliken's semi-empirical formula

$$E_{HH} = A_{HH} I S^2 / 2(1 - S^2)$$

to evaluate the HH repulsion energies of both the staggered and eclipsed forms of ethane. They chose a value of 0.95 for A_{HH} instead of the value 0.65 used by Mulliken. They neglected the *trans* interactions in the staggered form. Under these assumptions the estimated value for the rotation barrier of ethane is 2.85 kcal/mole in good agreement with experiment. Rotation barriers calculated in the above way for several other molecules were found to be in good agreement with experiment.

It may be noted that neglect of HH (*trans*) interactions for the staggered form increases the Cignitti & Allen's value for the rotation barrier in ethane from 1.34 to 2.46 kcal/mole. This may be true on the basis of screening by the C–C bond.

The origin of the barrier hindering internal rotation in ethane was recently discussed by Parr and others (89–92).

Bond Energy Terms in Alkenes and Alkynes

In most of the work of the past it was assumed that the C–H and C–C bond energies were constant in alkanes, alkenes, and alkynes. Several authors have shown in recent years that the C–H and C–C bond energies vary with the hybridization of the carbon orbitals. The empirical and semi-empirical methods to obtain the bond energy terms in alkenes, alkynes, and substituted paraffins are considered in this section.

Overlap integral as a measure of bond strength.—Maccoll (93) and later Mulliken (76) have pointed out that the overlap integral may be taken as a measure of the strength of the bond. Bloor & Gartside (58) assumed that the overlap integral is simply proportional to the bond energy and calculated the strengths of various types of CC bonds. Mulliken (94) suggested the use of overlap population as a measure of bond strength but this proved not to be very satisfactory. Another quantity tried by Mulliken (68) is the product of the overlap integral (S) and the mean ionization potential ($\bar{I}$) of the two atomic orbitals which form the bond. Recently, Peters (59) evaluated the quantity ($S\bar{I}$) for a number of bonds and showed that there was some improvement as compared with the overlap integral alone, but the correlation between this quantity and bond strength is still very poor. According to Peters, both the overlap integral and the overlap population may be useful in determining the strengths of a set of similar bonds, but these lack the necessary flexibility for dealing with the general bond. Both these schemes ignore the polarity of the bond.

Peters also showed that the potential energy of the overlap charge in the fields of the two atoms, which form a localized chemical bond, is a good guide to the strength of the bond. Like the overlap integral, however, it underestimates the strength of a strongly polar bond.

Mulliken (95) discussed the significance of LCAO-MO population analysis in a recent article.

Prediction of bond energies on the basis of bond lengths.—Glockler (49) has shown that the enthalpies of atomization of the paraffins can be consistently interpreted on the basis of the following expressions for the energies of the CC and C–H bonds:

$$E_{CC} = ar_{CC}^{-1} + br_{CC}^{-2} + cr_{CC}^{-3} + dr_{CC}^{-4}$$

$$E_{C-H} = A + Br_{C-H}^{-1}$$

where a, b, c, d, A and B are constants and r is the bond length.

Felchenfeld (50) also used the same procedure but used the following expressions for the energies of the CC and C–H bonds:

$$E_{CC} = k_1 r_{CC}^{-3}$$

$$E_{C-H} = k_2 r_{CH}^{-3}$$

Dewar & Schmeising (51) adopted three parameter relations of the form:

$$E = a_1 r_1^{-1} + a_2 r_2^{-2} + a_3 r_3^{-3}$$

for both carbon–carbon and carbon–hydrogen bonds.

According to Dewar & Schmeising (51), a relation between E and r should be such that E is finite at $r=0$ and vanishes as $r \to \infty$ and satisfies everywhere the following conditions

$$\frac{\partial E}{\partial r} < 0; \qquad \frac{\partial^2 E}{\partial r^2} > 0; \qquad E > 0 \ (r > 0)$$

None of the equations so far considered satisfy the above conditions. Dewar & Schmeising (51) showed that the tractrix function,

$$r = 1/b[a \log(a + (a^2 - E^2)^{1/2}) - a \log E - (a^2 - E^2)^{1/2}]$$

fulfills the necessary conditions.

Bernstein (54) questions the validity of the above mentioned bond energy-length relations on the basis that a monotonic energy-length curve cannot be expected to cover all types of CC bonds. However, for C–H bonds he proposes to use an equation of the form:

$$E = ar^2 + b$$

Skinner & Pilcher (44) also criticized the Dewar & Schmeising scheme on the ground that the tractrix equations for CC and for CH bonds in particular are extremely sensitive to experimental uncertainties in available bond-length data. According to Long (96) the intrinsic or *realistic* (R_{ij}) terms rather than the conventional bond energy terms should be related to bond lengths. Bernstein (97) has shown that the nonbonded repulsions also affect the bond lengths. Consequently bond energies predicted on the basis of bond lengths alone may not be of great use.

Relation between bond energy and average C–H stretching frequency.—Bernstein (98) has also shown that the average of the squares of the C–H stretching frequencies of an equivalent set of C–H bonds is to a high order of accuracy given by

$$\langle \nu_{C-H}^2 \rangle = k/4\pi^2 c^2 \mu$$

where the symbols have their usual meaning. The C–H bond energies when plotted versus the root-mean-square average ν_{C-H} fall on a smooth curve for C–H bonds in acetylene, ethylene, methane, and formaldehyde. The point for the C–H radical also falls on the same curve.

Sutherland's relation.—In 1938 Sutherland (57) showed that for diatomic molecules a simple potential function of the form

$$U = ar^{-m} - br^{-n}$$

where a, b, m, and n are constants and r is the internuclear distance, leads to the following important result

$$k_e r_e^2 / D_e = mn$$

Sutherland showed that the product mn did not vary much for a number of diatomic molecules. In 1939 Fox & Martin (56) applied this relation to CC bond energies and showed that the product mn is constant. They also showed that Sutherland's function resembles the Morse function in the immediate neighborhood of the equilibrium position ($r = r_e$), but agreement elsewhere can only be approximate. Varshni (99) reviewed a number of potential functions for diatomic molecules and showed that a number of other functions also lead to the Sutherland's relation. Somayajulu (55) used Sutherland's relation to derive bond energies for various bonds. The product mn for various types of bonds between two atoms X_i and X_j may be called the Sutherland's constant and denoted by S_{ij}. The S_{ij} values can be determined accurately from the molecular data for the diatomic molecules. The chief difficulty in extending this relation to bonds in polyatomic molecules concerns the force constants. The force constants derived on the basis of a force-field model with proper allowance for the nonbonded repulsion may be used. It is instructive to study Baughan's (100) comments on various force-field models. The bond energies obtained on the basis of S_{ij} values for diatomic molecules and the force constants based on valence force-field models for polyatomic molecules have been found to lead to bond energies which are accurate to within a few kcal.

BOILING POINT CORRELATIONS

Several attempts have so far been made to derive empirical and semi-empirical correlations between the normal boiling point of organic substances and the structures of their molecules. Such empirical correlations however, omit an essential step, namely the intermolecular forces, since the normal boiling point and other macroscopic bulk properties of liquids are determined by the force-fields acting between molecules. The relations between intermolecular forces and the bulk properties are imperfectly understood, and one has to turn to statistical thermodynamics for their elucidation. This is a task in itself, and so far there is not available a rigorous theoretical relation which is thermodynamically sound for the precise calculation (within $\pm 0.01°$ C) of the normal boiling point. We therefore have to settle for such empirical calculations whenever we desire precise correlation procedures for purposes of structural interpolation and extrapolation, though at times they may add little to our basic understanding.

Mention may be made of the attempts by Somayajulu & Palit (101) who used a relationship of the type for the normal boiling point

$$T_b = a(\Sigma Z)^p + b$$

where a, b, and p are constants for a given homologous series; ΣZ is the sum of the atomic numbers of the atoms in the molecules; and $p = \frac{1}{2}$ for n-alkyl fluorides, chlorides, bromides, iodides, acids, amines, aldehydes, ketones, n-alkyl benzenes, n-alkyl cyclohexanes, and n-alkyl cyclohex-1-enes, $p = \frac{1}{3}$ for n-alkenes, and n-alkynes, and $p = 1$ for n-alcohols, n-alkyl ethers, n-alkyl acetates, and n-alkanethiols.

We shall discuss six additional empirical and semi-empirical relations that have been proposed for the normal boiling point of homologous series. Each of these in turn will be examined by correlating the well established values for the normal boiling points of the normal alkanes. For n-alkanes, Cox (102) used a logarithmic relation for T_b (° K) as a function of molecular weight M, which is given by:

$$\log T_b = 1.07575 + 0.949128 \log M - 0.101 \log^2 M$$

while Francis (103) developed a polynomial in m to evaluate t_b:

$$t_b = -76.4 + 36.4m - 1.047m^2 + 0.016m^3 - 270m^{-1} + 216m^{-2}$$

Tatevskii et al. (48) proposed the following equation for the normal boiling point

$$t_b = \alpha - \beta/(\gamma + m)$$

where α, β, and γ are constants for n-alkanes with the following values:

$$\alpha = 782.179, \quad \beta = 16{,}401.98 \quad \text{and} \quad \gamma = 17.982.$$

Li et al. (104) used the relation

$$t_b = t_o + c \log (m - 1) + f(m) \qquad 10.$$

where t_b is the boiling point of a given homologue in ° C, t_o and c are constants for a given series, and f (m) is a complex function of the number of carbon atoms m with two additional constants. The formula is applicable for $m > 2$ and was the one used in the preparation of the American Petroleum Institute Research Project 44 tables and Manufacturing Chemists Association Research Project tables (105).

The contribution of each CH_2 group to molal refraction and thus to polarizability is additive, to a very good approximation, and is like the contribution to volume. This is, however, not the case with the cohesion energy and boiling point (106). If we assume that the hydrocarbons in the liquid are present mainly in the coiled configurations, then the inwardly directed parts of the chain will not contribute to the intermolecular interactions. Aten (107), therefore, took the cohesion energy to be proportional to the nearly spherical surface of such a coil, which will increase as $m^{2/3}$ with the number of CH_2 groups. Kurata & Isida (108) empirically found that the reciprocal of the boiling point of a given n-alkane is inversely proportional to $m^{2/3}$, thus supplementing Aten's empirical approach. Using T_b values from the American Petroleum Institute Research Project 44 tables, Kurata & Isida's expression for the normal boiling point of a n-alkane, fitted by the method of least squares, is given by:

$$1/T_b = 5.4364 \times 10^{-4} + 7.8665 \times 10^{-3} m^{-2/3}$$

Kreglewski & Zwolinski (109) following Kurata & Isida's approach have found that certain properties (y) of n-alkanes, such as normal boiling point, heat of vaporization, and the Antoine constant B, follow the simple law:

$$\log (y_\infty - y) = a - bm^{2/3}$$

where y_∞ is the value of a given property y for $m = \infty$ and a and b are constants. It is interesting to note from their work that n-alkyl compounds have the same value of T_∞ as n-alkanes. The equation of Tatevskii et al. also leads to a similar value for T_∞ for n-alkanes. Kreglewski & Zwolinski's equation involves only two constants and gives excellent fit for n-alkanes for $m \geq 3$ and for n-alkyl compounds for $m \geq 6$.

All of the above six procedures for correlating normal boiling points for homologous series are compared in Table IX using the n-alkanes as a basis. The deviations calculated by the Cox and Francis relations are not random. Among the remaining four relations, the average deviations favor the Li et al. equation. Although Kreglewski & Zwolinski's [K-Z] equation appears to be

TABLE IX

NORMAL BOILING POINT OF *n*-ALKANES

m	t_{obs} (° C)	Deviations (t°(calc)−t°(obs))					
		Cox	Francis	Tatevskii	Kurata & Isida	Li et al.	Kreglewski & Zwolinski
1	−161.49						
2	− 88.63		−0.03				
3	− 42.07	(−0.52)	−0.12		+0.12		−0.16
4	−0.50	−0.15	−0.03	+0.04	+0.18	(−0.87)	+0.08
5[a]	+36.074	−0.074	−0.004	−0.054	+0.006	−0.071	+0.109
6	68.740	−0.030	−0.020	−0.250	−0.122	+0.089	+0.163
7	98.427	−0.047	−0.007	−0.177	−0.289	+0.053	+0.136
8	125.665	−0.075	−0.055	−0.035	−0.448	−0.022	+0.060
9	150.798	−0.038	−0.078	+0.102	−0.533	−0.056	−0.005
10	174.123	+0.097	−0.063	+0.167	−0.534	−0.052	−0.051
11	195.890	+0.032	−0.040	+0.130	−0.463	−0.037	−0.092
12	216.278	+0.742	+0.002	−0.038	−0.311	−0.014	−0.119
13	235.434	+1.066	+0.086	−0.314	−0.072	+0.015	−0.126
14	253.515	+1.575	+0.195	−0.735	+0.221	+0.014	−0.143
15	270.614	+2.166	+0.366	−1.284	+0.580	−0.005	−0.156
16	286.793	+2.877	+0.677	−1.913	+1.031	−0.014	−0.137
17	302.15	+3.38	+1.14	−2.63	+1.55	−0.03	−0.11
[a] Average deviation beyond m=5)		0.94	0.21	0.60	0.47	0.04	0.11

slightly less accurate than the former, it is simple to use. The analysis further shows that for extrapolations to the higher members of the homologous series, only the Li et al. and the K-Z equations are to be preferred.

The unavailability of high purity samples and the associated difficulties of experimental measurements have thwarted the prediction of physical properties of high molecular weight hydrocarbons ($m > 20$). The following correlation scheme may be used. Assuming the correctness of the Antoine equation for the higher molecular weight hydrocarbons,

$$\log P = A - B/(C + t(°C))$$

we have developed the following power series expressions for the Antoine constants B and C in terms of the normal boiling points using the known API Research Project 44 values for the C_3 to C_{17} *n*-alkanes:

$$B = 3.53813T_b - 9.77736 \times 10^{-5}T_b^2 - 6.66695 \times 10^{-7}T_b^3$$

$$C = 273.15 - 4.49159 \times 10^{-2}T_b - 2.68408 \times 10^{-4}T_b^2 - 5.18608 \times 10^{-8}T_b^3$$

To apply these expressions, we need to predict the normal boiling point, for then A, B and C are known. For this purpose, the Li et al. and the K-Z equations were chosen, and the predicted boiling points at 5 *mm*, 10 *mm*, and 760 *mm* Hg compared with available API Research Project 42 values (110) in Table X.

TABLE X

BOILING POINTS (° C) OF *n*-ALKANES (C_{20} TO C_{100})

m	At 5 mm Hg			At 10 mm Hg			At 760 mm Hg	
	Obs. (110)	Calc.[a]	Calc.[b]	Obs. (110)	Calc.[a]	Calc.[b]	Li et al.	K-Z[d]
20	181.0	181.2	181.3	196.5	196.6	196.8	343.8	344.0
23	212.5	211.2	211.6	228.0	227.3	227.7	380.1	380.6
24	220.5	220.5	220.9	238.0	236.8	237.2	391.3	391.8
26	240.0	238.0	238.6	256.5	254.6	255.3	412.2	412.9
28	255.5	254.4	255.1	272.0	271.3	272.1	431.6	432.5
32	284.0	284.1	284.9	299.5	301.6	302.4	466.6	467.5
35	300.0	304.2	304.9	314.5	322.0	322.7	490.0	490.8
36	311.0	310.5	311.1	326.5	328.4	329.0	497.4	498.0
40		334.3	334.3		352.6	352.6	524.8	524.8
50		385.1	382.0		404.0	401.0	582.6	579.2
60		426.8	418.7		446.3	438.1	629.5	620.5
70		462.4	447.6		482.3	467.3	668.9	652.5
80		493.4	470.5		513.6	490.5	702.9	677.9
90		520.9	489.1		541.3	509.3	732.8	698.2
94[c]		531.1	495.5		551.6	515.7	743.8	705.2
100		545.6	504.2		566.3	524.5	759.5	714.7

[a] Normal boiling point from Li et al.'s equation was used to calculate Antoine constants and hence boiling points at 5 mm and 10 mm Hg.

[b] Similar calculations performed using Kreglewski & Zwolinski's equation for normal boiling point.

[c] *n*-Tetranonacontane synthesized by Dixon.

[d] Kreglewski & Zwolinski.

It should be noted that the Li et al. and K–Z equations lead to practically the same values for *n*-alkanes up to $m=40$. Beyond $m=40$, the K-Z equation yields lower values reaching a limit of 1078° K for $m=\infty$. Unfortunately the boiling points for higher paraffinic hydrocarbons are not available for comparison.

Boiling point of polar liquids.—Assuming additivity for the enthalpy and entropy contributions, the transition temperature may be expressed as

$$T_t = \sum_i (\Delta H_i/\Delta S_i)$$

To this approximation, the boiling point T_b of a polar liquid may be treated (111) as the sum of contributions from the London dispersion forces T_L and the dipole forces T_D, so that

$$T_b = T_L + T_D$$

For the boiling point, T_{RX}, of a polar *n*-alkyl compound, Kreglewski & Zwolinski (112) use a comparative scheme assuming that the above two contributions may be represented as functions of the boiling point T_{RH} of the corresponding *n*-alkane. Their formula for the boiling point is

$$T_{RX} = \alpha T_{RH} + \beta T_{RX}^{-\gamma}$$

where α, β, and γ are constants for a given homologous series. This equation was found to be applicable for a number of homologous series where $m \geq 2$, except for *n*-alkanoic acids.

The following modification of the Kreglewski–Zwolinski equation

$$\log (T_{RX} - T_{RH}) = a - bm^{2/3}$$

was tested for a large number of *n*-alkyl compounds including the *n*-alkanoic acids where $m \geq 3$. The results and average deviations are given in Table XI.

TABLE XI

VALUES OF a AND b IN THE EQUATION, $\log (T_{RX} - T_{RH}) = a - bm^{2/3}$

Compound	a	b	T_b(calc.) − T_b(obs.) Av. dev.
1-Fluoroalkanes	2.018	0.2003	0.21
1-Chloroalkanes	2.147	0.0989	0.44
1-Bromoalkanes	2.235	0.0895	0.42
1-Iodoalkanes	2.321	0.0805	0.82
Primary *n*-alkyl amines	2.166	0.1085	1.36
n-Alcohols	2.461	0.1546	0.59
n-Alkanoic acids	2.490	0.1086	0.78
n-Alkanethiols	2.239	0.0955	0.61
2-Thiaalkanes	2.287	0.1195	0.72

For the calculation of dipole contribution T_D, to the boiling point, some approximate methods are available (111). We may, however, suggest a different method for the estimation of T_D based on a method for the estimation of T_L. For simple molecules like HF, T_L may be calculated as the arithmetic mean of the boiling points of H_2 and F_2. If we call one half of the boiling point of H_2 the atomic boiling point of H and similarly for F, we may write

$$T_L = \sum T_a$$

where T_a is the atomic boiling point and the summation is over all atoms in the molecule. This would be the contribution due to the covalent character of HF. The second part, i.e., T_D, represents the contribution from the ionic character of HF and is calculated by subtracting T_L from the known normal boiling point. For a few simple molecules we have estimated T_D using this procedure and found that it is approximately given by the expression

$$T_D = a(4\pi N\mu^2/9k)$$

where a is a constant which was found to be 10^{-2} for a number of simple compounds. Thus,

$$T_b = \sum T_a + a(4\pi N\mu^2/9k)$$

Substituting the appropriate values in the above equation, we obtain the dipole moment in terms of T_D.

$$\mu = 0.128\sqrt{T_D}$$

Calculated values of μ for various compounds using the above formula are shown in Table XII along with the observed values (113). In most cases the agreement is quite satisfactory.

TABLE XII

Dipole Moments of Some Molecules Calculated from Boiling Point Data

Compound	T_b° K	T_L° K	T_D° K	$\mu_{calc.}$	$\mu_{obs.}$
HF	292.67	52.69	239.98	1.98	1.94
HCl	188.12	129.74	58.38	0.98	1.08
HBr	206.43	176.14	30.29	0.70	0.78
HI	237.55	238.96	—	—	0.38
FBr	293.15	208.45	84.70	1.18	1.3
FCl	183.15	162.05	21.10	0.59	0.88
ICl	370.95	348.32	22.63	0.61	1.
O_3	161.85	135.24	26.61	0.66	0.6
H_2O	373.15	65.46	307.69	2.25	1.87
H_2S	212.81	110.10	102.71	1.29	1.10
H_2Se	231.15	140.12	91.03	1.22	
NH_3	239.72	69.25	170.47	1.67	1.47

Some papers of general interest include those by Somayajulu & Palit (114), who showed that the boiling point in a given series of halides may be treated as a function of either the atomic surface area or the Van der Waals volume of the halogen. Seifer & Smolenskii (115) derived a formula for the normal boiling point of monoalkyl alkanes representing the structural formula of an alkane as a tree. They have extended this approach to calculate other thermodynamic properties of monoalkyl alkanes (116). Anderson (117) extended Kinney's (118) equation to bivalent sulfur compounds, trivalent organophosphorus compounds, and trivalent organoarsenic compounds. Flory, Orwoll & Vrij (119) derived an equation of state for *n*-paraffins (C_5 to C_∞). Correlations for the prediction of normal boiling points of branched hydrocarbons were reviewed by Greenshields & Rossini (69).

RECENT THERMODYNAMIC DATA

The comprehensive bibliography of abstracted and published papers on chemical thermodynamics and thermochemistry together with the Substance-Property Index continues to be published in the *Bulletin of Thermodynamics and Thermochemistry* prepared under the auspices of the Commission on Thermodynamics and Thermochemistry of the Section of Physical Chemistry of the International Union of Pure and Applied Chemistry. The comprehensive bibliography and substance-property index covering the literature for the period 1 December, 1962, to 1 December, 1963, appeared in the IUPAC Bulletin No. 7 published in April 1964. The bibliography and substance-property index for the period 1 December, 1963, to 1 December, 1964, covered by this review will appear in the IUPAC Bulletin No. 8 to be published early in 1965. The preparation and indexing of the bibliography continues to be a joint effort. The work on organic compounds of carbon with elements of finding numbers 1 through 22 [elements preceding carbon in the Standard Order of Arrangement (8, 105)] is conducted by the Chemical Thermodynamic Properties Center, Department of Chemistry, Texas A & M University (under the direction of Zwolinski), and the work on inorganic and organometallic compounds by the Thermochemistry Section, National Bureau of Standards (under the direction of Wagman and Evans). Attention should be called to a new appendix in the 1964 IUPAC Bulletin No. 7 containing a list of 47 books, monographs, and data compilations of interest to thermodynamicists.

The amount of information on thermodynamic data which is currently being generated in the open literature is truly remarkable. The number of journal references in the 1963 IUPAC Bulletin of Thermodynamics and Thermochemistry No. 6 was 1282. The 1964 Bulletin No. 7 listed 1819 journal references in the bibliography. It is readily seen why the practice of reporting a comprehensive bibliography and substance-property index in the Annual Reviews of Physical Chemistry was discontinued after the appearance of Volume 13 in 1962.

In closing this review we would like to discuss briefly some additional topics of current interest.

Noble gas compounds.—Ever since the discovery of compound formation between xenon and platinum hexafluoride by Bartlett & Lohmann (120) in 1962, the existence of about twenty other noble gas compounds has been established, and experimental evidence has been obtained for a few more compounds (121). Following are the compounds established to-date: $XePtF_6$, $Xe(PtF_6)_2$, $XeRuF_6$, $XeRhF_6$, $XePuF_6$, XeF_2, XeF_4, XeF_6, XeO_3, $XeOF_4$, H_4XeO_6, Ag_4XeO_6, Na_4XeO_6, $Xe(OH)_6$, Na_6XeO_6, K_6XeO_6, Ba_3XeO_6, $XeF_2 \cdot 2TaF_5$, $XeF_2 \cdot 2SbF_5$, KrF_2, KrF_4, and RnF_4. Experimental evidence has been obtained for the existence of the following compounds: $XeOF_2$, $XeOF_3$, XeO_2F_2, XeO_2F_4, and XeO_3F_2.

For information on the chemistry of noble gases, we refer the reader to the review articles by Bartlet (122), Hyman (123), and Chernick (124) and to the following recent books: *Noble Gas Compounds*, edited by Hyman(125) summarizing a symposium held at the Argonne National Laboratories in 1963, and *Noble Gases and Their Compounds* (Macmillan Company, New York, 1964) by Moody & Thomas (121). The currently available thermodynamic data for some of the noble gas fluorides are presented in Table XIII. It is fairly evident that additional precise thermodynamic data is badly needed.

Biological thermodynamic data.—The application of modern thermochemical and thermodynamic techniques to isolated substances important in

TABLE XIII

THERMODYNAMIC DATA FOR NOBLE GAS FLUORIDES

Compound	Form at room temperature	Melting point ° C	Boiling point ° C	Vapor pressure mm of Hg at 25° C	ΔH°_f(g; 25° C) (kcal/mole)	ΔH° Subl.) (kcal/mole)
XeF_2	Colorless crystal	140		3.8	-37 ± 10	12.3
XeF_4	Colorless crystal	100		3	-53 ± 5	15.3
XeF_6	Colorless crystal	46	76	30	−83	8.7
KrF_2	Colorless crystal				+5.3	
KrF_4	Colorless crystal	60 (dec.)		115	−0.5 to −53.0	8.8

biological systems as also to intact biological systems is still in its infancy. Wadsö presented an excellent review on biochemical calorimetry at the IUPAC Symposium on Thermodynamics and Thermochemistry held in Lund, Sweden (5), and also discussed thermodynamic data in biochemistry in the 1964 Gordon Conference on Critical Tables. Alekseeva & Ponomarev (126) have reported standard enthalpies of formation for several peptides. Brandts (127) studied the thermodynamics of the reversible denaturation of chymotrypsinogen. Wilhoit and his associates have been engaged in thermochemical studies of isolated biochemically important compounds and have reported enthalpies of combustion for several solid organic acids involved in the Krebs cycle (18, 128). Hutchens & Stout at the University of Chicago are continuing low-temperature calorimetric studies on levo amino acids (129–131). Heats of ionization are most important in unravelling energy changes in biological systems. Wadsö et al. have reported enthalpies of ionization for 11 sulfur compounds (132).

At the 19th annual Calorimetry Conference held in Washington, D. C. during October 1964, an afternoon session was devoted to bioenergetics. For his Huffman Memorial Lecture, J. M. Sturtevant discussed heats of biochemical reactions. T. H. Benzinger reported on key biothermodynamic data for ATP and DPNH.

ACKNOWLEDGMENTS

We wish to express our thanks to Mesdames Lesley J. Clague, Constance W. Haas, and June M. Husbands for their assistance in the preparation of the literature sources and to Professor Randolph C. Wilhoit for his helpful suggestions.

LITERATURE CITED

1. Jaynes, E. T., in *Statistical Physics*, **3**, 182–218 (W. A. Benjamin, Inc., New York, 1962); see also *Am. J. Phys.* (In press)
2. Tisza, L., *Rev. Mod. Phys.*, **35**, 151 (1963)
3. Rossini, F. D., Gucker, F. T., Jr., Johnston, H. L., Pauling, L., and Vinal, G. W., *J. Am. Chem. Soc.*, **74**, 2699 (1952)
4. *Natl. Bur. Std. (U.S.) Tech., News Bull.*, **47**, No. 10 (1963)
5. Plenary Lecture, in *Thermodynamics and Thermochemistry, IUPAC Lund Symp.* (Butterworths, London, 1964)
6. Rossini, F. D., *Proc. Intern. Conf. Nuclidic Masses, 2nd, Vienna, 1963* 99–100
7. Rossini, F. D. *Pure, Appl. Chem.*, 9 (In press)
8. NRC Office of Critical Tables *Natl. Acad. Sci.—Natl. Res. Council. Publ.* 976 (1962)
9. Standard Reference Data Committee, *Natl. Bur. Std. (U.S.) Tech. Note* **194 (1963)**
10. Wagman, D. D., *Natl. Bur. Std. (U.S.)* (Personal communication)
11. Good, W. D., Lacina, J. L., and McCullough, J. P., *J. Phys. Chem.*, **65**, 2229 (1961)
12. Scott, D. W., Good, W. D., and Waddington, G., *J. Am. Chem. Soc.*, **77**, 245 (1955)
13. Armstrong, G. T., *Experimental Thermochemistry*, **2**, 129–45 (Interscience, New York, 1962)
14. Wise, S. S., Margrave, J. L., Feder, H. M., and Hubbard, W. N., *J. Phys. Chem.* **66**, 381 (1962)
15. Wise, S. S., *The Heats of Formation of Some Inorganic Compounds by Fluorine Bomb Calorimetry* (Doctoral thesis, Univ. Wisconsin, Madison, Wis., 1962)
16. Good, W. D., *J. Phys. Chem.*, **66**, 380 (1962)
17. Pilcher, G., and Sutton, L. E., *Phil. Trans. Roy. Soc. London, Ser A*, **248**, 23 (1955)
18. Wilhoit, R. C., and Shiao, D., *J. Chem. Eng. Data*, **9**, 595 (1964)
19. Hubbard, W. N., Frow, F. R., and Waddington, G., *J. Phys. Chem.*, **65**, 1326 (1961)
20. Good, W. D., Lacina, J. L., Scott, D. W., and McCullough, J. P., *J. Phys. Chem.*, **66**, 1529 (1962)
21. Good, W. D., Scott, D. W., and Waddington, G., *J. Phys. Chem.*, **60**, 1080 (1956)
22. Hubbard, W. N., Katz, C., and Waddington, G., *J. Phys. Chem.*, 58 142 (1954)
23. Cayley, A., *Brit. Assoc. Advan. Sci. Rept.*, 257–305 (1875); *Collected Mathematical Papers* (Cambridge, 1889–98)
24. König, D., *Theorie der Endlichen und Unendlichen Graphen* (Akademie-Verlag, Leipzig, 1936; Chelsea, N. Y., 1950)
25. Berge, C., *The Theory of Graphs and its Applications* (Transl., Doig, Alison, Wiley, New York, 1962)
26. Henze, H. R., and Blair, C. M., *J. Am. Chem. Soc.*, **53**, 3077 (1931)
27. Iri, M., *RAAG Research Notes*, 2nd Ser., No. 35 (1957); Iri, M., *RAAG Memoirs*, **2**, 565 (1958); Kondo, K., and Iri, M., *RAAG Memoirs*, **2**, 42 (1958)
28. Henze, H. R., and Blair, C. M., *J. Am. Chem. Soc.*, **53**, 3042 (1931); **55**, 680 (1933); **56**, 157 (1934)
29. Blair, C. M., and Henze, H. R., *J. Am. Chem. Soc.*, **54**, 1098, 1538 (1932)
30. Pólya, G., *Compt. Rend.*, **201**, 1167 (1935); *Acta Mathematica*, **68**, 145 (1937)
31. Pólya, G. *Helv. Chem. Acta*, **19**, 22 (1936); *Z. Krist.*, **93**, 415 (1936)
32. Hill, T. L., *J. Chem. Phys.*, **11**, 294 (1943)
33. Hill, T. L., *J. Phys. Chem.*, **47**, 253 (1943)
34. Hill, T. L., *ibid.*, 413
35. Kimball, G. E., *J. Chem. Phys.*, **8**, 188 (1940)
36. Lunn, A. C., and Senior, J. K., *J. Phys., Chem.*, **33**, 1027 (1929)
37. Marchi, L. E., Fernelius, W. C., and McReynolds, J. P., *J. Am. Chem. Soc.*, **65**, 329 (1943)
38. Senior, J. K., *J. Chem. Phys.*, **19**, 865 (1951)
39. Senior, J. K., *Am. J. Math.*, **73**, 663 (1951)
40. Van Wazer, J. R., *Am. Scientist*, **50**, 450 (1962); Moedritzer, K., and Van Wazer, J. R., *J. Am. Chem. Soc.*, **86**, 802 (1964); Van Wazer, J. R., Moedritzer, K., and Matula, D. W., *J. Am. Chem. Soc.*, **86**, 807 (1964); Rausch, M. D., Van Wazer, J. R., and Moedritzer, K., *J. Am. Chem. Soc.*, **86**, 814 (1964); Matula,

D. W., Groenweghe, L. C. D., and Van Wazer, J. R., *J. Chem. Phys.*, **41,** 3105 (1964); Van Wazer, J. R., and Moedritzer, K., *J. Chem. Phys.*, **41,** 3122 (1964)
41. Dayhoff, M. O., Lippincott, E. R., and Eck, R. V., *Science*, **146,** 1461 (1964)
42. Cottrell, T. L., *The Strengths of Chemical Bonds*, 2/e, 103–22 (Butterworths, London, 1958)
43. Allen, T. L., *J. Chem. Phys.*, **31,** 1039 (1959)
44. Skinner, H. A., and Pilcher, G., *Quart. Rev.*, **17,** 264 (1963)
45. *Selected Values of Properties of Hydrocarbons and Related Compounds*, (*Am. Petrol. Inst. Res. Proj. 44*, Chem. Thermodyn. Properties Center, Texas A&M Univ. Coll. Sta. Tex., loose-leaf data sheets, 1964)
46. Fajans, K., *Ber. Deut. Chem. Ges.*, **53**(B), 643 (1920)
47. Laidler, K. J., *Can. J. Chem.*, **34,** 626 (1956)
48. Tatevskii, V. M., Benderskii, V. A., and Yarovoi, S. S., *Rules and Methods for Calculating the Physicochemical Properties of Paraffinic Hydrocarbons* (Transl.: Mullins, B. P., Pergamon, Oxford, 1961)
49. Glockler, G., *J. Chem. Phys.*, **21,** 1242, 1249 (1953); *J. Phys. Chem.*, **61,** 31 (1957); **62,** 1049 (1958)
50. Felchenfeld, H., *J. Phys. Chem.*, **61,** 1133 (1957)
51. Dewar, M. J. S., and Schmeising, H. N., *Tetrahedron*, **5,** 166 (1959); **11,** 96 (1960)
52. Mackle, H., and O'Hare, P. A. G., *Trans. Faraday Soc.*, **57,** 1521, 2119 (1961)
53. Cox, J. D., *Tetrahedron*, **18,** 1337 (1962)
54. Bernstein, H. J., *Trans. Faraday Soc.*, **58,** 2285 (1962)
55. Somayajulu, G. R., *J. Chem. Phys.*, **34,** 1449 (1961); *Univ. Calif. Rad. Lab.*, Rept. No. 9590 (1961)
56. Fox, J. J., and Martin, A. E., *J. Chem. Soc.*, **884** (1939)
57. Sutherland, G. B. B. M., *Proc. Indian Acad. Sci., Sect. A*, **8,** 341 (1938)
58. Bloor, J. E., and Gartside, S., *Nature*, **184, 1313** (1959)
59. Peters, D., *Trans. Faraday Soc.*, **60,** 1193 (1964); *J. Chem. Soc.*, 2003, 2015 (1963); 4017 (1964)
60. Klopman, G., *J. Am. Chem. Soc.*, **86,** 1463, 4550 (1964)
61. Ferreira, R., *J. Phys. Chem.*, **68,** 2240 (1964)
62. Zahn, C. T., *J. Chem. Phys.*, **2,** 671 (1934)
63. Dewar, M. J. S., and Petit, R., *J. Chem. Soc.*, 1625 (1954)
64. Brown, R. D., *J. Chem. Soc.*, 2615 (1953)
65. Skinner, H. A., *J. Chem. Soc.*, 4396 (1962)
66. Tatevskii, V. M., and Papulov, Yu. G., *Russ. J. Phys. Chem.* (*English Transl.*), **34,** 115, 231, 335 (1960); **36,** 93 (1962)
67. Allen, T. L., and Shull, H., *J. Chem. Phys.*, **35,** 1644 (1961)
68. Mulliken, R. S., *J. Phys. Chem.*, **56,** 295 (1952)
69. Greenshields, J. B., and Rossini, F. D., *J. Phys. Chem.*, **62,** 271 (1958)
70. Platt, J. R., *J. Chem. Phys.*, **15,** 419 (1947); *J. Phys. Chem.* **56,** 328 (1952)
71. Souders, M., Matthews, C. S., and Hurd, C. O., *Ind. Eng. Chem.*, **41,** 1048 (1949)
72. McCullough, J. P., and Good, W. D., *J. Phys. Chem.*, **65,** 1430 (1961)
73. Pilcher, G., Pell, A. S., and Coleman, D. J., *Trans. Faraday Soc.*, **60,** 499 (1964)
74. *Landolt-Börnstein Zahlenwerte und Functionen, 6 Auflage, II Band, 4 Teil* (Springer, Berlin, 1961)
75. Linnett, J. W., *The Electronic Structure of Molecules* (Wiley, New York, 1964)
76. Mulliken, R. S., Rieke, C. A., Orloff, D., and Orloff, H., *J. Chem. Phys.*, **17,** 1248 (1949); Mulliken, R. S., *J. Chem. Phys.*, **19,** 900 (1951)
77. Mulliken, R. S., *J. Chem. Phys.*, **2,** 782 (1934)
78. Hinze, J. A., and Jaffe, H. H., *J. Am. Chem. Soc.*, **84,** 540 (1962)
79. Pedley, J. B., *Trans. Faraday Soc.*, **57,** 1492 (1961); **58,** 23 (1962)
80. Lide, D. R., Jr., *J. Chem. Phys.*, **33,** 1514, 1519 (1960)
81. Coulson, C. A., *Valence*, 194–96 (Oxford Univ. Press, London, 1952)
82. Mizushima, S., and Shimanouchi, T., *Advances in Quantum Physics*, **II,** 103 (1947); Mizushima, S., *Structure of Molecules and Internal Rotation* (Academic Press, New York, 1954)
83. Searcy, A. W., *J. Chem. Phys.*, **28,** 1237 (1958); Searcy, A. W., *ibid.*, **31,** 1 (1959); Parsons, A. E., and Searcy, A. W., *J. Chem. Phys.* **30,** 1635 (1959)

84. Hoffmann, R., *J. Chem. Phys.*, **39,** 1397 (1963)
85. Pitzer, K. S., *Discussions Faraday Soc.*, **10,** 66, 124 (1951)
86. Pitzer, R. M., and Lipscomb, W. N., *J. Chem. Phys.*, **39,** 1995 (1963)
87. Cignitti, M., and Allen, T. L., *J. Phys. Chem.*, **68,** 1292 (1964)
88. Simmons, H. E., and Williams, J. K., *J. Am. Chem. Soc.*, **86,** 3222 (1964)
89. Karplus, M., and Parr, R. G., *J. Chem. Phys.*, **38,** 1547 (1963)
90. Parr, R. G., *J. Chem. Phys.*, **40,** 3726 (1964); Kim, H. J., and Parr, R. G., *J. Chem. Phys.*, **41,** 2892 (1964); Wyatt, R. E., and Parr, R. G., *J. Chem. Phys.* **41,** 3262 (1964)
91. Kirtman, B., *J. Chem. Phys.*, **41,** 775, 3262 (1964)
92. Ruedenberg, K., *J. Chem. Phys.*, **41,** 588 (1964)
93. Maccoll, A., *Trans. Faraday Soc.*, **46,** 359 (1950)
94. Mulliken, R. S., *J. Chem. Phys.*, **23,** 1833, 1841, 2338, 2343 (1955)
95. Mulliken, R. S., *ibid.*, **41,** 3428 (1964)
96. Long, L. H., *Quart. Rev.*, **7,** 134 (1953)
97. Bernstein, H. J., *Trans. Faraday Soc.*, **57,** 466 (1961)
98. Bernstein, H. J., *Spectrochim. Acta*, **18,** 161 (1962)
99. Varshni, Y. P., *Rev. Mod. Phys.*, **29,** 664 (1957)
100. Baughan, E. C., *Quart. Rev.*, **7,** 103 (1953)
101. Somayajulu, G. R., and Palit, S. R., *J. Chem. Soc.*, 2540 (1957)
102. Cox, E. R., *Ind. Eng. Chem.*, **27,** 1423 (1935)
103. Francis, A. W., *Ind. Eng. Chem.*, **35,** 442 (1943)
104. Li, K., Arnett, R. L., Epstein, M. B., Ries, R. B., Bitler, L. P., Lynch, J. M., and Rossini, F. D., *J. Phys. Chem.*, **60,** 1400 (1956)
105. *Selected Values of Properties of Chemical Compounds* (Manufacturing Chemists Assoc. Res. Proj. Chem. Thermodyn. Properties Center, Texas A&M Univ., Coll. Sta., Tex., loose-leaf data sheets, 1964)
106. Ketelaar, J. A. A., *Chemical Constitution* (Elsevier, New York, 1953)
107. Aten, A. H. W., Jr., *J. Chem. Phys.*, **5,** 260, 264, 598 (1937)
108. Kurata, M., and Isida, S., *J. Chem. Phys.*, **23,** 1126 (1955)
109. Kreglewski, A., and Zwolinski, B. J., *J. Phys. Chem.*, **65,** 1050 (1961)
110. *Properties of Hydrocarbons of High Molecular Weight* (Am. Petrol. Inst. Res. Proj. 42, Pennsylvania State Univ., University Park, Pa., 1962)
111. Böttcher, C. J. F., *Theory of Electric Polarization* (Elsevier, New York, 1952)
112. Kreglewski, A., and Zwolinski, B. J., *Roczniki Chem.*, **35,** 1041 (1961) (In Polish)
113. McClellan, A. L., *Tables of Experimental Dipole Moments* (W. H. Freeman & Co., San Francisco, 1963)
114. Somayajulu, G. R., and Palit, S. R., *J. Chem. Soc.*, 4837 (1957)
115. Seifer, A. L., and Smolenskii, E. A., *Zh. Fiz. Khim.*, **37,** 2657 (1963)
116. Seifer, A. L., and Smolenskii, E. A., *ibid.*, **38,** 202 (1964)
117. Anderson, H. H., *J. Chem. Eng. Data*, **9,** 272, 448, 592 (1964)
118. Kinney, C. R., *J. Am. Chem. Soc.*, **60,** 3032 (1938)
119. Flory, P. J., Orwoll, R. A., and Vrij, A., *J. Am. Chem. Soc.*, **86,** 3507 (1964)
120. Bartlett, N., and Lohmann, D. H., *Proc. Chem. Soc.*, 115 (1962)
121. Moody, G. J., and Thomas, J. D. R., *Noble Gases and Their Compounds* (Macmillan, New York, 1964)
122. Bartlet, N., *Intern. Sci. Technol.*, September, 56–66 (1964)
123. Hyman, H. H., *Science*, **145,** 773 (1964)
124. Chernick, G. L., *Record Chem. Prog.* (*Kresge-Hooker Sci. Lib.*), **24**(3) (1964)
125. Hyman, H. H., Ed., *Noble Gas Compounds* (Univ. Chicago Press, Chicago, 1963)
126. Alekseeva, T. A., and Ponomarev, V. V., *Zh. Fiz. Khim.*, **38,** 1337 (1964)
127. Brandts, J. F., *J. Am. Chem. Soc.*, **86,** 4291 (1964)
128. Wilhoit, R. C., and Lei, I., *J. Chem. Eng. Data*, **10** (In press)
129. Hutchens, J. O., Figlio, K. M., and Granito, S. M., *J. Biol. Chem.*, **238,** 1419 (1963)
130. Hutchens, J. O., Cole, A. G., Robie, R. A., and Stout, J. W., *J. Biol. Chem.*, **238,** 2407 (1963)
131. Hutchens, J. O., Cole, A. G., and Stout, J. W., *J. Biol. Chem.*, **239,** 591 (1964)
132. Irving, R. J., Nelander, L., and Wadsö, I., *Acta Chem. Scand.*, **18,** 769 (1964)

SHOCK WAVES[1,2]

By S. H. Bauer

Department of Chemistry, Cornell University, Ithaca, New York

Introduction

This review covers reports which are of interest to chemical kineticists on the properties of gases heated to moderately high temperatures by shock waves. It is a partial summary of the English language literature from the middle of 1963 to the end of 1964. The shock tube as a reaction vessel has been accepted by chemists as a useful tool for the investigation of the rates of fast reactions. Its obvious utility for the preparation of samples of gases at high temperatures has not been overlooked; it is being used in spectroscopic investigations, as a source for molecular beams, and in experiments designed to provide estimates of selected thermodynamic parameters. No data on shock heating via explosives of condensed phases are included, nor studies of gaseous systems raised to such high temperatures that these become plasmas. It is worth noting that the classification of research reports on the basis of the tool being exploited has merit when the problems thus studied are similar. As use of the new tool expands and its applicability to diverse situations is extended, emphasis shifts to the classification of papers on the basis of the problems solved, and away from the techniques used. This stage is being approached rapidly with regard to the study of gases in shock tubes. In contrast, the application of shock techniques to the investigation of liquids and solids for the measurement of properties such as electrical conductivity, equations of state, in the megabar pressure range, etc., has been limited (1).

The obvious useful features of heating with shock waves are the high temperatures obtainable under strictly homogeneous conditions and the rapid uniform heating of the samples studied. In properly designed experiments, quenching rates from 0.5 to 5° K per μsec are attainable in contrast to heating rates of the order 100° K per μsec. The adjustment in the distribution of molecular kinetic energy of translation from the ambient gas temperature ahead of the shock to the high temperature in the shocked sample requires 5 to 10 collisions, depending on the Mach number; indeed, the transition cannot be described merely as a sequence of layers with successively higher temperatures. Nevertheless, the shock front thickness is narrow enough to permit a separation in time of several subsequent relaxation processes. Attention is thus focused on the mechanics of energy transfer and on the possibility of detecting effects caused by excitation to selected states. During

[1] The survey of literature pertaining to this review was concluded in January 1965.

[2] This work was supported by NASA, Grant No. NsG 116–61, and the AFOSR, Grant No. AF49(638)-716. I wish to thank those who sent me manuscripts prior to their publication and Mr. David Lewis for his aid with the survey of literature.

the past two years, exploration of shock front structures has continued. There has also been considerable interest in the causes for and extent of departure of shock wave behavior from ideal one-dimensional theory, which limits the clean separation of events as they progress past the observation windows.

Since a shock tube is a gas dynamic device, its mode of operation is limited to densities at which the mean free path is of the order 10^{-2} or less of the dimensions of the apparatus. There can be only a modest unscrambling of molecular events. Chemical kineticists who are searching for devices in which the probabilities for transitions between specified states can be individually measured should pin their hopes on crossed beam techniques and on selective photoexcitation processes. In fact, the results derived from other experiments are necessary for the interpretation of shock tube data on complex molecules. For example, photochemical and thermal production of free radicals are now well-established procedures; specific reaction rate constants obtained for radicals will find confirmation in the analysis of shock tube kinetics data. After a flush period of intense activity with flash photolysis, the range of experimentation for chemical as well as spectroscopic studies is being extended into the vacuum ultraviolet as well as into the infrared regions (via matrix isolation techniques). Electrical discharges, the use of plasmas as a source of reactive molecular fragments, and the intense heating of samples in arc image furnaces and by focused laser beams have not yet produced sufficiently controlled conditions to permit quantitative studies of chemical kinetics. Meanwhile, investigation of ion molecule reactions in mass spectrometer sources shows a potential for generating reproducible data which unfortunately are not readily related to conventional rate constants.

For a qualitative introduction to chemical kinetics studies in shock tubes, the reader is referred to a review by Bauer (2); for basic theory and a summary of much experimental data, his attention is directed to two excellent books published during the past year by Greene & Toennies (3) and by Clark & McChesney (4). Other publications dealt with related problems on energy transfer during collisions (5, 6), combustion (7), and the structures of electronically excited species in the gas phase (8). Several bibliographies on shock tubes and of their application to physical and chemical reactions have also been assembled (9). Interest in shock tube investigations is increasing as indicated by the approximately half-dozen meetings held during 1964 in which reports on application were emphasized (10).

Shock Tube Operation and Technique

There are a number of causes for nonuniform flows in shock tubes, but the principal one is the presence of a boundary layer in the region behind the shock front. Depending on the shock speed, the roughness of the inner surface of the tube, and the presence of other perturbations, flow in the boundary layer will be laminar or turbulent or may pass from the former to the latter. The departure from ideal flow will be large and particularly disturbing at low pressures. A tracer method for studying the structure of the boundary layer was described by Chen & Emrich (11), who used suspended particles and dark field oscillatory spark microphotography. The velocity profiles they recorded show significant deviations from those expected on the basis of current boundary layer theory. The duration of laminar flow in the boundary layer (12) was measured in several gases for $1.5 \leq M \leq 3.5$. This flow time is critical when one wishes to measure quantitatively the spectroscopic behavior of the heated gas, as viewed through the cooler boundary layer. Small perturbations of shock flow have been analyzed by Mirels & Mullen (13), while Brocher (14) developed a theory applicable both to the laminar and turbulent cases in which he considered the influence of viscosity on the hot flow length and test time in shock tubes. The reduction of test time in low pressure shock tubes due to a laminar boundary layer has been analytically investigated by Mirels (15).

The reflected shock trajectory departs from that predicted on the basis of ideal theory to a greater extent than does the incident shock trajectory. A significant factor is the heat loss to the end wall which generates a growing thermal layer of cooled denser gas. To supply the material for this layer, mass motion toward the wall is induced in the heated gas thus perturbing the reflected shock wave (16). The theory for the interaction between this reflected wave and the laminar boundary layer, which may lead under certain conditions to boundary layer separation and shock bifurcation, has been developed a number of years ago by Mark and by Byron & Rott. Recently, Center (17) described a simple experimental procedure which can be used to detect this phenomenon. As the length to diameter ratio of a shock tube is increased and the initial pressure in the low pressure section is reduced, such that the preshock mean free path in the gas is larger than 10^{-3} of the tube diameter, wall-boundary layer effects become pronounced; curvature of the shock front results. De Boer (18) extended the analysis previously presented by Hartunian and developed approximate expressions for the curvature of shock fronts when followed by relaxation zones.

For chemical kinetics, departures from ideal behavior are particularly troublesome because these introduce ambiguities in the shock temperature. A thermometer with a time constant of 1 μsec with which one could measure gas temperatures to $\pm 5°$ K would be an ideal device. Under limited conditions one may record sodium line reversal temperatures which check the calculated values to about 50° K, but questions have been raised as to what temperature is being measured. In argon, the indicated temperature is considerably lower than the translational temperature even though the emission intensity (presumed caused by impurities) rises rapidly at the shock front (19). The low efficiency for translation-electronic energy exchange is consistent with previously measured low quenching cross-section for $Na(^2P)$ by $Ar(<10^{-20}\ cm^2)$. However, after 1 per cent N_2 or CO was added, a gradual rise in NaD emission was observed indicative of a relaxation process, and the final measured temperature agreed with that calculated to $\pm 35°$ K. Recently

Hurle (20) again demonstrated that in normal shocks in pure nitrogen T_{Na} equals T_{vib}. For reflected shocks in 0.5 per cent CO(Ar) at about $M = 3.3$, the expected vibrational relaxation zone was observed. The temperature behind the shock front, corrected for relaxation, was lower than that calculated, but the departure decreased with time ($\Delta T/\Delta t \approx 0.13 \pm 0.04°$ K/μsec) and eventually exceeded the theoretical value (21). Napier et al. (22) found that line reversal temperatures based on the (0,0) Swan band of C were slightly higher but within experimental error of those deduced from NaD line reversal.

The development of electrical shock tubes and wave engines for chemical processing is continuing (23), but their utility for kinetics studies is still limited by difficulties in obtaining reproducible conditions. It is pleasant to report progress in the development and use of new configurations and devices in shock tube technology. By using a liquid nitrogen-cooled ruby laser with a Kerr-cell shutter, Wachsler reduced the exposure time required to photograph the density profile at a shock front with a Mach-Zehnder interferometer to 20 nanoseconds (24). It is believed that the exposure time can be reduced to the width of a single emission spike (2–5 nsec), but curvature of the shock front and boundary layer perturbations will limit the effective resolution to about 40 nanoseconds. To facilitate studies of gases at low densities, several tubes with internal diameters in the range 40 to 60 cm have been put into operation. Lin & Fyfe (25) measured shock front curvature and found it to be independent of the shock Mach number but to vary inversely as the square root of the initial pressure. The curvature (the distance measured parallel to the tube axis from the shock front at the center to that at the tube wall) proved to be, for a 24-inch tube, $\delta(\text{mm}) = 1.75\ p^{-1/2}$ (p measured in mm Hg). To diagnose the shock front structure for investigations at initial pressures of 1 mm or less, one must place windows well within the boundary layer and close enough to the tube axis to minimize the magnitude of the correction for the curvature effect which must be made on the observations. A schematic of such a test section for spectrophotometric studies in the vacuum ultraviolet and for measurement of microwave reflection intensities is given by Wray (26). This tube has been instrumented by Camac (27) for determining gas density profiles at the shock front from the intensity of scattering of a collimated electron beam (accelerating voltage $\approx 10^5$).

Continued progress is being made in developing a shock tube-driven molecular-atomic beam of high intensity, known composition, and high kinetic energy (28). Skinner & Fetz (29) reported the production of a beam of N_2, 2.4×10^{18} molecules cm^{-2} sec^{-1}, having an energy 1.2 eV. However, painfully slow progress is being made in solving the sampling problem for a rapid-scan mass spectrometer coupled to a shock tube. That the integrity of the shocked gas sample is maintained by any of the slit geometries used or proposed to date has yet to be demonstrated. It now appears doubtful that a mass of shocked gas can be extracted from a shock tube and injected into the ion source while maintaining complete fidelity of composition. The question is whether the changes introduced by the slit system can be calculated, or measured, so that proper account can be taken of the altered composition. Work with a time-of-flight mass spectrometer is continuing in Kistiakowsky's laboratory. Diesen & Felmlee (93) used such an instrument for the study of the dissociation rates of chlorine, N_2H_4, and N_2F_4. They bled in the reflected shock-heated gas to the mass spectrometer ion source through a pinhole, 0.002 inches diameter in a thin plate (0.030 inches). Skinner (28a) pointed out that the thickness of the thermal boundary layer at the end of the shock tube, that is, of the cooler gas layer processed by collisions with the end plate, grows as $(\text{time})^{1/2}$, but the spherical region from which gas flows through the sampling hole has a radius which grows as $(\text{time})^{1/3}$. Thus, the region from which flow originates is soon entirely within the thermal boundary layer. The coupling section now under construction in our laboratory is a compressed version of the molecular beam slit system developed by Skinner.

Experiments in shock tubes which appear as extensions of the laboratory style of conventional kineticists can be done with a single pulse device. In such a shock tube (*a*) The sample is very rapidly raised to a calculable and constant temperature and is then rapidly quenched after a measured time by an expansion wave. This is easily achieved in any conventional shock tube. (*b*) The reflected shocks and expansion waves are damped so that the sample is not reheated or mixed. To achieve the latter, two designs have been developed. In the first, an expansion wave is generated by puncturing a diaphragm which separates the downstream end of the high pressure section from a large evacuated tank (30). For proper operation it is necessary that two diaphragms open fully, in sequence, with the correct time delay; this has not enhanced the popularity of the design. A simpler device is to connect a relatively small tank with a T-joint just past the low pressure side of the diaphragm (31).

The quench step (*a*) effectively stops all reactions with an activation energy greater than about 20 kcal/mole, but reactive fragments continue to react until depleted. Since a complete analysis can then be made of the products by conventional techniques, one can by inference deduce the complex set of reactions which did take place. A correction for the finite cooling rate in such a shock tube has been derived (32). As in all shock tube studies, the precise temperature to which the sample has been exposed is not known. It appears that the average reflected shock temperature, over regions up to 10 cm from the end, as estimated from the products obtained during an exposure of 0.5 to 1.0 msec are 10 to 35° K lower than computed (21, 33, 34). To circumvent this difficulty and errors caused by other perturbations, Tsang (35) showed that relative rates and activation energies can be obtained with great precision by measuring ratios of products from two parallel but non-interacting processes which take place concurrently.

Bradley & Tuffnell (36) constructed a tube in which a photolyzed gas sample is moved rapidly past observation ports by a shock wave; changes in the gas are followed spectrophotometrically. They tested their apparatus by photolyzing phosgene and oxalyl chloride, but only preliminary data were presented. An interesting application of shock tube technology was described by Goldstein (37). He showed that it was possible to study the condensation of a low vapor pressure material on the inner surface of the shock tube. Condensation takes place because of the isothermal compression of the gas in the thermal boundary layer, when initial conditions are adjusted so that the low vapor pressure material attains supersaturation. This occurs on a time scale of the order of microseconds. The condensation of water was shown to depend sensitively on the condition of the glass surface. For a clean hydrophilic surface, condensation appears initially in the form of droplets, whereas on a hydrophobic surface, condensation appears in the form of a continuous film.

It is instructive to consider the various gas dynamic devices available for quenching a sample which had been allowed to attain equilibrium under shock conditions, so that its composition was known either from direct measurements or from thermochemical calculations, and to observe the depletion of molecular fragments due to recombination or reaction with the major constituents. The recombination rate of chlorine atoms was followed in a shock tube in which the expansion chamber was located a few tube diameters from the shock tube end plate (38). A cylindrical diaphragm lying flush with the inside diameter of the tube was burst at a preset time, to initiate an expansion wave. Observations indicated that this technique provided an order of magnitude greater expansion rate ($-dT/dt \approx 5°/\mu$sec) than that attainable in the single pulse tube developed at Cornell Aeronautical Laboratory. The chlorine recombination rate was followed spectrophotometrically on the gas which remained in the shock tube; no attempt was made to capture the expanded gas for subsequent analysis. In a conventional shock tube, a sample heated in the reflected shock region can be rapidly cooled ($-dT/dt \approx 1°$ K/μsec) by an expansion wave generated at the interaction between the contact surface and the reflected shock, provided sound speeds across this interface are properly adjusted. In Cornell laboratories (39), the rate of depletion of CN radicals was measured in this manner following the attainment of equilibrium in a reflected shock propagated in a mixture 0.5 to 2 per cent of C_2N_2 with argon. The rate of decline in CN was found to be first order in the concentration of that radical, which indicates that the depletion is attributed to the growth of $(CN)_m$ polymers ($m \approx 3$–6). This is in agreement with data on the kinetics of disappearance of CN radicals obtained by another technique (40).

The most rapid quench rates are generated in expanding nozzle flows. The rates of several very rapid reactions were measured using sources close to room temperature. A recent paper deals with the reaction between nitric oxide and ozone (41). Of present interest are the nozzle flows of gas samples exposed to reflected shocks. Analytic solutions are not available for these nonequilibrium flows, when the chemical relaxation times are comparable to the characteristic flow times. However, exact numerical solutions have been obtained (42). Analyses of these solutions showed the unexpected fact that the entropy increment resulting from the nonequilibrium freezing was quite small compared with the total entropy change, for a wide range of flow conditions. This was confirmed (43) by an analytic investigation of a model for a single dissociating gas. To conclude this section on techniques, mention is made of the large variety of computer programs now available for general use in connection with shock tube investigations (44). These have helped substantially with the design of experiments so that optimum operating conditions can be established. Furthermore, programs permit the more rapid interpretation of complex kinetics studies and the analysis of nonequilibrium nozzle flows.

Intermolecular Energy Transfer

The shock front structure encompasses nearly all the features of shock waves which are of interest to chemical kineticists. The "front" may be treated as the filling of a sandwich between two slugs of gas each at equilibrium, one at the initial temperature and density, with the other at the final shock temperature and density. The profiles of molecular density and translational velocity distribution as functions of the distance parallel to the direction of shock propagation differ, but they do follow the same general shape. Recent reports of experimental studies and theoretical analyses at several levels of sophistication have contributed to the understanding of these profiles. A molecular model, viewed in a coordinate system moving with the shock, suggests that the Maxwell distribution in translational speeds of the inrushing molecules (initial temperature) becomes smoothly converted to another Maxwell distribution at the shock temperature. Because of the persistence of velocities, one may anticipate a required average collision number of about 2. That this transition, to an adequate approximation could be represented by a bimodal form $f=a(x)F(T_1)+[1-a(x)]\ F(T_2)$ is a pleasant surprise. Further, because the inrushing molecules carry directed momentum, most of which must be randomized, one may anticipate that a number of collisions greater than 2 would be needed to attain the steady state condition, perhaps of the order of 10. Since diffusion is involved, the transition region in density should be somewhat greater than that for molecular speeds. Finally, for all but the monatomic gases the efficiency per collision for the redistribution of energy among translation, rotation, and vibration must be considered.

Rotational relaxation times for many molecular species have been estimated from ultrasonic absorption or dispersion data, and from the dimensionless ratio: $f=\lambda M/\eta C_v$, which relates the thermal conductivity (λ), molecular weight (M), viscosity coefficient (η), and heat capacity (C_v). A recent set of values for Z_r, the required mean number of collisions per molecule, using collision cross sections derived from viscosity measurements are: CH_4, 9.4; CF_4, 3.0; SF_6, 2.5; C_2H_4, 2.4; C_2H_6, 4.0; O_2, 12; N_2, 7.3; CO_2, 2.4; C_2H_2, 1.8; N_2O, 2. (45, 46). These were found to be independent of the temperature. The relative efficiencies for spherical top molecules are in the sequence predicted by Widom (47). Holmes, Jones & Pusat (47a) pointed out that sound dispersion data for rotational relaxation times must be properly corrected for viscothermal processes. Z_r so determined are in reasonable agreement with theoretical estimates for ethylene (≈ 2), cyclopropane (5 ± 1), methane (12), and tetradeuteromethane (7). Thus, it follows that hydrogen and deuterium are the only known gases which require Z_r (300-200) of a magnitude large enough to distinguish rotation from translation and viscosity relaxation. Eventon (47b) estimated the efficiency for rotational-translation energy transfer from CN in the excited $B^2\Sigma$ state, by following emission intensities in the 3875 Å band. Rates of energy transfer were measured for rotational separations as great as 10; transitions with $\Delta J > 1$ take place with high probability. Approximately every collision leads to a change in rotational energy state, as in NO.

The varied approaches to the calculation of shock front structure may be judged from the following reports (48–52). For a given shock Mach number, the different profiles for directed velocity, T_{trans}, T_{rot}, T_{vib}, and density are clearly illustrated in Reference 50. The failure of the Navier-Stokes solutions for high Mach numbers is shown in Reference 51, in that a model based on the Boltzmann equation shows a long slow rise in the temperature profile ahead of the position of maximum slope, at the shock front. Muckenfuss (52) extended the bimodal formulation of Mott-Smith, and calculated mean collision numbers suffered by molecules in traversing the density zones for argon (Lennard-Jones potential) and for Maxwell molecules. For $M > 2.5$, these calculated collision numbers are ≈ 9.2 for L-J type, and ≈ 8.5 for Maxwell type.

Experimental investigations bear out the above analyses. With the optical reflectivity technique for exploring shock front structure in argon, for Mach numbers below 5, Hornig and co-workers (53) showed that the extended bimodal representation (52) is in closer agreement with the observed maximum-slope density thickness than that based on the Navier-Stokes equations. A reciprocal thickness, measured in units of the mean free path (undisturbed gas), rises rapidly to a maximum of 0.31 at $M = 3.5$ and decreases very gradually for higher shock speeds. In nitrogen, complete rotational relaxation occurs within the shock front; in fact, the measured shock front thicknesses in nitrogen are less than in argon. At higher shock speeds, measurements were made by Camac (54) in argon and nitrogen for $3 < M < 9$ with the electron beam technique, for initial pressures of 15 to 240 μ Hg. Over a small range in shock speeds where his runs overlap those of Linzer & Hornig, Camac's reciprocal lengths are about 30 per cent lower and thus fall directly on the bimodal curve. Here also the rotational relaxation region for nitrogen falls fully within the temperature-density profile. A question which can be answered only by experiment is whether in the shock front preceding a detonation rotational equilibrium is attained. Hornig (55) concluded that for mixtures ($H_2 + 3\ O_2$) in shock waves with strengths and final temperatures equivalent to those in detonations, translational and rotational equilibrium is reached in the initial compression region. Rotational temperatures for OH measured spectroscopically ($^2\Sigma^+ \rightarrow {}^2\Pi$) in rarefaction waves following detonations in mixtures of $C_2H_2+H_2+2\ O_2$ did agree with the theoretical gas temperatures when account was taken of rotation-vibration interaction (56).

Several ultrasonic absorption studies of gases at or near room temperature are pertinent to the following discussion of vibrational relaxation. Current techniques have extended the range of such measurements to 1000 Mc/sec atm. Characteristic times as short as a nanosecond-atmosphere can be estimated to abut 10 per cent. Reported (57) relaxation times at 1 atm for propane, propylene, and ethane are $(1.56\pm.08)\times10^{-8}$, $(1.48\pm0.08)\times10^{-9}$, $(1.12\pm0.03)\times10^{-8}$, and $(0.71\pm0.06)\times10^{-9}$ sec, respectively, with the latter two values belonging to ethane. These results are consistent with conclusions reached by Lambert more than a decade ago (58), that in polyatomic molecules the cross sections for energy transfer between translation and vibration are 10^{-1} to 10^{-2} of dimensions deduced from viscosity data, and, that the interchange of vibrational energy between the different modes within a molecule is usually very rapid. Ultrasonic dispersion appears at lower frequencies for small or rigid molecules, and in rare cases one finds more than one relaxation time. Sound absorption and velocity measurements on mixtures of O_2 and of CO_2 with H_2, D_2, and He at temperatures up to 300° C confirmed the observations based on shock tube studies that low mass colliders are highly efficient for the de-excitation of vibrational energies (59, 60). The extreme dependence of collisional vibrational relaxation time on the reduced mass of the collision partners was also demonstrated in a study of the quenching of vibrational fluorescence of very highly purified CO. Resonance excitation was sufficient to raise the gas initially at room temperature to $T_{vib}=993°$ K, while T_{rot} remained at 286° K (61). Finally, the efficient near resonance transfer of vibrational energy from N_2 to CO_2 provides a selective excitation mechanism and thus permits the operation of a laser system (62):

$$N_2(v = 1) + CO_2(00^\circ 0) \rightarrow N_2(v = 0) + CO_2(00^\circ 1)$$

$$CO_2(00^\circ 1) \rightarrow CO_2(00^\circ 0) + h\nu; \qquad \omega = 960.99\ \text{cm}^{-1}$$

At moderate or low shock speeds, the vibrational relaxation zone is distinct from the shock front so that the associated relaxation times can be measured as a function of shock Mach number. Millikan & White (63) assembled a large group of data for the vibrational relaxation times of diatomic molecules and deduced an empirical equation correlating values of $p\tau_v$ (atm sec) with the temperature, reduced mass of the colliding pair (μ), and the characteristic vibrational temperature ($\theta=ch\omega/k$) of the oscillator.

$$\log_{10}(p\tau_v) = (5.0\times10^{-4})\mu^{1/2}\theta^{4/3}[T^{-1/3} - 0.015\mu^{1/4}] - 8.00.$$

Although the comparative efficiency of energy transfer during collisions with light molecules is thus clearly indicated, this equation does not hold if the relaxing molecule is hydrogen or deuterium, nor can it be used for polyatomic gases, when these are relaxing or when they are the collision partners.

TABLE I

VIBRATIONAL RELAXATION TIMES (CO AND O_2)[a]

$CO + CO$	$p^b\tau_v =$	60×10^{-6}, atm-sec	$T(°K) = 2000$
Ar	$=$	350	2000
N_2	$=$	640	2000
H_2	$=$	0.7	2000
$CO + He$	$\log_{10}(p\tau_v) =$	$38\ T^{-1/3} - 8.3$	$500 < T(°K) < 1500$
Ne	$=$	$61.6\ T^{-1/3} - 9.2$	$1400 < T(°K) < 3000$
Kr	$=$	$81.2\ T^{-1/3} - 9.7$	$2100 < T(°K) < 7000$
N_2	$=$	$102\ T^{-1/3} - 11.24$	$1900 < T(°K) < 5400$
$O_2 + O_2$	$(p\tau_v) =$	$1.49 \times 10^{-10} \exp(133\ T^{-1/3})$	$600 < T(°K) < 2600$
He	$=$	$5.3 \times 10^{-9} \exp(60\ T^{-1/3})$	$400 < T(°K) < 1600$
H_2	$=$	$10^{-8} \exp(36\ T^{-1/3})$	
D_2	$\log_{10}(p\tau_v) =$	$27.3\ T^{-1/3} - 8.35$	$300 < T(°K) < 850$
CH_4	$=$	$18\ T^{-1/3} - 8.7$	$450 < T(°K) < 1300$
C_2H_2	$=$	$7.5\ T^{-1/3} - 7.3$	$450 < T(°K) < 1300$
C_2H_4	$=$	-6.92	independent of T

$$O_2 + Ar: \frac{1}{\tau_v} = nC_1T^{1/6}\,[1 - \exp(-228T)]\,\exp(-C/T)^{1/3}$$

$$C_1 = 1.2 \times 10^{-7};\ C = 1.04 \times 10^7;\ n \text{ molecules/cm}^3$$

[a] Camac, M., *J. Chem. Phys.*, **34,** 448 (1961).
[b] p refers to pressure (atm) of colliding species.

The measured values and those calculated from the equation generally agree within 50 per cent, but in the case of O_2-Ar at 1000°, the experimental time exceeds the calculated one by a factor of 5. The above empirical relation also fails for the halogen acids (64). Satisfactory measurements of the vibrational relaxation times for H_2 and D_2 are yet to be made. At about 1400° K in a 2/1 mixture of D_2 in Ar, measurement of the density behind a shock wave indicated a relaxation time of (4 ± 2) μsec atm, reduced on the assumption that the effective collisions were due to D_2/D_2 (65). Corresponding measurements on H_2/Ar mixtures led to values less than the experimental resolving time (≈ 1 μsec atm) (65a).

In a series of three papers Millikan and co-workers described shock tube measurements of the vibrational relaxation of carbon monoxide with various colliders, studied over an extended temperature range (66); see Table I. By following the emission intensities at the fundamental and first overtone frequencies, they demonstrated that the rate of populating the $v = 2$ level by successive unit quantum jumps was more than ten times faster than the direct excitation $\Delta v = +2$. In a mixture of CO and N_2 they observed the rate of vibrational excitation of CO by its infrared emission at 2000 cm^{-1} and of $(N_2 + CO)$ by measuring interferometrically the density profile of the shocked gas. Over the temperature range 3200° to 4200° K the two relaxation times were equal to within 20 per cent, whereas at 3600° K, pure N_2 has a relaxation time four times greater than CO. This shows that the intermolecule transfer of vibrational energy

$$CO(v = 1) + N_2(v = 0) \rightarrow CO(v = 0) + N_2(v = 1)$$

is faster than the rate of excitation of CO.

From the same laboratory White & Millikan reported another set of very interesting results on the vibrational relaxation of oxygen (67), estimated interferometrically. For combusting mixtures (with H_2 and D_2), shock tube data below 1500° K were combined with ultrasonic measurements at room temperature; thus the equations listed in Table I were derived. The efficiencies of helium and deuterium are very nearly equal both in magnitude and in temperature coefficient. Above 1500° K the induction zone merges into the shock front.

The vibrational relaxation of oxygen containing a small percentage of methane, acetylene, or ethylene was studied by shock tube interferometry from about 450° to 1300° K (67d). The relaxation zone appears to be unaffected by chemical reaction so that it was possible to measure the relaxation times for oxygen as a function of temperature. These appear to be shorter than when oxygen was being relaxed by hydrogen or deuterium, as given in Table I. For temperatures somewhat above 1300° K, an exothermic reaction results in a nonlaminar flow behind a nonplanar shock. At still higher temperatures, the relaxation zone becomes laminar again, with the vibrational excitation of oxygen accelerated by concurrent chemical reactions in the induction zone. From the relative magnitudes of relaxation times for the pure hydrocarbons and for the oxygen-hydrocarbon mixtures, White concluded that collisions with oxygen are more effective in the excitation of methane and ethylene than are self collisions, so that the excitation in oxygen occurs through vibrational exchange with the most easily excited molecule present in the mixture.

The relaxation sequence for carbon dioxide at low densities was investigated by Camac (68) over the temperature range 2000° to 6000° K. Density profiles at the shock front were determined by the electron scattering technique, and concurrently the rate of vibrational excitation of the ν_3 mode was determined from the infrared emission at the 4.3 μ band. As in the case of nitrogen the rotational relaxation was fully coupled with translation. Of particular interest is the conclusion that all four vibrational modes relax at the same rate since the gas density and the infrared profiles have the same shape. However, the emission intensity measures only the population in the first excited level of the ν_3 mode, which contributes only 20 per cent to the density change caused by vibration. When Camac combined his results with ultrasonic data he found that the relaxation times for carbon dioxide was well represented by the expression

$$p\tau_v = \exp\,[36.5T^{-1/3} - 3.9], \qquad \text{atm } \mu\text{sec}$$

over the temperature range $250° K < T < 6000° K$. Estimates of the dissociation rate of CO_2 were obtained from the decline in the infrared emission intensity after it passed through the maximum value. From the intensity of infrared emission, Camac estimated that dissociation occurs at shock speeds greater than 3.3mm/μsec; at 4mm/μsec and higher, the vibrational relaxation zone merges with the shock front, and only the dissociation relaxation could be followed.

The sequence of relaxation zones in nitrogen shocked to moderately high Mach numbers was followed by measuring the relative emission intensities with a three channel monochromator; resolution ≈5 Å (69). The $N_2^+(1-)$ $\Delta v = +1$ sequence at 4200 Å was used [two channels in the (0, 1) band gave the rotational temperature, while the ratio of the (2, 3) and (0, 1) bands gave the vibrational temperature]. The electronic temperature was estimated from the magnitude of the total molecular radiation [in the N_2 (1+) system, 5500–11,000 Å] normalized at the limiting value, since he assumed that the observations were recorded until full equilibrium was reached. As expected, the rotational temperature, which also measures the translational temperature, declines while the vibrational temperature rises to the equilibrium value. At the shortest measurable time (0.1 μsec) the electronic temperature is somewhat above T_{eq}; it attains the equilibrium value in 1.9 μsec, simultaneously with T_{vib} and T_{rot}. Interpretation of the emission sequence from shocked nitrogen and the mechanism of excitation of the gas to the $N_2(B)$ and $N_2^+(B)$ states are far from clear. Until an adequate formulation is presented it is advisable to avoid making loose statements.

From their study of vibrational relaxation in air, White & Millikan (70) concluded that the initial rapid relaxation attributed to the oxygen, was slightly larger than expected for the pure gas, and the subsequent vibrational relaxation of the nitrogen, for runs below 3000° K, was considerably faster than in the pure gas—by an order of magnitude at 1600° K. This is presumed to be brought about by vibrational energy exchange with the nearly equilibrated oxygen. Further, these authors suggest that the induction time for the formation of NO behind shock waves in air may result from the required preliminary vibrational relaxation of one or both of the reactants. (See below for discussions of the NO reaction, and of $HX + D_2$ exchange rates.) The vibrational relaxation in air shocked to very high temperatures was discussed by Wray (See 10d, Abstr. 4).

It is well known that the usually measured vibrational relaxation time at unit density is merely proportional to the inverse of the transition probability for de-excitation (k_{10}) from $(v=1) \rightarrow (v=0)$; also, for simple harmonic oscillators,

$$k_{v,v-1} = k_{10} \cdot v; \qquad \tau_{v,v-1} = \tau_{10}/v$$

For shock heated NO, Roth (71) deduced from the time sequence of peak emission intensities of the $\gamma(1, 1)$ and $\gamma(0, 0)$ bands (i.e., for the $A^2\Sigma^+$ state): $\tau_{10}/\tau_{21} \approx 1.9$, in agreement with theory. However, in HI (diluted 90 per cent with Ar) heated to temperatures in the range 1400 to 2400° K, adjustment of the populations of three excited states ($v=1, 2, 3$) to their respective equilibrium values followed the identical time dependence (72):

$$(1 - n_v/n_{v\infty}) = \exp\left[-k(\mathrm{HI})_o t\right]$$

with $k=4\times10^{-14}$, cm³ molecule⁻¹ sec⁻¹, at 1500° K; $k=8\times10^{-14}$ at 2400° K.

A comparison of theoretical and experimental vibrational relaxation times was undertaken by Parker (73) for the halogens, nitrogen, oxygen, and carbon monoxide. Reasonably good agreement was obtained for all gases except oxygen. Also, Rapp (74) reviewed and compared the three generally used methods for calculating transition probabilities for vibrational energy exchange in molecular collisions: (*a*) wave mechanical treatment using the method of distorted waves; (*b*) a semi-classical time dependent perturbation procedure in which the perturbation energy is obtained from the classical collision trajectory; and (*c*) a purely classical calculation of the energy transferred to an oscillator. The effect of attractive terms in intermolecular potential on the probability of vibrational energy transfer was analyzed by Shin (75). The strong collision model for the transfer of energy in systems of oscillators was treated again by Hoare (76) and by Ivanov & Sayasov (77). An excellent current summary of the theory was prepared by Takayanagi (47b). Mies analyzed the effect of anharmonicity on the probability of vibrational energy transfer (77a).

One of the most intriguing effects regarding vibrational excitation and de-excitation is the demonstration that in some cases vibration-rotation energy transfer plays a significant role. Cottrell and coworkers (78) proposed a simple model for the relative efficiencies of rotation-vibration energy exchange for a number of hydrides and deuterides of group IV and V elements. The measured ratios of relaxation times by ultrasonic absorption are in good agreement with this theory and are clearly in contradiction to the relative values predicted on the basis of the Landau-Teller theory (Table II). Even more striking was the observation by Millikan & Osburg (79) that parahydrogen was more than twice as effective in quenching infrared emission from vibrationally excited CO as was orthohydrogen. They estimated that around room temperature: $p\tau_{\mathrm{CO}}(\mathrm{para})=6.5\times10^{-5}$ atm sec, and $p\tau_{\mathrm{CO}}(\mathrm{ortho})=1.43\times10^{-4}$ atm sec. The difference decreases with rising temperature and becomes negligible at about 450° K. It is evident, however, that Cottrell's model calls attention too ne aspect of a complex problem. Highly specific interactions introduce apparent anomalies, as for NH_3 and ND_3 ($Z_{10}<15$; $Z_{10}\approx90$, respectively) (80). The relative mean collision numbers for C_2H_4 and C_2D_4 are as expected from Landau-Teller theory ($Z_{10}=820$, 390, respectively) (81); the relative efficiencies of H_2O and D_2O as catalysts for the transfer of vibrational energy cannot be accounted for on the basis of a single mechanism (82).

The question whether radiation generated at the shock front effectively pre-excites the gas downstream has been contemplated by many shock tube investigators but no definitive theoretical or experimental magnitudes have

TABLE II

RATIOS OF REDUCED RELAXATION TIMES

Pairs	τ_{10} Predicted		τ_{10} Observed
	vib-rot	*vib-trans*	
CD_4/CH_4	1.7	0.58	1.7_1
SiD_4/SiH_4	(1.5)[a]	0.25	1.5_1
PD_3/PH_3	1.5	0.13	1.5_2
AsD_3/AsH_3	1.2	0.04	0.6
AsH_3/PH_3	≈1	150	1.5

[a] ( ) denote an interpolated value.

been obtained. There are many possibilities, such as photoelectric emission by the walls of the shock tube and by minute concentrations of impurities with low work functions, photoexcitation in the visible and ultraviolet regions by resonance absorption, etc. Lapp (83) estimated that in the infrared such an effect is not large enough to significantly alter the vibrational relaxation time in CO, but it may be detectable for strongly absorbing species such as CO_2, if one carefully selected the experimental conditions.

Temperatures estimated from sodium line reversal were measured by Hurle, Russo & Hall for the nonequilibrium flow of shock heated nitrogen in a supersonic expansion (84). The gas was heated and compressed to reservoir temperatures and pressures at the range 2800° to 4600° K and 24 to 82 atm, and allowed to expand through a 15° axially symmetric nozzle attached to the end of the shock tube. The indicated temperatures at several locations along the flow axis were much lower than expected, and led to a relaxation time for vibrational de-excitation one fifteenth as long as that inferred from measurements behind normal shock waves (20). The reason for this difference is still elusive. Holbeche (84a) found that in oxygen, shocked to temperatures sufficiently elevated to produce up to 1 per cent dissociation, the vibrational relaxation time measured in an unsteady expansion via the Na line reversal technique was the same as that reported for normal shocks (84b). The possibility that there are several mechanisms for energy exchange processes (85) should not be dismissed until it has been established that the gases used were of very high purity. From available collision cross sections Hurle (86) estimated thermally averaged probabilities for the electron de-excitation of the first vibrational level in nitrogen. The probability increases from 10^{-4} at 1000° K to 10^{-2} at 5000° K, and is of a magnitude such that at 1000° K a concentration of electrons of one part in three million would decrease the vibrational relaxation time in nitrogen by one half. However, there are no indications that levels of electron concentration present in the above experiments were sufficiently high to have significantly influenced the relaxation rates.

Finally, attention is called to experiments on collisional de-excitation probabilities of chemically activated, relatively large molecules by inert gases. One example studied was the sec-butyl radical, excited to an excess of 40 kcal/mole (87). Depending on the model used in reducing the data, the average amount of energy transferred per collision is (1.3–2.5) kcal/mole for H_2, D_2, N_2, and the rare gases; with CH_3F, CH_3Cl, SF_6, and butene, the average energy transferred is greater than 9 kcal/mole, while with CO_2 and CH_4 an intermediate magnitude was obtained. Chemically excited cyclopropane, which carries almost 100 kcal/mole, loses between 10 to 14 kcal/collision with ethylene (at 298° K and 423° K). Nitrogen and argon are less efficient while helium is the least efficient of the gases tested; 67 per cent of the collisions appear to be elastic, while the inelastic encounters carry away on the average 6 kcal per collision (88). From these examples, it is evident that the mechanism for vibrational energy transfer of highly excited polyatomic molecules differs substantially from the process for the transfer of vibrational energy in diatoms excited to low vibrational levels. It is possible that the rather highly excited multi-atom molecules form short-lived complexes with their collision partners, which have lifetimes comparable to two to three typical vibrations. These are long enough to permit the efficient transfer of energy into the vibrational mode of the catalyst molecule, or into that vibrational mode of the complex which becomes the coordinate of separation for the pair.

For diatomic molecules, it is difficult to design experiments with which to measure the relative reactivities of highly excited vibrational states. One of the major problems is the preparation of sufficiently high concentrations of the diatoms in states with high v. This can be accomplished chemically by generating the molecules to be studied via highly exothermic reactions (89), but the quantitative measurement of the relative populations remains a problem. Another procedure is based on selective photoelectric excitation. Iodine molecules excited to $B^2\Pi_{o+u}$, $v \approx 15$, transfer vibrational energy rather efficiently to the rare gases and to a number of diatoms, approximately one quantum per eight kinetic theory collisions (90). This is similar to the efficient energy transfer deduced for I_2, in the $v = 25$ state.

Shock Front Structure Due to Diatom Dissociation and Recombination

At moderate temperatures, vibrational relaxation precedes dissociation as a distinct step. These processes can then be studied separately. However, at sufficiently high Mach numbers the two processes merge at the shock front. When vibrational relaxation is rapid the equilibrium populations appear to be maintained; analysis of the dissociation data is straightforward, although one may still question whether the upper vibrational levels are filled to their equilibrium populations. At very high temperatures the coupling between the vibration and dissociation negates this simplifying point of view.

It is disconcerting that one must still hesitate to quote the rate of dissociation of molecular hydrogen to within a factor of 10; also, there are significant differences between the measured activation energies. Another attempt is currently under way (91) in which the rate of production of hydrogen atoms behind an incident shock wave in H_2/Ar mixtures is being followed by atomic absorption spectrophotometry (at 1215.7 Å, Lyman α). Due to the very high absorption coefficient at this wavelength, it will be possible to extend the dissociation measurements to temperatures as low as 1700° K in dilute mixtures.

By comparing the rate of dissociation of bromine in back reflected shocks (1500°–1900° K) with those observed in the incident shock region, Johnson & Britton (34) found temperatures in the back reflected shock were $35 \pm 5°$ K lower than those calculated from one dimensional theory. Their study of the rate of dissociation of fluorine completes the list of homonuclear dihalogens for which rate data are available. The dissociation, in the presence of argon, over the temperature range 1300° to 1600° K was followed spectrophotometrically (92). For 5 per cent mixtures, using incident shocks, $\log_{10} k_D$ equals $[9.85 - 6520/T]$ (concentrations in moles/l); less reliable values were obtained for higher fluorine concentrations. The measured activation energy is about 7 kcal below the spectroscopic value, approximately as was observed for bromine, chlorine, and molecular hydrogen. Surprisingly, the rate of dissociation was found to be lower by a factor of 10 than was expected by analogy with the other halogens; no adequate explanation for this has been offered.

After an inauspicious beginning in which two sets of values, differing by a factor of 10, were cited for the rate of dissociation of chlorine, evidence is accumulating in favor of the lower value. Mass spectrometric analysis of the time dependent concentrations in Cl_2/Ar mixtures in back reflected shocks (93) checked an earlier spectrophotometric measurement of the rate (94). However, two sets of experiments performed independently (95, 96), in which the same spectrophotometric procedure was used, agreed with each other:

$$k_D = 8.90 \times 10^{13} \exp(-48{,}300/RT), \qquad \text{cm}^3\ \text{mole}^{-1}\ \text{sec}^{-1}.$$

A third determination is based on the intensity of light emission during the radiative recombination of chlorine atoms, which thus provides a measure of the instantaneous chlorine atom concentration. The measured activation energy at 2000° K is 48 kcal/mole, and the absolute rate agreed with the value cited above (97). As a collider, Cl_2 is only slightly more effective than argon. Since the spectroscopic dissociation energy $D_o = 57.080$ kcal/mole, one may write for the dissociation rate constant,

$$k_D = A(D_o/RT)^m \exp(-D_o/RT), \qquad \text{with } m = 1.85$$

Bader & Ogryzlo (98) have recently discussed the origin of luminescence from the recombination of chlorine atoms. It is interesting to note that no experiments have been reported on the dissociation of mixed diatoms. Clearly, dissociation of a species such as XY is more complex kinetically than that of X_2, because of the presence of chains; ClF would be further complicated by the possible generation of ClF_3. An analogous situation arose in the study of the dissociation of ClCN (see below).

The most extensive studies of a diatom dissociation rate have been made on oxygen. Because experiments were performed over a very wide range of temperatures and densities, it was possible to observe the effect on the shock front structure of the coupling between vibration and dissociation. Up to temperatures $\approx 7500°$ K, the dissociation rate constant for collisions with Ar is given by (99):

$$k_D = 1.80 \times 10^{18} T^{-0.98} \exp(-118{,}000/RT), \qquad \text{cm}^3\ \text{mole}^{-1}\ \text{sec}^{-1}$$

Vibrational relaxation appears to be well ahead of and distinct from the dissociation process. However, closer inspection of the data suggests that at temperatures as low as 4000° K the dissociation rate is sufficiently rapid to disturb the Boltzmann distribution among the upper vibrational levels. The effect becomes more marked as the temperature is raised. Above 11,000° K the observed rates are lower than those predicted from a least squares fit of points taken between 5000 to 11,000° K (26). In more recent studies of oxygen (100), shocked to $4 < M < 21$ at low initial pressures (15–30 μ Hg), the observed shock front structures could be qualitatively classified into three types: $4 < M < 10$, translation and rotation are relaxed within the shock front. This is followed by a vibrational relaxation zone and ultimately by dissociation. For $M < 14$, vibrational relaxation region enters the shock front and is completely engulfed by it at the upper Mach value. For $M > 16$, dissociation also begins at the shock front. The reciprocal shock front thickness for $4 < M < 18$ falls slightly below the bimodal curve, but there is considerable scatter in the experimental points. However, the identification of a measure of a thickness with a magnitude derived from theory is not clear. Since a quantitative molecular description of translational relaxation at the shock front has yet to be presented, it is not surprising that no adequate description is available for the more complex sequence of events which occurs in a diatomic gas.

Several years ago there was a flurry of speculation regarding the basis of the inverse temperature dependent pre-exponential term in the expression for the dissociation rate constants. As is well known, the inverse temperature dependence of the pre-exponential factor must be inserted only if one insists that the exponent in the exponential factor is the spectroscopically determined dissociation energy rather than the E_A deduced from a plot of ln k_{exp} versus $1/T_{transl}$. Due to vibration-dissociation coupling the "equivalent" vibrational temperature is less than T_{tr+rot} by a magnitude which increases appreciably with the shock Mach number, so that the lesser than expected slope may be due to use of an improper temperature scale.

Recombination rate constants calculated from the measured k_D, using the thermodynamic equilibrium constant, reflect the inverse pre-exponential temperature dependence. Several authors argued that since the upper vibrational levels were filled below their Boltzmann population, calculated for T_{tr+rot}, the relation $K_{eq}=k_D/k_R$ did not hold. There are now several reports on direct measurements of the recombination rates for oxygen and hydrogen at shock tube temperatures which test this point. Wray (101) followed the three-body recombination of oxygen atoms (argon catalyst) by monitoring the absorption of 1270 Å radiation and thus estimated the rate of production of O_2 molecules. He generated an excess of atoms (temperature range 1340°–2920° K) by shock heating dilute ozone-argon mixtures. Values for k_R were found to be compatible with those computed from k_D, within the precision of the measurements. The same conclusion was reached by Wilson, who used a different technique (102). Recently the recombination reaction was studied again by Kiefer & Lutz (103) with an excess of atoms produced by the pyrolysis of ozone, but in the presence of a high concentration of O_2. They used X-ray densitometry to follow the course of the exothermic recombination over the temperature range 1500° to 2800° K, and were able to demonstrate that the rate was truly second order in atom concentration.

$$k_R(O_2 \text{ catalyst}) = (6.81 \pm 0.54) \times 10^8 (2200/T)^{1.22\pm 0.53}, \qquad \text{liter}^2\ \text{mole}^{-2}\ \text{sec}^{-1}$$

This is in excellent agreement with the rates derived from the dissociation of O_2 using the same apparatus (104).

Preliminary values for the high temperature recombination rate constant for hydrogen atoms were estimated by Russo, Hall & Lordi (105) from nozzle-wall static pressure distributions in supersonic nozzle expansion flows of shock heated H_2/Ar mixtures. Back reflected shocks were used to produce the samples at reservoir temperatures of 6000° K and pressures of 28, 57, and 112 atm. The mean values (accurate to a factor of 2.5) check the corresponding recombination rate constants deduced from shock dissociation data. More precise values were deduced by Schott & Bird (106) from their measured rate of disappearance of OH radicals in shock heated mixtures of H_2 (2–8 per cent), O_2 (0.5–1 per cent), and Ar. Due to the very rapid atom abstraction reactions, equilibrium is maintained between the hydroxyl and hydrogen atoms, so that from the measured decline in OH one may estimate the rate of decline of the H atoms. They found

$$H + H + M \xrightarrow{k_1} H_2 + M$$

$$H + OH + M \xrightarrow{k_2} H_2O + M$$

$$k_1 = (6 \pm 1) \times 10^8\ \text{liter}^2\ \text{mole}^{-2}\ \text{sec}^{-1}\ \text{(in mixtures used)}; \qquad k_2/k_1 = 10 \pm 5$$

Estimation of the relative contributions of H_2, H, and H_2O as third bodies permitted the deduction of $k_1(\text{Ar})[(4\pm 2)\times 10^8\ \text{liter}^2\ \text{mole}^{-2}\ \text{sec}^{-1}]$, again in agreement with values deduced from the dissociation rate data to within the experimental error. To measure the rate of recombination of chlorine atoms Jacobs & Giedt (107) used the exploding collar technique (38) for a series of runs over the temperature range 1600° to 2100° K. The recombination rate constants derived from these experiments were found to equal those deduced from the dissociation of chlorine behind incident shock waves (95).

The three-body recombination rates of atoms at around room temperature have been studied for many years. A variety of techniques were developed for the production of appreciable quantities of atoms and for following the decline in their population as a function of the ambient gas, the surfaces of the vessels, and of the temperature. Due to the discordant values reported by the numerous investigators, presumably because of difficulties in estimating correctly the atom concentrations and of controlling the composition of the ambient gas, no useful comparison can yet be made of recombination rates at room temperature with those deduced from shock tube experiments or from flame studies. Wray (108) indicated that his shock tube data on oxygen with argon as the catalyst did check values for k_D as reported at 300° K, but no agreement was found when the highly catalytic atomic and molecular oxygen were used as catalysts. For the latter, the 300° K value extrapolated from about 5000° K proved to be about 40 times greater than that measured. The relative efficiencies of O and O_2 are not well established at low temperatures due to unknown magnitude of contribution from an alternate path for the recombination: $O+O_2+M \rightarrow O_3+M$; $O+O_3 \rightarrow 2\ O_2$. The role of the third body for oxygen atom recombination at around room temperature was discussed by Morgan & Schiff (109) who studied the rate of the reaction $O+O+N_2 \rightarrow O_2+N_2$ in an O_2-free system. Recombination rates at 300° K and 350° K for $N+N+N_2$; $N+O+N_2$; and $H+H+H_2$ have been remeasured recently (110) as were $Br+Br+Br_2$ and $Br+Br+Ar$ (111). The recombination rate of H atoms over the temperature range 200° to 350° K was measured by Thrush & Larkin (112) using a calorimetric probe. Their results agree with the values deduced from shock tube dissociation rate constants, if one assumes an overall T^{-1} dependence for k_R.

An excellent summary of three-body reactions at around room temperature which are of particular interest to the complex chemistry of the upper atmosphere was prepared by Barth (113). He assembled several reported values for each of six recombination reactions. In his table values are included for pairs of nitrogen atoms, pairs of oxygen atoms, NO+N, etc. In another table he collected data for five chemiluminescent reactions which may also require three-body collisions. Attention is also called to papers published in the Discussions of the Faraday Society, No. 37, on Chemical Reactions in the Atmosphere.

The above three-body recombinations of atoms at low temperatures may be briefly compared with the much more rapid bimolecular association reactions of small radicals. Association is then followed by a deactivation collision with the ambient gas. For example, a second order rate constant for $2\ CF_2 \rightarrow C_2F_4$ was measured by Dalby (114). He generated the radicals by flash photolysis of C_2F_4 and C_3F_6, and followed their decline spectrophotometrically in the absorption region 2340 to 2350 Å.

$$k_2 = 1.3 \times 10^8 (T/300)^{1/2} \exp(-1200/RT), \qquad \text{liter mole}^{-1}\ \text{sec}^{-1}$$

Note that this recombination rate increases with rising temperature in contrast to the similar association rate for CF_3. In another study (115), the net rate of recombination of methyl radicals was found to be sensitive to the composition of the third body, i.e., on its efficiency for deactivation.

Vibration-Dissociation Coupling and Diatom Dissociation Rates

During the past two years, several reports were published in which previously proposed theories for bimolecular rates of dissociation of diatomic molecules, or the equivalent rates of two-atom recombination in the presence of a third-body, were amended or extended. In the first category, Shuler & Weiss (116) treated the problem of dissociation in terms of the mean first passage time for the collisional excitation of diatomic oscillators to an absorbing barrier. The new aspects considered were the dependence of the mean first passage time on (*a*) the initial (arbitrary) distribution of molecules over their energy level system; (*b*) the height of the barrier; and (*c*) allowing transitions with a distribution in Δv such that $\Delta v \geq 1$. The latter appears to be the crucial factor in that on the basis of this model, with Δv restricted ± 1, the predicted rates of dissociation were too low by several orders of magnitude. If for sufficiently hard collisions transition probabilities for $\Delta v \gg 1$ are assumed to be about equal to $\Delta v = \pm 1$, correct orders of magnitude may be derived from this model. It is not clear whether this extension accounts for a pre-exponential factor with an inverse temperature dependence.

The classical treatment for diatom dissociation due to sufficiently energetic bimolecular collisions is referred to as the "available energy model," in which it is assumed that dissociation occurs when for the collision complex the total energy incorporated in an appropriate number of squared terms (in the Hamiltonian) exceeds the dissociation energy. This model leads directly to an inverse T dependence in the pre-exponential term but requires *ad hoc* assumptions regarding the selection of the squared terms. Hansen (117) suggested that one consider only the energy available in rotation and vibration of the diatom after the collision, i.e., $m = 4$. This neglects the specific catalytic effects of the colliders as noted in all cases studied so far. However, his suggestion leads to a correct order of magnitude estimate for a number of diatomic and for a few triatomic molecules for which dissociation rates are known.

The least arbitrary treatment is the one proposed by Wigner and developed by Keck (118), in which the least upper bound for the three-body recombination rate may be estimated by calculating the flux of reacting systems (as specified by a representative point for the three atoms in phase space) across a selected surface, and into a region of phase space which may be clearly identified as the diatom plus catalyst molecule. This treatment predicts an inverse T dependence for the recombination step. Woznick (119) extended this treatment to the case where the interaction between the catalyst and the recombining atoms is strongly attractive. An inverse T dependence for recombination rates for atoms or radicals can be deduced in a formal manner by treating the association either as a two or three step process involving a highly unstable complex:

$$A + A \rightleftharpoons A_2^*; \qquad A_2^* + M \rightarrow A_2 + M \qquad 1.$$

$$A + M \rightleftharpoons AM^*; \qquad AM^* + M \rightarrow AM + M; \qquad AM + A \rightarrow A_2 + M \qquad 2.$$

The lifetime of A_2^* or of AM^* (if M is an atom or a rigid diatom) is no longer than the collisional passage time. However, predictions which agree with experiment can thus be made (120). Using mechanism 2, it is possible to estimate (121) the catalytic efficiencies of large molecules by assigning lifetimes to AM^* deduced from kinetic data on unimolecular reactions.

The crucial role of vibration-dissociation coupling has been referred to several times. Specifically, the vibration relaxation equation with concurrent dissociation has three terms in the right member (122).

$$\frac{d\epsilon(T_v)}{dt} = \frac{\epsilon(T_t) - \epsilon(T_v)}{\tau_v} - \left[\frac{\bar{E}(T_t, T_v) - \epsilon(T_v)}{[x]}\right]\left(\frac{d[x]}{dt}\right)_f + \left[\frac{\bar{E}(T_t, T_t) - \epsilon(T_v)}{[x]}\right]\left(\frac{d[x]}{dt}\right)_r$$

The left member is the rate at which the energy per oscillator increases; $\epsilon(T)$ is the average energy per oscillator calculated at the temperature T; $\bar{E}(T_t, T_v)$ is the average vibrational energy lost per dissociation; $[x]$ is the concentration of diatoms with the corresponding rates for the forward (dissociation) and reverse (association) steps; and $\bar{E}(T_t, T_t)$ is the average energy per oscillator gained per recombination. To evaluate the $\bar{E}$ terms, one must introduce specific assumptions for the relative probabilities for dissociation from the individual vibrational states. The set of coupled equations for the time dependence of the diatom concentration in each vibrational state can be solved numerically, given the transition probabilities for the collisional transfer between vibrational levels and to the dissociated state. Such systems of equations have been solved by Treanor & Marrone (123). If one assumes that the distribution among the levels is approximately that of Boltzmann, calculations show that the effective T_v remains below T_t for a period which is very long compared with the measured vibration relaxation time, so that the measured dissociation rates are lower than those expected were T_v equal to T_t. The magnitude of the effect depends on the specific assumptions made regarding the relative probabilities for dissociation from levels below the uppermost state. If one assumes equal probabilities from all levels, an approximate T^{-1} dependence is deduced for oxygen shocked to temperatures in the range 4000 to 8000° K—preferential dependence from the upper levels gives a numerically larger power. This model also predicts an "incubation time" such as was reported by Wray (100).

The simplifying assumption that the population of the vibrational levels is approximately Boltzmannian is not essential. The concentration time dependence for an 18-level model has been calculated by Treanor assuming selective probabilities for dissociation and simple harmonic oscillator transition probabilities for translation-vibration exchange. Of course, one should not overlook the limitations of this approach. These essential transition probabilities (between vibrational levels and to the dissociated states) must be inserted in the program, based on theoretical estimates or deduced from specifically designed experiments. Neither is available for the highly excited vibrational levels, nor has adequate account been taken of the relatively rapid energy exchange between levels $X_2^{(v)}+X_2^{(w)} \leftrightarrows X_2^{(v+1)}+X_2^{(w-1)}$, which forces the system toward a local equilibrium. An analysis in which only overall rate constants appear was presented by Heims (124). It is based on moment equations for this coupled system; no numerical results have been published. The degree of departure of the vibrational level population from that expected for statistical equilibrium depends on the dissociation rate, i.e., on $\exp(-D_0/kT_v)$. Treanor's calculation indicates that the inverse temperature coefficient is not primarily a consequence of the non-Boltzmann feature of the model. The original basis for the argument that $(k_R)_{obs} \neq (k_D)_{obs}/K_{eq}$ no longer exists. Nevertheless, the question remains whether the departure from a Boltzmann population does invalidate the relation between the forward and reverse reaction rate constants. Rice pointed out (125) that the distorted population affects both rate constants by the same factor so that their ratio remains at K_{eq}. Snider (126) carefully analyzed the nonequilibrium relaxation processes, starting with master equations for a unimolecular reaction (such as an isomerization) and for a bimolecular dissociation. In the former case, the phenomenological equation $-dA/dt=dB/dt =k_f(A)-k_r(B)$ will always be valid after the initial transient has passed; and then $K_{eq}=k_f/k_r$ even though these differ from k_f^{eq} and k_r^{eq}. For the process $A_2+M \leftrightarrows 2A+M$ additional relations must be satisfied between the rate constants and the atom concentrations so that

$$-dA_2/dt = \frac{(dA/dt)}{2} = k_D(A_2)(M) - k_R(A)^2(M)$$

When the latter holds, K_{eq} equals $(k_D/k_R)_{obs}$. There are subtle but essential differences between the phenomenological rate constants, the specific equilibrium reaction rate constants, and the sums of transition probabilities between the states identified with A_2 and those identified with 2A (127)

Excitation and Ionization in Shocked Gases

Even though one of the strongest sensory impressions of shock propagation in a gas is the motion of the luminous front, the kinetics of excitation by collisions with atoms and molecules have received relatively little attention. Meaningful quantitative values for electronic excitation cross-sections are not easy to obtain from shock tube experiments because it is difficult to reduce the concentrations of impurities to a level such that they no longer dominate the rate processes. Measurements on the excitation of sulfur dioxide to a low-lying electronic state have been reported by Levitt & Sheen

(128). In a number of respects these are similar to a previously reported study of NO_2. Mixtures of 10 per cent SO_2 in Ar, shock heated to about 4300° K, produced a continuum emission over 3300 to 4400 Å. Since it has been established that the weak absorption by SO_2 at 3800 Å is due to the transition ($^3B_1 \leftarrow {}^1A_1$), these authors suggest that the emission they observed is due to thermally excited molecules in the $^3B_1$ state. The emission intensity rose to a maximum at the shock front (resolving time ≈ 1 μsec) and fell exponentially to zero. At the maximum, the absolute intensity per unit wave length at 4380 Å is given by: $dI/d\lambda = 10^{32.82} \exp(-73{,}200/RT)$ photons sec^{-1} $mole^{-1}$ cm^{-1}. As in NO_2, the intensity was shown to be first order in the emitter and zero order in argon. This indicates that the population of the upper state ($^3B_1$ is 73.5 kcal above $^1A_1$) is controlled by collisional activation and deactivation: $SO_2(^1A_1) + M \leftrightarrows SO_2(^3B_1) + M$.

Studies of ionization processes have been more extensive. Three types of species have been investigated: an alkali metal vapor, the rare gases, and the components of air. Haught (129) shock heated cesium vapor (in Ar) and from the intensity of the emitted radiation measured the thermal ionization rate as a function of the temperature and density of the sample. Only by exercising great care in outgassing the tube, by using very pure argon and pretreating the tube with cesium vapor to "getter" the less volatile impurities, was he able to obtain meaningful data. Total shock emission intensities were photographed with a drum camera; a delayed radiation overshoot was observed. These observations can be interpreted in terms of a two-stage mechanism. Immediately past the shock front, only atom-atom collisions occur, and since these have small cross-sections for either excitation or ionization, a low level of luminosity is produced. When the degree of ionization reaches some appreciable value (of the order 10^{-3}), atom-electron collisions take over control of the rates of excitation and ionization, leading first to the establishing of population equilibrium with respect to the allowed excited states. Further, suppose that under these conditions the rate of attaining equilibrium with respect to ionization is slower than that for excitation. As ionization equilibrium is approached, the temperature drops, and the population of the excited states also decreases, thus producing a radiation overshoot. However, details of the kinetics such as the dependence of the delay times on the cesium density and the temperature, etc., cannot be fully accounted for. This type of mechanism has been postulated to explain several overshoot phenomena—in particular, the overshoot of $N_2^+(1^-)$ emission in shock heated nitrogen. In that case, cooling results from delayed dissociation.

Ionization processes in the noble gases are more complex, and it appears that the mechanism for xenon differs to some extent from that for argon and krypton, although unequivical statements cannot be made due to the lack of full control over impurities. Most investigators agree that the observed activation energies for ionization coincide, within experimental error, with the first excitation potential rather than with the ionization potential. As in cesium, ionization equilibrium is attained via a two step process, of which the first is excitation of the atom by a heavy particle collision to states in the vicinity of the lowest lying electronic state above ground level. Harwell & Jahn (130) followed the early stages of ionization in these three inert gases (shock Mach numbers 7–10) with a transverse microwave beam. They deduced the following activation energies for ionization: Ar, 11.9 ± 0.5 eV; Kr, 10.4 ± 0.5 eV; Xe, 8.6 ± 0.5 eV, values consistent with the above mechanism.

In their second paper on xenon, Kornegay & Johnston (131) measured the rate of ionization by microwave absorption from samples heated in back reflected shocks. They corrected their data for the production of electrons by impurities and extraneous emission from the tube walls. For the mechanism:

$$\mathrm{Xe} + \mathrm{M} \underset{b}{\overset{a}{\rightleftarrows}} \mathrm{Xe}^* + \mathrm{M}; \qquad \mathrm{Xe}^* + \mathrm{M} \overset{c}{\rightarrow} \mathrm{Xe}^+ + \mathrm{e}^- + \mathrm{M}$$

the measured rate constants were: $k_1 = ac/(b+c)$ and $k_2 = b+c$. They reported that neither hydrogen nor oxygen (added impurities) were particularly efficient for exciting or ionizing xenon, but that in the presence of oxygen values for k_2 were lower than for pure xenon, possibly because of the dissociative quenching reaction $\mathrm{Xe}^* + \mathrm{O}_2 \rightarrow \mathrm{Xe} + 2\mathrm{O}$. By adding helium they demonstrated again that He-Xe cross-sections for excitation were about 20 times that for Xe-Xe. On the basis of a consistency test proposed by Alpher & White for the weight factor of the excited state (Xe*), Kornegay & Johnston concluded that the above mechanism was adequate for the case of pure xenon or in the presence of slight amounts of added hydrogen or oxygen, wherein Xe* refers to the multiplicity of five states located 8.3 eV above ground level. However, when neon or helium was added, the deduced multiplicity ranged from 10^2 to 10^3, indicating a large number of excitations were operating in parallel. They proposed the additional step:

$$\mathrm{Xe}^* + M \underset{e}{\overset{d}{\rightleftarrows}} \mathrm{Xe}^{**} + M \overset{f}{\rightarrow} \mathrm{Xe}^+ + e^- + M$$

The measured activation energy for ionization of Krypton is about 10 eV (IP = 14 eV).

In a lengthy paper, Hacker & Bloomberg (132) described their studies of the early ionization processes in shock heated xenon and arrived at a much more complex description of events. They made strenuous efforts to work in a scrupulously clean system. Their stainless steel shock tube was degassed at 500° under high vacuum. The reported leak rate was less than 10^{-3} μ/min; very pure gases were used in the driver as well as in the driven sections. The extent of ionization was measured from reflected microwave intensities (X-band). Also, spectra of the shock front were recorded. In back reflected shocks they found lines from XeI and XeII as well as a few silicon lines, at 2881 Å and 2520 Å. Time resolved spectrograms of the shock front showed emission due to C_2 as the only spectroscopic impurity. Surprisingly their measured activation energy was 1.6 ± 0.2 eV, which can be accounted for only on the assumption that electrons are produced via $Xe^* + Xe \rightarrow Xe_2^+ + e^-$, or by $Xe^* + C_2 \rightarrow XeC_2^+ + e^-$. They were forced to conclude that the initial excitation of the xenon was due to the diffusion of radiation ahead of the shock front, and that the production of luminosity at the shock front was due to radiative recombination of charged molecule ions with free electrons.

A still higher level of complexity was encountered in the study of ionization of air in strong shock waves (133). To account for the distribution of the various species, both neutral and charged, in the relaxing shock region, one must assume that ionization processes are coupled with vibration and dissociation relaxation. Lin & Teare concluded that for shock speeds up to 10 cm/μsec, ionization occurs predominantly from ionizing collisions between atoms. Then in decreasing order of importance, photo-ionization, electron-atom collisions, atom-molecule collisions, and finally molecule-molecule collisions contribute to the total ionization. Magnitudes of the rate constants required for the computer program, set up for the multiplicity of concurrent reactions, were obtained from other shock tube studies or were extrapolated from low temperature experiments. The computed electron density profiles were then compared with the electron density profiles obtained from a number of experiments. They were in good agreement in the steeply rising portion of the shock, but the experimentally observed distributions declined less rapidly behind the peak than did the predicted ones. This may be attributed to the slight deceleration of the shock wave under experimental conditions.

Stein and co-workers (134) measured the electron depletion rate in air at about 2900° K, produced by an unsteady expansion from a reflected shock heated sample at 3600° K. Their data indicate that the dissociative recombination $NO^+ + e^- \rightarrow N + O$ was the dominant electron depleting step. Their recombination coefficient $\alpha = (4 \pm 2) \times 10^8$ cm^3 sec^{-1} is in agreement with a value reported by Lin et al., at 5000° K, if a $T^{-3/2}$ dependence is assumed.

For diagnostic measurements of shocked gases, spectroscopic techniques have been extensively exploited. Here, brief mention is made of the use of shock tube heating for the solution of spectroscopic problems. The content and references of a dozen recent publications are listed in Table III. Berry & Reinmann (148) recorded the absorption spectra of gaseous halide ions produced by shock heating alkali metal salts, and determined the electron affinities of the halogen atoms (i.e., the *IP* of the negative ions) from the low energy threshold of their absorption spectra. In the case of F^-, the photo-detachment threshold was used. His results are summarized in Table IV. Allen & Textoris (149) report to have obtained evidence for the existence of N^- from the intensity of equilibrium continuum radiation in the vicinity of 5000 Å of shock heated nitrogen. An analysis of radiation cooling of a gas heated by a strong shock wave was presented by Sevastyanenko & Yakubov (150); they found good agreement between previously measured cooling

TABLE III

SPECTROSCOPIC STUDIES IN SHOCK TUBES

Substance	T range (°K)	Spectroscopic range	Observations (general)	Remarks	Reference
C_6H_6 C_6F_6	690–1900 (i.s.)[a]	2000–2900 Å	Integrated absorption coefficient (fnc T) obtained prior to pyrolysis—also kinetics of pyrolysis	f dependence on T indicates strong vibration—electronic interactions. Theory adequate for C_6F_6 but not for C_6H_6. Mechanism for pyrolysis proposed	(135)
Ar, Kr, Xe	5000–15,000 (i.s.)	visible	Dependence of intensity of continuum emission on T and ρ electrons	Good correlation with Kramers-Unsöld theory. Free-bound electron transitions primary source of continuum	(136)
air	4000–9000	4500–13,000 Å	Analysis of shock tube data. Estimate transition moments	Moments as function of T; deduced Z^2 values for free-free Kramers radiation in the infrared	(137)
air	1000–8000	2000–5000 Å	Absorption coefficients for six band systems, for diatoms	Developed digital computer programs based on molecular constants taken from various sources. Preserved spectra and line intensities on magnetic tape (atlas)	(138)
O_2,N_2	2600–6600 (i.s.)	UV and IR	Emission intensities; error analysis for estimated f	Discrepancy between calculated and observed intensity distribution in IR may be from transitions between excited states of NO	(139)
N_2, air	6000–9000 (r.s.)[a]	2μ–8μ	Absolute emission intensities; deduced f for various species	Continuum radiation caused by free-free scattering of electrons, mostly from neutral atoms and molecules	(140)
90% N_2–10% CO_2	8000 (r.s.)	2300–12,000 Å	Relative emission intensities from Venus-like atmosphere	CN most prominent radiator. Estimated transition probabilities of violet and red systems	(141)

TABLE III—(*continued*)

Substance	T range (K°)	Spectroscopic range	Observations (general)	Remarks	Reference
H_2O	3100–3500 (r.s.)	3063 Å	Absorption coefficient and f for OH($^2\Sigma-^2\Pi$)	$f=(3.9\pm0.9)\times10^{-3}$; compared this measurement with others	(142)
NO	1000–2500 (i.s.)	2.7μ and 5.29μ	Emission intensities for fundamental and harmonic	Measurements on carefully purified gas; checked shock conditions interferometrically. Vibration relaxation time <1 μsec	(143)
NO	1400–2400 (r.s.)	2.7μ and 5.29μ	Integrated band intensities independent of T for fund; overtone shows slight variation	T dependence in agreement with theory based on s.h.o. approximation. Integrated intensities extrapolated to STP $\alpha^0(5.3\mu)=76\pm7$ cm^{-2} atm^{-1}; $\alpha^0(2.7\mu)=2.8\pm0.5$	(144)
CO_2(Ar)	1500–3500 (i.s.)	4.25μ ($\Delta\lambda=0.043\mu$)	Absorption coefficient and emissivity	Deduced absorption coefficient in agreement with statistical model for pressure broadened lines	(145)
CO_2(Ar)	1200–3000 (r.s.)	2.7μ	Absolute integrated emission intensity—as function of T	Experimental integrated intensity shows strong T dependence, in reasonable agreement with theory based on s.h.o. wave functions. Must include higher order terms for electric dipole moments	(146)
CO_2 and H_2O	600–2400, shocks and burners	4.40μ CO_2 2.85μ H_2O	Absorption reduced to band model parameters; evaluated their T dependence	Random band model parameters evaluated to permit *in situ* quantitative analysis for components in gaseous mixtures	(147)

[a] i.s.—incident shocks; r.s.—reflected shocks.

TABLE IV

ELECTRON AFFINITIES OF THE HALOGENS FROM ABSORPTION SPECTRA (148)

Element	Electron affinity (eV)	Statistically weighted average energy difference of neutral atom (^2P) and negative ion (^1S)
F	3.448 ± 0.005	3.464
Cl	3.613 ± 0.003	3.649
Br	3.363 ± 0.003	3.515
I	3.063 ± 0.003	3.377

rates and their theoretical values. Penkin (151) assembled many results on oscillator strengths for atomic spectra. He compared data obtained with a King furnace and relative emission intensities of atomic lines from shock heated gases.

REACTIONS WHICH PASS THROUGH FOUR CENTER TRANSITION STATES

Development work on single-pulse shock tubes is beginning to pay dividends. Current practice permits chemical kineticists to treat this device as a short-duration, high-temperature reactor, which has the special feature that it insures homogeneous reaction conditions. Note that vibration-relaxation times for the polyatomic systems studied are generally less than one microsecond, whereas dwell times are greater than 500 μsec. Although full control of operating conditions is not yet available, one can measure relative rates precisely. On an absolute basis, reproducibility is often limited by adsorption of reactants and products on the walls of the metal gas handling system and of the shock tube, and by contamination with greases, pump oils, and minute diaphragm fragments. Accuracy is severely limited by the lack of precise knowledge of the temperature and by perturbations of the shock waves because of the irreproducibilities in diaphragm rupture and opening times. On the positive side, a relatively large range in concentrations of reactants can be studied, and small extent of reaction can be followed if one exploits the full sensitivity of current vapor phase chromatography analytical techniques.

Two types of four-center reactions have been successfully investigated using the single-pulse shock tube. Studies of the atom exchange rates $HX + D_2 \rightarrow DX + HD$, for $X = H$, Cl, HS, NH_2, CH_3, and C_2H as a function of composition and temperature led to the uncovering of the role of vibrational excitation in such metathesis reactions (152). These deuterium substitutions take place at temperatures between 1200° to 1700° K, with dwell times of about 800 μsec. Except for C_2H_2, the observed dependence of the rate on the concentrations of the reactants and the total densities is consistent with the following mechanism:

$$D_2^{(o)} + M_i \underset{k_{\epsilon i}}{\overset{k_{-\epsilon i}}{\rightleftarrows}} D_2^{(v)} + M_i \qquad M_i = \text{Ar}, \quad i = 1$$

$$D_2^{(v)} + HX \xrightarrow[k_x]{} DX + HD^{(v-1)} \qquad \begin{aligned} &= D_2, \quad i = 2 \\ &= HX, \quad i = 3 \end{aligned}$$

Then,

$$\frac{d\text{HD}}{dt} = k_{\epsilon 1}(D_2)(\text{Ar}) \frac{1 + \beta x + \gamma y}{1 + \Gamma}$$

$$x \equiv \text{HX}/\text{Ar} \qquad y \equiv D_2/\text{Ar} \qquad z \equiv D_2/\text{HX}$$

$$\alpha \equiv k_x/k_{-\epsilon 1}; \qquad \beta \equiv k_{\epsilon 3}/k_{\epsilon 1} = k_{-\epsilon 3}/k_{-\epsilon 1}; \qquad \gamma \equiv \frac{k_{\ 2}}{k_{\epsilon 1}} = k_{-\epsilon 2}/k_{-\epsilon 1}$$

$$\Gamma \equiv \frac{1}{\alpha}\left[\frac{1}{x} + \beta + \gamma z\right]$$

The vibrational relaxation time for $D_2(p\tau_v \approx 1\ \mu\text{sec atm})$ is sufficiently short to allow attainment of a steady state concentration of $D_2^{(v)}$; (v) is the lowest vibrational level which reacts at an appreciable rate with HX. Thus, for methane (153), $\alpha \approx 300$; $\beta \approx 1$; $\gamma \approx 100$; $E_A = 53.6 \pm 2.2$ kcal/mole, which indicates that $v \approx 7[k_{\epsilon 1} = 4.5 \times 10^{10} \exp(-53{,}600/RT)$ liter mole^{-1} sec^{-1}]. For NH_3, $\alpha \approx 500$; both β and γ appear to be quite small and may be neglected within the precision of the experiment; $E_A \approx 38$ kcal/mole, so that $v \approx 5$ (154). In both these cases

$$HX^{(v)} + D_2 \rightarrow HD + DX^{(v-1)}$$

contribute a negligible amount to the total exchange rate. The data for HCl and H_2S have not yet been fully reduced.

The model system for a metathesis reaction in the gas phase is $H_2 + D_2 \rightarrow 2\ HD$. Because of symmetry in the reactant, the rate of production of HD must be represented by the sum of two terms similar to the above, which involve, respectively, $D_2^{(v)} + H_2 \rightarrow 2\ HD$ and $H_2^{(v)} + D_2 \rightarrow 2\ HD$. The second contributes 0.35 as much as the first; $E_A = 42.43 \pm 1.6$ kcal/mole (155). One may readily demonstrate that the atom displacement mechanism is incompatible with the data under the conditions which were used for all the above reactions. From the known dissociation rates of D_2 and HX exchange rates via the mechanism:

$$D_2 + M \rightarrow 2D + M; \qquad D + HX \rightarrow HD + X$$

are too low by factors of 10^4–10^8, depending on the postulated chain length. It had been assumed that in general bimolecular processes the activation energy may be distributed over all the degrees of freedom of the collision complex. However, the mechanism deduced for the above exchange reactions requires that in this type of bimolecular process a much more stringent condition be fulfilled. There is a low probability for reaction between molecules in their ground or low-lying vibrational states even though their relative kinetic energy along the line of centers exceed the minimum required. However, the probability for exchange is high, irrespective of their relative kinetic energies, provided the D_2 (or H_2) is excited to its 5 to 8th vibrational level. Hard collisions of large amounts of momentum normal to the direction of separation are not effective, whereas large vibrational amplitudes which provide momentum along the direction of separation are effective.

The mechanism for exchange with acetylene (156) follows a different route. Here the rates were estimated from the intensity of infrared emission at selected wavelengths, characteristic of C_2H_2, C_2HD, and C_2D_2. The rate of appearance of C_2D_2 was half that of C_2HD, and their sum was equal to the rate of disappearance of the acetylene. The activation energy (over the temperature range 1300°–1665° K) is 32.5 kcal/mole, while the orders of the reaction were found to be unity for D_2, 0.24 for C_2H_2, and zero for the argon. This suggested a mechanism in which the addition of deuterium to acteylene generates a loose complex, which may be readily dissociated by collision with another acetylene. However, the loose complex may rearrange to form a tight complex of symmetry V_d, which resembles an excited ethylene. Reversal of the tight to the loose complex occurs with statistical redistribution of the H and D atoms, taken one from each carbon.

There are two other cases where high specificity to the mode of excitation has been demonstrated. In the dissociation of diatomic molecules the probability for dissociation appears to be highest for the topmost vibrational levels and declines exponentially with increasing distance below the dissociation limit (122). Also, in the ion-atom reaction $He + H_2^+ \rightarrow HeH^+ + H$ ($\Delta H° \approx 1.1$ eV) the rate appears to be independent of the relative kinetic energy of the reacting species and is appreciable only for H_2^+ in vibrational states above the 5th level (157). There is a possibility that the photosynthesis of CH_3I from CH_4 and I_2 by irradiation with 1849 Å occurs via atomic exchange between CH_4 and electronically excited I_2^*, as proposed by Harris & Willard (158).

It is instructive to contrast the above mechanism for metathesis with the extensively studied three-center transition states which play the dominant role in atom abstraction reactions. It is now generally accepted that the three-center systems have the lowest energy when in the linear configuration at the saddle point, i.e., at the kinetic transition state. It is presumed that the activation energy is provided by the relative kinetic energy of the colliding pair. There is no such information available for the four-center systems since reaction occurs only if the complex so generated is in a relatively high vibrational state. Further, it is now accepted that for highly exothermic abstraction reactions some of the bond energy appears as vibrational excitation of the newly formed bond, although the extent of such excitation depends on the detailed shapes of the potential energy surfaces (159, 160). We have not yet studied a highly exothermic four-center system under controlled conditions to ascertain whether a parallel effect is present.

The second group of reports concerns dissociation reactions, in which four-center transition states represent the highest energy configuration in the process of splitting out an HX from compounds of the type R_1-CH_2X (X = halogen, OH, SH). In his first paper, Tsang (33) measured the rates of dehydrohalogenation and dehydration on an absolute basis. He demonstrated that under shock conditions these reactions were strictly unimolecular, except for H_2S loss from the thioalcohols. For the latter he quenched the free radical path by adding a sufficient quantity of propylene to the RSH/Ar mixtures. These measurements permitted him to estimate accurately the activation energies for the corresponding dissociations. Tsang then demonstrated (35) that relative rates and activation energies can be measured with high precision provided the two reactions do not perturb one another, as in these unimolecular reactions at their high pressure limits. The irreproducible perturbations still inherent in shock tube operation and errors due to approximate reaction temperatures cancel practically completely. The results of these studies are summarized in Table V. The rate constants available in the literature in May 1963 for the unimolecular elimination of HX from the alkylhalides in the gas phase were critically analyzed by Benson & Bose (161). They call attention to a number of interesting regularities and argue that the data are compatible with the assumption of a very polar four-center transition state, as was proposed by Maccoll.

In his most recent report Tsang (162) presented high temperature data (900°–1150° K) on retro-Diels-Alder dissociations of cyclohexenes. These rate constants and others reported for related cyclobutanes are summarized in Table VI. He points to dissimilarities in the pre-exponential factors for the decylization of cyclohexenes and cyclobutanes. The exciting question remains to be answered: What factors favor ring opening as against ring fragmentation?

Isomerization and Fragmentation Reactions

In this section more than a dozen reactions are listed, the kinetics of which have been studied recently. In each case the shock tube technique was indicated by the need for high temperatures and the attainment of strictly homogeneous reaction conditions. Those reactions which proved to be unimolecular were readily interpreted. Since conveniently measured half times for current shock tube practice range from 10 to 10^3 μsec, operating conditions must be adjusted so that $20 \leq E_A/RT_{shock} \leq 25$, where E_A is the activation energy for the unimolecular step.

The rates of the *cis-trans* isomerization of butene-2 was measured in a single-pulse shock tube over the temperature range 1000 to 1250° K (31). The object of this investigation was to test the operation of the "magic hole" design. The high pressure limit rate constant fell slightly above the Arrhenius curve extrapolated from low temperature investigations, but a straight line drawn between the high and low temperature data gives the most reliable activation energy:

$$k_u^{\infty}{}_{(cis)} = 3.5 \times 10^{14} \exp(-65{,}000/RT), \text{ sec}^{-1}$$

TABLE V

SHOCK TUBE DATA ON DEHYDROHALOGENATIONS, ETC. (33, 35)

Reaction	log A	E_A (cal/mole)	T(°K)	Remarks
t-buCl→HCl+i−C_4H_8	13.9	46,200	750–950	Unaffected by added propylene
t-bu-Br→HBr+i−C_4H_8	13.5	41,500	700–900	Unaffected by added propylene
t-bu-OH→H_2O+i−C_4H_8	13.4	61,600	1050–1300	For $T>1300°$ K generated side products >1%: allene, propylene, isobutane, acetone
t-bu-SH→H_2S+i−C_4H_8	13.3	55,000	950–1230	Radical reaction concurrent with unimolecular decomposition. Radicals fully inhibited in 3/1 mixture of propylene/mercaptan
t-buCl→HCl+i−C_4H_8	13.74	44,690		Relative to iprBr→HBr+C_3H_6, for which log A = 13.62; E_A = 47,800 cal/mole
t-buBr→HBr+i−C_4H_8	13.87	41,490		
EtBr→HBr+C_2H_4	13.19	53,700	740–940	
EtCl→HCl+C_2H_4	13.16	56,460	820–1000	Measured relative to iprBr. A variety of combinations were run and data tested for mutual consistency
iprCl→HCl+C_3H_6	13.64	51,100	750–950	
iprI→HI+C_3H_6	13.67	45,065	680–850	
t-buI→HI+i−C_4H_8	13.73	38,080	650–760	

TABLE VI

RETRO-DIELS-ALDER AND RELATED REACTIONS

Reaction	log A	E_A (cal/mole)	T(° K)	Remarks	Ref.
cyclohexene$\rightarrow C_2H_4+(1,3)C_4H_6$	15.02	66,700	900–1150	Measured relative to iprCl	(162)
4-me-cyclo $C_6 \rightarrow C_3H_6+(1,3)C_4H_6$	15.13	66,600	900–1150	very slight amounts of C_2H_4 and CH_4	(162)
4-vinyl-cyclo $C_6 \rightarrow 2(1,3)C_4H_6$	15.20	62,000	900–1050	clean reaction—one product	(162)
$2\ C_4H_6 \rightarrow C_8H_{12}$	11.14	26,800	800	conc. in moles/cm³	(163)
$C_2H_4+(1,3)\ C_4H_6 \rightarrow C_6H_{10}$	10.45	27,500	800	conc. in moles/cm₃	(163)
$C_4H_8 \rightarrow 2\ C_2H_4$	15.6	62,500	693–541	Fall-off begins at 20 mm Hg	(164)
$C_4F_8 \rightarrow 2\ C_2F_4$	16.3	74,300	630–1200	Obtain C_3F_6 byproduct for $T>1400°$	(165)
Me-$C_4H_7 \rightarrow C_2H_4+C_3H_6$	15.4	61,200	683–723	Unimolecular down to 7 mm Hg	(166)
C \\ C-$C_4H_7 \rightarrow C_2H_4$+isoprene // C	14.64	51,030	578–622	Parallel isomerization to 1 me-cyclohexene	(167)

The thermal isomerization of methyl isocyanide (168) was studied in the same tube over the temperature range 710 to 860° K. For this reaction the rates measured under shock conditions fell below those extrapolated for the high pressure limit (8 atm) as deduced from the low temperature data (169). The shock tube data check the values derived from low temperature studies for the beginning of the fall-off region, calculated for $n=7$ (effective number of oscillators) on the assumption that the relative efficiency of excitation of CH_3NC in collisions with argon was 0.05 compared with CH_3NC collisions:

$$k_u^\infty = 10^{13.6} \exp(-38{,}400/RT), \text{ sec}^{-1}$$

Schofield, Tsang & Bauer (170) followed the thermal decomposition of ClCN (incident shock region; 2000°–2800° K) by absorption spectroscopy at the (0, 0) band head of the $B^2\Sigma^+ \leftarrow X^2\Sigma^+$ transition in CN. Although the first reaction is the bimolecular dissociation of the chloro-cyanogen by hard collisions with argon, the mechanism subsequently becomes involved because of atom abstractions and recombination steps; in all, six elementary reactions must be considered. Since the CN concentration time dependence had a characteristic form (a fast initial rise followed by a flatter portion, another fast rise and a final slow approach to equilibrium) the values for several ratios of rate constants can be estimated by trial curve fitting, using a digital computer. These data were analyzed on the basis of 99.4 kcal/mole for the heat of formation of CN. The directly measured quantity is: $k_1/k_2 = 3.1\times10^3 \exp(-85{,}500/RT)$, where

$$ClCN + Ar \underset{k_1}{\rightarrow} CN + Cl + Ar$$

$$ClCN + CN \underset{k_2}{\rightarrow} C_2N_2 + Cl$$

For the best overall fit the rate constant $k_4 = 10^{14.23} \exp(-100{,}400/RT)$, cm^3 mole^{-1} sec^{-1} for the reaction:

$$C_2N_2 + Ar \underset{k_4}{\rightarrow} 2CN + Ar$$

which is in agreement with our previously determined value. An order of magnitude estimate for $k_2 = 1.1\times10^{10} \exp(-6{,}000/RT)$ cm^3 mole^{-1} sec^{-1}, so that k_1 is $3.4\times10^{13} \exp(-91{,}500/RT)$ cm^3 mole^{-1} sec^{-1}, for the bimolecular dissociation of ClCN.

The thermal dissociation of N_2F_4 has been investigated in several laboratories. Modica & Hornig (171) followed the primary step by recording the characteristic absorption of NF_2 at 2600 Å. They checked the previously determined equilibrium constant and measured the dissociation rate over the temperature range 350 to 400° K for the N_2F_4 diluted with argon, helium, and SF_6 (using argon as the driver). The rate was total order two, being unit order with respect to the N_2F_4 and the diluent gas; the activation energy is 15.2 kcal/mole. Expressed in terms of the thermochemical dissociation energy ($D_0 = 18.4$ kcal/mole)

$$k_{(Ar)} = 3.0 \times 10^8 T^{1/2}(D_o/RT)^{4.0} \exp(-D_o/RT), \text{ l mole}^{-1} \text{ sec}^{-1}$$

This suggests that approximately eight squared terms contributed to the storage of the energy during the bimolecular collision. Relative collision efficiencies can be expressed as:

$$k_{He}/k_{Ar} = 1.2 \pm 0.1; \quad k_{SF_6}/k_{Ar} = 1.7 \pm 0.2; \quad k_{N_2F_4}/k_{Ar} = 1.8 \pm .02$$

Over a considerably higher temperature range (2200°–3000° K) the products of pyrolysis of N_2F_4 in the reflected shock region were determined via time resolved mass spectrometry (172); NF_2, NF, N_2, and F were the only species detected. These semiquantitative results can be interpreted on the basis of reactions:

$$NF_2 + M \xrightarrow{k_1} NF + F + M$$

$$2NF \xrightarrow{k_2} N_2F_2{}^* \xrightarrow{k_3} N_2 + 2F$$

which follow the almost instantaneous dissociation of the tetrafluoride into NF_2 radicals at these high temperatures. A preliminary value for k_2 is 2.5×10^{13}, cm³ mole⁻¹ sec⁻¹, independent of the temperature over the range investigated; k_1 is presumably known but not quoted by the author.

The unimolecular decomposition rate of chloropicrin (CCl_3NO_2) was measured by estimating the concentration of NO_2 spectrophotometrically at 4250 Å (interference filter band width: 60 Å) (173). Quantitative data were recorded over the temperature range 630 to 770° K, and additional qualitative observations were made over the range 1000 to 1300° K. At 790° K and above, a decline in absorption intensity indicated that after passing through a peak value the concentration of NO_2 was depleted by decomposition or reaction with the CCl_3. The deduced unimolecular rate constant for the primary step

$$k_u^{\infty}(630° - 770°K) = 10^{12.6} \exp(-30{,}900/RT), \text{ sec}^{-1}$$

is presumed to be due to C–N bond cleavage. However, there are difficulties. The pyrolysis of CCl_3NO_2 was investigated by Steacie & Smith (174) about 25 years ago who followed the reaction manometrically over the temperature range 400° to 450° K. They proposed the following stoichiometry on the basis of final product analyses

$$CCl_3NO_2 \rightarrow COCl_2 + NOCl \leftrightarrows COCl_2 + NO + 1/2Cl_2$$

for which they reported: $k_u^{\infty} = 10^{15.7} \exp(-37{,}700/RT)$, sec⁻¹. If the primary step under both sets of conditions is indeed the breaking of the C–N bond, the considerably lower activation energy found at about 700° K can only be explained by assuming that with rising temperature there is a significant concurrent increased contribution to the reaction coordinate by the vibrational modes, as in $ClNO_2$ (175). Mass spectrometric analysis showed that over the higher temperature range the products of reaction are: NO_2Cl, NOCl, NO, $COCl_2$, C_2Cl_4, CCl_2, and Cl_2 (176). A similar study was made of the dissociation of CH_3NO_2 and the consequent rate of decomposition of the NO_2, estimated from the decline in emission intensity at 4350 Å and 5100 Å

(177). At temperatures 1400 to 1800° K, the lifetime of the nitrogen dioxide is very short, presumably due to reaction with methyl radicals:

$$CH_3 + NO_2 \rightarrow CH_3O + NO$$
$$\qquad \longrightarrow CH_2O + HNO$$

Current interest in the high temperature chemistry of CO_2 is expressed by reports of a number of shock tube investigations, all of which are similar in approach but differ in the derived values and in their conclusions. Note that the primary dissociation to $CO(^1\Sigma)$ and $O(^3P)$ is endothermic by 126.7 kcal/mole. Jost and coworkers (178) followed the rate of disappearance of CO_2, highly diluted with Ar or N_2 (0.2–0.7 per cent), over the temperature range 2800° to 4400° K, in reflected shocks. They recorded the infrared emission intensity at 4.3 μ, and established that the rate was unit order both with respect to the carbon dioxide and the ambient gas. Their measured E_A was 98.8 ± 2.6 kcal/mole, with pre-exponential factors $10^{14.7}$ (cm^3 $mole^{-1}$ sec^{-1}) for Ar and $10^{14.9}$ for N_2. Brabbs, Belles & Zlatarich (179) confirmed the rate law for this dissociation using a single-pulse shock tube. They worked at higher concentration levels (3–7 per cent) over a narrower temperature range (2557°–3000° K). Also, they followed the decline in CO_2 concentration from the intensity of emission in the blue continuum $[CO_2(^3\Pi) \rightarrow CO_2(^1\Sigma) + h\nu]$ and found the activation energy to be 86 kcal/mole with a pre-exponential factor of $3 \times 10^{11} T^{1/2}$. Steinberg (180) and Davies (181) measured the dissociation rate of CO_2 over the temperature range 3300° to 6000° K, using incident shocks, by recording emission intensities at 4.3 μ and 2.7 μ. For Ar and N_2 diluents they agree that $E_A = 73 \pm 2$ kcal/mole, with pre-exponential factors $(1.2–2.9) \times 10^{11} T^{1/2}$ (cm^3 $mole^{-1}$ sec^{-1}), the lower values in N_2 diluent, while the higher ones in Ar diluent. An estimate of the dissociation relaxation time for CO_2 in reflected shocks was also given by Bazhenova (182).

Although the extremes of the reported activation energies differ by 26 kcal, all the experimenters agree that the kinetically deduced value is considerably less than the thermochemical value (by 28–51 kcal/mole). Davies suggested that the data can be interpreted in terms of a direct dissociation step (which is spin forbidden) by writing $k_{(Ar)} = 5.85 \times 10^{10}(D_o/RT)^{4.38}$ $\exp(-D_o/RT)$, with $D_o = 126.7$ kcal, a form analogous to expressions reported for diatom dissociations. The larger power dependence on D_o/RT is consistent with the squared terms available in the (tri-atom+noble atom) encounter. There is also the possibility, proposed by Brabbs et al., that carbon dioxide dissociates via a two step process:

$$CO_2(^1\Sigma) + Ar \rightarrow CO_2(^3\Pi) + Ar \xrightarrow{\text{fast}} CO + O + Ar.$$

Further spectroscopic as well as shock tube investigations are needed.

The thermal decomposition of carbon bisulfide appears to be similar to some extent to the mechanism proposed for the decomposition of carbon dioxide (183). The pyrolysis of CS_2 in argon was followed by photographic absorption spectroscopy. A short duration spark was triggered at preset times after passage of the shock front past the observation windows. Shock temperatures ranged from 2250° to 3350° K. The rate of appearance of CS was estimated photometrically, and for several runs the rates of disappearance of CS_2 and of S_2 were also determined. The rate was found to obey the relation

$$-\frac{d(CS_2)}{dt} = k_{Ar}(CS_2)(Ar)\left[1 + \frac{k_{CS_2}}{k_{Ar}}\frac{(CS_2)}{(Ar)}\right]$$

$$k_{Ar} = 10^{15.9} \exp(-81{,}800/RT), \qquad \text{cm}^3\ \text{mole}^{-1}\ \text{sec}^{-1}; \qquad k_{CS_2}/k_{Ar} \approx 20$$

It is probable that the measured value of k_{Ar} is twice that for the indicated dissociation (183a):

$$CS_2 + Ar \rightarrow CS_2^* + Ar$$
$$CS_2^* \longrightarrow CS + S$$

because the second step could be followed by the very fast reaction: $CS_2 + S \rightarrow CS + S_2$. The authors concluded that CS_2^* most likely represents an excited electronic state, possibly $^1B_2$ or $^1A_2$. Since the direct decomposition of CS_2 to ground state CS+S requires 96 kcal/mole ($= D_o$) the bimolecular rate constant was written in the form

$$k_{Ar} = 0.1(4.5 \times 10^{14})\left[\frac{D_o}{RT}\right]^3 \frac{1}{3!} \exp(-D_o/RT), \qquad \text{cm}^3\ \text{mole}^{-1}\ \text{sec}^{-1}$$

where 4.5×10^{14} is the estimated collision number at 2850° K.

Sulfur dioxide is another triatomic molecule, the decomposition of which follows a pattern similar to those of CO_2 and CS_2 but there are additional complications. The rate of disappearance of shock heated SO_2 in argon mixtures was estimated from the extent of absorption by SO, recorded photographically (184). Data taken over the temperature range 3000 to 4000° K are best explained on the basis of a net second order rate law with unit dependence on the SO_2 and Ar. The apparent activation energy is approximately 56 kcal/mole, when the concentration of SO_2 is 4 per cent to 32 per cent. At very low concentrations (0.1 per cent) Wagner (185) found that $E_A \approx 110$ kcal/mole, but this value decreased as the concentration of SO_2 increased, except for very high temperature runs when E_A remained at the high value for all concentrations. The proposed mechanism is:

$$SO_2 + Ar \underset{1}{\overset{2}{\leftrightarrows}} SO_2^* + Ar$$

$$SO_2^* + SO_2 \underset{3}{\rightarrow} SO_3 + SO$$

$$SO_2^* + Ar \underset{4}{\rightarrow} SO + O + Ar$$

$$SO_2 + Ar \underset{5}{\rightarrow} SO + O + Ar$$

where SO_2^* may be the isomer SOO. At very low SO_2 concentrations, reaction 3 is unimportant,

$$\frac{d(SO)}{dt} = \left(k_5 + \frac{k_1}{k_2}k_4\right)(SO_2)(Ar)$$

The high activation energy corresponds to k_5. When the SO_2 concentration is raised, reaction 3 provides an alternate path for the production of SO and provides competition with reaction 2 for SO_2^*, so that the measured E_A decreases. At sufficiently high concentrations, the following occurs:

$$\frac{d(\mathrm{SO})}{dt} \rightarrow k_1(\mathrm{SO_2})(\mathrm{Ar})$$

The intensity of thermal emission by NO_2 at 4380 Å and 4050 Å, when shocked at high dilution in argon to 1800° to 2200° K, may be used to measure the depletion rate of that species. The excited electronic state, populated by collisions with argon, is depopulated both by quenching and dissociation: $NO_2^* + Ar \rightarrow NO + O + Ar$. The emission intensities from the excited vibrational levels just below the predissociation limit are lower by a factor of about 3 than those calculated according to a quasi-Boltzmann distribution, but the levels well below the dissociation limit were filled to a Boltzmann limit. Levitt (186) interpreted this observation as evidence for excessive depopulation of the uppermost levels caused by dissociation. The minimum excitation energy to the (*) state is about 49 kcal/mole.

From data obtained on the decomposition of nitrous oxide in reflected shock waves Jost and coworkers (178) found that the rate of reaction was unit order both in the N_2O and Ar. The bimolecular rate constant is given by $k_{Ar} = 10^{15} \exp(-61{,}000/RT)$, cm^3 mole^{-1} sec^{-1}. Few additional details are available. Fishburne & Edse (187) extended the study of N_2O by testing the effect of added oxygen and nitric oxide (incident shocks). Changes in concentration of N_2O were measured from its infrared emission and ultraviolet absorption at Hg 2537 Å. With argon they found that the measured activation energy increased with the N_2O concentration from 49.5 kcal/mole at 0.13×10^{-4} mole per/cm^3 to 55.7 kcal/mole at 1.5×10^{-4} mole/cm^3. These data can be accounted for in terms of a unimolecular mechanism operative in the low pressure region, where the rate is limited by collisional activation.

$$\mathrm{N_2O + Ar} \underset{k_a}{\overset{k_{-a}}{\rightleftarrows}} \mathrm{N_2O^* + Ar}; \qquad \mathrm{N_2O^*} \xrightarrow{k_1} \mathrm{N_2 + O}$$

$$k_a = 2.8 \times 10^{11} T^{1/2} \exp(-41{,}700/RT), \qquad \mathrm{cm^3\ mole^{-1}\ sec^{-1}}$$

It is presumed that N_2O^* refers to an excited $^3\Sigma$ or $^3\Pi$ state, so that the product oxygen atom may be generated in its ^3P ground state. Were the dissociation to take place from the ground ($^1\Sigma$) state, a higher activation energy would be required to produce the lowest singlet state of oxygen (^1D). In the presence of molecular oxygen the rate of decomposition is slightly higher than in argon. In the presence of nitric oxide, rate constants were found to be equal to those obtained at the lower temperatures by Kaufman & Kelso (188).

Palmer summarized the recent work on the pyrolysis of methane and acetylene (189) and helped untangle previously contradictory conclusions. Of three single-pulse shock tube studies of methane shock heated in argon, only that of Skinner & Ruehrwein (190) gave an activation energy above 100 kcal/mole, as expected for the primary rupture of a C–H bond. When the shock tube data were combined with rate constants evaluated from low temperature pyrolyses, activation energies 85 to 93 kcal/mole were deduced. There is no longer any doubt that the decomposition in fixed volume and hot tube flow reactors were affected by heterogeneous contributions. In 1963 the kinetics of deposition of film carbon on the central tube of an annular flow reactor gave a new set of points for the CH_4 pyrolysis in the temperature range 1180° to 1400° K, which when combined with the shock tube data unequivocally led to an activation energy of 101 ± 2 kcal/mole (191). Kondratiev (192) reported on the rate of thermal cracking of methane by adiabatic compression and expansion. From the $(H_2)/(CH_4)$ in the residual gases, he estimated that $k_u^{\infty} = 10^{15} \exp(-103{,}000/RT)$, sec^{-1}, on the assumption that the mechanism is

$$CH_4 \leftrightarrows CH_4^* \rightarrow CH_3 + H; \qquad H + CH_4 \rightarrow CH_3 + H_2; \qquad 2CH_3 \rightarrow C_2H_6$$

Thus, Skinner & Ruehrwein, Kondratiev, and Palmer & Hirt agree with each other, but these data are not sufficient to establish whether the primary dissociation step gives CH_3+H (102 kcal) or CH_2+H_2 (103 kcal). One possible attack on this problem would be to shock a low concentration of CD_4 in a several-fold dilution with C_2H_4, followed by a rapid quench wave. If CD_2 is the primary product, appreciable concentrations of 1,1-dideuterocyclopropane should result; on the other hand, if CD_3 is the primary product, a variety of H/D substituted C_1, C_2, C_3 products should be generated. Concurrent pyrolysis of the ethylene will give only acetylene plus hydrogen in a clean unimolecular step.

The mechanism of pyrolysis of acetylene is not fully established, but Palmer & Dormish (193) proposed a possible mechanism into which both the low temperature data, the high temperature shock tube results, and hot flow reactor observations may be fitted. The difficulty is that although in both temperature regimes the reaction is second order with respect to acetylene, and the deduced activation energies are approximately equal (50 kcal/mole), the rate at high temperatures is about 10^{-2} of that expected from extrapolation of the low temperature data. It thus appears that at low temperatures pyrolysis proceeds via a chain. However, the chain terminating step has a steeper temperature coefficient than the chain propagating step, so that under shock tube conditions the chain length is essentially unity. These authors propose that the first step is excitation of the C_2H_2 to a triplet state in a spin-forbidden transition. New data obtained with a hot tube flow reactor were also reported by Cullis & Franklin (194, 194a).

Although about half a dozen reports on shock tube pyrolysis of ammonia have been presented, only a glimmer of the mechanism is beginning to appear. After improving his technique, Jacobs (195) repeated his study of the decomposition rate of NH_3 (temperature range 2080°–3070° K) by monitoring the infrared emission at the 2.7 to 3.2 μ fundamental region on the assumption that the emission was due entirely to vibrationally excited NH_3, and not at all to products of decomposition such as NH_2 and NH. He deduced the rate law:

$$-d(NH_3)/dt = k(NH_3)^{3/2}(Ar)^{1/2}$$

$$k = 2.5 \times 10^{13} \exp(-77{,}700/RT), \quad \text{liter mole}^{-1}\ \text{sec}^{-1}$$

Michel & Wagner (196) worked with very low concentrations of ammonia and measured its rate of pyrolysis by light absorption at 2300 Å (the extinction coefficient is high enough at 2000° K). Over the temperature range 2100° to 2900° K they found the rate to be first order in the NH_3 and in the Ar, indicating that the decomposition occurs in the low pressure limiting region: $k=10^{15.6} \exp(-79{,}500/RT)$, $cm^3\ mole^{-1}\ sec^{-1}$. Large quantities of NH radical were concurrently produced.

In the following papers the decomposition of ammonia was inferred from the emission intensity of NH. Cann and Kash (197) monitored the radiation at 3360 Å from NH_3-Xe mixtures, heated to shock temperatures 2900° to 9600° K. They recorded the intensity at the (0, 0) and (1, 1) band heads for the $A\ ^3\Pi \rightarrow X^3\Sigma^-$ transition. The production of the NH in the $A^3\Pi$ state was found to be dependent on the square of the initial ammonia concentration:

$$[NH(A\,^3\Pi)]/dt = k_2(NH_3)_0^2, \text{ with } k_2 = 2.4 \times 10^{10} \exp(-107{,}200/RT), \quad cm^3\ mole^{-1}\ sec^{1}$$

This suggests that the process followed is not that of Jacobs & Wagner. The relation between the time history of NH emission and the initial decomposition step of the NH_3 was discussed by Avery and co-workers (198 a, b).

The high temperature fate of another N/H compound has been extensively explored. It was demonstrated by Michel & Wagner (196) that the stoichiometry of the pyrolysis of hydrazine changes with the temperature; it may be expressed by the equation:

$$N_2H_4 = NH_3 + (2 - 3\nu/2)H_2 + (1 - \nu/2)N_2$$

such that $\nu=1$ for temperatures 900° to 1200° K and changes to $\nu=\frac{1}{2}$ for the range 1500° to 1600° K. The rate of disappearance of hydrazine was measured spectrophotometrically behind incident shock waves while the ammonia thus produced was estimated from its absorption at 2300 Å, when reheated by the reflected wave to $T_5>2000°$ K. Also, the concentration of NH was determined from its absorption at 3360 Å. Since no NH was detected below 1400° K they argue that this radical is not essential for chain propagation. Further, above 1400° K, NH did appear after an induction period which decreased with rising temperature. Under these conditions the decline in NH concentration appeared to follow a first order rate law. Around 2000° K no ammonia could be detected even though the latter does not decompose rapidly at that temperature. It appears therefore that two decomposition mechanisms are operative. The absorption traces of hydrazine show an induction period in the lower temperature regime, and the half life depends on the hydrazine density; these suggest a chain mechanism. For $T>1300°$ K, the induction period disappears and the rate law becomes first order in N_2H_4. There is a slight pressure dependence which indicates that the decomposition is close to the high pressure limit $\log_{10}k_u^{\infty} \approx 12.8-52{,}000/2.3\ RT$, sec^{-1} for $\rho_{total}=7.5\times10^{-5}$ mole/cm³.

Diesen & Felmlee (93) explored directly the species present during hydrazine pyrolysis by coupling a time-of-flight mass spectrometer to a shock tube. Over the temperature range 1200° to 2500° K and reflected shock pressures of 0.04 to 0.25 atm (from mixtures of 0.5–1.0 per cent hydrazine in argon), mass spectra taken at 25 μsec intervals showed only the undecomposed N_2H_4 and the products NH_3, NH_2, N_2, and H_2. The possible presence of N_2H_3 and of N_2H_2 must be inferred from mechanistic arguments. All investigators agree that the primary step is the rupture of the N–N bond in a unimolecular process; however, most of the experiments were performed in the second order fall-off region. For the subsequent steps Diesen proposed:

$$2\,NH_2 \xrightarrow{k_3} NH_3 + NH; \qquad k_3\,(2000°K) = 2.5 \times 10^{13}, \quad cm^3\,mole^{-1}\,sec^{-1}$$

$$2\,NH \xrightarrow{k_{4a}} N_2H_2^* \xrightarrow{k_{4b}} N_2 + H_2; \qquad k_4\,(2000°K) \approx 10^4, \quad cm^3\,mole^{-1}\,sec^{-1}$$

He admits that at lower temperatures N_2H_3 must be included. This radical could be produced via $NH_2 + N_2H_4 \rightarrow N_2H_3 + NH_3$, and reacts: $N_2H_3 + NH_2 \rightarrow NH_3 + N_2 + H_2$.

A very careful single-pulse shock tube study of the pyrolysis of hydrazine was undertaken by Palmer, Knox & McHale (10a, 199). They used highly purified reactant at about 1 mole per cent in Ar, shocked to temperatures in the range 970° to 1120° K (reflected shock pressures 2.3–8.6 atm). Under these conditions $\nu = 1$. By combining their data with those of Michel & Wagner (196) they deduced the high pressure limit rate constant: $k_u^\infty = 10^{13}$ exp $(-54{,}000/RT)$, sec^{-1}. If one assumes that six oscillators effectively contribute to the storage of energy in the activation step, one may then predict an apparent activation energy of 36 kcal/mole for the low pressure region at about 2000° K, which is consistent with flame studies. They found that hydrazine, which had aged under nitrogen for several weeks after purification, decomposed faster at the lower temperatures, according to: $k = 10^9$ exp $(-34{,}300/RT)$, sec^{-1}. The reason for this aging effect is not yet known. Under single-pulse shock tube conditions the chain length appears to be very short; indeed the following nonchain reaction sequence is adequate.

$$NH_2 + N_2H_4 \rightarrow NH_3 + N_2H_3$$
$$NH_2 + N_2H_3 \rightarrow NH_3 + N_2 + H_2$$

Avery & Bradley monitored the decomposition of hydrazine by the emission from the $(^3\Pi \rightarrow {}^3\Sigma^-)$ transition of NH (200). They generated shock temperatures 1560° to 3470° K. To deduce a self consistent explanation which was compatible with the maximum value for emission intensity, and with the dependence of the decay rate on temperature, they argued: (*a*) that the NH is generated via:

$$NH_2 + N_2H_4 \rightarrow NH_3 + N_2H_3; \qquad N_2H_3 \rightarrow NH_2 + NH$$

(*b*) that NH $(^3\Pi)$ is produced by resonance energy transfer:

$$NH + N_2^* \rightarrow NH^* + N_2$$

The emission intensity follows the growth of NH; (*c*) that the rate of formation of NH* is given by

$$d(\mathrm{NH}^*)/dt = k_f(\mathrm{N_2H_4})(\mathrm{Ar})$$

where k_f has an activation energy of 54.2 kcal/mole; and (*d*) that the maximum concentration of NH_2^* (at the peak of intensity) is proportional to the product (N_2H_4) exp $(-40{,}300/RT)$.

The gas phase decomposition of hydrazine (HY), hydrazine-water mixtures, unsymmetrical dimethyl hydrazine (UDMH), and of monomethyl hydrazine (MMH) were studied in a hot tube flow reactor (201) over the temperature range 800° to 1000° K. The relative rates for decomposition were: UDMH > MMH > HY. In the presence of appreciable amounts of water (25–50 per cent), the decomposition rate of HY is reduced by a factor of about 10, independent of the amount of water present. A mechanism proposed by the Göttingen group (196) for the decomposition of HY accounts for the flow reactor data obtained with the dry material. Effective first order rate constants were also evaluated for UDMH and MMH, but since the mechanisms for these decompositions are more involved and less understood than those for HY the significance of the apparent activation energy is far from clear.

Oxidation Reactions as Studied in Shock Tubes

Shock waves in mixtures which are undergoing oxidation may show special features, because generally such reactions are highly exothermic. If sufficient heat is released in the reaction zone, turbulence develops; that is, flow in the reaction zone is no longer one dimensional. When the shock velocity is less than the detonation velocity, the latter determines the gas dynamics for the system. Of course, this can be circumvented by using high dilution of the reactants with a nonreacting gas. By controlling the degree of dilution one can also gradually approach those conditions which show departures from one dimensional flow and thus study spinning detonations over a range of densities and temperatures. Because of their inherent complexity combustion problems cannot be solved solely through the use of any one technique. Indeed, properly to interpret shock tube data one must make use of the large amount of information which has been accumulated over decades, covering studies of low temperature oxidations, flame phenomena, and low density experiments with molecular fragments.

The initial applications of shock tubes to the study of combustion reactions were attempts to determine explosion limits and minimum temperatures for the initiation of detonations. With the improvement of techniques, these can now be measured reproducibly. And since it is possible to record with high sensitivity the time dependence of concentrations of specific intermediates such as OH, C_2, CH, etc., the duration of induction times can be reliably estimated. Samples have been extracted from combusting mixtures under shock conditions for analysis by mass spectrometry. Such data regarding products of pyrolysis proved helpful by suggesting possible reaction intermediates. More information will be obtained with single-pulse shock tubes, in which combustion reactions may be quenched at a sequence of stages for the analysis of products.

The kinetics of "nitrogen fixation" were reinvestigated by Camac, Feinberg & Ryan (202). They measured the time dependence of the intensity of characteristic infrared emission (5.0–5.5 μ) by the NO generated in air which had been shock heated to 2300° to 4500° K. Since the number of reactions which can and do occur is large, these authors tested previously proposed reaction rate constants (203) by comparing their measured intensities with those calculated by a computer program set up to solve the simultaneous equations. Acceptable agreement (20–25 per cent) was found when the following reactions were included: (*a*) direct bimolecular exchange:

$$N_2 + O_2 \rightarrow 2NO$$

(*b*) Zeldovitch chain:

$$O_2 + M \rightarrow 2O + M; \qquad O + N_2 \rightarrow NO + N; \qquad N + O_2 \rightarrow NO + O$$

(*c*) coupling of vibrational excitation to atom exchange:

$$O_2 + N_2^{(v)} \rightarrow 2NO; \qquad O + N_2^{(v)} \rightarrow NO + N$$

(*d*) all concurrent association and dissociation reactions for O_2, N_2, NO, as well as

$$N + O \leftrightarrows NO^+ + e^-.$$

For sufficiently strong shocks the NO emission passes through an overshoot. Because of the high rate constants for the Zeldovitch chain, the reaction $N_2 + O_2 = 2\ NO$ rapidly attains equilibrium during the early stages of the reaction, when the O_2 has dissociated only slightly. As molecular oxygen dissociates to its equilibrium level, the concentration of NO decreases.

The rate and mechanism of the CO oxidation received further study. Due to the exothermicity of this reaction, mixtures in a range of compositions of CO and O_2 explode when raised to critical temperatures. However, the explosion limits for such mixtures cannot be reproducibly measured unless the gases are thoroughly dried. When so treated these mixtures were found to be considerably less explosive than when slight amounts of water were present. For the very dry mixtures the upper explosion limits were found to be surface dependent and to occur at considerably higher temperatures than were heretofore reported (204).

White & Moore (205) shock heated mixtures of CO and O_2 in the ratios 1/1 to 8/1 to Mach numbers $5.3 < M < 7.4$, and recorded interferometrically gas densities at and past the shock front. Both reactants attained vibrational equilibrium before the exothermic reaction was initiated. The relaxation times observed are significantly lower than that in pure CO. The authors therefore concluded that the vibrational exchange between CO and O_2 is the main path for the vibrational relaxation of the carbon monoxide. Transverse shocks behind the relaxation zone are clearly present, as expected for highly exothermic reactions. The additions of small amounts of hydrogen produced a narrowing of the reaction zone due to the additional heat release and the more complex turbulent structure.

Investigation of the combustion of hydrogen is continuing at an accelerated rate. About a half-dozen shock tube studies and several computational programs were reported. Of all the possible elementary steps which had been listed long ago the four to six important reactions have now been identified and their rate constants estimated. Because of the length of the branching chain which dominates the first part of the reaction, no experiments have yet sufficiently clearly identified the chain initiating step; $H_2+O_2 \rightarrow 2\ OH$ is favored. Although the course of the reaction seems clear for $T>1100°$ K, further studies in the lower temperature regime, particularly for lean mixtures, are necessary to ascertain the role of the peroxide reactions.

A significant contribution to the unscrambling of the H_2-O_2 reaction was the application of shock tube techniques to measurement of the dependence of the induction time for the appearance of OH on the oxygen concentration and temperature (206). It was then established that following initiation, the branching chain sequence attains local equilibrium:

$$OH + H_2 \xrightarrow{2} H_2O + H; \qquad H + O_2 \xrightarrow{3} OH + O; \qquad O + H_2 \xrightarrow{4} OH + H$$

The buildup of the free radical concentration during the induction period (t_i), defined as the time required for the system to generate the lowest detectable OH concentration (approximately 10^{-6} moles/liter), involves essentially endothermic steps, which could now be distinguished from the subsequent exothermic reactions. Initially the OH concentration rises exponentially because chain branching is much faster than chain breaking by recombination. Reactions 2 and 4 are substantially faster than 3 so that the net stoichiometry is: $3\ H_2+O_2=2\ H_2O+2\ H$. The reaction rates increase exponentially with time: $d(H)/dt \doteq 2k_4(O_2)(H)$. The concentrations of H, O, and OH are approximately related, via the steady state condition:

$$(OH)/(H) \doteq 2k_4(O_2)/k_2(H_2); \qquad (O)/(H) \doteq k_4(O_2)/k_3(H_2)$$

Past the maximum, the OH declines slowly according to: $-d(OH)/dt \doteq k(OH)^2(M)$. Only in that region do the exothermic reactions begin to contribute due to recombinations such as:

$$H + H + M \rightarrow H_2 + M; \qquad H + OH + M \rightarrow H_2O + M$$

Schott & Bird (106) have extended these studies, again on the basis of OH concentration measured by absorption at the 3063 Å band. They followed the rate of depletion of OH past the maximum in hydrogen rich mixtures highly diluted in argon, shocked to temperatures of the range 1480° to 1970° K. The three-body recombination rate constants were cited above.

Induction times for OH absorption and emission were measured for shock heated mixtures of hydrogen and oxygen, highly diluted with argon (1400° to 2500° K), to ascertain the dependence of $[(O_2)t_i]$ on the oxygen and hydrogen concentrations as well as on the reaction temperature (207). It appears that the rates of both reactions 3 and 4 are of comparable importance; that for oxygen, vibrational equilibrium is attained before appreciable reaction occurs (except for possible slight effects when deuterium is used in place of hydrogen), and that the induction time does depend on the concentrations of the oxygen and hydrogen. These follow from the above mechanism provided the rate equations are solved simultaneously without introducing the steady state approximation. They used the following values [liter mole^{-1} sec^{-1}]:

$$k_1 = 10^{11} \exp(-70{,}000/RT); \qquad k_3 = 10^{12} \exp(-16{,}900/RT)$$

$$k_2 = 10^{10.8} \exp(-5{,}900/RT); \qquad k_4 = 10^{10} \exp(-7{,}700/RT)$$

The difficulty with the steady state approximation was emphasized by Belles & Lauver (208), in an extensive machine calculation for the hydrogen-oxygen reaction, for $T > 1000°$ K and for various hydrogen/oxygen ratios. They found that at constant temperature $[(O_2)t_i]$ should increase by a factor of 100 as the H_2/O_2 ratio decreases from 24 to 0.0075. There is no need to introduce a slow vibrational relaxation for oxygen to account for this dependence. Brokaw (209) found approximate analytic solutions for these rate equations including also:

$$H + O_2 + M \xrightarrow{5} HO_2 + M$$

$$HO_2 + H_2 \xrightarrow{6} H_2O_2 + H$$

Reaction 5 followed by 6 becomes important at low temperatures and high pressures. These analytic soltuions are involved expressions of rate constants and concentrations, but deductions based on these check the full machine calculations to within a few per cent. A brute force test of the adequacy of the above selected steps was made by integrating 15 simultaneous equations of possible reactions using literature values or reasonable estimates for the values of the rate constants (210). The conclusion reached is that at 1800° K the contribution of reaction 5 is negligible. Indeed, no significant effect could be detected by change in the magnitude of k_5, unless it was made unreasonably large.

The temperature range 890° to 1350° K was investigated by Miyama & Takeyama (211). They worked with H_2/O_2 mixtures highly diluted with argon, in the reflected shock region, and measured two induction times: that for the appearance of OH and that for a rapid pressure rise which correlates with the exothermic steps in the reaction. They checked the activation energy associated with the OH induction time as given by Schott & Kinsey (18.9 ± 0.9 kcal/mole), and concluded that reaction 3 is rate determining for OH production under all conditions. However, with regard to the overall reaction, as measured by the induction period for the pressure rise, reaction 3 is rate controlling at the higher temperatures (above 1100° K) but reaction 6 is rate controlling below 1100° K.

A shock tube study of undiluted hydrogen-oxygen mixtures showed that the initial course of the reaction follows the path outlined above for mixtures with dilute reactants (212). For this system White & Moore measured the density in the region of the shock front interferometrically ($3.5 < M < 6.5$). The expected turbulence in the reaction zone was observed. Both the oxygen and hydrogen attained vibrational equilibrium before the beginning of the exothermic reaction. The induction time (t_*) determined in these experiments corresponds to the onset of the exothermic reaction; it is given by $\log_{10} [(O_2)^{1/2}(H_2)^{1/2}t_*] = 3100/T - 9.8$, for temperatures in the range 1100° to 2200° K. In the above equation, concentrations are in moles/liter. This equation holds for hydrogen/oxygen ratios in the range of 0.075 to 24. Effective recombination rate constants for (2 H+M) and (2 O+M) were found to be about a factor of two greater than those deduced from other experiments.

A very interesting report on the mechanism of excitation of OH to the $^2\Sigma^+$ state, emission from which produces the characteristic luminescence, was published by Belles & Lauver (213). They measured the induction period for emission and the emission intensity for mixtures of 5 per cent hydrogen-95 per cent air shock heated to temperatures 1000° to 1900° K. It had been previously established by Kaskan (214) that the intensity of emission was proportional to the third power of the ground state concentration of OH. However, this merely requires that OH* be generated by a three-body process but does not require that the OH radicals be directly involved as reactants, since local equilibrium of the bimolecular steps is rapidly attained in flames. Belles & Lauver reached the conclusion that the excited state is generated chemically, via $H + O_2 + H_2 \xrightarrow{7} H_2O + OH^*$; $k_7 \approx 2 \times 10^5$, liter^2 mole^2 sec^{-1}; also that OH* is effectively quenched by collisions with water molecules only.

Attention is called to several recent reports on the oxidation of hydrocarbons. Although mechanisms have been proposed for the early stages in the combustion of methane and acetylene, much of the background information depends on data obtained from other techniques. Conclusions based on shock tube studies are tentative; hence only a brief discussion of these observations is given in this review. All investigators agree that the identification of the intermediate molecular species is not complete. The stable products generated in adiabatically compressed hydrocarbon-air mixtures (temperatures up to 700° K) were determined as a function of reaction time, 1 to 10 msec (215). The gaseous products were quenched after being maintained for a specified time under compression by rupturing a diaphragm which separated the compressor from a sampling vessel, which was filled to about 100 mm Hg with nitrogen or helium. Among the large number of products observed were acetone, $C_1 - C_8$ aldehydes, olefins, CO, etc.

The combustion of methane and acetylene is under investigation by Kistiakowsky and coworkers (216). Mass spectrometric analyses were made of samples extracted from the back reflected shock region under shock conditions; chemiluminescence of CH and C_2 were measured, induction periods for the appearance of chemiluminescence and of chemionization were determined. The duration of the induction period (t_*) was found to be independent of the methane concentration, and is given by

$$\log_{10} [(O_2)^{1/2}(Ar)^{1/2}t_*] = 56{,}000/4.5T - 13.18$$

where concentrations are in moles/liter. The authors suggest that t_* is determined by the dissociation rate of molecular oxygen in collision with the argon. The appearance of OH in this reaction was also followed by absorption spectrophotometry at 3067 Å. As in the H_2+O_2 system, the induction time for OH and for the sudden pressure increase due to the onset of exothermic steps were measured (217). The author finds that the rate determining step for OH formation in lean mixtures is

$$CH_3 + O_2 \rightarrow CH_2O + OH; \quad E_A \approx 20 \text{ kcal/mole}$$

The rate determining step for methane oxidation appears to be

$$CH_2O + O_2 \rightarrow HCO + HO_2; \quad E_A \approx 32 \text{ kcal/mole}$$

The final topic concerns shock tube studies of acetylene and oxygen, highly diluted with argon. In Professor Kistiakowsky's laboratory, chemiluminescence by CH and C_2 and chemionization ($C_3H_3^+$) were measured in mixtures under shock conditions; transient species from the reflected shock region were analyzed mass spectrometrically (216). Induction time (t_*) for the minimum detectable signals and the time constant (τ) for the exponential rise in the concentration of these species were determined. The latter is given by

$$\log_{10} [(O_2)\tau] = 17{,}000/4.58T - 14.29 \text{ (concentrations in moles/cm}^3\text{)}; t_* \approx 8\tau.$$

The first step of the following sequence appears to be rate controlling:

$$H + O_2 \rightarrow OH + O; \quad k = 1/[4\tau_{ion}(O_2)]$$
$$O + C_2H_2 \rightarrow OH + C_2H$$
$$OH + C_2H_2 \rightarrow H_2O + C_2H$$
$$C_2H + O_2 \rightarrow 2\,CO + H$$
$$C_2H + O_2 \rightarrow CO_2 + CH$$

Diacetylene was observed in acetylene rich mixtures, possibly generated per $C_2H+C_2H_2 \rightarrow C_4H_2+H$. Stubbeman & Gardiner (218) followed this oxidation via ultraviolet and visible chemiluminescence and by OH absorption spectroscopy. The OH appearance was delayed relative to the chemiluminescence, which is due to either $C_2H+O \rightarrow CH(A^2\Delta)+CO$, or $C_2H+O_2 \rightarrow CH(A^2\Delta)+CO_2$.

In conclusion, attention is called to one paper and two notes on problems relating to the destruction of acetylene by oxygen in flames (219).

LITERATURE CITED

1. Duvall, G. E., and Fowles, G. R., *High Pressure Physics and Chemistry*, **2**, 209 (Bradley, R. S., Ed., Academic Press, New York, 1963)
2. Bauer, S. H., *Science*, **141**, 867 (1963)
3. Greene, E. F., and Toennies, J. P., *Chemical Reaction in Shock Waves* (Edward Arnold, London, 1964)
4. Clarke, J. F., and McChesney, M., *The Dynamics of Real Gases* (Butterworths, Washington, 1964)
5. Hasted, J. B., *Physics of Atomic Collisions* (Butterworths, Washington, 1964)
6. Stoops, R., Ed., *Solvay Conf. Energy Transfer in Gases* (Interscience New York, 1963)

7a. Minkoff, G. J., *Chemistry of Combustion Reactions* (Butterworths, London, 1962)

7b. Sokolik, A. S., *Self-ignition, Flame and Detonation in Gases* (OTS Washington, D. C., 1963)

7c. Fenimore, C. P., *Chemistry in Premixed Flames* (Pergamon, Oxford, 1964)

7d. *Intern. Symp. Combust., 9th* (Academic Press, New York, 1963)

8a. *Discussions Faraday Soc.*, **35** (1963)

8b. Index in appendix of *Progress in*

Reaction Kinetics (Porter, G., Ed., Macmillan, New York, 1964)
9a. *ASTIA, Rept. ARB 3045, SHOCK TUBES* (ASTIA, Arlington Hall Station, Arlington, Va., Oct. 62)
9b. *AVCO-Everett Res. Lab., AMP93* (Everett, Mass., 1962)
9c. Melnick, A. M., *U. S. At. Energy Comm. Doc., R and D Rept. MLM-1189 and Suppl.* (September 1964)
10a. *Meeting Fundamental Phenomena in Hypersonic Flow, Cornell Aeronautical Lab., Buffalo, N.Y., 1964*
10b. *Intern. Symp. Combust., 10th, Cambridge Univ., 1964*
10c. *Lockheed Symp., Stanford, Calif. 1964*
10d. *Symp. Chem. Reactions Shock Tubes, Durham, N.C., 1964* (Reported by Greene, E. F., in *Physics Today*, **17**, (9), 52, 1964)
11. Chen, C. J., and Emrich, R. J., *Phys. Fluids*, **6**, 1 (1963)
12. Breeze, J. C., and Ferriso, C. C., *Phys. Fluids*, **7**, 1071 (1964)
13. Mirels, H., and Mullen, J. F., *Phys. Fluids*, **7**, 1208 (1964)
14a. Brocher, E. F., *Phys. Fluids*, **7**, 347 (1964)
14b. Rumpel, W., and Chapin, S., *Bull. Am. Phys. Soc.*, **9**, 583 (1964)
15. Mirels, H., *Phys. Fluids*, **6**, 1201 (1963)
16. Sturtevant, B., and Slachmuylders, E., *Phys. Fluids*, **7**, 1201 (1964)
17. Center, R. E., *Phys. Fluids*, **6**, 307 (1963)
18. De Boer, P. C. T., *Phys. Fluids*, 6, 962 (1963)
19. Tsuchiya, S., and Kuratani, K., *Bull. Chem. Soc., Japan*, **37**, 2487 (1964)
20. Hurle, I. R., *J. Chem. Phys.*, **41**, 3911 (1964)
21. Tsuchiya, S., and Kuratani, K. (Submitted to *J. Chem. Phys.*)
22. Napier, D. H., Nettleton, M., Simonson, J. R., and Thackeray, D. P. C., *AIAA J.*, **2**, 1136 (1964)
23. Lauer, J. L., and James, R. L. (See ref 10d, Abstr. 2); (Private communication)
24. Wachsler, E., Cornell Univ. (Private communication)
25. Lin, S. C., and Fyfe, W. I., *Phys. Fluids*, **4**, 238 (1961)
26. Wray, K. L., *J. Chem. Phys.*, **37**, 1254 (1962)
27. Camac, M., *AIAA Aerospace Sciences Meeting, New York, 1964* (Preprint 64-35, AVCO-Everett Research Lab., Everett, Mass., Rept. 172, Dec. 1963)
28a. Skinner, G. T., *Phys. Fluids*, **4**, 1172 (1961); (Private communications)
28b. General Motors Defense Res. Lab., TR 63–223, p. 8–2 (April 1964)
29. Skinner, G. T., and Fetz, B. H., *Intern. Symp. Rarefied Gas Dynamics, 4th Toronto, 1964*
30. Glick, H. S., Squire, W., and Hertzberg, A., *Proc. Intern. Symp. Combust., 5th, New York, 1955*, 393 (1956)
31. Lifshitz, A., Bauer, S. H., and Resler, E. L., Jr., *J. Chem. Phys.*, **38**, 2056 (1963)
32. Tschuikow-Roux, E., and Marte, J. E., *JPL Space Program Summary, No. 37–28*, **IV**, 112 (Jet Propulsion Laboratory, Aug. 1964)
33. Tsang, W., *J. Chem. Phys.*, **40**, 1498 (1964)
34. Johnson, C. D., and Britton, D., *J. Chem. Phys.*, **38**, 1455 (1963)
35. Tsang, W., *J. Chem. Phys.*, **40**, 1171 (1964)
35a. Tsang, W., *ibid.*, **41**, 2487 (1964)
36. Bradley, J. N., and Tuffnell, R., *Proc. Roy. Soc. (London), Ser. A*, **280**, 198 (1964)
37. Goldstein, R., *J. Chem. Phys.*, **40**, 2793 (1964)
38. Jacobs, T. A., Hartunian, R. A., Giedt, R. R., and Wilkins, R., *Phys. Fluids*, **6**, 972 (1963)
39. Bauer, S. H., and Watt, W. S. (Unpublished)
40. Paul, D. E., and Dalby, F. W., *J. Chem. Phys.*, **37**, 592 (1962)
41. Marte, J. E., Tschuikow-Roux, E., and Ford, H. W., *J. Chem. Phys.*, **39**, 3277 (1963)
42a. Hall, J. G., Eschenroeder, A. Q., and Marrone, P. V., *J. Aero. Sci.*, **29**, 1038 (1962)
42b. Eschenroeder, A. Q., Boyer, D. W., and Hall, J. G., *Phys. Fluids*, **5**, 615 (1962)
43. Eschenroeder, A. Q., *Phys. Fluids*, **6**, 1408 (1963)
44. Horton, T. E., *JPL Tech. Rept. No. 32-660* (Jet Propulsion Laboratory, Nov. 1964)
45. O'Neal, C., Jr., and Brokaw, R. S., *Phys. Fluids*, **6**, 1675 (1963)
46. Holmes, R., Jones, G. R., and Lawrence, R., *J. Chem. Phys.*, **41**, 2955 (1964)
47. Widom, B., *J. Chem. Phys.*, **32**, 913 (1960); Takayanagi, K., *Progr. Theoret. Phys. (Kyoto), Suppl.*, 25 (1963)
47a. Holmes, R., Jones, G. R., and Pusat, N., *Trans. Faraday Soc.*, **60**, 1220 (1964)
47b. Eventon, K. M., *Bull. Am. Phys. Soc.*, **10**, 191 (1965)

48. Chahine, M. T., *Symp. Rarefied Gas Dynamics, 3rd, Paris, 1962*, 260 (Laurmann, J. A., Ed., Academic Press, New York, 1963)
49. Haviland, J. K. (See ref. 48, p. 274)
50. Scala, S. M., and Talbot, L. (See ref. 48, p. 297)
51. Liepmann, H. W., Narasimha, R., and Chahine, M. T., *Phys. Fluids*, **5**, 1313 (1962)
52. Muckenfuss, C., *Phys. Fluids*, **5**, 1325 (1962)
53a. Linzer, M., and Hornig, D. F., *Phys. Fluids*, **6**, 1661 (1963)
53b. Schwartz, L. M., and Hornig, D. F., *Phys. Fluids*, **6**, 1669 (1963)
54. Camac, M., *Phys. Fluids*, **7**, 1076 (1964)
55. Hornig, D. F. (See ref. 6, p. 311)
56. Miyama, H., and Takeyama, T., *J. Chem. Phys.*, **39**, 851 (1963); Miyama, H., *Combust. Flame*, **6**, 319 (1962)
57. Holmes, R., Jones, G. R., and Pusat, N., *J. Chem. Phys.*, **41**, 2512 (1964)
58. Lambert, J. D., and Rowlinson, J. S., *Proc. Roy Soc. (London), Ser. A*, **204**, 424 (1951); (For a study of resonant energy transfer refer to Lambert, J. D., and Rowlinson, J. S., *ibid.*, **282**, 380 (1964)
59. Shields, F. D., and Lee, K. P., *J. Chem. Phys.*, **40**, 737 (1964)
60. Winter, T. G., *J. Chem. Phys.*, **38**, 2761 (1963)
61. Millikan, R. C., *J. Chem. Phys.*, **38**, 2855 (1963)
62. Patel, C. K. N., *Phys. Rev. Letters*, **13**, 617 (1964)
63. Millikan, R. C., and White, D. R., *J. Chem. Phys.*, **39**, 3209 (1963)
64. Borrell, P., and Hornig, D. F., *J. Chem. Phys.* (In press)
65. White, D. R. (Private communication)
65a. Added in proof: Vibrational relaxation times for H_2 and D_2 have been measured (J. Kiefer, Los Alamos Research Laboratory)
66a. Hooker, W. J., and Millikan, R. C., *J. Chem. Phys.*, **38**, 214 (1963)
66b. Millikan, R. C., *J. Chem. Phys.*, **40**, 2594 (1964)
66c. Millikan, R. C., and White, D. R., *J. Chem. Phys.*, **39**, 98 (1963)
67a. White, D. R., and Millikan, R. C., *ibid.*, 1803; **42**, 447 (1965)
67b. White, D. R., and Millikan, R. C., *ibid.*, 1807
67c. White, D. R., and Millikan, R. C., *ibid.*, 2107
67d. Millikan, R. C., and White, D. R. (See ref. 10d, Abstr. 14); *J. Chem. Phys.*, **42**, 2028 (1965)
68. Camac, M., *AVCO-Everett Res. Lab., Rept. 194* (Everett, Mass., Oct. 1964)
69. Allen, R. A., *AVCO-Everett Res. Lab., Rept. 186*, (Everett, Mass., Aug. 1964)
70. White, D. R., and Millikan, R. C., *AIAA J.*, **2**, 1844 (1964)
71. Roth, W., *J. Chem. Phys.*, **34**, 2204 (1961)
72. Chow, C. C., and Greene, E. F. (See ref. 10d, Abstr. 13)
73. Parker, J. G., *J. Chem. Phys.*, **41**, 1600 (1964)
74. Rapp, D., *J. Chem. Phys.*, **40**, 2813 (1964)
75. Shin, H. K., *J. Chem. Phys.*, **41**, 2864 (1964); *Can. J. Chem.*, **42**, 2351 (1964)
76. Hoare, M., *J. Chem. Phys.*, **41**, 2356 (1964)
77. Ivanov, G. K., and Sayasov, Yu. S., *Soviet Physics "Doklady" (English transl.)*, **9**, 171 (1964)
77a. Mies, F. H., *J. Chem. Phys.*, **40**, 523 (1964)
78. Cottrell, T. L., Dobbie, R. C., McLain, J., and Read, A. W., *Trans. Faraday Soc.*, **60**, 241 (1964)
79. Millikan, R. C., and Osburg, L. A., *J. Chem. Phys.*, **41**, 2196 (1964)
80. Cottrell, T. L. (See ref. 6, p. 523)
81. Hudson, G. H., McCoubrey, J. C., and Ubbelohde, A. R., *Proc. Roy. Soc. (London), Ser. A*, **264**, 289 (1961)
82. McCoubrey, J. C., Milward, R. C., and Ubbelohde, A. R., *Proc. Roy. Soc. (London), Ser. A*, **269**, 456 (1962)
83. Lapp, M., *General Electric Rept. No. 63-RL-3531 G* (December 1963)
84. Hurle, I. R., Russo, A. L., and Hall, J. G., *J. Chem. Phys.*, **40**, 2076 (1964)
84a. Holbeche, T. A., *Nature*, **203**, 476 (1964)
84b. Camac, M., *J. Chem. Phys.*, **34**, 448 (1961)
85. Bauer, S. H., and Tsang, S. C., *Phys. Fluids*, **6**, 182 (1963)
86. Hurle, I. R., *J. Chem. Phys.*, **41**, 3592 (1964)
87. Kohlmaier, G. H., and Rabinovitch, B. S., *J. Chem. Phys.*, **38**, 1709 (1963)
88. Simons, J. W., Rabinovitch, B. S., and Setser, D. W., *J. Chem. Phys.*, **41**, 800 (1964)
89a. Kane, R. A., McGarvey, J. J., and McGrath, W. D., *J. Chem. Phys.*, **39**, 840 (1963)
89b. Fitzsimmons, R. V., and Bair, E. J., *J. Chem. Phys.*, **40**, 451 (1964)

89c. Airey, J. R., Getty, R. R., Polanyi, J. C., and Snelling, D. R., *J. Chem. Phys.*, **41**, 3255 (1964)
90. Brown, R. L., and Klemperer, W., *J. Chem. Phys.*, **41**, 3072 (1964)
91. Myerson, A. L., Thompson, H. M., and Joseph, P. J., *Cornell Aeronautical Laboratory Rept. No. AD-1689-A-3* (May 1964)
92. Johnson, C. D., and Britton, D., *J. Chem. Phys.*, **68**, 3032 (1964)
93. Diesen, R. W., and Felmlee, W. J., *J. Chem. Phys.*, **39**, 2115, 2121 (1963)
94. Hiraoka, H., and Hardwick, R., *J. Chem. Phys.*, **36**, 1715 (1962)
95. Jacobs, T. A., and Giedt, R. R., *J. Chem. Phys.*, **39**, 749 (1963)
96. van Thiel, M., Seery, D. C., and Britton, D., *J. Chem. Phys.* (In press)
97. Palmer, H. B. (Private communication)
98. Bader, L. W., and Ogryzlo, E. A., *J. Chem. Phys.*, **41**, 2926 (1964)
99. Anderson, O. L., *United Aircraft Corp.*, Res. Rept. R-1828-1, (August 1961); Camac, M., and Vaughan, A., *J. Chem. Phys.*, **34**, 460 (1961)
100. Wray, K. L., and Freeman, T. S., *J. Chem. Phys.*, **40**, 2785 (1964)
101. Wray, K. L., *J. Chem. Phys.*, **38**, 1518 (1963); (See also ref. 10b, Abstr. 29)
102. Wilson, J., *AGARDograph 68: High Temperature Aspects of Hypersonic Flow*, 471 (Pergamon Press, New York, 1963)
103. Kiefer, J. H., and Lutz, R. W., *J. Chem. Phys.*, **42**, 1709 (1965)
104. Rink, J. P., Knight, H. T., and Duff, R. E., *J. Chem. Phys.*, **34**, 1942 (1961)
105. Russo, A. L., Hall, J. G., and Lordi, J. A., *CAL Rept. No. AD-1679-A-4* (Cornell Aeronautical Laboratory, June 1964)
106. Schott, G. L., and Bird, P. F., *J. Chem. Phys.*, **41**, 2869 (1964)
107. Jacobs, T. A., and Giedt, R. R. (See ref. 10d, Abstr. 5)
108. Wray, K. L. (See ref. 10b, Abstr. 29)
109. Morgan, J. E., and Schiff, H. I., *J. Chem. Phys.*, **38**, 1495 (1963)
110. Kretschmer, C. B., and Peterson, H. L., *J. Chem. Phys.*, **39**, 1772 (1963)
111. Basila, M. R., and Strong, R. L., *J. Phys. Chem.*, **67**, 521 (1963)
112. Thrush, B. A., and Larkin, F. S. (See ref. 10b, Abstr. 16); *Discussions Faraday Soc.*, **37**, 112 (1964)
113. Barth, C. A., *Ann. Geophys.*, **20**, 182 (1964)
114. Dalby, F. W., *J. Chem. Phys.*, **41**, 2297 (1964)
115. Toby, S., and Weiss, B. H., *J. Phys. Chem.*, **68**, 2492 (1964)
116. Shuler, K. E., and Weiss, G. H., *J. Chem. Phys.*, **38**, 505 (1963)
117. Hansen, C. F., *General Motors Aerospace Operations Dept.* Final Rept. Contract NAS7-217 (February 1964)
118a. Keck, J. C., *Discussions Faraday Soc.*, **33**, 173 (1962)
118b. Woznick, B. J., and Keck, J. C., *J. Chem. Phys.*, **40**, 1166 (1964)
119. Woznick, B. J., *AVCO-Everett Res. Lab., Rept. No. 199* (Everett, Mass., Nov. 1964)
120. Eusuf, M., and Laidler, K. J., *Trans. Faraday Soc.*, **59**, 2750 (1963)
121. Benson, S. W., *J. Chem. Phys.*, **38**, 2285 (1963)
122. Treanor, C. E., and Marrone, P. V., *Phys. Fluids*, **5**, 1002 (1962)
123. Marrone, P. V., and Treanor, C. E., *Phys. Fluids*, **6**, 1215 (1963)
124. Heims, S. P., *J. Chem. Phys.*, **38**, 603 (1963)
125. Rice, O. K., *J. Phys. Chem.*, **67**, 6, 1733 (1963); (See ref. 6, p. 17)
126. Snider, N. S. (Submitted to *J. Chem. Phys.*)
127. Widom, B., *Science* (In press); Pyun, C. W., and Ross, J., *J. Chem. Phys.*, **40**, 2572 (1964); Carrington, T., *J. Chem. Phys.*, **35**, 807 (1961)
128. Levitt, B. P., and Sheen, D. B., *J. Chem. Phys.*, **41**, 584 (1964)
129. Haught, A. F., *Phys. Fluids*, **5**, 1337 (1962)
130. Harwell, K. E., and Jahn, R. G., *Phys. Fluids*, **7**, 214 (1964)
131. Kornegay, W. M., and Johnston, H. S., *J. Chem. Phys.*, **38**, 2242 (1963)
132. Hacker, D. S., and Bloomberg, H., *J. Chem. Phys.*, **39**, 3263 (1963)
133. Lin, S. C., Neal, R. A., and Fyfe, W. I., *Phys. Fluids*, **5**, 1633 (1962); Lin, S. C., and Teare, J. D., *Phys. Fluids*, **6**, 355 (1963)
134. Stein, R. P., Scheibe, M., Syverson, M. W., Shaw, T. M., and Gunton, R. C., *Phys. Fluids*, **7**, 1641 (1964)
135. Bauer, S. H., and Aten, C. F., *J. Chem. Phys.*, **39**, 1253 (1963)
136. Alpher, R. A., and White, D. R., *Phys. Fluids*, **7**, 1239 (1964)
137. Keck, J. C., Allen, R. A., and Taylor, R. L., *J. Quant. Spectr. Radiative Transfer*, **3**, 335 (1963)
138. Churchill, D. R., Hagstrom, S. A., and

Landshoff, R. K. M., *J. Quant. Spectr. Radiative Transfer*, **4**, 291 (1964)
139. Wurster, W. H., *J. Quant. Spectr. Radiative Transfer*, **3**, 355 (1963)
140. Taylor, R. L., *J. Chem. Phys.*, **39**, 2354 (1963)
141. Fairbairn, A. R., *AIAAJ.*, **2**, 1004 (1964)
142. Watson, R., *J. Quant. Spectr. Radiative Transfer*, **4**, 1 (1964)
143. Fukuda, K., *J. Chem. Phys.*, **42**, 521 (1965)
144. Breeze, J. C., and Ferriso, C. C., *J. Chem. Phys.*, **41**, 3420 (1964)
145. Sultzmann, K. G. P., *J. Quant. Spectr. Radiative Transfer*, **4**, 375 (1964)
146. Breeze, J. C., and Ferriso, C. C., *J. Chem. Phys.*, **39**, 2619 (1963)
147. Penzias, G. J., and Mclay, G. J., *Report to NASA*, *CR-54002: W and STR-28*, Warner and Swasey, Control Instrument Div., Flushing, N. Y. (See ref. 10b Abstr. 123)
148. Berry, R. S., and Reinmann, C. W., *J. Chem. Phys.*, **38**, 1540 (1963)
149. Allen, R. A., and Textoris, A., *J. Chem. Phys.*, **40**, 3445 (1964)
150. Sevastyanenko, V. G., and Yakubov, I. T., *Opt. Spectr.* (*USSR*) (*English transl.*), **16**, 1 (1964)
151. Penkin, N. P., *J. Quant. Spectr. Radiative Transfer*, **4**, 41 (1964)
152. Bauer, S. H., and Resler, E. L., Jr., *Science*, **146**, 1045 (1964)
153. Bauer, S. H., Borrell, P., Watt, W., and Lewis, D. (In preparation)
154. Lifshitz, A., Lifshitz, C., and Bauer, S. H., *J. Am. Chem. Soc.*, **87**, 143 (1965)
155. Ossa, E., and Bauer, S. H. (In preparation)
156. Kuratani, K., and Bauer, S. H., *J. Am. Chem. Soc.*, **87**, 150 (1965)
157. Von Kock, H., and Friedman, L., *J. Chem. Phys.*, **38**, 1115 (1963); Moran, T. F., and Friedman, L., *J. Chem. Phys.*, **39**, 249 (1963); Weingartshofer, A., and Clarke, E. M., *Phys. Rev. Letters*, **12**, 591 (1964)
158. Harris, G. M., and Willard, J. E., *J. Am. Chem. Soc.*, **76**, 4678 (1954)
159. Polanyi, J. C. (See ref. 6, p. 526); Airey, J. R., Getty, R. R., Polanyi, J. C., and Snelling, D. R., *J. Chem. Phys.*, **41**, 3255 (1964)
160. Wall, F. T., and Porter, R. N., *J. Chem. Phys.*, **36**, 3256 (1963); Bunker, D. L., and Blais, N. C., *J. Chem. Phys.*, **41**, 2377 (1964); Karplus, M., Porter, R. N., and Sharma, R. D., *J. Chem. Phys.*, **40**, 2033 (1964)
161. Benson, S. W., and Bose, A. N., *J. Chem. Phys.*, **39**, 3463 (1963)
162. Tsang, W., *J. Chem. Phys.*, **42**, 1805 (1965)
163. Rawley, D., and Steiner, H., *Discussions Faraday Soc.*, **10**, 198 (1951)
164. Genaux, C. T., and Walters, W. D., *J. Am. Chem. Soc.*, **75**, 6196 (1953)
165. Lifshitz, A., Carroll, H., and Bauer, S. H., *J. Chem. Phys.*, **39**, 1661 (1963)
166. Das, M. H., and Walters, W. D., *Z. Physik. Chem.* (*Frankfurt*), **15**, 22 (1958)
167. Ellis, R. J., and Frey, H. M., *Trans. Faraday Soc.*, **59**, 2076 (1963)
168. Lifshitz, A., Carroll, H., and Bauer, S. H., *J. Am. Chem. Soc.*, **86**, 1488 (1964)
169. Schneider, F. W., and Rabinovitch, B. S., *J. Am. Chem. Soc.*, **84**, 4215 (1962); **85**, 2365 (1963); Rabinovitch, B. S., Gilderson, P. W., and Schneider, F. W., *J. Am. Chem. Soc.*, **87**, 158 (1965)
170. Schofield, D., Tsang, W., and Bauer, S. H., *J. Chem. Phys.*, **42**, 2132 (1965)
171. Modica, A. P., and Hornig, D. F., *Princeton Univ., Tech. Rept. No. 4* to ONR (NR 357-275) (October 1963); (See Ref. 10d, Abstr. 21); (See also forthcoming paper by Brown and Darwent, *J. Chem. Phys.*, **42**, 2158 (1965)
172. Diesen, R. W., *J. Chem. Phys.*, **41**, 3256 (1964); (See ref. 10d, Abstr. 10)
173. Lapidus, G., and Hiraoka, H. (Submitted to *J. Chem. Phys.*)
174. Steacie, E. W. R., and Smith, W. McF., *Can. J. Res.*, **16B**, 222 (1939)
175. Hiraoka, H., and Hardwick, R., *J. Chem. Phys.*, **36**, 2164 (1962)
176. Farmer, J. B., Henderson, I. H. S., Lossing, F. P., and Marsden, D. G. H., *J. Chem. Phys.*, **24**, 348 (1956)
177. Hiraoka, H., and Hardwick, R., *J. Chem. Phys.*, **39**, 2361 (1963); (See also ref. 128, 186)
178. Jost, W., Michel, K. W., Olschewski, H. A., Troe, J., and Wagner, H. G. (See ref. 10d, Abstr. 12)
179. Brabbs, T. A., Belles, F. E., and Zlatarich, S. A., *J. Chem. Phys.*, **38**, 1939 (1963)
180. Steinberg, M., *General Motors Defense Res. Lab.—TR64-49* (September 1964)
181. Davies, W. O., *J. Chem. Phys.*, **41**, 1846 (1964)

182. Bazhenova, T. W., *Soviet Physics "Doklady" (English transl.)*, **7,** 783 (1963)
183. Gaydon, A. G., Kimbell, G. H., and Palmer, H. B., *Proc. Roy. Soc. (London), Ser. A*, **279,** 313 (1964)
183a. Wagner, H. G. (Private communication): Pyrolysis of CS_2 at very low concentrations in Ar gave E_A-82 kcal/mole, and a rate constant half that reported in ref. 183
184. Gaydon, A. G., Kimbell, G. H., and Palmer, H. B., *Proc. Roy. Soc. (London), Ser. A*, **276,** 461 (1963); (Private communication from Palmer, H. B.)
185. Wagner, H. G. (Quoted by Palmer)
186. Levitt, B. P., *Trans. Faraday Soc.*, **59,** 59 (1963); (See ref. 10d, Abstr. 15)
187. Fishburne, E. S., and Edse, R., *J. Chem. Phys.*, **41,** 1297 (1964)
188. Kaufman, F., and Kelso, J. R., *J. Chem. Phys.*, **23,** 602 (1955)
189. Palmer, H. B. (To be published, see ref. 10a)
190. Skinner, G. B., and Ruehrwein, R. A., *J. Phys. Chem.*, **63,** 1736 (1959)
191. Palmer, H. B., and Hirt, T. J., *J. Phys. Chem.*, **67,** 709 (1963)
192. Kondratiev, V. N. (See ref. 10b, Abstr. 44)
193. Palmer, H. B., and Dormish, F. L., *J. Phys. Chem.*, **68,** 1553 (1964)
194. Cullis, C. F., and Franklin, N. H., *Proc. Roy. Soc. (London), Ser. A*, **280,** 139 (1964)
194a. Asaba, T., Yoneka, K., and Hikita, T., *Intern. Chem. Eng.*, **4,** 340 (1964) Studies the pyrolyses of C_2H_6
195. Jacobs, T. A., *J. Phys. Chem.*, **67,** 665 (1963)
196. Michel, K. W., and Wagner, H. G. (See ref. 10b, Abstr. 48)
197. Cann, M. W. P., and Kash, S. W., *J. Chem. Phys.*, **41,** 3055 (1964)
198. Avery, H. E., Bradley, J. N., and Tuffnell, R., *Trans. Faraday Soc.*, **60,** 335 (1964); Avery, H. E., and Bradley, J. N., *Trans. Faraday Soc.*, **60,** 857 (1964)
199. McHale, E. T., Knox, B. E., and Palmer, H. B. (See ref. 10b, Abstr. 47); (Private communication)
200. Avery, H. E., and Bradley, J. N., *Trans. Faraday Soc.*, **60,** 850 (1964)
201. Eberstein, I. J., and Glassman, I. (See ref. 10b, Abstr. 49)
202. Camac, M., and Feinberg, R. M., and Ryan, R. J. (See ref. 10d, Abstr. 11)
203. Wray, K. L., *Progress in Astronautics and Rocketry: Hypersonic Flow Research*, 181 (Riddell, F. R., Ed., Academic Press, New York, 1962)
204. Dickens, P. G., Dove, J. E., and Linnett, J. W., *Trans. Faraday Soc.*, **60,** 539 (1964)
205. White, D. R., and Moore, G. E., *General Electric Rept. No. 64-RL-3654C* (1964)
206. Schott, G. L., and Kinsey, J. C., *J. Chem. Phys.*, **29,** 1177 (1958); Schott, G. L., *J. Chem. Phys.*, **32,** 710 (1960)
207. Asaba, T., Gardiner, W. C., and Stubbeman, R. F. (See ref. 10b, Abstr. 38); (Private communications from Gardiner, W. C.)
208. Belles, F. E., and Lauver, M. R. (See ref. 10b, Abstr. 40)
209. Brokaw, R. S. (See Ref. 10b, Abstr. 41)
210. Hirsch, E., and Ryason, P. R., *J. Chem. Phys.*, **40,** 2050 (1964)
211. Miyama, H., and Takeyama, T., *J. Chem. Phys.*, **41,** 2287 (1964)
212. White, D. R., and Moore, G. E., *General Electric Rept. No. 64-RL-3654C* (1964); (See ref. 10b, Abstr. 78)
213. Belles, F. E., and Lauver, M. R., *J. Chem. Phys.*, **40,** 415 (1964)
214. Kaskan, W. E., *J. Chem. Phys.*, **31,** 944 (1959)
215. Martinengo, A., Melczer, J., and Schlimone, E. (See ref. 10b, Abstr. 45)
216. Glass, G. P., Kistiakowsky, G. B., Michael, J. V., and Niki, H., *J. Chem. Phys.*, **42,** 608 (1965)
217. Miyama, H., and Takeyama, T., *J. Chem. Phys.*, **40,** 2049 (1964); *Bull. Chem. Soc. Japan*, **38,** 37 (1965)
218. Stubbeman, R. F., and Gardiner, W. C., Jr., *J. Chem. Phys.*, **40,** 1771 (1964); *J. Phys. Chem.*, **68,** 3169 (1964)
219. Gardiner, W. C., *J. Chem. Phys.*, **40,** 2410 (1964); Fenimore, C. P., *J. Chem. Phys.*, **40,** 2411 (1964); Fenimore, C. P., and Jones, G. W., *J. Chem. Phys.*, **41,** 1887 (1964)

SOLUTIONS OF ELECTROLYTES[1]

By R. W. Stoughton

Chemistry Division, Oak Ridge National Laboratory,[2] Oak Ridge, Tennessee

Well over a thousand articles pertinent to this subject matter have appeared during the past year, making it mandatory that only a small fraction of the total be commented on. In recent years there has been a substantial increase in work at elevated temperatures and in D_2O solutions. Measurements of ion association (complex formation) continue unabated.

A very comprehensive *Encyclopedia of Electrochemistry* (1) has been published which includes a new table of electrode potentials and their temperature derivatives in aqueous solutions as well as emf series in fused salts versus temperature in several different media. The extended abstracts of a symposium in Toronto on transport and equilibrium properties in electrolytic solutions have been published (2), as have the papers presented at an international symposium on electrolytes, held in Trieste in 1959 (3).

WATER

A new set of steam tables, *NEL Steam Tables 1964*, has been compiled and prepared by Bain (4) presenting the physical properties of water and steam in the ranges 0 to 800° and 0 to 1000 bars. The tables are in conformity with values and tolerances agreed upon internationally in October 1963.

Mayinger (5) has measured the viscosity of water and steam to 700° and 800 atm.

Dielectric Constant

As pointed out by Scatchard (6) two years ago, any precise measurements of the dielectric constant of water are most newsworthy because of the importance of that quantity in the Debye-Hückel limiting law. The same may be said for D_2O in view of the increasing interest in this solvent and in comparisons with H_2O. Unfortunately, no new values have come to my attention although the interest in knowing this property over wide ranges of temperature and pressure has continued to increase. The spread of values is about 0.2 per cent at 0° and 0.05 per cent at 40° (6), while at 100° it is about 1 per cent (7, 8, 9, 10). Malmberg & Maryott (7) showed values of $(\partial \ln D/\partial T)_p$ to be nearly constant at $-(4.55 \pm 0.02) \times 10^{-3}$ from 0° to 100°; Wyman & Engalls (10) showed values of $-(4.59 \pm .08) \times 10^{-3}$ over this range, with that of largest magnitude being at 0°; while Åkerlöf & Oshry's equations of D versus temperature gave higher values of about $-(4.63 \pm .03) \times 10^{-3}$. This partial derivative is of interest in calculating enthalpies and entropies.

[1] The survey of literature was completed in December 1964.

[2] Operated by the Union Carbide Corporation for the United States Atomic Energy Commission.

Between 100 and 370° the only reliable measurements of D appear to be those of Åkerlöf & Oshry made at saturation pressure, although previously unpublished results of T. E. Gier and H. S. Young at 200, 250, 301 and 350° at pressures to 2000 bars have been reported by Lawson & Hughes (11). Measurements in the temperature range 377 to 393° at densities of 0.2 to 0.4 gm/cm^3, corresponding to a maximum pressure of 300 atm, have been reported by Fogo, Benson & Copeland (12). Quist & Marshall (13) have calculated values of the dielectric constant to 800° and to pressures corresponding to a water density of unity by the use of the Kirkwood equation for the dielectric constant of polar liquids (14, 15)

$$(2D+1)(D-1)/9D = (4\pi Nd/3W)(\alpha + \mu^2 g/3kT) \qquad 1.$$

where N is Avogadro's number, d is the density of the liquid, W is the molecular weight, α is the polarizability, μ is the dipole moment, g is the Kirkwood correlation factor; k and T have their usual significance. The quantity $\mu^2 g$ was fitted by the method of least squares to a simple empirical function of d and T by using literature data from 0 to 250°. The resulting equation was then used with Equation 1 to evaluate D at any desired density to 1.0 and temperature to 800° by using a high-speed computer. Franck (16) had previously made similar calculations using graphical techniques; in general, Franck's values were somewhat higher than those reported by Quist & Marshall (13). The experimental results of Gier and Young (11) were found (13) to be inconsistent with expectation in a number of instances; for example, when the experimental values at 250° were plotted against density, the extrapolated value of D at unit density was greater than the experimental value at 200°.

With increasing work being done in D_2O solutions, the difference in dielectric constant between H_2O and D_2O over a wide temperature range attains greater importance. Malmberg (17) measured the dielectric constant of D_2O from 0° to 100° and compared the difference $D(H_2O)-D(D_2O)$ (7, 17) with results of Wyman & Ingalls (10). The two sets of results showed close agreement at 0°, i.e., differences of about 0.28 (7, 17) versus 0.25 (10), with the discrepancy increasing with increasing temperature. At 100°, the two values were about 0.44 (7, 17) and 0.29 (10); the difference based on D_2O results of Malmberg (17) and H_2O results of Åkerlöf & Oshry (8) is only 0.11. Lietzke & Stoughton (18) show that activity coefficients of HCl in H_2O are larger than those of DCl in D_2O in the temperature range 25 to 225°. The two sets of activity coefficients are closer together at 90° than at 25 or 200°, suggesting that the dielectric constants may be closer together at around 100° than at 25 or 200°. The dielectric constant of H_2O was used (18) for both media above 100° since no measured values for D_2O were available.

Structure

A recent book by Kavanau, *Water and Solute-Water Interactions* (19), describes current knowledge and speculation concerning the structure of solid and liquid H_2O; ion-water interactions and their effects on the structure of the solvent; and interactions of water with nucleic acids and proteins. It appears to me that, whether or not the authors so state, many of the various theories and findings are consistent with a model of water (at any instant) composed of (*a*) a perturbed "ice-like" lattice, plus (*b*) interstitial molecules with fewer constraints and hence greater freedom of movement. (Additional components or variations within each category may well exist.)

Danford & Levy's (20, 21) X-ray diffraction measurements at 25° are in general agreement with those of other previous workers but give more detail. Their data are interpretable on the basis of a lattice structure which is like, although somewhat expanded and warped from, that of ice and with a ratio of framework to interstitial molecules of 4.0. More recently, Danford & Levy (22) compared X-ray diffraction measurements at 4° with those at 25° and found no significant differences in the oxygen spacings or in the ratio r of framework to interstitial oxygens. Narten & Levy (23) compared D_2O with H_2O at 4° and again found no significant difference in spacings or ratio r.

Marchi & Eyring have applied their Significant Structure Theory of Liquids (24) to water (25) with rather remarkable success. The procedure was to express a partition function for water in terms of those for the two assumed components: (*a*) "ice-like" molecules (i.e., those hydrogen bonded to their neighbors); and (*b*) freely rotating monomers. From this partition function, the various thermodynamic quantities were calculated over a temperature range of 0 to about 250°. The maximum errors in free energy and entropy were both about 1 per cent. The boiling point was obtained "almost exactly." The maximum error in molal volume was 0.3 ml; that no minimum was obtained near 4° is not surprising since the minimum volume is only a little over 0.01 per cent less than the value at 0°, and the existence of the minimum probably depends on very subtle differences in changes of processes with temperature. Perhaps the introduction of an appropriate temperature-dependent restriction on the "freely rotating" molecules would help the theory.

Mysels (26) finds very little excess turbidity, due to structural heterogeneities, above that caused by pressure fluctuations in measurements of light scattering by water. He interprets these results as being inconsistent with a model consisting of "iceberg-like clusters" with "fluid" water molecules in between (27, 28, 29), while they are more nearly consistent with any of the models involving a more homogeneous structure of "honeycomb networks of cages" (20, 25, 29, 30, 31, 32).

Choppin & Buijs (33, 34) and Yamatera, Fitzpatrick & Gordon (35) have studied the near infrared spectra of water from the visible to about 1.3 μ as a function of temperature (about 0° to 90°) and of added electrolytes. They interpret the large band between 1.1 and 1.25 μ as the $(\nu_1+\nu_2+\nu_3)$ band, since it has a wavelength intermediate between the corresponding (and more definitively known) bands of the gas (1.136 μ) and solid (1.252 μ). This assignment is consistent with that of previous investigators. They (33, 34, 35) assume three different kinds of water molecules and assign bands corresponding to each as follows: (*a*) those with neither hydrogen bonded to adjacent molecules, 1.16 μ (8630 cm^{-1}); (*b*) those with one hydrogen bonded to a neighbor molecule, 1.20 μ (8330 cm^{-1}); and (*c*) those with both hydrogens bonded to neighbors, 1.25 μ (7990 cm^{-1}). Whether or not the above interpretation is correct, there is no question of the existence of the bands at 8630 and 8330 cm^{-1} (33–37). Yamatera, Fitzpatrick & Gordon (35) present experimental evidence for the existence of the band at 7990 cm^{-1} in the temperature range 1° to 50° in pure water and evidence for its enhancement in the presence of "structure-making" ions. Choppin & Buijs assumed such a band to exist but present no evidence for it; in fact, it cannot be seen in their 21° spectrum of water presented in their Figure 1 (33). Other workers (36, 37) failed to find evidence for its existence in pure water although they have a number of spectra in the appropriate wavelength and temperature ranges. In pure water, the maximum in the 8330 cm^{-1} band exceeds that of the 8630 cm^{-1} band below 43°, while the reverse is true above this temperature (36, 37). Below 43°, the peak in the 8330 cm^{-1} band shifts toward that in ice (7990 cm^{-1}) attaining a value of 8196 cm^{-1} for super-cooled water at $-15°$ with an apparently linear-continuing shift of 5.6 $cm^{-1}/°C$. Above 120°, the maximum in the 8630 cm^{-1} band shifts linearly (0.23 $cm^{-1}/°C$) toward that of the dilute gas at 25° (8807 cm^{-1}); the maximum is at 8674 cm^{-1} at 250° (37). Thus, perhaps, the two bands with their temperature and phase-change shifts could "connect" the gas and solid spectra without the need of a third band.

Choppin & Buijs (33, 34) and Yamatera, Fitzpatrick & Gordon (35) on the basis of the above assignments of bands find general agreement with temperature and ion effects. The "unbonded" band at 8630 cm^{-1} grows in intensity with increasing temperature at the expense of the 8330 cm^{-1} band and presumably at the expense of a 7990 cm^{-1} band. These authors agree on the effects of anions: with increasing "structure-breaking" (or decreasing "structure-making") abilities they are $Cl^- < NO_3^- < Br^- < I^- < ClO_4^-$. For the alkali ions, the order of increasing "structure-breaking" (or decreasing "structure-making") effects is $Cs^+ < Li^+ < Na^+ < K^+$ as found by Choppin & Buijs (34) and $Li^+ < Na^+ < K^+ < Cs^+$ as found by Yamatera, Fitzpatrick & Gordon (35).

In general, the corresponding near infrared spectrum of D_2O is similar to that of H_2O but at a higher wavelength (or lower frequency) by a factor of about 1.35. Temperature and solute effects parallel those observed for H_2O (35, 37).

Goldstein & Penner (38, 39) have studied the infrared spectrum of water from 2200 to 8000 cm^{-1} and from 27° to 209°. At 209° there is considerable difference between the vapor and liquid spectra, although they do tend to resemble each other more the higher the temperature (37, 38, 39). Yukhnevich (40) has calculated the effects of hydrogen bonds on the vibrational spectra of H_2O and HDO, and the results were related to experimental measurements.

Bhandari & Sisodia (41) have attempted to apply the concept of free volume to study the temperature variation of dielectric relaxation time in water. Gurikov (42) has compared the structure of water as deduced from inelastic neutron-scattering measurements with that of ice I. He discusses (43) the mechanism of self-diffusion in water in terms of a lively exchange of molecules in the ice-like skeleton with those in voids.

There is some evidence indicating that D_2O shows stronger bonding at lower temperatures than H_2O and that more hydrogen bonds are broken with increasing temperature in D_2O (19). Density data versus temperature may be interpretable on such a basis, on the assumption that the greater the molar volume the more "ice-like" (the more hydrogen bonds) in the system. On this basis, the density measurements of Goldblatt (44) would indicate that T_2O shows still more hydrogen bonding and still greater changes with temperature. The ratios of molar volumes, $V(T_2O)/V(H_2O)$ and $V(D_2O)/V(H_2O)$, are 1.00710 and 1.00555 at 5° and 1.00305 and 1.00235 at 55°, respectively.

Smith (45) has reviewed relatively recent work on the hydrogen bond in aqueous and nonaqueous media. The various methods used are described along with the type of information derived from each. Only a few references are given.

Ion-Solvent Interactions

Proton spin-lattice relaxation times versus concentration of aqueous alkali halides were measured by Fabricand et al. (46). The results are interpreted in terms of three different types of sites: those near anions, those near cations, and those not near either. They come to the not surprising conclusion that the configurations of water molecules around Li^+, Na^+, and F^- are more stable than that of pure water, while the configurations around K^+, Rb^+, Cs^+, Cl^-, Br^-, and I^- are less stable than that of pure water. The degree of stability decreases with increasing ionic radius. All the ions in the second group may be said to show "negative" hydration in the terminology of Samoilov (47), i.e., water molecules in their immediate vicinity show greater mobility than in pure water. This effect correlates with the "structure-breaking" properties described by others, as opposed to "structure-making" properties of Li^+, Na^+, and F^-.

Buslaeva (48) reports good agreement between the average time that H_2O molecules remain near ions in aqueous solutions as calculated from the nuclear magnetic resonance measurements of Swift & Connick (49), and as calculated on the basis of Samoilov's theory (50).

Stabilization of the structure of water by addition of MeOH has been studied by a dielectric constant method in the range 0° to 60° (51) and by the use of nuclear magnetic resonance techniques (52). Valiev (53) has interpreted quadrupole relaxation times in terms of ion-H_2O interactions. A value of -1.02 kcal/mole was found for the difference between H_2O-Br^- and H_2O-H_2O interaction energies.

Balicheva & Andreev (54) have studied the infrared spectra of the hydrates $MSO_4 \cdot 7H_2O$, $MSO_4 \cdot 6H_2O$ and $M(ClO_4)_2 \cdot 6H_2O$ (M = Mg, Mn, Fe, Co, Ni, Zn) and of $Ca(ClO_4)_2 \cdot 6H_2O$ and $Fe(ClO_4)_3 \cdot 6H_2O$ in aqueous solution and in single crystals. They interpret the shift in the frequency of the O-H stretch with respect to those for H_2O liquid and gas in terms of solvent-ion interactions. Hester & Plane (55) report a Raman band at around 400 cm^{-1} in near-saturated solutions of bivalent and trivalent metal nitrates, sulfates and perchlorates after the lines due to the anions and water were subtracted. These lines were ascribed to aquo-metal ion complexes involving a significant degree of electron sharing in the metal-O bond. All the bivalent cations except Cu gave lines in the narrow region 360–400 cm^{-1}, consistent with the formation of hexacoordinated complexes. Kruglyak (56) has reviewed work on ions having external completed *sp*-shells and explains the results of many observations in terms of donor-acceptor bond representations.

Diamond (57) has suggested a new type of ion pair formation for electrolytes in which both ions are "hydrophobic," e.g., R_4NI (R = Me to Bu). He suggested that the surrounding water in order to minimize the disturbance to its own structure will force the cation-anion pair together, thus decreasing the activity coefficient. The extent of such ion pair formation should be greater the larger the unhydrated ions, in contradistinction to the Bjerrum electrostatic type of ion-pairing. More recently, Lindenbaum & Boyd (58) have made a more thorough and systematic study of the osmotic coefficients of the symmetrical tetra-alkyl ammonium halides versus concentration at 25°. They find that the bromides at low concentrations lie in the same order as the iodides, which are in opposite order to the chlorides and which lie below the Debye-Hückel limiting slope. Both chlorides and bromides lie above the limiting slope with the bromides closer to it. At higher concentrations there is considerable overlapping of curves. They interpret the observations in terms of water structure, ion-solvent interactions and micelle formation.

For ions of the same type, a linear relation exists between the values of apparent volumes and compressibilities according to Litvinenko (59, 60). The concept of "apparent incompressible volume" of an ion is introduced, and the difference in cation-solvent and anion-solvent interactions are discussed. Artemchenko (61) has suggested that the difference between the apparent molal volume of an ion and its crystal volume is a function of the difference between its ionic radius and its limiting radius of hydration which in turn may be evaluated from solution density measurements. Wen & Saito (62) have measured the apparent molal volumes of five symmetrical tetra-alkyl ammonium bromides in water over wide concentration ranges and have discussed the results in terms of water structure changes.

TRANSPORT PROPERTIES

Theory

Friedman (63, 64) has treated the conductance problem by use of a cluster expansion method similar to that of Mayer (65, 66, 67) in equilibrium theory. In the low concentration limit this treatment may be considered to provide a theoretical "confirmation" of Onsager's relaxation term of the limiting law (68, 69) of conductance versus concentration in a manner analogous to Mayer's cluster theory "confirmation" (65) of the Debye-Hückel limiting law for the activity coefficient (70, 71). Strel'tsova (72) has derived an equation for the electrical conductance of strong electrolyte solutions following the work of Bogolyubov (73) who had presented a rederivation of the Ursell-Mayer (74) method of obtaining the equation of state of a non-ideal gas in terms of "cluster integrals." Bogolyubov had generalized the results for the case of coulomb interaction and suggested that they also hold for the liquid state. Mayer also had indicated this possibility (74).

Aqueous Solutions

Hensel & Franck (75) have measured the equivalent conductance Λ of 0.01 *m* LiCl, KCl, CsCl, CsI, K_2SO_4, Et_4NI from 45 to 135° and from 1 to 8000 bars. With the exception of LiCl at 45°, all the values of Λ decreased monotonically with increasing pressure. The decrease was greater the larger the ions. At 8000 bars, the equivalent conductances of CsI and Et_4NI were between 60 and 70 per cent of their values at one bar, while the corresponding values for LiCl were 80 to 90 per cent. The "Walden product" using Λ at 0.01 *m* was not constant. Quist et al. (76) have measured the conductance of 0.0005, 0.0022, and 0.005 *m* aqueous potassium sulfate solutions from 25 to 800° and from 1 to 4000 bars. The values of Λ at 25° show relatively little change with pressure. At 107° a decrease is seen, in agreement with the results of Hensel & Franck (75). Above about 300° there is a large increase with increasing pressures at the lower pressures. When Λ is plotted as a function of solution density, at each concentration and temperature (above 300°) it goes through a maximum of some 800 to 1000 cm^2/ohm-equivalent at densities between 0.4 and 0.8 g/cm^3, after which it decreases. At densities below 0.8 g/cm^3, considerable ion association and hydrolysis appear to occur. Between 0.8 and 1.0 g/cm^3 at temperatures below 300° K_2SO_4 appears to behave more like a "strong" electrolyte consisting of the ions K^+, KSO_4^-, and SO_4^{--}. This general behavior was previously found by Franck (77, 78, 79) to hold for KCl and was explainable in terms of two conflicting effects with increasing pressure: the increasing dielectric constant which enhances ionic dissociation and increasing viscosity which decreases ionic mobilities.

Horne & Courant (80) have measured the specific conductance of 0.1 *m* KCl between $-1°$ and $+9°$. With the use of the Arrhenius equation they conclude that the activation energy for electrical conduction has a maximum between 2° and 4°, which may be pertinent to water structure considerations.

By using results from the literature for the limiting conductance of the sodium and hydrogen ion and for the dissociation constant of water versus temperature, Marsh & Stokes (81) derive equations for the limiting conductance of hydroxide ion and the specific conductance of water for the range 15° to 75°. When the latter is evaluated at 100° they find 7.96×10^{-7}/ohm-cm.

Passeron (82) suggests the following mechanism for deviations from Stokes' law for an ion moving under the influence of an applied electric field: as the ion moves through the solvent, nearby molecules of the latter rotate around it in an oriented position thereby storing up energy which is subsequently given up in thermal agitation. By the use of two adjustable parameters, the calculated correction factors agree with the experimental values for the smaller symmetrical tetra-alkyl ammonium ions to within 10 per cent.

Briere (83) has found that the increase in conductivity of highly purified water with applied electric field up to breakdown (10^6 V/cm) is in good agreement with the theory given by Plumley (84), but in disagreement with the equation of Onsager (85) which indicates a much smaller variation than that observed. Results with ethanol led to the same conclusion.

Spivey & Snell (86) have measured the conductance of $PrNMe_3^+Br^-$, $HOCH_2CH_2NMe_3Cl^-$, and $[Me_3NCH_2CONHMe]^+Cl^-$ in dilute aqueous solutions at 25°. The conductances of the last two cations and of $[Me_2N(EtOH)_2]^+$ are not significantly less than (actually somewhat greater than) those of tetra-alkyl ammonium ions when compared at the same total number of atoms other than hydrogen. They (86) conclude that either the $-OH$ and $-CO\cdot NH\cdot Me$ groups are not hydrogen bonded to water or else that there exist mechanisms to compensate for their presence. For example, as suggested by Frank & Evans (87), Me_3NPr^+ might be surrounded by solvent with a larger local viscosity than the bulk value because of the structure-promoting effects of the nonpolar groups. Spivey & Snell (86) cite evidence indicating that hydration may be expected for groups of the type studied, and, hence, a number of mechanisms may be required to explain all observations.

Smith & Dismukes (88) obtained cation transference numbers in aqueous 0.1 N sodium chloride from 54° to 125° by using an electrometric method. Pfeifer & Guenther (89) found a diffusion coefficient of water molecules of 2.95×10^{-5} cm²/sec in the presence of $Cr(CN)_6^{-3}$ at 20° compared to 1.8×10^{-5} cm²/sec in pure water, by proton resonance measurements. Good (90) discusses the fluidity of aqueous solutions of alkali and ammonium halides in terms of the parameters of the Arrhenius and Eyring equations (91) and of the structural properties of water.

Lengyel et al. (92) have measured the viscosities of solutions of sodium chloride and iodide, of potassium chloride, bromide, and idodide from 10° to 40°, also the viscosities of solutions of potassium, rubidium and cesium fluorides, lithium chloride, and rubidium and cesium chlorides from 15° to 35° to near saturation. Marcinkowsky, Nelson & Kraus (93) continue studies of diffusion coefficients D of ions by using a radiometric porous-frit method developed by Nelson & Kraus (94, 95). The method seems to be fast, simple and of general applicability giving values of D to within a few per cent. They find that values of $D^* = D\eta/\eta_0$ for K^+ and Br^-, where η/η_0 is the ratio of viscosities of solution to pure solvent, is much more nearly constant than are values of D. In the case of Zn(II) in chloride media there is a rapid change in D^* between 0.1 and 2 M HCl, where the Zn(II)-Cl^- species are changing rapidly with concentration. Beyond about 4 M where Zn(II) presumably exists principally as negative complexes there is little further change.

Nonaqueous and Mixed Solvents

Coplan & Fuoss (96) find a limiting conductance λ_0 of 36.74±0.11 for the triamylbutylammonium ion in methanol at 25°. This value is well within 1 per cent of that of tetraphenylboride ion, λ_0 (BPh_4^-) = 36.50±0.05. Hence, it is suggested that the electrolyte (Am_3BuN^+)(BPh_4^-) be used as a reference, $\lambda_0^+ = \lambda_0^- = \Lambda_0/2$, to establish single ion conductances in methanol and perhaps other nonaqueous solvents.

Fabry & Fuoss (97, 98) have measured conductances of lithium-7 chloride and rubidium iodide in dioxane-water mixtures with values of D in the ranges 13 to 62 and 13.5 to 78.5, respectively. The Stokes radius of the iodide ion was linear with $1/D$ over the whole range of D (including the water point); that of lithium ion was linear for values of D above about 50. Lithium chloride of all the alkali chlorides shows the most nearly linear relation of the logarithm of the association constant versus $1/D$, and the least association for values of D below about 40. Atkinson & Petrucci (99) measured the conductance of manganese (II) sulfate and manganese (II) m-benzenedisulfonate in acetone-water mixtures at 25° and analyzed the data by the Fuoss-Onsager theory (100). The variation of the association constant K_A and the Walden product $\Lambda_0\eta_0$ with dielectric constant were much closer to those observed in methanol-water mixtures than to those in dioxane-water media.

EQUILIBRIUM PROPERTIES

Theory

A previous approximate integration of the Poisson-Boltzmann equation at low but finite concentrations by Fuoss & Onsager (101) led to a chemical potential for symmetrical electrolytes expressible as a superposition of two major terms: (a) the usual square root term resulting from long-range electrostatic forces, i.e., for distances of the order of $1/\kappa$, where κ has its value given in the Debye-Hückel theory; and (b) a term linear in concentration due to short range interactions, i.e., applicable at distances of the order of ionic diameters. The value $\beta/2$, the Bjerrum radius, was chosen as the distance from the central ion for matching the two functions (*a*) and (*b*); the value β in terms of the charge ϵ of either ion (in a symmetrical electrolyte) and the Boltzmann constant k is given by ϵ^2/DkT. Recently Skinner & Fuoss (102) have investigated the effects of varying the distance of matching and find that the chemical potential is not sensitive to the value chosen for solvents of higher dielectric constants. They also noted that in extending the approximate solution (101) to terms linear and higher in concentration the general effect was to decrease the value of the activity coefficient. Thus, the often observed minimum in activity coefficients with increasing concentration cannot derive from the Poisson-Boltzmann equation. Rather qualitatively different theoretical sources are required for high concentration properties, as has been suggested before (103). Onsager (104) has computed the lowest order correction term to the Poisson-Boltzmann equation for unsymmetrical electrolytes. The dominant part of the correction is proportional to the first power in concentration, with a coefficient depending only on the Coulomb interaction. A few typographical errors were noted: in Equation 7, n_i should be replaced by η_1; in the following equation η_2, by η_1; and in Equation 8, n_2 by η_2.

Redlich & Meyer (105) have written a review article on molal volumes of electrolytes with 76 references. Although few people would question the validity of the Debye-Hückel limiting law for the logarithm of the activity coefficient, many authors have ignored the corresponding limiting law for the apparent molal volume of the electrolyte ϕV which in terms of its value at infinite dilution ϕV° and the molar concentration c is given by Equation 2.

$$\phi V = \phi V^0 + \mathcal{S}(0.5 \sum_i \nu_i Z_i^2)^{1.5} c^{0.5}, \qquad 2.$$

The slope $\mathcal{S}$ can be expressed in terms of Avagadro's number N, the charge of a univalent cation ϵ, the gas constant R, and the compressibility δ and pressure variation of the dielectric constant of the solvent:

$$\mathcal{S} = N^2\epsilon^3(8\pi/1000D^3RT)^{0.5}((\partial \ln D/\partial P)_T - \delta/3) \qquad 3.$$

Equations 2 and 3 may be derived by expressing the total free energy in terms of those of the solvent and solute where the Debye-Hückel limiting law is used for the activity coefficient of the latter. The above equations follow by taking the partial derivative with respect to pressure and comparing the result with the equation defining apparent molal volumes (106). By using the measurements of Owen and co-workers (107) for the variation of D with pressure and temperature and, in a private communication, values for the compressibility, Redlich & Meyer (105) give 1.867 as the value of $\mathcal{S}$ at 25°. This value seems to be well established at least to about 3 significant figures. The calculated value appears to decrease to about 1.44 at 0° and increase to about 2.77 at 70°. However, the measured values available, which require dilute solution density measurements to one part in 10^6 to 10^7, appear to decrease with increasing temperature. For higher concentrations, Redlich & Meyer (105) suggest adding a linear term in concentration to the right-hand side of Equation 2 with an empirically determined coefficient. From literature data, a value for $\mathcal{S}$ of 20.7 is estimated for ethanol solutions at 20° and of 13.0 for methanol at 25°.

Scatchard (6) mentioned that the tables of Loeb, Overbeek & Wiersema (108) may be used, though not very conveniently, to calculate the chemical potential or activity coefficient of an ion. Since I and others have had trouble doing so, a recent paper by Scatchard (109) showing how this is accomplished is welcomed. Some of the results were compared (109) with activity coefficients calculated by Guggenheim (110) for 2-2 electrolytes in water by numerical integration of the Poisson-Boltzmann equation; very good agreement was observed.

Glueckauf (111) has extended the Debye-Hückel theory and equation to cover changes of the dielectric constant in the vicinity of an ion. His treatment gives rise to two additional terms in the equation of the logarithm of the activity coefficient which are insignificant for small values of κa. At higher values of κa, these terms increase with concentration, temperature, and ionic charge. Hence, while they amount to only 0.5 per cent for a 1-1 electrolyte at 1 *M* and 25°, typical corrections at 325° are 12 per cent at 0.5 *M* for a 1-1 electrolyte and twice this value at 0.13 *M* for a 2-2 electrolyte.

Yalamov & Malaev (112) have investigated the effect of discreteness of surface charge on the electrostatic potential of macroions in solutions of electrolytes at low concentration. Yukhnovskii & Nekrot (113) have derived general formulas for the free energy of mixtures of ions with dipole molecules. Virial coefficients of the free energy and binary distribution functions were obtained in the form of Ursell-Mayer theory for molecular systems.

Hümbelin (114) has come to some "remarkable conclusions" in his discussion on the cluster theory of electrolyte solutions.

General

A new edition of *Stability Constants of Metal-Ion Complexes* has appeared (115). It was compiled by Sillén & Martell and includes both inorganic and organic ligands in the same volume. The field of metal ion-ligand complexes has grown so large that it is most useful to have such a comprehensive compilation available.

McKay (116) has emphasized the importance of giving full details on the composition of ionic media when reporting values of stability quotients of metal-ion complexes, and not just the ionic strength. Despite Lewis & Randall's pointing out as early as 1921 (117) that "activity coefficients are independent of the nature of the medium only at very low ionic strength," some workers still misuse the ionic-strength concept.

Wu, Smith & Young (118, 119) have extended the studies of Young's mixture rule (120) which may be expressed as follows: such thermodynamic properties as apparent molal volumes, excess molal free energies, heats of mixing, etc., of strong electrolytes are to a first approximation functions only of ionic strength at higher concentrations. This rule may be considered as a revision of Lewis & Randall's ionic-strength principle. The rule has been applied so far chiefly to 1-1 electrolytes. Its potential importance suggests that it should be tested as widely as possible. Perhaps it should be borne in mind that for mixtures of the same or complementary valence types the ionic strength is proportional to the total concentration of ions; this is not true for mixtures of other valence types. It has been shown (118) that the heat of mixing of Cl^- and Br^- is nearly the same whether the mixing be done in the presence of Li^+ or of Na^+, or of K^+, suggesting that the interactions between a pair of the same charge sign are typically affected relatively little by the common ion. Friedman (66) has shown that the cluster theory of Mayer is in accord with this general observation. The thermochemical relation that the sum of the two heats of cross mixing is approximately equal to the sum of the four heats of common ion mixing (in the system of four electrolytes M_iX_j; $i, j = 1, 2$) has been applied to four systems, with the observation that the rule works remarkably well (119). This may be illustrated as follows where Δ represents heat of mixing at constant ionic strength.

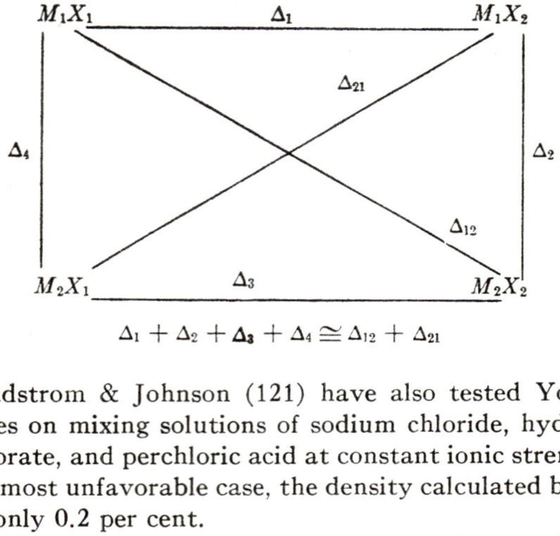

$$\Delta_1 + \Delta_2 + \Delta_3 + \Delta_4 \cong \Delta_{12} + \Delta_{21}$$

Wirth, Lindstrom & Johnson (121) have also tested Young's rule for volume changes on mixing solutions of sodium chloride, hydrochloric acid, sodium perchlorate, and perchloric acid at constant ionic strengths, 1.00 and 4.17 *m*. In the most unfavorable case, the density calculated by Young's rule is in error by only 0.2 per cent.

An interesting series of papers by Cobble and co-workers (122–128) presents tables of partial molal ionic standard entropies $\bar{S}°$ in aqueous solutions, to 200° in some cases (124); and tables including partial molal ionic standard heat capacities at 25°, and average values over the ranges 25 to upper limits of 60°, 100°, 150°, and 200° (125). Values of $\bar{C}_p°$ for the electrolytes sodium perrhenate (122), perrhenic acid (122), cesium chloride (127), and gadolinium chloride (128) were evaluated from 0° to 100° from measurements of integral heats of solution. An interesting correlation is noted between ionic entropies at 25° and at each of the higher temperatures, 60°, 100° and 150°: a value for $\bar{S}_{H^+}°$ could be found at each temperature such that all $\bar{S}_{25}°$ versus $\bar{S}_t°$ values for each class of ions lie on a straight line. The classes were simple cations, the halides plus OH^-, oxygenated anions, and oxygenated anions containing hydrogen. The values of $\bar{S}_{H^+}°$ allowing these linear relations (124) to hold were −5.0, −2.5, 2.0, and 6.5 cal/mole-degree [or gibbs/mole (129)] at 25°, 60°, 100°, and 150°, respectively. The authors' procedure (123, 126), for which they use the misleading title "third-law" method, involves expressing the free energy change ΔF versus temperature in a manner consistent with reasonable or known variations in heat capacity changes ΔC_p versus T. They prefer to use averaged ΔC_p values (where the temperature variation is not known) obtained or estimated from equilibrium data. An alternative, approximately equivalent method is to express ΔF in a manner consistent with $\Delta C_p=$const. (+linear term), i.e., $\Delta F = A_1 + A_2T + A_3T \ln T(+A_4T^2)$, and then evaluate the A parameters by fitting the data. Either procedure may be expected to give a more reasonable account of the data than representing the free-energy change as a power series in T. Ahluwalia & Cobble (123) recommend adoption of $E° = 0.222384$ V for the Ag,AgCl electrode and 0.268155 V for the calomel electrode at 25° as a result of their study. While their treatment in general is satisfactory, it is doubtful if the emf values of any cells are known to better than a few hundredths of a mV. They further conclude that the Ag,AgCl electrode is unreliable above 80°, although Cobble's (126) Figure 1 indicates it should be satisfactory to 125°. A retreatment of the measured values for the hydrogen versus Ag,AgCl electrode between 25° and 225° (18), in line with the above discussion, indicated that the implied cell is reliable to about 200°.

It is customary to assign a zero value to certain thermodynamic properties of hydrogen ion as a convention in order to evaluate the properties of other aqueous ions from those of electrolytes. In order to get a physically more meaningful assignment, Noyes (130) suggests dividing any partial molal property into the value for a neutral species of the same size plus an electrostatic contribution. The latter he expresses as a power series in $1/r$, where r is the radius of the ion. By suitably arranging terms, estimating the values for neutral species and extrapolating to $(1/r) = 0$, "absolute" values for each of several properties for H^+ may be obtained. The treatment yields a standard partial molal volume for H^+ of −1.5 ml/mole. Thus the hydrated proton appears to contribute only 1.5 ml/mole of the total 23.3 ml/mole electrostriction associated with the self-ionization of water.

Stokes (131) has compared experimental free energies of hydration of ions of noble gas structure with values calculated from an electrostatic model. He used the van der Waals radii calculated from those of the iso-electronic noble gases by the quantum mechanical scaling principle for the radii in the gas phase. In water, he used the crystal radii with the assumption that the effective dielectric constant went from 5 at the ion to 78 over one or two layers (each 2.8 Å thick), depending on the charge. He obtained very good agreement.

Desnoyers & Conway (132) have expressed logarithms of experimental activity coefficients as the sums of three contributions: (*a*) coulombic, (*b*) hydration, and (*c*) "mutual salting-out effect." They find for the alkali halides that if they subtract the latter two contributions, using Robinson & Stokes' (133) values for hydration in the concentration range 0.1 to 1.0 M, the remainder is linear in $M^{1/3}$, in support of the "diffuse lattice" treatment of Frank & Thompson (134). When they (132) used Glueckauf's theory (135) for evaluating the hydration contribution, the "cube-root law" was not obeyed. Pan & Chen (136) have presented evidence on mixtures in favor of the "cube-root law."

H_2O Solutions

At or near room temperature.—Lanier (137) has measured the activity coefficient of sodium chloride to 6 m in water as well as to lower molalities in mixed solvents by means of a commercial cation-sensitive glass versus a Ag,AgCl electrode. In water he notes that out to about 1.5 m there is agreement to within 0.4 per cent among various investigators (133, 138, 139); at 3 m two sets of values (133, 139) are about 1 per cent lower than the other two (137, 138); while above 4 m there is a 2 per cent discrepancy. Vlasov (140) has measured vapor pressures of sodium chloride solutions compared to that of water from 2 m to saturation at 45°. He calculated activity coefficients normalized to Robinson's value (141) at 2 m. The difference between two literature values and Vlasov's showed a maximum deviation of about 0.8 per cent in the range 2 to 6.3 m.

The activity coefficient of HCl (142) has been measured from 20° to 50° and 0.005 to 9.25 m and that of HI (143) from 0° to 50° and 0.005 to 0.9 m by emf measurements. Recent standard enthalpies of solution of gaseous HCl and HBr at 25° were $-17{,}875$, and $-20{,}350$ cal/mole, respectively, with a standard deviation of ± 7 as measured calorimetrically (144). Stephenson, Hopkins & Wulff (145) report 1070 ± 50 cal/mole for the standard enthalpy of solution of lithium fluoride obtained from calorimetric measurements. By using this value, they obtain 3.0 ± 0.4 gibbs/mole for the standard entropy of the lithium ion, compared to 2.5 ± 0.3 as calculated from literature measurements on the chloride. By the use of isopiestic measurements and literature data, activity coefficients have been reported for potassium fluoride from 2 m to saturation (146), of erbium and yttrium chlorides from 0.1 m to saturation (147), and of nitric acid from 0 to 100 per cent (148).

Plane and co-workers (149–151) have continued their studies of interionic association of metal nitrates, sulfates, and perchlorates by Raman spectral methods. With the criterion that strong association distorts the symmetry of oxyanions, it appears that nitrate complexes are much more common than sulfate complexes. In order to reconcile these findings with other evidence in favor of considerable association in metal sulfate solutions, they conclude that metal-sulfate complexes are "solvent separated," i.e., at least one water molecule is "trapped" between the metal ion and the sulfate ion. This interpretation is consistent with Eigen's findings from chemical relaxation spectra (152). Perhaps the water is reluctant to give up the sulfate ion completely in view of the apparent enhancement in solvent "structure" when sulfate ions are formed in solution (153–156). Klanberg, Hunt & Dodgen (157) found evidence for perchlorate complexing by Mn(II) in nuclear magnetic resonance studies, and Vdovenko, Romanov & Shcherbakov (158) claim complexing of it by U(IV) from proton relaxation time measurements. The reported formation quotient (158) for $UClO_4^{+3}$ was a factor of 40 smaller than that for UCl^{+3}, and a factor of about 400 smaller than that for USO_4^{++} in perchloric acid media. Rechnitz & Zamochnik (159) conclude from spectrophotometric and kinetic studies that Ag(II) is complexed by ClO_4^-. Lee & Wilmshurst (160), as a result of Raman spectra studies, suggest ion pair formation in concentrated solutions of $LiNO_3$, $AgNO_3$, $LiClO_4$, K_2CO_3, Li_2SO_4, and $(NH_4)_2SO_4$, although according to their graphs the difference between dilute and concentrated solutions is least obvious in the case of $LiClO_4$. Prue (161) has reviewed the 2-2 sulfates including the nature of ion-ion and ion-solvent interactions. Sullivan (162) reported the formation of a 1:1 complex between Np(V) and Cr(III) in perchlorate media from spectrophotometric measurements. Vasil'ev & Mukhina (163) have studied the complexes between Fe(III) and SCN^- spectrophotometrically in $NaSCN$-$NaNO_3$-HNO_3 media. Aveston & Johnson (164) find evidence for the presence of a hexatantalate species in solutions of tantalum in strong KOH. Raman spectra indicate a similar structure in solution and in crystalline $K_8(Ta_6O_{19}) \cdot 16\ H_2O$.

By using a conductance method Insley, Parfitt & Smith (165) have shown that the constant for formation of undissociated AgCl from the ions is less than 150, in contrast to values of 2000 and 1100 obtained from solubility measurements. As a result of spectrophotometric, pH, conductivity, and ion-exchange studies, Harvey, Redfern & Salmon (166) conclude that iodic acid condenses (to species containing more than one iodine) to a negligible degree at most in solutions below 0.1 *M*, in contradistinction to claims of some other investigators. Farrer & Rossotti (167) have studied potentiometrically the association between H^+ and F^- to 1 *M* total fluoride and 3 *M* total ionic strength. They report equilibrium quotients for the reactions $H^+ + F^- = HF$ and $HF + F^- = HF_2^-$ as a function of medium at 25°. They found no evidence for $H^+ - F^-$ complexes higher than HF_2^-. Horne and coworkers (168, 169) have studied the effect of pressure on dissociation of HSO_4^- and $FeCl^{+2}$.

Several studies have been made in strong acid mixed-electrolyte media involving the evaluation of Hammett and related acidity and single ion functions (170–174). Leifer & Hoegfeldt (174) calculated values of ion hydration and concluded that the species $H_9O_4^+$ has a marked stability in concentrated hydrochloric acid solutions.

Robinson & Bower (175) have measured values of the isopiestic ratio of NaCl-$BaCl_2$ mixtures compared to NaCl as a reference solution. They find that the ratio depends linearly on the ion fraction of $BaCl_2$, which simplifies considerably the work required to evaluate the activity coefficients. They also find that for this system Harned's rule (133), which states that the logarithm of the activity coefficient of one component in a two-electrolyte mixture varies linearly with the molality of the second component at constant total ionic strength, may be amended to advantage by changing the restriction to constant total concentration of ions. The simple relations obtaining in the NaCl-$BaCl_2$ mixtures do not hold in mixtures of NaCl-KCl or KCl-$BaCl_2$ (176).

Rush & Johnson (177) present a method for calculating activity coefficients of each of two solutes with a common ion in a three-component solution from ultracentrifugation measurements. They illustrate the method using the system HCl-$BaCl_2$-H_2O which they studied to an ionic strength of about 4.3.

Thermodynamic properties of the following aqueous mixtures have been measured at 25°: heats of mixing of KCl-NaCl and LiCl-NaCl at constant total molality to near saturation (178); HCl activities in HCl-KCl mixtures to a maximum ionic strength of 5.0 (179); isopiestic concentrations for the systems NaCl-LiCl, NaCl-KCl, NaCl-CsCl, NaCl-NH_4Cl (180), NaCl-RbCl (181), NaCl-$NaNO_3$, NaCl-NaBr, and NH_4Cl-NH_4Br (182); volume and heat capacity changes in the systems KCl-KBr (183) and KCl-KNO_3 (184). The HCl activity coefficient in HCl-KCl mixtures is reported (179) not to obey the same mixture rules as does HCl in mixtures with NaCl or RbCl.

High temperature studies.—Ryzhenko (185) has calculated the bisulfate acid constant K_2 from conductance measurements; he finds $pK_2 = 3.04$ (100°), 3.90 (156°), 4.87 (218°). These values are within 4 per cent of those reported by Lietzke & Stoughton (155). He (185) has calculated conditions of precipitation in the system Ca^{++}, $Ba^{++} || SO_4^{2-}$, CO_3^{2-}. Lietzke & Stoughton (186) have determined the activity coefficient of HBr in HBr-KBr mixtures to 150° and to a total ionic strength I of 3.8 from emf measurements, although the values above $I = 1.0$ are questionable. The activity coefficient of KBr, γKBr, in the mixtures was calculated. Harned's rule (133) was obeyed for γHBr but not for γ KBr. Marshall, Slusher & Jones (187) have measured the solubility of calcium sulfate in aqueous sodium chloride solutions to 200°. They find that the molality solubility product follows a single parameter equation of the Debye-Hückel type to an ionic strength of about 1.0 at 60° and to about 4.0 at 125° and above. Thus the work described here (153–156, 185–187) indicates that the Debye-Hückel equation is as useful to at least 225° as it is near room temperature, as well as being useful to concentrations far beyond its range of theoretical validity. The decrease in entropy on getting certain divalent ions in solution was shown to become greater the higher the temperature, at least to 200° (153–156, 187).

The vapor pressures of 1, 2, and 3 *m* NaCl solutions to 270° measured by Gardner, Jones & de Nordwall (188) provided for the first time an isopiestic standard above 100°. The results showed that above about 100° plots of the osmotic coefficient versus *m* were concave downward, and increasingly so with increasing temperature in the measured concentration range, contrary to behavior at lower temperatures. Recently Soldano, Bien & Secoy (189) have measured isopiestic ratios of NaCl, LiCl, KCl, CsCl, $BaCl_2$, Na_2SO_4, and $MgSO_4$ in the general concentration range 0.75 to 6 *m* at 151° and find this same behavior for all the electrolytes in the range of measurement. Soldano & Meek (190) found that a plot of isopiestic ratios (with respect to NaCl) versus temperature at any concentration of electrolyte gave a straight line within apparent experimental error. Further, if lines at several concentrations for any electrolyte were extrapolated, they tended to converge (meet) at about 160°. This behavior was observed for LiCl, KCl, $BaCl_2$, and $MgSO_4$.

Ravich & Borovaya have measured the solubilities of lithium sulfate to 388° and 1000 kg/cm^2 (191) and of sodium carbonate to 500° and 2400 kg/cm^2 (192). The solubility of magnesium hydroxide in water vapor was measured in the ranges 270 to 460° and 240 to 350 atm (193). The important factor determining the solubility at higher temperatures is the density of the water. The system H_2O-NaCl has been studied to 700° and 1240 bars (194). Densities and compressibilities of NaOH solutions were measured to 350° and to 49 weight per cent NaOH (195). The extrapolated apparent molal volume of NaOH increased from −13.0 at 0° to −5.2 at 70° after which it decreased to −320.0 at 350°. Gavrish & Galinker (196) have studied complexes of Ag(I) and Cu(I) with halide ions by solubility measurements at 300°.

D_2O Solutions

Gary, Bates & Robinson (197) have measured the standard emf of the cell Pt-$D_2(g)$ | DCl in D_2O | AgCl, Ag from 5 to 55°. At 25° they find it to be 4.31 mV lower (on the mole fraction scale) than that of the corresponding protium system. Lietzke & Stoughton (18) report standard emf values for the same cell and activity coefficients of DCl in D_2O to 1.0 *m* from 25 to 225°. When $E°$ values were expressed on a moles-solute/55.51-moles-solvent basis they found the difference ($E°_{HCl}$-$E°_{DCl}$) increased monotonically from 7.4 mV at 25° to 25.7 mV at 225°. The difference at 25° is nearly 3 mV greater than that found by Gary, Bates & Robinson (197); the reasons for this discrepancy are not known.

Kerwin (198) has measured the osmotic coefficients of a number of alkali halides in D_2O and compared them with the values in H_2O. For the chlorides, he finds a greater spread in D_2O at a given concentration (moles-solute/55.51-moles-solvent) in going from LiCl to CsCl than in H_2O. The values for KCl are nearly the same in the two solvents. He explains these and other differences in terms of greater order ("more structure") in D_2O than in H_2O.

Salomaa, Schaleger & Long (199) have measured the ionization constants of chloracetic acid, ammonia, and NH_4^+ in H_2O-D_2O mixtures by using a glass electrode. On extrapolation to pure D_2O they report K_H/K_D values of 2.69, 2.89, and 4.06, respectively. Tobias & Yasuda (200) have measured the acid dissociation quotients of the aquotriethyltin-(IV) and first and second dissociation quotients of aquodimethyltin(IV) ion in H_2O and D_2O media, $3M$ in ClO_4^-. They find log (K_H/K_D) values between 0.4 and 0.7 for all three. They suggest that these large isotope effects may result from the inability of the aquo cations to hydrogen bond effectively with the solvent. Gold & Lowe (201) have studied the dissociation of water and of acetic acid in mixtures of H_2O and D_2O with dioxane.

Hepler (202) has calculated the free energy of transfer of alkali chlorides from H_2O to D_2O solution by means of a Born-type equation in which dielectric saturation of the solvent near the ion is taken into account. This treatment gives values in the range of those experimentally measured.

Sheindlin & Gorbunova (203) report the measured values of enthalpy of D_2O in the ranges 250 to 500 kg/cm² and 350 to 460°. Narten (204) calculated the thermodynamic functions for the reaction $H_2O + D_2O = 2HDO$ in the gas and liquid state from recent data, although he did not take into account the zero point energy differences. He has now corrected (205) the previous values.

Nonaqueous and Mixed Solvents

Paabo & Robinson (206) have measured the standard potential of the Ag,AgCl electrode in water-mannitol solutions at 25°. Schwabe, Urlass & Ferse (207) measured emf values for the cell Pt,$H_2(g)$|HBr(m)|Hg_2Br_2,Hg in ethanol-water mixtures versus HBr concentration and mole fraction of H_2O in the solvent. A plot of $(E_m°\text{-}E_m°(H_2O))$ was linear in mole fraction of H_2O over a large range in composition. Kundu & Das (208) report standard potentials for Ag,AgCl and Ag,AgBr electrodes and activity coefficients of HCl and HBr at low concentrations in ethylene glycol and propylene glycol at 30°. Shkodin et al. (209) have measured emf values for the cell Pt,$H_2(g)$| HCl(m)|AgCl,Ag in methanol-dioxane mixtures. Activity coefficients and solvation energies are presented. Activity coefficients of sodium and cesium chloride were measured (210) in HCONHMe and $HCONH_2$ at 0.01 to 0.7 M. Enthalpies of solution of $CoCl_2$ and $NiCl_2$ in MeOH and of Mg in MeOH-HCl have been reported (211); from these and previous data ionic entropies were evaluated for six bivalent cations.

Schug & Dadgar (212) have measured values of the equilibrium quotients between mercury amalgams and methanol-water solutions containing Cl^- or Br^- for the reactions $Na(Hg) + (1/Z)M^{z+}$ (Sol.) $= Na^+$ (Sol.) $+ (1/Z)M(Hg)$, where Z is the charge of the cation and M = K, Rb, Cs, or Sr. They used a direct equilibration and analysis method and claim their results are more accurate than those of previous emf measurements. They also claim that, contrary to previous work, the Debye-Hückel limiting law describes the behavior (within 5 per cent) of NaCl and KCl in 90 per cent and of NaCl in 100 per cent methanol. Ivanova & Aleksandrov (213) have measured the activity coefficients of cesium iodide in methanol and of cadmium chloride in 1-butanol by emf measurements.

Kraus, Raridon & Baldwin (214) have measured activity coefficients of sodium and potassium chloride and barium nitrate in saturated water mixtures of glycol, glycerol, and their acetates at 25° by solubility measurements. Biktimirov (215, 216) reports solubilities of cesium nitrate in aqueous mixtures of alcohols, ketones, and complex ethers. The velocity of sound, density, and compressibility of solutions of five electrolytes in formamide were measured (217).

The papers presented at a metal-ammonia solutions colloquium in Lille, France in 1963 have recently been published (218). After designing experiments to avoid troubles from two liquid phases, Hallada & Jolly (219) found the absorptivities of Ca-NH_3 solutions at $-45°$ and between 10,000 and 21,000 Å to be twice those of alkali metal solutions within experimental error (estimated at ± 4 per cent).

Straub & Yung (220) apparently have prepared anionic complexes of Mn(II), Co(II), Ni(II), Cu(II), and UO_2(II) containing perchlorate as the only ligand in methylene chloride solutions.

LITERATURE CITED

1. Hampel, C. A., Ed., *The Encyclopedia of Electrochemistry* (Reinhold, New York, 1206 pp., 1964)
2. *Symp. Electrochem. Soc., Toronto, 1964*, 2 (Electrochemical Society, Inc., New York, 205 pp., 1964)
3. Pesce, B., Ed., *Proc. Intern. Symp. Electrolytes, Trieste, 1959* (Pergamon, New York, 360 pp., 1962)
4. Bain, R. W., Ed., *NEL Steam Tables 1964* (For Her Majesty's Stationery Office, Pickering & Inglis, Glasgow, 147 pp., 1964)
5. Mayinger, F., *Intern. J. Heat Mass Transfer*, **5**, 807 (1962)
6. Scatchard, G., *Ann. Rev. Phys. Chem.*, **14**, 161 (1963)
7. Malmberg, C. G., and Maryott, A. A., *J. Res. Natl. Bur. Std.*, **56**, 1 (1956)
8. Åkerlöf, G. C., and Oshry, H. I., *J. Am. Chem. Soc.*, **72**, 2844 (1950)
9. Lattey, R. T., Gatty, O., and Davies, W. G., *Phil. Mag.*, 12, 1019 (1931)
10. Wyman, J., Jr., and Engalls, E. N., *J. Am. Chem. Soc.*, **60**, 1182 (1938)
11. Lawson, A. W., and Hughes, A. J., in *High Pressure Physics and Chemistry*, **I**, Chap. 4.iv (Bradley, R. S., Ed., Academic Press, New York, 1963)
12. Fogo, J. K., Benson, S. W., and Copeland, C. S., *J. Chem. Phys.*, **22**, 209 (1954)
13. Quist, A. S., and Marshall, W. L. *J. Phys. Chem.* (In press)
14. Kirkwood, J. G., *J. Chem. Phys.*, **7**, 911 (1939)
15. Oster, G., and Kirkwood, J. G., *J. Chem. Phys.*, **11**, 175 (1943)
16. Franck, E. U., *Z. Physik. Chem. (Frankfurt)*, **8**, 107 (1956)
17. Malmberg, C. G., *J. Res. Natl. Bur. Std.*, **60**, 609 (1958)
18. Lietzke, M. H., and Stoughton, R. W., *J. Phys. Chem.*, **68**, 3043 (1964)
19. Kavanau, J. L., *Water and Solute-Water Interactions* (Holden-Day, San Francisco, 101 pp., 1964)
20. Danford, M. D., and Levy, H. A., *J. Am. Chem. Soc.*, **84**, 3965 (1962)
21. Danford, M. D., and Levy, H. A., *The Structure of Water and 1* M *NH_4F Solution at Room Temperature by X-Ray Diffraction* (145th Natl.

Meeting, Am. Chem. Soc., New York, **1963**)
22. Danford, M. D., and Levy, H. A. (Private communication)
23. Narten, A. H., and Levy, H. A. (Private communication)
24. Eyring, H., and Marchi, R. P., *J. Chem. Educ.*, **40**, 562 (1963)
25. Marchi, R. P., and Eyring, H., *J. Phys. Chem.*, **68**, 221 (1964)
26. Mysels, K. J., *J. Am. Chem. Soc.*, **86**, 3503 (1964)
27. Nemethy, G., and Scheraga, H. A., *J. Chem. Phys.*, **36**, 3382 (1962)
28. Nemethy, G., and Scheraga, H. A., *ibid.*, **41**, 680 (1964)
29. Frank, H. S., and Wen, W. Y., *Discussions Faraday Soc.*, **24**, 133 (1957)
30. Pauling, L., *The Nature of the Chemical Bond*, 472 (3rd Ed., Cornell Univ. Press, Ithaca, N. Y., 1960)
31. Frank, H. S., and Quist, A. S., *J. Chem. Phys.*, **34**, 604 (1961)
32. Samoilov, O. Ya., *Zh. Fiz. Khim.*, **20**, 1411 (1946)
33. Choppin, G. R., and Buijs, K., *J. Chem. Phys.*, **39**, 2035 (1963)
34. Choppin, G. R., and Buijs, K., *ibid.*, 2042
35. Yamatera, H., Fitzpatrick, B., and Gordon, G., *J. Mol. Spectr.*, **14**, 268 (1964)
36. Luck, W., *Ber. Bunsenges.*, **67**, 186 (1963)
37. Waggener, W. C., Weinberger, A. J., and Stoughton, R. W., *The Absorption Spectrum of H_2O and D_2O in the Near Infrared Region as a Function of Temperature from −20 to 250°C* (149th Natl. Meeting, Am. Chem. Soc., Detroit, Mich., 1965)
38. Goldstein, R., and Penner, S. S., *J. Quant. Spectr. Radiative Transfer*, **4**, 359 (1964)
39. Goldstein, R., and Penner, S. S., *ibid.*, 441
40. Yukhnevich, G. V., *Optika i Spektroskopiya, Akad. Nauk SSSR, Otd. Fiz.-Mat. Nauk Sb. Statei*, **2**, 223 (1963)
41. Bhandari, R. C., and Sisodia, M. L., *Indian J. Pure Appl. Phys.*, **2**, 132 (1964)
42. Gurikov, Yu. V., *Zh. Strukt. Khim.*, **4**, 824 (1963)
43. Gurikov, Yu. V., *ibid.*, **5**, 188 (1964)
44. Goldblatt, M., *J. Phys. Chem.*, **68**, 147 (1964)
45. Smith, J. W., *Sci. Progr. (London)*, **52**, 97 (1964)
46. Fabricand, B. P., Goldberg, S. S., Leifer, R., and Ungar, S. G., *Mol. Phys.*, **7**, 425 (1963–64)
47. Samoilov, O. Ya., *Discussions Faraday Soc.*, **24**, **141** (1957)
48. Buslaeva, M. N., *Zh. Fiz. Khim.*, **38**, 529 (1964)
49. Swift, F. J., and Connick, R. E., *J. Chem. Phys.*, **37**, 307 (1962)
50. Samoilov, O. Ya., *Zh. Fiz. Khim.*, **33**, 1147 (1959)
51. Yastremskii, P. S., and Samoilov, O. Ya., *Zh. Strukt. Khim.*, **4**, 844 (1963)
52. Matyash, I. V., and Yashkichev, V. I., *Zh. Strukt. Khim.*, **5**, 13 (1964)
53. Valiev, K. A., *Zh. Strukt. Khim.*, **5**, 517 (1964)
54. Balicheva, T. G., and Andreev, S. N., *Zh. Strukt. Khim.*, **5**, 29 (1964)
55. Hester, R. E., and Plane, R. A., *Inorg. Chem.*, **3**, 768 (1964)
56. Kruglyak, Yu. A., *Uch. Zap., Khar'kovsk. Gos. Univ.*, 133, *Tr. Khim. Fak. i Nauchn.-Issled. Inst. Khim.* **19**, 172 (1963)
57. Diamond, R. M., *J. Phys. Chem.*, **67**, 2513 (1963)
58. Lindenbaum, S., and Boyd, G. E., *J. Phys. Chem.*, **68**, 911 (1964)
59. Litvinenko, I. V., *Ukr. Fiz. Zh.*, **9**, 405 (1964)
60. Litvinenko, I. V., *Zh. Strukt. Khim.*, **4**, 830 (1963)
61. Artemchenko, A. I., *Zh. Fiz. Khim.*, **38**, 1601 (1964)
62. Wen, Wen-Yang, and Saito, S., *J. Phys. Chem.*, **68**, 2639 (1964)
63. Friedman, H. L., *Physica*, **30**, 509 (1964)
64. Friedman, H. L., *ibid.*, 537
65. Mayer, J. E., *J. Chem. Phys.*, **18**, 1426 (1950)
66. Friedman, H. L., *Ionic Solution Theory* (Interscience, New York, 265 pp., 1962)
67. Jepsen, D. W., and Friedman, H. L., *J. Chem. Phys.*, **38**, 846 (1963)
68. Onsager, L., *Physik. Z.*, **27**, 388 (1926)
69. Onsager, L., *Physik. Z.*, **28**, 277 (1927)
70. Debye, P., and Hückel, E., *Physik. Z.*, **24**, 185 (1923)
71. Debye, P., *Physik. Z.*, **25**, 97 (1924)
72. Strel'tsova, O. O., *Dopovidi Akad. Nauk Ukr. RSR*, **1963**, 1468

73. Bogolyubov, N., *J. Phys. (USSR)*, **10,** 257 (1946)
74. Mayer, J. E., and Montroll, E. M., *J. Chem. Phys.*, **9,** 2 (1941)
75. Hensel, F., and Franck, E. U., *Z. Naturforsch.*, **19**(a), 127 (1964)
76. Quist, A. S., Franck, E. U., Jolley, H. R., and Marshall, W. L., *J. Phys. Chem.*, **67,** 2453 (1963)
77. Franck, E. U., *Z. Physik. Chem. (Frankfurt)*, **8,** 92 (1956)
78. Franck, E. U., *ibid.*, 107
79. Franck, E. U., *Angew. Chem.*, **73,** 309 (1961)
80. Horne, R. A., and Courant, R. A., *J. Phys. Chem.*, **68,** 1258 (1964)
81. Marsh, K. N., and Stokes, R. H., *Australian J. Chem.*, **17,** 740 (1964)
82. Passeron, E. J., *J. Phys. Chem.*, **68,** 2728 (1964)
83. Briere, G. B., *Brit. J. Appl. Phys.*, **15,** 413 (1964)
84. Plumley, H. J., *Phys. Rev.*, **59,** 200 (1941)
85. Onsager, L., *J. Chem. Phys.*, **2,** 599 (1934)
86. Spivey, H. O., and Snell, F. M., *J. Phys. Chem.*, **68,** 2126 (1964)
87. Frank, H. S., and Evans, M. W., *J. Chem. Phys.*, **13,** 507 (1945)
88. Smith, J. E., Jr., and Dismukes, E. B., *J. Phys. Chem.*, **68,** 1603 (1964)
89. Pfeifer, G., and Guenther, K., *Zh. Strukt. Khim.*, **5,** 193 (1964)
90. Good, W., *Electrochim. Acta*, **9,** 203 (1964)
91. Glasstone, S., Laidler, K. J., and Eyring, H., *The Theory of Rate Processes*, **1,** 484 (McGraw, New York, 611 pp., 1941)
92. Lengyel, S., Tamas, J., Giber, J., and Holderith, J., *Acta Chim. Acad. Sci. Hung.*, **40,** 125 (1964)
93. Marcinkowsky, A. E., Nelson, F., and Kraus, K. A., *J. Phys. Chem.*, **69,** 303 (1965)
94. Nelson, F., and Kraus, K. A., *Production and Use of Short-Lived Isotopes from Reactors*, **1,** 191 (IAEA, Vienna 1962–63)
95. Nelson, F., *J. Polymer Sci.*, **40,** 563 (1959)
96. Coplan, M. A., and Fuoss, R. M., *J. Phys. Chem.*, **68,** 1177 (1964)
97. Fabry, T. L., and Fuoss, R. M. *J. Phys. Chem.*, **68,** 971 (1964)
98. Fabry, T. L., and Fuoss, R. M., *ibid.*, 974
99. Atkinson, G., and Petrucci, S., *J. Am. Chem. Soc.*, **86,** 7 (1964)
100. Fuoss, R. M., and Accascina, F., *Electrolytic Conductance* (Interscience, New York, 279 pp., 1959)
101. Fuoss, R. M., and Onsager, L., *Proc. Natl. Acad. Sci. U. S.*, **47,** 818 (1961)
102. Skinner, J. F., and Fuoss, R. M., *J. Am. Chem. Soc.*, **86,** 3423 (1964)
103. Onsager, L., *Chem. Rev.*, **13,** 73 (1933)
104. Onsager, L., *J. Am. Chem. Soc.*, **86,** 3421 (1964)
105. Redlich, O., and Meyer, D. M., *Chem. Rev.*, **64,** 221 (1964)
106. Redlich, O., *J. Phys. Chem.*, **44,** 619 (1940)
107. Owen, B. B., Miller, R. C., Milner, C. E., and Cogan, H. L., *J. Phys. Chem.*, **65,** 2065 (1961)
108. Loeb, A. L., Overbeek, J. Th. G., and Wiersema, P. H., *The Electrical Double Layer Around a Spherical Colloid Particle* (Mass. Inst. Technol. Press, Cambridge, Mass., 375 pp., 1961)
109. Scatchard, G., *Falkenhagen Sonderdruck (East German) Z. Physik. Chem.* (In press)
110. Guggenheim, E. A., *Trans. Faraday Soc.*, **56,** 1152 (1960)
111. Glueckauf, E., *Trans. Faraday Soc.*, **60,** 776 (1964)
112. Yalamov, Yu. I., and Malaev, A. M., *Zh. Fiz. Khim.*, **38,** 963 (1964)
113. Yuknovskii, I. R., and Nekrot, A. A., *Ukr. Fiz. Zh.*, **9,** 365 (1964)
114. Hümbelin, R., *Electrochim. Acta*, **9,** 685 (1964)
115. *Stability Constants of Metal-Ion Complexes, Spec. Publ. 17*, Sec. I: *Inorganic Ligands*, Sillén, L. G., Ed.; Sec. II: *Organic Ligands*, Martell, A. E., Ed. (Chem. Soc., London, 12 pp. use of tables, 754 pp. tables, 1964)
116. McKay, H. A. C., *J. Inorg. Nucl. Chem.*, **26,** 1767 (1964)
117. Lewis, G. N., and Randall, M., *J. Am. Chem. Soc.*, **43,** 1112 (1921)
118. Wu, Y. C., Smith, M. B., and Young, T. F., *J. Phys. Chem.*, **69,** 1868 (1965)
119. Wu, Y. C., Smith, M. B., and Young, T. F., *ibid.*, 1873
120. Young, T. F., and Smith, M. B., *J. Phys. Chem.*, **58,** 716 (1954)
121. Wirth, H. E., Lindstrom, R. E., and Johnson, J. N., *J. Phys. Chem.*, **67,** 2339 (1963)

122. Ahluwalia, J. C., and Cobble, J. W., *J. Am. Chem. Soc.*, **86,** 5377 (1964)
123. Ahluwalia, J. C., and Cobble, J. W., *ibid.*, 5381
124. Criss, C. M., and Cobble, J. W., *J. Am. Chem. Soc.*, **86,** 5385 (1964)
125. Criss, C. M., and Cobble, J. W., *ibid.*, 5390
126. Cobble, J. W., *J. Am. Chem. Soc.*, **86,** 5394 (1964)
127. Mitchell, R. E., and Cobble, J. W., *J. Am. Chem. Soc.*, **86,** 5401 (1964)
128. Jekel, E. C., Criss, C. M., and Cobble, J. W., *J. Am. Chem. Soc.*, **86,** 5404 (1964)
129. Giauque, W. F., Hornung, E. W., Kunzler, J. E., and Rubin, T. R., *J. Am. Chem. Soc.*, **82,** 62 (1960)
130. Noyes, R. M., *J. Am. Chem. Soc.*, **86,** 971 (1964)
131. Stokes, R. H., *J. Am. Chem. Soc.*, **86,** 979 (1964)
132. Desnoyers, J. E., and Conway, B. E., *J. Phys. Chem.*, **68,** 2305 (1964)
133. Robinson, R. A., and Stokes, R. H., *Electrolyte Solutions*, 438, 442, Chap. 9, Appendix 8.10, 2nd Ed. (Butterworths, London, 1959)
134. Frank, H. S., and Thompson, P. T., in *The Structure of Electrolyte Solutions*, Chap. 8 (Hamer, W. J., Ed., Wiley, New York, 441 pp., 1959)
135. Glueckauf, E., *Trans. Faraday Soc.*, **51,** 1235 (1955)
136. Pan, Hui-Chen, and Chen, Nien I., *Hua Hsueh Hsueh Pao*, **30,** 221 (1964)
137. Lanier, R. D., *Activity Coefficients of NaCl in Water-Organic Media by Cation-Sensitive Glass Electrodes* (149th Meeting, Am. Chem. Soc., Detroit, Mich., 1965)
138. Harned, H. S., and Nims, L. F., *J. Am. Chem. Soc.*, **54,** 423 (1932)
139. Scatchard, G., Hamer, W. J., and Wood, S. E., *J. Am. Chem. Soc.*, **60,** 3061 (1938)
140. Vlasov, Yu. G., *Zh. Fiz. Khim.*, **37,** 2586 (1963)
141. Robinson, R. A., *Trans. Faraday Soc.*, **35,** 1222 (1939)
142. Faita, G., and Mussini, T., *J. Chem. Eng. Data*, **9,** 332 (1964)
143. Hetzer, H. B., Robinson, R. A., and Bates, R. G., *J. Phys. Chem.*, **68,** 1929 (1964)
144. Vanderzee, C. E., and Nutter, J. D., *J. Phys. Chem.*, **67,** 2521 (1963)
145. Stephenson, C. C., Hopkins, H. P., and Wulff, C. A., *J. Phys. Chem.*, **68,** 1427 (1964)
146. Tamas, J., and Kosza, G., *Magy. Kem. Folyoirat*, **70,** 148 (1964)
147. Petheram, H. H., and Spedding, F. H., *U. S. At. Energy Comm.*, **IS-770,** 69 pp. (1963)
148. Davis, W., Jr., and de Bruin, H. J., *J. Inorg. Nucl. Chem.*, **26,** 1069 (1964)
149. Hester, R. E., Plane, R. A., and Walrafen, G. E., *J. Chem. Phys.*, **38,** 249 (1963)
150. Hester, R. E., and Plane, R. A., *J. Chem. Phys.*, **40,** 411 (1964)
151. Hester, R. E., and Plane, R. A., *Inorg. Chem.*, **3,** 769 (1964)
152. Eigen, M., *Z. Elektrochem.*, **64,** 115 (1960)
153. Lietzke, M. H., and Stoughton, R. W., *J. Phys. Chem.*, **64,** 133 (1960)
154. Lietzke, M. H., and Stoughton, R. W., *ibid.*, 816
155. Lietzke, M. H., and Stoughton, R. W., *ibid.*, **65,** 2247 (1961)
156. Lietzke, M. H., and Stoughton, R. W., *ibid.*, **67,** 652 (1963)
157. Klanberg, F., Hunt, J. P., and Dodgen, H. W., *Inorg. Chem.*, **2,** 139 (1963)
158. Vdovenko, V. M., Romanov, G. A., and Shcherbakov, V. A., *Radiokhimiya*, **5,** 664 (1963)
159. Rechnitz, G. A., and Zamochnik, S. B., *Talanta*, **11,** 713 (1964)
160. Lee, H., and Wilmshurst, J. K., *Australian J. Chem.*, **17,** 943 (1964)
161. Prue, J. E., *Experientia Suppl.*, **9,** 157 (1964)
162. Sullivan, J. C., *Inorg. Chem.*, **3,** 315 (1964)
163. Vasil'ev, V. P., and Mukhina, P. S., *Zh. Neorgan. Khim.*, **9,** 1134 (1964)
164. Aveston, J., and Johnson, J. S., *Inorg. Chem.*, **3,** 1051 (1964)
165. Insley, M. J., Parfitt, G. D., and Smith, A. L., *J. Phys. Chem.*, **68,** 2372 (1964)
166. Harvey, J. F., Redfern, J. P., and Salmon, J. E., *J. Inorg. Nucl. Chem.*, **26,** 1326 (1964)
167. Farrer, H. N., and Rossotti, F. J. C., *J. Inorg. Nucl. Chem.*, **26,** 1959 (1964)
168. Horne, R. A., Courant, R. A., and Frysinger, G. R., *J. Chem. Soc.*, **1964,** 1515
169. Horne, R. A., Myers, B. R., and Frysinger, G. R., *Inorg. Chem.*, **3,** 452 (1964)

170. Yates, K., *Can. J. Chem.*, **42,** 1239 (1964)
171. Ojeda, M., and Wyatt, P. A. H., *J. Phys. Chem.*, **68,** 1857 (1964)
172. Lee, D. G., and Stewart, R., *Can. J. Chem.*, **42,** 487 (1964)
173. Leifer, L., and Hoegfeldt, E., *Acta Chem. Scand.*, **17,** 338 (1963)
174. Leifer, L., and Hoegfeldt, E., *U.S. At. Energy Comm.*, CONF-*387-1*, 12 pp. (1964)
175. Robinson, R. A., and Bower, V. E., *J. Res. Natl. Bur. Std.*, **69**(A), **19** (1965)
176. Robinson, R. A. (Private communication)
177. Rush, R. M., and Johnson, J. S., *J. Phys. Chem.*, **68,** 2321 (1964)
178. Stern, J. H., and Anderson, C. W., *J. Phys. Chem.*, **68,** 2528 (1964)
179. Storonkin, A. V., Lagunov, M. D., and Prokof'eva, R. V., *Zh. Fiz. Khim.*, **38,** 509 (1964)
180. Kirgintsev, A. N., and Luk'yanov, A. V., *Zh. Fiz. Khim.*, **37,** 2773 (1963)
181. Kirgintsev, A. N., and Luk'yanov, A. V., *ibid.*, **38,** 1291 (1964)
182. Kirgintsev, A. N., and Luk'yanov, A. V., *ibid.*, 1603
183. Stakhanova, M. S., Epikhin, Yu. A., and Karapet'yants, M. Kh., *Zh. Fiz. Khim.*, **37,** 2570 (1963)
184. Epikhin, Yu. A., Stakhanova, M. S., and Karapet'yants, M. Kh., *Zh. Fiz. Khim.*, **38,** 692 (1964)
185. Ryzhenko, B. N., *Geokhimiya, 1964* (1), 23
186. Lietzke, M. H., and Stoughton, R. W., *J. Phys. Chem.*, **67,** 2573 (1963)
187. Marshall, W. L., Slusher, R., and Jones, E. V., *J. Chem. Eng. Data.* **9,** 187 (1964)
188. Gardner, E. R., Jones, P. J., and de Nordwall, H. J., *Trans. Faraday Soc.*, **59,** 1994 (1963)
189. Soldano, B. A., Bien, P. B., and Secoy, C. H., *U. S. At. Energy Comm.*, ORNL-*3591*, 99 (1964)
190. Soldano, B. A., and Meek, M., *U. S. At. Energy Comm.*, ORNL-*3417*, **94** (1963)
191. Ravich, M. I., and Borovaya, F. E., *Dokl. Akad. Nauk SSSR*, **155,** 1375 (1964)
192. Ravich, M. I., and Borovaya, F. E., *ibid.*, **156,** 894 (1964)
193. Martynova, O. I., Serov, E. P., and Smirnov, O. K., *Izv. Akad. Nauk SSSR, Energ. i Transp., 1963*(4), 555
194. Sourirajan, S., and Kennedy, G. C., *Am. J. Sci.*, **260,** 115 (1962)
195. Dibrov, I. A., Mashovets, V. P., and Matveeva, R. P., *Zh. Prikl. Khim.*, **37,** 29 (1964)
196. Gavrish, M. L., and Galinker, I. S., *Zh. Neorgan. Khim.*, **9,** 1289 (1964)
197. Gary, R., Bates, R. G., and Robinson, R. A., *J. Phys. Chem.*, **68,** 1186 (1964)
198. Kerwin, R. E., *The Osmotic and Activity Coefficients of Some Alkali Halides in Heavy Water at 25°C* (Doctoral thesis, Univ. Pittsburgh, 1964)
199. Salomaa, P., Schaleger, L. L., and Long, F. A., *J. Phys. Chem.*, **68,** 410 (1964)
200. Tobias, R. S., and Yasuda, M., *J. Phys. Chem.*, **68,** 1820 (1964)
201. Gold, V., and Lowe, B. M., *Pure Appl. Chem.*, **8,** 273 (1964)
202. Hepler, L. G., *Australian J. Chem.*, **17,** 587 (1964)
203. Sheindlin, A. E., and Gorbunova, N. I., *Teplofiz. Vysokik Temperatur, Akad. Nauk SSSR*, **2,** (3), 484 (1964)
204. Narten, A. H., *J. Chem. Phys.*, **41,** 1318 (1964)
205. Narten, A. H., *ibid.*, **42,** 814 (1964)
206. Paabo, M., and Robinson, R. A., *J. Phys. Chem.*, **67,** 2861 (1963)
207. Schwabe, K., Urlass, R., and Ferse, A., *Ber. Bunsenges. Physik. Chem.*, **68,** 46 (1964)
208. Kundu, K. K., and Das, M. N., *J. Chem. Eng. Data*, **9,** 87 (1964)
209. Shkodin, A. M., Sogoyan, T. P., Karkuzaki, L. I., and Kozynyuk, L. I., *Ukr. Khim. Zh.*, **30,** 237 (1964)
210. Povarov, Yu. M., Kessler, Yu. M., Gorbanev, A. I., and Safonova, I. V., *Dokl. Akad. Nauk SSSR*, **155,** 1411 (1964)
211. Jakuszewski, B., and Taniewska-Osinska, S., *Zeszyty Nauk. Uniw. Lodz*, Ser. II, **13,** 137 (1962)
212. Schug, K., and Dadgar, A., *J. Phys. Chem.*, **68,** 106 (1964)
213. Ivanova, E. F., and Aleksandrov, V. V., *Zh. Fiz. Khim.*, **38,** 878 (1964)
214. Kraus, K. A., Raridon, R. J., and Baldwin, W. H., *J. Am. Chem. Soc.*, **86,** 2571 (1964)

215. Biktimirov, R. S., *Zh. Fiz. Khim.*, **37**, 2356 (1963)

216. Biktimirov, R. S., *ibid.*, 2573

217. Mikhailov, I. G., Rozina, M. V., and Shutilov, V. A., *Akust. Zh.*, **10**, 13 (1964)

218. Lepoutre, G., and Sienko, M. J., Eds., *Metal-Ammonium Solutions. Physiochem. Properties, Colloq. Weyl, Lille, June 1963*, (Lille, Univ. Catholique; distributed by W. A. Benjamin, New York, 317 pp., 1964)

219. Hallada, C. J., and Jolly, W. L., *Inorg. Chem.*, **2**, 1076 (1963)

220. Straub, D. K., and Yung, Ying-Mae, *Perchlorate Complexes of* Mn^{2+}, Co^{2+}, Ni^{2+}, Cu^{2+}, *and* UO_2^{2+} (148th Natl. Meeting, Am. Chem. Soc., Chicago, 1964)

PHYSICAL ORGANIC CHEMISTRY: Π-COMPLEXES AS INTERMEDIATES IN ORGANIC REACTIONS[1]

BY MICHAEL J. S. DEWAR AND ALAN P. MARCHAND

Department of Chemistry, The University of Texas, Austin, 12, Texas

INTRODUCTION

Ever since the classic studies of Longuet-Higgins on the structure of diborane (1), the importance of many-center bonding in "electron-deficient" molecules has become increasingly apparent. During the last 20 years, a parallel, though less drastic, development has taken place in organic chemistry, and a number of reactions are now thought to involve intermediates with three-center bonds. Carbon differs from boron, however, in that the types of three-center bond it can form seem to be quite limited; most, if not all, can be regarded as π-complexes (2–6) formed by dative sharing of the π-electrons of unsaturated carbon skeletons with the empty orbitals of acceptor groups. For various reasons this has not been clearly recognized by organic chemists, although structures of this type are now very well known in inorganic chemistry. In view of the importance and topical interest of the subject, we have devoted the whole of this review to it in the hope of placing things in a better perspective.

MESOMERIC CARBONIUM IONS

The stability of conjugated carbonium ions (allyl, benzyl, tropylium, etc.) has long been recognized and the factors involved in their stabilization are well understood. Similar considerations apply to analogues of these ions in which one π-bond is stretched by interpolation of a methylene group [cf. homotropylium (7)]. Here we shall confine our attention to carbonium ions which are stabilized by other types of delocalization. Ions of this type have been loosely described as "nonclassical carbonium ions"; the problem of their structure and nomenclature is discussed below. Most of this article will be concerned only with the latest developments, since the chemistry of such ions has been reviewed (8, 9) recently.

Do nonclassical carbonium ions exist as stable intermediates?—Brown has for many years consistently opposed the view that "nonclassical carbonium ions" may exist as stable intermediates in chemical reactions, and his arguments have now appeared in concrete form (10–19).

The evidence for the participation of nonclassical carbonium ions in certain chemical reactions falls into four main categories. In descending order of cogency these are: (a) the abnormal steric course followed by the reactions

[1] The survey of literature pertaining to this review was concluded in December 1964.

in question, combined with very high stereospecificity; (b) the formation of products which are inconsistent with the intermediacy of classical carbonium ions; (c) the abnormal stability of certain of these ions, as indicated by their isolation in the form of stable salts; and (d) the abnormally high rates of reactions in which certain of these ions are postulated to be intermediates. Brown's arguments are mainly directed at evidence from the fourth and least cogent group.

As regards stereochemistry, Brown offers only three unconvincing arguments:

First, Brown & Bell have shown that when various norbornyl cations are trapped with borohydride, the products are formed with exclusive *exo* addition of hydrogen. Thus I[2] gave II, but no III. Moreover none of the secondary isomer IV was formed. Brown & Bell took this as evidence that the ion is

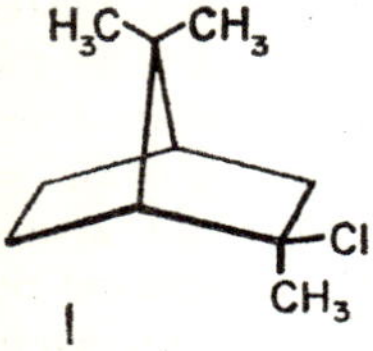

I

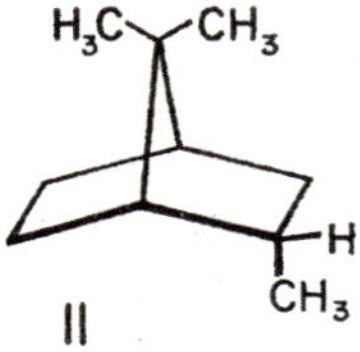

II

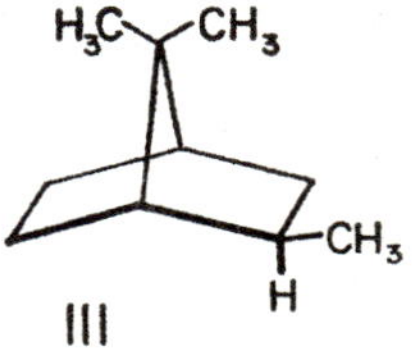

III

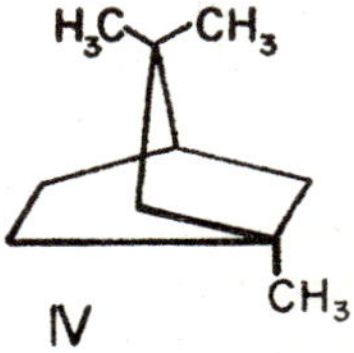

IV

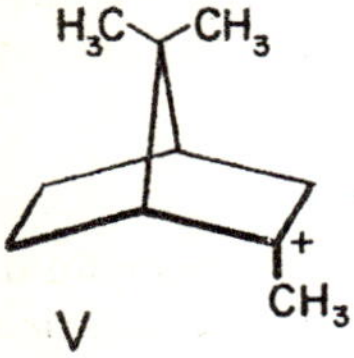

V

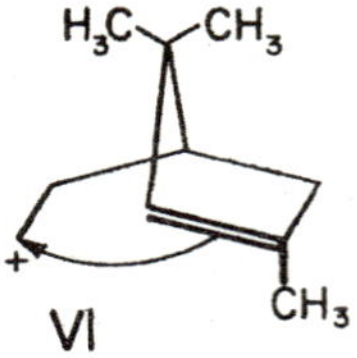

VI

VII
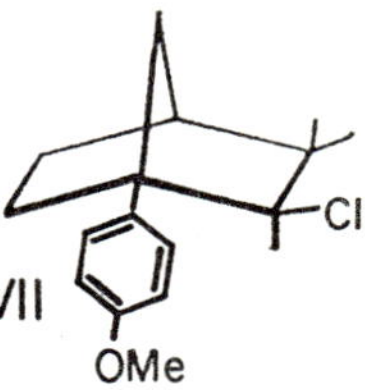

VIII
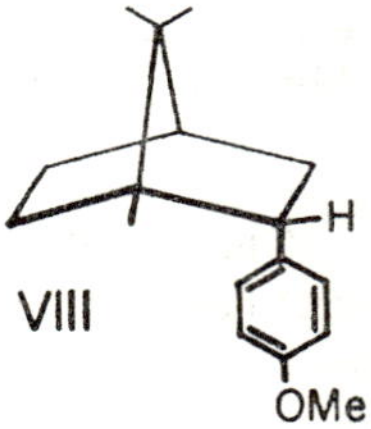

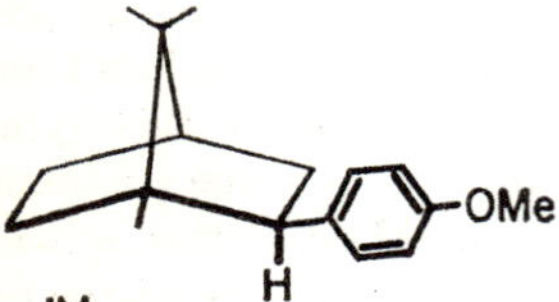

IX

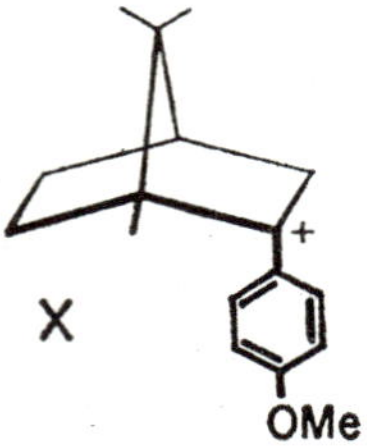

X

XI

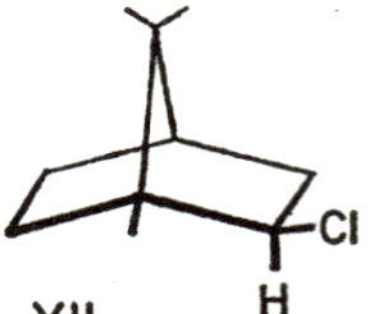

XII

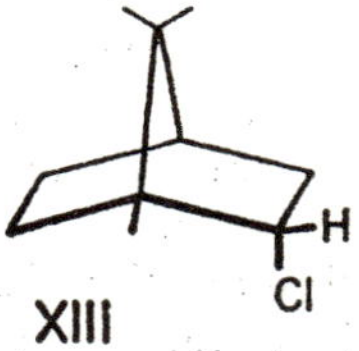

XIII

[2] These Roman numerals refer to the similarly designated structural formula.

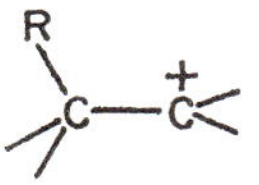

XIV

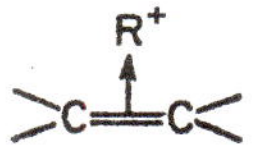

XV

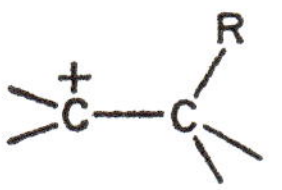

XVI

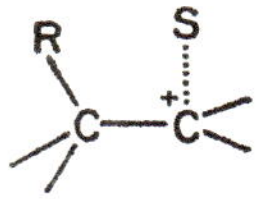

XVII

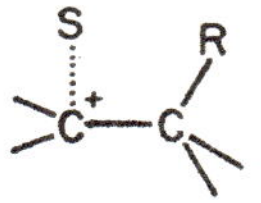

XVIII

the tertiary carbonium ion V, and that this reacts stereospecifically with borohydride. It is difficult to see the relevance of this evidence to Brown's thesis; for the results are precisely what would be expected if the intermediate ion were the π-complex VI. The preferential reaction of this at the tertiary carbon would of course be precisely analogous to the corresponding reactions of other π-complexes (e.g., olefine—Br^+ complexes), where similar discrimination between secondary and tertiary carbon atoms is responsible for Markownikov addition.

Second, Bell & Brown have shown that similar reduction of 1-*p*-anisylcamphene hydrochloride (VII) gives a mixture of *endo* (VIII) and *exo* (IX) 2-*p*-anisylbornanes in the ratio 6.5:1. There is reason (20) for believing that this reaction does involve a classical intermediate, the 2-*p*-anisylbornyl ion X; Bell & Brown take their results as further evidence that such ions combine with anions by preferential *exo* attack. However, this evidence is also entirely irrelevant for two reasons: (*a*), it is a far cry from the 6.5:1 *exo/endo* ratio observed in the reduction of VII to the essentially complete stereospecificity of solvolysis reactions of isobornyl chloride or I, for example. Bell & Brown's evidence, if anything, provides further support for the intervention of nonclassical intermediates in the latter reactions. (*b*) the stereospecificity of the rearrangement of camphene hydrochloride (XI) to isobornyl chloride (XII) is the more striking, in that bornyl chloride (XIII) is thermodynamically more stable than XII; Bell & Brown's evidence would be relevant to this only if IX had been shown to be more stable than VIII. This has not been done, and indeed all analogy suggests that VIII should be the more stable isomer.

Third, Brown has tried to explain the observed stereospecificity of these reactions by his "windshield wiper" mechanism. This supposes that in an equilibrating pair of isomeric classical ions XIV and XVI, the migrating group R oscillates to and fro so rapidly that it hinders attack by solvent and also inhibits free rotation about the bond linking the two adjacent carbon atoms. This suggestion seems unsound. Consider the energetics that must be postulated for such a situation to arise. We must suppose that the migration of R from one extreme position to another via the intermediate π-complex (XV) must occur in times of the order of a molecular vibration; the barrier to migration must then be $\ll kT$. At the same time we must suppose XV to be less stable than XIV or XVI; otherwise the system would adopt this nonclassical configuration. The energy of the system must therefore show a very flat maximum between the two extremes XIV and XVI. It would be a coincidence if this happened once; however, Brown supposes this is the situation wherever nonclassical ions have been postulated! Furthermore, the system will remain in this oscillating condition only if its free energy is less than that of solvated classical ion XVII or XVIII (S = solvent.) Since the energy of the oscillating ion is essentially identical with that of the individual classical structures XIV or XVI, the change in free energy in localizing R can arise only from the change in entropy due to restriction of its motion. This change has to be less than the free energy of solvation of a classical carbonium ion! A simple calculation shows this assumption to be unrealistic.

Similar difficulties arise in all the other analogous reactions where stereospecificity has been observed; Cram (21) has emphasized this for reactions involving phenonium ion intermediates, i.e., π-complex ions with apical aryl groups.

Brown does not consider arguments based on the formation of unexpected products in solvolysis reactions. A classic example of this is the formation of tricyclene in reactions thought to involve intermediate "bornyl" cation; this can be explained easily if the ions are π-complexes (e.g., VI). If the intermediate were a carbonium ion, the formation of tricyclene would involve a kind of electrophilic substitution for which no parallels exist. Recently two further cogent arguments of this kind have appeared. Winstein & Baker (22) have observed that the acetolysis of *threo*-3-*p*-anisyl-2-butyl *p*-bromobenzenesulfonate [XIX] in acetic acid is not only completely stereospecific, giving *threo* acetate containing less than 0.1 per cent of *erythro* isomer, but it also gives as the sole isolable byproduct the terminal olefine XX. If the intermediate in this reaction was a classical carbonium ion, the predominant olefine would be XXI. The formation of XX can be explained in terms of a synchronous deprotonation-rearrangement of the π-complex XXII.

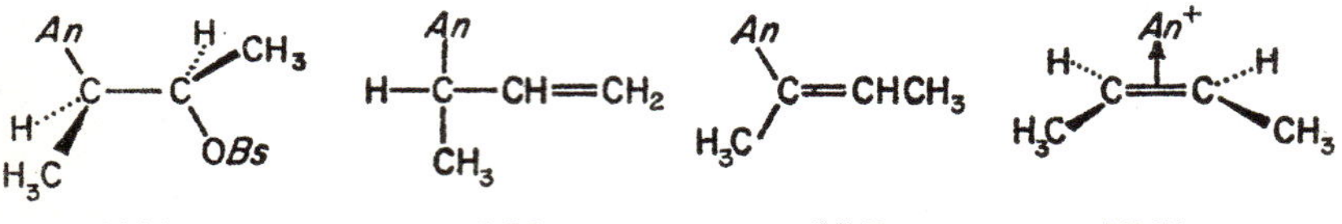

XIX XX XXI XXII

where *An* equals *p*-anisyl and *Bs* equals *p*-bromobenzenesulfonyl.

Collins et al. (23) have shown by elegant tracer methods that the acid-catalyzed dehydration of the dihydroxyphenylnorbornane (XXIII) gives the ketone (XXIV) in which there has been an inversion of configuration, the asterisked carbon is still attached to phenyl, and internal migration of the asterisked hydrogen has occurred. They point out that while these facts cannot be explained in terms of classical carbonium ions as intermediates, they are easily intelligible if the intermediates are π-complex ions.

Brown has not yet offered an explanation for our third line of evidence; the existence of certain π-complex ions as stable salts in cases where analogy

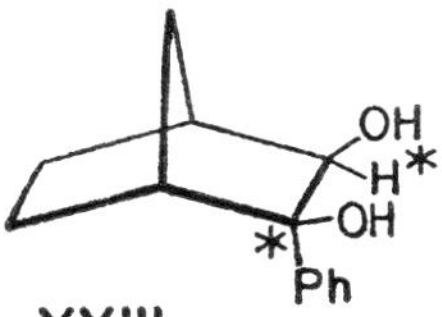

XXIII

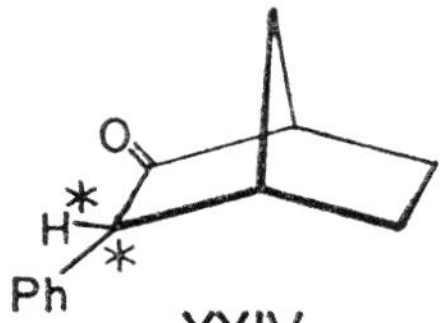

XXIV

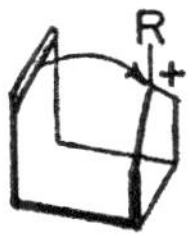

XXV

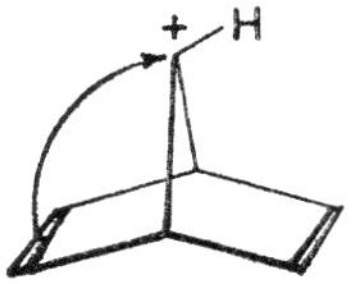

XXVI

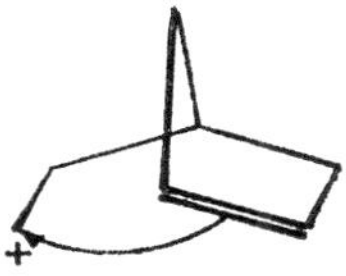

XXVII

suggests that the corresponding classical ions would be unstable. The first example of this kind was the isolation by Leal & Pettit (24) of the ion XXV ($R = Ph$) as its perchlorate, and their demonstration that a similar stable ion is formed when $R = CH_3$; the spectral properties of these ions were different from those expected for simple carbonium ions. Recently, Story et al. (25) have prepared the perchlorate of the norbornadiene ion XXVI, while Schleyer (26) has obtained salts of the norbornyl ion XXVII; the structures of these will be discussed below—here we stress only that the corresponding classical secondary carbonium ions could not have been obtained as stable entities under the conditions used.

Brown's (10) main attack on nonclassical carbonium ions derives from rate studies. First, he claims that in cases where *endo/exo* isomers react at different rates, it is the faster that is normal; the slower rate—e.g., of solvolysis of *endo* than of *exo* norbornyl esters—he attributes to steric effects. His argument is based on comparison of rates of these reactions with those of corresponding cyclopentyl derivatives. This comparison, however, is meaningless since it neglects the role of angle strain—a surprising circumstance since Brown himself first drew attention (27) to the importance of this as a factor in determining reaction rates! The trouble with arguments based on comparisons of rates with those of model compounds lies in the difficulty of selecting models. The only alternative course would be to estimate the rates of solvolysis of alkyl derivatives on the assumption that the intermediates are classical carbonium ions, allowing for all possible factors (angle strain, conformational strain, etc.) that might influence the rate. Recently, Schleyer (28b) has made an ingenious move in this direction using empirical methods (28a) to estimate the various relevant factors. He has developed an equation from which one can calculate the rates of solvolysis of secondary alkyl *p*-toleuenesulfonates, in cases where there is no reason to suspect the interaction of nonclassical intermediates. The calculated and experimental rates agree to within a factor of 10. No esters solvolyze more slowly than calculated; the deviations in rate are always positive and they occur only in cases where there is evidence for nonclassical intermediates. Thus the rate calculated for the *endo* norbornyl ester agrees very closely with experiment, while that for the *exo* ester is two orders of magnitude too small. While the approach used by Schleyer is frankly empirical and may be open to criticism, his results show that the observed rates are at least consistent with current ideas.

A second line of evidence put forward by Brown is based on comparisons of the rates of borohydride reductions of ketones with rates of solvolysis of esters of the corresponding secondary alcohols. Each of these reactions involves a change in geometry of the carbon atom at the reaction center from planar to tetrahedral, or vice versa, and in the case of the ketone, the structure is of course known to be classical. Brown claims that the rates of the reactions run parallel, even for solvolysis of esters when nonclassical ions have been proposed as intermediates. From this he concludes that the ions must, like the ketones, be classical. However, Schleyer claims (28b) that the relations reported by Brown depended on careful selection of data; he states that when the rates of borohydride reduction of ketones are plotted against the rates of solvolysis of esters of the corresponding secondary alcohols for all cases where data are available, the result is a scatter diagram with no indication of any correlation whatsoever.

A third argument rests on the small effect of $\pm E$ or $-E$ groups in the 1-position of bornane on the hydrolysis of *exo*-2-esters. Thus 1-*p*-anisylcamphene hydrochloride (VII) solvolyses more slowly than camphene hydrochloride itself. However this argument is based on a misunderstanding of theoretical principles, as will become clear in the following section. Compound VII would not be expected to solvolyse more rapidly than camphene hydrochloride, even if the intermediate is a π-complex.

One must conclude that Brown's case fails. However, it has had a healthy effect in promoting controversy and experimental work; this after all is an important function in science, since otherwise fallacies, once accepted, could be perpetuated indefinitely.

Structure and nomenclature of nonclassical carbonium ions.—The idea that nonclassical carbonium ions might occur as stable reaction intermediates was first proposed by Wilson (29a) to explain the rearrangement of XI into XII. Wilson, in effect, supposed that a carbonium ion, instead of existing as one of two possible classical isomers XXVIII or XXIX, might exist in an intermediate state represented as a resonance hybrid of the two. Stated in this way the idea carried little conviction and was ignored; indeed Wilson had difficulty (29b) in publishing it, even as a single sentence in his paper.

A radical change in the situation took place when the problem was considered in terms of molecular orbital theory (3, 5, 6). Here the intermediate ion can be represented as a dative π-complex XXX formed by an olefine acting as a π-donor, and the ion R^+ as an acceptor through its empty atomic orbital. It is of course implicit in this formulation that the olefine donor may be distorted out of its usual coplanar geometry in the π-complex, and that if

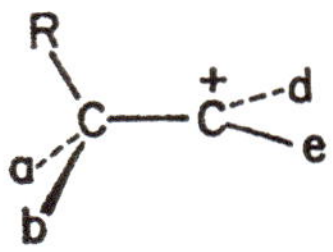

XXVIII

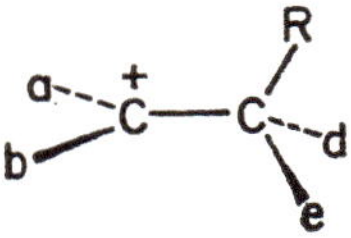

XXIX

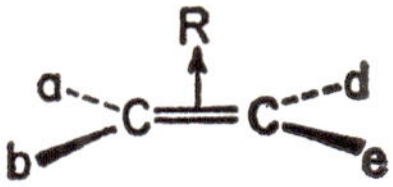

XXX

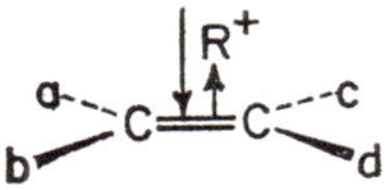

XXXI

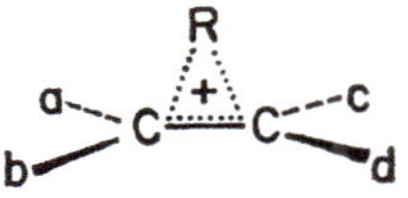

XXXII

R^+ is a carbonium ion, it may have a geometry intermediate between an ideal tetrahedral structure and the planar structure that it would have in the free state. A further immediate consequence of this formulation is the possibility of back coordination (5) in cases where the acceptor R^+ has π-electrons, or unshared *p*- or *d*-electrons; this can be conveniently represented (5) by the symbolism indicated in XXXI.

Olefines are known to act as π-donors in inorganic chemistry, and a wide range of such metal-olefine complexes have been prepared. The term π-complex has been rather generally adopted for compounds of this type, and they are commonly represented by the arrow symbolism of XXX, XXXI.

These investigations, together with the curious stereospecificity of the camphene hydrochloride rearrangement, left little doubt that Wilson's suggestion was correct, and since that time many other analogous cases have been discovered (8, 9).

In spite of this parallel, many organic chemists have preferred to use an alternative notation and terminology for ions such as XXX, describing them as "bridged ions" and designating them by a dotted line symbolism as in XXXII. Two arguments are commonly put forward in defence of the dotted line notation. First, it is argued that since carbonium ions are isoelectronic with boron compounds, and since boron can form compounds containing many different kinds of three-center bonds, one should adopt a symbolism and terminology for "nonclassical carbonium ions" flexible enough to accomodate any types that may turn up in the future. Secondly, it is argued that the olefine moiety in XXX must certainly be distorted from coplanarity, so that XXX fails to provide a satisfactory representation of the ion. Neither of these arguments seems convincing. In the first place all the types of nonclassical ions that have so far been discovered can, with one exception, be represented as π-complexes. Secondly, the distortion of the π-complex from its ideal geometry does not alter its properties in any qualitative way; the situation is no different from that in the trityl ion Ph_3C^+, where the rings are certainly twisted out of coplanarity, but where one still regards the ion as a conjugated system.

The fact that boron forms three-center bonds so readily, while C^+ forms them in only special cases, can be attributed in part to the fact that compounds of the latter type are charged and to the smaller size of the carbon valence orbitals. In a three-center bond, the orbitals cannot overlap efficiently with each other if they are too small. This difficulty is avoided in a π-complex, since most of the strength of the bond derives from the overlap of the p-atomic orbitals on the two basal carbon atoms; the apical atom can be attached much more weakly. Consider for example the change in bond energy when XXVIII (R = alkyl) isomerizes to XXX. Two C–C single bonds (total bond energy ~165 kcal/mole) are replaced by a complex bond whose energy can be written as a sum of a C=C bond energy (~145 kcal/mole) plus the energy of the dative bond formed by it to R^+. Evidently this second contribution need not be large for the two structures to be comparable in energy. Our picture of the π-complex then envisages strong bonding between the basal carbon atoms but only weak bonding to the apical atom or group *R*. Since the bond to R is weak, it can be distorted at relatively little cost in energy; thus a distortion which would halve the energy of a C–C σ-bond would cost 40 kcal/mole, while a similar distortion of the π-complex might not cost much more than 10–15 kcal/mole. This is why π-complex ions are often favored in strained systems; in such π-complexes the strain can be relieved at little cost by deforming the weak dative bond.

The π-complex notation has the advantage of indicating this situation very clearly. The dotted line notation gives no indication of stereochemistry or bonding until it has been translated into orbital terms. Moreover since the dotted line notation has no immediate physical interpretation, there is a temptation to extend it indiscriminately; some very strange structures have appeared in print for this reason (8). The existence of stable π-complexes must not be taken as a license for the indiscriminate postulation of other types of many-center bonding; as yet only one other such structure has appeared in carbonium ion chemistry, and that in a very special situation. In view of the dominant role of π-complex ions, and in view of the widespread importance of π-complexes in inorganic chemistry, it seems extremely unfortunate that a common nomenclature and symbolism has not been used for such structures in both fields.

One last point should be mentioned. Some time ago the term π-complex was used (30) to describe the weak hydrogen-bonded complexes formed by aromatic hydrocarbons with hydrogen chloride. Unfortunately the term has since been used for analogous charge transfer complexes. This has given the impression that π-complexes are weak. This is not the case. In a true π-complex the π-donor and acceptor are linked by a covalent bond, not by weak forces of van der Waals type. In Mulliken's terminology the term π-complex should be restricted to inner complexes in which a π-system acts as the donor.

Specific ions: The 7-norbornadienyl ion.—The alternative structures XXXIII and XXXIV have been suggested (31) for the 7-norbornadienyl ion (XXVI). Both of these look most attractive in dotted line garb but neither will stand up to close scrutiny. Structure XXXIII would have two valence electrons in an antibonding molecular orbital; no cases are known where such structures are stable in neutral or cationic systems. Structure XXXIV bears the same relation to XXVI that the transition state for a S_N2 reaction bears to the reactants; it can indeed be regarded as the transition state for interconversion of two isomeric π-complexes XXVI. On this basis alone one

XXXIII

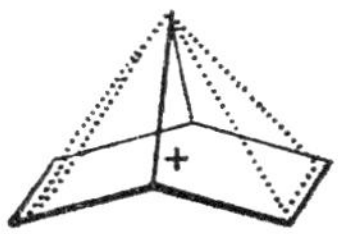

XXXIV

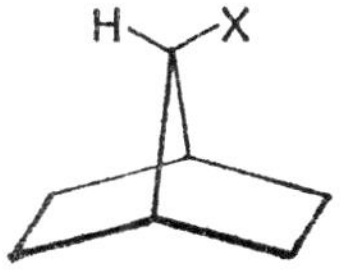

XXXV

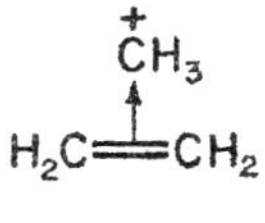

XXXVI

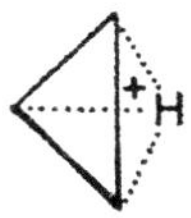

XXXVII

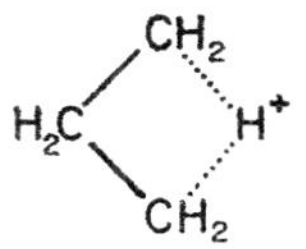

XXXVIII

would not expect XXXIV to be stable, and here the situation is made even worse by the fact that the apical carbon atom in XXXIV must have sp^2 hybridization. This would introduce severe angle strain in comparison with XXVI, where the apical carbon is formally quadricovalent and therefore more or less tetrahedral. The magnitude of this effect is indicated by the fact that 7-norbornyl esters (XXXV) solvolyse about 10^8 times slower than normal secondary alkyl analogues (2, 20). It is not surprising therefore that the NMR spectrum of the ion shows it to have an unsymmetrical structure (25).

Protonated cyclopropane.—Three structures have been considered for protonated cyclopropanes which occur as intermediates in several reactions; the π-complex XXXIV (6); the "dotted line" structure XXXVII (32); and XXXVIII (33), with a three-center bond analogous to those in diborane (1). Structure XXXVII can be ruled out on stereochemical and electronic grounds. Structure XXXVI accounts better (6) for the Markownikov orientation (34) in the ring opening of substituted cyclopropanes, and for the formation (35, 36) of cyclopropane in reactions thought to involve the n-Pr^+ ion. It also accounts for the lack of stereospecificity (37), in the conversion of tricyclene to *exo* norbornyl acetate-*d* by $DOAc/D_2SO_4$. On the other hand, tracer studies by Reutov & Shatkina (38) and Baird & Aboderin (39, 40) have shown that the hydration of cyclopropane to *n*-propanol cannot involve a single ion of type XXXVI. However all their results can be accommodated if the ion XXXVI can undergo rapid isomerisation by proton shifts; conversion of XXXIX to XL involves only a small displacement of atoms. LaLonde & Tobias (41) have also argued for intermediates of the type XXXVIII on the grounds that the acid-catalyzed solvolysis of XLI gives exclusively *trans*-1-acetoxy-2-methylcycloheptane XLII; however exactly the same result would follow if the intermediate were the π-complex XLIII. There is therefore no evidence as yet for the existence of stable proton-bridged ions analogous to XXXVIII.

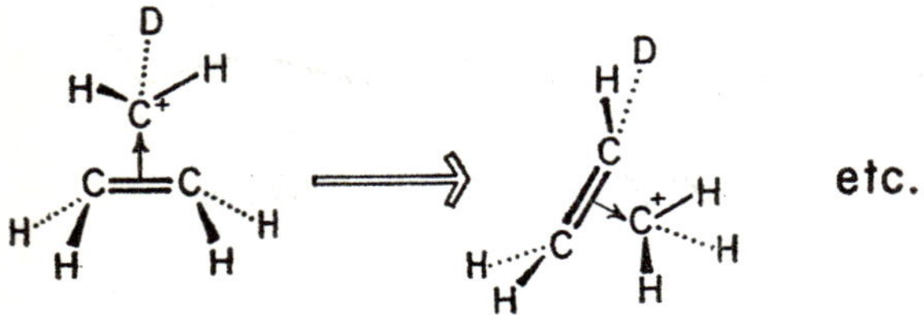

XXXIX XL

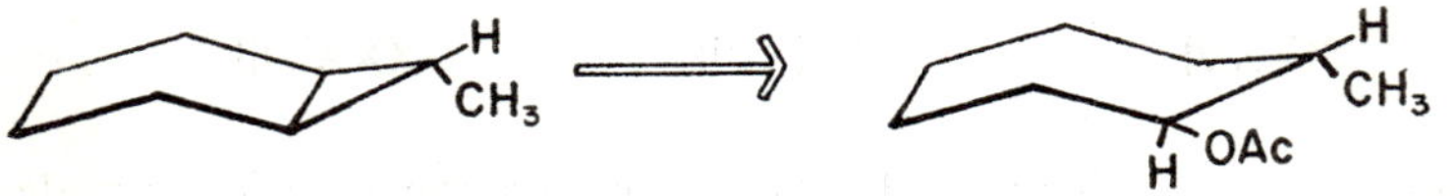

XLI XLII

The arylnorbornyl system.—Schleyer, Kleinfelder & Richey (20) found by NMR spectroscopy that the 1,2-*bis*-*p*-anisylnorbornyl ion XLIII (*An* = *p*–anisyl) exists as two isomeric ions rather than a single π-complex (XLIV). This is not surprising. The large resonance interactions in ions of the type RCH_2^+ or $RCH{=}CH{-}CH_2^+$ where R is an $\pm E$ or $-E$ group, are due (42) to the fact that the empty orbitals in CH_3^+ or $CH_2{=}CH{-}CH_2^+$ are nonbonding. The empty orbitals in the π-complex XXX, like those in cyclopropenium, are strongly antibonding; consequently one would not expect significant resonance interactions between substituents such as phenyl or *p*-anisyl with the three-center bond—this is borne out by the inefficiency (43) of phenyl groups in stabilizing the analogous cyclopropenium system. Substituents of this type should therefore very strongly displace the equilibrium between a classical carbonium ion and a π-complex in favor of the former; the ions XLIII are probably classical *p*-anisylcarbonium ions. A similar situation arises in the case of the phenylethyl system; whereas ions containing a single

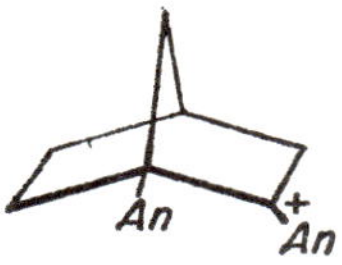

XLIII

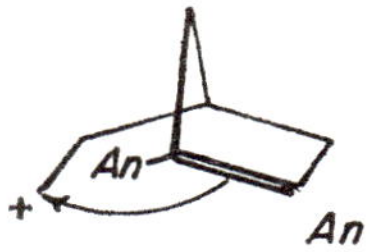

XLIV

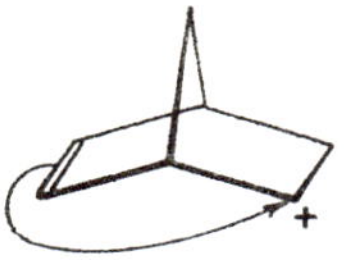

XLV

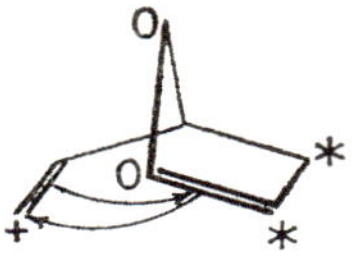

XLVI

phenyl group commonly exist as stable π-complexes (21) (phenonium ions), ions with several phenyl groups prefer classical carbonium structures (44).

The dehydronorbornyl system.—Dehydronorbornyl esters show signs of solvolysing via an intermediate nonclassical ion (8, 9), the conventional dotted line formulation of which is equivalent in π-complex notation to XLV. This formulation was proposed from analogy with ions such as XXVI, or cholesteryl (45), where such structures seem undoubtedly to be present; however, here the evidence seems to suggest that the ion has the alternative symmetrical structure XLVI, analogous to that of the bornyl ion but further stabilized by back-coordination. This can perhaps be seen more clearly in the equivalent representation XLVII.

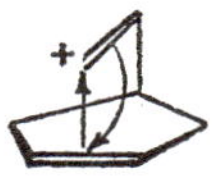

XLVII

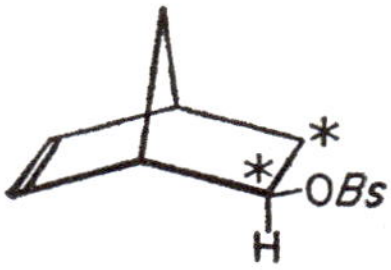

XLVIII

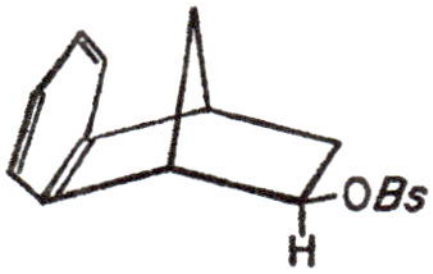

XLIX

Two lines of evidence support this conclusion. First, Roberts, Lee & Saunders (46) observed carbon scrambling in the solvolysis of *exo*norbornenyl *p*-bromobenzenesulfonate (XLVIII) labeled with C^{14} in the asterisked positions; this would not have happened had the intermediate ion been XLV whereas in XLVI the asterisked and circled positions are equivalent. Secondly, Giddings & Dirlam (47) have observed that the benzo derivative (XLIX) of XLVIII racemises during solvolysis; solvolysis via a benzo derivative of XLV would have led to retention of configuration, while the benzo derivatives of XLVI (or XLVII) have a plane of symmetry.

The norbornyl system.—Schleyer (26) has studied the NMR spectra of solutions containing the norbornyl cation. At room temperature this consists of a single line, indicating rapid equilibriation of all the hydrogen atoms, but at low temperatures the spectrum spreads into three bands in the intensity ratio 4:6:1, corresponding apparently to the protons marked respectively as *a*, *b*, and *c* in L. (See diagram page 332.) This result suggests that even at low temperatures there is a rapid equilibriation of the ions L, LI, and LII by proton transfer; this had been inferred previously on the basis of tracer studies (32) and explained in terms of the intervention of an isomeric nonclassical ion of type XXXVII. It seems more likely that the rearrangement is analogous to that postulated here for protonated cyclopropane (vide supra), no stable intermediates being involved. At higher temperatures a further migration of a proton is possible, from the back of L to the front, giving the isomeric ion LIII; this process is slower and can be inhibited by lowering the temperature.

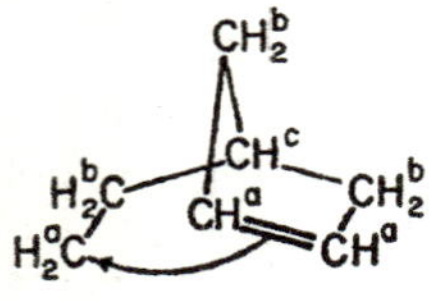

L

LI

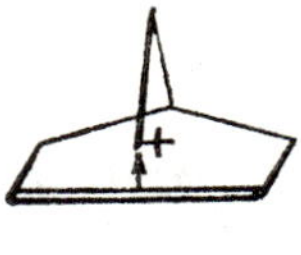

LII

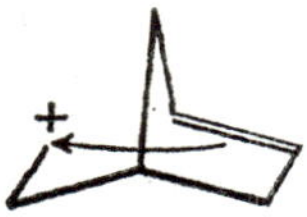

LIII

LIV

Here again there seems to be no need to postulate the intervention of stable intermediates other than L and its isomers. The rearrangement of L into LIII presumably takes place via LIV, but this may well be a transition state rather than a stable intermediate.

This conclusion seemed for a time to conflict with work by Corey et al. (48) on the deamination of *exo*-2-norbornylamine by nitrous acid. They claimed that their results could be explained only if the first intermediate is a classical norbornyl cation ion, requiring activation for rearrangement to the π-complex. Such a conclusion would be completely at variance with all other evidence concerning this system; Berson & Remanick (49) have now repeated this work with quite different results, and conclude that Corey's conclusions were incorrect.

The cyclopropylcarbinol system.—A common nonclassical intermediate A is involved in carbonium ion reactions of cyclopropylcarbinol, cyclobutanol, or allylcarbinol (50) derivatives, and after much study Roberts and his collaborators (51) have formulated this as an equilibriating mixture of three isomeric "bicyclobutonium" ions (LV). This formulation explains the ready formation of the ion from cyclobutene or butene derivatives—although less easily from cyclopropylcarbinol—and it also seems to explain the fact that cyclobutene derivatives are major products of reactions in which it is an intermediate. (See top diagram, page 333.)

This formulation becomes less convincing when one tries to translate the dotted line symbolism of LV into orbital terms. One possibility would be a coplanar homoallylic structure; this, however, would not give cyclobutane derivatives with nucleophiles since the interaction between the terminal carbon atoms in LV would be negligible. Similar difficulties apply to the π-

LV

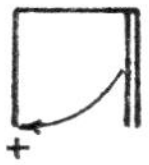
LVI

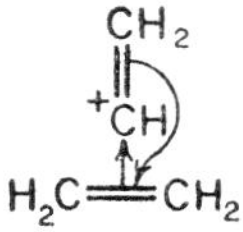

LVII

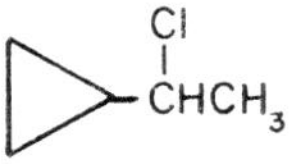

LVIII

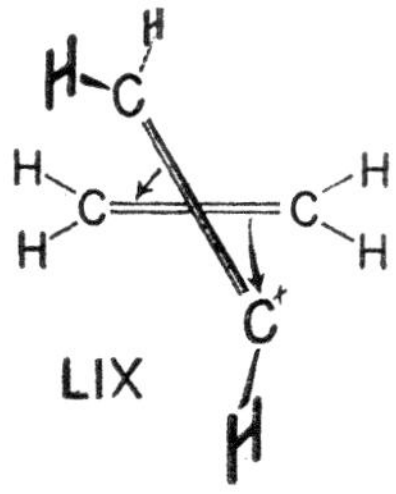

LIX

complex formation LVI, which, moreover, would be highly strained. A much more attractive formulation of A seems to be LVII, a π-complex with an apical vinyl group which would be stabilized by back-coordination—cf. the norbornenyl ion XLVII. Such a structure could be formed easily from cyclopropylcarbinol, cyclobutenol, or allylcarbinol derivatives; the only difficulty lies in explaining why it should so easily revert to derivatives of cyclobutene.

This difficulty arises only in the case of the unsubstituted C_4 systems; even the monomethyl derivative LVIII solvolyses "normally" to give cyclopropylethyl derivatives rather than methylcyclobutenes. A possible explanation would be that LVII structure is unlikely because of the positive charge that back-coordination places on the methylene of the apical vinyl group. As soon as we introduce a suitable substituent at this point, however, LVII becomes quite satisfactory. This argument suggests that A may be a distorted version of LVII, perhaps LIX. Such a structure could account for all the available evidence, including the "normal" reactions of compounds such as LVIII and also recent studies by Applequist & Landgrebe (52, 53).

This formulation would also account for the exceptional stabilizing effect of cyclopropyl groups on carbonium ions; Hart and his collaborators (54) have found this to be much greater than that of vinyl, and also to increase more or less uniformly with the number of cyclopropyl groups. This is indicated by the relative rates of solvolysis of mixed tertiary isopropyl/cyclopropyl carbinyl esters;

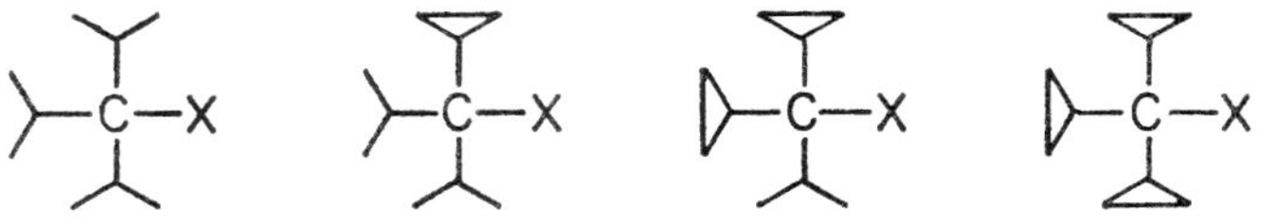

relative rate	1	246	23,500	~20,000,000

this result is difficult to explain in conventional terms. However the arguments given above suggest that the ion ▷–CH$^+$ should be stabilized by deformation into the π-complex LVII; this then provides a general mechanism for the stabilization of carbonium ions by cyclopropyl substituents. (See p. 334.)

Trishomocyclopropenyl and related systems.—Winstein et al. (55) have presented evidence for the intervention of a stable nonclassical ion as intermediate in the solvolysis of LXa; they formulated this as LXI, with a sym-

LX; a, *X* = tosyl
b, *X* = NH_2

LXI

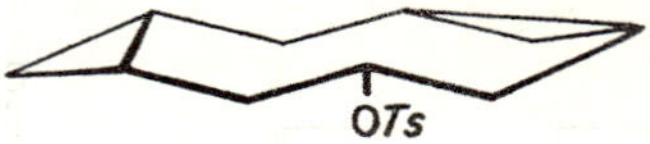

LXII

LXIII

metrical three-center bond formed by overlapping of sp^3 atomic orbitals of the three relevant carbon atoms. Recently Winstein et al. (56) have presented evidence for a similar intermediate in the solvolysis of LXII. However Corey & Dawson (57a) have found that the amine LXb deaminates with little rearrangement, an observation which suggests that the intermediate cannot be the symmetrical ion LXI. [Corey & Uda (57b) have presented a further argument in support of this conclusion, based on the failure of bridgehead phenyl groups to accelerate the solvolysis of LXa. This argument is, however, fallacious, being based on the mistaken idea that phenyl groups would mesomerically stabilize an ion such as LXI. From the viewpoint of each of the three relevant carbon atoms in LXI, the three-center bond is a normal covalent bond $C-C_2$. The fact that the other end of the bond involves a pair of carbon atoms instead of a single atom in no way alters its essential character. Consequently the three bridgehead carbon atoms in LXI are effectively saturated, being quadricovalent, and no resonance interactions with adjacent phenyl groups would be expected.] Now one can write an equivalent orbital description of a π-complex, the double bond of the olefine being represented as a combination of two bent "banana" bonds one of which is used to form a dative three-center bond to the apical atom. This suggests that the intermediate ion from LX may have the analogous structure LXIII, an internal coordination complex in which the donor is one of the bent σ-atomic orbitals of the cyclopropane ring. (Of course LXIII would be expected to equilibriate easily with two isomeric ions in which the three relevant carbon atoms interchange roles.)

If solvolysis of LXa does lead to a nonclassical ion, this cannot of course be a π-complex; however the argument above suggests that it could have a very closely similar structure. Thus this exception to our rule, that nonclassical carbonium ions are normally π-complexes, is at best a weak one.

Miscellaneous.—Erman (58) has studied the interconversions of various π-complex intermediates in the acid-catalysed reaction of phenol with camphene. Cram & Goldstein (59) report evidence for π-complex intermediates in the solvolysis of esters derived from *p*-phenylenepolymethylenes. Tanida, Tsuji & Ishitobi (60) have shown that a π-complex ion is formed in the solvolysis of 7-*anti* benzonorbornenyl *p*-bromobenzenesulfonate.

Electrophilic Addition

Evidence for π-complex intermediates can be obtained for reactions of solvolytic type only if the corresponding classical ions are unstable and so tend to rearrange to π-complexes during their formation (61). A much more powerful tool would be to study the intermediates in electrophilic addition to olefines; here π-complexes should be formed even if the isomeric carbonium ions are stable, provided that the π-complexes are more stable than the classical ions. Evidence of this kind has of course long been available in the case of ions formed from olefines with Br^+ or I^+; it has often been assumed that all such reactions would proceed in an analogous manner, with π-complex intermediates and with consequent stereo-specific *trans* addition.

Dewar & Fahey (62, 63) have now found that the polar additions of deuterium bromide to acenaphthylene, indene, and *cis* and *trans*-1-phenylpropene take place predominantly, though not exclusively, by *cis*-addition; they deduced from this that the intermediates could not be π-complexes, but most probably were classical carbonium ions.

In these cases the classical ions were given an unfair advantage over the π-complexes; for they would be selectively stabilized by resonance interactions with the aryl groups. It is entirely possible that π-complex ions might be the preferred intermediates in the case of simple olefines; this seems (63) to be the situation in the case of chlorine addition, where simple olefines (e.g., 2-butene) give exclusively *trans* adducts, while aryl olefines (e.g., stilbene, acenaphthylene) give mixtures in which the *cis*-adduct often predominates.

Fahey & Smith (64) have indeed found that hydrogen bromide adds *trans* to cyclohexene, no *cis* isomer being detected in the product. Hammond et al. (65, 66) had previously observed *trans* addition of hydrogen halides to 1,2-dimethylcycloolefines, but this evidence was less conclusive since the reactions might have been conformationally controlled. These results suggest that in nonpolar solvents, proton complexes of simple olefines may prove to be more stable than isomeric carbonium ions. However the solvolysis data

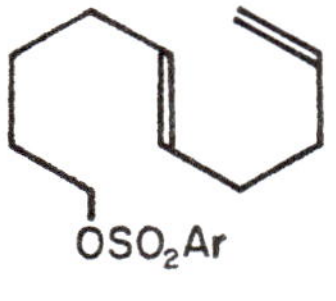

LXIV

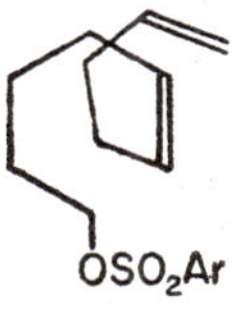

LXV

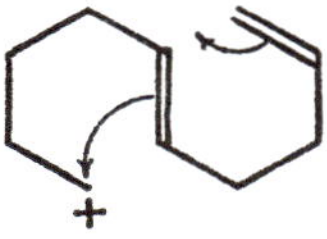

LXVI

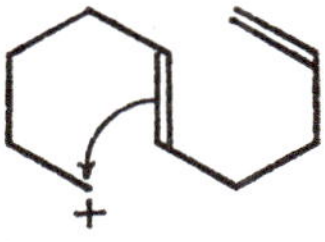

LXVII

show that even if this is the case, the isomeric classical ions are still stable entities, separated from the π-complexes by energy barriers.

Recent work by Johnson et al. (67, 68, 69) suggests that the same may be true for π-complexes in which the apical group (R in XV) is alkyl. Thus LXIV and LXV undergo cyclization on solvolysis to a mixture of products containing decalols and decalenes; LXIV gives derivatives of *trans*-decalin exclusively, LXV exclusively *cis*-decalin derivatives. The reactions take place by internal electrophilic attack, as indicated in LXVI; the stereochemistry of the products shows that addition across the actual bond takes place exclusively *trans*, implying that the intermediate is a π-complex, LXVII.

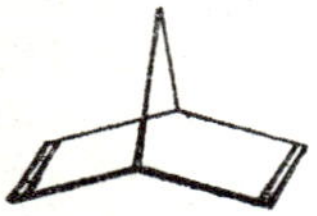

LXVIII

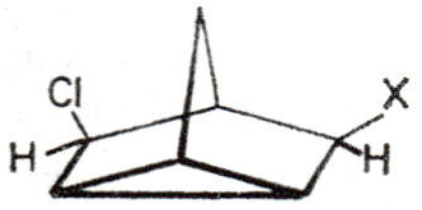

LXIX; a, =OAc
b, =NEt

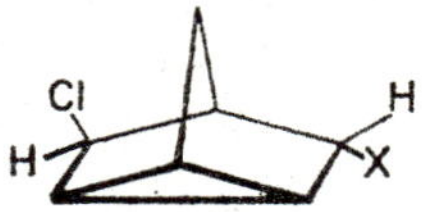

LXX

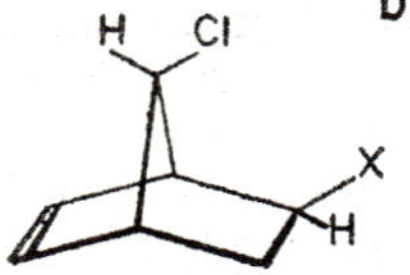

LXXI

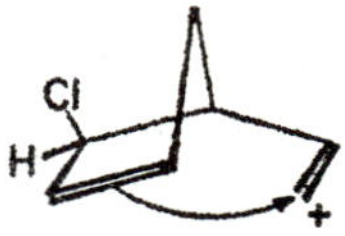

LXXII

In cases where solvolysis indicates that classical ions are unstable, addition reactions of corresponding olefines naturally take place via the same π-complex intermediates (2, 3). An interesting reaction of this kind has been reported by Neale & Whipple (70), the acid-catalyzed addition of diethyl-N-chloramine to norbornadiene (LXVIII) which gives comparable amounts of LXIXa, LXXa, LXIXb, LXXIb, and a little LXXIa. The reaction seemed to be ionic, involving an initial attack by Cl^+ derived from the ion Et_2NHCl^+. This would be expected to give the symmetrical π-complex LXXII as intermediate (cf., XLVII), and all the products could be derived from this. However two points are difficult to understand: first, the stereospecificity of the reaction (cf. the formation of LXIXb, but not LXXb); second, the formation of amines. Diethylamine should have undergone protonation extremely rapidly under the conditions used ($4M$ H_2SO_4) and so been prevented from reacting with LXXII.

Addition reactions involving primary attack by Br^+, I^+, or RS^+ (71) seem almost invariably to involve π-complex intermediates and to give *trans* adducts exclusively. A recent addition to the list of such reagents is iodine isocyanate (72).

Aromatic π-Complexes

Aromatic compounds can form π-complexes with metals; analogous organic π-complexes might therefore be expected (6) to play a role in chemistry. Several reactions are known when such compounds seem to be involved as intermediates.

Electrophilic substitution.—Olah and his collaborators (73, 74) have presented strong evidence that the rate-determining step in certain aromatic substitution reactions is the formation of a π-complex from the reagent and the aromatic substrate. In these cases different benzene derivatives react at comparable rates under conditions where there is a strong discrimination between the various positions; this indicates that the product-determining step is different from the rate-determining step. Since the product-determining step is almost certainly the formation of a benzenonium ion, the rate-determining step must be the formation of some intermediate that can subsequently rearrange to one or other of the possible isomeric benzenonium ions. This intermediate can only be reasonably formulated as a π-complex.

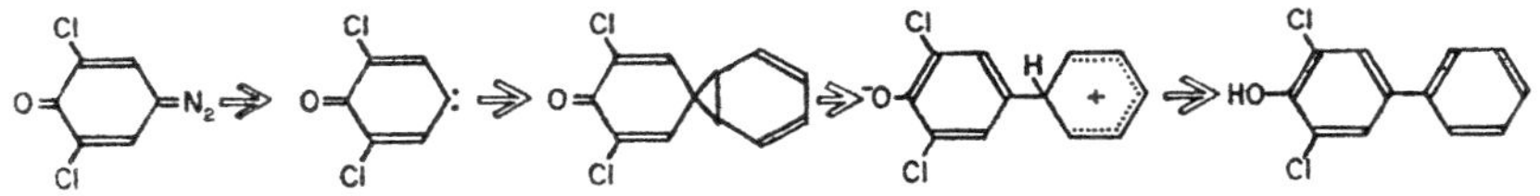

LXXIII LXXIV LXXV LXXVI LXXVII

A striking example of this has been found by Dewar & Narayanaswami (75). Decomposition of the diazooxide LXXIII in benzene derivatives gives derivatives of biphenyl LXXVII, through electrophilic substitution by the intermediate carbene LXXIV. While LXXIV shows high selectivity in choosing between the various positions in a given substrate (e.g., chlorobenzene gives <0.5 per cent of *m*-isomer), it reacts almost indiscriminately with substrates of different types (e.g., the rates of attack on anisole and methyl benzoate differ only threefold). Detailed studies suggested that the mechanism is as indicated above. The rate-determining step is the formation of the spiran or π-complex LXXV; this presumably is very exothermic and requires very little activation, so that the rates of reaction of different benzene derivatives differ little. The product-determining step is the conversion of LXXV to a benzenonium ion LXXVI; this step requires considerable activation, so that there is strong selectivity between the two isomers that can be formed from LXXV.

This kind of mechanism is not by any means general; in the majority of substitution reactions the rate-determining step is the formation of the intermediate arenonium ion. Formation of a π-complex can be rate-determining only in cases where the reaction is strongly exothermic; this can be so only with very "powerful" reagents, in particular ones whose π-complexes are stabilized by back-coordination (e.g., NO_2^+, LXXIV). It is of course possible that π-complexes may also be formed reversibly as intermediates, even in cases where the rate-determining step is the formation of the arenonium ion; in that case, however, they could have no effect on the course of the reaction and so would be kinetically and mechanistically irrelevant.

Rearrangements involving migrations from a side-chain into an aromatic nucleus.—A number of intramolecular reactions of this kind are known, e.g., the acid-catalyzed rearrangement of N-nitraniline to *o*- and *p*-nitraniline, that seem to involve π-complex intermediates [for a review, see (76)]. A recent example is the rearrangement of N-nitraniline in acid to give *o*- (52 per cent) and *p*- (31 per cent) nitro-IV-methylaniline, with no *m*-isomer, which has been studied by White, Lazdins & White (77). They interpreted their results in terms of a cage mechanism; this however seems extremely improbable since if free methylaniline had been formed at any stage in the reaction it would have been protonated immediately. In that case the products would have been very different [cf. the rearrangement of N-nitraniline, where it has been established (76) that free aniline cannot be formed as an intermediate]. There seems no reason to doubt that this reaction, like the rearrangement of N-nitraniline itself, takes place by the π-complex mechanism.

Benzidine rearrangement.—An important series of papers by Ingold and his collaborators (78) describes a detailed kinetic study of a number of benzidine and semidine rearrangements. Previous reactions of this type had, with one exception, been of second order with respect to acid; Ingold et al. have now found several examples of rearrangements (e.g., of α-hydrazonaphthalene) which are of first order with respect to acid, and even cases (e.g., 1-phenyl-2-α-naphthylhydrazine) where the order is intermediate and varies with acid concentration. Previous theories (76) had envisaged a variable acid dependence either through a mechanism where the product-determining step involves a proton attack on an intermediate π-complex, or on the assumption that it is the second conjugate acid of the hydrazine that rearranges, and that at high acid concentrations the hydrazine may be converted largely or completely to its monoacid salt. The new evidence shows that neither of these suggestions is correct; the kinetics of reactions with intermediate acid dependence indicate that they are taking place by two independent mechanisms, one of first order, the other of second order, with respect to acid. Studies of deuterium isotope effects moreover suggested that the first step is the reversible formation of the first or second conjugate acid of the hydrazine, this then rearranging in the rate-determining step. It was also shown that the uncatalyzed thermal rearrangements of several of these hydrazines gave products very similar to those formed in the acid-catalyzed reactions; this suggests that either the hydrazine, or its mono- or diacid salts, can rearrange, the specific rate increasing with increasing protonation, but the nature of the reaction remaining essentially the same.

The final paper of the series consists mostly of a discussion of the work of others in the field; this lacked the element of accuracy, however. [An interesting example is provided by their discussion of substituent effects. Having dismissed the "product rules" suggested by Dewar (6) as inconsistent with the facts, they then restate the same rules in a different and more complicated terminology and show them to be in full accord with experiment.] Their criticisms of the π-complex and cage mechanisms are in error and largely based on misinterpretation of the original papers. Kinetic evidence can throw no light on the nature of the intermediates in these reactions, and it is easy to write schemes involving π-complexes or caged radical pairs that are in full accord with their new data. The following scheme illustrates the interpretation in terms of π-complexes; here HA represents hydrazobenzene, π_0 the neutral π-complex LXXVIII, and π_1, π_2 the mono- and diacid salts of LXXVIII respectively, protonated on nitrogen.

$$HA \xrightarrow{\text{slow}} \pi_0 \longrightarrow \text{products}$$
$$HA + H^+ \rightleftarrows HA \cdot H^+$$
$$HA \cdot H^+ \xrightarrow{\text{slow}} \pi_1 \longrightarrow \text{products}$$
$$HA \cdot H^+ + H^+ \rightleftarrows HA \cdot 2H^+$$
$$HA \cdot 2H^+ \xrightarrow{\text{slow}} \pi_2 \longrightarrow \text{products}$$

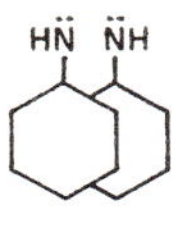

LXXVIII

A very similar scheme can be written in which the π-complexes are replaced by radical pairs.

Having refuted existing mechanisms for the reactions, Ingold et al. then propose their own "polar transition state" mechanism. The basis of this is their suggestion that the N–N bond undergoes heterolytic rupture, the diacid salt $ArNH_2^+NH_2^+Ar$ splitting into an amine $ArNH_2$ and a dication $ArNH_2^{++}$ which recombine to form the products. Now a change of this kind seems inherently improbable; all analogy suggests that homolytic cleavage to ion-radicals $ArNH_2^+$ should occur much more readily, as is postulated in the cage mechanism. Ingold et al. try to avoid this difficulty by supposing that $ArNH_2$ and $ArNH_2^{++}$ are not liberated in a free state but remain linked by some kind of bond. The obvious interpretation of this would be that they combine to form a π-complex (π_2 in equation 1). However Ingold et al. cannot allow this since they have already arbitrarily dismissed the π-complex mechanism, and one could no longer meaningfully distinguish between homolytic and heterolytic fission of the N–N bond if a π-complex were formed, for the same π-complex would be formed by union of two ion-radicals $ArNH_2^+$ as by dative bonding between $ArNH_2$ and $ArNH_2^{++}$. Ingold et al. are therefore forced to suppose that the aryl groups are linked by some special kind of interaction which they denote by a dotted line and describe as a "long predominantly electrovalent bond." There are no analogies for bonding of this type.

Certain mechanisms for the benzidine rearrangement have been postulated which include a reversible conversion of a diaryl-hydrazine into some intermediate (e.g., a π-complex). Hammond & Clovis (79) have devised an ingenious test of this hypothesis by studying the rearrangement of 3,3′-diaminohydrazobenzene (LXXIX) labelled with N^{15} in the hydrazine group. Any reversible formation of an intermediate in which the N–N bond had been broken would have led to scrambling of N^{15} between the amino and hydrazine groups; no such scrambling was observed. However kinetic studies indicated that the rearranging species is a monoprotonated form of LXXIX, the reaction being analogous to those benzidine rearrangments that are of first order with respect to acid (78); all current mechanisms predict an irreversible conversion of the diarylhydrazine into intermediates or products, or both, in such cases, so the behavior of LXXIX has no bearing on the possible reversibility of reactions which are of second order with respect to acid.

Although this investigation failed in its primary purpose, it nevertheless proved to be very illuminating. The rearrangement of LXXIX gives mainly the diaminobenzidine LXXX; this can be formed by linkage through carbon atoms that were either *ortho* or *para* to the hydrazine group in LXXIX. The N^{15} label enabled the proportions of *ortho* and *para* linkage to be established; the observed ratio showed that the reaction could not take place by the Cope mechanism suggested by Brownstein, Bunton & Hughes (76, 80). Moreover the *ortho*:*para* ratio varied with acid concentration, *para* coupling being favored at low acid concentrations. Since the rearrangement was accurately of inverse first order with respect to acid over the whole range, Hammond & Clovis concluded that the rate-determining and product-determining steps in the reaction must differ, and hence that the reaction must involve a stable intermediate.

This intermediate could well be the π-complex LXXXI. This would tend (76) to rearrange by coupling through the carbon atom *para* to the NH group. At higher acidities, however, LXXXI would be converted to a symmetrical salt in which the nitrogen atoms on each ring are equivalent and all memory of the original hydrazine would be lost. This would account for the observed increase in *ortho* coupling with increasing acidity.

A formally equivalent scheme could be written in terms of a cage mechanism, involving fission of a salt of LXXII into the radicals LXXXII and LXXXIII which then undergo primary recombination. The effect of increasing acidity on the products would then be attributed to protonation of LXXXIII to the symmetrical ion-radical LXXXII. Hammond & Clovis envisage a third possibility, that the π-complex LXXXI is formed first and

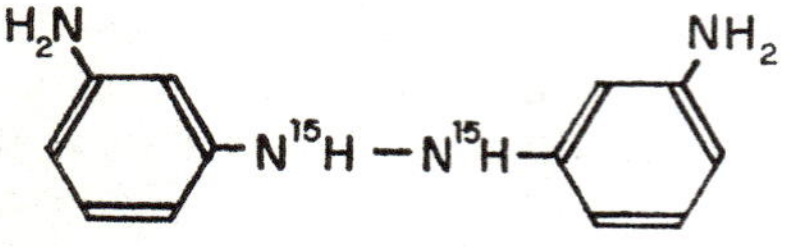

LXXIX

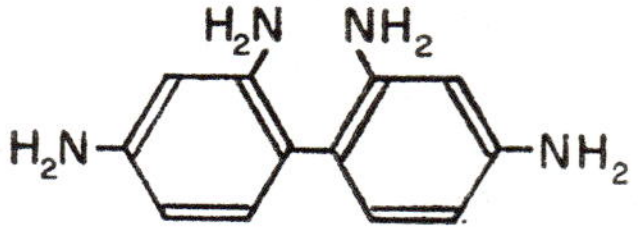

LXXX

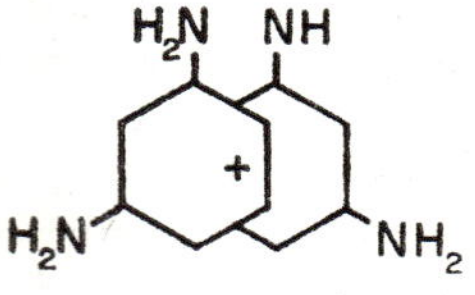

LXXXI

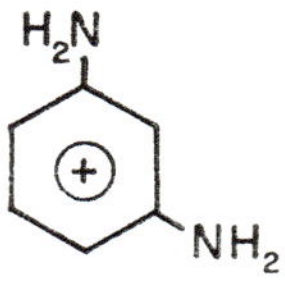

LXXXII

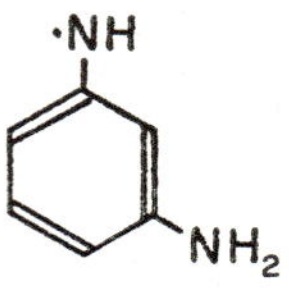

LXXXIII

then dissociates into a pair of radicals; this, however, seems an unnecessary complication since the products could certainly be formed directly from the π-complex.

It should perhaps be emphasized that there is still no definite evidence against the cage mechanism, although there are certainly strong arguments against it (76). At one time it seemed unreasonable to suppose that ion-radicals of like charge could be held together by a solvent cage against their mutual coulombic repulsion; however Hammond & Neuman (81) have discovered that the cation-radicals formed by photolysis of diacid salts of azo*bis*isobutyramidines do undergo primary recombination, to some extent at least, so this objection may be less forceful than had been supposed. However it still seems difficult to explain the essentially complete intramolecularity of the reaction ($>$99.9 per cent according to the most definitive test) in terms of a cage mechanism; the efficiency of primary recombination, even in the case of neutral radicals, never approaches 100 per cent. One might add that the primary recombination of the amine radicals studied by Hammond & Neuman might have been due to π-complex formation, rather than to a true cage effect.

Ferstandig (80) has suggested yet another structure for the benzidine intermediate, a sandwich-like double π-complex $(Ar \cdots H \cdots Ar)^+$ in which the rings are cemented together by a proton. This suggestion seems to have little to recommend it other than novelty. It offers no advantages over the π-complex mechanism, and it postulates a type of bond which is open to serious criticism and for which there are no analogies. [No π-complexes of aromatic donors with protons are known since these are invariably less stable than isomeric arenonium ions. Consequently the unsymmetrical hydrogen-bonded structure $(Ar \rightarrow H^+ \cdots Ar)$ could hardly be stable. On the other hand the symmetrical structure $(Ar \cdots H \cdots Ar)^+$ should be even worse (cf., structures XXXIV and XXVI for the 7-norbornadienyl ion).]

Benzidine-like rearrangements.—Compounds of the type Ar–X–Y–Ar, where X and Y are atoms electronically analogous to nitrogen (e.g., O, S), might be expected (76) to undergo rearrangement of the benzidine type. Only one definite example has so far been reported (82); the rearrangement of N-acetyl-O,N-diphenylhydroxylamine (LXXXIV) to LXXXV. (See p. 342.)

Quinamine rearrangement.—It has been known for some time that quinamines rearrange in presence of acid; recently Miller (83) reported a detailed study of these reactions.

Quinamines of the type LXXXVI, with the position *para* to nitrogen free, rearrange to give diphenyl ethers LXXXVII; quinamines with blocked *para* positions rearrange in an analogous manner to derivatives of biphenyl or diphenylamine. The reactions are of second order, first order with respect to quinamine, Q, and first order with respect to acid. Moreover since the rate of the reaction is affected much more by substituents in the benzenoid ring than in the quinonoid ring, it seems clear that the rearranging species QH^+ is a N-protonated quinamine rather than an O-protonated species. Further-

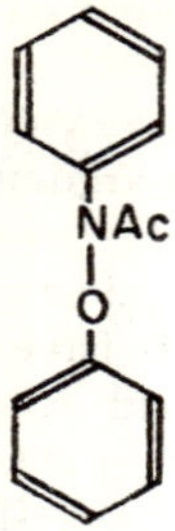

LXXXIV

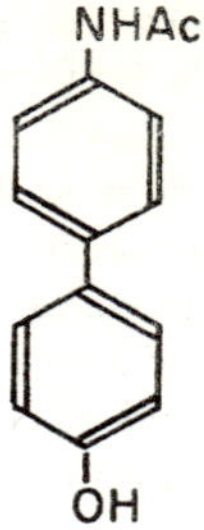

LXXXV

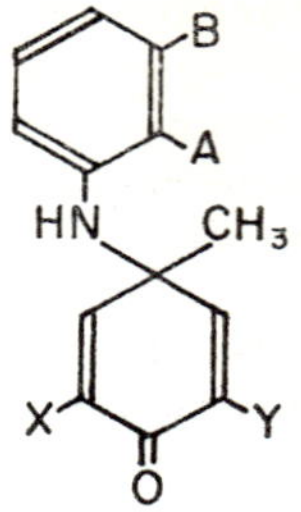

LXXXVI

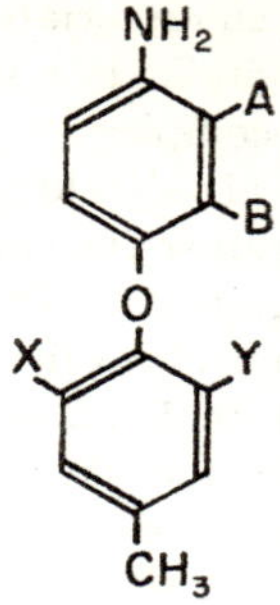

LXXXVII

more the reactions seem to be intramolecular, since mixtures of different quinamines rearrange independently, no cross products being formed.

If it can be assumed that QH^+ is formed reversibly and that the relative basicities of quinamines are the same as those of the corresponding anilines, the relative rates of rearrangement of different ions QH^+ can be estimated. Miller lists the following values for quinamines LXXXVI, with X=Y=Br:

A	B	relative rate
H	H	1
Br	H	3.9
CH_3	H	16.0
H	CH_3	6.1
H	CH_3CO	0.29

These results indicate that electron-releasing substituents in the benzenoid ring increase the rate of rearrangement of the ion QH^+, while electron-withdrawing substituents retard it. This in turn implies that the π-electron density in the benzenoid ring decreases in passing from QH^+ to the transition state.

There is a formal analogy between the quinamine and benzidine rearrangements in that the protonated quinamine LXXXVI could dissociate into the same pair of radicals LXXXIX that would be formed by homolytic cleavage of the N–O bond in the hydrazine analogue LXXXVIII. One can therefore write a series of mechanisms for the quinamine rearrangement that parallel those suggested for the benzidine rearrangement.
Miller concludes that his results are best explained in terms of a π-complex mechanism, the monoacid salt of LXXXVI rearranging to a π-complex formed by π-bonding of the radicals LXXXIX. (See p. 343.)

Structure of "sandwich" π-complexes.—It should be emphasized again that the bonding between the two rings in π-complexes such as LXXXI is

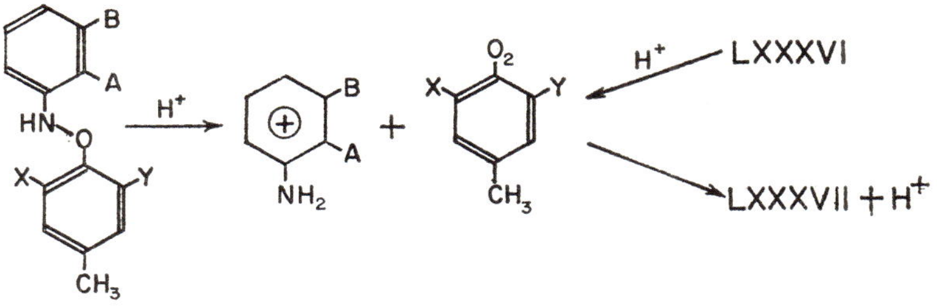

LXXXVIII LXXXIX

envisaged (76) as a definite covalent bond; π-complexes are not charge transfer complexes or other similar loose associations. It follows that the total number of classical covalent bonds in a π-complex is the same as in the starting material; this explains why rearrangements involving π-complex intermediates take place so very much more easily than do rearrangements of other kinds. At the same time the extended π-bond must be much weaker than a simple two-center π-bond; this again explains why rearrangements of this kind occur so readily in the case of diarylhydrazines, where the $\pi:\pi$ bond replaces a normal covalent bond (N–N) whose bond energy is quite low (60 kcal/mole). A different but analogous situation is seen in the case of the quinamine rearrangement; here the bond broken in the reaction (N–C) is a strong one, but this is compensated by the gain in resonance energy when the quinonoid ring becomes aromatic as a result of its rupture.

These considerations clearly indicate the lines that should be followed in searching for other rearrangements of this kind. Either one must choose systems where a weak bond is ruptured, or cases where bond rupture involves additional resonance stabilization.

Theoretical Calculations

Shortly after the π-complex theory was developed in qualitative terms, Dewar (84) published some simple Hückel calculations for the heats of formation of π-complexes. This approach was strongly criticized by Coulson, and after consultation, Coulson & Dewar published a paper (85) pointing out some of the dangers inherent in the use of the Hückel theory for calculations of this kind. Unfortunately this paper was apparently overlooked by other authors and a number of such calculations have since been reported, notably by Winstein and his collaborators (86), and more recently by Hoffmann (87). The use of the Hückel method for calculating resonance energies of conjugated systems is questionable (88); however its use to compare energies of classical and nonclassical ions of different geometries involves a much worse error. In an SCF treatment, the total energy of a set of electrons is not equal to the sum of their SCF orbital energies; for in these the interelectronic repulsions are counted twice over. The total electronic energy is equal to the sum of the orbital energies, less the corresponding interelectronic repulsions.

In chemistry, however, we are interested not so much in the electronic binding energy as in the total bond energy, i.e., the difference in energy between a molecule and atoms of which it is composed. In order to calculate this from the total binding energy of the electrons in the molecule, we must subtract not only the atomic binding energies but also the repulsions (core repulsion) between the nuclei in the molecule; i.e.,

$$E_B = \Sigma E_M - \Sigma E_O - E_R + E_{CR} \qquad 2.$$

Here E_M is the orbital energy of one of the electrons in the molecule, E_O its energy on an isolated atom, E_R is the mutual repulsion of the electrons in the molecule less their repulsion on isolated atoms, and E_{CR} is the core repulsion. If the molecule is neutral

$$E_R \sim E_{CR} \qquad 3.$$

In this case we may get a reasonable estimate of the total bonding energy in the molecule by neglecting both E_R and E_{CR}, which is what we do in the Hückel method. In the case of ions, however, this approximation breaks down completely. Now E_R and E_{CR} will differ by a quantity of the order of 100 kcal/mole. This still may not be serious if we are merely comparing systems of identical geometry, as when we are calculating resonance energies of π-systems; however it becomes extremely serious if we try to compare the total energies of structures of different geometry, e.g., π-complexes with classical carbonium ions.

Calculations of this kind must therefore be regarded as valueless except in a very qualitative sense. In order to achieve anything useful, it will be necessary to use a full Self Consistent Field Molecular Orbital treatment—and this is at present impracticable for the systems in which we are interested. Situations, of course, arise where exact theoretical treatments are impossible and approximate calculations are therefore justified and useful; however, the use of Hückel theory for ions of the type considered here is so grossly incorrect that no such justification can be accepted in this case.

LITERATURE CITED

1. Longuet-Higgins, H. C., *J. Chim. Phys.*, **46,** 268 (1949)
2. Dewar, M. J. S., *Nature*, **156,** 784 (1945)
3. Dewar, M. J. S., *J. Chem. Soc.*, 406 (1946)
4. Dewar, M. J. S., *ibid.*, 777
5. Dewar, M. J. S., *Bull. Soc. Chim.*, **18,** C71 (1951)
6. Dewar, M. J. S., *The Electronic Theory of Organic Chemistry* (Clarendon Press, Oxford, 1949)
7. von Rosenberg, J. L., Mahler, J. E., and Pettit, R., *J. Am. Chem. Soc.*, **84,** 2842 (1962)
8. Berson, J. A., in *Molecular Rearrangements*, **1** (de Mayo, P., Ed., Interscience, New York, 1963)
9. Capon, B., *Quart. Rev.* (*London*), **18,** 45 (1964)
10. Brown, H. C., *Chem. Soc.* (*London*), *Spec. Publ.*, **16,** 140 (1962)
11. Brown, H. C., and Chlovpek, F. J., *J. Am. Chem. Soc.*, **85,** 2322 (1963)
12. Brown, H. C., Chlovpek, F. J., and Rei, M.-H., *J. Am. Chem. Soc.*, **86,** 1246 (1964)
13. Brown, H. C., Chlovpek, F. J., and Rei, M.-H., *ibid.*, 1247
14. Brown, H. C., Chlovpek, F. J., and Rei, M.-H., *ibid.*, 1248

15. Brown, H. C., and Bell, H. M., *J. Am. Chem. Soc.*, **86,** 5003 (1964)
16. Brown, H. C., and Rei, M.-H., *J. Am. Chem. Soc.*, **86,** 5004 (1964)
17. Brown, H. C., and Bell, H. M., *J. Am. Chem. Soc.*, **86,** 5006 (1964)
18. Bell, H. M., and Brown, H. C., *ibid.*, 5007
19. Brown, H. C., and Rei, M.-H., *J. Am. Chem. Soc.*, **86,** 5008 (1964)
20. Schleyer, P. von R., Kleinfelder, D. C., and Richey, H. G., *J. Am. Chem. Soc.*, **85,** 479 (1963)
21. Cram, D. J., *J. Am. Chem. Soc.*, **86,** 3767 (1964)
22. Winstein, S., and Baker, R., *J. Am. Chem. Soc.*, **86,** 2071 (1964)
23. Collins, C. J., Cheema, Z. K., Werth, R. G., and Benjamin, B. M., *J. Am. Chem. Soc.*, **86,** 4913 (1964)
24. Leal, G., and Pettit, R., *J. Am. Chem. Soc.*, **81,** 3160 (1959)
25. Story, P. R., Snyder, L. C., Douglas, D. C., Anderson, E. W., and Kornegay, R. L., *J. Am. Chem. Soc.*, **85,** 3630 (1963)
26. Schleyer, P. von R., Watts, W. E., Fort, R. C., Jr., Comisarow, M. B., and Olah, G. A., *J. Am. Chem. Soc.*, **86,** 5679 (1964); Saunders, M., Schleyer, P. von R., and Olah, G. A., *J. Am. Chem. Soc.*, **86,** 5680 (1964)
27. Brown, H. C., Fletcher, R. S., and Johannesen, R. B., *J. Am. Chem. Soc.*, **74,** 1894 (1952)
28a. Foote, C. S., *J. Am. Chem. Soc.*, **86,** 1853 (1964)
28b. Schleyer, P. von R., *J. Am. Chem. Soc*, **86,** 1854 (1964); South Eastern Regional Meeting of A.C.S., Charleston, W. Va., Oct. 1964
29a. Nevell, T. P., DeSalas, E., and Wilson, C. L., *J. Chem. Soc.*, 1188 (1939)
29b. Wilson, C. L. (Personal communication)
30. Brown, H. C., and Brady, J. D., *J. Am. Chem. Soc.*, **74,** 3570 (1952)
31. Winstein, S., and Ordronneau, C., *J. Am. Chem. Soc.*, **82,** 2084 (1960)
32. Roberts, J. D., Lee, C. C., and Saunders, W. H., *J. Am. Chem. Soc.*, **76,** 4501 (1954)
33. Skell, P. S., and Starer, I., *J. Am. Chem. Soc.*, **84,** 3962 (1962)
34. LaLonde, R. T., and Forney, L. S., *J. Am. Chem. Soc.*, **85,** 3767, 3771 (1963)
35. Silver, M. S., *J. Am. Chem. Soc.*, **82,** 2971 (1960)
36. Skell, P. S., and Starer, I., *J. Am. Chem. Soc.*, **82,** 2971 (1960)
37. Nickon, A., and Hammons, J. H., *J. Am. Chem. Soc.*, **86,** 3322 (1964)
38. Reutov, O. A., and Shatkina, T. N., *Tetrahedron*, **18,** 237 (1962); *Izv. Akad. Nauk SSSR, Otd. Khim. Nauk*, **1963,** 195
39. Baird, R. L., and Aboderin, A. A., *J. Am. Chem. Soc.*, **86,** 252 (1964)
40. Baird, R. L., and Aboderin, A. A., *ibid.*, 2300
41. LaLonde, R. T., and Tobias, M. A., *J. Am. Chem. Soc.*, **86,** 4068 (1964)
42. Dewar, M. J. S., *Tetrahedron*, **19,** Suppl. 2, 89 (1963)
43. Breslow, R., Lockart, J., and Chang, H. W., *J. Am. Chem. Soc.*, **83,** 2375 (1961)
44. Bonner, W. A., and Collins, C. J., *J. Am. Chem. Soc.*, **78,** 5587 (1956)
45. Shoppee, C. W., and Ingold, C. K., *J. Chem. Soc.*, 1147 (1946); see Cohen, L. A., and Witkop, B., in Ref. 8, p. 1075
46. Roberts, J. D., Lee, C. C., and Saunders, W. H., *J. Am. Chem. Soc.*, **77,** 3034 (1955)
47. Giddings, W. P., and Dirlam, J., *J. Am. Chem. Soc.*, **85,** 3900 (1963)
48. Corey, E. J., Casanova, J., Vatakencherry, P. A., and Winter, R., *J. Am. Chem. Soc.*, **85,** 169 (1963)
49. Berson, J. A., and Remanick, A., *J. Am. Chem. Soc.*, **86,** 1749 (1964)
50. Servic, K. L., and Roberts, J. D., *J. Am. Chem. Soc.*, **86,** 3773 (1964)
51. Mazur, R. H., White, W. N., Semenov, D. A., Lee, C. C., Silver, M. S., and Roberts, J. D., *J. Am. Chem. Soc.*, **81,** 4390 (1959)
52. Landgrebe, J. A., and Applequist, D. E., *J. Am. Chem. Soc.*, **86,** 1536 (1964)
53. Applequist, D. E., and Landgrebe, J. A., *ibid.*, 1543
54. Hart, H., and Law, P. A., *J. Am. Chem. Soc.*, **86,** 1957 (1964)
55. Winstein, S., and Sonnenberg, J., *J. Am. Chem. Soc.*, **81,** 6524 (1959); **83,** 3235, 3244 (1961); Piccolini, R. J., and Winstein, S., *Tetrahedron*, **19,** Suppl 2, 423 (1963)
56. Winstein, S., Bruck, P., Radlick, P., and Baker, R., *J Am. Chem. Soc.*, **86,** 1867 (1964)
57a. Corey, E. J., and Dawson, R. L., *J. Am. Chem. Soc.*, **85,** 1782 (1963)
57b. Corey, E. J., and Uda, H., *J. Am. Chem. Soc.*, **85,** 1788 (1963)
58. Erman, W. F., *J. Am. Chem. Soc.*, **86,** 2887 (1964)
59. Cram, D. J., and Goldstein, M., *J. Am. Chem. Soc.*, **85,** 1063 (1963)

60. Tanida, H., Tsuji, T., and Ishitobi, H., *J. Am. Chem. Soc.*, **86**, 4904 (1964)
61. Dewar, M. J. S., *Ann. Rept. Progr. Chem.* (*Chem. Soc. London*), **48**, 118 (1951)
62. Dewar, M. J. S., and Fahey, R. C., *J. Am. Chem. Soc.*, **85**, 2245, 3645 (1963)
63. Dewar, M. J. S., and Fahey, R. C., *ibid.*, 2248
64. Fahey, R. C., and Smith, R. A., *J. Am. Chem. Soc.*, **86**, 5035 (1964)
65. Hammond, G. S., and Nevitt, T. D., *J. Am. Chem. Soc.*, **76**, 4121 (1954)
66. Hammond, G. S., and Collins, C. H., *J. Am. Chem. Soc.*, **82**, 4323 (1960)
67. Johnson, W. S., Bailey, D. M., Owyang, R., Bell, R. A., Jaques, B., and Crandall, J. K., *J. Am. Chem. Soc.*, **86**, 1959 (1964)
68. Johnson, W. S., Gray, S. L., Crandall, J. K., and Bailey, D. M., *J. Am. Chem. Soc.*, **86**, 1966 (1964)
69. Johnson, W. S., Lunn, W. H., and Fitzi, K., *J. Am. Chem. Soc.*, **86**, 1972, 2085 (1964)
70. Neale, R. S., and Whipple, E. B., *J. Am. Chem. Soc.*, **86**, 3130 (1964)
71. Prilezhaeva, E. N., and Shostakovskii, M. F., *Russ. Chem. Rev.* (*English transl.*), **32**, 399 (1963)
72. Hassner, A., and Heathcock, C. C., *Tetrahedron*, **20**, 1037 (1964) *Tetrahedron Letters*, 1125 (1964)
73. Olah, G. A., in *Friedel-Crafts and Related Reactions* (Olah, G. A., Ed., Interscience, New York, 1963)
74. Olah, G. A., Kuhn, S. J., Flood, S. H., and Hardie, B. A., *J. Am. Chem. Soc.*, **86**, 1039 (1964); Olah, G. A., Kuhn, S. J., Flood, S. H., and Hardie, B. A., *ibid.*, 1044; Olah, G. A., Flood, S. H., Kuhn, S. J., Moffett, M. E., and Overchuck, N. A., *J. Am. Chem. Soc.*, **86**, 1046 (1964); Olah, G. A., Kuhn, S. J., and Hardie, B. A., *J. Am. Chem. Soc.*, **86**, 1055 (1964); Olah, G. A., Flood, S. H., and Moffett, M. E., *J. Am. Chem. Soc.*, **86**, 1060, 1065 (1964)
75. Dewar, M. J. S., and Narayanaswami, K., *J. Am. Chem. Soc.*, **86**, 2422 (1964)
76. Dewar, M. J. S., in Ref. 8, p. 295
77. White, W. N., Lazdins, D., and White, H. S., *J. Am. Chem. Soc.*, **86**, 1517 (1964)
78. Banthorpe, D. V., Hughes, E. D., Ingold, C. K., Bramley, R., and Thomas, J. A., *J. Chem. Soc.*, 2864 (1964)
79. Hammond, G. S., and Clovis, J. S., *J. Org. Chem.*, **28**, 3283 (1963); Clovis, J. S., and Hammond, G. S., *ibid.*, 3290
80. Brownstein, S., Bunton, C. A., and Hughes, E. D., *Chem. Ind.* (*London*), 981 (1956)
81. Hammond, G. S., and Neuman, R. C., *J. Am. Chem. Soc.*, **85**, 1501 (1963)
82. Cox, J. R., and Dunn, M. F., *Tetrahedron Letters*, 985 (1963)
83. Miller, B., *J. Am. Chem. Soc.*, **86**, 1127, 1135 (1964)
84. Dewar, M. J. S., *Discussions Faraday Soc.*, **2**, 50 (1947)
85. Coulson, C. A., and Dewar, M. J. S., *Discussions Faraday Soc.*, **2**, 54 (1947)
86. Simonetta, M., and Winstein, S., *J. Am. Chem. Soc.*, **76**, 18 (1954); Piccolini, R. J., and Winstein, S., *Tetrahedron*, **19**, Suppl. 2, 423 (1963)
87. Hoffmann, R., *J. Am. Chem. Soc.*, **86**, 1259 (1964); *J. Chem. Phys.*, **40**, 2480 (1964)
88. Dewar, M. J. S., *Rev. Mod. Phys.*, **35**, 586 (1963)

RADIATION CHEMISTRY[1,2]

By Harold A. Schwarz
Chemistry Department, Brookhaven National Laboratory, Upton, Long Island, New York 11973

Introduction

Several books and symposia in and around radiation chemistry have appeared recently. Spinks & Woods (1) have published a textbook on radiation chemistry and Vereshinsky & Pikaev (2) have published a Russian text. Volume 6 (3) (radiation chemistry and radiation biology) and volume 7 (4) (radiation effects in metals and solids) of the series edited by Haissinsky have appeared. The Faraday Society Discussion on radiation chemistry (5), the proceedings of the conference on aqueous solutions, sponsored by the National Academy of Sciences and the National Research Council (6), and the proceedings of the 1962 Tihany symposium on radiation chemistry held in Budapest (7) have been published. There has been an English translation of a Russian book on hydrocarbon radiolysis (8), and the papers presented at a symposium on excitons sponsored by the Radiation Research Society have been published in one volume (9). Haissinsky & Magat have revised their compilation of radiolysis yields (10).

The literature of 1963 and 1964 is grouped in this review mainly by the types of processes involved (ionic, free radical, etc.). There has been considerable progress in understanding the characteristics of such processes, and diverse pieces of information can be more easily correlated by this division. However, the radiation chemistry of aqueous solutions is treated separately, as it is commonly recognized that aqueous radiation chemists live in a different (and better) world.

The research in radiation chemistry has expanded a great deal recently and two years have elapsed since it was last reviewed in this series. Consequently less than a quarter of the literature can be covered in this space. The author regrets that certain large areas such as the radiation chemistry of polymers and of gases have been left largely untouched.

Aqueous Solutions

Hydrogen and oxygen form a surprisingly large number of different compounds and ions. Three new ones have been recently reported as radiolysis products. Czapski & Bielski (11) report the formation of H_2O_3. They irradiated a flowing stream of oxygen-saturated water with an electron beam and mixed the irradiated solution with a reducing solution. A new intermediate was found which oxidized the ferrous ion and was suggested to be H_2O_3 on

[1] The survey of literature pertaining to this review was concluded in November 1964.

[2] Research performed under the auspices of the U. S. Atomic Energy Commission.

the basis of the observed yields. The H_2O_3 was found to decompose spontaneously in a pseudo-first order reaction catalyzed by acid and base. The maximum lifetime at room temperature was 2 sec at pH 2. The rate constant for ionization of H_2O_3 was found to be 9 sec^{-1}. The resulting HO_3^- ion was not stable, but assuming a rate constant of 10^{10} M^{-1} sec^{-1} for the reverse reaction, the pK of H_2O_3 is about 9. Benson (12) re-examined the thermochemical estimates for poly oxides and concluded that the properties of H_2O_3 noted by Czapski & Bielski are consistent with the present state of knowledge. Czapski & Dorfman (13) searched for the absorption spectrum of H_2O_3 by using the pulsed radiolysis technique, in which the light absorption of a solution is followed as a function of time after irradiation by a short (10^{-6} sec) intense (10^{10} rads/sec) pulse of high energy electrons from a linear accelerator. They did not find any spectrum which could be identified with H_2O_3, but did find two new absorptions in alkaline air-saturated water. Both absorptions exhibited pseudo-first order decay, the lifetime increasing with increasing base concentration. One absorption, at 254 mμ, had a lifetime of the order of seconds and seemed to be produced from the hydroperoxy radical ion O_2^-. Czapski & Dorfman suggested that the absorption may be due to either O_2^{-2} or O_4^{-2}, either of which could be formed by the bimolecular reaction of two O_2^- radicals. The second absorption had a maximum around 430 mμ and a lifetime of the order of milliseconds. The hydroxyl radical, or O^-, was found to be the precursor of the second species. The authors suggest that the absorbing species is O_3^- formed by the reaction of O^- with oxygen.

The absorption spectra of the HO_2 and O_2^- radicals have been observed by Baxendale (14) in flash photolysis experiments and by Czapski & Dorfman (13) in pulsed radiolysis experiments. The latter give the extinction coefficient of HO_2 at the maximum (230 mμ) as 1150 M^{-1} cm^{-1} and of O_2^- as 1060 at 240 mμ. A study of the optical density a few microseconds after the irradiation pulse as a function of pH gave the pK of HO_2 as 4.5 ± 0.2. Czapski & Bielski (11) have measured the pK to be 4.4 ± 0.4, and another study by Rabani, Mulac & Matheson (15) gives 4.45 ± 0.25.

The pK values of HO_2 (4.5 ± 0.2) and H_2O_3 (~ 9) can be combined with the known pK values for water (15.7) and hydrogen peroxide (11.7) and a recent measurement for the pK value of the hydroxyl radical (11.9 ± 0.2) by Rabani & Matheson (16) to yield the equation

$$pK = 19.0 - 7.1 \; O/H$$

where O/H is the ratio of oxygen atoms to hydrogen atoms in the compound. This dependence of pK on O/H is reminiscent of similar observations in inorganic acids (17). The relationship breaks down abruptly at $O/H = 0$, as the pK of the hydrogen atom is less than 12 (18).

Transient spectra and rate constants.—Many transient spectra have been produced by the reaction of the solvated electron with various solutes. Of particular interest is the report of Baxendale, Fielden & Keene (19) on optical absorptions near 300 mμ following reaction of the hydrated electron with

Cd^{+2}, Co^{+2}, Ni^{+2}, Zn^{+2}, Pb^{+2}, and Mn^{+2}. The absorbing species are believed to be the univalent ions.

Keene (20) has given a new determination of the absorption spectrum of the solvated electron. He measured the optical density as a function of time following the radiation pulse and extrapolated to zero time at each wavelength. Previous determinations, using the less accurate film density techniques, had indicated two peaks, the major one at 700 mμ and another at 520 mμ. Keene found a single absorption peak at 715 mμ with a long tail in the visible. The extinction coefficient at 300 mμ is still 7 per cent of the extinction coefficient at 715 mμ.

Absorption spectra of intermediates have been very useful for measuring absolute rate constants of reactions of the intermediates. The absorption spectrum of the electron has been a magnificent example of this application. So far, in the few years since discovery of the spectrum, approximately 200 rate constants for reactions of the electron with various solutes have been reported. It is obviously beyond the scope of this review to report on these measurements, and fortunately a thorough review of pulse radiolysis studies has been written by Dorfman & Matheson (21). Bimolecular combination reactions of radicals are of special interest in radiation chemistry as they are necessary for a meaningful test of the diffusion model, so a summary will be given here.

The predominant radicals in water radiolysis (in the absence of oxygen and below pH 10) are the hydrated electron, the hydroxyl radical, and the hydrogen atom. There are six possible combinations of these radicals, and rate constants have been reported for all of them (Table I). Reaction 1 is of interest because the ultimate product is hydrogen. Dorfman & Taub (23) demonstrated that the electron decayed by a second order process in dilute ethanol solutions. Chemical analysis of a similarly pulse-irradiated solution of C_2D_5OH indicated the product of the reaction was H_2, not HD as would be found if reaction 1 led to two hydrogen atoms. There is evidence that reaction 1, occurring in the spurs and tracks, is the principal source of the hydrogen yield in aqueous solution radiolysis (31). Reaction 2 is believed to produce hydrogen although corroboration is lacking.

The agreement among various workers on the perhydroxyl radical rate constants is excellent with the exception of a value of k_9 obtained from the flash photolysis of water. The rate constant for reaction 8 was computed from the data of Czapski & Bielski (11) who studied reactions 7, 8, and 9 from pH 2 to pH 7. They could not obtain reliable values for both k_8 and pK_{HO_2}, but the more recent and precise pK determinations (4.5 ± 0.2) (13, 15) allow k_8 to be estimated with a probable error of about 20 per cent.

Rate studies of perhydroxyl radical reactions with solutes, long neglected, have been appearing in the literature. Keene (33) has found the rate constant for oxidation of ferrous ion by HO_2 to be 7.3×10^5 M^{-1} sec^{-1}. Baxendale (14) has used the flash radiolysis technique to follow the reaction of HO_2 with copper ion and found a rate constant of 1.5×10^7 M^{-1} sec^{-1}.

TABLE I

Absolute Rate Constants of Radical Combination Reactions

Reaction	Rate constant, M^{-1} sec^{-1}[a]	Reference
1. $e_{aq}+e_{aq}\rightarrow(?)\rightarrow H_2$	0.45×10^{10}	(22)
	$<0.65\times10^{10}$	(23)
	0.55×10^{10}	(24)
2. $e_{aq}+H\rightarrow(?)\rightarrow H_2$	2.5×10^{10}	(24)
3. $e_{aq}+OH\rightarrow OH^-$	3.0×10^{10}	(24)
4. $H+H\rightarrow H_2$	1.5×10^{10}	(25)
	1.0×10^{10}	(26)
	1.4×10^{10}	(32)
5. $H+OH\rightarrow H_2O$	3.2×10^{10}	(27)
	4.5×10^{10}	(32)
6. $OH+OH\rightarrow H_2O_2$	0.4×10^{10}	(28)
	0.6×10^{10}	(27)
	0.7×10^{10}	(32)
7. $HO_2+HO_2\rightarrow H_2O_2+O_2$	2.4×10^{6}	(29)
	2.2×10^{6}	(11)
	2.7×10^{6}	(13)
	2.8×10^{6}[b]	(14)
8. $HO_2+O_2^-\xrightarrow{H_2O}H_2O_2+O_2$	4.4×10^{7}	[c]
9. $O_2^-+O_2^-\xrightarrow{H_2O}H_2O_2+O_2$	1.45×10^{7}	(30)
	1.5×10^{7}	(11)
	1.7×10^{7}	(13)
	1.2×10^{8}	(14)
10. $OH+HO_2\rightarrow H_2O_3$ (70%)[d]; $\rightarrow H_2O+O_2$ (30%)[d]	1.5×10^{10}	(27)
	1.6×10^{10}	(31)

[a] Rate constants for recombination of like species defined by $dc/dt=-2kc^2$.
[b] Corrected to $\epsilon_{HO_2}=1150$ at 230 mμ (12).
[c] Calculated from results of Ref. 10.
[d] Disproportionation factor from Ref. 10.

Scholes & Simic (34) have demonstrated the utility of nitrous oxide solutions for determining relative rate constants of the hydrated electron. The reaction with N_2O is

$$e_{aq} + N_2O \longrightarrow (N_2O)^- \xrightarrow{H_2O} N_2 + OH \qquad 11.$$

and relative rate constants of various solutes to N_2O are determined by the effect of the solutes in depressing the nitrogen yield. Absolute rate constants are available for 11 of the solutes studied by Scholes & Simic, and the agreement with the relative constants is of the order of ±30 per cent. Dainton (35) has pointed out a danger connected with the use of N_2O as a standard, however. The N_2O^- ion can sometimes transfer its charge to the second solute giving large apparent rate constant ratios.

Adams & Boag (36) have reported a rather unusual technique for determining relative rate constants of hydroxyl radicals. Hydroxyl radicals react with carbonate to form a species with an absorption at about 600 mμ which disappears in a second order reaction with a lifetime generally longer than 10^{-5} sec. Solutes which react with the hydroxyl radical lower the optical density observed immediately after the pulse. By following the initial optical density as a function of solute concentration, they determine rate constants relative to carbonate to be 1.6 for bromide, 3.6 for ethanol, 5.3 for glycerol, 15 for sulfite, 16 for thiocyanate, 26 for bisulfite, and 32 for nitrite. Ferradini & Koulkes-Pujo (37), in a study of acidic bromide-hydrogen solutions, have determined that the reaction of hydroxyl radicals with bromide ion is 800 times faster than the reaction with hydrogen. The reaction rate constant for the latter reaction is 4.5×10^7 M^{-1} sec^{-1}, and combining the ratios above would lead to the conclusion that the reaction rate constant for oxidation of nitrite is 7×10^{11} M^{-1} sec^{-1} which is clearly impossible. Either there is an acid effect on bromide oxidation [as is the case with chloride (38)] or the data of Adams & Boag and Ferradini & Koulkes-Pujo are incompatible.

Radical yields and the identification of the hydrogen atom.—Measurements of radical yields in water radiolysis, well-established five years ago, have progressed to the point of confusion. Allen discussed this topic recently (39). He gives the hydrated electron yield in neutral water as 2.8 per 100 eV, the hydrogen atom yield as 0.6, and molecular hydrogen as 0.45, or total reduction of 4.3 molecules per 100 eV. The comparable figure in 0.4 M H_2SO_4 is 4.5 indicating a rather small effect of acid on the overall decomposition. The hydroxyl radical yield in neutral water has been determined to be 2.3 and the hydrogen peroxide yield is 0.7, or total oxidation of only 3.7 molecules per 100 eV (increasing to 4.5 in acid solution). Assigning estimated errors to the yields, the material balance discrepancy in neutral solution is 0.6 ± 0.2, so that it does seem to be real.

The correspondence between the material balance discrepancy and the hydrogen atom yield has prompted Hayon (40) and Dainton & Watt (41) to suggest that the so-called hydrogen atom in neutral solutions is really some form of excited water molecule which reacts similarly to a hydrogen atom in some solutions and is quenched (41) or reacts as an electron (40) in others. The excited state proposals would seem to depend upon a reasonable experimental identification of the species. The only technique available for this at the moment is a comparison of relative rate constants determined for known hydrogen atoms and for the questionable ones. Czapski, Rabani & Stein (42) concluded the neutral solution species were hydrogen atoms on this basis as did Scholes & Simic (43). Anbar & Myerstein have questioned this conclusion (44). There are seven reasonably well-established cases in which relative rate constants have been measured for both neutral solution hydrogen atoms and known hydrogen atoms. The comparison is given in Table II. In most cases the ratios were not determined directly for both sources of hydrogen atoms, but the ratio given could be found simply as the product of two other ratios. The sources of known hydrogen atoms are: (*a*) the reaction of e_{aq} with H_3O^+ (25, 26, 42, 46, 50); (*b*) hydrogen atoms from microwave discharge in gaseous hydrogen (bubbled into solution) (49); and (*c*) hydrogen atoms from reaction 12 (48).

$$OH + H_2 \rightarrow H_2O + H \qquad 12.$$

The overall comparison gives the ratio between relative rates of H to "H" as 1.1 ±0.3.

TABLE II

Comparison of Rate Constant Ratios

Ratio	Neutral solution "H"	Reference	Known H atoms	Reference
$HCOO^-/Fe(CN)_6^{-3}$	3.7×10^{-2}	(43)	5.9×10^{-2}	(42)
$CH_3OH/Fe(CN)_6^{-3}$	4.1×10^{-4}	(43)	4.2×10^{-4}	(50)
$C_2H_5OH/Fe(CN)_6^{-3}$	3.7×10^{-3}	(43)	3.8×10^{-3}	(42)
$CH_3CHOHCH_3/Fe(CN)_6^{-3}$	1.2×10^{-2}	(43)	1.3×10^{-2}	(50)
CH_2ClCOO^-/OH^-	0.14	(43, 45)	0.22	(49)
H_2O_2/O_2	6.3×10^{-3}	(46)	$5\ \times10^{-3}$	(26, 25)
			2.2×10^{-3}	(48)
C_2H_5OH/O_2	2.6×10^{-3}	(46)	2.3×10^{-3}	(46)
	2.9×10^{-3}	(47)		

It might be argued that H and "H" are different species whose kinetics follow a free energy relationship, such as

$$\log k = A - \alpha\Delta F$$

and relative rate constants would be expected to be equal. Such a relationship would be expected among the alcohols, for instance, but their reactivities are compared to those of ferricyanide or oxygen. The latter two solutes are known to react with hydrogen atoms with diffusion control rates and could not follow a free-energy relationship. Furthermore, different reaction types are included in Table II (hydrogen abstraction, chlorine abstraction, e_{aq} formation, bond formation, and electron transfer) which would not be expected to follow a single free-energy relationship. The species formed in neutral solution must be a hydrogen atom.

Scholes (47) and Seddon (46) observe hydrogen atom kinetics in neutral solution in the presence of 2.5×10^{-4} *M* oxygen, which is incompatible with Hayon's proposal in which less than 10^{-4} *M* oxygen is sufficient to convert the species to a hydrated electron. Dainton & Watt (41) have suggested that the species might be a radical pair (H · · · OH) which would react with rate constants characteristic of hydrogen atoms but would generally lead to no net products, as the resulting radicals would combine to give water and the reactant. Reaction with carbon-hydrogen bonds would lead to hydrogen, however, which is the method used to measure hydrogen atoms. Such radical pairs have a very short lifetime, and it can be calculated from diffusion treatments of the problem that approximately one molar oxygen, for example, would be required to react with even half of the radical pairs. Seddon has found essentially complete reaction in 2.5×10^{-4} M oxygen.

Allen (39) accepted all the yield determinations at face value, noticed the pleasing agreement between reduction yields for water decomposition in acid solution and neutral solution (acid should have little effect other than to convert the hydrated electrons to hydrogen atoms), and proposed that the lack of material balance could be accounted for by assuming that some hitherto unnoticed oxidizing species is produced in water radiolysis. An oxygen atom yield of 0.3 per 100 eV, for instance, would be sufficient. Allen's hypothesis does offer explanations for several annoying, unexplained observations in the literature, but there is no solid evidence for another species.

Product yields from irradiated basic solutions still lead to conflicting mechanisms and radical yields. Draganic, Draganic & Kosanic (51) have made a thorough study of aerated oxalic acid solutions and conclude that the hydroxyl radical yield is only 2.0 per 100 eV in basic solution in agreement with Cheek & Linnenbom (52) and Cheek & Swinnerton (53). On the other hand, Dainton & Watt (41) and Jezowska-Trzebiatowska & Kalecinski (54) conclude that the hydroxyl radical yield is about 3 in strongly basic solutions. Haissinsky, Jove & Szymanski (55) find a small effect of pH on the hydrogen peroxide yield in contrast to other reports (52, 41).

The diffusion model.—Some new information has appeared on the effect of the LET (linear energy transfer or energy loss per unit distance along the track) of the ionizing radiation on yields. Carmo Anta, Mariano & Lima e Santos (56) have studied the hydrogen peroxide yield in 0.4 M H_2SO_4 irradiated with 5.3 MeV α particles (from Po^{210}). Hydrogen peroxide is believed to be produced by combination of hydroxyl radicals as they diffuse out of the track. The authors find the peroxide yield to be 1.34 molecules per 100 eV at low solute concentration, decreasing with increasing concentrations of chloride, bromide, and thallous ion (solutes which react with hydroxyl radicals). The shape of the yield versus concentration curve agrees well with the computed diffusion kinetic curve of Kuppermann & Belford; better, in fact, than was observed with low LET radiation. The relative effectiveness of chloride, bromide, and thallous ions are in the ratio of 1:1:16. The corresponding ratios for Co^{60} γ radiation are 1:6:14 (57), indicating a discrepancy for bromide ion.

Kochanny et al. (58) have measured the yields of ferrous sulfate oxidation and cerium sulfate reduction in the presence and absence of thallous ion for 10 to 22 MeV protons, all in 0.4 M H_2SO_4. This is a LET range somewhat below the capabilities of the other cyclotrons which have been used for radiation chemistry in the past. From these three yields, G_H, G_{OH}, G_{H_2}, and $G_{H_2O_2}$ can be determined (e_{aq} is all converted to hydrogen atoms at this acid concentration). They find that G_H and G_{OH} decrease with decreasing energy of the protons, but G_{H_2} and $G_{H_2O_2}$ remain relatively constant. The latter two yields are determined as the difference between large numbers, however, and perhaps the expected variation is obscured by error. Direct hydrogen yield measurements for 20 MeV deuterons have indicated a larger hydrogen yield than observed for γ rays and fast electrons.

The diffusion model faces at least two serious problems. One is that the evidence for the hydroxyl radical being the precursor of hydrogen peroxide (as is commonly assumed) is on shaky ground. At present, the only basis for such evidence would be a comparison of the relative effectiveness of solutes in lowering the peroxide yield with relative rate constants for hydroxyl radical reaction with the solutes. A few comparisons are possible in acid solution. The relative effectiveness of thallous ion and cerous ion agrees well with the rate constant ratio for hydroxyl radical oxidation of the two ions (59). Anbar & Thomas find the reaction of the hydroxyl radical with chloride ion to be catalyzed by hydrogen ion (38), and the effectiveness of chloride ion in lowering the peroxide yield increases with acid concentration (60). The situation is not as clear in neutral solution. Bromide ion is five times as effective as ethanol in lowering $G_{H_2O_2}$, while the hydroxyl radical rate constant ratio favors ethanol by a factor of 1.6 (61). Hughes & Willis (62) note that the effect of ferrocyanide in lowering $G_{H_2O_2}$ is the same in the presence and absence of 1 *M* methanol. In the last case, scavenging of the hydroxyl radical by methanol would be expected to predominate and ferrocyanide should have little effect. Clearly, more work is required to settle the question of the origin of the molecular peroxide.

The second problem confronting the diffusion model is the origin of the hydrogen atom yield. If electrons and hydroxyl radicals are present in the spurs then there must also be H_3O^+ present. Some hydrated electrons should react with H_3O^+ to form hydrogen atoms while diffusing out of the spurs. Kuppermann (63), using known rate constants, has shown that the yield should be about 1 for this reaction which agrees reasonably well with the observed yield of 0.6. This leaves no room for other modes of production of hydrogen atoms, however. Furthermore, the diffusion model would require that G_H depend on solute concentration. The extent of the dependence has not yet been predicted, but a simple model would suggest that the percentage effect of an electron scavenger on G_H should be about that of the same scavenger at half the concentration on G_{H_2}. Hayon (64) has looked for this effect and found none. His experiment was to measure $G(H_2)$ in solutions of 1 *M* isopropanol with 0.5 *M* KNO_3 or 1 *M* acetone present to react with the electron. The experiment is not conclusive, however, as the observed hydrogen yield is the sum of G_H and G_{H_2}, and knowledge of the variation of each quantity with alcohol as well as nitrate and acetone concentrations must be known. Hayon's study does suggest that the hydrogen atoms are not all formed by the diffusion process. This point is important and deserves further study.

Other studies.—Seddon & Sutton have introduced a brand new system to radiation chemistry: nitric oxide solutions (65). They find that the principal products are nitrite ($G=6.1$), nitrous oxide ($G=3.2$), hydrogen peroxide ($G=0.55$), and hydrogen ($G=0.38$). Apparently the electrons and hydrogen atoms react with NO to give NOH which reacts further with NO to give N_2O and OH. The hydroxyl radicals react with NO to give nitrite ion. The mechanism gives values of the radical yields of 3.15 for the sum of G_H and G_e and 2.9 for G_{OH}. The hydroxyl radical yield is considerably larger than previously reported values in neutral solution (discussed above) and does give material balance. The system is not specific for OH, however, and G_{OH} is perhaps better thought of as the sum of all oxidizing radicals.

Katakis & Allen (66) have studied the radiolysis of perchloric acid solutions at high concentrations. They find the acid to be inert to hydrogen atoms and hydroxyl radicals in agreement with other work (48); hence perchloric acid can be used to vary hydrogen ion concentration without introducing reactive anions. They also find a direct effect on perchlorate ion yielding chlorate and an unusual effect on the hydrogen yield which decreases linearly with perchloric acid concentration. The last observation could mean that ionization of ClO_4^- is rare compared to excitation, or that a termolecular reaction occurs in the spurs:

$$2H + ClO_4^- \rightarrow H_2O + ClO_3^-$$

Getoff (67) has summarized his research on the radiolysis of carbon dioxide solutions. Carbon dioxide is reduced by the hydrated electron forming the COOH radical (or its negative ion). Formaldehyde and formic acid are observed as products. The yields generally decrease with increasing dose and are strongly pH dependent indicating that secondary reactions are occurring.

Water vapor and ice.—Radical yields in water vapor have received new attention. Baxendale & Gilbert (68), confirming Firestone's result, find that G(HD) is 10.5 in mixtures of D_2O vapor and hydrogen. Other solutes in water vapor (methanol, ethanol, cyclohexane, ether, propane) gave hydrogen yields of about 8. They conclude the radical yield is 8 per 100 eV and that some other entity is responsible for the increased yield when hydrogen is present. They also find a molecular yield of hydrogen produced from water vapor of 0.48 molecules per 100 eV. Anderson, Knight & Winter (69) have also found a molecular yield of 0.47 in water-perdeutero benzene mixtures. The authors suggest as possible reactions, formation of hydride ion followed by reaction to give hydrogen [suggested by Platzman (70)], or direct production of hydrogen as is believed to occur in water photolysis at 1236 Å. The agreement between the gas phase molecular yield and the yield observed in liquid water (0.45 molecules per 100 eV) is probably just coincidence.

ESR studies of ice radiolysis have indicated that hydrogen atoms were formed in strongly acid ice but not in neutral ice. Flournoy, Baum & Siegel (71) have demonstrated that hydrogen atoms are formed in neutral ice but disappear rapidly above 50° K. They have studied the rate of disappearance as a function of temperature between 20° K and 50° K and find the kinetics to be complex. The rate of disappearance is large at first and slow after several minutes, but is also first order in initial hydrogen atom concentration. The results suggest either spur reactions or reaction of the atoms after escaping from traps of varying depth. Kevan, Moorthy & Weiss (72) have found hydrogen atoms in nearly neutral ices at 77° K in the presence of certain anions, principally $H_2PO_4^-$ and HSO_4^-. Solutes reactive towards electrons (N_2O, acetone, etc.) decrease the hydrogen atom ESR signal indicating that electrons are precursors of the atoms. Schulte-Frohlinde & Eiben (73) and Henricksen (74) have published more observations on electrons and hydroxyl radicals in strongly alkaline ice. Henricksen points out that the number of ESR centers depends on the physical state of the solution (glass or crystalline) and finds both electrons and hydrogen atoms in alkaline ices. The electron yield is considerably larger than the hydrogen atom yield.

Electron Energy Loss Spectra

The primary process in radiation chemistry is energy loss by charged particles to the medium. Considerable progress has been made in determining energy loss spectra of electrons in various media. At the present time direct application of the results to radiation chemistry is limited to the grosser features, such as estimating the average energy loss per primary event, making qualitative estimates of cross sections for formation of ions and various types of excited states, and estimating the distance traveled by a low energy electron before reaching thermal energy.

The selection rules for energy loss of electrons approach those for optical absorption at electron energies above a few hundred electron volts. Below 100 eV, exchange scattering becomes important and direct excitation of forbidden transitions is possible. Below about 5 eV (or the lowest excited state of the medium), energy loss must be by vibrational and rotational excitation.

Lassettre et al. (75) have published a series of papers on electron spectroscopy in gases for electrons with initial energies around 400 eV. The energy loss spectra are shown to be identical with optical spectra (upon integration over scattering angle). The authors give energy loss spectra for helium, hydrogen, methane, ethane, propane, cyclohexane, ethylene, water, nitrogen, carbon monoxide, and oxygen. Curiously, the spectra from the saturated hydrocarbons are very similar to one another.

Kuppermann & Ruff (76) have measured the energy loss of electrons scattered over large angles for electrons with initial energies below 100 eV and give spectra for helium, argon, hydrogen, and ethylene. They observe the same states as in ultraviolet spectroscopy; however, a few new peaks appear. The triplet state of ethylene at 4.6 eV was observed directly. The cross section for the triplet excitation decreased with increasing electron energy, becoming just a shoulder on the elastic scattering at 75 eV. Kuppermann (77) points out that even though triplet state formation is a resonant process, the widths of the resonances in terms of the initial electron energy are of the order of 50 eV and the cross sections can be comparable to allowed transitions.

Rauth & Simpson (78) have measured the average energy loss per primary event for 20 keV electrons passing through thin films (a few hundred Å) of polymers and metals. They measure the energy loss spectra as a function of scattering angle using a rather coarse energy interval and integrate over angle to obtain total scattering. The average energy losses are about 60 eV in Formvar, polystyrene, and carbon and increase with the average atomic number of the film. The average energy loss figure includes all losses (up to half the energy of the initial particle) and is thus not quite comparable to the energy per spur figure of 40 eV derived by Magee (79) on the basis of the results of Lassettre et al. on nitrogen and oxygen. Magee points out that energy losses greater than a few hundred electron volts result in electrons which are not much different in their properties from the primary electrons and are perhaps best thought of as delta rays which produce one or more other spurs.

The rate of energy loss of low energy electrons is crucial in estimating the distance between the cation and electron when the electron reaches thermal energy or is captured (i.e., the electron kinetic energy is reduced below the coulombic potential energy). Energy loss to electronic excitation predominates for electrons with kinetic energies greater than the lowest excited state of the medium. For subexcitation electrons it would appear that vibrational excitation by a mechanism of virtual negative ion state formation is extremely important. Chen (80) has made a comparison of cross sections computed on the basis of this model with experimental observations and finds good agreement between the two for vibrational excitation of nitrogen. The cross sections are large (several square angstroms) for electrons with energies between 2 and 3 eV. This corresponds to energy loss at a rate of about 10^6 eV cm^{-1} in liquid nitrogen which is sufficient to trap a low energy electron in a few tens of angstroms if maintained down to thermal energy. The vibrational-excitation cross section becomes very small below 1.5 eV, however. It is possible that rotational excitation is important at lower energies.

Ions and Ionic Reactions

There has been little information until quite recently on the fate of ions produced in liquids. It was expected that for liquids with low dielectric constants the great majority of the electrons would be captured in the field of their cationic partner, and ion pair neutralization would predominate. A small yield of freely diffusing ions would be expected from delta ray formation (79). Recent conductivity studies and chemical yield studies in hydrocarbons can be combined to make the beginning of a consistent model for ion production: In at least 95 per cent of the ionizations the electron is thermalized while still in the field of the parent ion, forming ion pairs with lifetimes sufficiently long to undergo ionic reactions with solutes present at concentrations below 1 *M*. The lifetime of the ion pair suggests that shallow traps exist for the electrons.

Conductivity studies in hydrocarbons.—Allen & Hummel (81) and Freeman (82, 83) have determined the yields of freely diffusing ions in hydrocarbons. The steady-state conductivity of hydrocarbons subjected to X or γ radiation depends on the rate of production of ions, the ion mobilities, and the rate of disappearance of ions. Freeman (82) measured the radiation-induced conductivity of several hydrocarbons, calculated the rate constant for ion recombination from the Debye equation, and assumed the ionic diffusion to be the same as the parent molecule. He finds the ion yields for *n*-hexane and cyclohexane to be 0.21 and 0.25 ion pairs per 100 eV, respectively. Allen & Hummel (81) measured the conductivity, the rate constant for ion recombination (by stopping the radiation and following the decay of conductivity), and the separate mobilities of the positive and negative ions (by producing a layer of ions near one electrode of a conductivity cell with a pulse of X radiation and following the drift time of the ions as a function of the applied voltage). They found the yield of ions in *n*-hexane to be 0.09 ion pairs per 100 eV. The ratio of the neutralization rate constant to the ion mobilities was 1.0×10^{-6} *V* cm in comparison with the corresponding quantity of the Debye equation, 0.96×10^{-6}, which is $4\pi\ \epsilon/\xi$ where ϵ is the charge on the electron and ξ is the dielectric constant. The reaction radius is large, about 260 Å, so that the agreement between experiment and theory, while gratifying, is expected.

The positive ion and negative ion mobilities were found to be 0.69×10^{-3} and 1.0×10^{-3} cm^2 V^{-1} sec^{-1} respectively (81), in reasonable agreement with earlier studies. Gazda (84) has found values of 0.61×10^{-3} and 0.95×10^{-3} cm^2 V^{-1} sec^{-1} respectively in *n*-hexane. Drift times in mobility experiments are of the order of a second so that the negative ion is unlikely to be the electron. Capture by impurities is highly probable in so long a time.

Jachym (85) has measured the mobilities of radiation-produced ions in cyclohexane as a function of temperature and finds activation energies of 3.8 and 3.0 kcal per mole for the positive and negative ions. Gazda (84) has measured the diffusion coefficients of the ions by allowing them to diffuse for a known time in the absence of a field. He finds, in cyclohexane, that the diffusion coefficients of the positive and negative ions are 1.43×10^{-5} and 2.75×10^{-5} cm^2 sec^{-1} in agreement with the Nernst-Einstein relation between diffusion coefficients and mobilities. Gazda points out that the diffusion constants are smaller by about half than the self diffusion constant of *n*-hexane, which aggravates the discrepancy between the ion yields of Freeman and Allen & Hummel. The scatter of the various yield determinations is of the order of ± 50 per cent so that perhaps the disagreement is not too serious at this stage of the art.

The measured ion yields increase with increasing applied field strength as some of the ions initially present as ion pairs are collected. Freeman (83) has treated the effect of applied voltage in terms of an Onsager model for the effect of field strength and finds reasonable agreement with experiment by taking an initial ion separation of 100 Å. He does not take into account a distribution of separations, however, so that the estimate is only order of magnitude.

Samuel et al. (86) have observed large pulses of current from conductivity cells filled with hydrocarbon and subjected to γ rays (with applied field strengths of the order of thousands of volts per cm). The current per pulse was about 10^4 times larger than could be accounted for by any reasonable ion yield, indicating electron multiplication. This is very unsettling as it implies that the rate of energy loss of electrons in the neighborhood of the ionization potential of the liquid is less than the applied field strength, i.e., less than 10^3 eV per cm, and electrons should rarely be captured by their cations. It is possible that the effect is spurious. Hummel (87) has tried to duplicate the effect in *n*-hexane without success.

Chemical yield studies in hydrocarbons.—Ion yields deduced from mechanism studies are in agreement with the measurements of Allen & Hummel and Freeman. Busler, Martin & Williams (88) have studied the inhibition by ammonia of the polymerization of cyclopentadiene and find the yield of chain starters to be 0.2 per 100 eV within a factor of two. A kinetic and mechanistic analysis indicated that the chain starters are freely diffusing ions. In a study of the radiation-induced chlorination of toluene by $SnCl_4$, Galimba, Wilson & Samuel (89) found a yield of 0.1 to 0.2 molecules per 100 eV for chlorotoluenes. There was no effect of temperature and the authors concluded that the reaction was ionic. Kaufman (90) studied the dimer fraction produced in the radiolysis of the straight chain butenes. The product distribution from butene-1 was found to be analogous to that observed from hexene-1. In particular, 3-methyl heptene, probably a product of ionic reaction, was produced with a yield of 0.6 compared to the yield of 0.6 molecules per 100 eV for 5-methyl undecene from hexene-1 radiolysis. Analogous products were found in *cis* and *trans* butene-2. The dimer yields from olefin radiolysis are all larger than the free ion yields and probably represent only part of the reactions which could reasonably be attributed to ions. Some of the positive ions must react with substrate before being neutralized by their original electron partner.

Van Dusen (91) has observed a yield of HD of 0.1 to 0.2 molecules per 100 eV from solutions of D_2O in cyclohexane, and Busler, Martin & Williams

(88) have found a similar yield from ND_3 solutions. They believe the mechanism to be proton transfer from positive ions to the bases followed by electron capture yielding deuterium atoms which then abstract hydrogen. A more precise study of ND_3 solutions by Williams (92) indicates that G(HD) is 0.08 molecules per 100 eV at low ND_3 concentration increasing roughly with the square root of the ammonia concentration. The yield of 0.08 is attributed to free positive ions and the increase with concentration is attributed to reaction with other positive ions before neutralization by their electron partner. This observation is of particular importance as it allows the author to estimate a median lifetime for the ion pair. Scavenging of ions would be about 50 per cent complete at 1 M ND_3 (by extrapolation), assuming a total ion yield of 3 ions per 100 eV. Williams assumes a rate constant for proton transfer to the ammonia of about 10^9 M^{-1} sec^{-1} (although it could be as large as 10^{10} according to Debye). The median lifetime is $\{k(ND_3)_{1/2}\}^{-1}$ or 10^{-9} to 10^{-10} sec which is much longer than estimated by Samuel & Magee (93) for a free electron (about 10^{-12} sec for an initial 100 Å separation) and suggests that some sort of electron trapping with trap depths of the order of 0.1 eV is involved.

Burton, Dillon & Rein (94) have observed a post-irradiation luminescence in 3-methylpentane glass at 77° K with lifetimes of the order of 10^3 sec which shows an activation energy of about 0.1 eV. The effect of additives indicates that ion neutralization is involved, and the authors attribute the effect to emitting states of products of recombination of electrons with their parent positive ions. The observed lifetime is larger by a factor of 10^6 or so than expected on the basis of the Samuel-Magee limit and a 0.1 eV activation energy. The authors attribute this factor to an entropy effect; however, at 77° K electrons are still in their parent ion field when separated by thousands of angstroms. A factor of 10^4 or so in lifetime is understandable on this basis alone, as the time required for ion pair neutralization increases with the cube of the separation distance (93, 92).

Cundall & Griffiths (95) have attributed the γ ray-induced *cis-trans* isomerization of butene-2 in dodecane solvent to positive ion transfer from solvent to olefin. Piperidine (a good base) reduces the yield of isomerization, while ferric chloride (an electron scavenger) increases the yield. The isomerization yield increases approximately with the square root of the butene concentration (similar to the observation of Williams).

Scholes & Simic (96) have attempted to capture the electron in cyclohexane radiolysis by reaction with N_2O. The corresponding reaction in water is rapid, leading to nitrogen formation. Hydrogen atoms do not react readily with N_2O. The yield of nitrogen is 4.3 molecules per 100 eV in 0.2 M N_2O, indicating that electron trapping is much more efficient than positive ion reaction if the mechanism is correct. It can be estimated that electron trapping is approximately 50 per cent complete in 0.03 M N_2O solutions. Since the cations and electrons exist for the same length of time and Williams indicates that positive ion scavenging is approximately 50 per cent complete in 1 *M* ammonia, the ratio of rate constant for the electron reacting with nitrous oxide and the positive ion reacting with ammonia would be 33. A large rate constant ratio is not unexpected, as the mobility of the electron is likely to be high and the reaction radius between the electron and nitrous oxide could be large (trap depths of 0.1 eV cannot confine the electron to a radius of less than 10 Å or so).

Hamill et al. (97), continuing earlier post-irradiation spectroscopic studies of ions in glasses, have observed both dissociative electron attachment and electron capture by solutes. Triphenylmethyl chloride and benzyl chloride in tetrahydrofuran glasses exhibit the former, while napthalene and biphenyl in 3 methylpentane, 2 methylpentene-1 glasses exhibit the latter. They note that in methyl tetrahydrofuran glasses the yield of benzyl radicals increases and the yield of trapped electrons decreases with increasing benzyl chloride concentration. Roberts & Hamill (98) have studied HCl formation in cyclohexane solutions of various chlorine-containing fluorocarbons. They show that $G(HCl)$ increases with solute concentration and that the relative efficiencies of the various solutes correlate with the energy range over which the solute can capture electrons and the appearance potentials for chloride ion formation obtained from mass spectral studies. The correlation with energy range was somewhat better than with appearance potential. Guarino & Hamill (99) have observed spectra in hydrocarbon glasses containing olefins, aromatics, or carbon tetrachloride, which they attribute to positive ions. A small amount of CCl_4 enhanced the olefin ion spectra, probably by attaching the electrons.

Solvated electrons in polar solvents.—The yield of solvated electrons in alcohols, as determined by chemical and pulse radiolysis techniques, is greater than the free ion yields in hydrocarbons. Taub et al. (100) found the electron absorption maximum in ethanol at 700 mμ. The extinction coefficient was determined to be 1.1×10^4 M^{-1} cm^{-1} at 546 mμ by comparing the electron optical density to the optical densities of the biphenylide ion and triphenyl methyl radical produced by reaction of biphenyl and triphenylcarbinol with the electron. (The latter two species have known spectra and extinction coefficients.) The yield of the electron was 1.0 per 100 eV. Adams & Sedgwick (101) determined the yield to be 1.02 in a study of the effect of solutes on the hydrogen yield. Taub et al. (100) have also observed the electron in methanol, propanol-1, and propanol-2. Adams, Baxendale & Boag (102) and Dainton et al. (103) also report the absorption spectrum of the electron in methanol with peaks at 620 and 650 mμ respectively. The latter authors have also observed the electron in methanol glass (peak at 533 mμ) and report that the absorption shows first-order decay. Baxendale & Mellows (104) have previously estimated the electron yield in methanol to be 1.9 per 100 eV from a study of ferric chloride reduction.

Ahrens, Suryanarayana & Willard (105) have shown that the conductivity of ammonia is increased by γ rays. Dainton et al. (106) have estimated an electron yield in ammonia of 0.3 to 0.4 per 100 eV on the basis of solute effects on yields. They also report that the molecular yields of hydrogen and hydrazine are larger than the free radical yields (H, e_s and NH_2) in contrast to the situation in water.

Reactions of ions.—With the exception of the electron, specific information concerning ionic intermediates and their reactions is mainly gleaned from mass spectrometric studies. A few new techniques for obtaining this information have appeared, however. Lindholm has developed a double mass spectrometer in which a selected ion from the first section can be retarded and reacted with a second molecule. The ions produced in the charge exchange process are then analyzed in the second mass spectrometer. The energy available to the second molecule depends upon the initial and final states of the parent ion and can be varied widely by selecting different parent ions in the first mass spectrometer. Of interest to radiation chemistry is the fact that the technique yields directly the products of a given charge exchange reaction. Lindholm & Wilmenius (107) have shown the application of their results to the xenon-sensitized radiolysis of methanol vapor. They find that the principal ion produced by charge exchange is CH_2OH^+ and, from the ion yields, can calculate hydrogen and ethylene glycol yields which agree well with experiment. Lindholm et al. have also given charge exchange data for ethanol, propane, and butane (108, 109).

Ausloos & Lias (110) have developed another useful charge exchange technique in which H_3^+, produced in hydrogen, is reacted with a trace amount of hydrocarbon. Selected ions can be produced this way (usually the proton transfer to the hydrocarbon). The fate of the positive ions is inferred from analysis of reaction products. Reaction of H_3^+ with methane yields CH_5^+ rather cleanly, but reaction with ethane gives $C_2H_5^+$. In general, the hydrocarbon ions tend to decompose to give a lower hydrocarbon and a carbonium ion. Ausloos, Lias & Gorden (111) have described similar proton transfers from CH_5^+ and $C_2H_5^+$ (produced in methane radiolysis with yields of 1.9 and 0.9 ions per 100 eV, respectively) and point out that in studies of methane radiolysis these ions will react with impurities or reaction products if the extent of conversion is not kept very small.

Decomposition of ions is believed to be responsible for a large portion of the fragmentation products in the gas phase. Borkowski & Ausloos (112) can account for more than 95 per cent of the propane produced in *n*-butane radiolysis by formation of $C_3H_7^+$ from the parent ion followed by hydride transfer. Decomposition of ions, or other species, is known to be much less important in the liquid phase. Kevan & Libby (113) have studied *n*-hexane radiolysis in the solid phase and find the trend from gas to liquid continued in the solid. The total yield of fragmentation products (which they attribute to ion decomposition) is given by the summation of all hydrocarbon yields below hexane. They find this sum to be 0.86 molecules per 100 eV compared to 1.7 in the liquid phase. Dewhurst found 7.3 fragmented molecules per 100 eV in the gas phase.

Free Radical Processes

The reactions of free radicals are known to be responsible for many of the products observed in radiation chemistry. This fact has been deduced from the change in yield of products of radiolysis in the presence of radical scavengers; indeed, iodine disappearance and iodide formation techniques are often used for measuring radical yields. Iodine is not specific for radicals, however, so that such measurements are not on firm ground. Recent work has been directed towards clear identification of the free radicals, measurement of their yields, studies of kinetics of free radical reactions, and determining the origin of the radicals.

ESR studies have been useful in determining which free radicals are produced, but until recently they have been confined to post-irradiation studies of solids and glasses. Now, Fessenden & Schuler (114) have published their extensive ESR studies of free radicals produced by electron irradiation of liquid hydrocarbons. Two obvious advantages of liquid phase studies are the narrow line widths obtainable (down to 0.13 gauss) and the applicability of the results to other radiolysis studies of liquids. The authors give α, β, and sometimes γ coupling constants for many primary, secondary, tertiary, and cyclic radicals, saturated and unsaturated, several deuterated radicals, and hydrogen and deuterium atoms. The main radical produced from *n*-alkanes is the parent radical (loss of a single hydrogen). Carbon-carbon bond breaking decreases with increasing length of the hydrocarbon chain, and branched alkanes show more fragmentation than *n*-alkanes. Hydrogen atoms were seen directly only in the radiolysis of methane and methane-ethane mixtures. There was evidence for production of hydrogen atoms in other hydrocarbons as addition of olefins always led to the corresponding hydrogen atom adduct.

Fessenden & Schuler also demonstrated the kinetic applications of the ESR technique in studies of vinyl radical addition to ethylene and the ethyl radical combination reaction. Fessenden (115) has developed two techniques for measuring radical lifetimes based on the use of intermittent radiation (analogous to rotating sector experiments). The ESR apparatus can be used with a long time constant to give average radical concentrations during the cycle, or used with time sampling to measure the radical concentration during a fixed interval (0.5 msec) of the radiation cycle. In the latter technique the signal from many cycles is added to produce a satisfactory signal to noise ratio. The rate constant for ethyl radical combination was found to be $1.3 \times 10^{10} \exp(-830/RT)$ M^{-1} sec^{-1}.

Gillis (116) has studied the radiolysis of liquid ethane and finds that the relative methyl and ethyl radical concentrations deduced from product analysis agree reasonably well with the radical concentrations reported by Fessenden & Schuler.

Holroyd & Klein (117) have developed a new and interesting chemical technique for estimating radical yields which is both more specific for radicals and less disruptive of the normal processes occurring in the system than are other chemical methods (such as iodine scavenging). A known radical labeled with C^{14}, R^*, is generated along with the other radicals R_i, R_j, in the irradiated system. At sufficiently high intensities the radicals, undergo random combination, and the ratio of yields of R^*R_i to R^*R_j represents the ratio of radical yields of R_i to R_j. The authors have developed two variations of the technique: the $C_2{}^{14}H_5$ radical is generated from $C_2{}^{14}H_4$, and the $C^{14}H_3$ radical is generated from $C^{14}H_3I$ added to the hydrocarbon or olefin in small amounts. They find that their yields compare well with radical yields determined by other techniques in the case of *n*-hexane (117). Their conclusions concerning radical formation in saturated hydrocarbons generally parallel the conclusions of Fessenden & Schuler. In olefin radiolysis they find that allylic radical yields are an order of magnitude larger than vinylic radicals where production of both is possible (118). The 1-methallylic radicals produced from *cis*- and *trans*-butene-2 exhibit greater than 80 per cent retention of structural configuration, indicating that little residual energy is left in the radical after formation. The alkyl radicals from olefin radiolysis show evidence of a formation process other than hydrogen atom addition, as *n*-propyl radicals (0.25 per 100 eV) and *n*-butyl radicals (0.5 per 100 eV) are produced from propylene and butene-1. Hydrogen atom addition produces *sec*-butyl radicals at 77° K (136).

Schuler & Kuntz (119) have measured methyl radical yields from a series of 36 hydrocarbons by measuring CH_3I^{131} production in the presence of radioactive iodine. The radical yield depends on structure according to the empirical rule

$$G(CH_3) = (n - 1)^{-2}[1.02a + 2.79b + 7.89c]$$

where $n-1$ is the number of carbon-carbon bonds (n is the number of carbon atoms) and a, b, and c are the number of methyl groups attached at carbons with 0, 1, and 2 other methyl groups respectively. The surprising feature is the inverse square dependence on n-1. A linear decrease might have been expected, but the square dependence is in qualitative agreement with the observation of Fessenden & Schuler from ESR studies that fragmentation decreases with increasing chain length.

Spectroscopic observation of radicals is becoming commonplace. Unfortunately, visible and ultraviolet absorption spectra, unlike ESR spectra, do not spell out the names of the radicals, Dorfman & Taub (23) have studied the pulse radiolysis of ethanol. They find an absorption with peaks at 290, 265, and 255 mμ which decays by a second order process. Essentially the same peaks were found in aqueous ethanol solutions and the authors attribute the peak to the α-ethanol radical. Dainton et al. (103) have observed a post-irradiation absorption spectrum in liquid and glassy methyltetrahydrofuran. The spectrum found in the liquid (pulse radiolysis technique) has a maximum at 270 mμ; the spectrum in the glass has a maximum at 250 mμ. Both decayed with second order kinetics. ESR examination of the glass gave a seven line spectrum which the authors interpret as the methyltetrahydrofuryl radical produced by removing the tertiary hydrogen. They also report an absorption in liquid methyl alcohol at 290 mμ. The glass phase methyl alcohol spectrum examined after γ irradiation does not have a peak at 290 mμ but rises to a peak at or below 200 mμ. The authors attribute both the liquid phase and glass phase spectra to the methanol radical CH_2OH; however the shift in maxima seems excessive. Hamill et al. (97) have observed several absorption spectra in hydrocarbon glasses containing aromatic hydrocarbon solutes. The difference in absorption spectra between 3-methylpentane glasses with and without 25 per cent 2-methylpentene-1 added are attributed to hydrogen atom adducts with the aromatic hydrocarbons. They give spectra for C_6H_7, $C_6H_6CH_3$, $C_6H_5C_6H_6$, and $C_{10}H_9$. For the latter they report peaks at 375, 580, 640, and 710 mμ. Dainton et al. (120) have looked at the pulse radiolysis of solutions of napthalene in tetrahydrofuran and in hexane. They report a strong absorption peak at 330 mμ which they attribute to the $C_{10}H_9$ radical. The spectra of Hamill et al. and of Dainton et al. would seem to be inconsistent. There is a smaller peak at about 370 mμ in the pulse radiolysis of napthalene-tetrahydrofuran solutions but the rate constant for disappearance of the 370 mμ peak is larger than for the 330 mμ peak (121).

The origin of the radicals.—Given the fact that the principal radical formed in almost all systems is the parent molecule less a hydrogen atom, it is very tempting to conclude that the radicals largely originate by simple homolytic breaking of carbon-hydrogen bonds. The yields of parent radicals from pentane, hexane, and heptane are between 4.0 and 4.6 radicals per 100 eV [secondary radicals predominate over primary radicals (117)]. The yield of scavengeable hydrogen from hydrocarbons is typically about 3 molecules per 100 eV (out of a total of 5 to 6), and a popular hypothesis holds that this represents the hydrogen atoms which, in the absence of scavenger, abstract hydrogen from the substrate. The yield of hydrogen which is not scavenged is called the "molecular" yield and according to various authors, is due to molecular elimination, ionic reactions, or hot hydrogen atom abstraction reactions. This simple model has run into much trouble lately, as discussed by Dyne, Smith & Stone (122). The bulk of the recent evidence does not support the existence of large quantities of hydrogen atoms.

Dyne, Denhartog & Smith (123) have amplified their studies of dilute solutions of perdeuterocyclohexane in hydrocarbons in which they find a molecular yield of D_2 as well as a yield of HD. The two yields vary with scavenger concentration in a parallel manner for many different scavengers, indicating that the scavengers are reacting with a common precursor which they believe to be an excited state, not hydrogen atoms. The effect is peculiar to the liquid phase as the molecular D_2 yield in gaseous cyclohexane is not affected by scavengers (124).

Molecular elimination of hydrogen is a prominent process in gaseous cyclohexane as well as in gaseous propane (125) and butane (126), so there is no reason to neglect the process in liquids. In propane, the effect of density is small. Ausloos, Lias & Sandoval (125) have studied the radiolysis of 1:1

propane-perdeuteropropane mixtures as a function of pressure and find the HD yield is 15 per cent of total hydrogen below one atmosphere increasing to only 20 per cent at 25 atmospheres.

Klots, Raef & Johnsen (127) indicate that the clean division of the hydrogen yield from liquid hydrocarbons into molecular and hydrogen atom yields is a snare and a delusion. The hydrogen atom portion of the yield was believed to follow regular competition kinetics in the presence of scavengers, but the authors show that the use of benzene, pyridine, acrylonitrile, carbon tetrachloride, tetrachloroethylene, and methyl methacrylate as scavengers give strongly curved competition plots with no well-defined intercepts (reciprocal of the hydrogen yield) or recognizable slopes.

Toma & Hamill (128) have obtained results which argue strongly for the participation of ions in the formation of hydrogen. The most telling piece of evidence was an experiment in which a dilute solution of cyclohexane-d_{12} in cyclopentane was irradiated with and without ethanol. They found that ethanol drastically reduced the yields of HD and D_2. This indicates that cationic precursors are involved, as ethanol is known to protonate readily. A re-examination of the effect of iodine on the hydrogen yield, taking into consideration ionic reactions, led them to suggest that the true yield of thermal hydrogen atoms is only about 0.5 per 100 eV instead of the more commonly accepted value of 2 to 3. They propose a hot atom mechanism for the production of the majority of the hydrogen, but such a large ratio of hot to thermal hydrogen atom reactions would require average initial energies for the atoms of several electron volts. Klots (129) has estimated that most of the excess energy available in an excited state of cyclohexane should go into internal energy instead of kinetic energy on the basis of quasi-equilibrium rate theory. Rosenberg & Wolfgang (130) have measured the reaction probability per collision for hot tritium atoms reacting with butane and point out that only 64 per cent of the hot reactions lead to HT formation (the rest being substitution reactions). The proportion would likely be somewhat less for hydrogen atoms.

On the other side of the question, there has been new evidence for the participation of hydrogen atoms in hydrocarbon radiolysis. Yang, Scott & Burr (131) have irradiated mixtures of tritium labeled cyclohexane and benzene and find excess tritium in the polymer (over that from the cyclohexyl portion). The excess tritium corresponds to several tritium atoms per benzene residue in the polymer. Cher, Hollingsworth & Browning (132) have examined the fate of cyclohexene-C^{14} in cyclohexane and find products characteristic of the cyclohexyl-C^{14} radical which are most easily explained on the basis of hydrogen atom addition. As mentioned earlier, Fessenden & Schuler (114) observed radicals in mixtures of normal and perdeutero hydrocarbons characteristic of hydrogen atom addition to olefins.

Thus the origin of the hydrogen and hence the free radicals in hydrocarbon radiolysis is still obscure. The work of Klots, Raef & Johnsen demonstrate that more than one precursor is involved in hydrogen formation as the scavenging curves for different solutes have different shapes and even cross each other for certain pairs of solutes (such as carbon tetrachloride and acrylonitrile). The work of Dyne, Denhartog & Smith demonstrates that the true molecular hydrogen in which both hydrogen atoms arise from the same molecule has a common precursor with the bimolecular hydrogen. These two facts would suggest that one of the precursors of hydrogen is a linear descendent of the other.

It is still conceivable that hydrogen atoms are involved in hydrogen formation. The unimolecular hydrogen could be formed in spur reactions between the hydrogen atom and the parent radical in competition with scavenging by the hydrocarbon to form H_2 and reaction with scavengers:

$$\begin{aligned} H + RH\cdot &\rightarrow H_2 + R \\ H + S &\rightarrow HS \end{aligned} \qquad 13.$$

Dyne (133) has discussed this possibility briefly but discarded it on the basis of the efficiency of scavengers.

There is evidence that reaction 13 in which RH· is the cyclohexyl radical may be important in cyclohexane radiolysis. Falconer & Burton (134) and Burns & Parry (135) have studied the α-particle radiolysis of cyclohexane. There is quantitative disagreement in the results, perhaps due to the different energies used ($\sim$4 MeV and 1.5 MeV, respectively), but qualitatively they find that the ratio of cyclohexene to bicyclohexyl increases with LET. Falconer & Burton, bearing in mind that $G(H_2)$ does not change much with LET, conclude that reaction 13 is the most probable reaction occurring in the spurs and tracks, with a yield of about 1 molecule per 100 eV. Independent studies of reaction 13 are not available, but Klein, Scheer & Kelley (136) have found that disproportionation predominates over combination for the reaction of hydrogen atoms with sec butyl radicals at 90° K, in agreement with reaction 13.

Rumfeldt & Armstrong have given convincing evidence for the participation of thermal hydrogen atoms as an ultimate intermediate in hydrogen formation from liquid hydrogen halides (137). The observed hydrogen yield is separable into two components, one of which is easily decreased by hydrogen atom scavengers. They obtain rate-constant ratios for hydrogen atoms reacting with either hydrogen chloride or hydrogen bromide in competition with chlorine, bromine or ethylene. These rate constant ratios agree with values in the literature for known hydrogen atoms (from photolysis experiments, etc.). Thermal hydrogen atoms account for one third of the hydrogen in liquid hydrogen chloride. The remaining hydrogen is scavenged less readily and the authors suggest it may be due to hot hydrogen atoms.

Schindler (138) has studied the γ radiolysis of methyl iodide-ethyl iodide mixtures and concludes that the hydrocarbon products are produced by radical combination in the spurs. (Homogenous combination cannot occur as iodine is the principal product of the radiolysis and scavenges all radicals which escape the spurs.) The propane yield goes through a maximum in the neighborhood of a 1:1 mixture while the butane yield increases monotonically with ethyl iodide concentration.

Energy Transfer and Excited State Effects

The term "energy transfer" is even more ill-defined than the term "direct effect" with which it is intimately connected. It is most usefully limited to effects observed at relatively large solute concentrations (a few per cent or more) which depart from linearity with electron fraction. In this description it is tacitly assumed that direct effects are proportional to electron fraction. Since this assumption is not on very firm ground, the departure from linearity with electron fraction should be quite large (an order of magnitude) before it is taken as evidence for energy transfer. Fessenden & Schuler (139) have observed energy transfer from ethane to ethylene in liquid mixtures irradiated with an electron beam. The ESR technique with normal and deuterated compounds was used to distinguish radicals produced from ethane (C_2H_5) from those produced from ethylene ($C_2H_5\cdot$ and $C_2H_3\cdot$). They found that vinyl radicals were produced as if the fraction of energy absorbed in the ethylene was ten times as great as expected from electron fraction.

Wakeford & Freeman (140) have studied the radiolysis of cyclohexene in the presence and absence of benzene, 1,3-cyclohexadiene, and 1,4-cyclohexadiene. They find the products of cyclohexene radiolysis to be depressed below the yields expected on an electron fraction basis alone. 1,3-cyclohexadiene was particularly effective and is also the only additive with an ionization potential below cyclohexene. Hatada, Takezai & Hirota (141) have studied the chain production of methyl cyclohexyl ketone in cyclohexene-acetaldehyde mixtures. The ketone yield decreases linearly with electron fraction of added benzene and cyclohexane, indicating that there is no energy transfer in this system.

Jones, Van Dusen & Theard (142) have made an interesting comparison of intermolecular and intramolecular energy transfer. They measured hydrogen yields from various alkylbenzenes and butane-benzene solutions. The hydrogen yields are assumed to be due to a contribution from the phenyl group (or benzene) and the alkyl group (or butane):

$$G(H_2) = g_{\text{phenyl}}x + g_{\text{alkyl}}(1 - x)$$

where x is the electron fraction of the phenyl or benzene and g_{phenyl} is taken as 0.038 molecules per 100 eV. The yield due to the alkyl group, g_{alkyl} is found to be 0.5 for a wide range of alkyl benzenes (toluene through octylbenzene) and is also 0.5 for solutions of butane in benzene. The yield g_{alkyl} is only one tenth as large as the hydrogen yields from alkanes, indicating the occurrence of both inter- and intramolecular energy transfer. Intermolecular energy transfer is not as efficient as intramolecular transfer, however, as the value of g_{alkyl} for butane dissolved in butylbenzene was 1.3 molecules per 100 eV compared to 0.5 for the butyl group itself.

A somewhat different facet of energy transfer has been reported by Rabe, Rabe & Allen (150) who studied the radiolysis of azoethane adsorbed on solids. They observed the usual transfer of energy to the azoethane but also found that the azoethane decomposition on magnesium oxide went through a maximum with increasing amounts of azoethane present. Apparently energy can be shared among several neighboring azoethane molecules so that none of them receives enough to decompose.

There is a general belief that reactions of excited states are important in understanding energy transfer effects, particularly for solutions in which benzene is one component. Recent studies of benzene solutions give strong evidence for the production of excited states. Dainton et al. (120) have observed the absorption spectrum of the napthalene triplet state in the pulse radiolysis of dilute solutions of napthalene in both benzene and cyclohexane solution.

Two other solute effects have been attributed to triplet state formation in benzene. Cundall & Griffiths (95) have measured the yield of *cis-trans* isomerization in solutions of butene-2 dissolved in benzene. The isomerization yield is concentration dependent below 0.1 *M*, resembling the solutions in dodecane which were discussed in terms of ionic effects earlier. The isomerization yield was larger in benzene, however, and ferric chloride and piperidine had little effect. The reaction was quenched by additives such as oxygen and napthalene, and the authors attribute the isomerization to the transfer of energy from the triplet state of benzene to butene-2. They report that the yield for conversion of *cis* to *trans* is 2.2 molecules/100 eV and for *trans* to *cis* is 1.8. If the effect is entirely due to triplet states, then the yield of triplets would be the sum of the two yields or 4 molecules per 100 eV. Cherniak, Collinson & Dainton (143) have studied the reduction of ferric chloride in benzene and find three distinct stages with respect to ferric chloride concentration. The lowest concentration stage, below 10^{-4} *M*, gave a yield of reduction of 0.86 molecules per 100 eV which is identifiable with earlier determinations of the free radical yield in benzene. The second stage reached a plateau below 10^{-3} *M*, the reduction yield for the second stage being 2.7 molecules per 100 eV. The authors attribute this stage to energy transfer from triplet benzene to ferric chloride. Since two ferrous chlorides are produced per triplet (one by dissociation and one by the resultant free radical) *G*(triplet) is 1.35 which is much smaller than the yield observed by Cundall & Griffiths. The lack of agreement could be attributable to (*a*) a contribution to the isomerization yield of Cundall & Griffiths from ionic processes or (*b*) inefficiency of ferric chloride dissociation following triplet transfer from benzene.

The third stage of ferric reduction in benzene, observed by Cherniak, Collinson & Dainton was very large. The yield of ferrous chloride increased from 4 at 10^{-3} *M* $FeCl_3$ to 10 for a 1.5×10^{-3} *M* solution with no indication that the upper limit had been reached. The authors attribute this stage to energy transfer from a singlet state of benzene (or the exciton state corresponding to the singlet).

Cherniak et al. point out the parallelism between their observations on the third stage and ferric reduction in crystalline benzene. Collinson, Conlay & Dainton (144) have studied the effect of phase on ferric chloride reduction. Certain solvents such as benzene, diphenyl ether, diphenyl methane, and phenatole give strongly concentration-dependent ferrous chloride yields. These solutions exhibit an abrupt increase in $G(FeCl_2)$ upon freezing superimposed on a general increase with decreasing temperature. Such characteristics are expected on the basis of exciton migration which will occur more readily in a crystal. Lower concentrations of ferric ion were required in the solid phase in agreement with the model.

Reactions of excited states have been invoked to explain some peculiar results observed for systems irradiated with high LET particles. The hydrogen yield from benzene increases with increasing LET but the increase seems to be due almost entirely to effects near the end of the particle tracks, i.e., when the energy of the helium ions has been reduced below a few MeV (145, 146). Burns & Reed (147) report similar behavior in biphenyl and orthoterphenyl. In the case of water radiolysis the increase in hydrogen yield is quite noticeable for low LET particles such as 20 MeV deuterons. Also, in benzene, biphenyl, and terphenyl, the yield for disappearance of solvent $G(-M)$ increases with increasing LET, which is not to be expected from radical reactions as the high radical concentrations in tracks should favor radical combination without involving more solvent. Burns & Reed believe that these effects suggest excited states of benzene are involved and that hydrogen production is attributed to reactions 14 and 15 occurring in the spurs and tracks.

$$RH^* + RH^* \rightarrow H_2 + \text{products} \qquad 14.$$

$$RH^* + RH \rightarrow 2RH \qquad 15.$$

Reaction 15 would serve to limit the lifetime of RH^* and hence decrease track effects at relatively low LET. Burns & Jones (148) have treated reactions 14 and 15 on the basis of a modified Ganguly-Magee model for diffusion and combination of species formed in a spur and find that with a reasonable choice of parameters they can obtain results in agreement with experiment. They must use large values of k_{15}, however, corresponding to a lifetime of RH^* in the neighborhood of 10^{-11} to 10^{-10} sec. The excited species RH^* cannot be any of those observed by Cherniak, Collinson & Dainton, as the low concentration of ferric chloride involved in the experiments of the latter authors (less than 10^{-2} *M*) indicate lifetimes for their species of at least 10^{-9} sec.

Schuler (145) suggested a novel mechanism for explaining the observation of large hydrogen yields in benzene only at very low particle energies.

Increased amounts of charge separation are produced by slow positive ions by charge exchange with the liquid

$$He^{+2} \xrightarrow{+e} He^{+}$$

Electrons are displaced large distances in solution by this process and may well lead to increased or different chemical effects.

Choi & Lichtin (149) have irradiated methanol with γ rays and with $B^{10}(n,\alpha)Li^7$ recoils in the presence and absence of oxygen. The hydrogen yield increases slightly, the formaldehyde yield increases, and the ethylene glycol yield decreases with increasing LET. The most startling observation concerns the effect of oxygen. The difference between hydrogen yields in the presence and absence of oxygen is 3.1 molecules per 100 eV for both γ radiation and for recoil radiation. If the oxygen effect was due to hydrogen atoms, it should decrease with increasing LET. The authors point out that any radical model will run into serious difficulty explaining this effect.

LITERATURE CITED

1. Spinks, J. W. T., and Woods, R. J., *An Introduction to Radiation Chemistry* (Wiley, New York, 477 pp., 1964)
2. Verechinsky, I. V., and Pikaev, A. K., *Introduction to Radiation Chemistry*, (Izd. AN CCCP, Moscow, 407 pp., 1963)
3. Haissinsky, M., Ed., *Actions Chimiques et Biologiques Des Radiations*, **6** (Masson, Paris, 194 pp., 1963)
4. Haissinsky, M., Ed., *ibid.*, **7** (291 pp., 1964)
5. Fundamental Processes in Radiation Chemistry, *Discussions Faraday Soc.*, **36**, 325 pp. (1963)
6. Glassner, A., and Hart, E. J., Eds., *Radiation Research*, Suppl. 4 (223 pp., 1964)
7. *Proc. Tihany Symp., Budapest, 1962*, 482 (1964)
8. *Radiolysis of Hydrocarbons* (Topchiev, A. V., Ed., Holroyd, R. A., English Ed., American Elsevier, New York, 323 pp., 1964)
9. Exciton Symposium, *Radiation Res.*, **20**, 53–158 (1963)
10. Haissinsky, M., and Magat, M., Eds., *Constantes Sellectionnees Rendements Radiolytique* (Pergamon, Paris, 217 pp., 1963)
11. Czapski, G., and Bielski, B. H. J., *J. Phys. Chem.*, **67**, 2180 (1963)
12. Benson, S., *J. Am. Chem. Soc.*, **86**, 3922 (1964)
13. Czapski, G., and Dorfman, L. M., *J. Phys. Chem.*, **68**, 1169 (1964)
14. Baxendale, J. H., *Radiation Res.*, **17**, 312 (1962)
15. Rabani, J., Mulac, W. A., and Matheson, M. S., *J. Phys. Chem.*, **69**, 53 (1965)
16. Rabani, J., and Matheson, M. S., *J. Am. Chem. Soc.*, **86**, 3175 (1964)
17. Cotton, F. A., and Wilkinson, G., in *Advanced Inorganic Chemistry*, 135 (Interscience, Wiley, New York, 1962)
18. Baxendale, J. H., *Radiation Res.*, Suppl. 4, 139 (1964)
19. Baxendale, J. H., Fielden, E. M., and Keene, J. P., *Proc. Chem. Soc.*, 242 (1963)
20. Keene, J. P., *Radiation Res.*, **22**, 1 (1964)
21. Dorfman, L. M., and Matheson, M. S., in *Prog. Reaction Kinetics*, **3** (In press)
22. Gordon, S., Hart, E. J., Matheson, M. S., Rabani, J., and Thomas, J. K., *Discussions Faraday Soc.*, **36**, 193 (1963)
23. Dorfman, L. M., and Taub, I. A., *J. Am. Chem. Soc.*, **85**, 2370 (1963)
24. Matheson, M. S., and Rabani, J., *J. Phys. Chem.*, **69**, 1324 (1965)
25. Schwarz, H. A., *J. Phys. Chem.*, **67**, 2827 (1963)
26. Sweet, J. P., and Thomas, J. K., *J. Phys. Chem.*, **68**, 1363 (1964)
27. Fricke, H., and Thomas, J. K., *Radiation Res.*, Suppl. 4, 35 (1964)
28. Schwarz, H. A., *J. Phys. Chem.*, **66**, 255 (1962)

29. Bielski, B. H. J., and Saito, E., *J. Phys. Chem.*, **66,** 2266 (1962)
30. Schmidt, K., *Z. Naturforsch.*, **16**(b), 206 (1961)
31. Schwarz, H. A., *Radiation Res.*, Suppl 4, 89 (1964)
32. Pikaev, A. K., *Dokl. Akad. Nauk SSSR*, **156,** 916 (1964)
33. Keene, J. P., *Radiation Res.*, **22,** 14 (1964)
34. Scholes, G., and Simic, M., *J. Phys. Chem.*, **68,** 1731 (1964)
35. Dainton, F. S., *Radiation Res.*, Suppl. 4, 71 (1964)
36. Adams, G. E., and Boag, J. W., *Proc. Chem. Soc.*, 112 (1964)
37. Ferradini, C., and Koulkes-Pujo, A. M., *J. Chim. Phys.*, **60,** 1310 (1963)
38. Anbar, M., and Thomas, J. K., *J. Phys. Chem.*, **68,** 3829 (1964)
39. Allen, A. O., *Radiation Res.*, Suppl. 4, 54 (1964)
40. Hayon, E., *Trans. Faraday Soc.*, **60,** 1059 (1964)
41. Dainton, F. S., and Watt, W. S., *Proc. Roy. Soc.* (*London*), *A*, **275,** 447 (1963)
42. Czapski, G., Rabani, J., and Stein, G., *Trans. Faraday Soc.*, **58,** 2160 (1962)
43. Scholes, G., and Simic, M., *J. Phys. Chem.*, **68,** 1738 (1964)
44. Anbar, M., and Myerstein, D., *J. Phys. Chem.*, **68,** 1713 (1964)
45. Nehari, S., and Rabani, J., *J. Phys. Chem.*, **67,** 1609 (1964)
46. Seddon, W. A. (Private communication)
47. Scholes, G., *Discussions Faraday Soc.*, **36,** 311 (1963)
48. Hochanadel, C. J., *Radiation Res.*, **17,** 286 (1962)
49. Jortner, J., and Rabani, J., *J. Phys. Chem.*, **66,** 2081 (1962)
50. Rabani, J., *J. Phys. Chem.*, **66,** 361 (1962)
51. Draganic, Z. D., Draganic, I. G., and Kosanic, M. M., *J. Phys. Chem.*, **68,** 2085 (1964)
52. Cheek, C. H., and Linnenbom, V. J., *J. Phys. Chem.*, **67,** 1856 (1963)
53. Cheek, C. H., and Swinnerton, J. W., *J. Phys. Chem.*, **68,** 1429 (1964)
54. Jezowska-Trzebiatowska, B., and Kalecinski, J., *Nukleonika*, **9,** 625 (1964)
55. Haissinsky, M., Jove, J., and Szymanski, W., *J. Chim. Phys.*, **61,** 572 (1964)
56. Carmo Anta, M., Mariano, M. H., and Lima e Santos, M., *J. Chim. Phys.*, **61,** 577 (1964)
57. Allen, A. O., in *The Radiation Chemistry of Water and Aqueous Solutions*, 61 (Van Nostrand, Princeton, N.J., 1961)
58. Kochanny, G. L., Jr., Timnick, A., Hochanadel, C. J., and Goodman, C. D., *Radiation Res.*, **19,** 462 (1963)
59. Sworski, T. J., *Radiation Res.*, **4,** 483 (1956)
60. Sworski, T. J., *ibid.*, **2,** 26 (1955)
61. Hummel, A., and Allen, A. O., *Radiation Res.*, **17,** 302 (1962)
62. Hughes, G., and Willis, C., *Discussions Faraday Soc.*, **36,** 223 (1963)
63. Kuppermann, A., *Radiation Res.*, Suppl. 4, 69 (1964)
64. Hayon, E., *J. Phys. Chem.*, **68,** 1242 (1964)
65. Seddon, W. A., and Sutton, H. C., *Trans. Faraday Soc.*, **59,** 2323 (1963)
66. Katakis, D., and Allen, A. O., *J. Phys. Chem.*, **68,** 3107 (1964)
67. Getoff, N., *Oesterr. Chemiker-Ztg.*, **65,** 143 (1964)
68. Baxendale, J. H., and Gilbert, G. P., *J. Am. Chem. Soc.*, **86,** 516 (1964)
69. Anderson, A. R., Knight, B., and Winter, J., *Nature*, **201,** 1026 (1964)
70. Platzman, R. L., *Astr. Intern. Cong. Radiation Res., Harrogate, Engl., August 1962*, 128
71. Flournoy, J. M., Baum, L. H., and Siegel, S., *J. Chem. Phys.*, **36,** 2229 (1962)
72. Kevan, L., Moorthy, P. N., and Weiss, J. J., *Nature*, **199,** 689 (1963)
73. Schulte-Frohlinde, D., and Eiben, K., *Z. Naturforsch.*, **18**(a), 99 (1963)
74. Henricksen, T., *Radiation Res.*, **23,** 63 (1964)
75. Lassettre, E. N., and Francis, S. A., *J. Chem. Phys.*, **40,** 1208 (1964); Lassettre, E. N., and Jones, E. A., *J. Chem. Phys.*, **40,** 1218, 1222 (1964); Lassettre, E. N., Berman, A. S., Silverman, S. M., and Krasnow, M. E., *J. Chem. Phys.*, **40,** 1232 (1964); Lassettre, E. N., Silverman, S. M., and Krasnow, M. E., *J. Chem. Phys.*, **40,** 1242, 1248, 1256, 1261 (1964); Silverman, S. M., and Lassettre, E. N., *J. Chem. Phys.*, **40,** 1265 (1964); Skerbele, A. M., and Lassettre, E. N., *J. Chem. Phys.*, **40,** 1271 (1964)
76. Kuppermann, A., and Ruff, L. M.,

Discussions Faraday Soc., **35,** 30 (1963)

77. Kuppermann, A., *Radiation Res.*, Suppl. 4, 15 (1964)
78. Rauth, A. M., and Simpson, J. A., *Radiation Res.*, **22,** 643 (1964)
79. Magee, J. L., *Ann. Rev. Phys. Chem.*, **12,** 389 (1961)
80. Chen, J. C. Y., *J. Chem. Phys.*, **40,** 3513 (1964)
81. Allen, A. O., and Hummel. A., *Discussions Faraday Soc.*, **36,** 95 (1963)
82. Freeman, G. R., *J. Chem. Phys.*, **39,** 988 (1963)
83. Freeman, G. R., *J. Chem. Phys.*, **39,** 1580 (1963)
84. Gazda, E., *Acta Phys. Polon.*, *Suppl.*, **24,** 209 (1963)
85. Jachym, B., *Acta Phys. Polon.*, *Suppl.*, **24,** 243 (1963)
86. Samuel, A. H., Halliday, F. O., Keast, A. K., and Taimuty, S. I., *Science*, **144,** 839 (1964)
87. Hummel, A. (Private communication)
88. Busler, W. R., Martin, D. H., and Williams, F., *Discussions Faraday Soc.*, **36,** 102 (1963)
89. Galimba, F. B., Wilson, W. E., and Samuel, A. H., *J. Phys. Chem.*, **68,** 647 (1964)
90. Kaufman, P. C., *J. Phys. Chem.*, **67,** 1671 (1963)
91. Van Dusen, W., *Discussions Faraday Soc.*, **36,** 260 (1963)
92. Williams, F., *J. Am. Chem. Soc.*, **86,** 3954 (1964)
93. Samuel, A. H., and Magee, J. L., *J. Chem. Phys.*, **21,** 1080 (1953)
94. Burton, M., Dillon, M. A., and Rein, R., *J. Chem. Phys.*, **41,** 2228 (1964)
95. Cundall, R. B., and Griffiths, P. A., *Discussions Faraday Soc.*, **36,** 111 (1963)
96. Scholes, G., and Simic, M., *Nature*, **202,** 895 (1964)
97. Hamill, W. H., Guarino, J. P., Ronayne, M. R., and Ward, J. A., *Discussions Faraday Soc.*, **36,** 169 (1963)
98. Roberts, J., and Hamill, W. H., *J. Phys. Chem.*, **67,** 2446 (1963)
99. Guarino, J. P., and Hamill, W. H., *J. Am. Chem. Soc.*, **86,** 777 (1964)
100. Taub, I. A., Harter, D. A., Sauer, M. C., and Dorfman, L. M., *J. Chem. Phys.*, **41,** 979 (1964)
101. Adams, G. E., and Sedgwick, R. D., *Trans. Faraday Soc.*, **60,** 865 (1964)
102. Adams, G. E., Baxendale, J. H., and Boag, J. W., *Proc. Roy. Soc.* (*London*), *A*, **277,** 549 (1964)
103. Dainton, F. S., Keene, J. P., Kemp, T. J., Salmon, G. A., and Teply, J., *Proc. Chem. Soc.*, **265** (1964)
104. Baxendale, J. H., and Mellows, F. W., *J. Am. Chem. Soc.*, **83,** 4720 (1964)
105. Ahrens, R. W., Suranarayana, B., and Willard, J. E., *J. Phys. Chem.*, **68,** 2947 (1964)
106. Dainton, F. S., Skworski, T., Smithies, D., and Wezranowski, E., *Trans. Faraday Soc.*, **60,** 1068 (1964)
107, Lindholm, E., and Wilmenius, P., *Arkiv Kemi*, **20,** 255 (1963)
108. von Koch, H., and Lindholm, E., *Arkiv Fysik*, **19,** 123 (1961
109. Petterson, E., and Lindholm, E., *Arkiv Fysik*, **24,** 49 (1963)
110. Ausloos, P., and Lias, S. G., *J. Chem. Phys.*, **40,** 3599 (1964)
111. Ausloos, P., Lias, S. G., and Gorden, R., Jr. *J. Chem. Phys.*, **39,** 3341 (1963)
112. Borkowski, R. P., and Ausloos, P., *J. Chem. Phys.*, **39,** 818 (1964)
113. Kevan, L., and Libby, W. F., *J. Chem. Phys.*, **39,** 1288 (1964)
114. Fessenden, R. W., and Schuler, R. H., *J. Chem. Phys.*, **39,** 2149 (1963)
115. Fessenden, R. W., *J. Phys. Chem.*, **68,** 1508 (1964)
116. Gillis, H. A., *J. Phys. Chem.*, **67,** 1399 (1963)
117. Holroyd, R. A., and Klein, G. W., *J. Am. Chem. Soc.*, **84,** 4000 (1962)
118. Holroyd, R. A., and Klein, G. W., *J. Phys. Chem.*, **69,** 194 (1965)
119. Schuler, R. H., and Kuntz, R. R., *J. Phys. Chem.*, **67,** 1004 (1963)
120. Dainton, F. S., Kemp, T. J., Salmon, G. A., and Keene, J. P., *Nature*, **203,** 1050 (1964)
121. Ho, S. K. (Private communication)
122. Dyne, P. J., Smith, D. R., and Stone, J. A., *Ann. Rev. Phys. Chem.*, **14,** 313 (1963)
123. Dyne, P. J., Denhartog, J., and Smith, D. R., *Discussions Faraday Soc.*, **36,** 135 (1963)
124. Blackford, J., and Dyne, P. J., *Can. J. Chem.*, **42,** 1165 (1964)
125. Ausloos, P., Lias, S. G., and Sandoval, I. B., *Discussions Faraday Soc.*, **36,** 66 (1963)
126. Spittler, E. G., Jordan, P., Dorfman, L. M., and Sauer, M. C., *J. Phys. Chem.*, **67,** 2235 (1963)
127. Klots, C. E., Raef, Y., and Johnsen, R. H., *J. Phys. Chem.*, **68,** 2040 (1964)
128. Toma, S. Z., and Hamill, W. H., *J. Am. Chem. Soc.*, **86,** 4761 (1964)

129. Klots, C. E., *J. Chem. Phys.*, **41**, 117 (1964)
130. Rosenberg, A. H., and Wolfgang, R., *J. Chem. Phys.*, **41**, 2159 (1964)
131. Yang, J. Y., Scott, B., and Burr, J. G., *J. Phys. Chem.*, **68**, 2014 (1964)
132. Cher, M., Hollingsworth, C. S., and Browning, B., *J. Chem. Phys.*, **41**, 2270 (1964)
133. Dyne, P. J., *J. Phys. Chem.*, **66**, 767 (1962)
134. Falconer, J. W., and Burton, M., *J. Phys. Chem.*, **67**, 1743 (1963)
135. Burns, W. G., and Parry, J. R., *Nature*, **201**, 814 (1964)
136. Klein, R., Scheer, M. D., and Kelley, R., *J. Phys. Chem.*, **68**, 598 (1964)
137. Rumfeldt, R. C., and Armstrong, D. A., *J. Phys. Chem.*, **68**, 761 (1964)
138. Schindler, R. N., *Radiochimica Acta*, **2**, 73 (1964)
139. Fessenden, R. W., and Schuler, R. H., *Discussions Faraday Soc.*, **36**, 147 (1963)
140. Wakeford, B. R., and Freeman, G. R., *J. Phys. Chem.*, **68**, 2992 (1964)
141. Hatada, M., Takezai, J., and Hirota, K., *Bull. Chem. Soc. Japan*, **37**, 166 (1964)
142. Jones, K. H., Van Dusen, W., and Theard, L. M., *Radiation Res.* **23**, 128 (1964)
143. Cherniak, E. A., Collinson, E., and Dainton, F. S., *Trans. Faraday Soc.*, **60**, 1408 (1964)
144. Collinson, E., Conlay, J. J., and Dainton, F. S., *Discussions Faraday Soc.*, **36**, 153 (1963)
145. Schuler, R. H., *Rept. Radiation Res. Lab., Mellon Inst.*, Pittsburgh, *April–June 1964*
146. Burns, W. G., *Trans. Faraday Soc.*, **58**, 961 (1962)
147. Burns, W. G., and Reed, C. R. V., *Trans. Faraday Soc.*, **59**, 101 (1964)
148. Burns, W. G., and Jones, J. D., *Trans. Faraday Soc.*, **60**, 2022 (1964)
149. Choi, S. U., and Lichtin, N. N., *J. Am. Chem. Soc.*, **86**, 3948 (1964)
150. Rabe, J. G., Rabe, B., and Allen, A. O., *J. Am. Chem. Soc.*, **86**, 3887 (1964)

MOLECULAR STRUCTURE[1,2]

By James A. Ibers[3]

Brookhaven National Laboratory, Upton, New York

Introduction

A very limited number of topics from the vast literature relevant to molecular structure have been selected for this review. In making this selection, the interests and prejudices of the author have been followed. These are probably typical of those of a structural chemist primarily interested in inorganic chemistry and the solid state. Yet, other considerations were involved in the selection. The molecular structures of noble-gas compounds are reviewed partly because this seems a propitious time to do so, and partly because of a sense of dissatisfaction that these compounds should so recently have been discovered, when the simple experiment of heating xenon and fluorine together to form xenon fluorides could have been performed successfully at almost any time over the past quarter century by a chemist with a little luck and a disregard for what he had been taught. Perhaps one should question frequently the underlying "principles" of his chosen field. Five-coordination in transition metal complexes is reviewed, in part to emphasize the great need for careful structural work not only in this area, but also in all of transition metal chemistry. The tremendous expansion of activity in preparative inorganic chemistry has not been followed by a careful characterization of many of the compounds prepared. Without such characterization, the reputable theoretician has little on which to build a theory which will be helpful in suggesting meaningful experiments to be performed. The review on isotope effects in hydrogen bonding hopefully will encourage interest in this area on the part of crystallographers, all of whom have the experimental tools at their disposal to make important contributions to our knowledge of hydrogen bonding. The concept of the metal–hydrogen bond in transition metal hydrides has undergone complete revision as the interpretations of earlier spectroscopic experiments have been questioned. This subject is reviewed both because of its fundamental and practical importance and also as a reminder that the subjectivity that is often responsible for the choice of an experiment can occasionally be extended to its interpretation.

Three topics regretfully omitted from this review deserve particular mention. Discussion of metal–metal bonding in transition metal complexes has been omitted, even though this is a field ripe for study by the crystallog-

[1] The survey of literature pertaining to this review was concluded in November 1964.

[2] Supported by the United States Atomic Energy Commission.

[3] Present address: Chemistry Department, Northwestern University, Evanston, Illinois 60201.

rapher and physical chemist interested in magnetism and is a field of fundamental importance to the theoretician interested in understanding and explaining chemical bonding. Fortunately, Lewis (1) has recently presented an excellent review of some aspects of metal–metal bonding. A review of the current status of charge density measurements has been omitted. This is perhaps just as well, for at a recent conference on the subject (2) the impression was conveyed that we are still far from determining experimentally the distributions of bonding electrons in simple as well as in complex substances. Finally, discussion of the exciting developments in the field of boron hydrides and related compounds, especially the carboranes, has been omitted. This has been done in part because of the appearance of Lipscomb's excellent book (3), and in part because a new book on boron chemistry (4) is in press.

Molecular Structures of Noble-Gas Compounds

A discussion of the molecular structures of noble-gas compounds is given here in part because these compounds were unknown when the last review on molecular structure in this series was written (5). Yet now that the initial activity has subsided and there is time for contemplation, the entire subject of noble-gas compounds emerges as one with philosophical as well as scientific implications. The striking facts are that these compounds are easy to prepare, at least in an impure state, and that they fit in well with the usual theories of chemical bonding. Beginning some seventy years ago with Berthelot (6) there have been reports of the formation of noble-gas compounds, and it seems incredible that in the process of refutation of these reports some true compounds would not have been produced. During this time also there have been a few predictions of the possibility of existence of such compounds (7, 8), and at least one serious attempt to prepare these (9). Pauling (7) predicted that XeF_6, KrF_6, and the salts of H_4XeO_6 would be stable and that XeF_8 might be stable. Yost & Kaye (9) attempted to prepare various noble-gas halogen compounds by the action of ultraviolet light and of a discharge through mixtures of the elements; such an experimental approach, with minor (but critical) changes, is one that is now known to yield such compounds. Although it is true that most of the best textbooks on inorganic chemistry (e.g., 10, 11) state that covalent compounds of the noble gases cannot exist, usually because their existence would be a violation of the octet rule, a few state otherwise; in fact Emeléus & Anderson (12) note that "the question as to whether the rare gases form compounds or not is still very open." Despite such statements, the idea seems to have developed that such compounds could not exist, and hence their discovery came in an indirect but brilliant series of experiments. Bartlett & Lohmann (13, 14) had prepared the compound $O_2^+PtF_6^-$ by oxidation of oxygen with platinum hexafluoride. Bartlett realized that O_2 and Xe have very nearly the same first ionization potential and reasoned that it might be possible to oxidize Xe with PtF_6. He was successful in doing this, although the stoichiometry of the $Xe \cdot PtF_6$ compound formed remains unsettled. His announcement

(15) of the first covalent compound of a noble gas immediately suggested to a group of workers at Argonne National Laboratory that it might be possible to prepare xenon fluorides by the direct reaction of the elements at elevated temperatures. That experiment (16), beautiful in its simplicity, yielded the first xenon fluoride, XeF_4, and led for the next two years to rushed experimental activity and to theoretical hindsight.

As a result of this activity, the structures of some of the xenon fluorides are known much more accurately than are those of any of the related interhalogen compounds, and the perxenate ion is better characterized structurally than the periodate ion. What is more, these noble-gas compounds fit in so well with current theories of chemical bonding, and with current notions of molecular structure, that no new concepts or changes in old ones have had to be introduced. It is not the purpose of this review to discuss these theories, but attention must be called to the marvelous review of the subject by Coulson (17) and to his remark: "It is no exaggeration to say that in principle almost everything described in this survey could have been said thirty years ago."

One of the interesting features of noble-gas research has been the way in which the compounds have been characterized. Although spectroscopic methods have played their usual important role in characterization, modern diffraction methods, both X-ray and neutron, in many instances (e.g., the perxenates and $XeF_2 \cdot XeF_4$) have supplanted the more traditional methods. Moreover, the electron diffraction method has been applied to only one of these compounds, even though many of the compounds are volatile and no less stable than others studied by electron diffraction in the past.

Xenon tetrafluoride.—The crystal and molecular structure of XeF_4 has been the subject of two independent X-ray studies (18–21) and one neutron diffraction study (22, 23). The three studies are in complete agreement. Xenon tetrafluoride in the solid state is required by crystallographic symmetry conditions to be planar and it is square within the limits of error of the studies. The two independent Xe–F distances, uncorrected for thermal motion, are 1.932 and 1.939 $\pm$ 0.002 Å (22). If one assumes that the F "rides" on the Xe, then the interatomic distance corrected for thermal motion is 1.95 $\pm$ 0.01 Å. Analysis of the thermal motion indicates that the molecule can be treated, at least in first approximation, as a rigid body and that the root-mean-square librational motions of the molecule are approximately 5 degrees about the molecular axes. In the crystal structure there are some surprisingly short nonbonded intermolecular F–Xe interactions of about 3.2 Å.

The square-planar structure of xenon tetrafluoride in the solid state is also consistent with the geometry deduced from spectroscopic studies of the vapor (24, 25, 26). Vibrational assignments have been made and force constants evaluated (25, 26). In particular, Pimentel & Spratley (27) have offered an explanation for the differences in force constants between XeF_4 and ICl_4^-. Finally, an electron diffraction study of XeF_4 has been carried out; in its revised form (28), it is in agreement with the structure derived from the solid-state and spectroscopic studies.

Xenon difluoride.—XeF_2 was surmised by Siegel & Gebert (29) to be linear and symmetric in the solid phase, but they were unable to determine the Xe—F distance accurately. The molecule was also found to be linear in the gas phase by Smith using spectroscopic techniques (25, 30). Levy & Agron (31), in a subsequent neutron diffraction study, established the Xe—F distance to be 1.984 ± 0.004 Å. (The molecule is required by crystallographic symmetry conditions to be linear and symmetric.) On the assumption of rigid-body motion, the Xe—F distance, corrected for the effect of the F riding on the Xe, is 2.00 ± 0.01 Å. This is significantly longer than the distance in XeF_4, and this increase is consistent with the differences in principal stretching constants of XeF_4 and XeF_2 of 3.02 and 2.82 md/Å (25). The closest intermolecular Xe · · · F distance is 3.41 Å.

Xenon difluoride·xenon tetrafluoride.—Originally, Burns (32) reported a unit cell for a crystalline xenon-fluorine compound and concluded, on the assumption that the compound was XeF_4, that a "high-density" form of XeF_4 existed. Subsequently (33), on the basis of a complete structure determination, the compound was found to be $XeF_2 \cdot XeF_4$. The structures of the separate parts are identical with those discussed above for XeF_4 and XeF_2. The close fluorine-to-xenon intermolecular bond distances range from 3.28 to 3.42 Å, about the same as in the parent compounds, and hence there is no evidence for forces holding the compound together that are different from those in the separate crystals.

It is amusing to speculate on the theories that would have been put forward to account for XeF_3, had analytical data confirmed the stoichiometry as XeF_3 prior to the structure determination!

Xenon hexafluoride. Compared with the lower fluorides of xenon, XeF_6 is extremely reactive; it cannot be contained in dry quartz. There has been much speculation about its structure. The basic theoretical problem centers on the possible influence of the lone pair of electrons. In the theories of Gillespie (34) and Allen (35, 36, 37) [which some workers feel do not explain the facts (17, 38)] XeF_6 should deviate from octahedral symmetry. Descriptions in terms of molecular orbital theory (17, 39) suggest octahedral symmetry.

The spectroscopic data on XeF_6 are not too helpful. Smith (25) suggests an assignment in which XeF_6 cannot possess a center of symmetry. If this be so then XeF_6 is the first known MF_6 compound that is not octahedral in the gas phase (40). Smith's basic argument rests on the assignment of a band at 520 cm^{-1} to XeF_6 and on a comparison of the infrared spectrum in the vapor state with a Raman study of the solid. There has been some discussion (41) of the possibility that, since XeF_6 is much more reactive than was thought originally, the band at 520 cm^{-1} may be caused by an impurity. Moreover, preliminary X-ray data (42) indicate that XeF_6 is a tetramer in the solid state, and this would preclude a meaningful comparison of the infrared data from the vapor with the Raman data from the solid.

Methods recently have been developed which allow XeF_6 to be produced in a very pure state (43), free from contamination by other xenon fluorides and by $XeOF_4$. What is now badly needed is not only a repetition of the spectroscopic studies, but also a reliable electron diffraction study on XeF_6. This should be possible, but precautions will have to be taken to insure that the sample does not come into contact with anything but pretreated metals. Since the experiment on the vapor would hopefully settle a theoretical point at issue, it seems well worth the effort.

Xenon oxytetrafluoride.—This compound is a colorless stable liquid (mp −28°) which is formed by the reaction of XeF_6 with either water or silica (44). Its structure in the solid phase is unknown, but spectroscopic studies of the vapor (25, 26) are consistent with C_{4v} symmetry for the molecule (square pyramidal). On the basis of other data for square pyramidal molecules (*vide infra*) one would guess that the O–Xe–F angles are around 100°.

Xenon trioxide.—The structure of this sensitive and powerfully explosive solid has been determined in an X-ray study (45, 46). XeO_3 is a trigonal pyramidal molecule with an average Xe–O distance of 1.76 Å and an average O–Xe–O bond angle of 103°. There are short intermolecular interactions of 2.86 Å between Xe and O. When dry the substance is about as powerful an explosive as TNT and since the products of the explosion are xenon and oxygen its use would be hard to detect.

Perxenate ion.—In striking confirmation of the prediction of Pauling (7), some of the various salts of perxenic acid, H_4XeO_6, are remarkably stable. A stable solid, sodium perxenate, was first prepared by hydrolysis of the xenon fluorides (47), but it was not possible to arrive at its formula. Two products of the hydrolysis of xenon hexafluoride by NaOH have now been identified by three-dimensional X-ray studies, a method that even today is a somewhat novel one in analytical chemistry. The two products are $Na_4XeO_6 \cdot 8H_2O$ (48, 49) and $Na_4XeO_6 \cdot 6H_2O$ (50, 51). Recently the compound $K_4XeO_6 \cdot 9H_2O$ (52) has been established by means of a complete structure determination. There are no basic differences in the perxenate geometry in these three compounds: the average Xe–O distances are 1.864 ± 0.012, 1.84 ± 0.02, and 1.86 ± 0.01 Å, and the shape in each case does not differ significantly from octahedral. It is interesting that this Xe–O distance is 0.10 Å longer than that reported for the Xe–O distance in XeO_3, just as the I–O distance in the very few periodates that have been studied (53, 54) is 0.06 to 0.11 Å longer than in a typical iodate (55).

Other noble-gas compounds.—Several other compounds of the noble gases are known, but details of their structures are not. The green, explosive gas XeO_4 has been prepared (56) and, on the basis of limited spectra, is thought to be tetrahedral. Two fluorides of krypton have been reported, namely

KrF_2 and KrF_4. Krypton difluoride has been prepared by several methods (57–60) and its spectrum is surprisingly close to that of XeF_2 (57). There is every reason to believe that KrF_2 is a linear symmetric molecule, although a preliminary report (61) suggests that it is not isostructural with XeF_2 in the solid state. The preparation of KrF_4 (62, 63) has been reported, but its existence is the subject of some dispute (60). Yet it is interesting to note that two molecular orbital calculations (39, 64) suggest that KrF_4 ought to be stable.

A radon fluoride has been prepared (65), but has not been well characterized.

Although the stoichiometry of the initial compound $Xe \cdot PtF_6$ is somewhat uncertain, a number of salts and possible adducts of noble-gas compounds are well characterized, although none structurally. These include $XeF_6 \cdot NaF$ (43); $XeF_6 \cdot BF_3$ and $XeF_6 \cdot AsF_5$ (66); $XeF_2 \cdot 2SbF_5$ and $XeF_2 \cdot 2TaF_5$ (67); and $XeF_6 \cdot CsF$ and $XeF_5 \cdot 2CsF$ (68).

Pimentel (8) suggested in 1951 that it might be possible to prepare compounds between noble gases and halogens and he recently (27) reiterated this prediction with regard to HeF_2. As of now, however, there have been no reports of the successful preparation of noble-gas compounds of Ar, Ne, or He.

Five-Coordination in Transition Metal Complexes

The difficulties alluded to earlier in predicting the ground-state configuration of xenon hexafluoride seem minimal compared with the difficulties of predicting the ground-state configurations and even the coordination numbers of metals in transition metal complexes. From time to time, certain panaceas for our theoretical ills have been put forward; the most recent examples are crystal- and ligand-field theory. Presumably our theoretical understanding of coordination has increased, and in part our failure to predict with much accuracy the bonding in transition metal complexes is traceable to our lack of good experimental data. This in turn is traceable in large measure to the inherent weakness of indirect spectroscopic, magnetic, and other physical measurements for the determination of coordination numbers and stereochemistry. To cite but two examples of this weakness, the ion $ReCl_4^-$ was predicted to be tetrahedral (69) in order to account for the available experimental data, particularly the diamagnetism of salts of the ion. The ion actually is $Re_3Cl_{12}^{3-}$, on the basis of two independent diffraction studies (70, 71). From two spectroscopic studies of the gas phase (72, 73), $HMn(CO)_5$ was required to have symmetry lower than C_{4v}. Yet this symmetry has now been elevated to C_{4v} (74) in a third spectroscopic study which was prompted by the discovery by diffraction methods (75, 76) that the molecule has C_{4v} symmetry in the solid state.

In the hope that further work, both experimental and theoretical, will be encouraged, we summarize here what little is known with certainty about five-coordinated transition metal complexes. We choose to discuss five-coordination for several reasons: there has been a reasonable amount of theoretical effort on the subject (77–82); reports of five-coordination are now frequent (83, 84, 85); and we wish to emphasize how little is really known. Since we are interested in the stereochemistry as well as the coordination number, and since this information is available only from diffraction studies, only the results of such studies are considered. We thus ignore the many credible accounts of five-coordination that are based on spectroscopic, magnetic, or other indirect measurements. Since five-coordination of the non-transition metals (Groups O through IVa, IIIb through VIIb) is well documented, particularly for compounds of P (86, 87), Se (88), Sn (89), Sb (90), and Ti (91), we do not discuss it here.

Group Va.—The vanadium in oxobis-acetylacetone vanadium (II) lies near the center of gravity of a distorted square pyramid (92). The axial vanadyl oxygen is 1.56 Å from V; the equatorial V–O distances are between 1.96 and 1.98 Å. A similar arrangement around V is found in the structure of V_2O_5 (93), with the axial V–O of length 1.54 Å, and the equatorial V–O lengths ranging from 1.77 to 2.02 Å. Yet a discussion of coordination in three-dimensional complexes, such as V_2O_5, is probably more difficult and less meaningful than in isolated complexes.

In an early electron diffraction study (94) of the chlorides and bromides of niobium and tantalum, arguments based on dipole moment measurements were used to assign a trigonal bipyramidal structure rather than a square pyramidal structure (with the metal in the basal plane). Clearly these substances should be reinvestigated, particularly since there is now interest in the details of the geometry, including the question of possible differences in equatorial and apical bond lengths. Niobium and tantalum chlorides are dimeric in the solid phase, and the metals are octahedrally coordinated (95).

Group VIa.—In another early electron diffraction study (96), $MoCl_5$ was assigned a trigonal bipyramidal structure. In the solid phase, $MoCl_5$ (97) is dimeric and the Mo octahedrally coordinated.

Group VIII.—Iron pentacarbonyl has been found to be a regular trigonal bipyramid, both in the gas phase (98) and in the solid phase (99, 100). The various interpretations of the infrared and Raman spectra (101, 102) are apparently now consistent with this stereochemistry, although at one time (103) it was thought that only a tetragonal pyramidal configuration was consistent with the spectral and dipole moment data.

Two iron porphyrins have been examined by diffraction methods. These are chlorohemin (104) and methoxyiron(III)mesoporphyrin-IX-dimethyl ester (105). In both of these compounds the iron is in a square pyramidal environment, with Fe–N distances of 2.06 Å. In both substances the iron lies approximately 0.5 Å above the basal toward the Cl in the first instance, (Fe–Cl = 2.22 Å) and toward an O in the second (Fe–O = 1.84 Å).

The single example of five-coordinated cobalt is the dimethyldithiocarbonatonitrosocobalt(II) (106, 107) in which the Co is toward the center of a square pyramid formed by the four basal sulfur atoms and the apical NO group. The oft-quoted Co–N–O angle of 139° might not survive a redetermination of the structure.

One example of five-coordinated nickel is also known. Nickel is in a distorted square pyramidal environment in $NiBr_2$triarsine (108) (triarsine is a tridentate arsenic ligand of formula $(CH_3)_2As(CH_2)\ As(CH_3)(CH_2)\ As(CH_3)_2$). The three As and one Br are basal, but are not coplanar. The basal Ni–Br distance is 2.37 Å, whereas the apical Ni–Br bond length is 2.69 Å.

The single example of five-coordinated ruthenium occurs in the compound dichlorotris-triphenylphosphine ruthenium (II) (109) in which the Ru lies about 0.4 Å above the basal plane formed by two *trans* chlorines and two phosphorus atoms toward an apical phosphorus. The configuration is that of a distorted square pyramid. The Ru–P apical bond is 0.2 Å shorter than the Ru–P basal bond. The existence of a d^6 five-coordinated compound is most surprising, for there is in general a ready formation of spin-paired octahedral complexes in this electronic configuration (79). A close examination of the structure reveals, on the assumption that the phenyl rings have D_{6h} symmetry and C–H distances of 1.08 Å, that a hydrogen, attached to the beta carbon of one of the phenyl rings, is 2.59 Å from the Ru in such a position as to complete the octahedral configuration about the metal. Thus, one can describe the configuration about Ru either as square pyramidal or as octahedral, depending upon how one feels about counting such metal–hydrogen interactions. This points out the fact that even from diffraction studies it is not always possible to assign the coordination number unambiguously. One might still wish to consider the Ru as five-coordinated, since the Ru–H distance is about what is expected from van der Waals radii. However, a number of "weak" metal–hydrogen interactions have been postulated to account for various spectroscopic anomalies in certain ferrocenyl alcohols (110), in protonated acylferrocenes (111), in cycloheptadienium complexes (112), and in the compound $(CH_2)_3(Mn(CO)_5)_2$ (113). Such evidence is extremely tenuous, and what is needed is a diffraction study of one of these compounds in the solid state that would provide geometrical proof of such an interaction and hopefully indicate a metal–hydrogen bond that is shorter than a van der Waals interaction. Alternatively, the existence of a metal–hydrogen interaction in the ruthenium compound in solution, as manifested by spectroscopic anomalies, would provide a sounder basis for postulates of such interactions. Unfortunately, the solution spectrum of the ruthenium compound has not been studied in detail.

The single example of five-coordinated rhodium occurs in the hydride of rhodium carbonyl tris-triphenylphosphine (114, 115). In this compound the Rh is in a trigonal bipyramidal configuration with the Rh about 0.34 Å above the basal plane of phosphorus atoms toward the CO group. Both the Rh–CO and Rh–H distances appear to be normal.

The tetradentate arsenic ligand QAS [tris-(*o*-diphenylarsinophenyl)arsine] coordinates with Pt in the compound tetraphenylboron platinum QAS iodide (116) to give a regular trigonal bipyramidal configuration to the Pt;

there are three As in the basal plane, and As and I at the apices. No other five-coordinated platinum compounds appear to have been reported.

Group IB.—Several five-coordinated copper compounds are known. Bis-salicylaldehyde isopropylenedi-imine copper monohydrate (117) has a square pyramidal configuration with the basal Cu–O bonds ranging in length from 1.78 to 1.95 Å and the apical Cu–OH_2 bond of 2.53 Å in length. The Cu lies 0.2 Å out of the basal plane toward the water molecule. The compound $Cu(NH_3)_4SO_4 \cdot H_2O$ (118) can similarly be thought of as an example of five-coordinated square pyramidal copper with Cu–N bonds in the basal plane 2.04 to 2.06 Å in length and an apical Cu–OH_2 bond length of 2.59 Å. There is a second Cu–OH_2 bond at 3.37 Å that completes the octahedral coordination; whether one counts this or not is a matter of choice. Two examples of Cu in trigonal bipyramidal configurations are known. In the compound $Cr(NH_3)_6CuCl_5$, the $CuCl_5^-$ ion is essentially a perfect trigonal bipyramid (119). In iodobis(2,2′bipyridyl)Cu(II) iodide (120) the basal plane is made up of two nitrogens from different bipyridyls and an iodine; the apical Cu–N bonds are of the same length as the basal Cu–N bonds.

There are several dimeric copper compounds known in which the copper is five-coordinated and invariably square pyramidal. In copper formate (121) the basal Cu–O distances range from 1.93 to 1.99 Å and the apical Cu–O (the O from a second formate) is 2.40 Å in length. A similar arrangement occurs in bis(N,N-di-*n*-propyldithiocarbonato)Cu(II) (122) where the basal Cu–S bonds are 2.32 Å in length and the apical Cu–S bond length is 2.71 Å. The Cu is 0.4 Å above the basal plane. Similarly in copper dimethylglyoxime (123) the Cu–N bonds range from 1.91 to 1.96 Å in length and the apical Cu–O bond length is 2.43 Å; in bis-salicylaldehyde ethylenedi-imine copper (124) the basal Cu–O bond lengths are 1.91 and 2.03 Å and the apical is 2.41 Å. There is a C completing the octahedral coordination at 3.84 Å, which may be counted if one prefers octahedral copper(II). Finally in acetylacetonato-*o*-hydroxyanilato-copper(II) one of the two nonequivalent copper environments is that of a distorted square pyramid with Cu–O and Cu–N basal bonds ranging from 1.91 to 2.00 Å and with an apical Cu–O bond length of 2.68 Å (125).

Group IIB.—One trigonal bipyramidal and two square pyramidal Zn compounds have been reported. In Zn(acetylacetone)$_2$monohydrate (126, 127), the square pyramidal arrangement is distorted[4]; the basal Zn–O bonds range from 2.01 to 2.04 Å in length, while the Zn–OH_2 apical bond length is 2.00 Å; the Zn lies 0.41 Å above the base. In bis-salicylaldehyde ethylenedi-imine zinc monohydrate (128) the basal plane contains two nitrogens and two oxygens (Zn–O = 1.94 Å, Zn–N = 2.08 Å). The apical Zn–OH_2 distance is 2.13 Å and the Zn is 0.34 Å above the basal plane. In Zn(terpyridyl)Cl_2 (129), the Zn is in a trigonal bipyramid with the basal plane composed of one

[4] The distortion is sufficient so that a distinction between square pyramidal (127) and trigonal bipyramidal (126) is largely a matter of taste.

nitrogen and two chlorines and with two nitrogens apical. The Zn–Cl distance is 2.3 Å and the Zn–N distance 2.2 Å. This compound is supposedly isomorphous with the Cu and Cd salts, but isomorphism, as judged by similarity of cell edges, is no guarantee that the compounds are isostructural.

In the compound Hg_2O_2NaI (130), the Hg lies in the base of a tetragonal pyramid composed of two *trans* oxygen atoms at 2.03 Å and two *trans* iodine atoms at 3.33 Å. An apical Hg–I bond of 3.35 Å completes the coordination. This is a three-dimensional structure with linkage through the O–I edges.

Insofar as we know, these are the only tabulated five-coordinated transition metal compounds for which there is diffraction evidence. [We have specifically ignored linkages which may be ascribed to π bonding, as in tetracarbonyl acrylonitrile iron (131). There seems no particularly compelling reason to talk about double-bond centers occupying vertices of specific coordination polyhedra.] In some instances, the evidence we have cited is inaccurate, and, unfortunately, in many of these studies the information is scanty and preliminary and has not been followed by a full paper. [It would be interesting to know the percentage of communications to the *Journal of the American Chemical Society* and to the *Proceedings of the Chemical Society* that are followed by full papers; in the case of *Physical Review Letters* for 1961 this appears to be less than 50 per cent (132).] Nyholm (79), in his remarkably informative Tilden lecture, emphasized that among the factors that may affect coordination number and geometry in transition metal complexes are: (*a*) the oxidation state of the metal (when this can be defined); (*b*) the nature of the ligand; (*c*) the particular metal atom involved; (*d*) the type of bond (especially if double bonding occurs); and (*e*) steric effects. Among the structures discussed above, there are 9 trigonal bipyramids, and 16 square pyramids (of which 7 have long apical bonds and 3 have short apical bonds). It is clear that since the number of possible factors that affect coordination is a significant fraction of the number of experimental results, nothing definitive can be said. Some of these factors have been favored on theoretical grounds (e.g., 81, 82), but obviously there are too few experimental facts to lend credence to any particular discussion in which some of these factors are emphasized at the expense of others. What is needed is additional careful diffraction work on five-coordinated or suspected five-coordinated compounds, and a casual reading of the recent literature will reveal any number of potentially interesting compounds (83, 84, 85).

Hydrogen Bonding

The investigation and interpretation of hydrogen bonding continues to occupy, and at times baffle, many investigators using a variety of techniques. The subject was reviewed in some depth three years ago in this series (5), and here an attempt will be made to include additional data and to discuss different aspects of this vast subject.

Strong hydrogen bonds in solids.—In simplest terms there are three possible potential functions that might be considered for the hydrogen-bonded system AHA. The first function is an asymmetric, double-minimum one, with one minimum substantially lower than the other; this potential function corresponds to the bonding system A–H···A and to an asymmetric hydrogen bond. (It also corresponds invariably to the system AHB.) Such bonds are easy to distinguish, both spectroscopically (133) and crystallographically (e.g., 134), and these will not be discussed further. The second possible potential function is symmetric, has a single minimum, and obviously corresponds to the truly symmetric hydrogen bond. The third possible potential function is a symmetric, double-minimum one. If the barrier is well above the level of the ground state, then the potential corresponds to bonding in which there is, relative to the time-scale of the experimental method, an equal distribution of A–H···A and A···H–A. Such a situation can be recognized by characteristic infrared (133) and proton magnetic resonance spectra (135). It is most easily recognized in the far infrared (136), if the intense transition between the symmetric and antisymmetric components of the split ground-state level can be detected. Now that commercial instruments are available that enable reliable spectra to be recorded down to 50 cm^{-1} or lower, the presence of a lower barrier with respect to the ground state should be detectable. Of course, in complex substances it is not always possible to interpret the far infrared region unambiguously, even with the use of deuterated analogues.

A simple determination of the A–A distance is not sufficient to enable distinction to be made between the double-minimum potential with low barrier (hereafter referred to simply as the double-minimum potential) and the single-minimum potential. In fact, in the range of OHO distances between 2.4 and 2.6 Å, there occur not only compounds whose infrared spectra are characteristic of these two potentials, but also compounds whose infrared spectra clearly indicate a high barrier. Thus, in this region the oft-quoted correlations of O–H stretching frequency with O–O distance (149–152) must break down. In Table I we present some data, obtained mainly by Speakman and co-workers and Hadži and co-workers, that illustrate this point; we also indicate the required crystallographic symmetry between the two A atoms in the AHA bond. The existence of this symmetry prevents the determination, by either X-ray or neutron diffraction methods alone, of the nature of the bonding, for the model A–H–A and the statistical distribution of A–H···A and A···H–A are equally consistent with the symmetry. Naturally, these two models should lead to a different distribution of scattering density, even though both must obey the symmetry operation. Yet the distribution of scattering density is also strongly affected by zero-point motions of the scatterers, and without some knowledge of such motions (perhaps from a different kind of experiment) it is impossible to separate the effects of motion and of hydrogen position. This has been pointed out recently in studies of the bifluoride ion (153, 154). In these studies, however, it was possible, from the neutron diffraction data, to derive the mean-square amplitudes of vibration of H (or D) and F along the F–F bonds. Although these amplitudes are a manifestation of all of the motions of the ion in the solid, and hence cannot be calculated with the necessary degree of accuracy, the difference between the amplitudes of H and F along the bond should depend only upon those motions in which H and F move differently, namely the symmetric and asymmetric stretching modes. Thus, on the assumption that the bond is symmetric, it is possible to compute, from a simple normal-coordinate treatment and the known spectroscopic frequencies, the expected difference in amplitudes of H and F along the bond and to compare this with the difference obtained from the diffraction studies as a function of the assumed half-H positions in the double-minimum model F–H···F and F···H–F. The agreement turns out to be very satisfactory for a symmetric FHF bond. This combination of diffraction and spectroscopic data is obviously a satisfying approach to the problem of determining the symmetry of hydrogen bonds, but its application is limited to very simple systems, unless one is willing to make many simplifying and dubious assumptions.

Investigation of the bonding in systems with very short AHA bonds is facilitated if no crystallographic symmetry is imposed on the bond. Surprisingly, few studies of such bonds have been made. Recently, Ellison & Levy (155) carried out a neutron diffraction study on potassium hydrogen chloromaleate. They find the O–O distance to be 2.403±0.003 Å, somewhat shorter than in the parent compound (Table I). Of more interest is the fact that in this compound there is no crystallographic symmetry imposed on the

TABLE I

DIFFRACTION AND SPECTROSCOPIC DATA ON SOME ACID SALTS OF MONOBASIC ACIDS

Compound	Symmetry at H site	Type[a]	O···O (Å)	ν_{OH}	ν_{OD}	References[c] Distance
KH di-*p*-hydroxybenzoate hydrate	$\bar{1}$	*SDM*	2.61, 2.69	1700	1380	(137)
KH dibenzoate	$\bar{1}$	*SDM*	2.51	1700	1390	(139)
NH_4(Rb, K)H disalicylate hydrate		*ADM*	2.45, 2.65, 2.76			(134)
RbH di-*o*-nitrobenzoate	$\bar{1}$	*SDM*	2.43 [6][b]			(140)
KH di-*p*-nitrobenzoate		*ADM*	2.492 [20]	2185		(140)
NaH diacetate	2	*SDM*	2.444 [10]	1700	1320	(141)
NH_4H dicinnamate	$\bar{1}$	*SDM*	2.51 [3]			(142)
K(NH_4, Rb)H di-*p*-chlorobenzoate	$\bar{1}$	*SDM*	2.457 [13]			(143)
RbH bisglycollate		*ADM*	2.53 [1]			(144)
KH di-trifluoroacetate	$\bar{1}$	*SDM*	2.435 [7]			(144)
CsH di-trifluoracetate	$\bar{1}$	*SDM*	2.38 [3]			(144)
di-*p*-chlorophenyl H phosphate	2	*SDM*	2.398 [18]			(145)
KH bisphenylacetate	$\bar{1}$	*SDM*	2.54	1715	1370	(146, 147)
KH maleate	m	*SDM*	2.437 [4]	1700	1420	(148)

[a] *SDM* = symmetric double-minimum (with the barrier possibly below ground-state level); *ADM* = asymmetric double-minimum.

[b] Numbers in brackets are estimated standard deviations in the least-significant digit(s) of the preceding entry.

[c] Reference to the spectra are (133, 138).

ion and hence the structure can be refined directly without the usual singularity problems (153) inherent in the analysis when symmetry is imposed. The refinement leads to a centered distribution of scattering density for the hydrogen nucleus. Naturally, this result does not preclude a statistical distribution of O–H···O and O···H–O; but since the separate halves of the molecule to which the oxygens are attached are chemically and crystallographically different, this result seems most consistent with a symmetric bond. Ellison & Levy, in a manner similar to that discussed above (153, 154), also employ arguments based on differences of motion along the bond as calculated from spectroscopic data and as derived from the neutron diffraction data and conclude that the bond is probably truly centered.

We have indicated that in general it is extremely difficult, if not impossible, on the basis of diffraction or spectroscopic data of the type discussed so far, to distinguish the single-minimum potential function from the double-minimum function. Yet it is clearly of interest to determine the potential functions in various strong hydrogen-bonding systems, for example in those depicted in Table I. How can this be done? A possible way is through the study of both the OHO and ODO systems, as discussed in the next section.

Isotope effects in hydrogen bonding in solids.—In principle, the substitution of D for H might provide a sensitive probe of the hydrogen-bond potential function. Yet very few definitive experiments have been carried out.

The first study of the effects of isotopic substitution on hydrogen bonds in solids was made by Robertson & Ubbelohde (156) and was followed by further studies by Ubbelohde and co-workers. These data, along with a few more recent entries, are presented in Table II. Several points are worth noting. The most important is that the differences in lengths of the hydrogen bond on isotopic substitution are derived from the observed changes in lattice constants of the solid. In essence, it is assumed that the isotope effects are localized in the bonds of interest, rather than being distributed over the entire crystal. This assumption is difficult to justify. A second point is that the obvious trend of the data of Table II is toward a small effect for very strong or very weak bonds, but a sizable effect in the region of intermediate bonds.

Qualitative explanations for the limiting cases of very weak and very strong hydrogen bonds are possible. For very weak bonds the A–H···A mean length probably exceeds that of the A–D···A length because A–A coupling can be ignored and because the anharmonic nature of the potential function leads to a mean A–H bond length that is longer than the mean A–D bond length. A simple calculation, based on guessed-at values for the anharmonicity coefficient in solids (165), suggests an effect of the order of 0.005 Å. For very short, presumably symmetric bonds the explanation also involves anharmonic effects, this time through a coupling of the two stretching modes; the direction of the effect will depend upon the sign of the coupling constant. This problem has been discussed in detail recently (158).

TABLE II

ISOTOPE EFFECTS ON HYDROGEN BOND LENGTHS IN SOLIDS

Crystal	r_{AA}	$(r_D - r_H)$	Reference
KHF_2	2.277 [6][a]	≤ -0.0024	(154, 157)
$NaHF_2$	2.264 [3]	−0.0046 [5]	(153, 158)
$NH_4H_2PO_4$	2.49	0.010	(159)
Oxalic acid dihydrate	2.49	0.040_5	(156)
$HCrO_2$	2.49	0.06	(160, 161)
KH_2PO_4	2.48	0.0097	(157)
KH_2AsO_4	2.49	0.0080	(162)
Acetylene dicarboxylic acid dihydrate	2.56	0.034	(163)
Succinic acid	2.64	≈ 0.018	(156)
Resorcinol (α and β)	2.70, 2.75	No measurable effect	(156)
Ice (0°)	2.76	0.001_5	(164)
Urea	2.99	Slightly negative	(157)

[a] See Table I for explanation.

One important point is that the isotope effect, as derived from diffraction data, provides a means of evaluating a coupling constant that would be very difficult to obtain from spectroscopic data. In this way, a possible potential function for the stretching region in the bifluoride ion was derived.

The explanation of the large isotope effects, with the deuterium bond longer than the hydrogen bond, in bonds of intermediate strength has been the subject of much discussion and some controversy (as have these effects in general) (166–170). Before his untimely death, Rundle (171) offered a revised qualitative explanation of the effect that was based in part on unpublished calculations of simulated potential functions of Boardman (see 171) and in part on the observation (160) that in the chromous acid system the spectroscopic data are consistent with a symmetric O–H–O bond (length 2.49 Å) but with an asymmetric O–D···O bond (length 2.55 Å). These data have been corroborated recently (172), and it has been shown that a similar change in potential function occurs in the $HCoO_2$–$DCoO_2$ system. Rundle argues that as the lowest hydrogen level approaches the top of the barrier, the difference between D and H distributions becomes marked, so that the hydrogen has considerably more density than deuterium in the center of the double well. Hence, the oxygens should be attracted and the O–H–O bond should be shorter. This qualitative explanation, given in more detail in Rundle's paper, should be amenable to a more extensive calculational study in which one solves for the energy levels and distributions of two asymmetric potential functions as their minima are brought closer together. Such calculations have not yet been made. An important consequence of Rundle's model is that the isotope effect should provide a more general means of distinguishing a single-minimum potential function from a double-minimum one with low barrier. If a pronounced isotope effect is present, then there is a double well with sufficient barrier to cause a split in the distribution function of the deuterium. If there is none, then the well has effectively a single minimum. One must detect a significant difference in OHO and ODO bond lengths, and hence one must determine the individual bond lengths to about 0.005 Å, a level easily within today's experimental techniques. Thus far, mainly because of Rundle's death, such studies have not been carried out.

Miscellaneous topics in hydrogen bonding.—In a very interesting paper, Baur (173) has shown from evaluation of the Madelung constants for different orientations of water molecules in crystalline hydrates, that the positions where the hydrogen atoms show least energy are very close to those determined experimentally. This appears to be true for severely bent (174) and bifurcated hydrogen bonds (175), as well as for one case where no hydrogen bonding occurs (174). This result supports the view that these bonds are essentially electrostatic.

There appear to have been no new developments in the explanation of anomalously short C–H, N–H, and O–H distances, found on occasion in accurate X-ray studies, since Hamilton's review (5). However, it is clear that as the quality of X-ray data improves, it should be possible to carry out meaningful refinements of the hydrogen atom parameters and perhaps to compare the results with those obtained in neutron diffraction studies. Jensen & Sundaralingam (176, 177) have shown that, at least in a few compounds, the thermal parameters of the hydrogen atoms determined by refinement appear to be less than those on the atoms to which they are bound. This physically untenable situation is ascribed to an incorrect model for the scattering, specifically to the fact that the scattering factor for the hydrogen is taken to be that of the free spherical atom, whereas for a bonded atom it would appear, on the basis of these results, that the electron charge cloud is more localized. What bearing, if any, these results on bonding effects will have on the explanation of the anomalous distances remains to be seen.

Metal–Hydrogen Bond in Transition Metal Hydrides

The nature of the metal–hydrogen bond has received renewed attention in the last few years, primarily as a result of the discovery in the last ten years of remarkable new classes of molecular hydrides (178). Prior to 1955, no molecular hydrides were known apart from the very unstable metal carbonyl hydrides. In 1955, Wilkinson & Birmingham (179) prepared π-$(C_5H_5)_2ReH$, a compound sensitive to air but stable to moisture; in 1957, Chatt, Duncanson & Shaw (180) prepared $HPtCl(P(C_2H_5)_3)_2$, a compound that is air-stable. A variety of other classes of stable molecular hydrides have now been prepared.

Of the carbonyl hydrides, only those of cobalt, $HCo(CO)_4$, and iron, $H_2Fe(CO)_4$, have been studied to any extent. These volatile substances decompose at about −20° C. Ewens & Lister (98) studied the molecular structures of these compounds by electron diffraction. They concluded that in both compounds the metal was tetrahedrally coordinated by carbons of the carbonyl groups, but they failed to test all reasonable models for the structures, in particular the *cis*-octahedral model for the iron compound. Since they found no evidence for a stereochemical influence of the hydride hydrogens, they concluded that the hydrogens must be attached to the oxygens of the carbonyl groups. Yet there was reasonable chemical evidence that the hydrogens in such compounds are attached to the metal. In fact, Hieber (181) thought it likely that the hydrogen atoms were imbedded as protons in the electronic core of the metal atoms, the groups H_2Fe and HCo functioning as pseudometal atoms of atomic number 28. This view received at least some support when Wilkinson & Birmingham (178) for the first time proved by proton magnetic resonance techniques that the hydrogen was indeed attached to the metal in a molecular hydride, in this case π-$(C_5H_5)_2ReH$.

The first of a series of spectroscopic studies of the metal carbonyl hydrides was made in 1955 (182), and, in 1956, the first metal–hydrogen vibration was identified in $HCo(CO)_4$ (183). In these studies, C_{3v} symmetry was assigned to $HCo(CO)_4$. [Stammreich et al. (184) have pointed out that although these spectra are compatible with C_{3v} symmetry for $HCo(CO)_4$, T_d symmetry cannot be ruled out, as it can in the case of the iso-electronic ion $HFe(CO)_4^-$.] There ensued a controversy between Cotton and Edgell (73, 183, 185–189) which finally resulted in Edgell's (189) proving that the 703 cm^{-1} band in $HCo(CO)_4$ is a Co–H bending vibration, rather than a Co–H stretching vibration as contended by Cotton (73, 186), and Cotton's (187) showing that the LCAO calculation of Edgell & Gallup (185) was incorrect. Cotton (187), in his LCAO calculation of the Co–H bond in $HCo(CO)_4$, concluded that the Co–H distance must be less than 1.2 Å. This gave added support to Hieber's view of the hydrogen buried in the metal orbitals, a view already accepted by Cotton & Wilkinson (186) as the explanation for the very large shifts to high fields of the proton resonance signals from these hydrides. Additional support for the short metal–hydrogen distance was obtained in a broad-line nuclear magnetic resonance study of $H_2Fe(CO)_4$ (190). The second moment of the proton resonance was found to be about 8 G^2; and, on the assumption that intermolecular contributions to this moment could be neglected, the H–H intramolecular distance was found to be 1.88 ± 0.05 Å. For the derivation of an Fe–H distance from this H–H distance, the geometry of the molecule must be known; the various assumed geometries are discussed in detail in the original paper (190). It is concluded that the Fe–H distance is about 1.1 Å, of the same order as, if not actually shorter than, the covalent radius of the iron atom. Additional apparent support for the buried hydrogen came with the discovery of $HMn(CO)_5$ (191), a substance considerably more stable than either the iron or cobalt carbonyl hydrides. In physical properties, $HMn(CO)_5$ resembles

$Fe(CO)_5$ very closely, and, since $Fe(CO)_5$ was known to be a trigonal bipyramid (98), it was concluded that $HMn(CO)_5$ was also a trigonal bipyramid, a geometry again consistent with the hydrogen exerting no stereochemical influence. This geometry was supported in two independent infrared studies of the spectra of manganese carbonyl hydride and deuteride (72, 73). From the number of bands observed in the carbonyl stretching region, it was concluded that the symmetry of the molecule could not be as high as C_{4v}, but could be either C_s or C_{2v}.

This history has been cited at some length to indicate that the conclusions that the hydrogen in these compounds exerts no stereochemical influence and that it is buried in the metal orbitals are based on interpretations of a number of independent spectroscopic experiments. The fact that these conclusions now appear to be extremely doubtful is thus all the more remarkable.

As soon as diffraction studies were carried out on some members of the newer, more stable classes of molecular hydrides, it became evident that the hydrogen does in fact exert a profound influence on the stereochemistry. In the first such study (192), that of the structure of $HPtBr(P(C_2H_5)_3)_2$, it was found that the Pt is arranged at the center of a square, with Br and *trans*-P atoms occupying three of the four corners. The basic features of this geometry have been confirmed recently (193) in a much more accurate study of the structure of $HPtCl(P(C_6H_5)_2(C_2H_5))_2$. Similarly, in a study of the crystal structure of $HOsBr(CO)(P(C_6H_5)_3)_3$ (194) Orioli & Vaska showed that the Br, C (of the CO), and three P atoms are situated at five of the six vertices of a distorted octahedron. The positions of the hydrogen atoms were not determined in these studies. Finally, in an X-ray study of $HRh(CO)(P(C_6H_5)_3)_3$ (114, 115), it was shown that the Rh is near the center of a trigonal bipyramid, with three P atoms in the base, the C of the CO at one apex, and the H at the other. The Rh–H distance was found to be a normal covalent length of 1.60 ± 0.12 Å (115). In a subsequent neutron diffraction study (195) of the remarkable hydride K_2ReH_9, the average Re–H distance was found to be 1.68 ± 0.01 Å.

It could be argued, of course, that these molecular hydrides are somehow different from the parent carbonyl hydrides. Yet in a recent X-ray study of the structure of $HMn(CO)_5$ (75, 76) it was shown that the $Mn(CO)_5$ portion of the molecule does not deviate significantly from C_{4v} symmetry; that is, the CO groups occupy five of the six vertices of an octahedron. (The position of the hydrogen atom was not located.) This symmetry is very different and is higher than that thought possible from the spectroscopic studies of $HMn(CO_5)$ in the gas phase (72, 73), and it was suggested that something must be wrong with the interpretation of the spectra. Huggins & Kaesz (74) have now re-investigated the spectra and conclude that they are indeed compatible with C_{4v} symmetry for the molecule. Apparently, previous investigators mis-assigned CO stretching bands. The results of the X-ray study of $HMn(CO)_5$ also suggest what might be wrong with the interpretation of the broad-line magnetic resonance study on $H_2Fe(CO)_4$ (190). The molecules of $HMn(CO)_5$ pack in such a way that the octahedral sites of the molecules not occupied by carbonyls are in close contact. Hence, there is very probably a relatively close H–H interaction. If it is assumed that the molecule has C_{4v} symmetry, then a van der Waals H–H contact of 2.2 to 2.4 Å leads to a Mn–H distance of 1.6 to 1.5 Å. More important, the possibility of this close intermolecular interaction in $HMn(CO)_5$ makes untenable the assumption made in the magnetic resonance study of $H_2Fe(CO)_4$ that intermolecular interactions could be neglected. In fact, such intermolecular interactions undoubtedly contribute significantly to the large second moment and eliminate the need to postulate a short H–H intramolecular contact.

The remaining evidence that could be cited for the buried hydrogen is the large chemical shift of the proton. However, calculations by Lipscomb and co-workers (196, 197) have shown that it is possible to account for these high-field shifts with normal geometries and with normal covalent metal–hydrogen distances. Quite recently, Buckingham & Stephens (198) have shown in a detailed calculation that it is possible to account for the observed shifts with abnormal geometry! i.e., with buried hydrogens. Yet they comment (in a note added in proof) that metal–hydrogen distances up to 1.6 Å in length could be tolerated in their calculations. Clearly, the differences in these various molecular orbital calculations indicate that they do not provide a sensitive means of determining the metal–hydrogen distance, and one must be satisfied that the high-field shifts can be accounted for with normal geometry.

Despite the seemingly impressive evidence of a few years ago, there are at this time no convincing experimental data to support the view that the hydrogen in transition metal hydrides is buried in the metal orbitals. Rather, all evidence now points to the hydrogen exerting a profound influence on the stereochemistry and being situated at a normal covalent distance from the metal, although the metal–hydrogen bond length has not yet been established directly for a parent carbonyl hydride.

ACKNOWLEDGMENT

It is a pleasure to acknowledge the help I received from a number of colleagues who sent me relevant information prior to publication.

LITERATURE CITED

1. Lewis, J., *Pure Appl. Chem.*, **10**, 11 (1965)
2. Weiss, R. J., *Physics Today*, **18**, No. 4, 43 (1965)
3. Lipscomb, W. N., *Boron Hydrides* (Benjamin, New York, 1963)
4. Muetterties, E. L., Ed., *The Chemistry of Boron* (Wiley, New York, 1965)
5. Hamilton, W. C., *Ann. Rev. Phys. Chem.*, **13**, 19 (1962)
6. Berthelot, M., *Compt. Rend.*, **120**, 581 (1895)
7. Pauling, L., *J. Am. Chem. Soc.*, **55**, 1895 (1933)
8. Pimentel, G. C., *J. Chem. Phys.*, **19**, 446 (1951)
9. Yost, D. M., and Kaye, A. L., *J. Am. Chem. Soc.*, **55**, 3890 (1933)
10. Thorne, P. C. L., and Roberts, E. R., *Fritz Ephraim Inorganic Chemistry*, 6th ed., 65 (Interscience, New York, 1954)
11. Cotton, F. A., and Wilkinson, G., *Advanced Inorganic Chemistry*, 145 (Interscience, New York, 1962)
12. Emeléus, H. J., and Anderson, J. S., *Modern Aspects of Inorganic Chemistry*, 3rd ed., 435 (Van Nostrand, New York, 1960)
13. Bartlett, N., and Lohmann, D. H., *Proc. Chem. Soc.*, 115 (1962)
14. Bartlett, N., and Lohmann, D. H., *J. Chem. Soc.*, 5253 (1962)
15. Bartlett, N., *Proc. Chem. Soc.*, 218 (1962)
16. Claassen, H. H., Selig, H., and Malm, J. G., *J. Am. Chem. Soc.*, **84**, 3593 (1962)
17. Coulson, C. A., *J. Chem. Soc.*, 1442 (1964)
18. Ibers, J. A., and Hamilton, W. C., *Science*, **139**, 106 (1963)
19. Templeton, D. H., Zalkin, A., Forrester, J. D., and Williamson, S. M., *J. Am. Chem. Soc.*, **85**, 242 (1963)
20. Hamilton, W. C., and Ibers, J. A., in *Noble-Gas Compounds*, 195 (Hyman, H. H., Ed., Univ. Chicago Press, Chicago, 1963)
21. Templeton, D. H., Zalkin, A., Forrester, J. D., and Williamson, S. M., in *Noble-Gas Compounds*, 203 (See ref. 20)
22. Burns, J. H., Agron, P. A., and Levy, H. A., in *Noble-Gas Compounds*, 211 (See ref. 20)
23. Burns, J. H., Agron, P. A., and Levy, H. A., *Science*, **139**, 1208 (1963)
24. Chernick, C. L., Claassen, H. H., Fields, P. R., Hyman, H. H., Malm, J. G., Manning, W. M., Matheson, M. S., Quarterman, L. A., Schreiner F., Selig, H .H., Sheft, I., Siegel, S., Sloth, E. N., Stein, L., Studier, M. H., Weeks, J. L., and Zirin, M. H., *Science*, **138**, 136 (1962))
25. Smith, D. F., in *Noble-Gas Compounds*, 295 (See ref. 20)
26. Claassen, H. H., Chernick, C. L., and Malm, J. G., *J. Am. Chem. Soc.*, **85**, 1927 (1963)
27. Pimentel, G. C., and Spratley, R. D., *J. Am. Chem. Soc.*, **85**, 826 (1963)
28. Bohn, R. K., Katada, K., Martinez, J. V., and Bauer, S. H., in *Noble-Gas Compounds*, 238 (See ref. 20)
29. Siegel, S., and Gebert, E., *J. Am. Chem. Soc.*, **85**, 240 (1963)
30. Smith, D. F., *J. Chem. Phys.*, **38**, 270 (1963)
31. Levy, H. A., and Agron, P. A., *J. Am. Chem. Soc.*, **85**, 241 (1963)
32. Burns, J. H., *J. Phys. Chem.*, **67**, 536 (1963)
33. Burns, J. H., Ellison, R. D., and Levy, H. A., in *Noble-Gas Compounds*, 226 (See ref. 20)
34. Gillespie, R. J., in *Noble-Gas Compounds*, 333 (See ref. 20)
35. Allen, L. C., and Horrocks, W. D., Jr., *J. Am. Chem. Soc.*, **84**, 4344 (1962)
36. Allen, L. C., *Science*, **138**, 892 (1962)
37. Allen, L. C., in *Noble-Gas Compounds*, 317 (See ref. 20)
38. Jortner, J., Wilson, E. G., and Rice, S. A., in *Noble-Gas Compounds*, 358 (See ref. 20)
39. Lohr, L. L., Jr., and Lipscomb, W. N., in *Noble Gas Compounds*, 347 (See ref. 20)
40. Weinstock, B., Claassen, H. H., and Chernick, C. L., *J. Chem. Phys.*, **38**, 1470 (1963)
41. Claassen, H. H., in *Noble-Gas Compounds*, 304 (See ref. 20)
42. Agron, P. A., and Levy, H. A., *Oak Ridge Natl. Lab. Chem. Div. Ann. Progr. Rept. No. ORNL-3679*, 108 (20 June 1964)
43. Sheft, I., Spittler, T. M., and Martin, F. H., *Science*, **145**, 701 (1964)
44. Chernick, C. L., Claassen, H. H., Malm, J. G., and Plurien, P. L., in *Noble-Gas Compounds*, 106 (See ref. 20)
45. Templeton, D. H., Zalkin, A., Forrester, J. D., and Williamson, S. M., *J. Am. Chem. Soc.*, **85**, 817 (1963)
46. Templeton, D. H., Zalkin, A., Forrester, J. D., and Williamson, S. M.,

in *Noble-Gas Compounds*, 229 (See ref. 20)
47. Malm, J. G., Holt, B. D., and Bane, R. W., in *Noble-Gas Compounds*, 167 (See ref. 20)
48. Hamilton, W. C., Ibers, J. A., and MacKenzie, D. R., *Science*, **141**, 532 (1963)
49. Ibers, J. A., Hamilton, W. C., and MacKenzie, D. R., *Inorg. Chem.*, **3**, 1412 (1964)
50. Zalkin, A., Forrester, J. D., Templeton, D. H., Williamson, S. M., and Koch, C. W., *Science*, **142**, 501 (1963)
51. Zalkin, A., Forrester, J. D., and Templeton, D. H., *Inorg. Chem.*, **3**, 1417 (1964)
52. Zalkin, A., Forrester, J. D., Templeton, D. H., Williamson, S. M., and Koch, C. W., *J. Am. Chem. Soc.*, **86**, 3569 (1964)
53. Helmholz, L., *J. Am. Chem. Soc.*, **59**, 2036 (1937)
54. Peterson, E. R., and Peterson, S. W., *Abstr. H-10, Am. Cryst. Assoc. Meeting, Bozeman, Montana, 26–31 July 1964*
55. Ibers, J. A., *Acta Cryst.*, **9**, 225 (1956)
56. Selig, H., Claassen, H. H., Chernick, C. L., Malm, J. G., and Huston, J. L., *Science*, **143**, 1322 (1964)
57. Turner, J. J., and Pimentel, G. C., in *Noble-Gas Compounds*, 101 (See ref. 20)
58. Turner, J. J., and Pimentel, G. C., *Science*, **140**, 974 (1963)
59. MacKenzie, D. R., *Science*, **141**, 1171 (1963)
60. Schreiner, F., Malm, J. G., and Hindman, J. C., *J. Am. Chem. Soc.*, **87**, 25 (1965)
61. Siegel, S., and Gebert, E., *J. Am. Chem. Soc.*, **86**, 3896 (1964)
62. Grosse, A. V., Kirshenbaum, A. D., Streng, A. G., and Streng, L. V., *Science*, **139**, 1047 (1963)
63. Streng, A. G., Kirshenbaum, A. D., Streng, L. V., and Grosse, A. V., in *Noble-Gas Compounds*, 73 (See ref. 20)
64. Waters, J. H., and Gray, H. B., *J. Am. Chem. Soc.*, **85**, 825 (1963)
65. Fields, P. R., Stein, L., and Zirin, M. H., *J. Am. Chem. Soc.*, **84**, 4164 (1962)
66. Selig, H., *Science*, **144**, 537 (1964)
67. Edwards, A. J., Holloway, J. H., and Peacock, R. D., *Proc. Chem. Soc.*, 275 (1963)
68. Peacock, R. D., Selig, H., and Sheft, I., *Proc. Chem. Soc.*, 285 (1964)
69. Orgel, L. E., *Quelques Problèmes de Chimie Minérale, 10th Solvay Conf.*, 289 (Stoops, Brussels, 1956)
70. Robinson, W. T., Fergusson, J. E., and Penfold, B. R., *Proc. Chem. Soc.*, 116 (1963)
71. Bertrand, J. A., Cotton, F. A., and Dollase, W. A., *J. Am. Chem. Soc.*, **85**, 1349 (1963)
72. Wilson, W. E., *Z. Naturforsch.*, **13b**, 349 (1958)
73. Cotton, F. A., Down, J. L., and Wilkinson, G., *J. Chem. Soc.*, 833 (1959)
74. Huggins, D. K., and Kaesz, H. D., *J. Am. Chem. Soc.*, **86**, 2734 (1964)
75. La Placa, S. J., Ibers, J. A., and Hamilton, W. C., *J. Am. Chem. Soc.*, **86**, 2288 (1964)
76. La Placa, S. J., Hamilton, W. C., and Ibers, J. A., *Inorg. Chem.*, **3**, 1491 (1964)
77. Daudel, R., and Bucher, A., *J. Chim. Phys.*, **42**, 6 (1945)
78. Gillespie, R. J., and Nyholm, R. S., *Prog. Stereochem.*, **2**, 261 (1958)
79. Nyholm, R. S., *Proc. Chem. Soc.*, 273 (1961)
80. Gillespie, R. J., *J. Chem. Soc.*, 4672 (1963)
81. Gillespie, R. J., *J. Chem. Soc.*, 4679 (1963)
82. Ballhausen, C. J., and Gray, H. B., *Inorg. Chem.*, **2**, 426 (1963)
83. Jensen, K. A., Nygaard, B., and Pedersen, C. T., *Acta Chem. Scand.*, **17**, 1126 (1963)
84. Langford, C. H., Billig, E., Shupack, S. I., and Gray, H. B., *J. Am. Chem. Soc.*, **86**, 2958 (1964)
85. Benner, G. S., Hatfield, W. E., and Meek, D. W., *Inorg. Chem.*, **3**, 1544 (1964)
86. Holmes, R. R., *Abstr. 13A1, Intern. Coordination Chem. Conf., Vienna, Austria, 1964*
87. Muetterties, E. L., Mahler, W., and Schmutzler, R., *Inorg. Chem.*, **2**, 613 (1963)
88. Lindqvist, I., and Nahringbauer, G., *Acta Cryst.*, **12**, 638 (1959)
89. Hulme, R., *J. Chem. Soc.*, 1524 (1963)
90. Wells, A. F., *Z. Krist.*, **99**, 367 (1938)
91. Andersson, S., and Wadsley, A. D., *Nature*, **187**, 499 (1960)
92. Dodge, R. P., Templeton, D. H., and Zalkin, A., *J. Chem. Phys.*, **35**, 55 (1961)
93. Byström, A., Wilhelmi, K. A., and Brotzen, O., *Acta Chem. Scand.*, **4**, 1119 (1950)
94. Skinner, H. A., and Sutton, L. E.,

Trans. Faraday Soc., **36**, 668 (1940)
95. Zalkin, A., and Sands, D. E., *Acta Cryst.*, **11**, 615 (1958)
96. Ewens, R. V. G., and Lister, M. W., *Trans. Faraday Soc.*, **34**, 1358 (1938)
97. Sands, D. E., and Zalkin, A., *Acta Cryst.*, **12**, 723 (1959)
98. Ewens, R. V. G., and Lister, M. W., *Trans. Faraday Soc.*, **35**, 681 (1939)
99. Hanson, A. W., *Acta Cryst.*, **15**, 930 (1962)
100. Donohue, J., and Caron, A., *Acta Cryst.*, **17**, 663 (1964)
101. Stammreich, H., Sala, O., and Tavares Y., *J. Chem. Phys.*, **30**, 856 (1959)
102. Pistorius, C. W. F. T., and Haarhoff, P. C., *J. Chem. Phys.*, **31**, 1439 (1959)
103. O'Dwyer, M. F., *J. Mol. Spectr.*, **2**, 144 (1958)
104. Koenig, D. F., *Acta Cryst.*, **18**, 663 (1965)
105. Hamor, M. J., Hamor, T. A., Hoard, J. L., and Caughey, W. S., *Abstr. K-12, Am. Cryst. Assoc. Meeting, Bozeman, Montana, 26–31 July 1964*
106. Alderman, P. R. H., and Owston, P. G., *Nature*, **178**, 1071 (1956)
107. Alderman, P. R. H., Owston, P. G., and Rowe, J. M., *J. Chem. Soc.*, 668 (1962)
108. Mair, G. A., Powell, H. M., and Henn, D. E., *Proc. Chem. Soc.*, 415 (1960)
109. La Placa, S. J., and Ibers, J. A., *Inorg. Chem.*, **4**, 778 (1965)
110. Trifan, D. S., and Bacskai, R., *J. Am. Chem. Soc.*, **82**, 5010 (1960)
111. Rubalcava, H. E., and Thomson, J. B., *Spectrochim. Acta*, **18**, 449 (1962)
112. Davison, A., McFarlane, W., Pratt, L., and Wilkinson, G., *J. Chem. Soc.*, 4821 (1962)
113. King, R. B., *J. Am. Chem. Soc.*, **85**, 1922 (1963)
114. La Placa, S. J., and Ibers, J. A., *J. Am. Chem. Soc.*, **85**, 3501 (1963)
115. La Placa, S. J., and Ibers, J. A., *Acta Cryst.*, **18**, 511 (1965)
116. Mair, G. A., Powell, H. M., and Venanzi, L. M., *Proc. Chem. Soc.*, 170 (1961)
117. Llewellyn, F. J., and Waters, T. N., *J. Chem. Soc.*, 2639 (1960)
118. Mazzi, F., *Acta Cryst.*, **8**, 137 (1955)
119. Mori, M., Saito, Y., and Watanabe, T., *Bull. Chem. Soc. Japan*, **34**, 295 (1961)
120. Barclay, G. A., and Kennard, C. H. L., *Nature*, **192**, 425 (1961)
121. Barclay, G. A., and Kennard, C. H. L., *J. Chem. Soc.*, 3289 (1961)
122. Pignedoli, A., and Peyronel, G., *Gazz. Chim. Ital.*, **92**, 745 (1962)
123. Frasson, E., Bardi, R., and Bezzi, S., *Acta Cryst.*, **12**, 201 (1959)
124. Hall, D., and Waters, T. N., *J. Chem. Soc.*, 2644 (1960)
125. Barclay, G. A., Harris, C. M., Hoskins, B. F., and Kokot, E., *Proc. Chem. Soc.*, 264 (1961)
126. Lippert, E. L., and Truter, M. R., *J. Chem. Soc.*, 4996 (1960)
127. Montgomery, H., and Lingafelter, E. C., *Acta Cryst.*, **16**, 748 (1963)
128. Hall, D., and Moore, F. H., *Proc. Chem. Soc.*, 256 (1960)
129. Corbridge, D. E. C., and Cox, E. G., *J. Chem. Soc.*, 594 (1956)
130. Aurivillius, K., *Acta Chem. Scand.*, **14**, 2196 (1960)
131. Luxmoore, A. R., and Truter, M. R., *Proc. Chem. Soc.*, 466 (1961)
132. *Physics Today*, **17**, No. 9, 92 (1964)
133. Blinc, R., Hadži, D., and Novak, A., *Z. Elektrochem.*, **64**, 567 (1960)
134. Downie, T. C., and Speakman, J. C., *J. Chem. Soc.*, 787 (1954)
135. Blinc, R., and Hadži, D., *Spectrochim. Acta*, **16**, 853 (1960)
136. Hadži, D., *J. Chem. Phys.*, **34**, 1445 (1961)
137. Skinner, J. M., and Speakman, J. C., *J. Chem. Soc.*, 185 (1951)
138. Hadži, D., and Novak, A., *Nuovo Cimento*, **2**, Suppl. No. 3, 715 (1955)
139. Skinner, J. M., Stewart, G. M. D., and Speakman, J. C., *J. Chem. Soc.*, 180 (1954)
140. Shrivastava, H. N., and Speakman, J. C., *J. Chem. Soc.*, 1151 (1961)
141. Speakman, J. C., and Mills, H. H., *J. Chem. Soc.*, 1164 (1961)
142. Bryan, R. F., Mills, H. H., and Speakman, J. C., *J. Chem. Soc.*, 4350 (1963)
143. Mills, H. H., and Speakman, J. C., *J. Chem. Soc.*, 4355 (1963)
144. Golič, L., and Speakman, J. C., *J. Chem. Soc.*, 2521, 2530 (1965)
145. Calleri, M., and Speakman, J. C., *Acta Cryst.*, **17**, 1097 (1964)
146. Speakman, J. C., *J. Chem. Soc.*, 3357 (1949)
147. Bacon, G. E., and Curry, N. A., *Acta Cryst.*, **10**, 524 (1957)
148. Darlow, S. F., and Cochran, W., *Acta Cryst.*, **14**, 1250 (1961)
149. Rundle, R. E., and Parasol, M., *J. Chem. Phys.*, **20**, 1487 (1952)
150. Lord, R. C., and Merrifield, R. E., *J. Chem. Phys.*, **21**, 166 (1953)

151. Nakamoto, K., Margoshes, M., and Rundle, R. E., *J. Am. Chem. Soc.*, **77,** 6480 (1955)
152. Pimentel, G. C., and Sederholm, C. H., *J. Chem. Phys.*, **24,** 639 (1956)
153. McGaw, B. L., and Ibers, J. A., *J. Chem. Phys.*, **39,** 2677 (1963)
154. Ibers, J. A., *J. Chem. Phys.*, **40,** 402 (1964)
155. Ellison, R. D., and Levy, H. A., *Acta Cryst.* (In press, 1965)
156. Robertson, J. M., and Ubbelohde, A. R., *Proc. Roy. Soc. (London), A*, **170,** 222 (1939)
157. Ubbelohde, A. R., *Proc. Roy. Soc. (London), A*, **173,** 417 (1939)
158. Ibers, J. A., *J. Chem. Phys.*, **41,** 25 (1964)
159. Ubbelohde, A. R., and Woodward, I., *Proc. Roy. Soc. (London), A*, **179,** 399 (1942)
160. Snyder, R. G., and Ibers, J. A., *J. Chem. Phys.*, **36,** 1356 (1962)
161. Hamilton, W. C., and Ibers, J. A., *Acta Cryst.*, **16,** 1209 (1963)
162. Dickson, D. H. W., and Ubbelohde, A. R., *Acta Cryst.*, **3,** 6 (1950)
163. Gallagher, K. J., Ubbelohde, A. R., and Woodward, I., *Proc. Roy. Soc., (London), A*, **222,** 195 (1954)
164. Megaw, H. D., *Nature*, **134,** 900 (1934)
165. Ibers, J. A., *Acta Cryst.*, **12,** 251 (1959)
166. Bernal, J. D., and Tamm, G., *Nature*, **135,** 229 (1935)
167. Nordman, C. E., and Lipscomb, W. N., *J. Chem. Phys.*, **19,** 1422, (1951)
168. Rundle, R. E., *J. Chem. Phys.*, **21,** 937 (1953)
169. Ubbelohde, A. R., and Gallagher, K. J., *Acta Cryst.*, **8,** 71 (1955)
170. Gallagher, K. J., in *Hydrogen Bonding*, 45 (Hadži, D., Ed., Pergamon, London, 1959)
171. Rundle, R. E., *J. Phys.*, **25,** 487 (1964)
172. Benoit, A., *Spectrochim. Acta*, **19,** 2011 (1963)
173. Baur, W. H., *Acta Cryst.* (In press, 1965)
174. Baur, W. H., *Acta Cryst.*, **17,** 863 (1964)
175. Padmanabhan, V. M., Busing, W. R., and Levy, H. A., *Abst. 4.19, Intern. Union Cryst., Sixth Intern. Congr., Rome, Italy, 1963*
176. Jensen, L. H., and Sundaralingam, M., *Science*, **145,** 1185 (1964)
177. Sundaralingam, M., and Jensen, L. H., *Acta Cryst.* (In press, 1965)
178. Chatt, J., *Proc. Chem. Soc.*, 318 (1962)
179. Wilkinson, G., and Birmingham, J. M. *J. Am. Chem. Soc.*, **77,** 3421 (1955)
180. Chatt, J., Duncanson, L. A., and Shaw, B. L., *Proc. Chem. Soc.*, 343 (1957)
181. Hieber, W., *Die Chemie*, **55,** 24 (1942)
182. Friedel, R. A., Wender, I., Shufler, S. L., and Sternberg, H. W., *J. Am. Chem. Soc.*, **77, 3951** (1955)
183. Edgell, W. F., Magee, C., and Gallup, G., *J. Am. Chem. Soc.*, **78,** 4185 (1956)
184. Stammreich, H., Kawai, K., Tavares, Y., Krumholz, P., Behmoiras, J., and Bril, S., *J. Chem. Phys.*, **32,** 1482 (1960)
185. Edgell, W. F., and Gallup, G., *J. Am. Chem. Soc.*, **78,** 4188 (1956)
186. Cotton, F. A., and Wilkinson, G., *Chem. Ind. (London)*, 1305 (1956)
187. Cotton, F. A., *J. Am. Chem. Soc.*, **80,** 4425 (1958)
188. Edgell, W. F., Asato, G., Wilson, W., and Angell, C., *J. Am. Chem. Soc.*, **81,** 2022 (1959)
189. Edgell, W. F., and Summitt, R., *J. Am. Chem. Soc.*, **83,** 1772 (1961)
190. Bishop, E. O., Down, J. L., Emtage, P. R., Richards, R. E., and Wilkinson, G., *J. Chem. Soc.*, 2484 (1959)
191. Hieber, W., and Wagner, G., *Z. Naturforsch.*, **13b,** 339 (1958)
192. Owston, P. G., Partridge, J. M., and Rowe, J. M., *Acta Cryst.*, **13,** 246 (1960)
193. Eisenberg, R., and Ibers, J. A., *Inorg. Chem.*, **4,** 773 (1965)
194. Orioli, P. L., and Vaska, L., *Proc. Chem. Soc.*, 333 (1962)
195. Abrahams, S. C., Ginsberg, A. P., and Knox, K., *Inorg. Chem.*, **3,** 558 (1964)
196. Stevens, R. M., Kern, C. W., and Lipscomb, W. N., *J. Chem. Phys.*, **37,** 279 (1962)
197. Lohr, L. L., Jr., and Lipscomb, W. N., *Inorg. Chem.*, **3,** 22 (1964)
198. Buckingham, A. D., and Stephens, P. J., *J. Chem. Soc.*, 2747 (1964)

GAS KINETICS[1,2]

By S. W. Benson

Department of Thermochemistry and Kinetics, Stanford Research Institute, Menlo Park, California

AND

W. B. DeMore

Jet Propulsion Laboratory, California Institute of Technology, Pasadena, California

Introduction

The "literature explosion" has been making its impact as strongly felt in chemical kinetics as elsewhere in science. As a result, the suspension of the customary *Annual Review* for gas phase kinetics during "editorial" 1963 and 1964 created for the reviewers the forbidding prospect of almost 1500 manuscripts. This was a catastrophic situation whose recurrence might prove fatal for the series.

Because of the impossibility of dealing with this large quantity of work in any small space we have taken rather extreme liberties in compression and deletion of material. The present review covers journals appearing 1 January 1962 through 15 October 1964. We have omitted most data which were not quantitative and most data not capable of reasonably direct interpretation. Thus, the fields of combustion, flames, most complex pyrolyses, and oxidations have been almost entirely neglected. However, many of these are already the concern of specialized reviews and journals. Photochemistry, being covered in special reviews, is touched on only insofar as it yields direct rate data. The relatively new fields of ion reactions in gases and molecular beam kinetics are totally omitted. This is certainly not a rational, and hopefully only a temporary, divorce. Recent reviews of these fields have, however, appeared.

For more complex reasons, the specialties of energy transfer and isotope effect have been treated very superficially as has the field of radiation and "hot" radical kinetics. A decreasing but not inappreciable amount of work continues to be done in the formal mathematics of complex differential equations of kinetics. We have felt that this is no longer of sufficient interest to warrant inclusion. Our feelings are the same in regard to analyses of coupled hydrodynamics and kinetics occurring in subsonic and sonic flows—and the related problems of departures from "steady state" kinetics. A discussion of

[1] The survey of literature pertaining to this review was concluded October 1964.

[2] This paper presents the results of one phase of research carried out at the Jet Propulsion Laboratory, California Institute of Technology, under Contract No. NAS 7-100, sponsored by the National Aeronautics and Space Administration.

newer experimental techniques, while not without interest, seemed also too detailed for the present review. There have been a large number of books, reviews, symposia, informal meetings, and conferences in kinetics which will be touched upon only as they concern particular areas. There is clearly not adequate space to discuss these even briefly.

In the course of our searches, some interesting facts emerged which may be of general interest. Ninety per cent of the literature of homogeneous, gas phase kinetics is covered within the confines of about eight journals, and most of these are published in English. Since kinetics represents less than 10 per cent of the volume of these journals, a very compelling argument can be made for an International Journal of Gas Phase Kinetics. Gas Phase Kinetics and Photochemistry might be even more appealing. Who will be the first to sponsor it?

There is a striking need for common standards in proposing mechanisms and in presenting data. With rare exceptions it seems to us that rate constants are useful only where stoichiometry is known. This implies some concern with mass balances and while much current literature recognizes this, it is often glaringly absent.

Rate constants are generally presented in one of three forms, the Arrhenius form, $Ae^{-E/RT}$; the collisional form $A'T^{1/2}e^{-E'/RT}$, and the transition-state theory, $(kT/h)e^{-E^{\ddagger}/RT}e^{\Delta S^{\ddagger}/R}$. It would be pleasant to have at least one universal form, and we would plead that whatever form is used, the Arrhenius parameters be given in addition. This is experimentally justified on the grounds that no one has yet demonstrated curvature in Arrhenius plots of gas phase data. Also, since rate constants are frequently transferable between gas phase and solution, it would be nice to agree on a universal set of concentration units. In this case we would hope for moles/liter.

We should also like to propose an innovation. Our own experience has led us to represent equilibrium constants and rate constants in the form: $k(\text{or } K) = 10^{\log A}10^{-E/\theta} = 10^{\log A - E/\theta}$ where $\theta = 2.303$ RT in kcal/mole. This leads to much brevity and efficiency in expression and makes it easier to do mental arithmetic with rate constants and comparative rates. It leads to a very simple "pK" type language.

Finally, we should like to suggest some common criteria for rate constants; namely, that frequency factors for first- and second-order reactions be scrutinized for "reasonableness." It is not too difficult to estimate entropy changes for chemical reactions to within ± 3 Gibb/mole and, thus, the measurement of an A-factor for a reaction can be used to deduce the A-factor for the reverse reaction. Both should be "reasonable." Similarly, in proposing steps in a reaction mechanism, workers should avoid transition states with large strain or very low entropy unless no other path is available for reaction. Quite a large number of implausible reactions have survived in the literature with no more supporting evidence than an apparent inadequacy of a proposed mechanism. Our understanding of the details of step reactions has reached the point where such circumstantial evidence is surely inadequate and it is hoped that authors will exercise extreme parsimony in their proposals for new or "unusual" step reactions.

Elementary Atom Reactions

A number of recent reviews have appeared on atom reactions. Some excellent ones are by Kaufman (219) on O-atom reactions, by Cvetanovic (105) on addition of atoms to olefins, by Volman (391) on reactions in photochemically activated systems containing O_2 and H_2, by Avramenko & Kolesnikova (7) on reactions involving OH and O, and by Fettis & Knox (139) on halogen atom reactions.

Recombination.—Measurements of the homogeneous rates of recombination of atoms are extremely difficult to make, particularly at high temperatures where the equilibrium favors dissociation. Shock tube measurements give rates of dissociation:

$$M + X_2 \rightleftarrows M + 2X \qquad 1,2.$$

$$K_{eq} = k_1/k_2 \qquad 3.$$

and rate constants for dissociation are deduced from the thermodynamic relation $k_2 = k_1/K_{eq}$. This has generated quite some controversy because of the lack of attainment of a Maxwell distribution of energy states (48, 62, 300, 301, 320, 321), but the present consensus of opinion (48, 62, 320, 321) based on detailed balancing favors validity of Equation 3.

The measurement of E_2, the temperature coefficient of k_2, from shock tube measurements of E_1 is fraught with serious errors since E_2 is a very small number and E_1 is a large number, of the order of the bond dissociation energy, with a large error, easily of the order of E_2. Since shock tube measurements of k_1 by different authors show spreads of as much as a power of 10, it is difficult to place much faith in values of E_2 deduced in this fashion.

Recent measurements of atom recombination rates are summarized in Table I. Although agreement is fair in some cases one notes that the rates reported by different workers for Cl recombination vary by a factor of more than 1000. The reported rates for O+O recombination at high (322, 403) and low (274) temperatures are consistent with a temperature dependence of the order of T^{-1}. Previous measurements of this rate at high and low temperatures could not be correlated without postulating an unexpectedly strong temperature dependence (219).

The problem of the negative activation energy observed for recombination has received a great deal of theoretical and experimental investigation. Extending an old theory of Rice, Benson & Fueno (48) have shown a dependence of the order of T^{-1} is to be expected if excited molecules X_2^* lose or gain energy in units of about kT. The T^{-1} dependence then arises from the increasing chance of redissociation of X_2^* with increasing temperature. This is made reasonable by a machine calculation of the energy exchange of highly excited X_2^* in linear collisions (41).

Discussions of the mechanisms of atom recombination have centered

TABLE I

ATOM RECOMBINATION RATES[a]

Reaction	Method	Temp. °K	Log k ($lit^2/mole^2$-sec) (third body in parentheses)	Reference
$O+O+M \rightarrow O_2+M$	N+NO reaction, air afterglow	300	9.0(N_2) 8.5(Ar, He) 9.2(N_2O) 9.5(CO_2, SF_6)	Morgan & Schiff (274)
$O+O+M \rightarrow O_2+M$	Shock tube pyrolysis O_3, O_2 absorption at 1270A	2000	7.4(Ar)	Wray (403)
$O+O+M \rightarrow O_2+M$	Shock tube pyrolysis O_2, X-ray densitometry	4000	7.3(Ar) 7.6(Kr) 8.1(Xe)	Rink (322)
$O+O+M \rightarrow O_2+M$	Flow discharge, ESR	300	8.9(O_2)	Marshall (261)
$H+H+M \rightarrow H_2+M$	H_2–O_2 flame, Na emission, D exchange	1072	9.5(H_2) 9.3(N_2, H_2O)	Dixon-Lewis, Sutton & Williams (116)
$H+H+M \rightarrow H_2+M$	Flow discharge, Wrede gauge	300	9.95(H_2)	Kretschmer & Peterson (241)
$H+H+M \rightarrow H_2+M$	Shock tube dissociation, UV absorption	3500	8.3(Ar) 9.3(H_2) 10.2(H)	Patch (287)
$H+H+M \rightarrow H_2+M$	Flow discharge, ESR	300	10.1(H2)	Marshall (261)
$D+D+M \rightarrow D_2+M$	Shock tube dissociation, X-ray densitometry	3500	8.5(D_2) 9.3(D) 8.3(Ar, Kr)	Rink (323)
$N+N+M \rightarrow N_2+M$	Flow discharge, NO titration, and afterglow decay	300	8.9(N_2)	Kretschmer & Peterson (241)
$N+N+M \rightarrow N_2+M$	Flow discharge, ESR	300	9.5(N_2)	Marshall (261)
$N+N+M \rightarrow N_2+M$	Flow discharge, NO and C_2H_4 titration	293	9.3(N_2)	Avramenko & Krasnen'kov (11)
$N+O+M \rightarrow NO+M$	Static discharge, Wrede gauge and afterglow decay	350	9.5(N_2)	Kretschmer & Peterson (241)
$I+I+M \rightarrow I_2+M$	Flash photolysis	333	9.3(Ne) $E_a = -1.5$ kcal/mole 13.5(NO) $E_a = -2.2$ kcal/mole	Porter, Szabo & Townsend (296)
$Br+Br+M \rightarrow Br_2+M$	Flash photolysis	300	9.4(Ar) 10.7(Br_2)	Basila & Strong (21)
$Cl+Cl+M \rightarrow Cl_2+M$	Flow discharge, thermocouple probe	300	8.5(Cl_2)	Linnett & Booth(250)
$Cl+Cl+M \rightarrow Cl_2+M$	Shock tube, spectrophotometric measurements	1600	9.4(Ar)	Hiraoka & Hardwick (199)
$Cl+Cl+M \rightarrow Cl_2+M$	Shock tube, absorption spectrophotometry	2000	8.1(Cl_2)	Jacobs & Giedt (208)
$Cl+Cl+M \rightarrow Cl_2+M$	Flow discharge, isothermal probe	313	10.5(Cl_2) 9.6(Ar) 9.5(He)	Bader & Ogryzlo (16)
$Cl+Cl+M \rightarrow Cl_2+M$	Photochlorination, rotating sector	313	11.8(Cl_2)	Ayscough et al. (12)
$Cl+Cl+M \rightarrow Cl_2+M$	Flow discharge, NOCl titration, light emission	300	10.3(Cl_2)	Hutton (207)

[a] Rate constants defined by $d(X_2)/dt = k(X)^2(M)$.

around two processes, the energy-transfer mechanism and the radical-molecule complex mechanism (32, 91, 134, 295). The energy-transfer mechanism can be represented by:

$$2R \rightleftarrows R_2^* \qquad 4, 5.$$

(R = atom or radical, M = third body)

$$R_2^* + M \rightleftarrows R_2 + M \qquad 6.$$

(also includes $R + M \rightleftarrows RM^*$ as initial step)

The radical-molecule complex mechanism is usually given as:

$$R + M \rightleftarrows RM^* \qquad 7.$$

$$RM^* + M \rightarrow RM + M \qquad 8.$$

$$R + RM \rightarrow R_2 + M \qquad 9.$$

The former mechanism predominates when the binding between R and M is less than RT, since the complex RM cannot then be deactivated to a long-lived state (134). However, when binding occurs (by charge transfer in most of the cases studied), the complex mechanism is necessary to explain the very high recombination rates which are observed, as in the $I+I+I_2 \rightarrow 2I_2$ reaction. In this connection, it is noteworthy that NO has now displaced I_2 as the most effective third body for I recombination, being about twenty times more effective than I_2 and 10^4 times more effective than neon (296). The suggested mechanism is:

$$I + NO \rightleftarrows NOI \qquad 10, 11.$$

$$I + NOI \rightarrow NO + I_2 \qquad 12.$$

$$NOI + NOI \rightarrow 2NO + I_2 \qquad 13.$$

However, the thermodynamic parameters deduced for $K_{10,11} = k_{10}/k_{11}$ are not consistent with the observed activation energy and it may be that the authors have over-simplified their kinetic treatment. At low NO (≤ 0.4 torr) they assume NOI at equilibrium and neglect Equation 13. If we assume a steady state for NOI and make both Equations 10 and 11 pressure dependent, the rate law becomes (M = Ne):

$$\begin{aligned} d(I_2)/dt &= k_{10}k_{12}(M)(NO)(I)^2/[k_{11}(M) + k_{12}(I)] \\ &= k_{12}K_{10,11}(NO)(I)^2/[1 + K_{12,11}(I)/(M)] \end{aligned} \qquad 14.$$

If $k_{11}(M)$ and $k_{12}(I)$ are comparable, which seems quite likely, the observed low-negative, overall activation energy seems quite understandable since the effect of $K_{12,11}$ in the denominator would tend to cancel part or more of the temperature dependence of $K_{10,11}$ in the numerator. The quantity $\Delta H_{10,11}$ should be about -11 kcal/mole and $\Delta S_{10,11}$ about -26 cal/mole-°K.

At very high NO(>200 torr), (I) reaches a steady state and the overall rate should be:

$$d(I_2)/dt = k_{13}(NOI)^2 + k_{12}k_{11}(NOI)^2(M)/[k_{10}(M)(NO) + k_{12}(NOI)] \qquad 15.$$

$$d(I_2)/dt = k_{13}(NOI)^2[1 + K_{12,13}K_{11,10}]/[(NO) + K_{12,10}(NOI)/(M)] \qquad 16.$$

where none of the terms is negligible.

Nitric oxide also catalyzes the recombination of H atoms through formation of an HNO intermediate (70, 92). The mechanism seems to be analogous to that proposed above.

As may be seen from Equation 14, the apparent termolecular rate constant for recombination may be a function of atom concentration when recombination occurs by the complex mechanism. Hutton (207) has recently suggested a dependence of this type on (Cl) in the recombination of Cl atoms in the presence of Cl_2. To account for his data, which imply a decrease in $k(M = Cl_2)$ with increasing (Cl), he suggests Cl_3 as an intermediate. However, the interpretation depends on knowing accurately the 6 to 40 per cent contribution of wall reaction.

Christie (91) has called attention to the fact that at very low $(I_2)/(M)$ concentrations the apparent termolecular rate constant for recombination falls off by as much as a factor of two from its extrapolated value. She suggests that this may be because of improper allowance for vibrationally excited I_2^* which are only slowly relaxed by the rare gases and may have smaller extinction coefficients than ground state I_2. This may be a significant problem for flash photolysis experiments where ground state extinction coefficients are normally used.

Mavroyannis & Winkler (267) report a lower limit of log k (lit^2/mole^2-sec) = 8.7 (298° K) for the recombination $H+N+M \rightarrow NH+M$ in the pressure range 2.5 to 4.5 torr. No evidence of product formation could be found upon mixing N atoms with H_2 at temperatures as high as 350° C. This is not surprising considering that the reaction $N+H_2 \rightarrow NH+H$ is about 24-kcal/mole endothermic. If it did occur, it would certainly be followed by the very fast step $N+NH \rightarrow N_2+H$.

Addition.—The 40-year-old problem of energy chains started by vibrationally excited molecules seems now to have been settled for O_3. New measurements of the quantum yields of O_3 decomposition in red light (83) confirm earlier estimates (39) for the relative rates of $O+O_3 \rightarrow 2O_2^*$ and $O+O_2+M \rightarrow O_3+M$ as well as the relative efficiencies of different third bodies M and the absence of an energy chain from vibrationally excited O_2^* (24, 29, 340).

Recent quantitative data on atom addition rates are listed in Table II. A significant development is that previous discrepancies in the rate of $O+O_2+M \rightarrow O_3+M$ between flow-discharge measurements and results obtained by O_3 decomposition have been somewhat resolved by the findings that long-lived metastable species ($^1\Delta$ or $^1\Sigma$) decompose O_3 in discharge systems (96, 220, 265). The rate given by Kaufman & Kelso for this reaction (see Table II) which agrees closely with the results from static O_3 decomposition, was obtained by a nondischarge flow method designed to eliminate effects of metastable species.

The reaction of atomic oxygen with olefins is another case where results obtained by discharge methods can be compared with those from a different technique. Elias & Schiff (123) and Elias (122) produced O atoms by a dis-

TABLE II

ATOM ADDITION REACTIONS

Reaction	Method	Log k^a (liter/mole-sec)	Reference
$O+O_2+O_2 \rightarrow O_3+O_2$	Flow, O from O_3 pyrolysis, air afterglow, room temperature	8.4	Kaufman & Kelso(220)
$O+O_2+N_2 \rightarrow O_3+N_2$	Shock tube, O_3 pyrolysis	$6.97+1.7/\theta$ 8.2(300° K)	Jones & Davidson (215)
$O+NO+N_2 \rightarrow NO_2+N_2$	Mass spectrometer (300 to 500° K)	$9.16+1.93/\theta$ 10.57(298° K)	Klein & Herron (228)
$O+NO+O_2 \rightarrow NO_2+O_2$	Flow discharge, NO_2 titration (300° K)	10.46	Westenberg & DeHaas (395)
$O+NO+O_2 \rightarrow NO_2+O_2$	Flow discharge, air afterglow (212 to 315° K)	$8.9+1.8/\theta$ 10.2(300° K)	Clyne & Thrush (93)
$O+CO \rightarrow CO_2$	Stirred-flow discharge, NO_2 titration (300 to 530° K)	$7 \quad -4.0/\theta$	Mahan & Solo (256)
$O+CO \rightarrow CO_2+h\nu$	Stirred-flow discharge	$8^b \quad -9.5/\theta$	Mahan & Solo (256)
$O+CO \rightarrow CO_2+h\nu$ (O_2 carrier)	Flow discharge (200 to 300° K)	$3.8-3.7/\theta$	Clyne & Thrush (93)
$O+NO \rightarrow NO_2+h\nu$ (O_2 carrier)	Flow discharge (200 to 300° K)	$3.5+1.5/\theta$	Clyne & Thrush (93)
$O^*+O_2 \rightarrow O^*O+O$	Mass spectrometer	8.78	Herron & Klein (196)
$O^*+NO \rightarrow NO^*+O$	Mass spectrometer	9.04	Herron & Klein (196)
$O^*+NO_2 \rightarrow NO_2^*+O$	Mass spectrometer	9.62	Herron & Klein (196)
$H+NO+H_2 \rightarrow HNO+H_2$	Flow discharge, isothermal probe (298° K)	10.04	Simonaitis (348)
$H+NO+H_2 \rightarrow HNO+H_2$	Flow discharge, HNO emission (293° K)	10.17 ($E_a=-0.6$ kcal/mole)	Clyne & Thrush (92)
$H+O_2+Ar \rightarrow HO_2+Ar$	Flow discharge, isothermal probe—data combined with explosion limit studies (250 to 800° K)	$8.67+1.6/\theta$	Clyne & Thrush (95)

* Indicates labeled O.
[a] $\theta=2.3RT$.
[b] A-factor uncertain.

charge through molecular oxygen and also by an N_2 discharge using the "titration" reaction $N+NO \rightarrow N_2+O$. Cvetanovic and co-workers, on the other hand, have carried out a detailed study of these reactions by the method of mercury-photosensitized decomposition of N_2O, which gives mainly $N_2+O(^3P)$. [For a review, see (105).] Relative rates at 25°C for a series of olefins (and *n*-butane) are nearly identical using the two methods, but Arrhenius parameters are not in agreement, the discrepancy being small but outside the apparent combined experimental errors. The reason for the differing results is not known.

Both the mercury photosensitized decomposition of N_2O and the reaction $N(^4S)+NO\ (^2\pi) \rightarrow N_2+O$ involve a short-lived triplet-excited state of N_2O

with about 112±1 kcal/mole energy above the ground state. (The approximate equality of excitation energies is fortuitous.) Although the spin-conservation rule favors production of $O(^3P)$ in each case, it should be borne in mind that intersystem crossing between triplet and singlet states is not rigorously forbidden for N_2O. The most recent estimation (164), based on analysis of data from the thermal decomposition of N_2O, suggests that the singlet-triplet crossing is forbidden by a factor of about 10^{-1} or 10^{-2}. Decomposition to give the products $N_2+O(^1D)$ in the present cases is exothermic by about 30 kcal/mole and almost certainly occurs to some extent. A figure of 0.5 to 5 per cent $O(^1D)$ is a reasonable guess. An assessment of the kinetic importance of a small proportion of $O(^1D)$ requires more information on the deactivation rate than is presently available; but in cases where $O(^3P)$ reactions are slow, as with the alkanes, the higher expected reactivity of $O(^1D)$ could be important.

A flow-discharge technique has also been used by Avramenko, Kolesnikova & Savinova (9) to study the reaction of atomic oxygen with olefins. However, the rate constants are calculated by a quite indirect method based on product analysis. The reported results do not agree with those of Cvetanovic nor Elias and show unexpected trends in activation energies for a series of olefins. A similar method was used to study the reaction of atomic oxygen with methane and ethane (8) and with *n*-butane (10). A conclusion from this work, which is at variance with those of other workers, is that the initial step is not H abstraction to give OH. It is assumed, instead, that the initial step involves both valences of O to give CH_2O or H_2O in the reaction with methane, and leads to a C–C bond rupture in the reaction with higher alkanes. The reported rates are summarized below, along with those of Azatyan, Nalbandyan & Tsui (13, 14, 15). Measurements by the latter group were made by determining the effect of small quantities of hydrocarbon on the first combustion limit of H_2–O_2 and CO–O_2 mixtures.

Reaction	*Log k(liter/mole-sec)*	*Reference*
$O+CH_4$	$10.75-8.7/\theta$	Azatyan, Nalbandyan & Tsui (15)
$O+CH_4$	$10.3-7.35/\theta$	Avramenko, Kolesnikova & Kuznetsova (8)
$O+C_2H_6$	$10.9-7.5/\theta$	Azatyan, Nalbandyan & Tsui (13)
$O+C_2H_6$	$9.73-5.2/\theta$	Avramenko, Kolesnikova & Kuznetsova (8)
$O+C_3H_8$	$11.05-6.2/\theta$	Azatyan, Nalbandyan & Tsui (14)
$O+n-C_4H_{10}$	$10.9-4.2/\theta$	Azatyan, Nalbandyan & Tsui (14)
$O+n-C_4H_{10}$	$9.9-4.1/\theta$	Avramenko, Kolesnikova & Savinova (10)

Fontijn, Meyer & Schiff (150) have measured the absolute rate and spectral distribution (3875 to 14,000 Å) for the chemiluminescent reaction $O+NO\rightarrow NO_2\dagger\rightarrow NO_2+h\nu$. This reaction may be used as a standard to measure rates of other light-emitting reactions. Reeves, Harteck & Chace (316) have extended, in large vessels, the data to 3μ and shown that $(h\nu)$ is strictly second-order, independent of added gases. At the highest pressure (151) of 4 torr, there is no evidence of the expected collisional quenching which raises some rather interesting problems for the excited states of NO_2.

Klein & Herron (196, 228) have measured the exchange rates of atomic oxygen with O_2, NO, and NO_2 (1 to 3 torr) by a mass-spectrometer method (see Table II). The results confirm the a priori calculations of Benson (31) which indicated that these reactions would lead to rapid isotopic equilibration. The exchange results also have significance in regard to the overall reactions, which Klein & Herron represent as follows:

$$O + XO \underset{18}{\overset{17}{\rightleftarrows}} (OXO)^* + M \overset{19}{\rightarrow} XO_2 + M \qquad 17, 18, 19, 20.$$

$$\overset{20}{\rightarrow} X + O_2$$

$$(XO = O_2, NO, \text{ or } NO_2)$$

The measured exchange rates show that step 17 for NO is only twice as fast as step 17 for O_2, thus proving that the slowness of the reaction $O+O_2+M \rightarrow O_3+M$ (by a factor of about 10^2) relative to $O+NO+M \rightarrow NO_2+M$ is mainly attributed to a lower k_{19}/k_{18} ratio in the former case. Klein & Herron assign equal values of k_{18} for O_3^* and NO_2^* and thus attribute all of the difference in overall reaction rate to a lower rate of collisional deactivation of O_3^*. However it would be more consistent with the RRK theory (26) to expect that O_3^* would have a shorter lifetime $(1/k_{18})$ than NO_2^*.

The reaction of atomic oxygen with NO_2 may give stabilized NO_3 or $NO+O_2$, and the NO_3^* intermediate may be either symmetric NO_3 or OONO. Klein & Herron assume a common intermediate for both reaction paths and are thereby led to deduce $k_{18} \approx k_{20}$ for this reaction. The latter result is highly improbable for reasons previously given (31) and therefore constitutes good evidence that the two reaction paths actually involve different intermediates. Another problem, apart from the foregoing, is that Klein & Herron have employed a low $(\sim 10^{-5})$ collisional deactivation efficiency for NO_3^* which then requires postulation of an unreasonably long ($10^{-4.7}$ sec) lifetime for NO_3^* in order to account for the observed overall rates.

Dixon-Lewis & Williams (117) have shown that HO_2 is an important radical species in H_2–O_2 flames at 1072°K. The mode of formation is by the reaction $H+O_2+M \rightarrow HO_2+M$, which is competitive with the chain branching step $H+O_2 \rightarrow OH+O$ under these conditions. Evidence is given that the major path of HO_2 disappearance is by the reaction $H+HO_2 \rightarrow OH+OH$. This could be the source of excited OH during the induction period behind incident shocks in H_2+O_2 mixtures (23).

Transfer reactions.—Table III summarizes Arrhenius parameters for reactions of the type $H+RH \rightarrow H_2+R$. For the alkanes, most of the A-factors are close to 10^{11} liter/mole-sec, indicating a collision efficiency of about 0.1. The two values reported for methane (169, 210), obtained by quite different methods, are in poor agreement and also are inconsistent with the moderately

TABLE III

REACTIONS OF THE TYPE $H+RH \rightarrow H_2+R$

RH	Method	Log k^a (liter/mole-sec)	Reference
CH_4	Flow discharge, product analysis (500 to 787° K)	$8.78-7.4/\theta$	Jamieson & Brown (210)
CH_4	Effect of added CH_4 on lower H_2–O_2 explosion limit (675 to 825° K)	$11.52-15.1/\theta$	Gorban & Nalbandyan (169)
C_2H_6	Statistical analysis of existing data (300 to 1100° K)	$11.117-9.7/\theta$	Baldwin & Melvin (18)
C_2H_6	Effect of added C_2H_6 on lower H_2–O_2 explosion limit (675 to 825° K)	$11.53-12.2/\theta$	Gorban & Nalbandyan (169)
C_3H_8	H_2+O_2 inhibition, combined with low temperature data (300 to 800° K)	$10.81-7.8/\theta$	Baldwin (17)
C_3H_8	Flow, H_2 pyrolysis on W filament, isothermal probe, product analysis (368 to 443° K)	$11.12-8.2/\theta$	Kazmi, Diefendorf & LeRoy (221)
C_3H_8	Effect of added C_3H_8 on lower H_2–O_2 explosion limit (675 to 825° K)	$10.80-8.8/\theta$	Gorban & Nalbandyan (169)
n-C_4H_{10}	H_2+O_2 inhibition, combined with low temperature data (300 to 800° K)	$10.9-7.5/\theta$	Baldwin & Walker (19)
n-C_4H_{10}	Rate measured relative to $H+C_3H_8$	$10.82-7.1/\theta$	Kazmi & LeRoy (222)
i-C_4H_{10}	H_2+O_2 inhibition, combined with low-temperature data (300 to 800° K)	$11.0-6.8/\theta$	Baldwin & Walker (19)
i-C_4H_{10}	Effect of added C_4H_{10} on lower H_2–O_2 explosion limit (675 to 825° K)	$10.76-5.1/\theta$	Gorban & Nalbandyan (169)

[a] $\theta=2.3RT$.

well-established Arrhenius parameters for the reverse reaction $CH_3+H_2 \rightarrow CH_4+H$, for which log A (liter/mole-sec) $=8.7$ and $E_a=10.0$ kcal/mole (398). From the latter data, and taking known or estimated values of $S°$ for the species involved (27), it is found that log A should be about 10 and E_a about 9 to 10 kcal/mole for abstraction from methane.

Considerable data involving transfer reactions of the iodine atom have been obtained by Benson and co-workers (45, 47, 146, 185, 277, 283, 369, 370) in studies of the dehydrogenation of alkanes by I_2, the reduction of iodides by HI, and the iodine-catalyzed positional isomerization of olefins. Data for these and other halogen-transfer reactions are listed in Table IV.

Despite some occasional scatter, it appears that the majority of atom-abstraction reactions by atoms from molecules have A-factors in the range $10^{11 \pm 0.5}$ liter/mole-sec (139). A great deal of theoretical work (140, 141, 168, 234, 235, 401) has been undertaken to predict the small differences that do occur in terms of transition-state theory. However, it is doubtful if the calculations can do much better than the simple choice given above. A very naive model can give adequate justification for the above results.

TABLE IV

TRANSFER REACTIONS OF HALOGEN ATOMS

R	E_a (kcal/mole)	Log A (liter/mole-sec)	Reference
		$I+RI \rightarrow R+I_2$	
CH_3	20.5	11.4	Flowers & Benson (146)
CH_3	19.2	10.7	Boyd et al. (61)
CF_3	17.6	10.8	Boyd et al. (61)
C_2H_5	17.1	11.01	Hartley & Benson (185)
$t-C_4H_7$	13.0	10.71	Teranishi & Benson (369)
CH_3CO	14.6	10.9	O'Neal & Benson (283)
		$I+RH \rightarrow R+HI$	
CH_3	32.4	11.6	Flowers & Benson (146)
C_2H_5	27.9	11.22	Hartley & Benson (185)
i-C_3H_9	25.0	11.22	Nangia & Benson (277)
		$I+RH \rightarrow R+HI$	
t-C_4H_9	21.4	10.88	Teranishi & Benson (369)
Methallyl	13.2	9.83	Benson, Bose & Nangia (47)
CH_3CO	15.7	10.3	O'Neal & Benson (283)
H	33.5	11.1	Sullivan (364)
D (from D_2)	34.5	11.0	Sullivan (365)
		$Cl+RH \rightarrow R+HCl$	
CH_2Cl	3.3	10.53	Knox (233)
$CHCl_2$	3.0	10.43	Knox (233)
CCl_3	3.35	9.84	Knox (233)
		$Br+RH \rightarrow R+HBr$	
CH_2OH	6.55	8.9	Buckley & Whittle (68)
CF_3	23.5	10.46	Corbett, Tarr & Whittle (98)

For the reaction

$$H + RH \underset{2}{\overset{1}{\rightleftarrows}} (R\text{-}H\text{-}H)^{\ddagger}, \qquad K_{1,2} = k_1/k_2 \text{ and } \log (A_1/A_2) = \Delta S^{\ddagger}/4.575.$$

If we assume that S(R-H-H)$-S$(R-H) is just given by the difference in symmetry, spin, and an extra internal rotation, these amount to about R1n6 +R1n2+1.4 eu for most hydrocarbons. Thus, $\Delta S_p^{\ddagger} = -S°(\mathrm{H})+6.4 = -21.0$ eu. Corrected to standard states of mole/liter, this becomes $\Delta S_c^{\ddagger} = -12.6$ eu so that if A_2 has the expected value of $10^{13.5}$ sec^{-1} then $A_1 \sim 10^{10.7}$ liter/mole-sec. The uncertainty in this value is less than a factor of two, due consideration being given to symmetry numbers for specific examples.

For reactions of heavier atoms X such as Br or Cl or I, the entropy of (R-H-X)‡ is most easily approximated by taking $S°$(R-X) and adding to it corrections for spin, symmetry, about 3 eu for internal rotation, and about 2 eu$\pm$1 for extra bending and about 1 eu for increased moments of inertia.

Then setting $A_2 = 10^{13}$ sec^{-1} we find again that $A_1 \sim 10^{11}$ liter/mole-sec within a factor of 3. Similar analysis for H-abstraction by CH_3 leads to a value of $10^{8.5}$ liter/mole-sec within a factor of 2. This is for a relatively loose complex with nearly free internal rotation and bending modes near 400 cm^{-1}. But even a "tight" complex would not lower this by more than a factor of 4. This makes inherently suspicious some of the very low A-factors which have been reported for H-atom abstraction (Table VIII) from alcohols and amines. Also suspect are the values for H abstraction by Br from CH_3OH (Table IV).

Data for other atom-transfer reactions are listed in Table V. A series of studies of N-, O-, and H-atom reactions with species such as O_3, NO_2 and NO has been carried out by Phillips & Schiff (290, 291, 293), using a mass spectrometer coupled with a fast-flow discharge system. In the course of this work, evidence was found that the reactions $N + NO \rightarrow N_2^* + O$ and $H + O_3 \rightarrow OH^* + O_2$ give vibrationally excited products which are able to dissociate O_3. The average energy of N_2^* was found to be 20 ± 4 kcal/mole by means of a calorimetric probe (273). The rate of the reaction $N_2^* + O_3 \rightarrow O_2 + O + N_2$ was given as $3.2 \pm 0.7 \times 10^8$ liter/mole-sec (291). By contrast, vibrationally

TABLE V

ATOM TRANSFER REACTIONS

Reaction	Method	Log k[a] (liter/mole-sec)	Reference
$N + O_3 \rightarrow NO + O_2$	Flow discharge, mass spectrometer (298° K)	8.5	Phillips & Schiff (290)
$N + NO \rightarrow N_2 + O$	Flow discharge, mass spectrometer (298° K)	10.1	Phillips & Schiff (290)
$N + O_2 \rightarrow NO + O$	Static discharge, Wrede gauge, NO emission (350° K)	4.7	Kretschmer & Peterson (241)
$H + NO_2 \rightarrow OH + NO$	Flow discharge, mass spectrometer (298° K)	10.5	Phillips & Schiff (293)
$H + O_3 \rightarrow OH + O_2$	Flow discharge, mass spectrometer 298° K)	10.2	Phillips & Schiff (293)
$H + O_2 \rightarrow OH + O$	Statistical treatment of existing data (300 to 1650° K)	$11.31 - 16.49/\theta$	Baldwin & Melvin (18)
$H + O_2 \rightarrow OH + O$	Burning velocity of H_2–O_2 flames	$11.82 - 17.4/\theta$	Lovachev (253)
$H + HNO \rightarrow H_2 + NO$	Addition of NO to H_2–O_2 flame (1600 to 2000° K)	9.78	Bulewicz & Sugden (70)
$O + NO_2 \rightarrow O_2 + NO$	Flow discharge, mass spectrometer (298° K)	9.2	Phillips & Schiff (290)
$O + NO_2 \rightarrow O_2 + NO$	Mass spectrometer (278 to 374° K)	$10.29 - 1.06/\theta$ 9.5(298° K)	Klein & Herron (228)
$O + O_3 \rightarrow O_2 + O_2$	Flow discharge, mass spectrometer (298° K)	7.2	Phillips & Schiff (290)
$O + OH \rightarrow H + O_2$	Flow discharge, NO titration, air afterglow (265 to 293° K)	10.5 ($E_a \sim 0$)	Clyne & Thrush (94)
$O + H_2 \rightarrow OH + H$	Flow discharge, NO_2 titration, air afterglow (409 to 733° K)	$10.5 - 9.5/\theta$	Clyne & Thrush (94)

[a] $\theta = 2.3RT$.

excited O_2 from the reaction of O (^3P) with O_3 and NO_2 does not decompose O_3 (292).

The overall reaction of H with NO_2 has the stoichimetry $2H+3NO_2 \rightarrow 3NO+O_2+H_2O$ and is sufficiently rapid to be considered for use in H-atom titration (293) in the absence of molecular hydrogen.

In another mass spectrometer study of atom reactions, Klein & Herron (228) used an Eltenton-type reactor capable of monitoring relative O-atom concentrations. The measurements could be put on an absolute basis by calibrating the O-atom flow by a titration method. With this apparatus the previously unreported activation energy for the reaction $O+NO_2 \rightarrow NO+O_2$ (1.06 ± 0.2 kcal/mole) was measured. At 298°K the reported rate constant agrees very closely with the previous value of Ford & Endow (152), obtained using a photolytic method, and is about a factor of 2 higher than that of Phillips & Schiff (290).

The use of ESR to measure O- and N-atom concentrations receives some support from the finding of Westenberg & DeHaas (395) that concentrations so obtained, using O_2 as an internal standard, agree with measurements by the techniques of NO_2 and NO titration. The ESR method was used also by Marshall (261) to measure O, N, and H recombination rates (see Table I), again with O_2 as an internal standard.

Electronically excited atoms.—Considerable interest has arisen in the chemistry of electronically excited atomic oxygen, largely because of the importance of these species in photochemical and atmospheric systems. DeMore & Raper (110, 311) used ultraviolet photolysis of O_3 to produce $O(^1D)$ for a study of the reactions $O(^1D)+N_2+M \rightarrow N_2O+M$ and $O(^1D)+CO+M \rightarrow CO_2+M$. These reactions are very slow in the gas phase and give low product yields even in liquid N_2 or CO at 77° K. Evidence was presented to show that the low yields do not indicate low reactivity of $O(^1D)$ toward N_2 and CO, but rather are attributed to predissociation of the initially formed adducts N_2O^* and CO_2^*, giving spin-disallowed $N_2+O(^3P)$ and $CO+O(^3P)$.

Katakis & Taube (218) found that isotopically labelled O_3 undergoes a photochemically induced exchange with CO_2 which is attributed to the reaction $O^*(^1D)+CO_2 \rightarrow CO_2^*+O(^1D)$, where the asterisk denotes labelled O. It was proposed that CO_3 is an intermediate and that this species can be stabilized at high pressures, as follows: $O(^1D)+CO_2+M \rightarrow CO_3+M$. Raper & DeMore (311) have suggested that an electronically excited (triplet) form of $CO_3\dagger$ is produced as an intermediate by reaction of O_2 ($^3\Sigma_u^+$) with CO in liquid CO–O_2 mixtures upon irradiation with 2537A light. The $CO_3\dagger$ formed in this manner may undergo unimolecular decomposition to give either $CO+O_2$ (the quenching reaction) or $CO_2+O(^3P)$ (which leads to the observed products CO_2+O_3). These reaction paths are spin-allowed and are of comparable exothermicity. Evidence was also found that $CO_3\dagger$ may react bimolecularly with O_2 or CO to give CO_2+O_2 or, in the latter case, $2CO_2$.

Yamazaki & Cvetanovic (405, 406) have shown that excited $O(^1D)$ or $O(^1S)$ produced by N_2O photolysis at 1849A is rapidly deactivated by CO_2

and Xe, probably by incipient chemical bonding, but is not efficiently deactivated by He, Ar, Kr, C_2F_6 and SF_6.

S(^{1}D) formed by the ultraviolet photolysis of OCS reacts with olefins to give the corresponding cyclic sulfides and inserts into the CH bonds of alkanes and cycloalkanes to give mercaptans (231, 232, 362). Evidence was found for collisional deactivation of S(^{1}D) to S(^{3}P) by the above reactants and by inert gases.

Atomic selenium also adds to olefins to give cyclic selenides (78). The selenium atoms are produced by flash photolysis of CSe_2 vapor and are believed to be in ^{3}P states.

Experiments by Tanaka & McNesby (368) indicate that H(^{2}P), generated by irradiation of atomic hydrogen with 1215Å light, interacts with N_2 to give NH_3 but not N_2H_4.

Elementary Radical Reactions

Recombination.—Theoretical aspects of radical recombination have been actively discussed, with particular emphasis on explaining the high *A*-factors which are observed (34, 64, 114, 266). It is generally agreed that the activated complex must be quite "loose," but the problem has been to account for the bonding under these circumstances. A complex made up of loosely bound particles is inconsistent with the directional properties of ordinary covalent bonds. Benson (34) has proposed that the rocking freedom of the complex is best accounted for by partial ionic bonding, as in the following structures for the case of two CH_3:

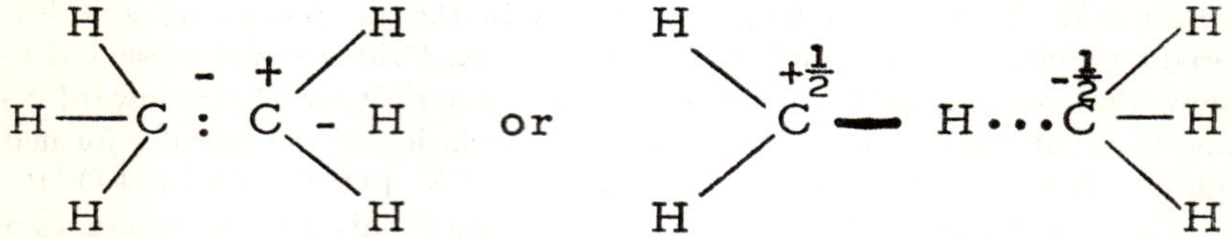

A related problem is the structure of the activated complex leading to disproportionation, and the question of whether recombination and disproportionation stem from a common transition state. Although the net result of disproportionation is abstraction of H by one radical from another, the high *A*-factors and small temperature dependence of disproportionation rates suggest that the process is not closely related to ordinary radical abstraction reactions. Some of the previously advanced theories (114, 227, 266) have envisioned initial formation of the excited dimer RR′ which may decompose unimolecularly to give alkane and alkene. However, this picture is inconsistent with the observed pressure dependence and other experimental observations (34). Participation of ionic states again seems quite probable (34) with structures such as the following for C_2H_5 disproportionation:

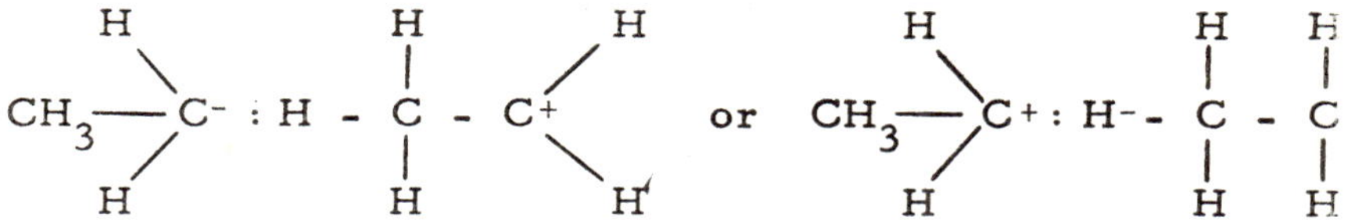

The rotating sector method has been used to measure absolute recombination rates for isopropyl (270) and *t*-butyl radicals (269). The results were log k(liter/mole-sec) = 10.8(isopropyl, 81 to 169° C) and log k(liter/mole-sec) = 9.5(*t*-butyl, 81 to 115° C).

Dalby (106) has reported $k = 10^{8.1}\ (T/300°\ K)^{1/2}\ 10^{-1.2/\theta}$ for the combination of CF_2 radicals. The radicals were produced by flash photolysis of tetrafluorethylene and perfluoropropylene, and radical concentrations were followed by absorption spectrophotometry.

March & Polanyi (258) have described a technique for determining absolute rates of radical dimerization and abstraction by pulsed photolysis. To calculate the rate constants it is necessary only to measure two ratios of average rates of formation of dimer and abstraction product at different (known) pulse times or quantum dosages, or both. Application of the method to acetone photolysis gave a mean value for log k(liter/mole-sec) = 10.37 for methyl-radical recombination, which was in good agreement with earlier values.

In addition to the foregoing absolute rate measurements, new experimental data have been accumulated on rate ratios for disproportionation and combination (Table VI). Most of the results are in fair agreement with previous measurements [for a summary, see (79)], the principal exception being data for the *t*-butyl radical. Using the symbol $\Delta(R, R')$, where H is transferred from R′ to R, the previous results are 4.6 (240) and 4.4 (53) for $\Delta(t\text{-}C_4H_9,\ t\text{-}C_4H_9)$, 1.7 for $\Delta(C_2H_5,\ t\text{-}C_4H_9)$ (57), and 1.2 for $\Delta(i\text{-}C_3H_7,\ t\text{-}C_4H_9)$ (57).

The question of temperature dependence is still somewhat in doubt, since the effects are small and difficult to measure. A negative temperature coefficient has been reported for k_{dis}/k_{comb} for ethyl radicals (−65 to 40° C) which can be represented by $E_{dis} - E_{comb} = -0.3$ kcal/mole or by $k_{dis}/k_{comb} \sim T^{-0.7}$ (114). A positive temperature coefficient for k_{dis}/k_{comb} was previously reported for *t*-butyl radicals (226), corresponding to $E_{dis} - E_{comb} = 1.3$ kcal/mole; but more recent work by Thynne (375) shows no temperature dependence for this ratio in the region of 78 to 186° C. Gunning & Stock (180) found no temperature dependence for k_{dis}/k_{comb} for cyclopentyl radicals in the range 26 to 250° C.

There has been continued interest in determining the cross combination ratio $k_{ab}/(k_{aa}k_{bb})^{1/2}$. Recent data are shown in Table VII. Cross-combination ratios for alkyl radicals fall remarkably near 2.0, as previously noted (227), but for other radicals they do not always do so (79).

TABLE VI

DISPROPORTIONATION-COMBINATION RATIO k_{dis}/k_{comb}

$$R + R' \xrightarrow{k_{dis}} RH + R'(-H)$$

$$R + R' \xrightarrow{k_{comb}} RR'$$

R	R'	Ratio k_{dis}/k_{comb}[a] (Reference in parentheses)
CH_3	C_2H_5	0.039 (177)
		0.055 (373)
CH_3	n-C_3H_7	0.041 (177)
		0.025 (178)
		0.095 (374)
CH_3	i-C_3H_7	0.195 (374)
CH_3	n-C_4H_9	0.15 (375)
CH_3	s-C_4H_9	0.08 (272)
CH_3	CH_3O	1.5 (378)
C_2H_5	C_2H_5	0.12 (326)
		0.14 (373)
C_2H_5	n-C_3H_7	0.081 (177)
C_2H_5	i-C_3H_7	0.43 (118)
C_2H_5	t-C_4H_9	0.48 (118)
n-C_3H_7	C_2H_5	0.058 (177)
n-C_3H_7	n-C_3H_7	0.141 (177)
		0.14 (374)
i-C_3H_7	C_2H_5	0.19 (118)
i-C_3H_7	i-C_3H_7	0.65 (374)
i-C_3H_7	t-C_4H_9	0.70 (118)
n-C_4H_9	n-C_4H_9	0.95 (375)
t-C_4H_9	C_2H_5	0.31 (118)
t-C_4H_9	i-C_3H_7	0.67 (118)
t-C_4H_9	t-C_4H_9	3.19 (118)
s-C_4H_9	s-C_4H_9	~1.2 (272)
cyclo-C_5H_9	*cyclo*-C_5H_9	1.0 (180)
C_2H_4Cl	C_2H_5	0.22 (329)
C_2H_4Cl	C_2H_4Cl	≤0.1 (330)
		≤0.05 (329)
CCl_3	C_2H_5	0.22 (329)
		0.24 (176)
CCl_3	C_2H_4Cl	≤0.11 (330)
		0.14 (329)
CH_3O	CH_3O	>60 (373)
		9.3 (190)
C_2H_5O	C_2H_5O	12 (191)
I	C_2H_5	0.33 (179)

[a] The figures under this heading are sometimes referred to as $\Delta(R, R')$.

TABLE VII

CROSS COMBINATION RATIOS $k_{ab}/(k_{aa}k_{bb})^{1/2}$

$$A+B\xrightarrow{k_{ab}} AB,\ 2A\xrightarrow{k_{aa}} A_2,\ 2B\xrightarrow{k_{bb}} B_2$$

A	B	$k_{ab}/(k_{aa}k_{bb})^{1/2}$	References
CH_3	C_2H_5	2.02	Thynne (373)
CH_3	n-C_3H_7	2.04	Thynne (374)
CH_3	n-C_3H_7	2.08	Grotewold & Kerr (177)
CH_3	CH_3COCH_2	1.0	Batt & Benson (22)
CH_3	i-C_3H_7	1.96	Thynne (374)
CH_3	n-C_4H_9	2.05	Thynne (375)
C_2H_5	i-C_3H_7	1.96	Dominguez, Kerr & Trotman-Dickenson (118)
C_2H_5	n-C_3H_7	1.93	Grotewold & Kerr (177)
C_2H_5	t-C_4H_9	1.9	Dominguez, Kerr & Trotman-Dickenson (118)
C_2H_5	CCl_3	2.0	Roquitte & Wijnen (329)
i-C_3H_7	t-C_4H_9	2.02	Dominguez, Kerr & Trotman-Dickenson (118)
C_2H_4Cl	CCl_3	2.5	Roquitte & Wijnen (330)

For CH_3 and n-C_3F_7 the ratio is given by (303) $0.743 \pm 0.046 \exp 1440 \pm 50/RT$ (300 to 579° K). This temperature dependence can be accounted for by a 2.9 kcal/mole barrier to n-C_3F_7 recombination, if it is assumed that the other recombinations have zero activation energies.

Addition.—Much of the work on the addition reactions of radicals has centered about special cases. One of these is the addition of CF_3 to olefins and benzene derivatives, which has been studied extensively (85, 115, 204, 239, 284, 289, 360, 396). The rate of addition of CF_3 to substituted benzenes is increased by electron donating groups and decreased by electron-withdrawing groups (85, 396). Although the major reaction is addition, abstraction of H occurs to some extent (85, 204). The mechanisms may involve direct abstraction or else addition followed by the disproportionation reaction $CF_3\cdot + CF_3C_6H_6\cdot \rightarrow CF_3H + CF_3C_6H_5$.

Brown & James (67) give the following Arrhenius parameters for addition of ethyl radicals to conjugated dienes, with heptene-1 included for comparison:

Substrate	E_a(kcal/mole)*	log A(liter/mole-sec)*
2,3-Dimethylbutadiene-1,3	4.5	8.5
Cyclohexadiene-1,3	5.2	8.4
Heptene-1	7.0	8.1

* Data recalculated using log A (recomb) = 11.0 and E_a (recomb) = 0.

The higher reactivity of the di-olefins is attributed mainly to a lower E_a, the A-factors being in the ratio of about 2 as would be predicted statistically.

Dominguez & Trotman-Dickenson (119) determined the following Arrhenius parameters for alkyl-radical addition to acetylene (100 to 200° C). The radicals were prepared by photolysis of the corresponding aldehydes. Methyl migration in the alkenyl radicals formed by isopropyl and t-butyl addition was postulated to account for the appearance of isomerization

R	E(kcal/mole)	log A(liter/mole-sec)
CH_3	7.7	8.4[a]
C_2H_5	7.0	8.0[b]
i-C_3H_7	6.9	8.2[b]
t-C_4H_9	5.3	7.7[b]

[a] log A recomb = 10.3.
[b] log A recomb = 11.0.

products (pentene-1 and 4-methyl pentene-1, respectively) at low pressures. An alternative mechanism not involving methyl migration can be proposed, however, in which the excited adduct vinyl radical ($\sim$35 kcal/mole excitation energy) undergoes intramolecular hydrogen abstraction ($E_a \sim 15$ kcal/mole) followed by intramolecular addition to the double bond ($E_a \sim 16$ kcal/mole):

$$[(CH_3)_2CH—CH{=}CH\cdot]^* \rightarrow CH_3CH(CH_2\cdot)CH{=}CH_2 \rightarrow$$

$$CH_3—\underset{\llcorner\!\!-\!-\!-\!\!\lrcorner}{CH—CH_2—CH}—CH_2\cdot \rightarrow CH_3—\dot{C}H—CH_2—CH{=}CH_2 \qquad 21.$$

The resulting cyclopropane intermediate may then undergo ring fission to give the precursor of pentene-1. (Similar excited molecule decompositions are discussed in references 35 and 109a.)

Miyoshi & Brinton (272) report that addition of CH_3 to propylene occurs mainly (about 90 per cent) at the terminal carbon with log $k/k_r^{1/2} = 3.7 - 8.8/\theta$, where k_r is the rate of methyl recombination. This rate is very similar to that previously found for addition of CH_3 to ethylene.

These addition reactions are extremely difficult to study quantitatively except in very special cases. In order to get sufficient product to analyze, the addition of the alkyl radical R to the olefin should be favorable relative to termination of 2R. But in this case the adduct radical R will probably also add to olefin and it will be difficult to make a quantitative analysis for product. This limits severely the composition range that can be employed. A further problem is the multitude of products to which adduct radical can transform by recombination or disproportionation (see further discussion on p. 416).

Alkyl-O_2 reactions have been studied by photo-oxidation of the corresponding azoalkanes coupled with long-path IR analysis (84, 112, 113, 363, 372) and by photo-oxidation of the alkyl iodides using mass spectrometer analysis (20, 138, 190, 191). Most of the work was near room temperature. The derived mechanisms are complex, but there is agreement that the major products of the radical oxidation are the corresponding aldehydes and alcohols plus lesser amounts of the alkyl hydroperoxide. The mechanisms presented show some similarities, but there are differences in emphasis. The groups led by Calvert and by Johnston consider that the important radicals present in their systems are (for CH_3 oxidation) CH_3, CH_3O_2, and CH_3O, and that the main products (for O_2-rich mixtures) are formed by disproportionation reactions among the latter two, such as

$$2CH_3O \rightarrow CH_3OH + CH_2O \qquad 22.$$

$$CH_3O + CH_3O_2 \rightarrow CH_3O_2H + CH_2O \qquad 23.$$

On the other hand, the Linnett group is of the opinion that CH_3, CH_3O_2, and OH are the principal radical species and that the major mechanism is

$$CH_3 + O_2 + M \rightarrow CH_3O_2 + M \qquad 24.$$

$$CH_3O_2 \rightarrow CH_2O + OH \qquad 25.$$

$$OH + CH_3O_2 \rightarrow CH_3OH + O_2 \qquad 26.$$

The very excellent studies of RO_2 radicals by Dingledy & Calvert (113) do not support reaction 25.

Dingledy & Calvert (113) report that the reaction $C_2H_5 + O_2 \rightarrow C_2H_5O_2$ is second order in the pressure range 0.48 to 10.3 cm at 22° C, with log k(liter/mole-sec) = 9.62.

By photolyzing mixtures of CH_3CHO with O_2^{32} and O_2^{36}, McDowell & Sifniades (255) have shown that the chain-terminating combination of peroxyacetyl radicals takes place with elimination of one oxygen atom from each radical. The proposed mechanism involves a four-center transition state giving O_2 elimination and diacetyl peroxide formation in one step:

$$(CH_3CO)OO\cdot + (CH_3CO)O^*O^* \rightarrow (CH_2CO)_2OO^* + OO^* \qquad 27.$$

However, the isotopic composition of the peroxide was not determined, and it seems more likely that the initial step would be formation of O_2 plus two acetoxy radicals, the latter recombining randomly.

Further evidence has been reported for multiple addition of CH_3 to NO at low temperatures (less than about 200° C). In the photolysis of CD_3NNCD_3 with added NO at room temperature, mass spectral analysis indicated formation of CD_3NO as the only product until NO was consumed (263, 264). Subsequently, $(CD_3)_2NOCD_3$ and C_2D_6 are formed by competition between addition and combination reactions of CD_3. It was estimated that log k(lit/mole-sec) ≥ 7.6 for the reaction $CD_3 + CD_3NO \rightarrow (CD_3)_2NO$. In cases where all NO is not used up, the products are CH_3NO and its dimer $(CH_3NO)_2$ (80). Hoare (202) also reported that up to three NO's were consumed per CH_3 radical produced in the photolysis of acetone at 200° C with

NO added in low concentrations. For the reaction $CH_3+NO+M(\text{acetone}) \rightarrow CH_3NO+M$, the rate $\log k(\text{liter}^2/\text{mole}^2\text{-sec}) = 11.5$ was deduced.

The situation is apparently different at higher temperatures. In the pyrolysis of tetraethyl lead-NO mixtures at 233 to 267° C, with NO in excess, ethyl radicals reacted with NO to give acetaldoxime which decomposed to CH_3CN and H_2O (297). No evidence for further addition of CH_3 to the adduct was found. The rate for the reaction $C_2H_5+NO \rightarrow C_2H_5NO$ was given as $\log k(\text{liter/mole-sec}) \geq 7.1$.

Nitric oxide adds to vinyl and substituted vinyl radicals in a manner which leads to scission of the double bond, as in the following proposed mechanism (344):

$$R_1R_2C{=}\dot{C}R_3 + NO \rightarrow [R_1R_2C{=}CR_3NO]^* \rightarrow \begin{bmatrix} R_1R_2C{-}CR_3 \\ | \quad \| \\ O{-}N \end{bmatrix} \rightarrow R_1R_2C{=}O + R_3CN \qquad 28$$

Abstraction.—Table VIII lists the rather large amount of new data on H abstraction by hydrocarbon radicals. Data for other classes of radical abstraction reactions are summarized in Table IX. Most of the *A*-factors for H abstraction by CH_3 fall near the expected value of $10^{8.5\pm0.5}$. However, a few are anomalously low. Among these are some of the data of Gray & Thynne (174), who used a deuterium labelling technique to obtain Arrhenius parameters for H abstraction by CH_3 from the CH_3 and NH_2 groups of methylamine. A similarly low *A*-factor was reported by Shannon & Harrison (342) for abstraction from the OH group of CH_3OH. The *A*-factors for abstraction from the NH_2 or OH groups are about 1.5 to 2 powers of 10 below those for abstraction from the CH_3 groups, although the *A*-factors found by Gray & Thynne for H abstraction from NH_3 and N_2H_4 are in the usual range, about 10^8 (175). There is no obvious reason why a CH_3 group attached to NH_2 should lower the *A*-factor more than does an NH_2 group. The low *A*-factors are difficult to reconcile on the basis of the transition-state model, since an abnormally rigid activated complex would be required. A possible source of error in these experiments lies in failure to take into account secondary reactions of radicals formed in the initial abstraction reaction. The major fate of these secondary radicals would be recombination or disproportionation with CH_3, the latter of which steps may lead to a CH_4 product labelled isotopically in the opposite sense from that formed in the initial reaction. The process may be illustrated as follows:

$$\text{initial:} \quad CH_3 + CH_3ND_2 \rightarrow \cdot CH_2ND_2 + CH_4 \qquad 29.$$

$$\text{secondary:} \quad CH_3 + \cdot CH_2ND_2 \rightarrow CH_2 = ND + CH_3D \qquad 30.$$

A similar possibility exists in the case of CH_3OH:

$$\text{initial:} \quad CH_3 + CH_3OD \rightarrow CH_3O + CH_3D \qquad 31.$$

$$\text{secondary:} \quad CH_3 + CH_3O \rightarrow CH_2O + CH_4 \qquad 32.$$

In view of the improbably low *A*-factors obtained and the possibilities for experimental error in studies of this type, the current results for CH_3NH_2 and CH_3OH must be considered suspect.

TABLE VIII

H-Atom Abstraction Reactions of Radicals

$$R + R'H \rightarrow RH + R' \qquad a$$
$$2R \rightarrow R_2 \qquad b$$

R	R'H***	$E_a - \frac{1}{2}E_b$	$\log(A_a/A_b^{\frac{1}{2}})$	$\log A_a$*	References
			A (liter/mole-sec)		
CH_3	CH_3NNCH_3	7.8	2.7	7.9	Gray & Thynne (174)
CH_3	CH_3ND_2	9.0	3.0	8.2	Gray & Thynne (174)
CH_3	$CH_3\mathit{ND}_2$	7.0	1.4	6.6	Gray & Thynne (174)
CH_3	$CD_3\mathit{NH}_2$	6.0	1.6	6.8	Gray & Thynne (174)
CH_3	CD_3NH_2	10.1	2.7	7.9	Gray & Thynne (174)
CH_3	CH_3NH_2	8.7	2.8	8.0	Gray & Thynne (174)
CH_3	$CH_3\mathit{NH}_2$	5.7	1.4	6.6	Gray & Thynne (174)
CH_3	NH_3	9.8	2.6	7.8	Gray & Thynne (175)
CH_3	ND_3	10.9	2.8	8.0	Gray & Thynne (175)
CH_3	N_2H_4	5.0	2.8	8.0	Gray & Thynne (175)
CH_3	CH_3OH	10.4	3.2	8.4	Shannon & Harrison (342)
CH_3	$CH_3O\mathit{H}$	6.4	1.1	6.3	Shannon & Harrison (342)
CH_3	CD_3OH	11.7	3.1	8.3	Shannon & Harrison (342)
CH_3	$\mathit{CH}_3OCOOCH_3$	8.9	2.1	7.3	Thynne & Gray (377)
CH_3	$CH_3\mathit{CH}_2OC_2H_5$	9.8	3.7	8.9	Long & Skirrow (252)
CH_3	$\mathit{H}COO(n\text{-}C_3H_7)$	7.3	1.9	7.1	Thynne (374)
CH_3	$\mathit{H}COO(i\text{-}C_3H_7)$	8.9	2.7	7.9	Thynne (374)
CH_3	$\mathit{H}COOCH_3$	9.0	2.7	7.9	Thynne (373)
CH_3	$\mathit{H}COOC_2H_5$	8.2	2.3	7.5	Thynne (373)
CH_3	$\mathit{H}COO(n\text{-}C_4H_9)$	8.2	2.4	7.6	Thynne (375)
CH_3	CH_3OOCH_3	10.0	4.4	9.6	Thynne & Gray (378)
CH_3	$p\text{-}\mathit{CH}_3C_6H_4F$	5.7	1.4	6.6	Wunderlich & Rebbert (404)
CH_3	$o\text{-}\mathit{CH}_3C_6H_4F$	6.0	1.6	7.8	Wunderlich & Rebbert (404)
CH_3	$m\text{-}\mathit{CH}_3C_6H_4F$	7.1	2.2	7.4	Wunderlich & Rebbert (404)
CH_3	$p\text{-}C_6H_4(\mathit{CH}_3)_2$	7.4	2.6	7.8	Sanders & Rebbert (333)
CH_3	$o\text{-}C_6H_4(\mathit{CH}_3)_2$	7.8	3.0	8.2	Sanders & Rebbert (333)
CH_3	$m\text{-}C_6H_4(\mathit{CH}_3)_2$	8.5	3.3	8.5	Sanders & Rebbert (333)
CH_3	$C_6H_5\mathit{CH}_3$	7.4	2.3	7.5	Burkley & Rebbert (74)
CH_3	$C_6H_5\mathit{CH}_2CH_3$	7.0	2.6	7.8	Burkley & Rebbert (74)
CH_3	$C_6H_5\mathit{CH}(CH_3)_2$	6.4	2.6	7.8	Burkley & Rebbert (74)
CH_3	$C_6H_5COCH_3$	7.4	2.5	7.7	Duncan & Trotman-Dickenson (120)
CH_3	HI	2.3	—	9.5**	Flowers & Benson (146)
C_2H_5	HI	1.1	—	8.9**	Hartley & Benson (185)
CH_3CO	HI	1.5	—	9.1**	O'Neal & Benson (283)
CD_3	$(CH_3)_3C\mathit{D}$	9.8	—	8.6**	Jackson, McNesby & Darwent (207b)
CD_3	$(\mathit{CH}_3)_3CD$	11.7	—	9.3**	Jackson, McNesby & Darwent (207b)
CD_3	$(CH_3)_3C\mathit{H}$	8.2	—	8.5**	Jackson, McNesby & Darwent (207b)
CD_3	cyclo-C_6H_{12}	9.7	—	—	Gordon & Smith (170)
CD_3	cyclo-C_7H_{14}	9.1	—	—	Gordon & Smith (170)
CD_3	cyclo-C_4H_8	10.3	—	—	Gordon, Smith & Drew (171)
CD_3	CF_3H	10.2	—	—	Pritchard & Thommarson (305)
CD_3	C_2F_5H	9.5	—	—	Pritchard & Thommarson (305)
CD_3	C_3F_7H	9.2	—	—	Pritchard & Thommarson (305)
C_2H_5	$C_2H_5COC_2H_5$	8.9	3.5	8.7	Thynne (373)
C_2H_5	$\mathit{H}COOC_2H_5$	7.8	2.4	7.6	Thynne (373)
$n\text{-}C_3H_7$	$\mathit{H}COO(n\text{-}C_3H_7)$	7.6	2.7	7.9	Thynne (374)
$i\text{-}C_3H_7$	$\mathit{H}COO(i\text{-}C_3H_7)$	6.6	1.9	7.1	Thynne (374)
$n\text{-}C_4H_9$	$\mathit{H}COO(n\text{-}C_4H_9)$	5.3	1.7	6.9	Thynne (375)
C_6H_5	H_2	6.5	2.5	7.7	Fielding & Pritchard (142)
C_6H_5	CH_4	7.5	2.7	7.9	Fielding & Pritchard (142)
C_6H_5	CH_4	11.1	3.4	8.6	Duncan & Trotman-Dickenson (120)
C_6H_5	CF_3H	5.2	2.7	7.9	Fielding & Pritchard (142)
C_6H_5	$CH_3COC_6H_5$	6.2	3.1	8.3	Duncan & Trotman-Dickenson (120)
C_6H_5	*cyclo*-C_3H_6	8.5	2.9	8.1	Duncan & Trotman-Dickenson (120)
C_6H_5	*iso*-C_4H_{10}	6.7	3.3	8.5	Duncan & Trotman-Dickenson (120)

* Calculated by assuming log A_b equal to 10.4.

** These figures are independent of the assumed log A_b.

*** The atom abstracted is italicized.

TABLE IX

TRANSFER REACTIONS OF RADICALS

$$R + R'X \rightarrow RX + R' \quad a$$
$$2R \rightarrow R_2 \quad b$$

R	R'X***	E_a-½E_b (kcal/mole)	log ($A_a/A_b^{½}$) (A liter/mole-sec)	log A*	References
CH_3	CCl_4	13.4	5.0	10.2	Tomkinson, Galvin & Pritchard (380)
CH_3	C_2Cl_6	10.2	3.2	8.4	Tomkinson, Galvin & Pritchard (380)
CH_3	CCl_3COCCl_3	9.8	4.1	9.3	Tomkinson, Galvin & Pritchard (380)
CH_3	*Cl*$COOCH_3$	6.3	—	—	Thynne & Gray (376)
CCl_3	H_2	11.3**	—	—	Hautecloque (187)
CCl_3	CCl_3Br	18.6**	—	—	Hautecloque (188)
CCl_3	CCl_3COCCl_3	7.0**	—	—	Hautecloque (189)
C_2F_5	H_2	12.4	4.7	9.9	Pritchard & Foote (302)
C_2F_5	D_2	12.6	4.0	9.2	Pritchard & Foote (302)
C_3F_7	H_2	12.1	4.4	9.6	Pritchard & Foote (302)
C_3F_7	D_2	14.0	4.5	9.7	Pritchard & Foote (302)
CF_3	$C\mathit{H}D_3$	12.7	3.33	8.5	Sharp & Johnston (343)
CF_3	$CH\mathit{D}_3$	10.5	2.86	8.1	Sharp & Johnston (343)
C_2F_5	$C_2F_5C\mathit{H}O$	4.5	2.0	7.2	Pritchard, Miller & Foote (304)
C_3F_7	$C_3F_7C\mathit{H}O$	4.0	1.8	7.0	Pritchard & Miller Foote (304)
CH_3O	*H*$COOCH_3$	8.2	—	—	Thynne & Gray (378)

* Calculated by assuming log A_b equal to 10.4

** See reference 380 for certain problems regarding CCl_3 abstraction reactions.

*** The atom abstracted is italicized.

Two groups have studied the rate of H abstraction by phenyl radicals from methane in order to determine $D(C_6H_5\text{–}H)$. Duncan & Trotman-Dickenson (120) used acetophenone photolysis to generate phenyl radicals and Fielding & Pritchard (142) used photolysis of diphenyl mercury. The results do not agree (see Table VIII), and there is still some disagreement about $D(CH_3\text{–}H)$. If one takes $D(CH_3\text{–}H) = 102$ kcal/mole, the values for $D(C_6H_5\text{–}H)$ are 100 kcal/mole (Duncan & Trotman-Dickenson) and 103.7 kcal/mole (Fielding & Pritchard). We feel the latter is the more reliable result, mass balances being very poor in the former experiment.

Thynne has developed a method involving H abstraction by CH_3 from alkyl formates as a new source of alkyl radicals (373–375). The mechanism is given as

$$CH_3 + HCOOR \rightarrow CH_4 + \cdot COOR \qquad 33.$$

$$\cdot COOR \rightarrow CO_2 + R \qquad 34.$$

The methyl radicals are generated by acetone photolysis (>3000 Å) at a wavelength where the formates do not absorb. The method has been criticized by Grotewold & Kerr (178) on the grounds that H abstraction from R in the alkyl formate may occur in some cases.

Activation energies for H abstraction by methyl-d_3 from CF_3H, C_2F_5H, and C_3F_7H were measured by Pritchard & Thommarson (305) to determine the CH bond energies in these compounds. The results (Table VIII) lead to $D(R_f\text{–}H) = 102 \pm 2$ kcal/mole for each compound. (The authors take $D(CH_3\text{–}H) = 103 \pm 1$ kcal/mole.) It is pointed out that activation energies for H abstraction by CF_3, C_2F_5, and C_3F_7 from a given substrate RH are roughly constant, with the exception of the substrates H_2 and D_2 (302, 304, 305).

McNesby et al. (207a) have carried out some very careful studies of abstraction reactions by CH_3 and CD_3 from specially heterolabelled hydrocarbons such as $CH_3CD_2CH_3$. While these reactions are subject to the criticisms mentioned on such heterolabelling experiments (p. 416), they do yield very convincing evidence (at high temperatures) on the relative rate constants for decomposition of radicals (e. g., H vs. CH_3 elimination) and also on the nonisomerization of simple alkyl radicals. Thus *i*-C_3H_7 does not isomerize to *n*-C_3H_7, and a lower limit to the E_a for the isomerization of about 45 kcal can be estimated.

A series of investigations has been carried out on the rates of H abstraction from the α-carbon of aromatic compounds by methyl radicals (74, 333, 404). The method used involved acetone photolysis in the presence of the aromatics at about 100 to 200° C. The results, listed in Table VIII, show an increase in activation energies and pre-exponential factors in the order *p*, *o*, *m* for xylenes and fluorotoluenes. Pearson & Szwarc (289) have claimed, however, that both the activation energies and *A*-factors are nearly constant in this series of reactions, with the apparent E_a differences attributed to experimental errors.

Cher (86) has found evidence that methyl radicals abstract from the ring at a rate 0.17 as fast as from the methyl group of toluene at 60° C in the gas phase. Ethylbenzene was a major product in the gas phase, along with a small amount of *o*-xylene, whereas liquid phase experiments yielded *o*-xylene as the major product, with a smaller quantity of ethylbenzene. Cher considers that gas phase abstraction does not involve permanent addition of CH_3 to the ring, but rather leads directly to a tolyl radical which then undergoes a metathetical reaction with toluene to form a benzyl radical and thus ultimately ethylbenzene. Since the predominant fate of radicals in this system is recombination, it seems more likely that the tolyl radicals would also recombine with CH_3 radicals to give xylenes. The low yield of xylenes relative to the liquid phase can be explained by the fact that the disproportionation reaction is favored in the liquid phase by a high stationary concentration of the cyclohexadienyl radical $CH_3C_6H_5CH_3\cdot$ due to the high concentration of toluene in the liquid relative to the gas. It can be estimated that, for $CH_3+C_6H_5CH_3 \leftrightarrows CH_3C_6H_5CH_3\cdot$, $K_{eq}=10^{-5}\,10^{8/\theta}$ liter/mole, so that $(CH_3C_6H_5CH_3\cdot)/(CH_3)=K_{eq}(C_6H_5CH_3)$. At 60°C($\theta=1.5$ kcal), $K_{eq}=2$ liter/mole and the ratio is about $10^{-2.2}$ in the gas and 22 in the liquid, since the corresponding toluene concentrations are about $10^{-2.5}$ moles/liter and 11 moles/liter.

Clyne, Thrush & Wayne (97) studied both the overall reaction and the chemiluminescent reaction between NO and O_3 in a flow system at pressures of 0.8 to 3.0 torr and at 216 to 322° K. The reported rate (see below) is in good agreement with the previous value of Johnston & Crosby (211) and also with the recent value of Marte, Tschuikow-Roux & Ford (262) who used a supersonic nozzle technique. Clyne, Thrush & Wayne showed that the overall reaction proceeds by two paths, one giving ground state NO_2 and having an activation energy of 2.4 kcal/mole, and the other giving electronically excited $NO_2^{\dagger}$ with an activation energy of 4.18 kcal/mole. The best evidence is that the *A*-factor for each reaction path is about $10^{8.7}$ liter/mole-sec, and that the path giving chemiluminescence contributes about 10 per cent to the overall reaction at room temperature. Spectral comparisons show that the emission is similar to the thermal emission of NO_2 at 1200° K and that these transitions involve the same electronic state of NO_2 as that of the O+NO afterglow and the NO_2 fluorescence excited by 4358 Å radiation.

Rate results from the above work and from other recent studies involving NO and NO_2 are summarized below.

Reaction	E_a (kcal/mole)	log *A* (*A*, liter/mole-sec)	Reference
$O_3+NO \rightarrow O_2+NO_2$	2.4	8.76	Clyne, Thrush & Wayne (97)
$O_3+NO \rightarrow O_2+NO_2$	2.5	9.08	Marte, Tschuikow-Roux & Ford 262)
$XeF_4+NO \rightarrow XeF_3+ONF$	7 to 9	—	Johnston & Woolfolk 214)
$XeF_2+NO \rightarrow XeF+ONF$	10 to 12	—	Johnston & Woolfolk (214)
$NO_2+NO_3F \rightarrow NO_2F+NO_3$	14.3	8.28	Viscido, Sicre & Schumacher (389)
$NO_2+F_2O \rightarrow NO_2F+FO$	14.5	8.12	DeStaricco, Sicre & Schumacher (111)
$NO_2+NO_2 \rightarrow 2NO+O_2$	26.9	9.6	Ashmore & Burnett (3)
$NO_2+NO_2 \rightarrow NO_3+NO$	23.6	9.7	Ashmore & Burnett (3)

A new study (165) of the reaction of NO with O_2 at 23° C in the parts per million range with N_2 added to a total pressure of 760 torr shows that the reaction under these conditions is accurately second order in NO concentration and first order in O_2 concentration. The observed rate constant, $10^{4.19}$ liter^2/mole^2-sec, is in good agreement with early results obtained at higher NO and O_2 concentrations, such as the work of Bodenstein (58). However, these results and those of Ashmore, Burnett & Tyler (5) show no evidence of certain complicating features of the reaction as found by Treacy & Daniels (381).

Unimolecular Reactions

Isomerizations.—Thermal reactions of small ring compounds have been extensively investigated, as shown by the large amount of data for this reaction type in Table X. Cyclopropane and cyclobutane compounds undergo structural isomerization to give olefins and may also undergo a geometric *cis-trans* isomerization. A third type of reaction is recyclization leading to ring enlargement, which can occur when a vinyl group is attached to the ring. An example of the latter type is the rearrangement of vinyl cyclopropane to cyclopentene (148, 394):

$$\overline{CH_2CH_2CH}CH{=}CH_2 \rightleftarrows [\cdot\dot{C}H_2CH_2\dot{C}HCH{=}CH_2$$

$$\rightleftarrows \cdot\dot{C}H_2CH_2CH{=}CH\dot{C}H_2\cdot] \rightarrow \overline{CH_2CH_2CH{=}CHCH_2} \qquad 35.$$

A similar reaction has been demonstrated in the case of isopropenylcyclobutane. which gives 1-methylcyclohexene in addition to ethylene and isoprene (125).

Most authors consider that the mechanism of these three reaction types involves a biradical intermediate, although there is still discussion on this point (30, 50, 161, 275, 341, 359).

$$\text{Ring} \rightleftarrows \text{Biradical} \rightarrow \text{Olefin} \qquad 36.$$

Activation energies for structural isomerization of cyclopropane and cyclobutane alkyl derivatives normally fall in the range 62 to 65 kcal/mole with the exception of the vinyl substituted derivatives. The latter decompose with activation energies about 12 to 15 kcal/mole lower because the biradical transition state is stabilized by allylic resonance (125, 126, 157, 159, 275). Resonance energy in the biradical also accounts for lower activation energies for decomposition of cyclobutane carbonyl derivatives (275, 408).

Methylenecyclobutane can form an allylic type radical upon ring opening but nevertheless decomposes with a normal activation energy of about 62 kcal/mole (65, 87, 275). An explanation which has been offered (275) is that allylic resonance is sterically hindered in this case, since one of the H atoms in the CH_2 radical of the opening ring is not in the same plane as the vinyl group.

Ring strain reduces activation energies for decomposition (156, 359) and in cases where rigid biradical intermediates are formed, as in the isomerizations of certain bicyclic hydrocarbons, the Arrhenius A-factors are lowered (156, 161).

The isomerizations of cyclobutenes (see Table X) and methylenecyclopropanes do not seem to fit any of the simple mechanisms, and Benson (34,

TABLE X

THERMAL ISOMERIZATION[a]

Reaction	E_a (kcal/mole)	Log A (sec^{-1})	Reference
$CH_2CH_2C(CH_3)CH{=}CH_2 \rightarrow CH_2(CH_2)_2CH{=}CCH_3$	49.35	14.11	Ellis & Frey (126)
$CH_2{=}C(CH_3)CHCH_2CH_2 \rightarrow CH_2(CH_2)_2C{=}CCH_3$	50.9	13.89	Frey & Marshall (159)
$CH_2CH_2CHCH{=}CH_2 \rightarrow$ overall	50.0	13.73	Wellington (394)
$\rightarrow CH_2(CH_2)_2CH{=}CH$	49.7	13.61	
$\rightarrow CH_2{=}CHCH_2CH{=}CH_2$	57.3	14.43	
$\rightarrow$ *trans* $CH_2{=}CHCH{=}CHCH_3$	53.6	13.00	
$\rightarrow$ *cis* $CH_2{=}CHCH{=}CHCH_3$	56.2	13.90	
cis $CH_3CHCH_2CHCH{=}CH_2 \rightarrow$ *cis* $CH_2{=}CHCH_2CH{=}CHCH_3$	31	~11	Ellis & Frey (127)
$(CH_3)_2CCH_2C(CH_3)_2 \rightarrow (CH_3)_2C{=}CHCH(CH_3)_2$	64.4	15.83	Frey & Marshall (158)
$CH_3CHCH_2CH_2 \rightarrow$ overall	62.4	14.61	Setser & Rabinovitch (341)
$CH_3CHCHDCHD \rightarrow$ overall	62.3	14.43	
$CH_3CHCH_2C{=}CH_2 \rightleftarrows CH_3C{=}CCH_2CH_2$ ($k_{forward}+k_{reverse}$)	40.4	14.26	Chesick (87)
cis $CH_3CHCH(CH_3)CHCH_3 \rightarrow$ *trans*	60.95	15.78	Frey & Marshall (160)
cis $CH_3CH_2CHCH_2CHCH_3 \rightarrow$ *trans*	58.87	15.08	Elliot & Frey (124)
cis $C_6H_5CHCH_2CHC_6H_5 \rightarrow$ *trans*	33.53	11.2	Rodewald & DePuy (324)
$CH_2CH_2CH_2CH_2CH_2CH_2 \rightarrow$ overall	60.71	15.36	Flowers & Frey (149)
$\rightarrow CH_2{=}CH_2 + CH_2{=}CHCH{=}CH_2$	70.8	16.8	
$(CH_3)_2CHCH(CH_2)_2CH_2 \rightarrow (CH_3)_2CHCH{=}CH_2 + CH_2{=}CH_2$	62.6	15.63	Zupan & Walters (407)
$CF_2CF_2CF_2CF_2 \rightarrow 2CF_2{=}CF_2$	74.3	16.0	Butler (76)
$\rightarrow C_3F_6 + CF_2$	87.0	17.2	

[a] Also includes some thermal decomposition data.

TABLE X—(*Continued*)

Reaction	E_a (kcal/mole)	Log A (sec^{-1})	Reference
$CClFCF_2CF_2CClF \rightarrow 2CClF{=}CF_2$	65.3	15.4	Atkinson & Stedman (6)
cis $CClFCF_2CF_2CClF \rightarrow$ *trans*	60.2	15.1	
trans $CClFCF_2CF_2CClF \rightarrow$ *cis*	60.2	14.9	
$CH_2{=}C(CH_3)CH(CH_2)_2CH_2 \rightarrow CH_2{=}CH_2 +$ $CH_2{=}C(CH_3)CH{=}CH_2$	51.03	14.64	Ellis & Frey (125)
$\rightarrow CH_3CH{=}CH(CH_2)_3CH_2$	51.03	14.53	
$CH_2(CH_2)_2CHCHO \rightarrow CH_2{=}CHCHO + CH_2{=}CH_2$	53.3	14.43	Roquitte & Walters (327)
$CH_2(CH_2)_2CHCOC_2H_5 \rightarrow CH_2{=}CHCOC_2H_5 + CH_2{=}CH_2$	54.2	14.50	Roquitte & Walters (328)
$CH_2(CH_2)_2CHCOOCH_3 \rightarrow CH_2{=}CHCOOCH_3 +$ $CH_2{=}CH_2$	57.3	14.85	Zupan & Walters (408)
$CH_3C{=}CHCH_2CH_2 \rightarrow CH_2{=}CHC(CH_3){=}CH_2$	35.10	13.79	Frey (154)
$CH_3C{=}C(CH_3)CH_2CH_2 \rightarrow CH_2{=}C(CH_3)C(CH_3){=}CH_2$	36.04	13.84	Frey (155)
$CH_3CHCH{=}CHCH_2 \rightarrow$ *trans* $CH_3CH{=}CHCH{=}CH_2$	31.55	13.53	Frey (157)
$CF{=}CFCF_2CF_2 \rightarrow CF_2{=}CFCF{=}CF_2$	47.08	14.12	Schlag & Peatman (335)
$CF_2{=}CFCF{=}CF_2 \rightarrow CF{=}CFCF_2CF_2$	35.38	12.03	
2-methyl(2.1.0)bicyclopentane$\rightarrow (k_{cis}+k_{trans})$	38.9	14.45	Chesick (88)
Bicyclo(2.1.0)pentane$\rightarrow CH{=}CH(CH_2)_2CH_2$	46.6	14.58	Halberstadt & Chesick (181)
Bicyclo(2.1.0)pentane$\rightarrow CH{=}CH(CH_2)_2CH_2$	45.6	14.1	Steel et al. (359)
$\rightarrow CH_2{=}CHCH_2CH{=}CH_2$	52.3	14.4	
Bicyclo(2.2.0)hexane $\rightarrow CH_2{=}CH(CH_2)_2CH{=}CH_2$	36.0	13.4	
Bicyclo(2.1.1)hexane $\rightarrow CH_2{=}CH(CH_2)_2CH{=}CH_2$	55.0	15.17	Srinivasan & Levi (357)
Bicyclo(3.1.0)hexane $\rightarrow CH{=}CH(CH_2)_3CH_2$	57.4	13.29	Frey & Smith (161)
$\rightarrow CH_3C{=}CH(CH_2)_2CH_2$	61.17	13.89	

TABLE X—(*Continued*)

Reaction	E_a (kcal/mole)	Log A (sec^{-1})	Reference
Quadricyclene→bicyclo(2.2.1)hepta-2, 5-diene	33.51	12.81	Frey (156)
Bicyclo(2.2.1)hepta-2, 5-diene→$HC\equiv CH+$ $CH=CHCH_2CH=CH$	50.19	14.62	Herndon & Lowry (195)
→$CH=CHCH=CHCH_2CH=CH$	50.61	14.68	
→$C_6H_5CH_3$	53.14	14.23	
Bicyclo(2.2.1)hepta-2, 5-diene→$HC\equiv CH+$ $CH=CHCH_2CH=CH$	50.5	14.7	Birely & Chesick (52)
→$CH=CHCH=CHCH_2CH=CH$	51.6	14.9	
Bicyclo(2.2.1)hepta-2, 5-diene→$HC\equiv CH+$ $CH=CHCH_2CH=CH$	49.8	14.3	Roquitte (325)
→$CH=CHCH=CHCH_2CH=CH$	49.8	13.94	
Bicyclo(2.2.1)hept-2-ene→$CH_2=CH_2+$ $CH=CHCH_2CH=CH$	42.75	13.78	Herndon, Cooper & Chambers (193)
1, 3-dimethylbicyclo(1.1.0)butane→ $CH_2=C(CH_3)C(CH_3)=CH_2$	43.3	14.5	Chesick (89)
Tricyclo(3.3.0.0^{2,6})octane-1, 5-cyclooctadiene→overall	55.9	15.51	Srinivasan & Levi (358)
$CH=CHCH=CHCH_2CH=CH$→$C_6H_5CH_3$	51.1	13.55	Klump & Chesick (230)
$CH=CHCH=CHCH_2CH=CH$→$C_6H_5CH_3$	52.1	13.90	Herndon & Lowry (195)
$CH_2=CHCH=CHCH=CH_2$→$CH_2CH=CHCH=CHCH_2$	29.9	11.8	Lewis & Steiner (245)
cis-$CH_3CH=CHCOOCH_3$→*trans* $CH_3CH=CHCOOCH_3$	57.8	13.2	Butler & Small (77)
cis-$CH_3CH=CHCN$→*trans* $CH_3CH=CHCN$	51.3	11.0	Butler & McAlpine (77)
CH_3NC→CH_3CN	38.4	13.6	Schneider & Rabinovitch (338)
CH_2CH_2O→CH_3CHO	57	14.5	Benson (35)

37) has proposed that ion-pair or semi-ion pair intermediates may play an important role in such reactions where the activation energies are abnormally low and one would expect steric inhibition of resonance.

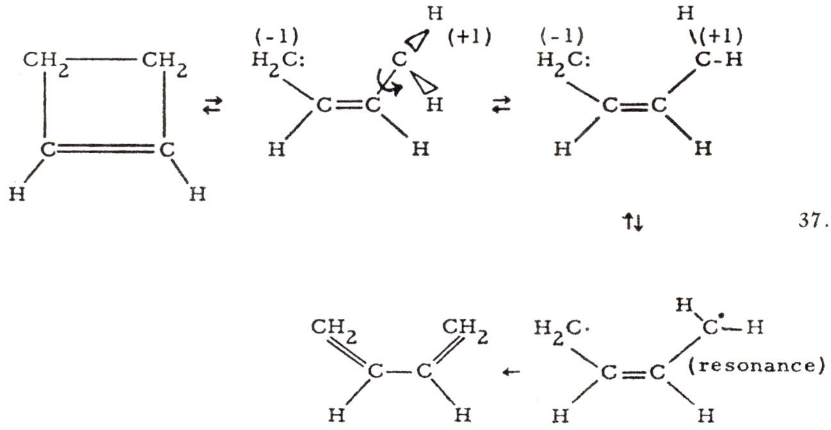

37.

If this is applied to the methylene cyclobutane case it would imply that the high activation energy comes from the splitting out of C_2H_4 from allene rather than from steric inhibition of resonance. An interesting labelling experiment could settle this point.

Perhaps one of the most unusual of the small ring pyrolyses involving the formation of a "hot" molecule is the isomerization of ethylene oxide to acetaldehyde. Benson (35) has shown that much of the discordant data on this system can be understood in terms of a biradical mechanism similar to that for the isoelectronic propylene formation from cyclopropane. In this case, however, the overall reaction is 28 kcal exothermic and this, added to the activation energy, results in a CH_3CHO^* molecule with $\sim$85 kcal of excess energy, more than enough to split the 78 kcal C–C bond. Recent experimental work (279) has confirmed this mechanism and demonstrated that up to 90 per cent yield of CH_3CHO can be obtained with adequate quenching of the hot CH_3CHO.

Fissions.—Considerable information on unimolecular reactions and energy transfer has been obtained by Rabinovitch and co-workers (102, 104, 137, 236–238, 242, 308, 309) in studies of the decomposition of chemically activated radicals. Vibrationally excited radicals are produced with a narrow spread of energy by addition of H atoms to the appropriate olefin at 25° C or −78° C. These nonequilibrium systems are well-defined and have the advantage of being quite sensitive to relative rates of decomposition and collisional stabilization, as opposed to thermal systems where details of interest are often masked by experimental uncertainties or, in the high pressure limit, by the fact that the unimolecular decomposition is a small perturbation on a near equilibrium system. The method used by the Rabinovitch group involves experimental determination of the pressure dependence of average decomposition rate constant k_a from the relation $k_a = \omega D/S$. The ratio D/S expresses the relative rates of decomposition and stabilization, as determined by product analysis, and ω is the calculated collision rate. The observed k_a values are compared with calculations based on a quantum statistical formulation of the Marcus-Rice model for unimolecular reaction rates (260). Good agreement is found for the dependence on excess energy and structure of the decomposing species provided that all vibrational and rotational degrees of freedom are taken as active. An important result in regard to energy transfer is that low pressure behavior of k_a shows evidence of multistage deactivation of excited radicals by monatomic and diatomic gases. Depending upon the model of transition probability employed, the amounts of energy removed per collision by these species fall in the range of about 1 to 3 kcal/mole. For the olefins and other polyatomic species as deactivating bodies the assumption of unit collisional efficiency appears to be valid. A recent and comprehensive review of these and other aspects of the chemical activation method is available (307).

Palmer & Hirt (285) measured the rate of methane pyrolysis at 875 to 1125° C by a flow method which used the rate of carbon deposition on a porcelain rod as an index of decomposition rate. The rate of the initial step was $\log k(\text{sec}^{-1}) = 14.1 - 101.0/\theta$. Combined with the higher temperature shock tube results of Skinner & Ruehrwein (353), the best value is $\log k = 14.58 - 103.0/\theta$. The results are consistent with either $CH_4 \rightarrow CH_3 + H$ or $CH_4 \rightarrow CH_2 + H_2$ as the initial step, and both may occur.

Quinn (306) studied the kinetics of methane formation in the pyrolysis of ethane at 61 to 202 torr and 564 to 608° C. By identifying the main source of CH_4 as being CH_3 from the primary dissociation of ethane, he concludes that the primary step is first order with rate $\log k(\text{sec}^{-1}) = 17.45 - 91.74/\theta$. This constitutes the first direct evidence for the very high-frequency factors, long predicted from CH_3 recombination, for the unimolecular decomposition of C_2H_6. The numbers, however, would imply that the C_2H_6 decomposition is in a fall-off region and this is not observed.

Thermal decomposition attracts a large number of workers and the trend of the Arrhenius parameters seems to be quite well established. The splitting of single bonds into two large radicals (Table XI) seems always to be accompanied by "abnormally" large A-factors, e.g., 10^{15} to 10^{18} sec^{-1}. The implications of this are quite clearly that a "loose" transition state is formed in which nearly free rotation exists between the radical portions (see discussion on p. 410). This can be taken now as a general rule. Splitting off of atoms seems to have smaller A-factors in the range of 10^{14} to 10^{15} sec^{-1}, again in agreement with a loose transition state.

The splitting off of stable molecules, on the other hand, which involves producing or shifting unsaturation, goes generally through tighter complexes and these have A-factors for 4 center reactions in the range of $10^{13 \pm .5}$ sec^{-1} (see Table XII).

TABLE XI

Unimolecular Reactions: Fission Into Radicals

Reaction	E_a (kcal/mole)	Log A (A sec^{-1})	Reference
$C_6H_5NHCH_3 \rightarrow C_6H_5NH + CH_3$	60.00	13.40	Esteban, Kerr & Trotman-Dickenson (133)
$(CH_3)_2NN{=}NN(CH_3)_2 \rightarrow 2(CH_3)_2N + N_2$	36.1	14.4	Gowenlock, Jones & Snelling (172)
$(C_2H_5)_2NN{=}NN(C_2H_5)_2 \rightarrow 2(C_2H_5)_2N + N_2$	33.0	13.7	Gowenlock, Jones & Snelling (172)
$CH_3(CH_2)_3N{=}N(CH_2)_3CH_3 \rightarrow 2CH_3(CH_2)_2CH_2 + N_2$	53.2	17.7	Blackham & Eatough (54)
$(CH_3)_2CHCH_2N{=}NCH_2CH(CH_3)_2 \rightarrow 2(CH_3)_2CHCH_2 + N_2$	49.0	16.2	Blackham & Eatough (54)
$CH_3CH_2CH(CH_3)N{=}NCH(CH_3)CH_2CH_3 \rightarrow 2CH_3CH_2CHCH_3 + N_2$	48.4	17.3	Blackham & Eatough (54)
$(CH_3)_3CN{=}NC(CH_3)_3 \rightarrow 2(CH_3)_3C + N_2$	43.0	17.2	Blackham & Eatough (54)
$CH_3N{=}NCH_3 \rightarrow 2CH_3 + N_2$	55.5	17.24	Forst & Rice (153)
$CF_3N{=}NCF_3 \rightarrow 2CF_3 + N_2$	55.20	16.17	Leventhal, Simonds & Steel (244)
$NH_2NH_2 \rightarrow 2NH_2$ (in fall-off region)	54.15	11.70	Kerr, Sekhar & Trotman-Dickenson (225)
$CH_3NHNH_2 \rightarrow CH_3NH + NH_2$	51.90	13.19	Kerr, Sekhar & Trotman-Dickenson (225)
$(CH_3)_2NNH_2 \rightarrow (CH_3)_2N + NH_2$	49.60	13.22	Kerr, Sekhar & Trotman-Dickenson (225)
$C_6H_5CH_2NH_2 \rightarrow C_6H_5CH_2 + NH_2$	59.80	13.00	Kerr, Sekhar & Trotman-Dickenson (225)
$C_6H_5CH_2NHCH_3 \rightarrow C_6H_5CH_2 + CH_3NH$	57.70	12.86	Kerr, Sekhar & Trotman-Dickenson (225)
$(CH_3)_3COOC(CH_3)_3 \rightarrow 2(CH_3)_3CO$	37.40	15.6	Batt & Benson (22)
$C_6H_5CH_2CH_3 \rightarrow C_6H_5CH_2 + CH_3$	70.10	14.6	Esteban, Kerr & Trotman-Dickenson (132)
$C_6H_5CH_2CH_2CH_3 \rightarrow C_6H_5CH_2 + CH_3CH_2$	68.60	14.9	Esteban, Kerr & Trotman-Dickenson (132)
$C_6H_5CH_2CH_2CH_2CH_3 \rightarrow C_6H_5CH_2 + CH_3CH_2CH_2$	67.20	14.5	Esteban, Kerr & Trotman-Dickenson (132)
$C_6H_5CH_3 \rightarrow C_6H_5CH_2 + H$	85.00	14.8	Price (299)
p-$C_6H_4(CH_3)_2 \rightarrow CH_3C_6H_4CH_2 + H$	76.00	14.0	Errede DeMaria (131)
$NO_3F \rightarrow NO_3F$	32.30	15.4	Viscido, Sicre & Schumacher (390)
$CF_2HCl \rightarrow CF_2 + HCl$	55.8	13.84	Edwards & Small (121)
$CF_2HCl \rightarrow CF_2 + HCl$	51.4	—	Gozzo & Patrick (173)
$(C_2H_5)_4Pb \rightarrow (C_2H_5)_3Pb + C_2H_5$	37.0	12.6	Pratt & Purnell (298)
$CH_3CO\cdot \rightarrow CH_3 + CO$	15.0	10.3	O'Neal & Benson (283)
$(CH_3)_3CCH_2\cdot \rightarrow i\text{-}C_4H_8 + CH_3$	32	12.5	Anderson & Benson (2)
$CH_2(CH_2)_3CH\cdot \rightarrow C_2H_4 + CH_2{=}CHCH_2$	36.9	—	Gunning & Stock (180)
	Second-Order Reactions (A, liter/mole-sec)		
$NOCl + M \rightarrow NO + Cl + M$, $M = NOCl$	35.4	13.8	Ashmore & Burnett (4)
$M = N_2$	35.8	13.3	
$M = NO$	36.0	13.5	
$M = H_2$	34.9	13.0	
$NO_2Cl + M \rightarrow NO_2 + Cl + M$, $M = NO_2$	29.2	14.2	Ashmore & Burnett (4)
$M = N_2$	29.8	14.1	
$CH_3CHO + M \rightarrow CH_3 + CHO + M$, $M = CH_3CHO$	73.8	17.0	Trenwith (382)
$CH_3CO + M \rightarrow CH_3 + CO + M$, $M = CH_3COCH_3$	12	11.5	O'Neal & Benson (283)

TABLE XII

UNIMOLECULAR REACTIONS: FISSIONS TO STABLE MOLECULAR PRODUCTS

Reaction	E_a (kcal/mole)	Log A (A sec^{-1})	Reference
$CH_2(CH_2)_4CHCl \rightarrow CH_2(CH_2)_3CH{=}CH + HCl$	50.2	13.88	Herndon, Henly & Sullivan (194)
$HCOF \rightarrow HF + CO$	10.0	2.3	Fischer & Buchanan (143)
$CH_2{=}CHCH_2COOH \rightarrow CH_3CH{=}CH_2 + CO_2$	39.3	11.3	Smith & Blau (355)
$CH_3CHClCH_2CH_3 \rightarrow C_4H_8$(mixture) $+ HCl$	50.75	14.07	Heydtmann & Rinck (198)
$(CH_3)_2C{-}N{=}N \rightarrow CH_3CH{=}CH_2 + N_2$(4 torr)	33.165	13.89	Frey & Stevens (162)
$CHF_2CH_2SiF_3 \rightarrow SiF_4 + CHF{=}CH_2$	32.72	12.27	Haszeldine, Robinson & Simmons (186)
$HO_2CCO_2H \rightarrow HCO_2H + CO_2$	30.0	11.9	Lapidus, Barton & Yankwich (243)
$CH_2ClCO_2C(CH_3)_3 \rightarrow CH_2ClCO_2H + (CH_3)_2C{=}CH_2$	38.11	13.09	Emovon (128)
$CHCl_2CO_2C(CH_3)_3 \rightarrow CHCl_2CO_2H + (CH_3)_2C{=}CH_2$	36.09	12.77	Emovon (128)
$CH_3CO_2CH(CH_3)CH_2CH{=}CH_2 \rightarrow CH_3CO_2H +$ Pentadienes	44.41	13.00	Emovon & Maccoll (130)
$CH_3CO_2CH(CH_3)CH_2COCH_3 \rightarrow CH_3CO_2H + CH_3CH{=}CHCOCH_3$	37.40	11.88	Emovon & Maccoll (130)
$CH_3CO_2C(CH_3)_2 \rightarrow CH_3CO_2H + CH_3CH{=}CH_2$	46.34	13.42	Emovon & Maccoll (129)
$CH_3CO_2CH(CH_3)CH_2CH_3 \rightarrow CH_3CO_2H + C_4H_8$	46.60	13.30	Emovon & Maccoll (129)
$CH_3CO_2C(CH_3)_3 \rightarrow CH_3CO_2H + C_4H_8$	40.02	13.15	Emovon & Maccoll (129)
$CH_3CO_2C(CH_3)_2C_2H_5 \rightarrow CH_3CO_2H + C_5H_{10}$	40.26	13.43	Emovon & Maccoll (129)
$CH_2(CH_2)_2CH{=}CH \rightarrow CH_2{=}CH_2 + CH_2{=}CHCH{=}CH_2$	66.0	15.2	Uchiyama, Tomioka & Amano (388)
$C_2H_5I \rightarrow C_2H_4 + HI$	50.0	13.36	Benson & Bose (44)
i-$C_3H_7I \rightarrow C_3H_6 + HI$	43.5	12.96	Teranishi & Benson (371)
i-$C_3H_7I \rightarrow C_3H_6 + HI$	48.0	14.49	Holmes & Maccoll (205)
s-$C_4H_9I \rightarrow C_4H_8 + HI$	47.9	15.2	Holmes & Maccoll (205)
t-$C_4H_9I \rightarrow i$-$C_4H_8 + HI$	36.4	12.52	Bose & Benson (60)

When such reactions involve internal ring formations, the A-factors are usually low, of the order of $10^{11.5\pm.5}$ sec^{-1} arising mainly from the loss of hindered internal rotations. This is evident in the Cope-type rearrangements such as the following, (55, 366) for which the A-factors are $10^{11.43}$ and $10^{12.0}$ sec^{-1}, respectively. One can assign roughly 2.5 entropy units per lost internal rotation which would lead to the value of $10^{11.5}$ sec^{-1}. Of the ester pyrolyses listed in Table XII only three fit this behavior. The other five seem to have A-factors too high by a factor of about 5 which might imply very strongly hindered rotation in these esters to start with (note that in each case we must compare the intrinsic or symmetry-corrected A-factors which would require dividing by 3 for CH_2CH_3 esters, 6 for i-Pr and 9 for t-Bu).

The data obtained from flow experiments are always subject to greater errors than static experiments conducted at lower temperatures, and, thus, much of the data of Table XI on fission into radicals obtained from such measurements are always somewhat more uncertain. They are nevertheless of considerable interest to workers concerned with measurements of bond dis-

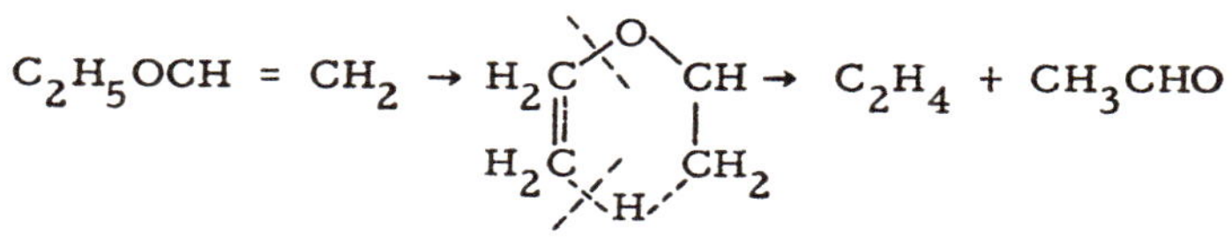

38.

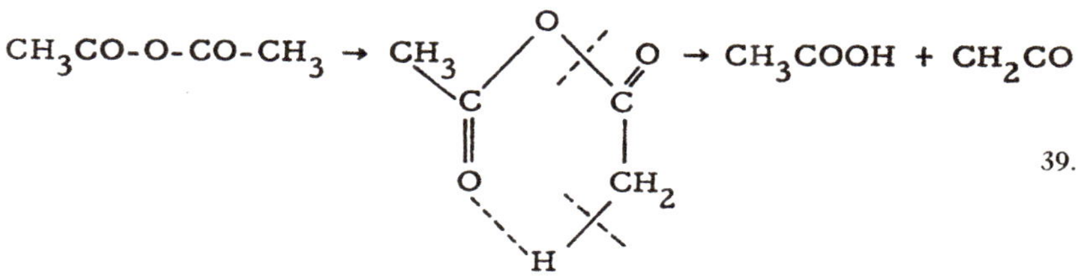

39.

sociation energies. In this respect it is nevertheless noteworthy that the newer data on benzyl radical formation favors values of the ϕCH_2–H bond dissociation energy nearer 84 kcal implying a resonance energy close to 14 kcal in the $\phi CH_2\cdot$ radical. This is also supported by Price's most recent analysis of the ϕCH_3 pyrolysis.

The *A*-factor for CHFO pyrolysis in Table XII is almost certainly a reflection of a wall reaction. No conceivable transition state could have an *A*-factor of $10^{2.3}$ sec^{-1} even if it were a forbidden transition.

One of the skeletons in the kineticist's closet was laid to rest in the case of di-tertiary butyl peroxide when Benson & Batt (Table XI) showed by direct measurement that exothermic heating was significant and accounted for the spread in Arrhenius parameters reported for this system.

A very interesting class of 3 center elimination reactions may now be coming to light in the studies of the pyroylsis of simple haloforms. Studies of $CHCl_3$ (345) had indicated that HCl and CCl_2 radicals are formed in the elementary act and this is now extended to CF_2 production from CF_2HCl (Table XII). CHF_3 (387) also seems to go to CF_2+HF in an elementary reversible act. However, recent data from the laboratory of one author (SWB) indicate that the quoted Arrhenius parameters for $CHCl_3$ are certainly both too low. The data would indicate linear structures for the CF_2 and CCl_2 ground state radicals. These observations focus considerable interest on elementary reactions such as $C_2H_6 \rightarrow CH_4+CH_2$ which have activation energies only 2 to 3 kcal higher than that for production of two CH_3 radicals but *A*-factors perhaps tenfold less. It certainly throws into the pyrolysis of hydrocarbons a new complicating feature which may be worthwhile exploring.

Kinetic data on unimolecular reactions measured in shock tubes are perhaps the most suspect of all. The data seem to scatter considerably and the Arrhenius parameters seem to be generally too low. One reason for this, which is perhaps true of many high-temperature flow studies, is that the reactions studied are not at their first order, high pressure limits. Tsang (Table XIII) has made probably the most precise use of the shock tube in his studies

TABLE XIII

SHOCK TUBE PYROLYSIS

Reaction	Method	Temp (° K)	Log k^a (liter/mole-sec)	Reference
$CS_2+Ar \rightarrow CS+S+Ar$	Absorption spectroscopy	2250–3350	$12.6-81.8/\theta$	Gaydon, Kimbell & Palmer (163)
$CO_2+Ar \rightarrow CO+O+Ar$	Infrared emission	3500–6000	$8.46+\log T/2 -74.5/\theta$	Davies (108)
$CO_2+N_2 \rightarrow CO+O+N_2$	Infrared emission	3500–6000	$8.39+\log T/2 -74.5/\theta$	Davies (108)
$CO_2+Ar \rightarrow CO+O+Ar$	Product analysis, light emission	2550–3000	$8.5+\log T/2 -86/\theta$	Brabbs, Belles & Zlatarich (63)
$N_2O+Ar \rightarrow N_2+O+Ar$	Absorption spectroscopy	1500–2500	$12-61/\theta$	Jost et al. (216)
$N_2O \rightarrow N_2+O$	Infrared emission Argon carrier	1500–2200	$10.4-55.7/\theta$	Fishburne & Edse (145)
$O_3+N_2 \rightarrow O_2+O+N_2$	Absorption spectroscopy (results combined with low temperature data)	689–863	$11.76-23.15/\theta$	Jones & Davidson (215)
$BrCN+Ar \rightarrow Br+CN+Ar$	Light emission	2500–7000	$9.3+\log T/2 -90.5/\theta$	Paterson & Greene (288)
$HNO_3+Ar \rightarrow HO+NO_2+Ar$	Absorption spectroscopy	800–1200	$12.2-30.6/\theta$	Harrison, Johnston Hardwick (183)
$NO_2+Ar \rightarrow NO+O+Ar$	Light emission	1800–2200	$12.8-65.9/\theta$	Hiraoka & Hardwick (201)
$NO_2Cl+Ar \rightarrow NO_2+Cl+Ar$	Absorption spectroscopy	600–1000	$10.4+\log T/2 -25.9/\theta$	Hiraoka & Hardwick (200)
$NO_2Cl+O_2 \rightarrow NO_2+Cl+O_2$	Absorption spectroscopy	600–1000	$10.6+\log T/2 -26.8/\theta$	Hiraoka & Hardwick (200)
$NO_2Cl+NO_2Cl \rightarrow NO_2+Cl+NO_2Cl$	Absorption spectroscopy	600–1000	$10.2+\log T/2 -23.6/\theta$	Hiraoka & Hardwick (200)
$C_2N_2+Ar \rightarrow 2CN+Ar$	Absorption spectroscopy	1700–2500	$\log(10^{9.52}/5.5.) +\log T/2+5.5 \log(125\times10^3/RT) -125/\theta$	Tsang, Bauer & Cowperthwaite (386)
$CF_2CF_2CF_2CF_2 \rightarrow 2C_2F_4$	Product analysis	1040–1200	$16.32-74.3/\theta$	Lifshitz, Carroll & Bauer (247)
cis Butene-2 → *trans* Butene-2	Product analysis	1000–1250	$14.55-65/\theta$	Lifshitz, Bauer & Resler (246)
t-$C_4H_9Cl \rightarrow i$-C_4H_8+HCl	Product analysis	750–950	$13.9-46.2/\theta$	Tsang (384)
t-$C_4H_9Br \rightarrow i$-C_4H_8+HBr	Product analysis	700–900	$13.5-41.5/\theta$	Tsang (384)
t-$C_4H_9OH \rightarrow i$-$C_4H_8+H_2O$	Product analysis	1050–1300	$13.4-61.6/\theta$	Tsang (384)
t-$C_4H_9SH \rightarrow i$-$C_4H_8+H_2S$	Product analysis	950–1230	$13.3-55.0/\theta$	Tsang (384)
t-$C_4H_9Cl \rightarrow i$-C_4H_8+HCl	Internal standard	715–845	$13.74-44.69/\theta$	Tsang (383)
t-$C_4H_9Br \rightarrow i$-C_4H_8+HBr	Internal standard	690–780	$13.87-41.49/\theta$	Tsang (383)

TABLE XIII—(*Continued*)

Reaction	Method	Temp (° K)	Log k^a (liter/mole-sec)	Reference
$C_2H_5Br \rightarrow C_2H_4 + HBr$	Internal standard	740–940	$13.19 - 53.7/\theta$	Tsang (385)
$C_2H_5Cl \rightarrow C_2H_4 + HCl$	Internal standard	820–1000	$13.16 - 56.46/\theta$	Tsang (385)
i-$C_3H_7Cl \rightarrow C_3H_6 + HCl$	Internal standard	750–950	$13.64 - 51.10/\theta$	Tsang (385)
i-$C_3H_7I \rightarrow C_3H_6 + HI$	Internal standard	680–850	$13.67 - 45.06/\theta$	Tsang (385)
t-$C_4H_9I \rightarrow i$-$C_4H_8 + HI$	Internal standard	650–760	$13.73 - 38.08/\theta$	Tsang (385)

[a] $\theta = 2.3$ RT.

of dehydrohalogenation reactions by using two independent reactions, one of which, with known kinetics, serves as an internal time and temperature-measuring device. By this device, many of the inherent uncertainties of the shock tube are eliminated. However, he does not avoid in this way problems arising from fall-off in rate constant at the higher temperatures since the two chosen reactions generally do not have the same fall-off characteristics. Since he chooses to use i-PrBr as a test substance and this is known to "fall-off" in rate at 3 torr (217, 371) at 600° K, one would expect pronounced effects in the range 700° K to 1000° K. Some of his rates are too low and this could be the explanation.

Molecule-Molecule Reactions

Relatively few rates of addition or disproportionation reactions between molecules are easily measurable, since for the first reaction type the equilibrium constants tend to favor dissociation, and for the second type the reactions are usually too endothermic, being the reverse of radical-radical reactions. Accordingly, few studies of these reactions have been reported. One of the cases where new data have been obtained is the addition of HI to a series of olefins, as studied by Benson and co-workers (see Table XIV). These results and others for HCl and Br addition (46) show interesting substituent effects. Each α-CH_3 substitution lowers the activation energy for the addition reaction by 5 to 7 kcal, and each βCH_2 substitution lowers the activation energy by 1 to 2 kcal. The kinetic directive effect (Markownikoff rule) of the CH_3 group in propylene is such that the n-PrI/i-PrI ratio is less than 10^{-3}, whereas the thermodynamic equilibrium ratio would be about $\frac{1}{4}$. A similar result holds for addition of HI to isobutene. The directive effect can be expressed as an activation energy difference of at least 7 kcal favoring the Markownikoff addition. The interpretation of these results (46) is discussed further under the section on Theory.

The reaction $F_2 + CO \rightarrow FCO + F$, which occurs rapidly at 15 to 45° C, is the rate-determining step in the overall reaction $F_2 + 2CO + O_2 \rightarrow (FCO)_2O_2$

(192). The fluorine atoms are believed to react with CO to give FCO, and the FCO radicals react further with O_2 leading to the peroxide. Sice products are F_2CO and CO_2, the former being favored by low O_2 concentration. Fluorine has two chemical effects on the thermal decomposition of NO_3F (347), one arising from a bimolecular reaction between F_2 and NO_3F (Table XIV), and a second from the rapid reaction of F_2 with NO_3 formed in the unimolecular dissociation of NO_3F: $NO_3F \rightarrow NO_3 + F$, $NO_3 + F_2 \rightarrow FO + NO_2$. The FO radicals then decompose a second NO_3F by the reaction $FO + NO_3F \rightarrow NO_2F + O_2 + F$.

TABLE XIV

MOLECULE-MOLECULE REACTIONS

Reaction	E_a	Log A (liter/mole-sec)	Reference
$HI + C_2H_4 \rightarrow C_2H_5I$	28.9	8.52	Benson & Bose (44)
$HI + C_3H_6 \rightarrow i\text{-}C_3H_7I$	23.4	7.89	Bose & Benson (59)
$HI + i\text{-}C_4H_8 \rightarrow t\text{-}C_4H_9I$	18.0	6.5	Bose & Benson (60)
HI + Butene-2 $\rightarrow$ sec-C_4H_9I	21.1	6.26	Nangia & Benson (278)
$2CF_2{=}CF_2 \rightarrow CF_2(CF_2)_2CF_2$	24.0	7.4	Butler (76)
$2CClFCF_2 \rightarrow CClF(CF_2)_2CClF$	26.6	7.33	Atkinson & Stedman (6)
$D_2 + I_2 \rightarrow 2DI$	41.5	11.1	Sullivan (365)
$2NOCl \rightarrow 2NO + Cl_2$	23.6	9.7	Ashmore & Burnett (4)
$F_2 + CO \rightarrow FCO + F$	13.5	8.67	Heras et al. (192)
$F_2 + NO_3F \rightarrow NO_2F + FO + F$	22.7	11.5	Sicre & Schumacher (347)

THEORY

Theory of homogeneous, gas phase reactions.—For over a century now, the kinetic-molecular theory of matter and the experimental study of quantitative chemical kinetics have enjoyed a close and fruitful relation. Just prior to and shortly after World War II it appeared that the theory was advanced quite far beyond the experimental work. The situation now appears righted and for the moment there are enough data to keep the theorists occupied. Unfortunately, not all the data is of sufficiently high accuracy to warrant complaints of surfeit. This lack of accuracy has been and will continue to be a source of confusion to theorists not equipped with the proper grains of salt.

Present theoretical work can be described as "applied." It has followed very closely along the lines laid down in the 1920's and 1930's, very little in the way of new approaches or models having been turned up. Two main fields of activity can be discerned. The first is on the theory of unimolecular reactions and now includes detailed studies of the behavior of these reactions in the nonequilibrium or "fall-off" region where the rate of formation of energized molecules is rate limiting. The second has to do with bimolecular reactions and the use of transition state theory to calculate Arrhenius A-

factors for these reactions from a priori models of the transition state. Closely allied to these efforts are the attempts to predict activation energies from semi-empirical models of the potential energy surface for a transition state.

Transition-state theory.—Certainly one of the most important developments of this century in chemical kinetic theory has been the transition-state theory of chemical reactions. This has been given its most significant expression by Henry Eyring and his co-workers (166) during the late 1930's. Applied to bimolecular abstraction reactions of the form $A+BC \rightleftarrows AB+C$, the theory assumes that a transition state complex $(A \cdots B \cdots C)^{\ddagger}$ is formed which is in "apparent" equilibrium with reactants. Then the rate constant for the forward reaction $k_f = \kappa\, kTK^{\ddagger}/h$ where κ, the transmission coefficient, is almost always taken as unity and $K^{\ddagger}$ is the equilibrium constant for formation of complex with one vibrational degree of freedom $\nu^{\ddagger}$ corresponding to the reaction coordinate factored out. Alternatively, if we choose to use the complete equilibrium constant, $K^{\#}$, we may write $k_f = \nu^{\ddagger}\, K^{\#}$ (28). Since $\nu^{\ddagger}$ is expected to be about 10^{13} sec^{-1} (within a factor of 3 for most systems) while $kT/h \sim 10^{13}$ sec^{-1} at 500° K, the two forms do not differ significantly. For either form, if one chooses a reasonable structure for $(A \cdots B \cdots C)^{\ddagger}$, it is in principle possible to calculate $K^{\ddagger}$ or $K^{\#}$ and thus k_f.

Early proponents of the theory were somewhat over-optimistic about the possibilities of a priori calculation of $K^{\ddagger}$ from such models, and the initially unified "absolute rate" approach has split into two approaches. The more successful of the two has confined itself to calculations of the pre-exponential or "A"-factor. The other approach, which is to calculate $E^{\ddagger}$ (the absolute value of the activation energy) from quantum mechanics, has been on the whole unsuccessful and few practioners remain. In its place have sprung up a number of empirical and semi-empirical approaches, mostly of uncertain utility and accuracy.

Symmetry numbers.—If one writes $K^{\ddagger}$ in terms of partition functions, then it can be separated into a product of factors $K_{\sigma}^{\ddagger} K_{\text{int}}^{\ddagger}$ where $K_{\sigma}^{\ddagger} = \Pi(\sigma \text{ reactants})/\sigma^{\ddagger}$ is the ratio of the products of symmetry numbers for the reactants to transition state (25). Generally this ratio also gives the number of distinct kinetic paths from reactants to products. However, Gold (167) has pointed out that this latter is not true if optical isomerism is involved and has proposed a special correction to σ for optical isomerism. If followed, his suggestion could lead to considerable confusion since proper partition functions already include contributions for optical isomerism. These contributions are simple and consist in multiplying the partion function by a number n_i corresponding to the number of energetically equivalent isomers. To maintain the spirit of the factoring of the partition function into intrinsic $K_{\text{int}}^{\ddagger}$ and nonintrinsic or symmetry features, it is better to redefine $K_{\sigma}^{\ddagger}$ as $\Pi(\sigma \text{ reactant}/n_i\text{-reactant})\ (n_i^{\ddagger}/\sigma^{\ddagger})$. This would accomplish the purpose pointed out by Gold without changing our conventional definition of partition functions. This also takes care of the same problem which has been discussed by Schlag (334) from the point of view of group theory. Schlag and

Rapp & Weston (314) in their considerations have made an erroneous application of these symmetry numbers to simple abstractions. The latter noted that for the exchange reactions $^{a}H+^{b}H-^{c}H \rightleftarrows ^{a}H-^{b}H+^{c}H$ and $H+^{b}D-^{c}D \rightleftarrows H-^{b}D+^{c}D$, $K_\sigma^\ddagger=1$ and 2, respectively, if the transition states are symmetrical. They then argued that this is unreasonable, since the number of paths is actually the same for both reactions and there should be no statistical differences between them. This argument is correct, however, only if the H atoms are distinct from each other and the D atoms are distinct also. But in this case $K_\sigma^\ddagger=1$ for both reactions. If H atoms are indistinguishable, then the first reaction is not really a reaction at all. What the authors really mean by their first reaction is $H+o\text{-}H_2 \rightleftarrows p\text{-}H_2+H$ and the $K_\sigma^\ddagger$ is adequately and properly given by including nuclear specific weights.

The above paradox leads Rapp & Weston to conclude that the transition state is never in true equilibrium with reactants and is never symmetrical along the reaction coordinate. This is an incorrect conclusion but there is not enough space here to discuss it at length.

Arrhenius A-*factors.*—Efforts to calculate A-factors from simple models have on the whole been quite successful to within a factor of about 3 (81, 90, 212, 315, 343). It is doubtful if much improvement can be expected. The reasons for this are about equally divided between uncertainties in the data and in our lack of knowledge of the vibrational frequencies of complex molecules. It is extremely difficult to measure activation energies of abstraction reactions to absolute accuracies of better than ± 0.6 kcal. This implies a related uncertainty in A of about a factor of 2.

During the past decade, Johnston and co-workers (81, 90, 212, 343) have been the most active in exploring models for predictions of A-factors and have devised some not-too-complicated schemes for assigning frequencies to transition state complexes. However, the known difficulties in making accurate assignments of low-frequency vibrations even to stable molecules make some of these efforts seem more complex than is warranted. As noted in some of the earlier sections, educated, empirical guesswork does almost as well in producing A-factors within the same limits. What is clearly required here is a much more direct approach to chemical bonding than is yet available from current quantum mechanics.

Machine calculations.—Ever since the pioneering calculations with computers by Wall and others (223, 251, 392), on the classical mechanical trajectories of the exchange reaction of $H+H_2$, an increasing number of workers have attempted to gain new insight into chemical reactions by such methods. Blais & Bunker (56, 71) have made very detailed calculations for reactions of alkali metal atoms with CH_3I and other exchange reactions. Some very interesting results concerning the effects of molecular weight on reaction cross section and energy distribution among the products have been obtained (72, 73, 294). It is also clear from these and other calculations, that normal coordinate analysis is hopelessly inadequate to describe a highly excited, small molecule.

Unimolecular reactions.—Wieder & Marcus have made the most extensive and detailed application of the Rice-Ramsperger-Kassel-Marcus theory (399) and have shown that it is in satisfactory agreement with all of the available data on unimolecular reactions. This revised theory, which incorporates the transition-state theory description of the unimolecular decomposition, was first elaborated in 1951 by Rice & Marcus (259, 260) and probably represents the best description to date of unimolecular reactions. The rival Slater Theory (354) which has stimulated much work and thought on the subject has been pretty well shown to be inadequate. It cannot account for the high-frequency factors found for many reactions and its main axiom, that the energy distribution among normal modes of a molecule is frozen between collisions, seems very untenable. This latter conclusion is one of the most important results to come out of the machine calculations on the motions of highly energized molecules. Bunker (72, 73) and Wilson and coworkers (66, 206) have shown that in small, highly energized molecules anharmonicity becomes very important and rotational-vibrational energy coupling is very large. In no case can a simple normal coordinate treatment describe the vibrational behavior.

Most of the other current work on unimolecular reactions has been directed towards improving various phases of the calculations of the rate constants for energy transfer and the density of energy states in highly excited molecules. The RRKM mechanism of an irreversible, unimolecular reaction is:

$$A + M \underset{d}{\overset{a}{\rightleftarrows}} A^* + M \qquad 40.$$

$$A^* \underset{c}{\overset{b}{\rightleftarrows}} A^{\ddagger} \qquad 41.$$

$$A^{\ddagger} \xrightarrow{r} \text{products} \qquad 42.$$

where A^* is a critically energized molecule not at the transition state geometry $A^{\ddagger}$.

At the high pressure limit where the first-order rate constant k_1 is independent of total pressure, we can write for k_1:

$$k_1 = \Sigma K^{\ddagger}(E)k_r(E) \qquad 43.$$

where $K^{\ddagger}(E)$ and $k_r(E)$ are both functions of energy E and the sum is over all energies. Note that this is not quite the same as the transition state formulation:

$$k_1 = \langle K^{\ddagger}\rangle\langle k_r\rangle \qquad 44.$$

At the low-pressure limit where activation is rate limiting and the overall reaction is second order, the second-order rate constant k_2 is:

$$k_2 = \Sigma k_a(E) = \Sigma \frac{k_a(E)}{k_d(E)} \cdot k_d(E) \qquad 45.$$

$$k_2 = \Sigma K^{\ddagger}(E)k_d(E) \qquad 46.$$

RRK theory gives the form for $k_r(E)$ and a number of authors have developed numerical methods for computing $K^{\ddagger}(E)$ for harmonic oscillators (249, 336, 397). These have been adapted to ion fragmentation in mass spectrometers (402) and Schlag, Sandsmark & Valence (337) have estimated corrections for anharmonicity.

Isotope effects.—Closely related to the problems of transition-state theory is the treatment of isotope effects. In bimolecular and unimolecular reactions, the ratio of rates of reactions of isotopically labeled species can be used to eliminate some of the guesswork in the frequency assignments to various transition-state models. It seems very doubtful if isotopic rate studies could ever be used to make unique assignments of transition-state geometries and frequencies because there are still too many unknowns involved. A number of books and chapters have appeared on isotope effects (51, 182, 268) so that we shall treat the subject only casually here.

Some of the most extensive studies of isotope effects in unimolecular reactions have come from Rabinovitch and co-workers (101, 103, 310, 349–352). It is now clear that secondary isotope effects can be as large or larger than primary isotope effects (276).

A great deal of work has been done on isotope effects in bimolecular exchange reactions. Since the only large effects are those involving H and D isotopes, most of the work involves compounds containing them. One of the most elusive problems in such cases has been the "tunnelling" correction. Because of its light mass, H can tunnel through small, thin barriers at low temperatures. However, the quantitative demonstration of such behavior, which in principle would come from a curvature of the (k_H/k_D)-Arrhenius plots, has eluded direct demonstration. Indirect theoretical evidence is much less convincing and represents the only method of demonstration. This is of concern in dealing with complex systems such as $CH_3 \cdots H{-}R$ abstraction since it is by no means clear that the reaction coordinate is truly the uncoupled movement of an H atom along the $C \cdots R$ axis. If the reaction coordinate is coupled to bending and stretching modes in R and CH_3 as is likely to be the case, then the reduced mass for this coordinate becomes a complex function and the very existence of tunnelling is in question (332).

Much work has gone into the problem of tunnelling without resolving the present situation greatly (90, 109, 229, 257, 332, 346, 379). One significant difficulty which has generally been ignored lies for small molecules in the coupling of rotation to vibration. This affects A-factors and geometry of the transition state (69).

Some very recent work by Weston et al. (229, 379) indicates that none of the a priori theories are in good agreement with data on the reaction of Br atoms with H_2-HD-D_2 mixtures or of Cl atoms with HCl-DCl mixtures. Salomon (332) has shown that all of the available data on $CH_3 + H{-}R$ or D–R abstraction can be fitted to a simple classical model with no tunnelling if the two new bending frequencies on the transition state are assumed to be very small. The one deviation he finds from the fit is for the reaction of CD_3

$+^{14}CD_4$ (100). However the Arrhenius parameters reported for this reaction, $A = 10^{9.6}$ 1/mole-sec and $E = 17.8$ kcal/mole are both impossibly high. No reasonable transition-state model can give so large an A-factor for this abstraction. (See discussion on pp. 407, 408.)

Theory of dissociation of diatomic molecules and atom recombination.—With the increased experimental work on dissociation of diatoms and recombination of atoms has come an extension of theoretical work in this area. Rice (317, 318) first pointed out the fact that to account for the high A-factors for dissociation of diatomic molecules it was necessary to assume that dissociation took place from the highest vibrational levels of a diatomic molecule within a band RT of the dissociation limit. This has now been extended to a complete treatment of such systems by Benson & Fueno (48). A similar approach had been taken earlier by Nikitin & Sokolov (280, 282) with different results and Rice (319, 320) has recently extended the discussion to consider slightly different assumptions and the masses of third bodies. These theories seem to give reasonably good agreement with data and the Benson-Fueno treatment gives the negative temperature coefficient for the recombination rate. It does not give a proper dependence on the mass of the third body and this has been somewhat rectified by independent machine calculations (42) of 3-body trajectories. Lacking in all theories to date is a proper accounting of the effects of upper electronic levels on the rates of dissociation and recombination. This is a difficult theoretical problem and very few experimental data are available with which to discuss it.

An alternate model, in principle more rigorous, has been developed by Keck (224) from an early formulation by Wigner (400). It gives an upper limit to the reaction rate although not in explicit form. Quite difficult integrations must be performed on computers by the Monte Carlo-type techniques. The theory ignores the nonequilibrium aspects of the problem and we feel that as a result it cannot be very useful. Light (248) has tried to develop a theory from over-simple collision assumptions. He also ignores the nonequilibrium aspects of the problem and chooses as a model a truncated harmonic oscillator. The apparent close agreement with experiment (average factor of $\sim$3) must be looked upon as fortuitous.

The fact that there is a serious departure from equilibrium occupation of the vibrational states of a diatomic molecule during dissociation has occasioned considerable controversy about whether or not the relation between forward- and reverse-rate constants holds under these nonequilibrium conditions, namely $k_f/k_r = K_{\text{equil}}$. Pritchard (300, 301) initiated the discussion in a paper on H_2 dissociation where he challenged the relation for the special case of H_2 and, by import, other diatomic molecules as well. This has been answered by Rice (320) and others (33, 107, 281) and recently in very elegant fashion by Boynton (62). The latter showed, using Laplace transforms, that only if the translational energy distribution is severely non-Maxwellian will the relationship likely be disturbed. This is assuredly not the case in studies to date.

One of the experimental difficulties with the data on dissociation and thus recombination can be the effects of small amounts of homogeneous, catalytic impurities. In the case of H atoms, as an example, Benson (32, 33) has shown that CH_4 can be an efficient catalyst above 1000° K (at less than 1 per cent mole fraction) by virtue of the rapid chain reaction:

$$H + CH_4 \rightarrow H_2 + CH_3 \qquad 47.$$

$$CH_3 + H \rightarrow CH_4 \qquad 48.$$

The two body $H+CH_3$ recombination is made possible by the long lifetime of nascent CH_4^*. In similar fashion, NO and NO_2 can act as a catalyst for the recombination of O atoms at lower temperatures as can CO, though less effectively. This is an aspect of such systems not very well investigated.

Energy transfer processes.—An important aspect of unimolecular reaction-rate theory and the dissociation of diatomic molecules is the rate at which translational energy can exchange with vibration. In view of recent surveys of this field (99, 197, 307, 367) we shall deal with the results very briefly. There has been a good deal of machine computations of vibrational relaxation for the first two levels of an oscillator on collision (1, 40, 42, 43, 184, 271, 286, 313), and, aside from chemical interactions, they all confirm the original view taken by Zener, Mott and Jackson and Landau and Teller. This gives a temperature dependence of the collision relaxation probability for the first level, P_{1-0}, predominantly of the form exp $(-A/T^{1/3})$. Harter, Alterman & Wilson (184) have shown that vibrational exchanges between two colliding oscillators do not take place readily.

The situation is quite different at the very highest vibrational levels where energy transfer seems to take place in small amounts of the order of RT or less per collisions (41). For polyatomic molecules Rabinovitch et al. (236, 237, 307) seem to find small energy transfer per collision with rare gases, but larger energy transfer with complex molecules.

Activation energies.—The first attempts at quantum-mechanical estimates of the activation energies of chemical reactions were made by London using very crude wave functions. Since then, efforts by Eyring, Polanyi and recent workers have tended towards more complex calculations on computers, but with relatively little success. In the past three years, a great deal of work has appeared on the $H+H_2$ potential surface with still some dispute as to the question of linear versus nonlinear transition state. The most recent work is semi-empirical (82) and gives a reasonably good activation energy 8.9 ± 1.2 kcal (obs 8.0 ± 0.5) for a linear complex. However, the significance of such an agreement seems very difficult to evaluate in view of the approximations employed.

More empirical approaches have been quite numerous but few have been very useful. Evans & Polanyi (135) suggested some time ago that for exothermic abstraction reactions there might be a relation between activation energies and heat of reaction of the type:

$$E = E_0 + \alpha(\Delta H) \qquad 49.$$

where α and E_0 are constants in a homologous series. While such a relation holds crudely for a few cases, e.g., Me+H−R, there are too many exceptions to it. As an example, the following reactions are all thermoneutral while their activation energies vary over almost the entire range of values for such reactions:

$$CH_3 + H\text{–}CH_3 \rightarrow \text{(exchange)} \quad (E = 14.5\text{ kcal}) \qquad 50.$$

$$CH_3 + H\text{–}OCH_3 \rightarrow CH_4 + OCH_3 \quad (E \sim 8\text{ kcal}) \qquad 51.$$

$$CH_3 + H\text{–}Cl \rightarrow CH_4 + Cl \; (E = 3.8\text{ kcal}) \qquad 52.$$

Since about 90 per cent of all reported exothermic radical abstraction reactions have activation energies of 8 ± 3 kcal, it is going to take a fairly sophisticated rule to be of any use. The relative insensitivity of E to ΔH can be seen in the following comparisons. All values in kcal/mole:

$$CH_3 + H\text{–}Cl \rightarrow CH_4 + Cl \quad (E = 3.8;\ \Delta H = 0) \qquad 53.$$

$$CH_3 + H\text{–}Br \rightarrow CH_4 + Br \quad (E = 3.2;\ \Delta H = -15) \qquad 54.$$

$$CH_3 + H\text{–}I \rightarrow CH_4 + I \quad (E = 1.5;\ \Delta H = -31) \qquad 55.$$

and in an organic sequence:

$$CH_3 + CH_3CH_3 \rightarrow C_2H_5^{\cdot} + CH_4 \quad (E = 10.9;\ \Delta H = -4) \qquad 56.$$

$$CH_3 + i\text{-}C_4H_{10} \rightarrow t\text{-Bu}^{\cdot} + CH_4 \quad (E = 8.1;\ \Delta H = -10) \qquad 57.$$

$$CH_3 + CH_3OCH_3 \rightarrow {}^{\cdot}CH_2OCH_3 + CH_4 \quad (E = 10.0;\ \Delta H = -9) \qquad 58.$$

$$CH_3 + CH_3COCH_3 \rightarrow {}^{\cdot}CH_2COCH_3 + CH_4 \quad (E = 10.2;\ \Delta H = -8) \qquad 59.$$

$$CH_3 + CH_3CH{=}CH_2 \rightarrow \text{allyl} + CH_4 \quad (E = 8.2;\ \Delta H = -18) \qquad 60.$$

$$CH_3 + \phi H = \phi^{\cdot} + CH_4 \quad (E = 9.7;\ \Delta H \sim 0) \qquad 61.$$

$$CH_3 + \phi CH_3 \rightarrow \phi CH_2^{\cdot} + CH_4 \quad (E = 8.8;\ \Delta H = -19) \qquad 62.$$

Starting from a simple intuitive idea that the sum of the "bond orders" is conserved in a chemical reaction, Johnston & Parr (213) have developed a somewhat elaborate, empirical calculation for predicting activation energies to within about ± 2 kcal. In view of the general uniformity of these values already noted, the utility and significance of the scheme are not yet clear.

An interesting modification of the Evans-Polanyi equation has been proposed by Spirin (356). He observed that the original London equation can be put in a form which suggests that the important correction terms to the first approximation correspond to coulombic interactions. On this basis, he was led to the empirical equation:

$$E = 0.66(0.75D_{AC} - Q)(1/P_A + 1/P_C) - 1.6 \qquad 63.$$

where in the abstraction reaction $A\cdot + B\text{–}C \rightarrow A\text{–}B + C\cdot$, $Q = -\Delta H$ is the heat liberated in the reaction, $D_{A\text{–}C}$ is, very curiously, the bond dissociation energy of A–C, and P_A and P_C are the molar refractions of A and C, estimated from empirical equations. For some 40 abstraction reactions the activation energies fit to within ± 2 kcal, and in only three cases does the maximum deviation amount to 5 kcal. This appears to be the most successful attempt yet at correlating these activation energies.

A completely different approach has been taken by Benson & Bose (46) in developing what appears to be an astonishingly simple model for predicting the activation energies of four center molecule-molecule reactions. Starting with data on the addition of hydrogen halides to substituted olefins, they showed that the marked effect of apparently innocuous substituents like CH_3 (Markownikoff Rule) could be quantitatively explained in terms of a "semi-ion pair" model in which the double bond is polarized to give formal charges of $\pm\frac{1}{2}$ at each C atom. Assuming a similar, parallel polarization for HX, namely $H^{+1/2}-X^{-1/2}$ and correcting for dipole-dipole attraction in the transition state they could calculate activation energies E for some 25 exothermic addition reactions covering a range of 16 to 40 kcal to within an average error of ± 3 kcal. The formula for E is:

$$D = \frac{\epsilon^2}{8}\left(\frac{r^2_{c-c}}{\alpha_{c-c}} + \frac{r^2_{H-X}}{\alpha_{H-X}} - \frac{3}{2}\,\frac{r_{c-c}r_{H-X}}{r_a^3}\right) \qquad 64.$$

where ϵ is the electronic charge, r_{c-c} and r_{H-X} are the respective bond lengths of olefin and HX, α_{c-c} and α_{H-X} their longitudinal polarizabilities and r_a is the mean separation distance between the C–C and H–X dipoles in the transition state. Recent calculations (49) have shown that the method can be extended very successfully to other reactions such as $I_2+C_2H_4$; H_2+I_2, etc. In addition, the precision has been improved to ± 1 kcal with a maximum deviation of ± 3 kcal.

The theoretical justification for such a classical approach may be made from the argument that in chemical reactions generally, new bonds cannot be formed between closed-shell molecules because no low-lying open orbitals are available for chemical bonding. On this basis, one would predict that in closed-shell molecule-molecule reactions the activation energy is just the minimum work required to open orbitals in each species. Polarization work seems to offer the lowest energy for such a promotion process and the semi-ion pair for an X–Y molecule corresponds to leaving one electron in the bond and localizing the other on Y(or X) thus forming X·Y· with formal charges of $+\frac{1}{2}$ on X and $-\frac{1}{2}$ on Y. X now has an open orbital and can receive an electron from a donor species.

Looked at from this point of view we can classify activation energies in exothermic chemical reactions as follows:

Reaction type	*Observed range of activation energies*
I. Molecule-molecule	20–60 kcal
II. Radical-molecule	0–15 kcal
III. Radical-radical	0 kcal

The closed shell molecule-molecule reactions have the highest activation energies because two promotions (i.e., polarization) must be accomplished. In radical-molecule reactions, only one promotion is needed, the radical having an open orbital. In radical-radical reactions, no promotions are required and the activation energy is zero, quite independent of the mode of reaction, i.e., addition or recombination. Ion-molecule reactions would be like radical-molecule reactions, but the large, long-range coulombic attraction would exceed the expected 8 ± 3 kcal of activation energy to give a net of zero as is observed.

On this basis, two center abstraction reactions would take the same polarization path, i.e.:

$$H\cdot + H\overset{\times}{\underset{\times}{}}H \rightleftarrows H\cdot H \times H\times \leftrightarrow \cdot H \times H \times H \rightleftarrows H \dot{\times} H + \times H \qquad 65.$$

Activation energies calculated for such abstraction reactions have the correct range of values, but, unfortunately, the models are too sensitive to distances to give meaningfully the small differences in observed activation energies.

Miscellany.—While heterogeneous catalysis is a widely studied, little-understood but important phenomenon, homogeneous catalysis in gases has been relatively obscure. Several recent studies indicate that this situation may not be merited. In the field of chain reactions, small molecules such as HCl and HBr can bring about a considerable acceleration of otherwise slow processes by substituting two fast steps with high A-factors and low activation energies for one slow step. This has been demonstrated now for HCl in the pyrolysis of ditertiary butyl peroxide (147), of CH_3OCH_3 (2), and of neopentene (38) and was, of course, known for some time in the oxidation of *i*-butane (331).

It now appears that the phenomenon can be used as an important diagnostic tool in the analysis of complex pyrolytic chains (36) where it can convert the normally fast radical-pyrolysis step into the rate-determining step. Thus, in neopentane pyrolysis, the normal chain is:

$$CH_3 + (CH_3)_4C \rightleftarrows CH_4 + \cdot CH_2-C(CH_3)_3 \quad \text{(slow)} \qquad 66.$$

$$\cdot CH_2 - C(CH_3)_3 \rightarrow CH_3 + i\text{-butene} \quad \text{(fast)} \qquad 67.$$

while with HCl it becomes:

$$\left[\begin{array}{l} CH_3 + HCl \leftrightarrows CH_4 + Cl \\ Cl\cdot + (CH_3)_4C \leftrightarrows HCl + \cdot CH_2-C(CH_3)_3 \end{array}\right] \quad \text{(fast)} \qquad \begin{array}{r} 68. \\ 69. \end{array}$$

$$\cdot CH_2-C(CH_3)_3 \rightarrow \cdot CH_3 + i\text{-butene} \quad \text{(slow)} \qquad 70.$$

Iodine has been known for some time to act as a catalyst for dehydrohalogenation reactions (60, 278), but its precise mode of action is still obscure. A very unusual example of catalysis at the molecular level seems to be provided from studies on the effects of HBr and HCl on the dehydration of *t*BuOH and the elimination of MeOH from *t*-BuOMe (136, 254, 361, 393). The data are still somewhat suspect because of possible wall effects, but they seem certainly worth pursuing.

In contrast to organic ions which are quite labile and rearrange easily, organic radicals in the gas phase seem very stable. In a number of cases which have been reported of group migration in radicals (144), most have involved conjugated unsaturation, biradicals, or else high energies ("hot" radicals). It is therefore of interest to note that the migration of CH_3 in the isomerization of CH_3NC to CH_3CN (338, 339), is a genuine migration with what must be an unusual transition state. It is difficult to write any normal structures for the pentavalent carbon in the transition state, and the most reasonable explanation for the reaction would appear to require ion pairs:

$$CH_3-N=C \rightleftarrows CH_3^+(N\equiv C)^- \rightleftarrows \overset{CH_3^+}{(N\equiv C)^-} \rightarrow N\equiv C-CH_3 \qquad 71.$$

Such interpretation is favored by the very large dipole moments already present in both reactants and products.

Another very interesting transition state arises in the pyrolysis of sym-trioxane (75, 203). The Arrhenius parameters preclude any radical process or biradical intermediate and the implication is that the ring falls apart into three CH_2O units in essentially one concerted step. ΔS for the overall reaction is about 80 Gibbs/mole, and the reverse termolecular combination of CH_2O to form sym-trioxane would have the unprecedentedly low A-factor of 10^{-2} 1^2/mole^2-sec!

LITERATURE CITED

1. Allen, R. T., and Feuer, P., *J. Chem. Phys.*, **40**, 2810 (1964)
2. Anderson, K. H., and Benson, S. W., *J. Chem. Phys.*, **39**, 1677 (1963)
3. Ashmore, P. G., and Burnett, M. G., *Trans. Faraday Soc.*, **58**, 253 (1962)
4. Ashmore, P. G., and Burnett, M. G., *ibid.*, 1801
5. Ashmore, P. G., Burnett, M. G., and Tyler, B. J., *Trans. Faraday Soc.*, **58**, 685 (1962)
6. Atkinson, B., and Stedman, M., *J. Chem. Soc.*, 512 (1962)
7. Avramenko, L. I., and Kolesnikova, R. V., *Advances in Photochemistry*, **2**, 25 (Interscience, New York, 1963)
8. Avramenko, L. I., Kolesnikova, R. V., and Kuznetsova, N. W., *Izvest. Akad. Nauk SSSR, Otd. Khim. Nauk*, 620 (1963)
9. Avramenko, L. I., Kolesnikova, R. V., and Savinova, G. I., *Izv. Akad. Nauk SSSR, Otd. Khim. Nauk*, **17**, 36 (1963)
10. Avramenko, L. I., Kolesnikova, R. V., and Savinova, G. I., *ibid.*, 976
11. Avramenko, L. I., and Krasnen'kov, V. M., *Izv. Akad. Nauk SSSR, Otd. Khim. Nauk*, 1196 (1963)
12. Ayscough, P. B., Cocker, A. K., Dainton, F. S., and Hirst, S., *Trans. Faraday Soc.*, **58**, 295 (1962)
13. Azatyan, V. V., Nalbandyan, A. B., and Tsui, M. Y., *Dokl. Akad. Nauk SSSR*, **147**, 361 (1962)
14. Azatyan, V. V., Nalbandyan, A. B., and Tsui, M. Y., *Dokl. Akad. Nauk Arm. SSSR*, **36**, 23 (1963)
15. Azatyan, V. V., Nalbandyan, A. B., and Tsui, M. Y., *Kinetika i Kataliz*, 201 (1964)
16. Bader, L. W., and Ogryzlo, E. A., *Nature*, **201**, 498 (1964)
17. Baldwin, R. R., *Trans. Faraday Soc.*, **60**, 527 (1964)
18. Baldwin, R. R., and Melvin, A., *J. Chem. Soc.*, 1785 (1964)
19. Baldwin, R. R., and Walker, R. W., *Trans. Faraday Soc.*, **60**, 1236 (1964)
20. Barber, M., Farren, J., and Linnett, J. W., *Proc. Roy. Soc.* (*London*), *Ser. A*, **274**, 306 (1963)
21. Basila, M. R., and Strong, R. C., *J. Phys. Chem.*, **67**, 521 (1963)
22. Batt, L., and Benson, S. W., *J. Chem. Phys.*, **36**, 895 (1962)
23. Belles, F. E., and Lauver, M. R., *J. Chem. Phys.*, **40**, 415 (1964)

24. Benson, S. W., *J. Chem. Phys.*, **26**, 1351 (1957)
25. Benson, S. W., *J. Am. Chem. Soc.*, **80**, 5151 (1958)
26. Benson, S. W., *Foundations of Chemical Kinetics*, 312 (McGraw, New York, 1960)
27. Benson, S. W., *ibid.*, 662
28. Benson, S. W., *ibid.*, Sect. XI-7, XII-4
29. Benson, S. W., *J. Chem. Phys.*, **33**, 939 (1960)
30. Benson, S. W., *ibid.*, **34**, 521 (1961)
31. Benson, S. W., *ibid.*, **38**, 1251 (1963)
32. Benson, S. W., *ibid.*, 2285
33. Benson, S. W., *Symp. Combust.*, *9th*, 760 (Academic Press, New York, 1963)
34. Benson, S. W., *Advances in Photochemistry*, **2**, 1 (Interscience, New York, 1964)
35. Benson, S. W., *J. Chem. Phys.*, **40**, 105 (1964)
36. Benson, S. W., *Ind. Eng. Chem.*, **56**, 18 (1964)
37. Benson, S. W., Paper presented at *Symp. Pyrolysis, Ottawa, 1964*
38. Benson, S. W., and Anderson, K. H., *J. Chem. Phys.*, **40**, 3747 (1964)
39. Benson, S. W., and Axworthy, A. E., Jr., *J. Chem. Phys.*, **26**, 1718 (1957)
40. Benson, S. W., and Berend, G., *J. Chem. Phys.*, **38**, 2777 (1963)
41. Benson, S. W., and Berend, G. C., *ibid.*, **40**, 1289 (1964)
42. Benson, S. W., Berend, G., and Wu, J., *J. Chem. Phys.*, **37**, 1386 (1962)
43. Benson, S. W., Berend, G., and Wu, J., *ibid.*, **38**, 25 (1963)
44. Benson, S. W., and Bose, A. N., *J. Chem. Phys.*, **37**, 2935 (1962)
45. Benson, S. W., and Bose, A. N., *J. Am. Chem. Soc.*, **85**, 1385 (1963)
46. Benson, S. W., and Bose, A. N., *J. Chem. Phys.*, **39**, 3463 (1963)
47. Benson, S. W., Bose, A. N., and Nangia P., *J. Am. Chem. Soc.*, **85**, 1388 (1963)
48. Benson, S. W., and Fueno, T., *J. Chem. Phys.*, **36**, 1597 (1962)
49. Benson, S. W., and Haugen, G. (Unpublished)
50. Benson, S. W., and Nangia, P., *J. Chem. Phys.*, **38**, 18 (1963)
51. Bigeleisen, J., and Wolfsberg, M., *Advan. Chem. Phys.*, **1**, 15 (1958)
52. Birely, J. H., and Chesick, J. P., *J. Phys. Chem.*, **66**, 568 (1962)
53. Birrell, R. N., and Trotman-Dickenson, A. F., *J. Chem. Soc.*, 4218 (1960)
54. Blackham, A. U., and Eatough, N. L., *J. Am. Chem. Soc.*, **84**, 2922 (1962)
55. Blades, A. T., and Murphy, G. W., *J. Am. Chem. Soc.*, **74**, 1039 (1952)
56. Blais, N. C., and Bunker, D. L., *J. Chem. Phys.*, **37**, 2713 (1962)
57. Boddy, P. J., and Robb, J. C., *Proc. Roy. Soc. (London), Ser. A*, **249**, 547 (1959)
58. Bodenstein, M., *Z. Physik. Chem. (Frankfurt)*, **24**, 183 (1918)
59. Bose, A. N., and Benson, S. W., *J. Chem. Phys.*, **37**, 1081 (1962)
60. Bose, A. N., and Benson, S. W., *ibid.*, **38**, 878 (1963)
61. Boyd, R. K., Downs, G. W., Gow, J. S., and Horrex, C., *J. Phys. Chem.*, **67**, 719 (1963)
62. Boynton, F. P., *J. Chem. Phys.*, **40**, 3124 (1964)
63. Brabbs, T. A., Belles, F. E., and Zlatarich, S. A., *J. Chem. Phys.*, **38**, 1939 (1963)
64. Bradley, J. N., and Rabinovitch, B. S., *J. Chem. Phys.*, **36**, 3498 (1962)
65. Brandaur, R. L., Short, B., and Kellner, S. M. E., *J. Phys. Chem.*, **65**, 2269 (1961)
66. Brauner, D. L., and Wilson, D. J., *J. Phys. Chem.*, **67**, 1134 (1963)
67. Brown, A. C. R., and James, D. C. L., *Proc. Chem. Soc.*, 81 (1962)
68. Buckley, E., and Whittle, E., *Trans. Faraday Soc.*, **57**, 536 (1962)
69. Buff, F. P., and Wilson, D. J. *J. Am. Chem. Soc.*, **84**, 4063 (1962)
70. Bulewicz, E. M., and Sugden, T. M., *Proc. Roy. Soc. (London), Ser. A*, **277**, 143 (1964)
71. Bunker, D. L., *Nature*, **194**, 1277 (1962)
72. Bunker, D. L., *J. Chem. Phys.*, **37**, 393 (1962)
73. Bunker, D. L., *ibid.*, **40**, 1946 (1964)
74. Burkley, I. B., and Rebbert, R. E., *J. Phys. Chem.*, **67**, 168 (1963)
75. Burnett, G., and Bell, R. P., *Trans. Faraday Soc.*, **34**, 420 (1938)
76. Butler, J. N., *J. Am. Chem. Soc.*, **84**, 1393 (1962)
77. Butler, J. N., and McAlpine, R. D., *Can. J. Chem.*, **41**, 2487 (1963); Butler, J. N., and Small, G. J., *ibid.*, 2492
78. Callear, A. B., and Tyerman, W. J. R., *Proc. Chem. Soc.*, 296 (1964)
79. Calvert, J. G., *Ann. Rev. Phys. Chem.*, **11**, 41 (1960)
80. Calvert, J. G., Thomas, S. S., and Hanst, P. L., *J. Am. Chem. Soc.* **82** 1 (1960)

81. Carmichael, H., and Johnston, H. S., *J. Chem. Phys.*, **41,** 1975 (1964)
82. Cashion, J. K., and Herschbach, D. R., *J. Chem. Phys.*, **40,** 2358 (1964)
83. Castellano, E., and Schumacher, H. J., *J. Chem. Phys.*, **36,** 2238 (1962)
84. Cerfontain, H., and Kutschke, K. O., *J. Am. Chem. Soc.*, **84,** 4017 (1962)
85. Charles, S. W., Pearson, J. T., and Whittle, E., *Trans. Faraday Soc.*, **59,** 1156 (1963)
86. Cher, M., *J. Phys. Chem.*, **68,** 1316 (1964)
87. Chesick, J. P., *J. Am. Chem. Soc.*, **85,** 2720 (1963); *J. Phys. Chem.*, **65,** 2170 (1961)
88. Chesick, J. P., *J. Am. Chem. Soc.*, **84,** 3250 (1962)
89. Chesick, J. P., *J. Phys. Chem.*, **68,** 2033 (1964)
90. Chiltz, Z., Eckling, R., Goldfinger, P., Huybrechts, G., Johnston, H. S., Meyers, L., and Verbeke, G., *J. Chem. Phys.*, **38,** 1053 (1963)
91. Christie, M. I., *J. Am. Chem. Soc.*, **84,** 4066 (1962)
92. Clyne, M. A. A., and Thrush, B. A., *Discussions Faraday Soc.*, **33,** 139 (1962)
93. Clyne, M. A. A., and Thrush, B. A., *Proc. Roy. Soc. (London), Ser. A*, **269,** 404 (1962)
94. Clyne, M. A. A., and Thrush, B. A., *ibid.*, **275,** 544 (1963)
95. Clyne, M. A. A., and Thrush, B. A., *ibid.*, 559
96. Clyne, M. A. A., Thrush, B. A., and Wayne, R. P., *Nature*, **199,** 1057 (1963)
97. Clyne, M. A. A., Thrush, B. A., and Wayne, R. P., *Trans. Faraday Soc.*, **60,** 359 (1964)
98. Corbett, P., Tarr, A. M., Whittle, E., *Trans. Faraday Soc.*, **59,** 1609 (1963)
99. Cottrell, T. L., and McCoubrey, J. C., *Molecular Energy Transfer in Gases* (Butterworths, London, 1961)
100. Creak, G. A., Dainton, F. S., and Ivin, K. J., *Trans. Faraday Soc.*, **58,** 326 (1962)
101. Current, J. H., and Rabinovitch, B. S., *Can. J. Chem.*, **40,** 557 (1962)
102. Current, J. H., and Rabinovitch, B. S., *J. Chem. Phys.* **38,** 783 (1963),
103. Current, J. H., and Rabinovitch, B. S., *ibid.*, 1967
104. Current, J. H., and Rabinovitch, B. S., *ibid.*, **40,** 2742 (1964)
105. Cvetanovic, R. J., *Advances in Photochemistry*, **1,** 115 (Interscience, New York, 1963)
106. Dalby, F. W., *J. Chem. Phys.*, **41,** 2297 (1964)
107. Davidson, N. (Unpublished)
108. Davies, W. O., *J. Chem. Phys.*, **41,** 1846 (1964)
109. Davy, J. A., Guerra, C. R., Johnston, H. S., Weaver, C. F., and Young, C. E., *J. Chem. Phys.*, **41,** 1517 (1964)
109a. DeMore, W., and Benson, S. W., *Advances in Photochemistry*, **2,** 246 (Interscience, New York, 1964)
110. DeMore, W., and Raper, O. F., *J. Chem. Phys.*, **37,** 2048 (1962)
111. DeStaricco, E. R., Sicre, J. E., and Schumacher, H. J., *Z. Physik Chem. (Frankfurt)*, **39,** 337 (1963)
112. Dever, D. F., and Calvert, J. G., *J. Am. Chem. Soc.*, **84,** 1362 (1962)
113. Dingledy, D. P., and Calvert, J. G., *J. Am. Chem. Soc.*, **85,** 856 (1963)
114. Dixon, P. S., Stefani, A. P., and Szwarc, M., *J. Am. Chem. Soc.*, **85,** 2551 (1963)
115. Dixon, P. S., and Szwarc, M., *Trans. Faraday Soc.*, **59,** 112 (1963)
116. Dixon-Lewis, G., Sutton, M. M., and Williams, A., *Discussions Faraday Soc.*, **33,** 205 (1962)
117. Dixon-Lewis, G., and Williams, A., *Nature*, **196,** 1309 (1962)
118. Dominguez, J. A. G., Kerr, J. A., and Trotman-Dickenson, A. F., *J. Chem. Soc.*, 3357 (1962)
119. Dominguez, J. A. G., and Trotman-Dickenson, A. F., *J. Chem. Soc.*, 940 (1962)
120. Duncan, F. J., and Trotman-Dickenson, A. F., *J. Chem. Soc.*, 4672 (1962)
121. Edwards, J. W., and Small, P. A., *Nature*, **202,** 1329 (1964)
122. Elias, L., *J. Chem. Phys.*, **38,** 989 (1963)
123. Elias, L., and Schiff, H. I., *Can. J. Chem.*, **38,** 1657 (1960)
124. Elliot, C. S., and Frey, H. M., *J. Chem. Soc.*, 900 (1964)
125. Ellis, R. J., and Frey, H. M., *Trans. Faraday Soc.*, **59,** 2076 (1963)
126. Ellis, R. J., and Frey, H. M., *J. Chem. Soc.*, 959 (1964)
127. Ellis, R. J., and Frey, H. M., *Proc. Chem. Soc.*, 221 (1964)
128. Emovon, E. U., *J. Chem. Soc.*, 1246 (1963)
129. Emovon, E. U., and Maccoll, A., *J. Chem. Soc.*, 335 (1962)
130. Emovon, E. U., and Maccoll, A., *ibid.*, 227 (1964)
131. Errede, A., and DeMaria, F., *J. Phys. Chem.*, **66,** 2664 (1962)

132. Esteban, G. L., Kerr, J. A., and Trotman-Dickenson, A. F., *J. Chem. Soc.*, 3873 (1963)
133. Esteban, G. L., Kerr, J. A., and Trotman-Dickenson, A. F., *J. Chem. Soc.*, 3879 (1963)
134. Eusuf, M., and Laidler, K. J., *Trans. Faraday Soc.*, **59**, 2750 (1963)
135. Evans, M. G., and Polanyi, M., *Trans. Faraday Soc.*, **34**, 11 (1938)
136. Failes, R. L., and Stimson, V. R., *J. Chem. Soc.*, 653 (1962)
137. Falconer, W. E., Rabinovitch, B. S., and Cvetanovic, R. J., *J. Chem. Phys.*, **39**, 40 (1963)
138. Farren, J., Gilbert, J. R., Linnett, J. W., and Read, I. A., *Trans. Faraday Soc.*, **60**, 740 (1964)
139. Fettis, G. C., and Knox, J. H., *Progress in Reaction Kinetics*, **2**, 1 (Pergamon, New York, 1964)
140. Fettis, G. C., Knox, J. H., and Trotman-Dickenson, A. F., *J. Chem. Soc.*, 1064 (1960)
141. Fettis, G. C., Knox, J. H., and Trotman-Dickenson, A. F., *ibid.*, 4177
142. Fielding, W., and Pritchard, H. O., *J. Phys. Chem.*, **66**, 821 (1962)
143. Fischer, G., and Buchanan, A. S., *Trans. Faraday Soc.*, **60**, 378 (1964)
144. Fish, A., *Quart. Rev.*, **18**, 243 (1964)
145. Fishburne, E. S., and Edse, R., *J. Chem. Phys.*, **41**, 1297 (1964)
146. Flowers, M. C., and Benson, S. W., *J. Chem. Phys.*, **38**, 882 (1963)
147. Flowers, M., Benson, S. W., and Batt, L., *J. Chem. Phys.*, **37**, 2662 (1962)
148. Flowers, M. C., and Frey, H. M., *J. Chem. Soc.*, 3547 (1961)
149. Flowers, M. C., and Frey, H. M., *ibid.*, 1689 (1962)
150. Fontijn, A., Meyer, C. B., and Schiff, H. I., *J. Chem. Phys.*, **40**, 64 (1964)
151. Fontijn, A., and Schiff, H. I., *Chemical Reactions in the Lower and Upper Atmosphere*, 239 (Interscience, New York, 1961)
152. Ford, H. W., and Endow, N., *J. Chem. Phys.*, **37**, 1156 (1957)
153. Forst, W., and Rice, O. K., *Can. J. Chem.*, **41**, 562 (1963)
154. Frey, H. M., *Trans. Faraday Soc.*, **58**, 957 (1962)
155. Frey, H. M., *ibid.*, **59**, 1619 (1963)
156. Frey, H. M., *J. Chem. Soc.*, 365 (1964)
157. Frey, H. M., *Trans. Faraday Soc.*, **60**, 83 (1964)
158. Frey, H. M., and Marshall, D. C., *J. Chem. Soc.*, 3052 (1962)
159. Frey, H. M., and Marshall, D. C., *ibid.*, 3981
160. Frey, H. M., and Marshall, D. C., *ibid.*, 5717 (1963)
161. Frey, H. M., and Smith, R. C., *Trans. Faraday Soc.*, **58**, 697 (1962)
162. Frey, H. M., and Stevens, I. D. R., *J. Chem. Soc.*, 3865 (1962)
163. Gaydon, A. G., Kimbell, G. H., Palmer, H. B., *Proc. Roy. Soc.* (*London*), *Ser. A.*, **279**, 313 (1964)
164. Gill, E. K., and Laidler, K. J., *Can. J. Chem.*, **36**, 1570 (1958)
165. Glasson, W. A., and Tuesday, C. S., *J. Am. Chem. Soc.*, **85**, 2901 (1963)
166. Glasstone, S., Laidler, K. J., and Eyring, H., *Theory of Rate Processes* (McGraw, New York, 1941)
167. Gold, V., *Trans. Faraday Soc.*, **60**, 738 (1964)
168. Goldfinger, P., and Martens, G., *Trans. Faraday Soc.*, **57**, 2220 (1961)
169. Gorban, N. I., and Nalbandyan, A. B., *Zh. Fiz. Khim.*, **36**, 1757 (1962)
170. Gordon, A. S., and Smith, S. R., *J. Phys. Chem.*, **66**, 521 (1962)
171. Gordon, A. S., Smith, S. R., and Drew, C. M., *J. Chem. Phys.*, **36**, 1824 (1962)
172. Gowenlock, B. G., Jones, P. P., and Snelling, D. R., *Can. J. Chem.*, **41**, 1911 (1963)
173. Gozzo, F., and Patrick, C. R., *Nature*, **202**, 80 (1964)
174. Gray, P., and Thynne, J. C. J., *Trans. Faraday Soc.*, **59**, 2275 (1963)
175. Gray, P., and Thynne, J. C. J., *ibid.*, **60**, 1047 (1964)
176. Gregory, J. E., and Wijnen, M. H. J., *J. Chem. Phys.*, **38**, 2925 (1963)
177. Grotewold, J., and Kerr, J. A., *J. Chem. Soc.*, 4337 (1963)
178. Grotewold, J., and Kerr, J. A., *ibid.*, 4342
179. Guercione, J. A., and Wijnen, M. H. J., *J. Chem. Phys.*, **38**, 1 (1963)
180. Gunning, H. E., and Stock, R. L., *Can. J. Chem.*, **42**, 357 (1964)
181. Halberstadt, M. L., and Chesick, J. P., *J. Am. Chem. Soc.*, **84**, 2688 (1962)
182. Halevi, E. A., *Advan. Phys. Org. Chem.*, **1**, 109 (1963)
183. Harrison, H., Johnston, H. S., and Hardwick, E. R., *J. Am. Chem. Soc.*, **84**, 2478 (1962)
184. Harter, R. J., Alterman, E. B., and Wilson, D. J., *J. Chem. Phys.*, **40**, 2137 (1964)
185. Hartley, D. B., and Benson, S. W., *J. Chem. Phys.*, **39**, 132 (1963)

186. Haszeldine, R. N., Robinson, P. J., and Simmons, R. F., *J. Chem. Soc.*, 1890 (1964)
187. Hautecloque, S., *Compt. Rend.*, **256**, 131 (1963)
188. Hautecloque, S., *ibid.*, 2601
189. Hautecloque, S., *ibid.*, **254**, 3671 (1962)
190. Heicklen, J., and Johnston, H. S., *J. Am. Chem. Soc.*, **84**, 4030 (1962)
191. Heicklen, J., and Johnston, H. A., *ibid.*, 4394
192. Heras, J. M., Arvia, A. J., Aymonino, P. J., and Schumacher, H. J., *Anales Asoc. Quim. Arg.*, **50**, 120 (1962)
193. Herndon, W. C., Cooper, W. B., and Chambers, M. J., *J. Phys. Chem.*, **68**, 2016 (1964)
194. Herndon, W. C., Henly, M. B., and Sullivan, J. M., *J. Phys. Chem.*, **67**, 2842 (1963)
195. Herndon, W. C., and Lowry, L. L., *J. Am. Chem. Soc.*, **86**, 1922 (1964)
196. Herron, J. T., and Klein, F. S., *J. Chem. Phys.*, **40**, 2731 (1964)
197. Herzfeld, K. F., and Litovitz, T. A., *Absorption and Dispersion of Ultrasonic Waves* (Academic Press, New York, 1959)
198. Heydtmann, H., and Rinck, G., *Z. Phys. Chem. (Frankfurt)*, **36**, 75 (1963)
199. Hiraoka, H., and Hardwick, R., *J. Chem. Phys.*, **36**, 1715 (1962)
200. Hiraoka, H., and Hardwick, R., *ibid.*, 2164
201. Hiraoka, H., and Hardwick, R., *ibid.*, **39**, 2361 (1963)
202. Hoare, D. E., *Can. J. Chem.*, **40**, 2012 (1962)
203. Hogg, W., McKinnon, D. M., Trotman-Dickenson, A. F., and Verbeke, G. J. O., *J. Chem. Soc.*, 1403 (1961)
204. Holmes, J. L., and Kutschke, K. O., *Trans. Faraday Soc.*, **58**, 333 (1962)
205. Holmes, J. L., and Maccoll, A., *J. Chem. Soc.*, 5919 (1963)
206. Hung, N. C., and Wilson, D. J., *J. Chem. Phys.*, **38**, 828 (1963)
207. Hutton, E., *Nature*, **203**, 835 (1964)
207a. Jackson, W. M., and McNesby, J. R., *J. Am. Chem. Soc.*, **83**, 4891 (1961); Jackson, W. M., and McNesby, J. R., *J. Chem. Phys.*, **36**, 2272 (1962); McNesby, J. R., and Jackson, W. M., *J. Chem. Phys.*, **38**, 692 (1963); Jackson, W. M., McNesby, J. R., and Darwent, B. DeB., *J. Chem. Phys.*, **37**, 2256 (1962)
207b. Jackson, W. M., McNesby, J. R., and Darwent, B. DeB., *J. Chem. Phys.*, **37**, 1610 (1962)
208. Jacobs, T. A., and Giedt, R. R., *J. Chem. Phys.*, **39**, 749 (1963)
209. Jaffe, S., and Ford, H. W. (Private communication)
210. Jamieson, J. W. S., and Brown, G. R., *Can. J. Chem.*, **42**, 1638 (1964)
211. Johnston, H. S., and Crosby, H. J., *J. Chem. Phys.*, **22**, 689 (1954)
212. Johnston, H. S., and Goldfinger, P., *J. Chem. Phys.*, **37**, 700 (1962)
213. Johnston, H. S., and Parr, C., *J. Am. Chem. Soc.*, **85**, 2544 (1963)
214. Johnston, H. S., and Woolfolk, R., *J. Chem. Phys.*, **41**, 269 (1964)
215. Jones, W. M., and Davidson, N., *J. Am. Chem. Soc.*, **84**, 2868 (1962)
216. Jost, W., Michel, K. W., Troe, J., and Wagner, H. G., *Z. Naturforsch.*, **19a**, 59 (1964)
217. Kale, M. N., and Maccoll, A., *J. Chem. Soc.*, 1513 (1964)
218. Katakis, D., and Taube, H., *J. Chem. Phys.*, **36**, 416 (1962)
219. Kaufman, F., *Progress in Reaction Kinetics*, **1**, 1 (Pergamon, New York, 1961)
220. Kaufman, F., and Kelso, J. R., *J. Chem. Phys.*, **40**, 1162 (1964)
221. Kazmi, H. A., Diefendorf, R. J., and LeRoy, D. J., *Can. J. Chem.*, **41**, 690 (1963)
222. Kazmi, H. A., and LeRoy, D. J., *Can. J. Chem.*, **42**, 1145 (1964)
223. Keck, J. C., *J. Chem. Phys.*, **29**, 410 (1958)
224. Keck, J. C., *Discussions Faraday Soc.*, **33**, 173 (1962)
225. Kerr, J. A., Sekhar, R. C., and Trotman-Dickenson, A. F., *J. Chem. Soc.*, 3217 (1963)
226. Kerr, J. A., and Trotman-Dickenson, A. F., *J. Chem. Soc*, 1602 (1960)
227. Kerr, J. A., and Trotman-Dickenson, A. F., *Progress in Reaction Kinetics*, **1**, 113 (Pergamon, New York, 1961)
228. Klein, F. S., and Herron, J. T., *J. Chem. Phys.*, **41**, 1285 (1964)
229. Klein, F. S., Persky, A., and Weston, R. E., Jr., *J. Chem. Phys.*, **41**, 1799 (1964)
230. Klump, K. N., and Chesick, J. P., *J. Am. Chem. Soc.*, **85**, 130 (1963)
231. Knight, A. R., Strausz, D. P., and Gunning, H. E., *J. Am. Chem. Soc.*, **85**, 1207 (1963)
232. Knight, A. R., Strausz, D. P., and Gunning, H. E., *ibid.*, 2349
233. Knox, J. H., *Trans. Faraday Soc.*, **58**, 275 (1962)

234. Knox, J. H., and Nelson, R. H., *Trans. Faraday Soc.*, **55**, 937 (1959)
235. Knox, J. H., and Trotman-Dickenson, A. F., *J. Phys. Chem.*, **60**, 1367 (1956)
236. Kohlmaier, G. H., and Rabinovitch, B. S., *J. Chem. Phys.*, **38**, 1692 (1963)
237. Kohlmaier, G. H., and Rabinovitch, B. S., *ibid.*, 1709
238. Kohlmaier, G. H., and Rabinovitch, B. S., *ibid.*, **39**, 490 (1963)
239. Komezawa, H., Stefani, A. P., and Szwarc, M., *J. Am. Chem. Soc.*, **85**, 2043 (1963)
240. Kraus, J. W., and Calvert, J. G., *J. Am. Chem. Soc.*, **79**, 5921 (1957)
241. Kretschmer, C. B., and Peterson, H. L., *J. Chem. Phys.*, **39**, 1772 (1963)
242. Kubin, R. F., Rabinovitch, B. S., and Harrington, R. E., *J. Chem. Phys.*, **37**, 937 (1962)
243. Lapidus, G., Barton, D., and Yankwich, P. E., *J. Phys. Chem.*, **68**, 1863 (1964)
244. Leventhal, E., Simonds, C. R., and Steel, C., *Can. J. Chem.*, **40**, 930 (1962)
245. Lewis, K. E., and Steiner, H., *J. Chem. Soc.*, 3080 (1964)
246. Lifshitz, A., Bauer, S. H., and Resler, E. L., Jr., *J. Chem. Phys.*, **38**, 2056 (1963)
247. Lifshitz, A., Carroll, H. F., and Bauer, S. H., *J. Chem. Phys.*, **39**, 1661 (1963)
248. Light, J. C., *J. Chem. Phys.*, **36**, 1016 (1962)
249. Lin, S. H., and Eyring, H., *J. Chem. Phys.*, **39**, 1577 (1963)
250. Linnett, J. W., and Booth, M. H., *Nature*, **199**, 1181 (1963)
251. Lippincott, E. R., and Leifer, A., *J. Chem. Phys.*, **28**, 769 (1958)
252. Long, J., and Skirrow, F., *Trans. Faraday Soc.*, **58**, 1403 (1962)
253. Lovachev, L. A., *Combust. Flame*, **7**, 388 (1963)
254. Maccoll, A., and Stimson, V. R., *J. Chem. Soc.*, 2836 (1960)
255. McDowell, C. A., and Sifniades, S., *Can. J. Chem.*, **41**, 300 (1963)
256. Mahan, B. H., and Solo, R. B., *J. Chem. Phys.*, **37**, 2669 (1962)
257. Majer, J. R., Capey, W. D., and Robb, J. C., *Nature*, **203**, 294 (1964)
258. March, R. E., and Polanyi, J. C., *Proc. Roy. Soc. (London), Ser. A*, **273**, 360 (1963)
259. Marcus, R. A., *J. Chem. Phys.*, **20**, 352 (1952)
260. Marcus, R. A., and Rice, O. K., *J. Phys. Colloid Chem.*, **55**, 894 (1951)
261. Marshall, T. C., *Phys. Fluids*, **5**, 743 (1962)
262. Marte, J. E., Tschuikow-Roux, E., and Ford, H. W., *J. Chem. Phys.*, **39**, 3277 (1963)
263. Maschke, A., Shapiro, B. S., and Lampe, F. W., *J. Am. Chem. Soc.*, **85**, 1876 (1963)
264. Maschke, A., Shapiro, B. S., and Lampe, F. W., *J. Am. Chem. Soc.*, **86**, 1929 (1964)
265. Mathias, A., and Schiff, H. I., *J. Chem. Phys.*, **40**, 3118 (1964)
266. Matsuoka, M., Dixon, P. S., Stefani, A. P., and Szwarc, M., *Proc. Chem. Soc.*, 304 (1962)
267. Mavroyannis, C., and Winkler, C. A., *Can. J. Chem.*, **40**, 240 (1962)
268. Melander, L., *Isotope Effects on Reaction Rates* (Ronald Press, New York, 1960)
269. Metcalf, E. L., *J. Chem. Soc.*, 3560 (1963)
270. Metcalf, E. L., and Trotman-Dickenson, A. F., *J. Chem. Soc.*, 4620 (1962)
271. Millikan, R. C., and White, D. R., *J. Chem. Phys.*, **39**, 3209 (1963)
272. Miyoshi, M., and Brinton, R. K., *J. Chem. Phys.*, **36**, 3019 (1962)
273. Morgan, J. E., and Schiff, H. I., *Can. J. Chem.*, **41**, 903 (1963)
274. Morgan, J. E., and Schiff, H. I., *J. Chem. Phys.*, **38**, 1495 (1963)
275. Nangia, N., and Benson, S. W., *J. Am. Chem. Soc.*, **84**, 3411 (1962)
276. Nangia, P. S., and Benson, S. W., *J. Chem. Phys.*, **38**, 18 (1963)
277. Nangia, P. S., and Benson, S. W., *J. Am. Chem. Phys.*, **86**, 2773 (1964)
278. Nangia, P. S., and Benson, S. W., *J. Chem. Phys.*, **41**, 530 (1964)
279. Neufeld, M. L., and Blades, A. T., *Can. J. Chem.*, **41**, 2956 (1963)
280. Nikitin, E. E., *Dokl. Akad. Nauk SSSR*, **121**, 991 (1958)
281. Nikitin, E. E., *Kinetika i Kataliz*, **3**, 830 (1962)
282. Nikitin, E. E., and Sokolov, N. D., *J. Chem. Phys.*, **31**, 1371 (1959)
283. O'Neal, H. E., and Benson, S. W., *J. Chem. Phys.*, **37**, 540 (1962)
284. Owen, G. E., Jr., Pearson, J. M., and Szwarc, M., *Trans. Faraday Soc.*, **60**, 564 (1964)
285. Palmer, H. B., and Hirt, T. J., *J. Phys. Chem.*, **67**, 709 (1963)
286. Parker, J. G., *J. Chem. Phys.*, **41**, 1600 (1964)

287. Patch, R. W., *J. Chem. Phys.*, **36,** 1919 (1962)
288. Paterson, W. L., and Green, E. F., *J. Chem. Phys.*, **36,** 1146 (1962)
289. Pearson, J. M., and Szwarc, M., *Trans. Faraday Soc.*, **60,** 553 (1964)
290. Phillips, L. F., and Schiff, H. I., *J. Chem. Phys.*, **36,** 1509 (1962)
291. Phillips, L. F., and Schiff, H. I., *ibid.*, 3283
292. Phillips, L. F., and Schiff, H. I., *ibid.*, **37,** 924 (1962)
293. Phillips, L. F., and Schiff, H. I., *ibid.*, 1233
294. Polanyi, J. C., and Rosner, S. O., *J. Chem. Phys.*, **38,** 1028 (1963)
295. Porter, G., *Discussions Faraday Soc.*, **33,** 198 (1962)
296. Porter, G., Szabo, Z. G., and Townsend, M. G., *Proc. Roy. Soc.* (*London*), *Ser. A*, **270,** 493 (1962)
297. Pratt, G. L., and Purnell, J. H., *Trans. Faraday Soc.*, **60,** 371 (1964)
298. Pratt, G. L., and Purnell, J. H., *ibid.*, 519
299. Price, S. J., *Can. J. Chem.*, **40,** 1310 (1962)
300. Pritchard, H. O., *J. Phys. Chem.*, **65,** 504 (1961)
301. Pritchard, H. O., *ibid.*, **66,** 2111 (1962)
302. Pritchard, G. O., and Foote, J. K., *J. Phys. Chem.*, **68,** 1016 (1964)
303. Pritchard, G. O., Hsia, Y. P., and Miller, G. H., *J. Am. Chem. Soc.*, **85,** 1568 (1963)
304. Pritchard, G. O., Miller, G. H., and Foote, J. K., *Can. J. Chem.*, **40,** 1830 (1962)
305. Pritchard, G. O., and Thommarson, R. L., *J. Phys. Chem.*, **68,** 568 (1964)
306. Quinn, C. P., *Proc. Roy. Soc.* (*London*), *Ser. A*, **275,** 190, (1963)
307. Rabinovitch, B. S., and Flowers, M. C., *Quart. Rev.*, **18,** 122 (1964)
308. Rabinovitch, B. S., Kubin, R. F., and Harrington, R. E., *J. Chem. Phys.*, **38,** 405 (1963)
309. Rabinovitch, B. S., and Pearson, M. J., *J. Chem. Phys.*, **41,** 280 (1964)
310. Rabinovitch, B. S., and Setser, D. W., *J. Am. Chem. Soc.*, **84,** 1765 (1962)
311. Raper, O. F., and DeMore, W. B., *J. Chem. Phys.*, **40,** 1047 (1964)
312. Raper, O. F., and DeMore, W. B., *ibid.*, 1053
313. Rapp, D., *J. Chem. Phys.*, **40,** 2813 (1964)
314. Rapp, D., and Weston, R. E., Jr., *J. Chem. Phys.*, **36,** 2807 (1962)
315. Ree, T. S., Ree, T., Eyring, H., and Fueno, T., *J. Chem. Phys.*, **36,** 281 (1962)
316. Reeves, R. R., Harteck, P., and Chace, W. H., *J. Chem. Phys.*, **41,** 764 (1964)
317. Rice, O. K., *J. Chem. Phys.*, **9,** 259 (1941)
318. Rice, O. K., *ibid.*, **21,** 750 (1953)
319. Rice, O. K., *J. Phys. Chem.*, **65,** 1972 (1961)
320. Rice, O. K., *ibid.*, **67,** 6 (1963)
321. Rice, O. K., *ibid.*, 1733
322. Rink, J. P., *J. Chem. Phys.*, **36,** 572 (1962)
323. Rink, J. P., *ibid.*, 1398
324. Rodewald, L. B., and DePuy, C. H., *Tetrahedron Letters*, **40,** 2951 (1964)
325. Roquitte, B. C., *Can. J. Chem.*, **42,** 2134 (1964)
326. Roquitte, B. C., and Futrell, J. H., *J. Chem. Phys.*, **37,** 378 (1962)
327. Roquitte, B. C., and Walters, W. D., *J. Am. Chem. Soc.*, **84,** 4049 (1962)
328. Roquitte, B. C., and Walters, W. D., *J. Phys. Chem.*, **68,** 1606 (1964)
329. Roquitte, B. C., and Wijnen, M. H. J., *J. Am. Chem. Soc.*, **85,** 2053 (1963)
330. Roquitte, B. C., and Wijnen, M. H. J., *J. Chem. Phys.*, **38,** 4 (1963)
331. Rust, F. F., and Vaughn, W. E., *Ind. Eng. Chem.*, **41,** 2595 (1949)
332. Salomon, M., *Can. J. Chem.*, **42,** 610 (1964)
333. Sanders, W. A., and Rebbert, R. E., *J. Phys. Chem.*, **67,** 170 (1963)
334. Schlag, E. W., *J. Chem. Phys.*, **38,** 2480 (1964)
335. Schlag, E. W., and Peatman, W. B., *J. Am. Chem. Soc.*, **86,** 1676 (1964)
336. Schlag, E. W., and Sandsmark, R. A., *J. Chem. Phys.*, **37,** 168 (1962)
337. Schlag, E. W., Sandsmark, R. A., and Valence, W. G., *J. Chem. Phys.*, **40,** 1461 (1964)
338. Schneider, F. W., and Rabinovitch, B. S., *J. Am. Chem. Soc.*, **84,** 4215 (1962)
339. Schneider, F. W., and Rabinovitch, B. S., *ibid.*, **85,** 2365 (1963)
340. Schumacher, H. J., *J. Chem. Phys.*, **33,** 938 (1960)
341. Setser, D. W., and Rabinovitch, B. S., *J. Am. Chem. Soc.*, **86,** 564 (1964)
342. Shannon, T. W., and Harrison, A. G., *Can. J. Chem.*, **41,** 2455 (1963)
343. Sharp, T. E., and Johnston, H. S., *J. Chem. Phys.*, **37,** 1541 (1962)
344. Sherwood, A. G., and Gunning, H. E., *J. Am. Chem. Soc.*, **85,** 3506 (1963)
345. Shilov, A. E., and Sabirova, R. D.,

J. Phys. Chem. (USSR), **34**, 408 (1960)
346. Shin, H. K., *J. Chem. Phys.*, **39**, 2934 (1963)
347. Sicre, J. E., and Schumacher, H. J., *Z. Phys. Chem. (Frankfurt)*, **32**, 355 (1962)
348. Simonaitis, R., *J. Phys. Chem.*, **67**, 2227 (1963)
349. Simons, J. W., and Rabinovitch, B. S., *J. Am. Chem. Soc.*, **85**, 1023 (1963)
350. Simons, J. W., and Rabinovitch, B. S., *J. Phys. Chem.*, **68**, 1322 (1964)
351. Simons, J. W., Rabinovitch, B. S., and Kubin, R. F., *J. Chem., Phys.*, **40**, 3343 (1964)
352. Simons, J. W., Setser, D. W., and Rabinovitch, B. S., *J. Am. Chem. Soc.*, **84**, 1758 (1962)
353. Skinner, G. B., and Ruehrwein, R. A., *J. Phys. Chem.*, **63**, 1736 (1959)
354. Slater, N. B., *Theory of Unimolecular Reactions* (Cornell Univ. Press, Ithaca, N. Y., 1959)
355. Smith, G. G., and Blau, S. E., *J. Phys. Chem.*, **68**, 1231 (1964)
356. Spirin, Y. L., *Zh. Fiz. Khim.* **36**, 1202 (1962)
357. Srinivasan, R., and Levi, A. A., *J. Am. Chem. Soc.*, **85**, 3363 (1963)
358. Srinivasan, R., and Levi, A. A., *ibid.*, **86**, 3756 (1964)
359. Steel, C., Zand, R., Hurwitz, P., and Cohen, S. G., *J. Am. Chem. Soc.*, **86**, 679 (1964)
360. Stefani, A. P., and Szwarc, M., *J. Am. Chem. Soc.*, **84**, 3661 (1962)
361. Stimson, V. R., and Watson, E. J., *J. Chem. Soc.*, 524 (1963)
362. Strausz, D. P., and Gunning, H. E., *J. Am. Chem. Soc.*, **84**, 4080 (1962)
363. Subbaratnam, N. R., and Calvert, J. G., *J. Am. Chem. Soc.*, **84**, 1113 (1962)
364. Sullivan, J. H., *J. Chem. Phys.*, **36**, 1925 (1962)
365. Sullivan, J. H., *ibid.*, **39**, 3001 (1963)
366. Szwarc, M., and Murawski, J., *Trans. Faraday Soc.*, **47**, 268 (1951)
367. Takayanagi, K., *Vibrational and Rotational Transition in Molecular Collision* (In press, 1964)
368. Tanaka, I., and McNesby, J. R., *J. Chem. Phys.*, **36**, 3170 (1962)
369. Teranishi, H., and Benson, S. W., *J. Am. Chem. Soc.*, **85**, 2887 (1963)
370. Teranishi, H., and Benson, S. W., *ibid.*, 2890
371. Teranishi, H., and Benson, S. W., *J. Chem. Phys.*, **40**, 2946 (1964)
372. Thomas, S. S., and Calvert, J. G., *J. Am. Chem. Soc.*, **84**, 4207 (1962)
373. Thynne, J. C. J., *Trans. Faraday Soc.*, **58**, 676 (1962)
374. Thynne, J. C. J., *ibid.*, 1394
375. Thynne, J. C. J., *ibid.*, 1533
376. Thynne, J. C. J., and Gray, P., *Proc. Chem. Soc.*, 141 (1962)
377. Thynne, J. C. J., and Gray, P., *Trans. Faraday Soc.*, **58**, 2403 (1962)
378. Thynne, J. C. J., and Gray, P., *ibid.*, **59**, 1149 (1963)
379. Timmons, R. B., and Weston, R. E., Jr., *J. Chem. Phys.*, **41**, 1654 (1964)
380. Tomkinson, D. M., Galvin, J. P., and Pritchard, H. O., *J. Phys. Chem.*, **68**, 541 (1964)
381. Treacy, J. C., and Daniels, F., *J. Am. Chem. Soc.*, **77**, 2033 (1955)
382. Trenwith, A. B., *J. Chem. Soc.*, 4426 (1963)
383. Tsang, W., *J. Chem. Phys.*, **40**, 1171 (1964)
384. Tsang, W., *ibid.*, 1498
385. Tsang, W., *ibid.*, **41**, 2487 (1964)
386. Tsang, W., Bauer, S. H., and Cowperthwaite, M., *J. Chem. Phys.*, **36**, 1768 (1962)
387. Tschuikow-Roux, E., *J. Chem. Phys.*, **42**, 2049 (1965)
388. Uchiyama, M., Tomioka, T., and Amano, A., *J. Phys. Chem.*, **68**, 1878 (1964)
389. Viscido, L., Sicre, J. E., and Schumacher, H. J., *Z. Phys. Chem. (Frankfurt)*, **32**, 182 (1962)
390. Viscido, L., Sicre, J. E., and Schumacher, H. J., *ibid.*, **33**, 206 (1962)
391. Volman, D. H., *Advances in Photochemistry*, **1**, 43 (Interscience, New York, 1963)
392. Wall, F. T., Hiller, L. A., Jr., and Mazur, J., *J. Chem. Phys.*, **29**, 255 (1958)
393. Watson, E. J., and Stimson, V. R., *J. Chem. Soc.*, 1392 (1961)
394. Wellington, C. A., *J. Phys. Chem.*, **66**, 1671 (1962)
395. Westenberg, A. A., and DeHaas, N., *J. Chem. Phys.*, **40**, 3087 (1964)
396. Whittemore, I. M., Stefani, A. P., and Szwarc, M., *J. Am. Chem. Soc.*, **84**, 3799 (1962)
397. Whitten, G. Z., and Rabinovitch, B. S., *J. Chem. Phys.*, **38**, 2466 (1963)
398. Whittle, E., and Steacie, E. W. R., *J. Chem. Phys.*, **21**, 993 (1953)
399. Wieder, G. M., and Marcus, R. A., *J. Chem. Phys.*, **37**, 1835 (1962)
400. Wigner, E., *J. Chem. Phys.*, **5**, 720 (1937)
401. Wilson, D. J., and Johnston, H. S., *J. Am. Chem. Soc.*, **79**, 29 (1957)

402. Wolfsberg, M., *J. Chem. Phys.*, **36**, 1072 (1962)
403. Wray, K. L., *J. Chem. Phys.*, **38**, 1518 (1963)
404. Wunderlich, F. J., and Rebbert, R. E., *J. Phys. Chem.*, **67**, 1382 (1963)
405. Yamazaki, H., and Cvetanovic, R. J., *J. Chem. Phys.*, **39**, 1902 (1963)
406. Yamazaki, H., and Cvetanovic, R. J., *J. Chem. Phys.*, **40**, 582 (1964)
407. Zupan, M., and Walters, W. D., *J. Phys. Chem.*, **67**, 1845 (1963)
408. Zupan, M., and Walters, W. D., *J. Am. Chem. Soc.*, **86**, 173 (1964)

THE QUANTUM THEORY OF VALENCE[1,2]

By Benjamin M. Gimarc and Robert G. Parr

Departments of Chemistry, Georgia Institute of Technology, Atlanta, Georgia, and The Johns Hopkins University, Baltimore, Maryland

Research in the theory of the electronic structure of molecules, which we may call valence theory (1), has become so extensive and varied that any review of the field must be selective rather than comprehensive. In this article we shall consider, in turn, accurate calculations on small molecules, semiempirical calculations on large molecules, basic theoretical problems, and applications to reaction rates and other questions of interest. We shall slight, among many topics, the theories of intermolecular forces, of Rydberg states of molecules, of magnetic properties of molecules, of molecular crystals, and of the interactions of light with matter.

Ab Initio Calculations

The fundamental equation to be solved in the theory of the electronic structure of a molecule is the Schrödinger equation for the electrons moving in a fixed nuclear frame

$$H\Psi = W\Psi \qquad 1.$$

Here H is the many-electron, nonrelativistic Hamiltonian, comprising electron kinetic energy, nuclear-nuclear repulsion energy, nuclear-electron attraction energy, and electron-electron repulsion energy parts; and Ψ is the antisymmetric electronic wavefunction for a particular stationary state of energy W.

The usual procedure for solving this equation is the variational method. For various antisymmetric functions Φ which approximate Ψ, one computes the expectation values

$$E = \int \Phi^* H \Phi d\tau \Big/ \int \Phi^* \Phi d\tau \qquad 2.$$

That Φ which minimizes E is the best approximation to the ground state eigenfunction Ψ_0, since by the variational theorem, $E \geq W_0$ for any Φ. By indefinitely enlarging the class of trial functions Φ, one can, by this procedure, come arbitrarily close to the true Ψ_0 and W_0.

The SCF method.—A convenient approximate Φ is the single determinant, or antisymmetrized product of spinorbitals λ_i, each containing one electron,

[1] The survey of the literature pertaining to this review was concluded in January 1965.

[2] Aided by grants to the Johns Hopkins University from the National Science Foundation and the Petroleum Research Fund of the American Chemical Society.

$$\Phi = \mathcal{A}[\lambda_1(1)\lambda_2(2) \cdots \lambda_N(N)] \quad 3.$$

The best such set λ_i are the Hartree-Fock or self-consistent-field (SCF) orbitals, but one must usually be satisfied with less than the best. Each λ_i is a spin factor α or β times a space factor ϕ_i; for molecules, the ϕ_i are the molecular orbitals (MO). The MO may be linearly composed from atomic orbitals (AO) centered on the various nuclei (or elsewhere),

$$\phi_i = \sum_p C_{ip}\chi_p \quad 4.$$

This is the LCAO approximation; the equations for determining the best wavefunctions of this form define the LCAO-SCF procedure of Roothaan (2). In Equation 4, the AO χ_p themselves may be linear combinations of other functions, and inclusion of enough terms will give convergence toward the SCF orbitals. The formulation is applicable to atoms.

A single molecular orbital describes the motion or distribution of one electron, depending on the other electrons only in an average way. It is useful to define the error in the SCF approximation as the correlation energy E_{corr},

$$E_{corr} = E_{exp} - E_{rel} - E_{HF} \quad 5.$$

where E_{exp} is the experimental energy, E_{HF} is the Hartree-Fock energy, and E_{rel} is the relativistic correction or difference between the exact eigenvalue W and E_{exp}; $W = E_{HF} + E_{corr}$. Experimental extrapolations sometimes must be carried out to obtain E_{exp}.

Relativistic corrections are appreciable for large atomic number Z and cannot be neglected. They can be calculated approximately by perturbation methods and empirical extrapolations. Clementi (3) and Hartmann & Clementi (4) have estimated E_{rel} for atoms with 2 to 18 electrons and atomic numbers 2 to 20 and in some cases to 36. The relativistic correction is apparently the sum of contributions from atomic subshells, with the $1s$ shell providing the dominant term; terms from higher subshells diminish rapidly. Relativistic corrections for molecules probably can be well approximated by the sum of corrections for constituent atoms.

The work of Clementi, Roothaan, Synek, McLean and others (5–19) has been providing very good approximations to the Hartree-Fock wavefunctions for atoms and positive and negative ions from $Z=2$ to 36. Herman & Skillman (19) have given, in a book less than an inch thick, the orbitals in numerical form for all neutral atoms from helium through lawrencium. Their calculations, however, were based on Slater's simplified version of the Hartree-Fock equations. The book includes discussion of computational procedures and contains actual FORTRAN programs.

From values of E_{exp}, E_{HF}, and E_{rel} for neutral atoms and positive ions, Clementi (20) and Allen, Clementi & Gladney (21) have constructed tables of atomic correlation energies by partitioning E_{corr} into contributions from closed electron shells, including intrashell and intershell corrections, from which one can estimate the correlation energies (and hence the experimental energies) of larger atoms and ions. In this way Clementi and associates (5–7) estimated atomic electron affinities and predicted the stability of several atomic negative ions. They found that certain second-row negative ions have excited states that are stable relative to the neutral-atom ground state.

Over the last several years many molecular wavefunctions in the LCAO-SCF approximation have become available. Three different types of atomic basis orbitals have been used for the formation of the molecular orbitals: (*a*) exponential Slater-type orbitals (STO): $\chi = r^n \exp(-\zeta r) Y_{lm}(\theta,\phi)$; (*b*) one-center orbitals or STO centered on or expanded about only a single center in the molecule; (*c*) Gaussian orbitals: $\chi = r^n \exp(-\alpha r^2) Y_{lm}(\theta,\phi)$. Tables I, II, and III list some of the molecules which have been treated with these three bases, with references.

The most accurate MO calculations to date are those with the STO basis. Calculations of this type have been performed by Clementi & McClean (22–33), Nesbet, Kahalas & Carlson (34–40), Karo & Allen (41, 42, 43), Ransil (44, 45), Lefebvre-Brion, Moser & Nesbet (46, 47), and others (48–79). A number of these wavefunctions and energies are very good approximations to the Hartree-Fock limits, the MO formed from extensive basis sets and all integrals evaluated by accurate methods. Much of the older work, however, involves only minimum basis set MO and approximation of the most difficult integrals, particularly the many-center electron repulsion integrals. For typical recent work on molecular integrals involving STO see the papers by Bonham, Peacher & Cox (80), Geller (81), and Wahl, Cade & Roothaan (82); Magnusson (83) has given a compact review.

One-center functions or Gaussian orbitals individually are poor approximations to the Hartree-Fock solutions for molecules, but the easy evaluation of integrals involving them and the availability of digital computers make it feasible to construct good molecular wavefunctions from many of them. One-center functions are particularly appropriate for molecules of the type AH_n (84, 85). Recent calculations by the one-center method include calculations by Bishop and collaborators (86–90), by Moccia (91, 92, 93), and others (94, 95, 96). Bishop (97) has suggested a method for improving one-center wavefunctions in the neighborhood of off-center nuclei, and Turner et al. (98) have examined the one-center approximation for the off-center hydrogen atom. Gaussian orbitals have recently been applied to molecules in the MO formulation by Nesbet (99), Krauss (100), Reeves & Harrison (101), Harrison (102, 103), and others (104, 105). Considering the success of the many-Gaussian wavefunctions, particularly the methane calculation of Krauss (100) and the HF calculation by Harrison (103), surprisingly few results have been published. However, extensive calculations by Krauss (107) have begun to appear.

Tables I to III are not exhaustive, although a complete and critical compilation of the MO-SCF literature along the lines of the bibliography of Allen & Karo (108) indeed would be valuable. In fact, good wavefunctions now are available for a wide range of molecules. Such calculations are useful

TABLE I

LCAO MO SCF CALCULATIONS USING SLATER ORBITAL BASES

Homonuclear diatomics	Reference	Diatomic hydrides	Reference
H_2	(64)	LiH	(36, 43, 44, 48)
He_2	(45)	BeH^+	(48)
Li_2	(44, 56, 57, 72)	BH	(44, 54)
Be_2	(44)	CH^+	(60)
B_2	(63)	CH	(49, 60)
C_2	(44)	NH	(44, 53, 60, 61)
N_2	(37, 44, 50, 69)	OH	(59, 60)
F_2	(44, 78)	OH^-	(74)
N_2^+, N_2^{++}	(69)	HF	(24, 34, 35, 41, 42, 44, 60)
Mn_2	(40)	HCl	(38, 77)
Other diatomics			
LiF	(33, 44)	CO	(37, 44, 46, 47, 67)
BF	(37, 44)	TiO	(39)
BeO	(76)		
Polyatomics			
BH_3	(62)	HCN	(31)
CH_2	(66)	CO_2	(30, 32)
CH_3	(51, 66)	COS	(25)
CH_4	(75, 79)	C_3	(23, 26)
NH_2	(53)	C_4	(22)
		N_3^-	(23, 29)
NH_3^+	(68)	NO_2^+	(23, 29)
NH_3	(53, 55, 58)	HF_2^-	(23, 28)
H_2O	(52)	C_2H_2	(30)
H_3O^+	(70, 71)	C_2H_6	(73)
H_4O^{++}	(74)	H_2CO	(65)
		C_2N_2	(27)

for interpretation of electronic spectra, and they yield theoretical values for such physical properties as spectroscopic constants, internuclear distances, and dipole and quadrupole moments. Agreement with experimental properties is generally good (37, 76, 78, 106).

Other methods.—The Hartree-Fock method is in error by the correlation energy E_{corr}. Other methods can in principle more closely approach the exact energy W.

In the method of configuration interaction (CI), the wavefunction is assumed to be a sum,

TABLE II

MO SCF CALCULATIONS USING ONE-CENTER BASES[a]

	Reference		Reference
H_2	(84, 87)	CH_4	(85, 87, 91, 96)
LiH, BeH, BH, CH, NH, OH	(90)	SiH_4	(91)
OH^-	(95)	NH_3	(89, 92)
HF	(89, 91, 95)	NH_4^+	(92, 96)
HS^-	(94)	NH_4	(88)
HCl	(93)	PH_3, PH_4^+	(92)
H_2O	(89, 93, 95)	BH_4^-	(96)
H_2S	(93, 94)		
H_3O^+	(95)		

[a] Strictly speaking, some of these calculations are not SCF calculations.

$$\Phi = \sum_i c_i \Phi_i \qquad 6.$$

the leading term of which may be the LCAO–SCF or SCF approximation to the wavefunction in question. The other terms in the sum correspond to "configurations" in which one or more electrons are excited into or occupy an excited MO (or AO for atoms). The energy is minimized with respect to the coefficients c_i. The most natural (although not necessarily the best) choice for ground state and excited MO are the SCF MO. Das & Wahl (109) report CI calculations using SCF–MO in H_2, Li_2, and F_2.

Using the CI approach, Browne (110) and Kato, Hayes & Duncan (111) have calculated potential energy curves and spectroscopic constants for several excited states of H_2.

Christoffersen, Hagstrom & Prosser (112), Christoffersen (113), and Conroy (114–118) have worked out accurate potential energy surfaces for H_3^+. They conclude that the most stable nuclear configuration for this molecular ion is the equilateral triangle. Christoffersen used a 12 term CI wavefunction

TABLE III

LCAO MO SCF CALCULATIONS INVOLVING GAUSSIAN ORBITAL BASES

	Reference		Reference
H_2	(105)	CH_4	(99, 100, 107)
O_2	(104)	NH_3	(101)
HF	(103)	NH_4^+	(100, 107)
H_3	(107)	H_2O	(107)
CH_2	(107)	H_3O^+	(107)
NH_2	(107)	BH_4^-	(107)

constructed from 1*s*, 2*s*, and 2*p* orbitals centered on the three protons. Conroy's 29-term wave function (114, 118), determined by his variance minimization procedure (see below), contained many r_{12} terms. Shull (119) and Linnett (120) comment on the bonding in H_3^+.

Bowen & Linnett (121) have studied the linear symmetric configurations of H_3^+, H_3, and H_3^-, using various types of wavefunctions constructed from 1*s* orbitals. Their results show that the simple nonpaired spatial orbitals of Hirst & Linnett (122) yield energies nearly as good as CI functions and much better than the molecular orbital and valence bond functions constructed from the same 1*s* basis set. Bradley (123) also has studied linear symmetric H_3. Preuss (124) is applying an "atom association" method to the same problem.

There is much interest in the H_3 system. For this molecule, nobody has yet published accurate potential energy surfaces obtained by a priori calculations, however. A rash of papers on H_3 may be expected (107).

Harris & Taylor (125, 126) have used flexible two-centered orbitals in elliptical coordinates for calculations on the atoms and diatomic molecules: H^-, He, HeH, He_2, Li, and LiH. Taylor & Gerhauser (127) have applied the same sort of many-parameter orbitals to the study of the excited states of H_2^-. Browne (128) has mixed elliptical orbitals with STO to construct wavefunctions for LiH^+ and LiH.

Stuart & Matsen (129) have studied HeH^+ by a one-center CI treatment. Preuss (130) has calculated the $^1\Sigma$ and $^3\Sigma$ potential energy curves for the two-electron diatomic systems H_2, HeH^+, He_2^{++}, and LiH^{++}.

The one-center idea is being developed as a perturbation method (131). Levine (132) has recently simplified some of the mathematics.

Csizmadia, Sutcliffe & Barnett (133) have made a group orbital study of the LiH molecule. They used separated pair functions to describe pairs of electrons in the lithium inner shell and in the lithium-hydrogen bond.

An extension of the one-center variational method, most appropriate for molecules of the AH_n type, is a two-center expansion method, appropriate for molecules like C_2H_6 or CH_3OH, of the H_mABH_n type. Hoyland (134) has applied this scheme to linear H_3^+ and H_3, with the end protons as the A and B expansion centers for elliptical-coordinate wavefunctions.

In another interesting paper, Hoyland (135) has investigated the basis sets of 1*s* Slater-type orbitals, exp $(-\zeta r)$, and spherical Gaussians, exp $(-\alpha r^2)$, of all possible orbital exponents centered at all possible points in space. His emphasis was on an increase in the number of expansion centers over previous work and careful minimization of all exponential parameters. From systematic calculations on H_2^+ and H_2, he concluded that the STO 1*s* basis is too slowly convergent and the 3- and 4-center integrals which arise take too long to compute. His Gaussian results, however, lead to a rather optimistic view. For H_2^+ the molecular energies he calculated are among the most accurate yet obtained over a wide range of internuclear separations. Earlier work along related lines includes that of Parr (136) and Whitten

(137). Preuss (138) has also urged independently the use of simple Gaussians scattered through space as a basis for molecular calculations.

Using the local energy method of Frost et al. (139, 140), Harriss & Frost (141) have calculated the energy and ground state wavefunction of H_2 at its equilibrium separation. They compare their results with the energy and coefficients determined variationally from a trial function containing exactly the same terms.

Goodisman (142) has suggested that numerical methods should be used to evaluate all molecular integrals. He tests Gauss quadrature molecular integral evaluation on H_2^+ and H_2. Bonaccorsi, Scrocco & Tomasi (143) have used AO with an angular dependent orbital exponent in an H_2 calculation, $\chi = \exp[-\zeta(1+\gamma \cos\theta)]$, where ζ and γ are adjustable parameters. This form allows an orbital on one nucleus to be distorted by another nucleus. Zung & Parr (144) have suggested using a 0*s* (zero-*s*) function of the form, $(1/r[\exp(-\alpha r) - \exp(-\beta r)]$, in atomic and molecular wavefunctions.

In two fine papers, Kolos and Wolniewicz (145, 146) have made the most accurate studies of H_2 to date. Using wavefunctions with as many as 80 terms in elliptical coordinates and including the interelectronic coordinate r_{12}, they first (145) have carefully studied the H_2 potential energy curve in the clamped nuclei (Born-Oppenheimer) approximation. They report binding energies for H_2, HD, and D_2, corrected for nuclear motions and relativity. These final theoretical binding energies still differ from experiment by about 5 cm^{-1}, even though the theoretical calculation should be good to less than 1 cm^{-1} in the energy. The discrepancy is ascribed to the adiabatic approximation. In their second paper (146) they report nonadiabatic calculations for H_2, D_2, and T_2, treated as four-particle problems employing 147-term wavefunctions to describe both electronic and nuclear motions. The calculated binding energies, with relativistic corrections, are too large by 0.6 cm^{-1} for the ground state. The discrepancy is larger than the experimental error, and it would increase, the authors argue, if a more accurate wavefunction were used.

Semiempirical Calculations

One of the significant successes of valence theory has been the development of simple semiempirical theories of the electronic structure of large molecules, from which molecular properties, such as electronic spectra, electron densities, bond orders, aromaticity, electron affinities, and ionization potentials, can be calculated, rationalized, and correlated through series of molecules. The number of applications of these theories is growing, in spite of their limitations. Almost all of these theories were developed within the framework of the molecular orbital method for handling the mobile or π electrons in the multiple bonds of planar organic molecules. Some or all of the integrals entering the formulas are treated as empirical parameters.

Theories of π electrons.—For π-electron systems, one begins by assuming the separation of the σ and π electrons (147). Only the π electrons are considered explicitly, but the effects of the other electrons enter through empirical parameters. The most commonly used of the semiempirical theories are the theory of Hückel (148) in which electron repulsions are included only in a primitive way, and the theories of Pariser & Parr (149) and Pople (150), in which electron repulsions are more correctly treated (PPP methods). Many variations, modifications, and combinations of these methods exist (151).

The simplest of the semiempirical treatments is the Hückel method. In the most elementary form of this method the wavefunctions are represented by unantisymmetrized products of molecular orbitals. the MO are taken to be linear combinations of AO as in Equation 4, and the Hamiltonian is assumed to be a sum of unspecified effective, one-electron operators $H_{eff}(\mu)$,

$$H = \sum_{\mu} H_{eff}(\mu) \qquad 7.$$

Preuss (152) has discussed the problem of expressing effective potentials, which are important in both the Hückel and free-electron MO theories; a discussion emphasizing the relation with more complete theories has been given by Del Re & Parr (153). The coefficients c_i in the MO are determined by the variation method, which leads to a set of simultaneous linear homogeneous equations:

$$\sum_{q} (H_{pq} - \epsilon_i S_{pq})c_{iq} = 0 \qquad 8.$$

for each p, where ϵ_i is the energy of an electron in MO ϕ_i, and the matrix elements H_{pq} and S_{pq} are defined by

$$H_{pq} = \int \chi_p H \chi_q d\tau, \qquad S_{pq} = \int \chi_p \chi_q d\tau \qquad 9.$$

Except perhaps for the overlap elements S_{pq}, the matrix elements are not evaluated mathematically, but treated semiempirically; one searches for parameter values which will correlate molecular properties through a series of molecules. Because of its simplicity and success the method continues to receive wide application (154–171).

A number of extensions and modifications of the Hückel method attempt to introduce the effect of the electron repulsions, at least explicitly excluded from the Hamiltonian, by making the Coulomb integral $\alpha_p = H_{pp}$ dependent on calculated charge densities q_p and the resonance integral $\beta_{pq} = H_{pq}$ dependent on bond orders p_{pq}. One such modification is the ω-technique of Streitwieser (154), in which new values α' are calculated iteratively from the old α by the formula

$$\alpha_p' = \alpha_p + (1 + q_p)\omega\beta \qquad 10.$$

where q_p is the charge density on atom p, ω is a dimensionless parameter, and β is an average resonance integral. This method has been used recently by Ettinger (157), Janssen & Sandstrom (158), and Moffat (159). Streitwieser,

Heller & Feldman (155) have tested an extension of the ω-technique which includes effects on α_p of charge densities on nearest-neighbor and next-nearest-neighbor atoms. Chatterjee & Ali (160) and Golebiewski & Nowakowski (161) propose modifications of β_{pq} values to include dependence on bond orders p_{pq}.

The Pariser-Parr treatment grew out of the more rigorous method of Goeppert-Mayer & Sklar (172). In their study of the electronic spectrum of benzene, Goeppert-Mayer & Sklar, after assuming separation of σ and π electrons, composed the π-electron wavefunctions from antisymmetrized products of LCAO molecular orbitals. Their Hamiltonian H was the sum, over all π electrons, of one-electron core Hamiltonians (which include kinetic energy operators and potential energy terms due to interaction of the π electrons with the nuclei and the σ electrons) plus the Coulombic potential terms arising from repulsions between π electrons:

$$H = \sum_{\mu} H_{\text{core}}(\mu) + \sum_{\mu<\nu} 1/r_{\mu\nu} \qquad 11.$$

Their method involves proceeding in a straightforward manner, to evaluate the integrals that appear by direct quadrature, and including configuration interaction as appropriate. The method is unwieldy and does not give good accuracy.

Pariser & Parr systematically examined the integrals which arise in the energy formulas of the Goeppert-Mayer & Sklar method. Some they calculated explicitly, others they related to experimental quantities, and many others they neglected entirely. Using generally the same approximations for integrals, Pople simplified the SCF equations of Roothaan. The Pariser-Parr method is to the Pople method as the CI method is to the LCAO-SCF method; that is, the first contains the second as a special case. The methods continue to receive wide application (173–191) and study (191–195). Many of the investigations include comparisons among the Pariser-Parr, Pople, and Hückel methods.

There is interest in justifying the semiempirically treated and neglected integrals. Ohno (192) has performed variational calculations, including variation of the Slater orbital exponential parameters, on the valence states of the C and N atoms and their positive and negative ions. From an inspection of the integrals involved he has given additional justification to the $\gamma_{pp}=I-A$ (valence state ionization potential minus electron affinity) correction to the one-center electron repulsion integral of the Pariser-Parr method. He also shows some rough but simple "semitheoretical" estimates of core resonance integrals, and he finds support for an earlier conjecture (193) that the Slater. effective charge for a carbon $2p\pi$ orbital in a molecule may be too large.

Weltin, Weber & Heilbronner (191) approximate the core resonance integral with the formula

$$\beta_{pq} = \langle p \mid H_{\text{core}} \mid q \rangle \approx k_2 p_{pq}^2 + k_1 p_{pq} + k_0 \qquad 12.$$

where the p_{pq} are the bond orders computed from the Hückel MO and the constants k_0, k_1, and k_2 are calibrated to make the results match a specific set of spectroscopic data. The dependence of β_{pq} on the p–q bond length R_{pq} enters through the R_{pq}-dependent bond orders. These authors also use a simplified method for handling electron repulsions. They apply their scheme to some conjugated molecules, including benzene, napthalene, anthracene, tetracene, phenanthrene, fulvene, and azulene.

In an important comprehensive paper, Koutecký, Hochman & Michl (190) have made a careful study of 45 nonalternant hydrocarbons, finding generally very good agreement with observed spectra.

Valence state data enter PPP calculations through approximations for the one-center repulsion integrals γ_{pp} and the one-center core integrals α_p. Adams & Miller (194) have investigated the result of using sets of valence state data from different sources. Schmeising & Miller (195) have described an SCF theory of localized electron sets. The theory may be useful for calculating π-electron localization energies. Simmons (196) presents a new evaluation of the basic parameters in the Pariser-Parr theory, and Koutecký and co-workers (197) discuss effects of CI on these parameters.

Scheibe & Brück (198) have stated that for a molecule the difference between the ionization potential and the first excited state of the same symmetry as the ground state is approximately the same value (around 3.8 eV) for many molecules. Hartmann (199) investigated Scheibe's rule and concluded that it could not be explained in terms of the simple Hückel MO theory. The rule has been re-examined recently by Kollaard & Colpe (200), who have obtained fair agreement with Scheibe's rule from PPP theory.

Rein & Ladik (201) use a semiempirical SCF-MO theory of PPP type to investigate the π-electron structure of the hydrogen-bonded guanine-cytosine nucleotide base pair. The base pair forms one of the simplest units of the structure of DNA. Rein & Harris (202) consider explicitly the σ electrons in the hydrogen bond region to determine the protonic potential function in the guanine-cytosine hydrogen bond. Their results agree with the notion that radiation-induced mutations are the result of tautomeric rearrangements in the base pair. Ladik & Appel (203), Yomosa (204), and Beleznay & Biczó (205) have studied the band structure of some polynucleotide models of DNA. Other recent MO investigations of molecules of biological interest include those by Yonezawa, Del Re & Pullman (206) and Fukui, Imamura & Nagata (207).

In the first volume of a new "advances in quantum chemistry" series, edited by Löwdin (208), there is an excellent review of π-electron theory by Lykos (209).

Theories of σ electrons.—The theory of σ electrons in saturated molecules has received relatively little attention, but this area is now the focus of much activity. Some of the older work has been reviewed by Klopman (210).

Hoffmann & Lipscomb (211–213) have developed a Hückel theory of saturated and unsaturated compounds of carbon, boron, and nitrogen, in which are considered all valence electrons in $2s$ and $2p$ atomic orbitals on B, C, or N and $1s$ orbitals on the hydrogens. The H_{pp} matrix elements (Coulomb integrals) are taken to be the valence state ionization potentials of the atoms p. The H_{pq} (resonance integrals) are approximated either by

$$H_{pq} = 0.5K(H_{pp} + H_{qq})S_{pq} \qquad 13.$$

which seems to work well for hydrocarbons, or by

$$H_{pq} = K'S_{pq} \qquad 14.$$

which is apparently better for the resonance integrals involving B or N. No off-diagonal matrix elements are neglected; the molecular geometry enters through the overlap integrals S_{pq}, which are computed from Slater orbitals. The remaining parameter, either K or K', is the calibration parameter. For the hydrocarbons K is chosen to reproduce the observed rotational barrier in ethane. By minimizing the Hückel energy with respect to the molecular conformation, given a set of bond lengths, Hoffmann has determined the relative stabilities of various molecular configurations. Hoffmann has applied his method to conformational analysis for a variety of molecules: normal and branched alkanes, cycloalkanes, and olefinic, acetylenic and aromatic hydrocarbons, the B–N isoelectronic analogues of various hydrocarbon series, and many carbonium ions.

The success and simplicity of Hoffmann's calculations make his method an attractive way for chemists to study the relative stabilities of reaction intermediates. His technique also might be used to predict the geometries of molecules in excited states. In many cases the configuration of the excited state of a reactant may determine the configuration of a product. Thus Burnelle (214) has explained the *trans*-addition to the carbon-carbon triple bond in acetylenic compounds as the result of the *trans*-bent configuration of excited acetylenic molecules.

Pople & Santry (215) have also been concerned with the simple molecular orbital theory of saturated hydrocarbons. They too have considered all valence electrons, using the $2s-2p$ hybrid atomic orbitals of carbon and the $1s$ orbitals of the hydrogens as their basis set. From a perturbation theory starting from completely localized bonds, they conclude that the total electronic energy (in the Hückel approximation) can be expressed as the sum of energies of electrons in localized bonds but with stabilizing corrections due to electron delocalizations, which themselves arise from the energy difference between the carbon $2s$ and $2p$ electrons and from bonding between atoms in the geminal and vicinal positions. The stabilization resulting from these delocalization corrections is considerable but it does not vary greatly among isomers. Apparently the most important delocalizations are those involving vicinal bonds (giving the C–C bonds some π character) and these are greater for the *trans* than for the *gauche* configuration.

Some time ago, Hall (216) calculated the first ionization potentials for saturated hydrocarbons by a method in which a set of localized orbitals were assumed for C–C and C–H bonds. Fueki (217) has recently extended this method to higher ionization potentials of these molecules.

The success of these very simple treatments of σ systems is bound to trigger attempts to obtain more quantitative results by inclusion of electron repulsion. The calculation on formaldehyde by Parks & Parr (218) was an early attempt of this kind.

More recently, Pohl, Rein & Appel (219) have presented a semi-empirical method for treating σ-electron systems based on the Pariser-Parr and Pople theories. They start from Pople's form of the Roothaan SCF equations and make systematic approximations of the core, electron repulsion, one-center Coulomb, core-core repulsion, and resonance integrals by methods drawn from π-electron theory. A single calibration parameter enters through the resonance integrals and this is adjusted to make calculated dipole moments agree with experiment. They studied the ground states of the diatomic hydrogen halides HX. Only the electron pair in a bonding orbital composed of a 1*s* AO on H and an *np* AO on the X atom were considered explicitly, the core electrons entering only through the empirical values chosen for the integrals. The calculated molecular properties were binding energies, equilibrium bond distances, dipole moments, and vibrational force constants. The results were in reasonable agreement with the observed values and qualitatively well followed the actual trends through the HX series. Klopman (220) has introduced a similar molecular orbital theory of saturated molecules. He has made an error in his formal theoretical development of the kinetic energy terms in the Hamiltonian, but his subsequent empirical development may be valid nevertheless; his results are good and provide grounds for optimism.

Other methods.—Ellison (221) has developed a theory of "diatomics in molecules" which is related to semiempirical valence bond theory. He partitions the polyatomic matrix elements between valence bond structures into parts corresponding to independent contributions from each diatomic and monatomic fragment. He evaluates these contributions in terms of overlaps between the polyatomic structures and experimental ground and excited state energies of the fragments. No exchange integrals, Coulomb integrals, or calibration parameters occur. He has applied this method to calculate the O–H and H–H bond energies and the O-atom promotional energy in H_2O (221). Potential energy surfaces have been studied and activation energies calculated (222, 223) for H_3, H_3^+, and the hydrogen halides H_2X and X_2H (224). The results are in encouraging agreement with experiment.

Julg (225) has developed an "improved LCAO method" in which he has modified the $1/r_{ij}$ electron repulsion terms in the π-electron Hamiltonian by multiplicative parameters, in order to reduce the effect of these terms; these yield electron repulsions that are too large (because of no electron correlation) when simple Slater AO are used in the calculation. In a recent paper (226), he attempts to justify his procedure. He and co-workers have most recently used the method to study the spectra of *p*-benzoquinone (227), tropone (228), furan (229), several acetylenes, butatriene, and N_2 (230), and C_3 (231). Operationally, this method does not appear to differ except in detail from the PPP methods. Neither does the "variable electronegativity" method of Brown & Heffernan (232), as for example recently applied to N_2O_4 by Harcourt (233).

Linnett and co-workers continue to study nonpaired spatial orbitals and interpret molecular structure in terms of double quartet diagrams (234, 235). Chong & Linnett (236) compare alternant molecular orbitals and nonpaired spatial orbitals for the allyl cation, radical, and anion. Empedocles & Linnett (237) show that the nonpaired spatial orbitals for benzene using only two variation parameters give a total energy only slightly above that of the 22-configuration calculation of Moskowitz & Barnett (238).

Valence bond theory has been applied by den Boer-Veenendaal & den Boer (239), Simonetta & Heilbronner (240), Goodwin (241), Brown (242) and others (243–245). Rutner & Bauer (246), Tinoco & Woody (247), and Basu (248) have used the free-electron MO model. Platt's electrostatic model has been used by Moran & Friedman (249). Maximum overlap orbitals have been considered by Veillard & Del Re (250) and by Certain, Watts & Goldstein (251).

Swalen & de Heer (252) have applied the many-parameter alternant molecular orbital method to a number of closed-shell rings. This method allows horizontal correlation of the electrons (by allowing each electron in a pair to occupy the opposite ends of a bond, say) while the split *p*-orbital (SPO) method of Dewar & Chung (253) introduces vertical correlation by forcing electrons of a pair into the opposite lobes of the *p*-orbital. Lowe's (254) recent investigation sought to determine the relative importance of horizontal and vertical correlation. His results for ethylene indicate that almost all of the energy improvement comes from horizontal correlation. This confirms conclusions concerning the SPO method of Saturno, Joy & Snyder (255).

Special Theoretical Topics

The foregoing account shows that many calculations on molecules have been carried out, by many methods. This does not mean, however, that basic theoretical questions do not remain, or that new and original approaches are not needed. In this section, we review some of the more novel, very recent theoretical developments.

The method of Conroy.—Conroy (114–118) has obtained impressive accuracy in calculations on H^-, He, H_2^+, HeH^+, Li, H_3^{++}, H_3^+, by quite unique methods.

Characteristic of Conroy's approach is that he takes great care to choose trial forms for wavefunctions that have the right behavior at singular points in the potential, and at large distances. The latter requires him to include in the wavefunctions themselves the energy itself, and to abandon the conventional variational procedure.

To determine wavefunctions, Conroy minimizes the energy variance, the integral

$$U^2 = \int (H\Phi - W\Phi)^2 d\tau \bigg/ \int \Phi^2 d\tau \qquad 15.$$

Here Φ is a linear variation function, $\Phi = \sum_i c_i \Phi_i$. As has been discussed by

Conroy (117), the procedure is similar to the minimization of the variance in the local energy, proposed by Frost et al. (139, 140). In fact, both methods lead to equations of the same form,

$$\sum_i c_i[G_{ij} - WV_{ij} + (W^2 - U^2)S_{ij}] = 0 \qquad 16.$$

where

$$G_{ij} = \int H\Phi_i H\Phi_j d\tau, \qquad V_{ij} = \int (\Phi_i H\Phi_j + \Phi_j H\Phi_i)d\tau, \qquad S_{ij} = \int \Phi_i\Phi_j d\tau \qquad 17.$$

Minimization of U^2 has certain advantages over the variation method. Minimization of E is sensitive to errors in the integrals, whereas the variance minimization is more stable with respect to these errors. Thus Conroy (116) is able to use a numerical approximation method (a Monte Carlo process) in calculating the difficult many-center integrals and integrals containing the r_{ij} coordinates. The technique may be generalized to integrals arising in more complicated molecules. Monte Carlo procedures for calculating complicated molecular integrals are interesting in their own right.

In his local energy method, Frost uses quantities of the type G_{ij}, V_{ij}, and S_{ij} of Equation 17, where the integral signs and volume elements are replaced by summation signs and weighting factors respectively. His matrix elements are sums of functions (the integrands) evaluated over a set of points (chosen so as to approximate integration) in configuration space. Since Conroy's quantities are integrals evaluated by a Monte Carlo method, they too are summations over sets of points.

A final important aspect of Conroy's scheme is an extrapolation method to obtain the true ground state energy W at zero energy variance U^2 (117).

The correlation problem.—The study of the correlation energy and ways to include correlation effects in wavefunctions are important current problems in quantum chemistry. An excellent two-part review of the correlation problem was given in 1959 by Löwdin (256) and Yoshizumi (257).

A discussion of the helium atom offers a good introduction to the problem. The two most used methods for including correlation effects in wavefunctions are configuration interaction (CI) and the Hylleraas r_{12} method. The space part of the CI wavefunction for the He ground state can be written

$$\Phi_{CI} = \sum_{nml} A_{nml}[f_n(r_1)f_m(r_2)]P_l(\cos\theta_{12}) \qquad 18.$$

Here the f's are radial functions, the brackets indicate symmetrization of the closed factor if $m \neq n$, and θ_{12} is the angle at the nucleus between $\vec{r}_1$ and $\vec{r}_2$. The space part of one form of r_{12} type wavefunction can be written

$$\Phi_H = \sum_{nml} B_{nml}[f_n(r_1)f_m(r_2)]r_{12}{}^l \qquad 19.$$

Wavefunctions containing the r_{12} coordinate have been very successful for helium. The simple correlation factor $g(1,2) = 1 + cr_{12}$ multiplied onto the HF function accounts for over 86 per cent of E_{corr}. Hylleraas' six-term wavefunction of 1929 gives a slightly better result than the 35-configuration function of Weiss (258). Attempts to apply the r_{12} method to more complicated atoms and molecules meet some very difficult integrals. Nevertheless, r_{12} wavefunctions have been used to calculate the energies of Li and Be, by Burke (259) and Szász (260), respectively.

Others, discouraged by the integrals, have followed the more tractable but slowly convergent CI approach. For the Be atom the 55-configuration function of Weiss (258) gives a somewhat better result than the r_{12} function of Szász (260). Brooks & Simpson (261) have recently reported the results of a variation-perturbation calculation on Be. Their energy is in error by 0.4 per cent, while there is an 0.04 per cent energy error in the CI result of Weiss.

From the law of cosines, $\cos\theta_{12}=(r_1^2+r_2^2-r_{12}^2)/(2r_1r_2)$, one can see that Φ_{CI} contains only even powers of r_{12} while Φ_H may contain r_{12} to any power. In order to accord with the $1/r_{12}$ singularity in the wave equation, the true solution Ψ must satisfy the correlation cusp condition

$$\left(\frac{1}{\Psi}\frac{\partial\Psi}{\partial r_{12}}\right)_{r12=0}=\frac{1}{2} \qquad \textbf{19a.}$$

For any truncated Φ_{CI}, on the other hand, $(\partial\Phi_{CI}/\partial r_{12})=0$ at $r_{12}=0$—no truncated CI function can satisfy the correlation cusp condition. The relatively slow convergence of the CI wavefunction has been commonly attributed to its failure to build up a correlation cusp with only a few terms (258). Scherr & Knight (262) have recently made additional comments on the rate of convergence of the CI approach.

Handler & Joy (263) have used a series of open-shell functions of the form $r_1^m r_2^n \exp(-a_{mn}r_1-b_{mn}r_2)$ in an examination of radial correlation. Pritchard (264) has studied open and closed shell correlated wavefunctions using correlation factors $g(1,2)=\exp(-\alpha r_{12})\Sigma c_k r_{12}^k$. Lester & Krauss (265) have considered Gaussian correlation factors of the form $g(1,2)=\Sigma b_k \exp(-\beta_k r_{12}^2)$, for H_2 and He-like atoms with SCF functions constructed from Gaussians. Because of the relative ease of these integrals compared to those which arise from the exponential basis functions, correlated Gaussians may be useful for more complicated atoms and molecules. Longstaff & Singer (266) have made a survey of Gaussian wavefunctions for helium.

Gimarc, Cooney & Parr (267) have recently tested a new type of correlation factor of the form $g(1,2)=1+ch(r_{12})$ where $h(r_{12})$ is a cut-off correlation variable such that $h(r_{12})=r_{12}$ for $0\leq r_{12}\leq D$ and $h(r_{12})=D$ for $D\leq r_{12}\leq\infty$, with c and D adjustable parameters. By calculating energies corresponding to systematically selected values of D, one finds that the most important region for electron correlation in the helium atom is $0.2\leq r_{12}\leq 2.0$ Bohr radii. The region $r_{12}<0.2$ B has very little effect on the energy, hence the failure of the CI wave-function to build up a sharp r_{12} cusp is not very important for the calculated energy.

Although the linear correlation factor $1+cr_{12}$ is successful for $1s^2$ electrons, one is by no means certain that the same procedure will work for $2s^2$ electrons. In fact, the Be atom results of Watson (268) indicate that in correlating the $2s$ electrons with each other a $2p^2$ (or r_{34}^2) term is more important. This can explain the rather disappointing Be result of Szász (260), who included only the linear r_{34} term for $2s$ shell correlation.

To avoid problems of slow convergence and difficult integrals, a fresh approach to the correlation problem is needed. One such attempt is that of Sinanoğlu (269, 270) who is developing a "many-electron theory of atoms and molecules" to calculate wavefunctions and energies to chemical accuracy. In this theory the exact wavefunction Ψ is written as $\Psi = \Phi_{HF} + \chi$, where Φ_{HF} is the Hartree-Fock solution and χ is the correction function, the terms of which account for one-, two-, and many-electron correlations. The two-electron correlations are by far the most important, and χ for closed-shell systems can be approximated by retaining only pairs and their unlinked clusters. Thus the problem of electron correlation is reduced to a set of independent variational problems, each involving only pairs of electrons or pair correlations. Lefebvre (271) suggests that the importance of the pair correlations is a direct result of the antisymmetry constraint imposed on the wavefunction.

A careful study of pair correlations, in which all integrals that arise are explicitly exhibited, has been made by Krauss & Weiss (272). Levine, Geller & Taylor (273) have studied the consequences for the spin properties of the total electronic wavefunction. Tuan & Sinanoğlu (274) have recently used the many-electron theory to calculate correlation energies of 3- and 4-electron atoms. Their result for the Be atom correlation energy is 99.87 per cent of the known value. Their results confirm the notion that the correlation of electrons of Be^{+2} is virtually the same as that of the $(1s)^2$ cores of Be^+ and Be. McKoy & Sinanoğlu (275) use the theory to show that the $1s^2 2s^2$ core correlations are not the same through the series of ground-state first-row atoms, however.

The many-electron theory has conceptual value, and the numerical results are of interest. Sinanoğlu & Tuan (274) argue that one of the strong points of the formulation is its ease (compared to the method of configuration interaction) of application and extension to more complicated systems. Yet in the Be case, Tuan & Sinanoğlu make use of Watson's 37-configuration Be wavefunction (268) to estimate the dynamical part of the Be $(2s^2)$ correlation energy, and the difficult cyclic integrals involving $r_{12}r_{13}/r_{23}$ occur even though pairs are formally considered separately (276–278).

There is need to examine the relations between Sinanoğlu's theory and the theory of separated electron pairs (279, 280), along the lines started by Edmiston (281). It is in part only a matter of taste whether one prefers to impose strong orthogonality constraints at the beginning or to proceed without strong orthogonality constraints but to neglect certain terms in the energy expressions at the end.

Kelly & Sesler (282) and Kelly (283) have been developing a theory of electron correlation which utilizes the systematic expansion of the true wavefunction in terms of multiple particle excitations from the Hartree-Fock solution as the zeroth order approximation. The theory is developed in second-quantization form. Yaris & Musher (284) and Yaris (285) give general discussions of the appearance of unlinked clusters in SCF-CI calculations.

Concerning the relation between the correlation energy of a molecule and the correlation energies of its constituent atoms, Clementi (286) and Carlson & Skancke (287) have argued that the correlation energy does not change as a molecule dissociates into the appropriate atomic fragments, with, in particular, an unchanged number of paired electrons. Thus the correlation energy of H_2 is different from the correlation energy of 2H, but very close to the correlation energy of $H^+ + H^-$.

Scherr & Knight (288) continue to generate the remarkable perturbation expansion of helium wavefunctions in inverse powers of Z. Stewart (289), Coulson (290), and Rosenfield & Konowalow (291) have recently applied an expansion technique to open shell or unrestricted Hartree-Fock functions. They show that half-integer as well as integer powers of Z enter the open shell expansion.

The simple product $\exp[-\zeta(r_1+r_2)]$ for the $1s^2$ configuration of helium is a fairly good approximation to the Hartree-Fock function. The electron repulsion part of the energy is $V_{ee}=(5/8)\zeta$. Correlation may be thought of as lowering the energy by decreasing V_{ee} to $(5/8)\ (\zeta-\zeta_0)$. Jørgensen (292) points out that ζ_0 is very nearly 137/1836, the fine structure constant divided by the ratio of proton and electron masses, and he proposes a set of empirical rules for approximating the ζ-independent part of the many-electron atom correlation energy.

Integral Hellmann-Feynman theorem.—If the electronic Hamiltonian operator for a molecule depends on some parameter λ, such as a nuclear coordinate, the energy of the molecule in one of its real eigenstates Ψ also depends on that parameter; and it is well-known that the change of energy with change in λ is governed by the Hellmann-Feynman theorem,

$$\partial W/\partial\lambda = \int \Psi(\partial H/\partial\lambda)\Psi d\tau \Big/ \int \Psi\Psi d\tau \qquad 20.$$

Among the consequences of this formula is the fact that the force on a nucleus can be computed by classical electrostatics, as the Coulomb force on a point charge due to the other nuclei plus the electron distribution. Among recent applications of the formula are the interesting papers of Bader (293, 294). In a recent collection of papers on molecular orbital theory (295), Hurley (296) has carefully reviewed the formula from the theoretical point of view.

For two different values of λ, say λ_x and λ_y, corresponding for example to two distinct conformations of one molecule, the energy difference $W_y - W_x$ could be obtained by integration of Equation 20:

$$\Delta W = \int_{\lambda_x}^{\lambda_y} (\partial W/\partial\lambda)d\lambda \qquad 21.$$

This may be called the integrated Hellmann-Feynman theorem.

The integral Hellmann-Feynman theorem is still another formula for this energy difference (297, 298). With Ψ_x and Ψ_y the real eigenfunctions for conformers X and Y, one may define the transition density between the two states,

$$\rho_{xy}(1) = N \int \Psi_x \Psi_y d\tau(2) \cdots d\tau(N) \Big/ \int \Psi_x \Psi_y d\tau(1) d\tau(2) \cdots d\tau(N) \qquad 22.$$

Then if $\Delta H = H_y - H_x = \Delta V_{nn} + \Delta V_{ne}$, with $\Delta V_{ne} = \Sigma_\mu H'(\mu)$, one has

$$\Delta W = \int \Psi_x \Delta H \Psi_y d\tau \Big/ \int \Psi_x \Psi_y d\tau \qquad 23.$$

or

$$\Delta W = \Delta V_{nn} + \int \rho_{xy}(1) H'(1) d\tau(1) \qquad 24.$$

These formulas, it should be noted, do not contain kinetic energy operators or electron-electron repulsion operators; their extreme simplicity suggests that the most careful examination be made of them. Application to the problem of barriers to internal rotation about single bonds has been initiated by Wyatt & Parr (299).

For approximate wavefunctions, there would be three different ways to compute the energy difference $W_y - W_x$. One could use the expectation value difference $E_y - E_x$, the integrated Hellmann-Feynman difference of Equations 20 and 21, or the integral Hellmann-Feynman difference of Equation 23. Which would be the best method is a most interesting question that merits investigation. Equation 23 is favored by one simple argument: a first-order error in the wavefunctions Ψ_x and Ψ_y will produce a first-order error in ΔW computed from Equation 23, but first-order with respect to the true ΔW. In contrast, errors in expectation values E_x and E_y will be second order, but second order with respect to these energies, not their difference.

The recent discussion by Richardson & Pack (300) of "partitioning of Hamiltonians and molecular energies" contains Equation 23 as a special case. Their calculations of dissociation energies from approximate molecular wavefunctions provide grounds for optimism that Equation 23 may be a generally very useful "bootstrap principle" in molecular energetics.

Density matrices and natural orbitals.—In an important 1963 paper, Coleman (301) brings up to date research on the question, "Can the wavefunction be eliminated from quantum mechanics and its role taken over, in the discussion of physical systems, by reduced density matrices?" Of less profound importance, but equally enticing from the operational point of view, is the question of whether the natural spinorbitals for a system can be determined directly. Recent contributions relevant for these questions are by Kutzelnigg (302), Kutzelnigg & Smith (303), Harriman (304), and Linderberg (305). Natural orbitals for He_2^{++} have been discussed by Shull & Prosser (306).

Miscellaneous theoretical developments.—Löwdin (307) has published a lucid set of papers on perturbation methods. Synek (308) discusses relativistic self-consistent-field theory. Blinder (309) proposes the direct calculation of energy eigenvalue spectra from evolution in time of nonstationary states. Wyatt & Parr (310) have made use of the invariance of Hartree-Fock wavefunctions to unitary transformations of the orbitals to reduce the coupling in the differential equations governing the effects of small one-electron perturbations on Hartree-Fock wavefunctions. Helliwell (311) has developed a method for treating atomic "cores," which may be useful for treating inner shells in molecules.

Frost (312) has proposed a method for the series solution to the wave equation which is a generalization of the method used so successfully on helium by Pekeris. Frost, Inokuti & Lowe (313) have applied the generalization to a number of three particle problems, and Frost, Harriss & Scargle (314) discuss the possible application to more complicated systems.

Gay (315), Johnson & Coulson (316), and Goodisman & Secrest (317), have studied the calculation of lower bounds to the energy. All have used methods related to that of Bazley & Fox (318). Upper and lower bounds in perturbation theory have been treated by Sando & Hirschfelder (319).

Joy, Schaad & Handler (320) outline a "minimum energy orthogonalization" method, that constitutes a generalization of the Roothaan LCAO-SCF procedure. Finally, one must mention two sets of interesting observations, with interesting general implications. Fraga & McClung (321), in the process of developing extensions and variations of the LCAO-SCF method, noticed a remarkable constancy of the ratio of electron repulsion energy to total energy for a series of atoms and negative ions (the value is about minus one third), and suggests that this ratio may be good for molecules as well; if an empirical rule of this kind were to hold up, combined with the virial theorem it would allow the separate estimate of V_{ee} and $V_{ne}+V_{nn}$ for a molecule, from its total energy.

On the other hand, in a consideration of possible reasons for the success of extended Hückel calculations on molecules, Boer, Newton & Lipscomb (322) have found, by examination of actual LCAO-SCF calculations, a remarkable cancellation, on going from atoms to a molecule, of the nuclear-nuclear repulsion V_{nn} against one half of the change in the kinetic energy plus nuclear-electron attraction energy.

Properties of Molecules

In the following we briefly review recent work in three important areas of application of valence theory: the determination of electron distributions in molecules, the explanation of barriers to internal rotation about single bonds in molecules, and the theory of rates of chemical reaction. As typical papers in quite different areas, which we shall not review otherwise, we mention in passing the calculation by Karplus (323) of the refractive index of the hydrogen molecule, the calculation by Lowe & Flygare (324) of the field gradient at the deuteron in formaldehyde, and the treatment of Pople (325) of σ-electron contributions to magnetic anisotropy in benzene.

Electron distribution.—It can be instructive in a study of chemical bonding to plot the electron density in contours over the nuclear geometry. The difference in electron densities between the molecule and the unperturbed valence-state separated atoms shows the shift in charge density which results from formation of a chemical bond.

Turner and associates (326) have calculated electron densities in methane using the one-center wavefunction of Bishop (87). They examined the question of the CH bond moment in methane. From their contour plots of the difference in molecular and separated atom electron charge densities one may infer that a considerable amount of charge from the C and some from the H is moved into the CH bond region during bonding. On several different definitions of the bond moment in this context, they calculated a moment which indicates C^+H^- polarity. They also calculated the octupole and hexadecapole moments for the molecule. Sinai (327) has recently used the methane wavefunctions of Krauss (100), constructed from Gaussian orbitals, and of Woznick (75), from Slater orbitals, in calculating the octupole moment. Values from the Gaussian and Slater orbitals agree. The result of Turner et al. is higher, but an over-estimate from the one-center method may be expected (98). King (328) has published an interesting study of the methane infrared intensities as related to its electronic wavefunction.

Kern & Karplus (329) have studied the electron charge density of hydrogen fluoride calculated from several different LCAO-SCF wavefunctions. They show contour maps of charge densities for the different wavefunctions, differences in charge densities between molecule and separated atoms, differences in charge densities between different wavefunctions, and differences between charge densities for a particular MO in different approximations. Burnelle (330) has drawn contour diagrams of the difference in electron charge densities between molecule and separated atoms for HCN and C_2N_2.

How rigidly do bonding electrons follow the motions of vibrating nuclei? Schrader & Karplus (331) have found that the assumption of incomplete orbital following appears to be necessary to obtain a consistent theoretical interpretation of the hyperfine splittings in the ESR spectrum of the methyl radical.

Barriers to internal rotation about single bonds.—Recently it has been argued (297, 299), from the integral Hellmann-Feynman theorem, that the 3 kcal/mole barrier to internal rotation in the ethane molecule is ascribable to proton-proton repulsions moderated by electron shielding effects originating near protons. Molecular wavefunctions composed from 1*s*, 2*s* and 2*p* orbitals on the carbons and 1*s* orbitals on the hydrogens are adequate to explain the effect.

Let the eclipsed form E be obtained from the staggered form S by 60° rotation of one end of the molecule. We may then insert into Equation 24 the Fourier expansions

$$\rho_{ES}(1) = \rho_0(r, \theta) + \rho_3(r, \theta) \cos 3\phi + \cdots \qquad 25.$$

$$H'(1) = A_3(r, \theta) \cos 3\phi + A_9(r, \theta) \cos 9\phi + \cdots \qquad 26.$$

where ϕ is the azimuthal angle. The result is the basic formula

$$\Delta W \cong \Delta V_{nn} + \pi \int \rho_3(r, \theta) A_3(r, \theta) d\tau_{r,\theta} \qquad 27.$$

The quantity ΔV_{nn} is 5 kcal (we assume CH_3 geometry unchanged on rotation); the second term must then be -2 kcal. This comes from the three-fold component of the electron (transition) density. That it is as low as -2 is not in disagreement with the more or less homopolar nature of CH bonds. The polarity of the bonds involves a mixture of ρ_0 and ρ_3, and hence is not directly related to the barrier.

This argument is supported by analysis (299) of the LCAO-SCF wavefunctions for the two forms computed by Pitzer & Lipscomb (73, 332). It may be regarded as a development of an approach due to Karplus & Parr (333).

Other recent treatments of the internal rotation problem include those of Cignitti & Allen (334), Simmons & Williams (335), and Ruedenberg (336). It seems certain that this strange fact of molecular structure will, at long last, soon be understood.

Reaction rates.—Studies of reaction rates have always been an important part of theoretical chemistry. The simplest exchange type-reaction $H+H_2 = H_2+H$ has been a subject of great concern up to the present time (337). One goal has been to calculate from quantum-mechanical principles the potential energy surfaces for the reacting and product species. According to the activated complex formulation of chemical kinetics, one can obtain from the potential energy surfaces the activation energy, the nuclear configuration of the activated complex, and the rate of the reaction. The chief difficulty is that the calculation of energy values for the many different configurations of the nuclei required to construct accurate energy surfaces is a lot of work, even for this simple reaction, even in this era of high speed computers. The best recent purely theoretical calculations on the surface are those of Shavitt (338), and these fall far short of high accuracy.

Ellison, Huff & Patel (222, 223), Cashion & Herschbach (339), and Porter & Karplus (340) have calculated potential energy surfaces for $H+H_2$ based on the London approximation or related modifications or extensions. The various integrals in the formulas are approximated using data from empirical or calculated potential energy curves, for the ground and first excited triplet states of H_2. The three investigations all gave activation energies in good agreement with the recent experimental result of Schulz & Le Roy (341), and indicate the linear structure as the most stable single arrangement for the activated complex. The surfaces contain a single saddle point which contains no wells or depressions, the simple form of surface one assumes lacking contrary evidence.

With the potential energy surfaces given, reaction kinetics can be considered from the viewpoint of the dynamics of collisions, following the lines set by Wall and co-workers (342, 343). The energy surface is used in the appropriate equations of motion to calculate total and differential reaction cross-sections, from which rate constants can be calculated by Monte Carlo averaging over initial conditions corresponding to experimental situations. Karplus, Porter & Sharma (344) recently have used this collision dynamics approach for the $H+H_2$ reaction. Their computed rate constant agrees well with experiment, and shows the correct dependence on temperature. An interesting result of their calculations is the prediction that slightly bent configurations are important in this reaction. This seems to agree with the experimental results of Datz & Taylor (345), who have studied $D+H_2$ using crossed molecular beams.

Assuming an energy surface of the type suggested by Blais & Bunker (346), Karplus & Raff (347) have investigated the collision dynamics of the reactive scattering of $K+CH_3I$ treating the methyl group as a single atom.

Scattering theory is receiving much application to systems comprising small numbers of nuclei and electrons; the methods employed in these calculations do not differ greatly from those used in ordinary valence theory calculations. The following listing is but an indication of problems now being studied. The scattering of electrons by H atoms has been investigated by Waddington (348), Kyle & Temkin (349), Omidvar (350), Sloan (351), and Kingston (352); electron scattering by He by Simpson & Fano (353), Lassettre & Jones (354), and LaBahn & Callaway (355); electron scattering by H_2^+ by Peek (356). Cheshire (357), Roth (358), and Helbig & Everhart (359) have treated electron capture in H^+-H or H^+-He collisions.

Conclusion

Over the last ten years the literature of quantum chemistry has grown immensely in volume and diversity. Many new books have appeared that supplement the older books of Pauling & Wilson; Hellmann; Eyring; Eyring, Walter & Kimball; Pitzer; and Coulson. Pauling (360) has published a new and revised edition of his classic discussion of chemical bonds. The three volumes by Slater (361, 362) on the quantum theory of atoms and molecules are the beginning of a series which is to include the theory of solids. The works of Kauzmann (363) and of Daudel, Lefebvre & Moser (364) both attempt breadth and depth in a single volume. Linnett (365, 366) has published two small monographs which provide a nice introduction to quantum chemistry. The books of Goodenough (367), Hartree (368), Jørgensen (369), Simpson (370), and Parr (151) offer personal reviews of selected topics. The books of Roberts (371) and Streitwieser (148) describe the Hückel theory and its applications in organic chemistry. In very recently published volumes, Murrell (372), King (373), and Sandorfy (374) discuss the quantum theory of electronic spectra. The books by Orgel (375), Griffith (376), Ballhausen (377), and Jørgensen (378) are concerned primarily with ligand field theory and the spectra and bonding of inorganic molecules and ions. Pullman & Pullman (379) describe quantum chemistry to biochemists and biochemistry to quantum chemists. Reprints of research papers by the Chicago group on conjugated molecules have been collected into two volumes by Platt and co-workers (380, 381).

Several mathematical monographs have been published by authors who are primarily chemists or physicists. These include books on group theory by Griffith (382), Cotton (383), McWeeny (384), and Tinkham (385), and a book on matrices by Hall (386). The three volume set of books edited by Bates (387) is intended to span the whole spectrum of topics referred to as quantum theory. A collection of articles has been produced in book form by Löwdin & Pullman (295) as a tribute to Robert S. Mulliken. The first issue of a new annual review series, *Advances in Quantum Chemistry*, is off the press (208). Of importance to quantum chemists is one volume of the series *Methods in Computational Physics*, edited by Alder, Fernbach & Rotenberg (388), which comprises articles detailing methods for the calculation of atomic and molecular electronic structures. The journal *Theoretica Chimica Acta*, founded by Hartmann in 1962, is devoted almost exclusively to quantum chemistry.

Significant meetings whose proceedings have been preserved are the Boulder Conference of June 1959 (389), the Sanibel Island symposium of January 1963 (390), and the paper symposium on quantum mechanics in organic chemistry sponsored by *Tetrahedron* in 1963 (391). A series of Gordon Research Conferences on theoretical chemistry has been started.

The quantum theory of valence is becoming a mature part of physical chemistry.

LITERATURE CITED

1. Parr, R. G., and Ellison, F. O., *Ann. Rev. Phys. Chem.*, **6**, 171–92 (1955)
2. Roothaan, C. C. J., *Rev. Mod. Phys.*, **23**, 69–89 (1951); **32**, 179–85 (1960)
3. Clementi, E., *J. Mol. Spectr.*, **12**, 18–22 (1964)
4. Hartmann, H., and Clementi, E., *Phys. Rev.*, **133**, A1295–99 (1964)
5. Clementi, E., and McLean, A. D., *Phys. Rev.*, **133**, A419–23 (1964)
6. Clementi, E., McLean, A. D., Raimondi, D. L., and Yoshimine, M., *Phys. Rev.*, **133**, A1274–79 (1964)
7. Clementi, E., *Phys. Rev.*, **135**, A980–84 (1964)
8. Clementi, E., *J. Chem. Phys.*, **38**, 996–1000, 1001–8 (1963); **40**, 1944–45 (1964); **41**, 295–302, 303–15 (1964)
9. Clementi, E., Roothaan, C. C. J., and Yoshimine, M., *Phys. Rev.*, **127**, 1618–20 (1964)
10. Roothaan, C. C. J., and Kelly, P. S., *Phys. Rev.*, **131**, 1177–82 (1963)
11. Roothaan, C. C. J., and Synek, M., *Phys. Rev.*, **133**, A1263–66 (1964)
12. Synek, M., *Phys. Rev.*, **131**, 1572–77 (1963); **133**, A961–64 (1964)
13. Synek, M., and Stungis, G. E., *J. Chem. Phys.*, **41**, 971–74 (1964); *Phys. Rev.*, **136**, A112–14 (1964)
14. Watson, R. E., *Phys. Rev.*, **119**, 1934–39 (1960)
15. Kelly, P. S., *Proc. Phys. Soc.* (*London*), **83**, 533–35 (1964)
16. Moses, H. A., and Russek, A., *Phys. Rev.*, **135**, A1547–49 (1964)
17. Snow, E. C., Canfield, J. M., and Waber, J. T., *Phys. Rev.*, **135**, A969–73 (1964)
18. Davidson, E. R., *J. Chem. Phys.*, **41**, 656–58 (1964)
19. Herman, F., and Skillman, S., *Atomic Structure Calculations* (Prentice-Hall, Englewood Cliffs, N. J., 424 pp., 1963)
20. Clementi, E., *J. Chem. Phys.*, **38**, 2248–56 (1963); **39**, 175–79 (1963)
21. Allen, L. C., Clementi, E., and Gladney, H. M., *Rev. Mod. Phys.*, **35**, 465–73 (1963)
22. Clementi, E., *J. Am. Chem. Soc.*, **83**, 4501–5 (1961)
23. Clementi, E., *J. Chem. Phys.*, **34**, 1468–69 (1961)
24. Clementi, E., *ibid.*, **36**, 33–44 (1962)
25. Clementi, E., *ibid.*, 750–52
26. Clementi, E., and McLean, A. D., *J. Chem. Phys.*, **36**, 45–47 (1962)
27. Clementi, E., and McLean, A. D., *ibid.*, 563–64
28. Clementi, E., and McLean, A. D., *ibid.*, 745–49
29. Clementi, E., and McLean, A. D., *ibid.*, **39**, 323–25 (1963)
30. McLean, A. D., *J. Chem. Phys*, **32**, 1595–97 (1960)
31. McLean, A. D., *ibid.*, **37**, 627–30 (1962)

32. McLean, A. D., *ibid.*, **38,** 1347–55 (1963)
33. McLean, A. D., *ibid.*, **39,** 2653–65 (1963)
34. Nesbet, R. K., *Rev. Mod. Phys.*, **32,** 272–73 (1960)
35. Nesbet, R. K., *J. Chem. Phys.*, **36,** 1518–33 (1962)
36. Kahalas, S. L., and Nesbet, R. K., *J. Chem. Phys.*, **39,** 529–44 (1963)
37. Nesbet, R. K., *J. Chem. Phys.*, **40,** 3619–33 (1964)
38. Nesbet, R. K., *ibid.*, **41,** 100–4 (1964)
39. Carlson, K. D., and Nesbet, R. K., *J. Chem. Phys.*, **41,** 1051–62 (1964)
40. Nesbet, R. K., *Phys. Rev.*, **135,** A460–65 (1964)
41. Karo, A. M., and Allen, L. C., *J. Am. Chem. Soc.*, **80,** 4496–99 (1958)
42. Karo, A. M., and Allen, L. C., *J. Chem. Phys.*, **31,** 968–77 (1959)
43. Karo, A. M., *J. Chem. Phys.*, **30,** 1241–45 (1959)
44. Ransil, B. J., *Rev. Mod. Phys.*, **32,** 239–44, 245–54 (1960)
45. Ransil, B. J., *J. Chem. Phys.*, **34,** 2109–18 (1961)
46. Lefebvre-Brion, H., and Moser, C. M., *J. Chem. Phys.*, **32,** 1194–99 (1960)
47. Lefebvre-Brion, H., Moser, C. M., and Nesbet, R. K., *J. Chem. Phys.*, **33,** 931–32 (1960); **34,** 1950–57; **35,** 1702–7 (1961)
48. Fischer, I., *Arkiv Fysik*, **5,** 349–76 (1952)
49. Higuchi, J., *J. Chem. Phys.*, **22,** 1339–48 (1954)
50. Scherr, C. W., *J. Chem. Phys.*, **23,** 569–78 (1955)
51. Higuchi, J., *J. Chem. Phys.*, **23,** 2197–98 (1955); **28,** 527–31 (1958)
52. Ellison, F. O., and Shull, H., *J. Chem. Phys.*, **23,** 2348–57 (1955)
53. Higuchi, J., *J. Chem. Phys.*, **24,** 535–45 (1956)
54. Sahni, R. C., *J. Chem. Phys.*, **25,** 332–36 (1956)
55. Kaplan, H., *J. Chem. Phys.*, **26,** 1704–13 (1957)
56. Ishiguro, E., Kayama, K., Kotani, M., and Mizuno, Y., *J. Phys. Soc. Japan*, **12,** 1355–85 (1957)
57. Faulkner, J. E., *J. Chem. Phys.*, **27,** 369–73 (1957)
58. Duncan, A. B. F., *J. Chem. Phys.*, **27,** 423–29 (1957)
59. Freeman, A. J., *J. Chem. Phys.*, **28,** 230–43 (1958)
60. Krauss, M., *J. Chem. Phys.*, **28,** 1021–26 (1958)
61. Boyd, M. E., *J. Chem. Phys.*, **29,** 108–15 (1958)
62. Clinton, W. L., and Rice, B., *J. Chem. Phys.*, **29,** 445–46 (1958)
63. Padgett, A. A., and Griffing, V., *J. Chem. Phys.*, **30,** 1286–91 (1959)
64. Kolos, W., and Roothaan, C. C. J., *Rev. Mod. Phys.*, **32,** 219–32 (1960)
65. Goodfriend, P. L., Birss, F. W., and Duncan, A. B. F., *Rev. Mod. Phys.*, **32,** 307–11 (1960)
66. Padgett, A., and Krauss, M., *J. Chem. Phys.*, **32,** 189–95 (1960)
67. Merryman, P., Moser, C. M., and Nesbet, R. K., *J. Chem. Phys.*, **32,** 631–32 (1960)
68. Lorquet, J. C., and Lefebvre-Brion, H., *J. Chim. Phys.*, **57,** 85–89 (1960)
69. Richardson, J. W., *J. Chem. Phys.*, **35,** 1829–39 (1961)
70. Grahn, R., *Arkiv Fysik*, **19,** 147–61 (1961)
71. Grahn, R., *ibid.*, **21,** 1–12 (1962)
72. Manneback, C., *Physica*, **29,** 769–83 (1963)
73. Pitzer, R. M., and Lipscomb, W. N., *J. Chem. Phys.*, **39,** 1995–2004 (1963)
74. Rosenfeld, J. L. J., *J. Chem. Phys.*, **40,** 384–89 (1964)
75. Woznick, B. J., *J. Chem. Phys.*, **40,** 2860–64 (1964)
76. Yoshimine, M., *J. Chem. Phys.*, **40,** 2970–76 (1964)
77. Scrocco, E., and Tomasi, J., *Theoret. Chim. Acta (Berlin)*, **2,** 386–92 (1964)
78. Wahl, A. C., *J. Chem. Phys.*, **41,** 2600–11 (1964)
79. Sinai, J. J., *J. Chem. Phys.*, **39,** 1575–77 (1963)
80. Bonham, R. A., Peacher, J. L., and Cox, H. L., Jr., *J. Chem. Phys.* **40,** 3083–86 (1964)
81. Geller, M., *J. Chem. Phys.*, **41,** 4006–7 (1964)
82. Wahl, A. C., Cade, P. E., and Roothaan, C. C. J., *J. Chem. Phys.*, **41,** 2578–99 (1964).
83. Magnusson, E. A., *Rev. Pure Appl. Chem.*, **14,** 57–66 (1964)
84. Joy, H. W., and Parr, R. G., *J. Chem. Phys.*, **28,** 448–53 (1958)
85. Saturno, A. F., and Parr, R. G., *J. Chem. Phys.*, **33,** 22–27 (1960)
86. Bishop, D. M., *Theoret. Chim. Acta*, **1,** 410–17 (1963)
87. Bishop, D. M., *Mol. Phys.*, **6,** 305–15 (1963)
88. Bishop, D. M., *J. Chem. Phys.*, **40,** 432–34 (1964)
89. Bishop, D. M., Hoyland, J. R., and Parr, R. G., *Mol. Phys.*, **6,** 467–76 (1963)

90. Bishop, D. M., and Hoyland, J. R., *Mol. Phys.*, **7**, 161–64 (1964)
91. Moccia, R., *J. Chem. Phys.*, **40**, 2164–76 (1964)
92. Moccia, R., *ibid.*, 2176–85
93. Moccia, R., *ibid.*, 2186–92
94. Banyard, K. E., and Hake, R. B., *J. Chem. Phys.*, **41**, 3221–26 (1964)
95. Gáspár, R., Tamássy-Lentei, I., and Kruglyak, Y., *J. Chem. Phys.*, **36**, 740–45 (1962)
96. Albasiny, E. L., and Cooper, J. R. A., *Proc. Phys. Soc. (London)*, **82**, 289–303, 801–2 (1963)
97. Bishop, D. M., *J. Chem. Phys.*, **40**, 1322–25 (1964)
98. Turner, A. G., Honig, B. H., Parr, R. G., and Hoyland, J. R., *J. Chem. Phys.*, **40**, 3216–20 (1964)
99. Nesbet, R. K., *J. Chem. Phys.*, **32**, 1114–22 (1960)
100. Krauss, M., *J. Chem. Phys.*, **38**, 546–65 (1963)
101. Reeves, C. M., and Harrison, M. C., *J. Chem. Phys.*, **39**, 11–17 (1963)
102. Harrison, M. C., *J. Chem. Phys.*, **41**, 495–98 (1964)
103. Harrison, M. C., *ibid.*, 499–502
104. Meckler, A., *J. Chem. Phys.*, **21**, 1750–62 (1953)
105. Kimball, G. E., and Neumark, G. F., *J. Chem. Phys.*, **26**, 1285–87 (1957)
106. McLean, A. D., *J. Chem. Phys.*, **40**, 243–44 (1964)
107. Krauss, M., *J. Res. Natl. Bur. Std.*, **68**(A), 635–44 (1964).
108. Allen, L. C., and Karo, A. M., *Rev. Mod. Phys.*, **32**, 275–85 (1960)
109. Das, G., and Wahl, A. C., *Bull. Am. Phys. Soc.*, **10**, 102 (1965)
110. Browne, J. C., *J. Chem. Phys.*, **40**, 43–46; **41**, 1583–86 (1964)
111. Kato, Y., Hayes, E. F., and Duncan, A. B. F., *J. Chem. Phys.*, **41**, 986–88 (1964)
112. Christoffersen, R. A., Hagstrom, S., and Prosser, F., *J. Chem. Phys.*, **40**, 236–37 (1964)
113. Christoffersen, R. A., *J. Chem. Phys.*, **41**, 960–71 (1964)
114. Conroy, H., *J. Chem. Phys.*, **40**, 603–4, 3121 (1964)
115. Conroy, H., *ibid.*, **41**, 1327–31 (1964)
116. Conroy, H., *ibid.*, 1331–35
117. Conroy, H., *ibid.*, 1336–40
118. Conroy, H., *ibid.*, 1341–51
119. Shull, H., *J. Am. Chem. Soc.*, **86**, 1469–74 (1964)
120. Linnett, J. W., *J. Am. Chem. Soc.*, **86**, 2519–20 (1964)
121. Bowen, H. C., and Linnett, J. W., *Trans. Faraday Soc.*, **60**, 1185–92 (1964)
122. Hirst, D. M., and Linnett, J. W., *J. Chem. Soc.*, 1035–47, 3844 (1962)
123. Bradley, J. N., *Trans. Faraday Soc.*, **60**, 1353–62 (1964)
124. Preuss, H., *Theoret. Chim. Acta (Berlin)*, **2**, 344–61, 362–70 (1964)
125. Harris, F. E., and Taylor, H. S., *Physica*, **30**, 105–12 (1964)
126. Taylor, H. S., and Harris, F. E., *Mol. Phys.*, **7**, 287–95 (1964)
127. Taylor, H. S., and Gerhauser, J., *J. Chem. Phys.*, **40**, 244–45 (1964)
128. Browne, J. C., *J. Chem. Phys.*, **41**, 3495–98 (1964)
129. Stuart, J. D., and Matsen, F. A., *J. Chem. Phys.*, **41**, 1646–50 (1964)
130. Preuss, H., *Mol. Phys.*, **8**, 233–51 (1964)
131. Hauk, P., Parr, R. G., and Hameka, H. F., *J. Chem. Phys.*, **39**, 2085–98 (1963)
132. Levine, I. N., *J. Chem. Phys.*, **40**, 3444–45; **41**, 2044–52 (1964)
133. Csizmadia, I. G., Sutcliffe, B. T., and Barnett, M. P., *Can. J. Chem.*, **42**, 1645–63 (1964)
134. Hoyland, J. R., *J. Chem. Phys.*, **41**, 1370–76 (1964)
135. Hoyland, J. R., *ibid.*, **40**, 3540–47 (1964)
136. Parr, R. G., *J. Chem. Phys.*, **26**, 428 (1957)
137. Whitten, J. L., *J. Chem. Phys.*, **39**, 349–52 (1963)
138. Preuss, H., *Mol. Phys.*, **8**, 157–68 (1964)
139. Frost, A. A., Kellogg, R. E., and Curtis, E. C., *Rev. Mod. Phys.*, **32**, 313–17 (1960)
140. Frost, A. A., Kellogg, R. E., Gimarc, B. M., and Scargle, J. D., *J. Chem. Phys.*, **35**, 827–31 (1961)
141. Harriss, D. K., and Frost, A. A., *J. Chem. Phys.*, **40**, 204–7 (1964)
142. Goodisman, J., *J. Chem. Phys.*, **41**, 3889–97 (1964)
143. Bonaccorsi, R., Scrocco, E., and Tomasi, J., *J. Chem. Phys.*, **41**, 3658–60 (1964)
144. Zung, J. T., and Parr, R. G., *J. Chem. Phys.*, **41**, 2888–92 (1964)
145. Kolos, W., and Wolniewicz, L., *J. Chem. Phys.*, **41**, 3663–73 (1964)
146. Kolos, W., and Wolniewicz, L., *ibid.*, 3674–78
147. Lykos, P. G., and Parr, R. G., *J. Chem. Phys.*, **24**, 1166–73 (1956); **25**, 1301 (1956)
148. Streitwieser, A., Jr., *Molecular Orbital Theory for Organic Chemists* (Wiley, New York, 489 pp., 1961)

149. Pariser, R., and Parr, R. G., *J. Chem. Phys.*, **21**, 466–71, 767–76 (1953)
150. Pople, J. A., *Trans. Faraday Soc.*, **49**, 1375–85 (1953)
151. Parr, R. G., *Quantum Theory of Molecular Electronic Structure* (W. A. Benjamin, New York, 510 pp., 1963)
152. Preuss, H., *Theoret. Chim. Acta (Berlin)*, **2**, 93–97, 98–101 (1964)
153. Del Re, G., and Parr, R. G., *Rev. Mod. Phys.*, **35**, 604–11 (1963)
154. Streitwieser, A., Jr., *J. Am. Chem. Soc.*, **82**, 4123–35 (1960)
155. Streitwieser, A., Jr., Heller, A., and Feldman, M., *J. Phys. Chem.*, **68**, 1224–26 (1964)
156. Smith, W. F., Jr., *Tetrahedron*, **20**, 671–85 (1964)
157. Ettinger, R., *Tetrahedron*, **20**, 1579–88 (1964)
158. Janssen, M. J., and Sandstrom, J., *Tetrahedron*, **20**, 2339–49 (1964)
159. Moffat, J. B., *Can. J. Chem.*, **42**, 1323–32 (1964)
160. Chatterjee, H., and Ali, M. A., *Tetrahedron*, **20**, 2829–34 (1964)
161. Golebiewski, A., and Nowakowski, J., *Acta Phys. Polon.*, **XXV**, 647–61 (1964)
162. Goodwin, T. H., and Morton-Blake, D. A., *Theoret. Chim. Acta (Berlin)*, **2**, 75–83 (1964)
163. Goodwin, T. H., and Morton-Blake, D. A., *J. Chem. Soc.*, 615–18 (1964)
164. Howe, J. A., *J. Chem. Phys.*, **40**, 3744–45 (1964)
165. Wagnière, G., *Theoret. Chim. Acta (Berlin)*, **2**, 281–84 (1964)
166. Malrieu, J.-P., and Pullman, B., *Theoret. Chim. Acta (Berlin)*, **2**, 293–301, 302–14 (1964)
167. Nakajima, T., Saijo, T., and Yamaguchi, H., *Tetrahedron*, **20**, 2119–24 (1964)
168. de Giambiagi, M. S., Giambiagi, M., and Ferreira, R., *J. Chim. Phys.*, **61**, 697–705 (1964)
169. Suzuki, H., and Mizuhashi, S., *J. Phys. Soc. Japan*, **19**, 724–38 (1964)
170. Godfrey, M., and Murrell, J. N., *Proc. Roy. Soc. (London), A*, **278**, 57–63, 64–70, 71–90 (1964)
171. Fukui, K., Kato, H., Yonezawa, T., and Okamura, S., *Bull. Chem. Soc. Japan*, **37**, 904–5 (1964)
172. Goeppert-Mayer, M., and Sklar, A. L., *J. Chem. Phys.*, **6**, 645–52 (1938)
173. Allinger, N. L., and Miller, M. A., *J. Am. Chem. Soc.*, **86**, 2811–19 (1964)
174. Basu, R., *Theoret. Chim. Acta (Berlin)*, **2**, 87–92, 215–18 (1964)
175. Bloor, J. E., Daykin, P. N., and Boltwood, P., *Can. J. Chem.*, **42**, 121–36 (1964)
176. Chalvet, O., Daudel, R., and Kaufman, J. J., *J. Phys. Chem.*, **68**, 490–503 (1964)
177. Donath, W. E., *J. Chem. Phys.*, **40**, 77–80; **41**, 3646 (1964)
178. Donath, W. E., *ibid.*, **41**, 626–31 (1964)
179. Forster, L. S., *J. Am. Chem. Soc.*, **86**, 3001–4 (1964)
180. Inuzuka, K., and Yokota, T., *Bull. Chem. Soc. Japan*, **37**, 1224–27 (1964)
181. Ito, H., and I'Haya, Y., *Theoret. Chim. Acta (Berlin)*, **2**, 247–57 (1964)
182. Kaufman, J. J., *J. Phys. Chem.*, **68**, 2648–51 (1964)
183. Mataga, N., *Bull. Chem. Soc. Japan*, **36**, 1607–18 (1963)
184. Moreau, C., and Serre, J., *Theoret. Chim. Acta (Berlin)*, **2**, 40–54 (1964)
185. Peacock, T. E., and Wilkinson, P. T., *Proc. Phys. Soc. (London)*, **83**, 355–59, 525–32 (1964)
186. Kwun, S.-I., Ree, T., and Eyring, H., *J. Chem. Phys.*, **40**, 3320–23 (1964)
187. Wachters, A. J. H., and Davies, D. W., *Tetrahedron*, **20**, 2841–49 (1964)
188. Gondo, Y., *J. Chem. Phys.*, **41**, 3928–38 (1964)
189. Azumi, T., Armstrong, A. T., and McGlynn, S. P., *J. Chem. Phys.*, **41**, 3839–52 (1964)
190. Koutecký, J., Hochman, P., and Michl, J., *J. Chem. Phys.*, **40**, 2439–56 (1964)
191. Weltin, E., Weber, J.-P., and Heilbronner, E., *Theoret. Chim. Acta (Berlin)*, **2**, 114–24 (1964)
192. Ohno, K., *Theoret. Chim. Acta (Berlin)*, **2**, 219–27 (1964)
193. Lykos, P. G., and Parr, R. G., *J. Chem. Phys.*, **28**, 361 (1958)
194. Adams, O. W., and Miller, R. L., *J. Chem. Phys.*, **41**, 2948–50 (1964)
195. Schmeising, H. N., and Miller, R. L., *J. Chem. Phys.*, **40**, 926–32 (1964)
196. Simmons, H. E., *J. Chem. Phys.*, **40**, 3554–62 (1964)
197. Koutecký, J., Čížek, J., Dubský, J., and Hlavaty, K., *Theoret. Chim. Acta (Berlin)*, **2**, 462–67 (1964)
198. Scheibe, G., and Brück, D., *Z. Elektrochem.*, **54**, 403–12 (1950); *Chem. Ber.*, **85**, 867–77 (1952); Scheibe, G., Kern, J., and Dörr, E., *Z. Elektrochem.*, **63**, 117–21 (1959); Scheibe, G., *Chimia (Aarau)*, **15**, 10–20 (1961)

199. Hartmann, H., *Z. Naturforsch*, **15** (*A*), 993–1003 (1960); *Pure Appl. Chem.*, **4**, 15–22 (1962); *Advan. Chem. Phys.*, **V**, 1–31 (1963)
200. Kollaard, U. H., and Colpa, J. P., *Mol. Phys.*, **8**, 295–97 (1964)
201. Rein, R., and Ladik, J., *J. Chem. Phys.*, **40**, 2466–70 (1964)
202. Rein, R., and Harris, F. E., *Science*, **146**, 649–50 (1964); *J. Chem. Phys.*, **41**, 3393–401 (1964)
203. Ladik, J., and Appel, K., *J. Chem. Phys.*, **40**, 2470–73 (1964)
204. Yomosa, S., *J. Phys. Soc. Japan*, **19**, 1718–30 (1964)
205. Beleznay, F., and Biczó, G., *J. Chem. Phys.*, **41**, 2351–55 (1964)
206. Yonezawa, T., Del Re, G., and Pullman, B., *Bull. Chem. Soc. Japan*, **37**, 985–89 (1964)
207. Fukui, K., Imamura, A., and Nagata, C., *Bull. Chem. Soc. Japan*, **36**, 1450–53 (1963)
208. Löwdin, P.-O., Ed., *Advances in Quantum Chemistry*, **1** (Academic Press, New York, 385 pp., 1964)
209. Lykos, P. G., in *Advances in Quantum Chemistry*, **1**, 171–201 (Academic Press, New York, 1964)
210. Klopman, G., *Tetrahedron*, *Suppl.*, **19**(2), 111–22 (1963)
211. Hoffmann, R., and Lipscomb, W. N., *J. Chem. Phys.*, **37**, 2872–83 (1962)
212. Hoffmann, R., *J. Chem. Phys.*, **39**, 1397–1412 (1963); **40**, 2047–48, 2474–80, 2480–88, 2745 (1964)
213. Hoffmann, R., *J. Am. Chem. Soc.*, **86**, 1259–61 (1964)
214. Burnelle, L., *Tetrahedron*, **20**, 2403–11 (1964)
215. Pople, J. A., and Santry, D. P., *Mol. Phys.*, **7**, 269–86 (1964)
216. Hall, G. G., *Proc. Roy. Soc.* (*London*), *A*, **205**, 541–52 (1951); **213**, 102–23 (1952)
217. Fueki, K., *J. Phys. Chem.*, **68**, 2656–60 (1964)
218. Parks, J. M., and Parr, R. G., *J. Chem. Phys.*, **32**, 1657–81 (1960)
219. Pohl, H. A., Rein, R., and Appel, K., *J. Chem. Phys.*, **41**, 3385–92 (1964)
220. Klopman, G., *J. Am. Chem. Soc.*, **86**, 1463–69, 4550–57 (1964)
221. Ellison, F. O., *J. Am. Chem. Soc.*, **85**, 3540–44 (1963)
222. Ellison, F. O., Huff, N. T., and Patel, J. C., *J. Am. Chem. Soc.*, **85**, 3544–47 (1963)
223. Ellison, F. O., *J. Chem. Phys.*, **41**, 2198–99 (1964)
224. Ellison, F. O., and Patel, J. C., *J. Am. Chem. Soc.*, **86**, 2115–19 (1964)
225. Julg, A., *J. Chim. Phys.*, **57**, 19–30 (1960)
226. Julg, A., *Theoret. Chim. Acta* (*Berlin*), **2**, 134–57 (1964)
227. Julg, A., Béry, J. C., and Bonnet, M., *Tetrahedron*, **20**, 2237–41 (1964)
228. Julg, A., and Bonnet, M., *Tetrahedron*, **20**, 2243–46 (1964)
229. Pujol, L., and Julg, A., *Theoret. Chim. Acta* (*Berlin*), **2**, 125–33 (1964)
230. Julg, A., and Pellegatti, A., *Theoret. Chim. Acta* (*Berlin*), **2**, 202–9, 396–402 (1964)
231. Julg, A., and Pellegatti, A., *ibid.*, 210–14
232. Brown, R. D., and Heffernan, M. L., *Trans. Faraday Soc.*, **54**, 757–64 (1958)
233. Harcourt, R. D., *Theoret. Chim. Acta* (*Berlin*), **2**, 437–52 (1964)
234. Linnett, J. W., and Rosenberg, R. M., *Tetrahedron*, **20**, 53–66 (1964)
235. Linnett, J. W., *Am. Scientist*, **52**, 459–75 (1964)
236. Chong, D. P., and Linnett, J. W., *Mol. Phys.*, **8**, 151–55 (1964)
237. Empedocles, P. B., and Linnett, J. W., *Proc. Roy. Soc.* (*London*), *A*, **282**, 166–77 (1964)
238. Moskowitz, J. W., and Barnett, M. P., *J. Chem. Phys.*, **39**, 1557–60 (1963)
239. den Boer-Veenedaal, P. C., and den Boer, D. H. W., *Tetrahedron*, **20**, 2157–61 (1964)
240. Simonetta, M., and Heilbronner, E., *Theoret. Chim. Acta* (*Berlin*), **2**, 228–46 (1964)
241. Goodwin, T. H., *Theoret. Chim. Acta* (*Berlin*), **2**, 315–23 (1964)
242. Brown, T. H., *J. Chem. Phys.*, **41**, 2223–27 (1964)
243. Jordan, P. C., *J. Chem. Phys.*, **41**, 1442–49 (1964)
244. Huber, L. M., and Thorson, W. R., *J. Chem. Phys.*, **41**, 1829–32 (1964)
245. Dixon, R. N., *Mol. Phys.*, **8**, 201–12 (1964)
246. Rutner, E., and Bauer, S. H., *J. Phys. Chem.*, **68**, 1768–75 (1964)
247. Tinoco, T., Jr., and Woody, R. W., *J. Chem. Phys.*, **40**, 160–65 (1964)
248. Basu, S., *J. Chem. Phys.*, **41**, 1453–54 (1964)
249. Moran, T. F., and Friedman, L., *J. Chem. Phys.*, **40**, 860–66 (1964)
250. Veillard, A., and Del Re, G., *Theoret. Chim. Acta* (*Berlin*), **2**, 55–62 (1964)
251. Certain, P. R., Watts, V. S., and Goldstein, J. H., *Theoret. Chim. Acta* (*Berlin*), **2**, 324–31 (1964)
252. Swalen, J. D., and de Heer, J., *J. Chem. Phys.*, **40**, 378–84 (1964)

253. Dewar, M. J. S., and Chung, A. L. H., *J. Chem. Phys.*, **39**, 1741–47 (1963)
254. Lowe, J. P., *J. Chem. Phys.*, **41**, 502–8 (1964)
255. Saturno, A. F., Joy, H. W., and Snyder, L. C., *J. Chem. Phys.*, **38**, 2494–99 (1963); Snyder, L. C., and Parr, R. G., **28**, 1250–51 (1958); **34**, 1661–65 (1961)
256. Löwdin, P.-O., *Advan. Chem. Phys.*, **2**, 207–322 (1959)
257. Yoshizumi, H., *Advan. Chem. Phys.*, **2**, 323–65 (1959)
258. Weiss, A. W., *Phys. Rev.*, **122**, 1826–36 (1961)
259. Burke, E. A., *Phys. Rev.*, **130**, 1871–76 (1963)
260. Szász, L., *Phys. Letters*, **3**, 263–64 (1963)
261. Brooks, F. L., and Simpson, W. T., *J. Chem. Phys.*, **41**, 909–10 (1964)
262. Scherr, C. W., and Knight, R. E., *J. Chem. Phys.*, **40**, 1777–78 (1964)
263. Handler, G. S., and Joy, H. W., *J. Chem. Phys.*, **40**, 2408 (1964)
264. Pritchard, H. O., *J. Chem. Phys.*, **40**, 1496–97 (1964)
265. Lester, W. A., Jr., and Krauss, M., *J. Chem. Phys.*, **41**, 1407–13 (1964)
266. Longstaff, J. V. L., and Singer, K., *Theoret. Chim. Acta (Berlin)*, **2**, 265–80 (1964)
267. Gimarc, B. M., Cooney, W. A., and Parr, R. G., *J. Chem. Phys.*, **42**, 21–25 (1965)
268. Watson, R. E., *Phys. Rev.*, **119**, 170–77 (1960)
269. Sinanoğlu, O., *Advan. Chem. Phys.*, **6**, 315–412 (1964)
270. Sinanoğlu, O., and Tuan, D. F., *Ann. Rev. Phys. Chem.*, **15**, 251–80, (1964)
271. Lefebvre, R., *J. Chem. Phys.*, **41**, 2957–58 (1964)
272. Krauss, M., and Weiss, A. W., *J. Chem. Phys.*, **40**, 80–85 (1964)
273. Levine, H. B., Geller, M., and Taylor, H. S., *J. Chem. Phys.*, **40**, 595–602 (1964)
274. Tuan, D. F., and Sinanoğlu, O., *J. Chem. Phys.*, **41**, 2677–88 (1964)
275. McKoy, V., and Sinanoğlu, O., *J. Chem. Phys.*, **41**, 2689–95 (1964)
276. Szász, L., *J. Chem. Phys.*, **35**, 1072–76 (1961)
277. Öhrn, Y., and Nordling, J., *J. Chem. Phys.*, **39**, 1864–71 (1963)
278. Hinze, J., and Pitzer, K. S., *J. Chem. Phys.*, **41**, 3484–87 (1964)
279. Hurley, A. C., Lennard-Jones, J. E., and Pople, J. A., *Proc. Roy. Soc. (London), A*, **220**, 446–55 (1953)
280. Parks, J. M., and Parr, R. G., *J. Chem. Phys.*, **28**, 335–45 (1958)
281. Edmiston, C., *J. Chem. Phys.*, **39**, 2394–95 (1963)
282. Kelly, H. P., and Sessler, A. M., *Phys. Rev.*, **132**, 2091–95 (1963)
283. Kelly, H. P., *Phys. Rev.*, **134**, A1450–53 (1964)
284. Yaris, R., and Musher, J. I., *J. Chem. Phys.*, **41**, 1701–4 (1964)
285. Yaris, R., *J. Chem. Phys.*, **41**, 2419–21 (1964)
286. Clementi, E., *J. Chem. Phys.*, **38**, 2780–82 (1963); **39**, 487–88 (1963)
287. Carlson, K. D., and Skancke, P. N., *J. Chem. Phys.*, **40**, 613–14 (1964)
288. Scherr, C. W., and Knight, R. E., *J. Chem. Phys.*, **40**, 3034–39 (1964)
289. Stewart, A. L., *Proc. Phys. Soc. (London)*, **83**, 1033–37 (1964)
290. Coulson, C. A., *Proc. Phys. Soc. (London)*, **84**, 511–16 (1964)
291. Rosenfeld, J. L. J., and Konowalow, D. D., *J. Chem. Phys.*, **41**, 3556–59 (1964)
292. Jørgensen, C. K., *Mol. Phys.*, **8**, 191–93 (1964)
293. Bader, R. F. W., *J. Am. Chem. Soc.*, **86**, 5070–75 (1964)
294. Bader, R. F. W., *Can. J. Chem.*, **42**, 1822–34 (1964)
295. Löwdin, P.-O., and Pullman, B., Eds., *Molecular Orbitals in Chemistry, Physics, and Biology: A Tribute to R. S. Mulliken* (Academic Press, New York, 578 pp., 1964)
296. Hurley, A. C. (Ref. 295, pp. 165–75)
297. Parr, R. G., *J. Chem. Phys.*, **40**, 3726 (1964)
298. Kim, H. J., and Parr, R. G., *J. Chem. Phys.*, **41**, 2892–97 (1964)
299. Wyatt, R. E., and Parr, R. G., *J. Chem. Phys.*, **41**, 3262–63 (1964)
300. Richardson, J. W., and Pack, A. K., *J. Chem. Phys.*, **41**, 897–98 (1964)
301. Coleman, A. J., *Rev. Mod. Phys.*, **35**, 668–87 (1963)
302. Kutzelnigg, W., *J. Chem. Phys.*, **40**, 3640–47 (1964)
303. Kutzelnigg, W., and Smith, V. H., Jr., *J. Chem. Phys.*, **41**, 896–97 (1964)
304. Harriman, J. E., *J. Chem. Phys.*, **40**, 2827–39 (1964)
305. Linderburg, J., *J. Mol. Spectr.*, **12**, 210–18 (1964)
306. Shull, H., and Prosser, F., *J. Chem. Phys.*, **40**, 233–35 (1964)
307. Löwdin, P.-O., *J. Mol. Spectr.*, **10**, 12–33 (1963); **13**, 326–37; **14**, 112–44 (1964)
308. Synek, M., *Phys. Rev.*, **136**, A1552–60 (1964)

309. Blinder, S. M., *J. Chem. Phys.*, **41,** 3412–16 (1964)
310. Wyatt, R. E., and Parr, R. G., *J. Chem. Phys.*, **41,** 514–20 (1964)
311. Helliwell, T. M., *Phys. Rev.*, **135,** A325–31 (1964)
312. Frost, A. A., *J. Chem. Phys.*, **41,** 478–82 (1964)
313. Frost, A. A., Inokuti, M., and Lowe, J. P., *J. Chem. Phys.*, **41,** 482–89 (1964)
314. Frost, A. A., Harriss, D. K., and Scargle, J. D., *J. Chem. Phys.*, **41,** 489–94 (1964)
315. Gay, J. G., *Phys. Rev.*, **135,** A1220–26 (1964)
316. Johnson, B. P., and Coulson, C. A., *Proc. Phys. Soc.* (*London*), **84,** 263–68 (1964)
317. Goodisman, J., and Secrest, D., *J. Chem. Phys.*, **41,** 3610–12 (1964)
318. Bazley, N., and Fox, D. W., *J. Math. Phys.*, **4,** 1147–53 (1963)
319. Sando, K. M., and Hirschfelder, J. O., *Proc. Natl. Acad. Sci. U. S.*, **52,** 434–38 (1964)
320. Joy, H. W., Schaad, L. J., and Handler, G. S., *J. Chem. Phys.*, **41,** 2026–31 (1964)
321. Fraga, S. *Theoret. Chim. Acta* (*Berlin*), **2,** 403–15 (1964); McClung, R. E. D., and Fraga, S., *Theoret. Chim. Acta* (*Berlin*), **2,** 416–21 (1964)
322. Boer, F. P., Newton, M. D., and Lipscomb, W. N., *Proc. Natl. Acad. Sci. U. S.*, **52,** 890–93 (1964)
323. Karplus, M., *J. Chem. Phys.*, **41,** 880–83 (1964)
324. Lowe, J. T., and Flygare, W. H., *J. Chem. Phys.*, **41,** 2153–58 (1964)
325. Pople, J. A., *J. Chem. Phys.*, **41,** 2559–60 (1964)
326. Turner, A. G., Saturno, A. F., Hauk, P., and Parr, R. G., *J. Chem. Phys.*, **40,** 1919–28 (1964)
327. Sinai, J. J., *J. Chem. Phys.*, **40,** 3596–98 (1964)
328. King, W. T., *J. Chem. Phys.*, **39,** 2141–46 (1963)
329. Kern, C. W., and Karplus, M., *J. Chem. Phys.*, **40,** 1374–89 (1964)
330. Burnelle, L., *Theoret. Chim. Acta* (*Berlin*), **2,** 177–80 (1964)
331. Schrader, D. M., and Karplus, M., *J. Chem. Phys.*, **40,** 1593–601 (1964)
332. Pitzer, R. M., *J. Chem. Phys.*, **41,** 2216–17 (1964)
333. Karplus, M., and Parr, R. G., *J. Chem. Phys.*, **38,** 1547–52 (1963)
334. Cignitti, M., and Allen, T. L., *J. Phys. Chem.*, **68,** 1292–97 (1964)
335. Simmons, H. E., and Williams, J. K., *J. Am. Chem. Soc.*, **86,** 3222–26 (1964)
336. Ruedenberg, K., *J. Chem. Phys.*, **41,** 588–89 (1964)
337. Davy, J. G., Guerra, C. R., Johnston, H. S., Weaver, C. F., and Young, C. E., *J. Chem. Phys.*, **41,** 1517–18 (1964)
338. Shavitt, I., *J. Chem. Phys.*, **31,** 1359–67 (1959)
339. Cashion, J. K., and Herschbach, D. R., *J. Chem. Phys.*, **40,** 2358–63; **41,** 2199–200 (1964)
340. Porter, R. N., and Karplus, M., *J. Chem. Phys.*, **40,** 1105–15 (1964)
341. Schulz, W. R., and Le Roy, D. J., *Can. J. Chem.*, **42,** 2480–87 (1964)
342. Wall, F. T., Hiller, L. A., Jr., and Mazur, J., *J. Chem. Phys.*, **29,** 255–63 (1958); **35,** 1284–89 (1961)
343. Wall, F. T., and Porter, R. N., *J. Chem. Phys.*, **39,** 3112–17 (1963)
344. Karplus, M., Porter, R. N., and Sharma, R. D., *J. Chem. Phys.*, **40,** 2033–34 (1964)
345. Datz, S., and Taylor, E. H., *J. Chem. Phys.*, **39,** 1896–97 (1963)
346. Blais, N. C., and Bunker, D. L., *J. Chem. Phys.*, **37,** 2713–20 (1962); **39,** 315–23 (1963)
347. Karplus, M., and Raff, L. M., *J. Chem. Phys.*, **41,** 1267–77 (1964)
348. Waddington, J. F., *Proc. Phys. Soc.* (*London*), **84,** 47–53 (1964)
349. Kyle, H. L., and Temkin, A., *Phys. Rev.*, **134,** A600–7 (1964)
350. Omidvar, K., *Phys. Rev.*, **133,** A970–85 (1964)
351. Sloan, I. H., *Proc. Roy. Soc.* (*London*), *A*, **281,** 151–63 (1964)
352. Kingston, A. E., *Phys. Rev.*, **135,** A1529–36, A1537–39 (1964)
353. Simpson, J. A., and Fano, U., *Phys. Rev. Letters*, **11,** 158–59 (1963)
354. Lassettre, E. N., and Jones, E. A., *J. Chem. Phys.*, **40,** 1218–21 (1964)
355. LaBahn, R. W., and Callaway, J., *Phys. Rev.*, **135,** A1539–46 (1964)
356. Peek, J. M., *Phys. Rev.*, **134,** A877–83 (1964)
357. Cheshire, I. M., *Proc. Phys. Soc.* (*London*), **84,** 89–98 (1964)
358. Roth, B., *Phys. Rev.*, **133,** A1257–62 (1964)
359. Helbig, H. F., and Everhart, E., *Phys. Rev.*, **136,** A674–79 (1964)
360. Pauling, L., *The Nature of the Chemical Bond*, 3/e (Cornell Univ. Press, Ithaca, N. Y., 644 pp., 1960)
361. Slater, J. C., *Quantum Theory of*

Atomic Structures, **I, II** (McGraw, New York, 439 pp., 502 pp., 1960)
362. Slater, J. C., *Quantum Theory of Molecules and Solids*, **I** (McGraw, New York, 485 pp., 1963)
363. Kauzmann, W., *Quantum Chemistry* (Academic Press, New York, 744 pp., 1957)
364. Daudel, R., Lefebvre, R., and Moser, C., *Quantum Chemistry* (Interscience, Wiley, New York, 572 pp., 1959)
365. Linnett, J. W., *Wave Mechanics and Valency* (Wiley, New York, 184 pp., 1960)
366. Linnett, J. W., *Electronic Structure of Molecules* (Wiley, New York, 167 pp., 1964)
367. Goodenough, J. B., *Magnetism and the Chemical Bond* (Interscience, Wiley, New York, 393 pp., 1963)
368. Hartree, D. R., *The Calculation of Atomic Structures* (Wiley, New York, 181 pp., 1957)
369. Jørgensen, C. K., *Orbitals in Atoms and Molecules* (Academic Press, New York, 162 pp., 1962)
370. Simpson, W. T., *Theories of Electrons in Molecules* (Prentice-Hall, Englewood Cliffs, N. J., 183 pp., 1962)
371. Roberts, J. D., *Notes on Molecular Orbital Calculations* (W. A. Benjamin, New York, 156 pp., 1961)
372. Murrell, J. N., *The Theory of the Electronic Spectra of Organic Molecules* (Methuen and Co., London, Wiley, New York, 328 pp., 1963)
373. King, G. W., *Spectroscopy and Molecular Structure* (Holt, Rinehart, and Winston, New York, 482 pp., 1964)
374. Sandorfy, C., *Electronic Spectra and Quantum Chemistry* (Prentice-Hall, Englewood Cliffs, N. J., 385 pp., 1964)
375. Orgel, L. E., *An Introduction to Transition-Metal Chemistry: Ligand-Field Theory* (Methuen and Co., London, Wiley, New York, 180 pp., 1960)
376. Griffith, J. S., *The Theory of Transition-Metal Ions* (Cambridge Univ. Press, Cambridge, 455 pp., 1961)
377. Ballhausen, C. J., *Introduction to Ligand Field Theory* (McGraw, New York, 298 pp., 1962)
378. Jørgensen, C. K., *Absorption Spectra and Chemical Bonding in Complexes* (Pergamon, Oxford, Macmillan, New York, 351 pp., 1962)
379. Pullman, B., and Pullman, A., *Quantum Biochemistry* (Interscience, Wiley, New York, 867 pp., 1963); Pullman, B., Ed., *Electronic Aspects of Biochemistry* (Academic Press, New York, London, 582 pp., 1964); Kasha, M., and Pullman, B., Eds., *Horizons in Biochemistry* (Academic Press, New York, London, 604 pp., 1962)
380. Platt, J. R. et al., *Systematics of the Electronic Spectra of Conjugated Molecules* (Wiley, New York, 388 pp., 1964)
381. Platt, J. R., Ruedenberg, K., Scherr, C. W., Ham, N. S., Labhart, H., and Lichten, W., *Free-Electron Theory of Conjugated Molecules* (Wiley, New York, 177 pp., 1964)
382. Griffith, J. S., *The Irreducible Tensor Method for Molecular Symmetry Groups* (Prentice-Hall, Englewood Cliffs, N. J., 134 pp., 1962)
383. Cotton, F. A., *Chemical Applications of Group Theory* (Interscience, Wiley, New York, 295 pp., 1963)
384. McWeeny, R., *Symmetry: An Introduction to Group Theory and its Applications* (Pergamon, Oxford, Macmillan, New York, 248 pp., 1963)
385. Tinkham, M., *Group Theory and Quantum Mechanics* (McGraw, New York, 340 pp., 1964)
386. Hall, G. G., *Matrices and Tensors, International Encyclopedia of Physical Chemistry and Chemical Physics* (Pergamon, Oxford, Macmillan, New York, 106 pp., 1963)
387. Bates, D. R., Ed., *Quantum Theory*, **I, II, III** (Academic Press, New York, 447 pp., 475 pp., 402 pp., 1962)
388. Alder, B., Fernbach, S., and Rotenberg, M., Eds., *Methods in Computational Physics*, **2** (Academic Press, New York, 271 pp., 1963)
389. Boulder Conf., June 1959, *Rev. Mod. Phys.*, **32** (2), 169–476
390. Sanibel Island Symp., January 1963, *Rev. Mod. Phys.*, **35** (3), 415–736
391. Quantum Chemistry Symp., 1963, *Tetrahedron, Suppl.*, **19**(2), 1–478 (1963)

MASS SPECTROMETRY[1]

By P. Bommer and K. Biemann

Department of Chemistry, Massachusetts Institute of Technology, Cambridge, Massachusetts

Introduction.—Since its discovery more than seven decades ago, mass spectrometry has been continuously involved directly or indirectly in studies of phenomena related to physical chemistry. In the last ten or twenty years, it has become a very important tool in certain fields of physical chemistry or chemical physics and a vast amount of work in these areas has been accomplished. Not having been reviewed before in the *Annual Review of Physical Chemistry*, it would be an impossible task to attempt a complete comprehensive discussion covering all the areas of interest to the physical chemist. Rather than to exhaustively enumerate all the papers that have been published in the past few years, a task undertaken by the review compilation of *Analytical Chemistry* every other year (1), an attempt will be made to delineate the development during the last decade in an area of mass spectrometry where this field has expanded most, and has led to the most fruitful developments as far as its application to chemical problems goes.

It is the evolvement of this technique to an almost indispensable tool of the organic chemist who has found it to be very valuable in the characterization and determination of the structure of both simple as well as complex organic molecules. Although this area is, in principle, not very close to the interest of the physical chemist in general, work in this field has provided an immense amount of data which by necessity were interpreted by the organic chemist with an almost crude empiricism with the only major advantage that it led, in most instances, to useful results. This work has been summarized at the various stages of development in a number of books by Beynon (2); Reed (3); Biemann (4); McLafferty (5); Budzikiewicz, Djerassi & Williams (6, 7, 8). These works contain detailed discussions of the results and the present state of the field as well as, taken together, a quite complex bibliography. In the past, the physical chemist and physicist have been mainly interested in the complete solution of electron impact phenomena concerning comparatively small, oligoatomic molecules, and they have attempted to understand their behavior on a quantitative basis. It would seem to be very fruitful, however, to subject the empirical correlations and results obtained by the organic chemist to a scrutinizing investigation which might hopefully put these rules on a physically better-founded basis. There is no doubt that in the course of such work much useful phenomena and quantitative correlations will be uncovered which, in addition, might have considerable bearing on problems only distantly related to the fragmentation of organic molecules under electron impact. An attempt shall thus be made to outline the present status of our knowledge of the behavior of organic molecules in the mass

[1] The survey of literature pertaining to this review was completed November 1964.

spectrometer using a number of representative papers in this field with ample cross reference to sources for a more complete bibliography rather than to try to enumerate all papers that have appeared in this area and cite them individually.

First, the present status of the purely theoretical views concerning the ionization and fragmentation of organic molecules upon impact of low-energy electrons (up to 100 eV) shall be outlined, followed by a brief discussion of conclusions that can be drawn from ionization and appearance-potential measurements concerning the mechanisms of fragmentation of organic molecules. These are areas in which conclusions of quantitative character are based on experimentally determined values of physical meaning. Finally, the present state of the much more empirical argumentation and correlations used by the organic chemist in the interpretation of more complex spectra on a rather qualitative basis will be outlined. It is hoped that this contrast will evoke some interest on the part of the reader to scrutinize those qualitative relationships and empirical rules and subject them to a more rigorous experimental test to prove or disprove their validity.

Theory of mass spectra.—A theory of mass spectra which could become of great interest was presented for the first time by Rosenstock et al. (9) in 1952. The principle of the statistical theory of mass spectra states that an ionized molecular ion, processing electronic and vibrational energy, does not dissociate immediately, but survives for a sufficient period of time to permit random distribution of the excitation energy among its internal degrees of freedom. The parent ions then dissociate unimolecularly into various fragment ions corresponding to the different transition states in which enough vibrational energy is concentrated in the proper mode in which decomposition takes place. The unimolecular rate constant k for such a decomposition is related to the excitation energy E by

$$k = \int_{o}^{E-\epsilon_o} \frac{1}{h} \frac{\rho^{+}(E, \epsilon_o, \epsilon_{+})}{\rho(E)} dE_t \qquad 1.$$

where $\rho^{+}(E, \epsilon_o, \epsilon_t)$ denotes the density of energy levels of the transition state, whereas E is the energy difference to minimum energy for decomposition; ϵ_o is the energy difference between the ground state of the transition complex and the ground state of the decomposing ion, and ϵ_t is the kinetic energy along the path of decomposition; $\rho(E)$ is the density of states of decomposing ions at energy E, h is Planck's constant and E-ϵ_o corresponds to the internal energy of the transition complex. Use of Equation 1 would require knowledge of the density functions ρ and ρ^{+}, and, therefore, the authors had to make several assumptions which led to the following modified equation for the rate of decomposition

$$k = \left(\frac{E - \epsilon_o}{E}\right)^{N-1} \frac{\prod_{i=1}^{N} \nu_i}{\prod_{j=1}^{N} \nu_j^{+}} \qquad 2.$$

where ν_i are the vibrational frequencies of the molecule ion, ν_i^+ are the vibrational frequencies of the activated complex and N is the number of loosely coupled oscillators. Rosenstock et al. (9) tested this theory on the mass spectrum of propane. Thus, choosing a set of reasonable parameters, the calculated spectrum agreed well with the experimental spectrum.

Friedman, Long & Wolfsberg (10) studied the mass spectra of methanol, ethanol, 1-propanol, and 2-propanol and had some success in explaining their fragmentation using the statistical theory. But they found, however, that in the threshold region lower values for the number of oscillators, N, had to be taken, and different frequency factors for different processes like carbon-carbon bond cleavage or OH-abstraction, etc., were required to achieve reasonable agreement with the experimental data.

Chupka (11) investigated *n*-propylamine, *n*-propanol, and methylethyl ketone by photoionization and found that Rosenstock's theory is qualitatively correct for the fragmentation processes, but that in the threshold region the rate constant changes much faster than the theory predicts. Only when using a much smaller value for N a good agreement with experimental data was reached.

A very careful study of all saturated paraffins from C_2 to C_6 plus *n*-heptane and *n*-octane was undertaken by Steiner, Giese & Inghram (12), and the experimental data were compared with the predictions of the statistical theory. Again, to get reasonable agreement between theory and experiment, the total number of free oscillators had to be reduced although different values for N had to be used for calculating different physical properties, e.g. (*a*) to calculate the effect of temperature on the mass spectra, N had to be made a free parameter; (*b*) to predict the energy excess needed to give a rate constant such that dissociation occurs within the ion chamber requires reducing of N; (*c*) to accommodate the tails of fragment and metastable curves as well as the intensity of metastable ions; and (*d*) the shifts in fragment curves resulting from changing the extraction require again a smaller value for N as long as reasonable values for E_0 or ν, or both, are used.

Several of the differences between experiment and theory could be traced in part to inadequate assumptions for the rate equation (13), and Vestal, Wahrhaftig & Johnston (14) showed the classical approximations used in Rosenstock's theory to be invalid at low energies. They developed an improved continuous approximation to numbers of states of a collection of harmonic oscillators of different frequencies and were able to remove many of the difficulties in the application of the statistical theory to a variety of problems in mass spectrometers.

A detailed review of this interaction between experiment and theory of mass spectra including the contributions of many other workers in this field is given by Rosenstock & Krauss (15, 16).

Several papers applying and discussing Rosenstock's theory have since appeared, most of them dealing with the improvement of the rate expressions. Wolfsberg (17) applied the theoretical formulation of the unimolecular decomposition of excited molecules developed by Marcus (18) to numerical calculation of rate constants in the threshold region. Some of the previous difficulties of the old theory could be removed in this new approach. Wieder & Marcus (19) showed that all vibrational modes participate in energy exchange in the activated molecule. They estimate the magnitude of the error introduced by using the classical theory and show that for complex molecules, especially at low temperatures, a quantum mechanical treatment should be employed. Monahan & Stanton (20) investigated the mass spectra of hydrocarbons such as acetylene, ethylene, benzene, neopentane, and *n*-heptane from high-electron energy (200 to 2400 eV) impact. They assume that the accessible electronic states in the molecular ion separate into noninteracting groups which lead to different fragmentation processes. Two of these groups could be detected experimentally, but only for one of them an energy randomization required by the statistical theory could be assumed. The ions belonging to the other group are formed mostly with measurable kinetic energy and do not fit the predictions of the statistical theory.

Dibeler & Rosenstock (21) studied the mass spectra of methane and deuteromethane to determine metastable transitions of the parent ion. They found that all metastable transitions except $CD_4^{\oplus} \rightarrow CD_3^{\oplus} + D$ were collision induced (including the process $CH_4^{\oplus} \rightarrow CH_3^{\oplus} + H$). This unusual isotope effect is difficult to explain with the statistical theory, especially because of the similarity of the shape of the ionization-efficiency curve and the energy difference of the parent and methyl-ionization curves. There is no explanation for the obviously different excitation-energy distribution function of $CH_4^{\oplus}$ as compared with $CD_4^{\oplus}$. The authors discuss several other explanations, particularly the influence of the discreteness of the vibrational levels which could lead to a discreteness in the possible values of the rate constants for unimolecular decomposition, but, on the other hand, the uncertainty principle should lead to a diffuseness of the level which would just remove this discreteness. That would be contrary to the experimental finding concerning the difference in dissociation of CD_4 versus CH_4 mentioned above. Another reason could be a nonequilibrium effect, but none of these explanations is satisfactory.

Using an improved quasiequilibrium theory of mass spectra, Vestal (22) however, recently found that a discontinuity in the rate expression is to be expected, and that the results of Dibeler & Rosenstock are entirely in accord with the theory.

Another approach to experimental verification of the statistical theory concerned the investigation (23) of the metastable peaks of the two competing reactions.

$$C_6H_{13}^{\oplus} \rightarrow C_4H_9^{\oplus} + C_2H_4 \quad m^* = 38.2$$

$$C_6H_{13}^{\oplus} \rightarrow C_3H_7^{\oplus} + C_3H_6 \quad m^* = 21.8$$

which occur in different molecules such as *n*-alkanes, *n*-hexyl bromide, and di-*n*-hexyl ether.

If the statistical theory were incorrect, one should detect a significant change in the ratio of the two metastable peaks, for it is very likely that the initial distribution of energy in the $C_6H_{13}^{\oplus}$ ion differs for different molecules. Although the results did not show any nonequilibrium effects at 39 eV, contrary to a slight dependence at 50 and 70 eV, there is still the possibility that the initial distribution in phase space immediately after formation by unimolecular decomposition is the same for different molecules.

Murad & Inghram (24) investigated the photoionization mass spectra of acetone, 2-butanone, 2-pentanone, 3-methyl-2-butanone, 2-hexanone, and 4-methyl-2-pentanone. Comparison of the experimental data with the modified statistical theory led to the following results: (*a*) temperature shifts for the first fragmentation process in the ketones agree, in contrast to alkanes, with the predictions of the statistical theory; (*b*) the shapes of the threshold agree with the prediction of the modified theory; and (*c*) comparison of the calculated and measured intensity of metastable ions does not yield information useful for a test of the theory.

Klots (25) found that, except for two relatively unimportant reactions, the statistical theory of mass spectra in its usual form as well as in a modified form encompasses the presently available experimental data (26) of kinetic energies of fragmentation in unimolecular dissociation.

Sharp & Rosenstock (27) derived general expressions for calculating Franck-Condon factors of polyatomic molecules making the assumption that the molecule consists of a set of harmonic oscillators, but ignoring effects of rotation and Fermi resonance. Further, the number of vibrational degrees of freedom in the initial and final states are assumed to be equal, thus excluding linear-to-bent transitions. An attempt was made to fit the photoionization data of acetylene of Dibeler & Reese (28) and a good first-order approximation was reached, but, to get a higher accuracy, better estimates of the vibrational frequencies and perhaps an experimental re-examination of the photoionization data would be desirable.

One of the fields of interest for the organic chemist in mass spectrometry is the method of determining ionization and appearance potentials. With help of such data, dissociation energies can be determined and, indirectly, fragmentation mechanisms can be predicted. Unfortunately, many difficulties have still to be overcome in regard to both experiment and in the interpretation of data. In the paper of Chupka (11), many of these problems are discussed, e.g., the effects of the kinetics of dissociation and of internal thermal energy on the ionization efficiency. But in spite of all these difficulties, some very useful results were obtained, e.g., Rylander, Meyerson & Grubb (29) in their investigation of the mass spectra of toluene where they determined the appearance potentials of the two reactions:

$$e + C_6H_5CH_3 \rightarrow C_6H_6^{\oplus} + CH_3 + 2e \qquad \text{A.P. } 13.8 \text{ eV} \qquad 1.$$

$$e + C_6H_4(CH_3)_2 \rightarrow CH_3C_6H_4^{\oplus} + CH_3 + 2e \qquad \textit{ortho } 11.73 \text{ eV} \qquad 2.$$

$$\textit{para } 11.78 \text{ eV}$$

$$\textit{meta } 11.57 \text{ eV}$$

The difference between the measured values of reactions 1 and 2 can be explained only by assuming the ion $CH_3C_6H_5^{\oplus}$ is not just a methylated benzene ion, but that it must have a completely different structure which in this case could be explained by a tropylium ion of the form

Even when exact appearance-potential measurements and thermal data are available, it is often still difficult to determine the actual structure of relatively simple ions. Wiberg, Bartley & Lossing (30) attempted to deduce the structure of the $C_3H_3^{\oplus}$ ion at mass 39 by consideration of the heat of formation of cyclopropene, methylacetylene, and allene, as well as the appearance potentials of the $C_3H_3^{\oplus}$ ion produced from these three compounds. These data indicate identical heat of formation ΔH_f, which leads to three possible conclusions: (*a*) the three $C_3H_3^{\oplus}$ species are of different structures and the identity of their heat of formation is accidental; (*b*) all three $C_3H_3^{\oplus}$ species are propargyl cations; and (*c*) all three are cyclopropenyl cations. Case (*a*) cannot be excluded on the basis of the data, and of the other two (*c*) is preferred because the ratio of $C_3D_3^{\oplus}$ versus $C_3D_2H^{\oplus}$ is 1:3 in the mass spectrum of trideuteriomethylacetylene (31, 32) indicating equivalence of all three carbon atoms before loss of hydrogen.

Ionization not involving electron impact.—Aside from the commonly used method of ionization by electron impact, several others were developed during the last few years. Von Koch & Lindholm (33) and Petterson & Lindholm (34), for instance, investigated the mass spectra of small molecules such as ethanol and propane produced upon charge exchange with slow positive ions. The mass spectra observed were, in general, similar to those obtained by electron impact.

In the last few years, ionization by field emission was developed, a technique which could become of importance as an alternative to electron impact mass spectrometry. Under certain conditions, it avoids fragmentation of the molecule which is important if one is mainly interested in the molecular weight and composition of a compound. Inghram & Gomer (35) investigated, for the first time, the ions created on the surface of a point of a field-emission microscope by mass spectrometry. They point out the simplicity of field-emission spectra compared with electron-impact spectra which may make them particularly useful for the analysis of mixtures.

A few years later, Beckey (36) developed a field-emission ion source and used it for the investigation of ion-molecule reactions of molecules with paired electrons (Ar, N_2, H_2O) and unpaired electrons (NO, ClO, etc.). In a second paper (37), he determined the spectra of saturated hydrocarbons with one to nine carbon atoms at different field strengths. At low field strengths mainly, the molecular ion is formed and the only fragment of importance was the $C_2H_5^{\oplus}$ ion. At higher field strength, the intensity of $CH_3^{\oplus}$ became of the same magnitude as the $C_2H_5^{\oplus}$ ion. At low field strengths, only dissociation of carbon-carbon bonds is observed, whereas at high field strengths, the dissociation of carbon-hydrogen bonds also becomes important. More recently, the basis of a theory of field emission mass spectrometry was discussed (38) and some preliminary criteria of a general theory of field-emission ionization were formulated. Robertson & Viney (39) described the use of the edge of a razor blade for field-emission ionization, and determined with this ion source the spectra of some ketones, olefins, alcohols, propane, and diethylether. With few exceptions, most of the fragments could be explained by simple dissociation processes, but some substances, such as methanol and ethanol, gave small peaks of a mass greater than the molecular ion. Advances in experimental techniques, applications, and theory of field-emission mass spectrometry have been reviewed by Beckey et al. (40). It includes the description of a combined electron and field-emission ion source and its application in the analysis of a gasoline spectra as well as in natural product chemistry. For example, the mass spectrum of *d*-ribose shows a very strong molecular ion contrary to the electron-impact spectra, and, even for neopentane, a molecular ion is observed.

For the measurement of appearance potentials, photoionization became of considerable interest in addition to electron-impact techniques, such as the commonly used retarding potential-difference method of Fox et al. (41). Lossing & Tanaka (42) had shown that photon impact can be used for the ionization of molecules such as propane and 1,3-butadiene. Later, Hürzeler, Inghram & Morrison (43) demonstrated the value of this method in determining appearance and ionization potentials, and improvements of photoionization sources have since permitted the measure of ionization potentials down to 500 Å. Dibeler & Reese (28) determined the ionization potentials of acetylene and deutero-acetylene with great accuracy using such an ion source. They found 11.406 ± 0.006 eV for acetylene, and 11.416 ± 0.006 eV for deutero-acetylene. It seems that in the near future much more reliable data of ionization and appearance potentials of organic molecules may become available.

Empirical interpretation of mass spectra.—The aspects discussed in the previous section were concerned with a quantitative correlation of the abundance of various fragment ions formed on electron impact from a particular organic molecule with physical parameters such as bond energies, ionization potentials, or first-order reaction rates, approaches that are, at least at the present time, largely limited to small and simple molecules. On the other hand, there is a vast amount of data available in the form of more qualitative observations which have been used with considerable success in the interpretation or prediction of mass spectra of much more complicated molecules. This situation is comparable to the one prevailing in organic chemistry in general that one in principle knows all the factors which concertedly govern the behavior of an organic molecule, and should thus be able to predict by calculation the best synthesis of a given molecule, but any synthesis devised today is still almost exclusively based upon empirical observations and experience and not on quantum mechanical calculations. Seemingly empirical rules based on at least qualitative, in many instances even quantitative, understanding of reaction mechanisms are providing the organic chemist today with a much better basis for the design of a workable synthetic route, and it is to a large extent the physical chemist who has produced this basis. In the much younger field of organic mass spectrometry it would seem that here again the physical chemist may be able to provide the much-needed fundamental data to put the empiricism of the organic chemist on better grounds.

One of the most widely used and most acknowledged factors governing the fragmentation path which organic molecules follow after loss of an electron is the stability of the electron deficient positively charged particle formed after cleavage of one or more bonds (2–8). One of the very first, and by now classical, examples is the investigation by Cummings & Bleakney (44) concerning the mass spectrum of methanol, more specifically the loss of hydrogen from the molecular ion to form the fragment of mass 31. Since there are two different types of hydrogen available in this molecule, namely those in the methyl group and the one at the hydroxyl group, two different processes could lead to an ion of that mass which would therefore have structure I or II.

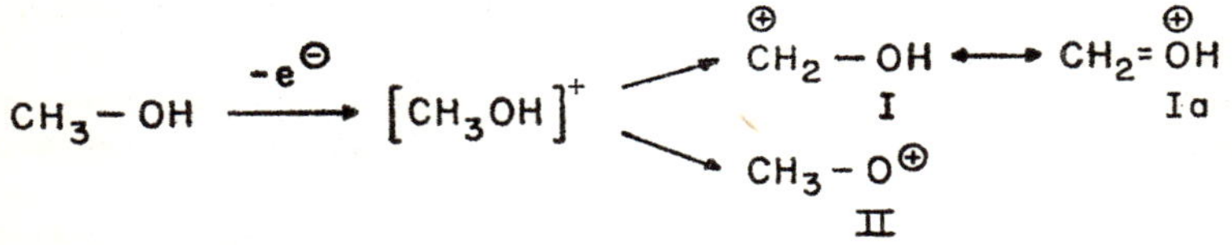

On the basis of the appearance potential (11.8±0.1 eV) of this ion and the known bond-dissociation energies of C–H and O–H as well as the ionization potential of methanol, neither structure I nor structure II could be the correct one for this ion; but when taking into account the magnitude estimated for the resonance energy between I and Ia, the observed value agreed quite well. Thus, the authors concluded (44) that the loss of hydrogen from the methyl group is very much favored over the loss of hydrogen from the hydroxyl group because in the former case a carbonium ion in resonance with an oxonium ion is formed, while the latter process would produce an ion (II) with electron-deficient oxygen atom, an energetically less favorable situation.

Since this first observation, a wealth of mass spectra have indicated that those fragments in which the positive charge can be delocalized by resonance are always much more abundant than those in which the positive charge would have to be confined to a single atom within the fragment. A generalization (45) of this view is found in what is generally referred to as "Stevenson's rule," namely, that upon cleavage of a bond the positive charge will reside on that particle which has the lower ionization potential. This is, of course, in most cases the particle where the loss of an electron leads to additional gain in energy either by resonance or a distribution of the charge attributed to electron-supplying substituents such as the inductive effect of alkyl substituents.

The large amount of experimental data accumulated over the last two or three decades concerning the magnitude of resonance and inductive effect and similar phenomena makes it possible to predict the degree of probability with which a certain bond in a given molecule might break. However, one has to keep in mind that these predictions are of a very qualitative nature and are more concerned with relative probabilities of two possible competing fragmentation processes rather than their absolute magnitude. Frequently, this suffices if one is only faced with a choice between a few possible structures on the basis of the mass spectrum. The underlying arguments, namely the above mentioned importance of the stability of the positive ion based on resonance and inductive effect, are rather indirect, particularly since most of the experimental data that led to the knowledge of the magnitude of these effects have been determined in solution. It is to be expected that these phenomena might be of a very different magnitude in the absence of the close proximity of other molecules, as is the case in the ion source of a mass spectrometer which is generally operated at pressures around 10^{-5} mm.

A typical example is the previously mentioned $C_7H_7^{\oplus}$ ion formed from many compounds containing a benzyl moiety (i.e., $C_6H_5CH_2$). Numerous chemical reactions are known in which the benyzl cation ($C_6H_5CH_2^{\oplus}$) is involved, while the work of Rylander, Meyerson & Grubb (29) showed that on electron impact in the gas phase the $C_7H_7^{\oplus}$ specie is best represented by a symmetrical ion such as the tropylium ion. The facility of this ring expansion is very much in contrast to the reactions occuring in solution (46) where the ring contraction from seven- to six-membered rings (i.e., tropone to benzaldehyde) is much more easily achieved than the reverse process, the expansion of a benzenoid system to a troponoid one. This difference may well be caused by the possibility of solvating the benzyl carbonium ion in solution and thus dissipating some of the asymmetry of this cation by participation of solvent molecules. This is impossible in the gas phase where ring expansion to the completely symmetrical tropylium ion is an energetically more favorable process.

A similar explanation might be offered for the unexpectedly large (in contrast to solution reactions) difference in the stabilizing effect of the free electron pair of nitrogen on a neighboring positive charge on carbon as compared with the stabilizing effect of oxygen. There are abundant examples (2–8) showing that a nitrogen atom facilitates fragmentation of the adjacent carbon-carbon bond much more than oxygen does. This is borne out most strikingly by compounds containing both an amino and a hydroxyl group at adjacent carbon atoms which are competing for the same type of fragmentation process. In all those cases, the nitrogen-containing ion is 10 to 100 times more abundant than the corresponding oxygenated ion and, similarly, some fragmentation processes that are dominant in certain nitrogen compounds are almost completely absent in the spectra of the oxygen analogs. It may well be that the less pronounced difference between the chemical reactivity of amino and hydroxy compounds in solution is in fact due to the greater basicity of the amino group which leads to more extensive protonation in the course of interaction with solvent thus making the free-electron pair less available for participation in resonance structures.

Various other concepts familiar to the physical organic chemist have been invoked as an explanation of certain regularities as well as anomalies in mass spectra. One of them is an earlier attempt by McLafferty (47) to correlate the cleavage of the phenyl–CO–bond in compounds containing the $R'-C_6H_4-CO-R$ grouping with the Hammet σ constant of various substituents R′. A reasonable correlation was found between the abundance of the ion arising from the cleavage of that bond and the σ-constant of R′ with only very few exceptions, one being chlorine as a substituent. This work seems to be the only one in which such a quantitative correlation has been accomplished with some success, but in spite of the number of years that have elapsed since then, this possibly quite fruitful approach has not yet been explored further.

In addition to the better known stabilizing effects due to direct resonance or the inductive effect mentioned above, other phenomena have been invoked to explain the unusual predominance of certain fragment ions. One of the more recent instances is the suggestion by Weininger, Mai & Thornton (48) of homoallylic participation of double bonds in the stabilization of carbonium ions. It had been observed that the mass spectrum of 1,2,3,6-tetrahydrophthalic anhydride (III) does not exhibit a peak for the molecular ion in the mass spectrum, but has a considerable tendency to lose carbon monoxide. The corresponding saturated anhydride (IV), while also not exhibiting a peak for the molecular ion, lacks the fragment caused by loss of carbon monoxide. It was suggested (48) that there must exist in the unsaturated system a driving force which leads to the expulsion of CO from V produced by the rupture of the bond between C_α and the anhydride oxygen, a fragmentation very common in anhydrides. The driving force suggested is the participation of the homoallylic double bond in the stabilization of the carbonium ion formed upon expulsion of the neutral molecule carbon monoxide from specie V. The resulting ion VI then undergoes hydrogen rearrangement to VII and loss of COOH to yield ion VIII of mass 79 which corresponds to a protonated benzene molecule.

This is an intriguing concept and deserves further investigations both concerning its correctness as well as its consequences. While a limited amount of isotope labeling was done, it had bearing only on the hydrogen migration in the later step (VI→VII) and not the homoallylic participation in the first. One could, for example, invoke the difference in geometry between a cyclohexane and cyclohexene ring as the reason for the apparent stability of the M–CO specie formed from III if the β-lactone (IX) would have a better chance to be formed from the unsaturated molecule (III) than from the saturated one IV (which loses C_2O_3, i.e., CO_2+CO directly). This would explain the experimental data just as well as the suggested homoallylic

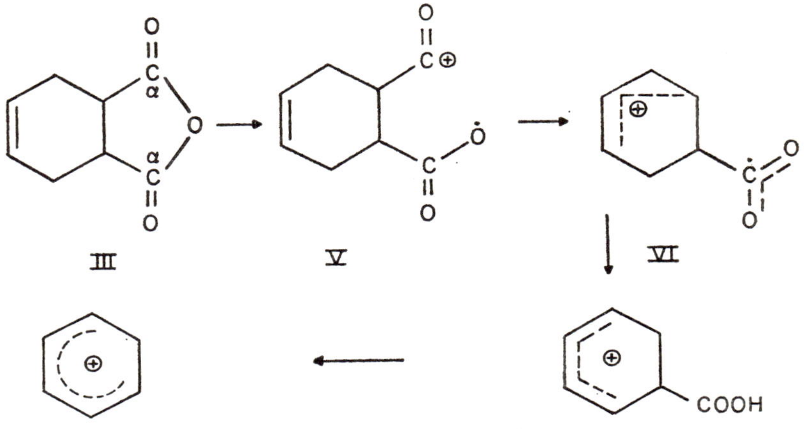

III V VI

VIII VII

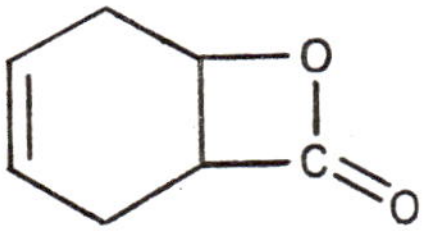

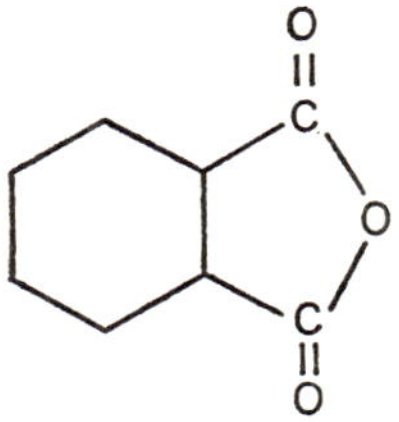

IX IV

participation. Careful and accurate appearance-potential measurements on the specie caused by loss of carbon monoxide would probably be the best means for distinguishing these two pathways. If such a participation is in fact involved, one should observe considerable variation in the magnitude of this effect upon change of ring size and steric relationship between the double bond and the carbon atom formerly bearing the initial positive charge.

A stabilizing effect distantly related to the one discussed above is borne out by various apparent anomalies in the mass spectra of compounds in which a functional group is attached to a reasonably long aliphatic chain. While unbranched hydrocarbons exhibit a rather uneventful mass spectrum in which peak height decreases smoothly from low to high mass, the incorporation of a substituent containing free electron pairs frequently gives rise to a certain irregularity of the abundance of fragments which are caused by cleavage of consecutive carbon-carbon bonds. For 1-chloro and 1-bromoalkanes (X) fragmentation of the bond between C–4 and C–5 is very much favored over the cleavage of any other carbon–carbon bond. This has been attributed by McLafferty (49) to cyclic chloronium or bromonium ions such as XI.

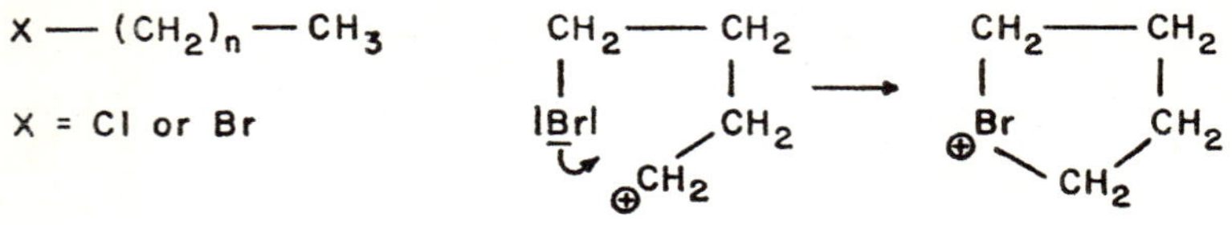

X XI

Similar phenomena are observed with nitriles (50) containing long unbranched chains except that fragments involving five and six carbon atoms of the chain can equally well be accommodated which might be due to the linear nature of the nitrile group. Fatty acids and their esters exhibit a characteristic periodicity with strong peaks for the mass of fragments containing 6, 10, 14, etc., carbon atoms in addition to the carboxyl or ester group (51). These phenomena indicate that the molecular ions produced in the ion source of the mass spectrometer have the tendency to "curl up" which may be nothing else than "internal solvation" of the center of the positive charge by the aliphatic chain for lack of any other solvating molecules. Such a phenomenon is further suggested by certain expulsion reactions (see below) in which a segment is lost from within the chain in a "*trans*-helical" reaction (51, 52).

A more vexing problem, from the mechanistic point of view, is the so-called rearrangement ions which are caused by a transfer of an atom or group taking place before or during fragmentation of the molecular ion or a fragment ion. It seems that in the vast majority of the cases it is a hydrogen that moves from one side in the ion to another, although recently the migration of phenyl groups have been proposed. Because of the large number of hydrogen atoms in any organic molecule, there is a multitude of reactions that could lead to a given fragment containing one hydrogen atom more than the number it would have had only simple bond cleavage occurred. In some of these instances, the origin of this hydrogen atom has been rigorously proven by deuterium labeling, as for example in the case of fatty esters (51, 53, 54) and alkyl benzenes (55). In those cases, the rearrangement involves the transfer of a hydrogen at the γ-position to the carbonyl oxygen or the aromatic ring by a mechanism that shall be discussed in detail later.

Hydrogen rearrangements seem to be particularly favored in those instances where cleavage of a carbon-carbon bond alone would lead to a poorly stabilized ion, while the simultaneous transfer of hydrogen leads to a much better ion which frequently corresponds to the molecular ion of a smaller molecule. For example, cleavage of the α, β-bond in an aliphatic ester (XI) and retention of the positive charge at the oxygen containing fragment (XII) would lead to an ion bearing a positive charge adjacent to a carbonyl carbon atom which, because of the polarization of this group, is already partly positive. On the other hand, if the γ-hydrogen rearranges to the carbonyl oxygen the resulting fragment (XIII) is the enol-form of an ester that has lost one electron, i.e., it corresponds to a molecular ion. Appearance potential measurements are in agreement with this proposal (56).

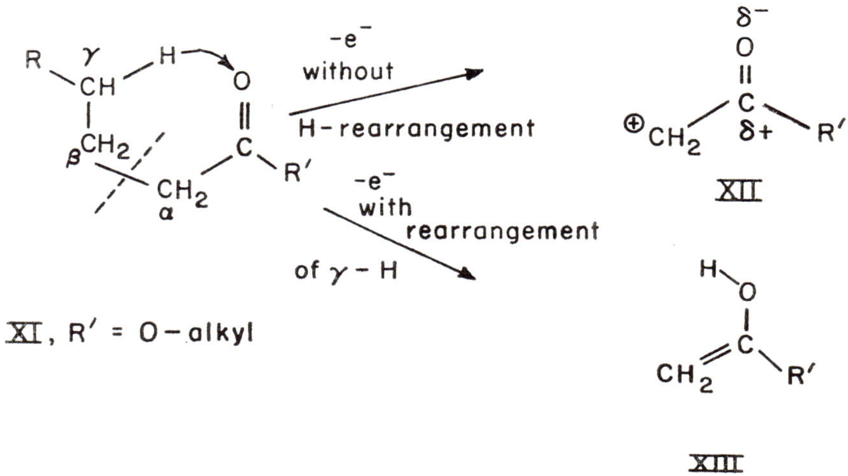

This type of rearrangement may be accompanied by the transfer of a second hydrogen to the oxygen-containing fragment, a frequently observed process in esters of alcohols containing two or more carbon atoms and involving the loss of the alcohol carbon chain. Its overall result is the following:

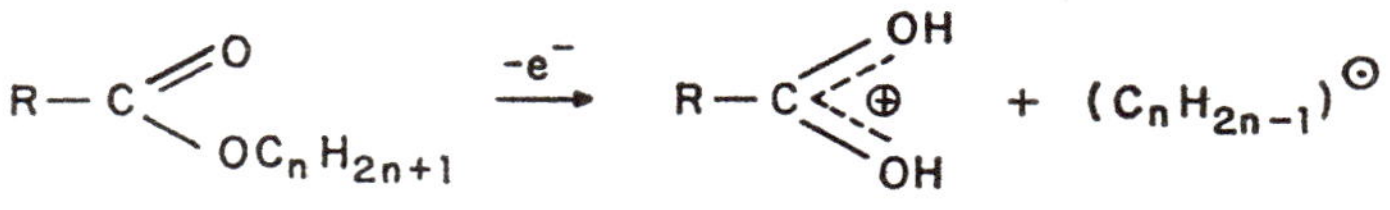

In contrast to the rearrangement discussed before where the identity of the migrating hydrogen atom has been established with considerable certainty, this reaction seems to be less clear. The driving force is obviously the formation of a protonated acid molecule which should represent a very stable positive ion. In an earlier attempt to gain insight into the mechanism, McLafferty & Hamming (57) investigated the mass spectra of deuterated sec-butyl acetates. It was concluded that C-1, C-3, and C-4 of the alcohol moiety together supply 80 per cent of the rearranged hydrogens of which C-3 supplies only 15 per cent. If the rearrangement were as clear-cut as the one discussed above involving the alkyl chain of the acid part, one would have expected that 20 per cent of the hydrogens in the first step come from C-3 and an additional amount in the second step. Whether this inconsistency is due to a prior rearrangement of the hydrogens in the alcohol moiety or to a strong isotope effect or to a lack of specificity of the reaction cannot be decided at present. The situation is further complicated by the recent finding by Harrison & Jones (58) that such "scrambling" seems to be involved in the formation of protonated propionic acid from *n*-propyl propionate while it is very specific in the case of isopropyl propionate where labeling experiments seem to indicate that it is one hydrogen of each of the two methyl groups that is transferred with the concomitant formation of an allyl radical.

Such double rearrangements of hydrogen may occur much more frequently than hitherto assumed and are often obscured by the fact that they may involve rearrangement of one hydrogen each in either direction involving no net gain of mass by the fragments produced. In the course of the very detailed work by Djerassi and his collaborators (6, 7, 8, 59) on the mass spectra of steroid ketones and related model compounds in which a wide variety of specifically deuterated analogs was employed, such double rearrangements had to be invoked frequently to explain the results. The mechanisms proposed are sometimes rather involved in order to account for the shifts in the mass spectra of the deuterated analogs, the only experimental data available. Careful and accurate appearance potential measurements as well as measurement of isotope effects should provide more conclusive evidence for the occurrence of such processes. Obviously this is more difficult in the case of molecules as complex as steroids, but should not present any great obstacles in the smaller bicyclic ketones.

While some of the hydrogen rearrangement reactions are in line with our intuitive expectations and even correspond to known chemical reactions, such as the pyrolytic elimination of an olefin from an ester or the photolysis of an alkyl ketone (60), others differ from the analogous chemical reaction. For example, the elimination of water from alcohols in solution leads always to the formation of an olefin in the course of a true 1,2-elimination, but this is not the case when water is lost from an alcohol on electron impact. Recently, Benz & Biemann (61) have shown that this elimination does not involve hydrogen from C-2, but to at least 90 per cent the hydrogen at C-4. This is shown by the fact that in a series of alcohols containing CD_2 in place of CH_2 at the various positions further and further removed from the hydroxyl group, it was 1-hexanol-4, 4-d_2 which lost almost exclusively HDO rather than H_2O. This apparent discrepancy between solution reaction and electron-impact reactions again illustrates the point made above that the presence or absence of solvent molecules may completely change the course of a reaction which superficially would seem to be the same. The chemical 1,2-elimination of water from alcohols is, of course, a bimolecular reaction involving participation of the solvent. For example, the protonation of the alcoholic hydroxyl group and the molecule of water formed cannot be considered as consisting of the hydroxyl group at C-1 and the hydrogen from C-2. In the gas phase, this type of reaction is, of course, impossible, and since the overall reaction is energetically favorable only if it involves the formation of an intact molecule of water rather than of a hydroxyl and a hydrogen radical, the hydrogen in question must come easily within bond distance of the oxygen. The hydrogen at C-4 which seems to be best suited for active participation in the transition state would involve a six-membered ring:

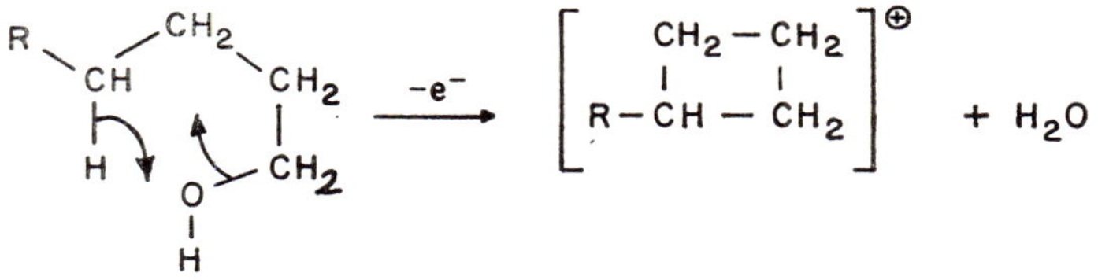

It would follow from this that the product ion is not an olefin as previously assumed but an alkyl cyclobutane. The data presently available and the lack of conclusive appearance-potential measurements do not permit such a definite conclusion.

A more specific case of these rearrangement phenomena are a variety of fragmentation processes generally called "expulsion reactions" in which part of the molecule is lost as a neutral molecule while the remainder forms a new bond. These are less startling in those cases where the two remaining parts were still connected by a bond, i.e., if the particle lost was part of a ring. A classical example is the elimination of carbon monoxide from anthraquinone which in two steps leads to biphenylene (62). But even in acyclic molecules, such expulsions with retention of the rest of the molecule is observed. There it must involve the formation of a new bond between the two remainders before or during cleavage of those bonds that previously connected them. A detailed mechanism has been proposed by Beynon (63) for the fragmentation of benzalacetophenone with elimination of CHO. More recently, Das, Funke & Bose (64) reported a number of instances of the reaction.

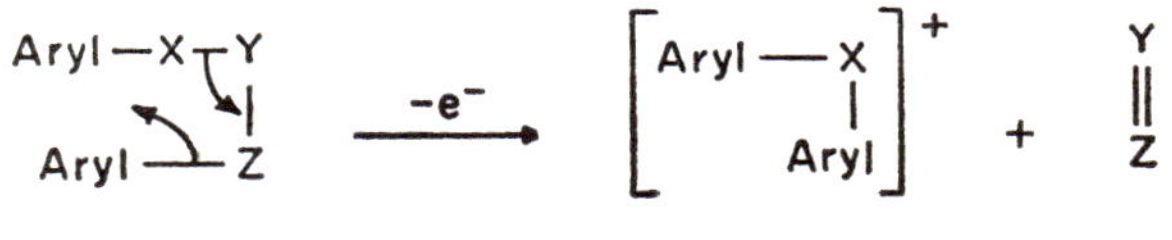

which seem to simply involve the migration of an aryl group, resulting in the loss of the Y=Z moiety.

While in these instances attraction of a positively charged aromatic ring by a neutral one (analogous to a π-complex) could possibly be invoked, an even more surprising expulsion reaction is one apparently occurring in aliphatic esters containing long alkyl groups. Stenhagen, Ryhage, and collaborators (51, 52) have shown by both substituent and isotope labeling that the first three carbon atoms following the ester group are eliminated in the course of the formation of a suspiciously intense ion nominally corresponding to the loss of C_3H_7. This had previously been assumed to be merely the loss of the last three carbon atoms of the chain, but deuterium introduced at the end of the chain, for example in the form of a CD_3-group was retained in the

M-43 peak, while the spectrum of methyl 4,4-dideuterioeicosanoate exhibited predominantly a peak caused by a loss of 45 and not 43 mass units. As mentioned earlier a "*trans*-helical" elimination was suggested to occur while the aliphatic chain is wound around the point of positive charge assumed to be at the carboxyl group.

It is apparent from the above discussions that the organic chemist has become used to interpreting the fragmentation of an organic molecule on electron impact in terms of mechanisms, and outlining them in a form employed when outlining chemical reactions, and to write structures for the various fragments similarly. This was done earlier in order to show that there is in fact some system and logic in such fragmentations which is the reason why different molecules exhibit different mass spectra. At the beginning of the use of mass spectrometry for the determination of the structure of organic molecules, it was realized that certain types of compounds have a tendency to fragment along certain pathways and that many of the most abundant fragments are clearly related to the structure of the original molecule. The fragmentation processes were written in considerable detail, and that part of the molecule that gave rise to a particular peak was drawn in terms of a more or less conventional structure deficient by one electron. This was done both for bookkeeping purposes and, also, to show that a certain peak in a mass spectrum was related to a particular part of the molecule because it increased in mass upon substitution of this area of the molecule by a functional group or a heavy isotope.

More recently, and with the availability of more and more mass spectrometers and, thus, more and more laboratories working in this field, this basic assumption seems sometimes to be overlooked and much too much is sometimes read into the fragmentation mechanisms drawn for complex molecules for the purpose of discussion of a mass spectrum. Papers appear more and more frequently in which the mass spectra of a few otherwise trivial compounds are being presented, followed by a discussion of a specifically delineated fragmentation mechanism leading to specific fragment ion structures without, however, containing any evidence and specific data aside from the fact that they look plausible or at least pretty. The fragmentation mechanisms themselves are frequently interpreted or proposed in a fashion that makes it appear that one really would know the minute details of these processes sometimes with a confidence which we do not even have in the much more fully investigated and more accessible area of solution chemistry. The situation is, of course, made very difficult because of the fact that the fragments escape direct investigation by other experimental techniques since they are formed at such a low level of concentration, and exist unaltered only until they strike the collector of the instrument. Nonetheless, recent proposals (65, 66, 67) have gone into extreme detail even to the physically indefensible point of telling which electron is removed in the ionization process and which electron moves in which fashion and in which direction during the process of fragmentation.

As an example, the previously mentioned transfer of hydrogen to the carbonyl oxygen of an ester group involving a six-membered transition state shall be discussed since it was the subject of a few recent controversial proposals. This mechanism had been proposed a few years ago by McLafferty (68), and in this original proposal arrows were drawn in a fashion (XIV) used to indicate movement of an electron pair in chemical reactions. In this case, it would also have meant the transfer of hydrogen with an intact electron pair to oxygen, but by no means was the statement made that this is actually believed to be the case. The same notation is used in Biemann's book (4) for a number of reasons, namely to distinguish the electron impact-induced reaction from chemical reactions passing through a similar transition state and because of the fact that the carbonyl group is most probably the bearer of the initial positive charge. It was, however, specifically stated (69) that such notations should by no means imply that it is, in fact, believed that the electrons follow in detail the path indicated by arrows.

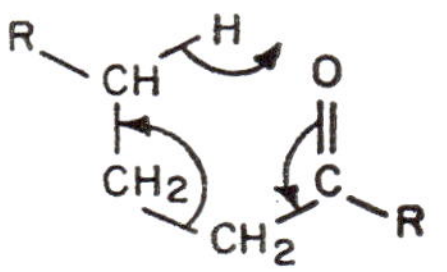

XIV

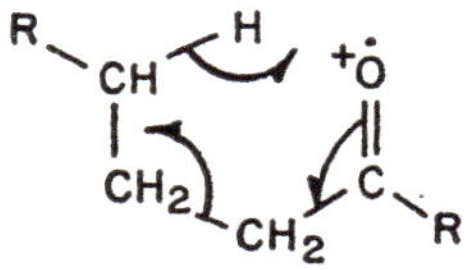

XV

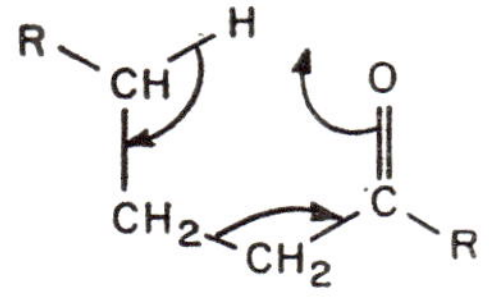

XVI

More recently, some much more detailed proposals have been made. Djerassi, for example, suggests (66) that the process actually involves the shift of a hydrogen radical, because a single electron is said to be missing at the carbonyl oxygen. In order to specifically show that this is thought to involve homolytic bond cleavage involving unpairing and re-pairing of electrons, a new symbol (fish hook) is proposed (XV). It is, thus, an instance in which the reader is lead to assume that one would really know not only which hydrogen moves (in this particular case, an experimentally proven fact), but also which electron moves.

In addition to this suggestion of a radical mechanism for the transfer of hydrogen, proton transfer XVI has been recently suggested by Spiteller & Spiteller-Friedmann (67), merely on the basis of its analogy to chemical reactions of this type. The argument involved a case in which it was believed that the positive charge is residing not at the carbonyl oxygen of the ester group involved in the fragmentation, but at a nitrogen atom located in another part of the molecule. This was specifically contrasted to the earlier proposals (68) which were interpreted as implying a hydride shift. It is this kind of misunderstanding of the degree of actuality implied in the drawing of fragmentation mechanisms as well as the unjustified distinction between two possible processes that may result in a misleading picture of the status of the actual knowledge concerning these phenomena. One has to keep in mind that even in chemical reactions it is very difficult to distinguish between hydride, hydrogen radical, and proton transfers even though many experimental techniques are available to test the system. A typical example is the chemical decarboxylation of β-keto acids which have been investigated in great detail by Swain et al. (70). To assert that the mere inspection of mass spectra would enable one to tell the difference between those three possibilities is completely unjustified. It is however, an intriguing problem worth being investigated. Conclusive results can be hoped for, however, only after a detailed investigation of the systems using all available techniques to first prove that it is exclusively this particular mechanism that is operative, which requires isotope labeling and appearance potential measurements, followed by the measurement of hydrogen isotope effects which would seem to be the only possible technique that would enable one to decide whether the rate-determining step involves the transfer of hydride, hydrogen, or proton.

For all these reasons an effort should be made to always clearly indicate the tenuousness of the implied fragmentation mechanisms when rationalizing a mass spectrum in terms of the structure of the compound. On the other hand, it would be very important to accumulate as much reliable data as possible to put fragmentation mechanisms on a firm basis. A number of techniques are available to achieve this, and these should be all employed, if possible, because each one furnishes a certain but different kind of information about the system: isotopic labeling, ionization and appearance-potential measurements, high resolution mass spectrometry, metastable peaks, and isotope effects.

Of these, specific labeling of the molecule with heavy isotopes, particularly deuterium, has been used most frequently to support or investigate fragmentation mechanisms. It is most conclusive an argument if specific and practically complete retention of the isotope is observed while an incomplete retention or "scrambling" of the isotopes should cause suspicion. Thus, clear-cut results such as those we mentioned above for the elimination of the side chain of fatty esters (51, 53, 54) or of water from alcohols (61) demonstrate at least which atoms are involved in the fragmentation while it still does not prove the detailed path it follows. In the case of the esters, the appearance potential measurements (56), in addition, favor the suggested proposals.

The vast majority of the labeling experiments with deuterium which have been reported often imply only that a particular region of the molecule is in fact involved. Because of the lesser tendency of carbon, nitrogen, and oxygen atoms to migrate, and the small number of the latter two generally occurring in a molecule the dangers inherent in isotope scrambling on electron impact is decreased and ^{15}N labeling has been used (71) in support of some fragmentation mechanisms proposed to occur in amino acids and their esters.

High resolution mass spectrometry (72) should always be employed when investigating a fragmentation mechanism to ascertain that the fragment in question in fact contains at least the elements it is proposed to have. With this technique, it is possible quite routinely to determine the mass of an ion with an accuracy (in the order of a millimass unit) which enables one to calculate the number and type of atoms constituting the particle. It is equally important to ascertain that the peak produced by the ion in question is in fact single specie and not a multiplet consisting of different ions of varying origin. If this were the case isotope-labeling experiments involving conventional mass spectra could be very misleading. High resolution mass spectrometry reduces to a certain extent, but by no means completely, the need for isotope labeling.

The importance of consideration of appearance potentials cannot be overemphasized because any mechanisms proposed which would require more energy than that corresponding to the appearance potential of the ion must be incorrect. The prime requirement is, however, an accurate and reliable determination of these and all other data required for the entire calculation, since methods and instrumentation used by various laboratories differ sufficiently to preclude the use of values obtained by others. Not infrequently crude values obtained with analytical instruments not specifically modified for the accurate determination of appearance potentials are being published and lead to confusion rather than to an enrichment of the literature.

Metastable peaks (73) are important since they relate the mass of a primary ion to that of a daughter ion formed in a one-step decomposition, a fact of great value in outlining the course of a fragmentation process.

It seems that no use has yet been made of isotope effects observed when either a light or a heavy isotope can take part in a fragmentation. In a few instances, isotope effects have been measured using molecules partially deuterated at a position known to be involved in a hydrogen rearrangement such as in the γ-position of butyric esters (54) and with butyramides (74) as well as with aromatic systems such as α'-monodeuterioethyl-benzyl ether (74).

Appearance potential measurements have been useful in many instances in the elucidation of fragmentation mechanisms such as the aforementioned formation of the tropylium ion from compounds containing a benzyl group or, for example, the more recent case of the specie formed on thermal decomposition of *para*-diiodobenzene which was suggested by Fisher & Lossing (75) to have an open structure rather than that of benzyne. High resolution spectra have also found use, for example, in some of the cases of expulsion reactions mentioned above as well as for the differentiation of fragments of the same nominal mass, for example, in the amino acid threonine where mass 75 can be and, in fact, is represented by two different parts of the molecule (76). In no case have all these techniques been exhaustively employed in the rigorous proof of the fragmentation mechanism; but with the availability of high resolution mass spectrometers in many laboratories, as well as the improved techniques for specific isotope labeling, it is expected that in the near future fragmentation mechanisms proposed for their own sake will be better supported by actual evidence than they have been in the past. None of the techniques by themselves can provide conclusive evidence even if the data were obtained under reliable conditions. For example, the fragmentation of a relatively simple molecule trimethylhydrazine was proposed (77) on the basis of the appearance potential of the various ions formed, but the conclusions were later shown to be incorrect (78) on the basis of the molecule's high resolution spectrum which indicated that the elemental composition of some of the fragments did not agree with those proposed.

Similarly, overestimation of the significance of the "structures" written for fragment ions should be avoided. For the sake of convenience a ground-state-like description of that part of the molecule believed to represent a given ion is generally used, and there is the danger that the reader might get the impression that such structures have been established with some degree of confidence. In most instances, however, this is not the case, and great care should be exercised to avoid such a misinterpretation by pointing out the tenuousness of such "structures" and the great difference in their meaning as compared with the conventional structures of intact molecules. Even the frequently cited argument (64) that because part of a mass spectrum resembles the mass spectrum of a certain compound, that, therefore, this fragment must have the structure of that compound, is completely unwarranted. Such a finding merely means that a fragment is formed that decomposes further to the same type of fragments to which the authentic compound also decomposes. But it could just as well be that the molecular ion of the authentic compound rearranges first to another structure which is, in fact, identical with the one formed from another larger molecule by fragmentation. Again, appearance-potential measurements should enable one to at least confirm or reject certain structures and this should always be done if a concrete structure is seriously proposed for a fragment.

LITERATURE CITED

1. Reese, R. M., and Harlee, F. N., *Anal. Chem.*, **36,** 278R (1964)
2. Beynon, J. H., *Mass Spectrometry and its Applications to Organic Chemistry* (Elsevier, Amsterdam, 640 pp., 1960)
3. Reed, R. I., *Ion Production by Electron Impact* (Academic Press, London, New York, 242 pp., 1962)
4. Biemann, K., *Mass Spectrometry* (McGraw, New York, 370 pp., 1962)
5. McLafferty, F. W., *Mass Spectrometry of Organic Ions* (Academic Press, London, New York, 730 pp., 1963)
6. Budzikiewicz, H., Djerassi, C., and Williams, D. H., *Interpretation of Mass Spectra of Organic Compounds* (Holden-Day, San Francisco, 271 pp., 1964)
7. Budzikiewicz, H., Djerassi, C., and Williams, D. H., *Structure Elucidation of Natural Products by Mass Spectrometry*, **I** (Holden-Day, San Francisco, 273 pp., 1964)
8. Budzikiewicz, H., Djerassi, C., and Williams, D. H., *ibid.*, **II,** 306 pp.
9. Rosenstock, H. M., Wallenstein, M. B., Wahrhaftig, A. L., and Eyring, H., *Proc. Natl. Acad. Sci. U. S.*, **38,** 667 (1952)
10. Friedman, L., Long, F. A., and Wolfsberg, *J. Chim. Phys.*, **27,** 613 (1957)
11. Chupka, W. A., *J. Chim. Phys.*, **30,** 191 (1959)
12. Steiner, B., Giese, C. F., and Inghram, M.G., *J. Chim. Phys.*, **34,** 189 (1961)
13. Chupka, W. A., and Kaminsky, M., *J. Chim. Phys.*, **35,** 1991 (1961)

14. Vestal, M., Wahrhaftig, A. L., and Johnston, W. H., *J. Chim. Phys.*, **37,** 1276 (1962)
15. Rosenstock, H. M., and Kraus, M., in *Advances in Mass Spectrometry*, **II,** 251–81 (Elliott, R. M., Ed., Pergamon, Oxford, 628 pp., 1963)
16. Rosenstock, H. M., and Kraus, M., in *Mass Spectrometry of Organic Ions*, Chap. 1, 2–64 (McLafferty, F. W., Ed., Academic Press, New York, London, 730 pp., 1963)
17. Wolfsberg, M., *J. Chim. Phys.*, **36,** 1072 (1962)
18. Marcus, R. A., *J. Chim. Phys.*, **20,** 352, 355, 359 (1952)
19. Wieder, G. M., and Marcus, R. A., *J. Chim. Phys.*, **37,** 1835 (1962)
20. Monahan, J. E., and Stanton, H. E., *J. Chim. Phys.*, **37,** 2654 (1962)
21. Dibeler, V. H., and Rosenstock, H. M., *J. Chim. Phys.*, **39,** 1326 (1963)
22. Vestal, M. L., *J. Chim. Phys.*, **41,** 3997 (1964)
23. Rosenstock, H. M., Dibeler, V. H., and Harlee, F. N., *J. Chim. Phys.*, **40,** 591 (1964)
24. Murad, E., and Inghram, M. G., *J. Chim. Phys.*, **40,** 3263 (1964)
25. Klots, C. E., *J. Chim. Phys.*, **41,** 117 (1965)
26. Taubert, R., *Z. Naturforsch.*, **19**(a), 484, 494 (1964)
27. Sharp, T. E., and Rosenstock, H. M., *J. Chim. Phys.*, **41,** 3453 (1964)
28. Dibeler, V. H., and Reese, R. M., *J. Chim. Phys.*, **40,** 2034 (1964)
29. Rylander, P. N., Meyerson, S., and Grubb, H. M., *J. Am. Chem. Soc.* **79,** 842 (1957)
30. Wiberg, K. B., Bartley, W. J., and Lossing, F. P., *J. Am. Chem. Soc.*, **84,** 3980 (1964)
31 Collin, J., and Lossing, F. P., *J. Am. Chem. Soc.*, **79,** 5848 (1957)
32. Collin, J., and Lossing, F. P., *ibid.*, **80,** 1568 (1958)
33. von Koch, H., and Lindholm, E., *Arkiv. Fisik*, **19,** 123 (1961)
34. Petterson, E., and Lindholm, E., *Arkiv. Fisik*, **24,** 49 (1963)
35. Inghram, M. G., and Gomer, R., *Z. Naturforsch.*, **10**(a), 863 (1955)
36. Beckey, H. D., *Z. Naturforsch.*, **14**(a), 712 (1959)
37. Beckey, H. D., *ibid.*, **17**(a), 1103 (1962)
38. Beckey, H. D., *ibid.*, **19**(a), 71 (1964)
39. Robertson, A. J. B., Viney, B. W., in *Advances in Mass Spectrometry*, **III** (In press)
40. Beckey, H. D., Knöppel, H., Metzinger, G., and Schulze, P., in *Advances in Mass Spectrometry*, **III** (In press)
41. Fox, R. E., Hickam, W. M., Grove, D. J., and Kjeldaas, T., *Rev. Sci. Instr.*, **84,** 559 (1951)
42. Lossing, F. P., and Tanaka, I., *J. Chim. Phys.*, **25,** 1031 (1956)
43. Hürzeler, H., Inghram, M. G., and Morrison, J. D., *J. Chim. Phys.*, **28,** 76 (1958)
44. Cummings, C. S., and Bleakney, W., *Phys. Rev.*, **58,** 787 (1940)
45. Stevenson, D. P., *Discussions Faraday Soc.*, **10,** 35 (1951)
46. Nozoe, T., in *Non-Benzoid Aromatic Compounds*, Chap, VII, 339–463 (Ginsburg, D., Ed., Interscience, New York, 543 pp., 1959)
47. McLafferty, F. W., *Anal. Chem.*, **31,** 477 (1959)
48. Weininger, S. J., Mai, V. T., and Thornton, E. R., *J. Am. Chem. Soc.*, **86,** 3737 (1964)
49. McLafferty, F. W., *Anal. Chem.*, **34,** 2 (1962)
50. McLafferty, F. W., *Anal. Chem.*, **34,** 26 (1963)
51. Dinh-Nguyen, Ng., Ryhage, R., Ställberg-Stenhagen, S., and Stenhagen, E., *Arkiv. Kemi*, **18,** 393 (1961); Ryhage, R., and Stenhagen, E., *Arkiv. Kemi*, **15,** 291 (1959)
52. For a comprehensive review see Stenhagen, E., and Ryhage, R., in *Mass Spectrometry of Organic Ions*, 399–452 (Academic Press, London, New York, 730 pp., 1963)
53. Biemann, K., *Mass Spectrometry*, 120 (McGraw, New York, 370 pp., 1962)
54. Williams, D. H., Budzikiewicz, H., and Djerassi, C., *J. Am. Chem. Soc.*, **86,** 284 (1964)
55. McCollum, J. D., and Meyerson, S., *J. Am. Chem. Soc.*, **81,** 4116 (1959)
56. Sharkey, A. G., and Hickam, W. M., *ASTM E-14, Conf. Mass Spectrometry, New York, May 1957*
57. McLafferty, F. W., and Hamming, M. C., *Chem. Ind. (London)*, **1958,** 1366.
58. Harrison, A. G., and Jones, E. G., *Can. J. Chem.* (In press)
59. See Karliner, J., Budzikiewicz, H., and Djerassi, C., *J. Am. Chem. Soc.*, **87,** 580 (1965) and earlier papers in this series.
60. Srinivasan, R., *J. Am. Chem. Soc.*, **81,** 5061 (1959)
61. Benz, H., and Biemann, K., *J. Am. Chem. Soc.*, **86,** 2375 (1964)
62. Beynon, J. H., Lester, G. R., and Wil-

liams, A. E., *J. Phys. Chim.*, **63,** 1861 (1959)
63. Beynon, J. H., *Mass Spectrometry and its Applications to Organic Chemistry*, 270 (Elsevier, Amsterdam, 640 pp., 1960)
64. Das, K. G., Funke, P. T., and Bose, A. K., *J. Am. Chem. Soc.*, **86,** 3729 (1964)
65. Shannon, J. S., *Tetrahedron Letters*, 801 (1963)
66. Budzikiewicz, H., Djerassi, C., and Williams, D. H., *Structure Elucidation of Natural Products by Mass Spectrometry*, **I,** 2 (Holden-Day, San Francisco, 273 pp., 1964)
67. Spiteller, G., and Spiteller-Friedmann, M., *Monatsh. Chem.*, **95,** 257 (1964)
68. Cummings, C. S., and Bleakney, W., *Phys. Rev.*, **58,** 787 (1940); McLafferty, F. W., *Anal. Chem.*, **31,** 82 (1959)
69. Biemann, K., *Mass Spectrometry*, 71 (McGraw, New York, 370 pp., 1962)
70. Swain, C. G., Bader, R. F. W., Esteve, R. M., Jr., and Griffin, R. N., *J. Am. Chem. Soc.*, **83,** 1945 (1961)
71. Biemann, K., Seibl, J., and Gapp, F., *J. Am. Chem. Soc.*, **83,** 3795 (1961)
72. See for example Beynon, J. H., in *Advances in Mass Spectrometry*, **I,** 328–54 (Waldron, J. D., Ed., Pergamon, Oxford, 704, 1959)
73. Hipple, J. A., and Condon, E. U., *Phys. Rev.*, **68,** 54 (1945) (See also discussions in ref. 2–6)
74. Smith, D., *Rearrangements in the Mass Spectrometer involving a Hydrogen Shift to a multiple Bond* (B. S. thesis, M.I.T., June, 1964)
75. Fisher, I. P., and Lossing, F. P., *J. Am. Chem. Soc.*, **85,** 1018 (1964)
76. Biemann, K., *Symp. Mass Spectrometry, 11th, San Francisco, May, 1963*
77. Dibeler, V. H., Franklin, J. L., and Reese, R. M., *J. Am. Chem. Soc.*, **81,** 68 (1959)
78. Beynon, J. H., Saunders, R. A., and Williams, A. E., *J. Am. Chem. Soc.*, **82,** 288 (1960)

CHEMICAL ASPECTS OF GENETICS[1]

By John R. Platt

Department of Biophysics, University of Chicago, Chicago, Illinois

The chemistry of genetics is primarily the chemistry and structure of the hereditary nucleic acid chains, DNA and RNA, and of the proteins whose structure they in turn control, and of the mechanism of this control. The problem of review is hard because this is the most information-rich and rapidly moving field of science today, and is already the most-reviewed and best-reviewed. Accounts of new developments and summaries appear regularly on the front pages of the *New York Times* and in the *Scientific American*, sometimes in advance of Journal publication, and anyone interested in the field will find it easy to become well read at this level. In 1964 also, three masterly reviews appeared which it would be presumptuous to amend or repeat: a 28-page article on "The New Genetics" by Beadle (1), a 20-page article on "Protein Synthesis" by Watson (2), and a 20-page article on "The Biochemistry of Genetics" by Crick (3). There are numerous up-to-date texts constantly appearing; an up-to-date background book by Asimov (4) is available in paperback; a particularly rich symposium collection on "Synthesis and Structure of Macromolecules" (5), and two fine collections of articles on nucleic acid research [Eds., Davidson & Cohn (6)] and on the composition, structure and function of proteins [Ed., Neurath (7)] have appeared within the year; and there is a lucid new monograph on the molecular biology of the viruses [Stent (8)].

The present article will therefore attempt only to list some of the recent high spots and controversies, and to give at the end some thoughts on the central chemical problems that must now be solved—the problem of the relation of protein structure to function, and the problem of the apparently general appearance of various kinds of "complementarity" in molecules with high-information sequences. It appears that concrete solutions to these problems could give us ways of predicting and creating new compounds, and new structures and catalysts, and might have an importance to chemistry perhaps comparable to the importance of valence theory a hundred years ago.

General Framework

In the chemistry of genetics the "central dogma" is that "DNA makes RNA makes protein." That is, the hereditary information that is passed from one cell or virus to its progeny or "daughters" is normally contained in the form of long chains of DNA, or deoxyribonucleic acid. (In some viruses the chains are made of RNA, or ribonucleic acid.) The chains have polymeric sugar-phosphate backbones, with specific sequences of nucleic acid bases attached to the ribose groups. As Figure 1 indicates, they can be copied, or "replicated", into daughter chains of the same kind. Short sections can also

[1] The survey of literature pertaining to this review was concluded January 1965.

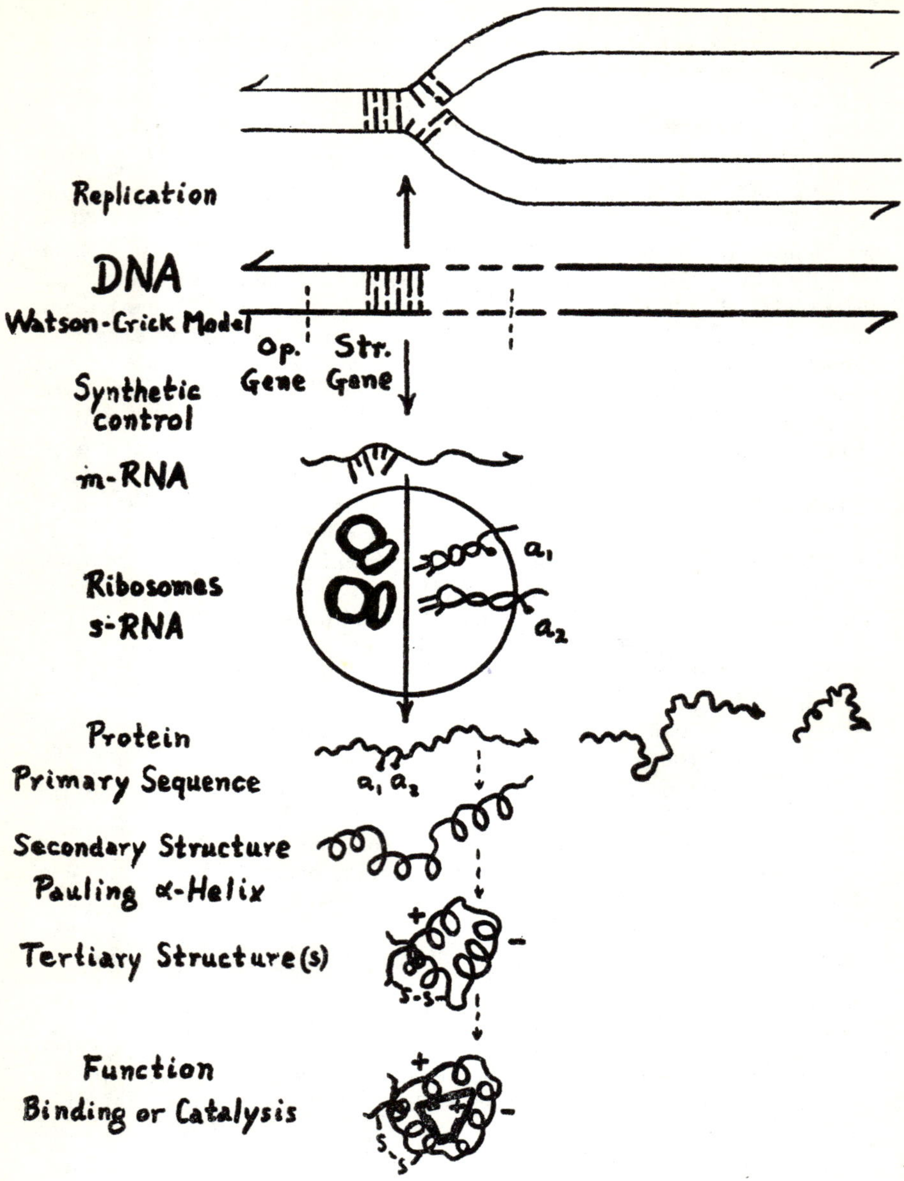

FIG. 1. Information transfer and structural transfer as DNA replicates or makes messenger-RNA which controls protein sequence, which in turn determines enzyme structure(s) and functions. Double helices rendered parallel for simplicity.

be copied, or "transcribed," into "messenger-RNA," or "mRNA," chains; these sections are commonly identified with the hereditary "genes" of classical genetics. Finally, the mRNA chains can be copied or, better, "translated," into the amino acid sequences of the polypeptide chains of enzymatic or other cellular proteins that do the chemical work of the cell. (The translation involves shorter chains of "transfer-RNA" or "soluble-RNA," or "sRNA," and a fourth form of RNA combined with protein in the ribosomes; but these forms of RNA, although copied from the DNA, do not seem to be copied themselves.) The copying or "reading" of the DNA or RNA chains to make other chains or to make protein is carried out by special enzymes for each kind of step, and is in many ways like reading an instruction book and carrying out the instructions, as Platt (9) pointed out in detail earlier.

There are four standard nucleic acid bases in DNA: two pyrimidines, thymine (T) and cytosine (C); and two purines, adenine (A) and guanine (G). In RNA, thymine (5-methyl-uracil) is replaced by uracil (U). The structure of DNA, determined by the classic work of Watson and Crick in 1953 using Wilkins' X-ray data, is a double-stranded helix in which the two strands are supposed to "point" in opposite directions; that is, the sequence . . . sugar(base)-3′-phosphate-5′-sugar(base)-3′-phosphate . . . has a polarity, and points "up" the helix on one strand and "down" it on the other. There are two slightly different geometries, "A" and "B," depending on the water content; but in both, the bases are on the "inside," approximately perpendicular to the helix axis, close-stacked with a spacing of about 3.4 Å, and with the pyrimidines (or purines) on one chain bonded across by hydrogen bonds to "complementary" purines (or pyrimidines) on the other. Thus T (or U) is bonded or "paired" to A, with two hyrdogen bonds; and C to G, with three hydrogen bonds. (Other bases occur in trace amounts in sRNA, but their role and geometry are not known.) There is single-stranded DNA in the bacteriophages ØX174 and S13, and single-stranded RNA in some viruses, but in both cases the chains make double-stranded "replicative forms" before replication. In mRNA, however, it appears that for translation the strand must remain single, with no "self-twisting" or "secondary structure."

It is helpful to have an idea of physical sizes. In the sRNA chains, there are about 80 bases; in mRNA chains, of the order of 1000, with wide variations depending on the size of the protein to be made. In the small DNA chain of ØX174, there are about 5000 bases; in the DNA chain of a T2 bacterial virus, about 250,000 base pairs. The molecular weight is about 600 daltons per base pair, so the molecular weight of the T2 DNA molecule is about 150 million. With 3.4 Å separation between the bases in the double helix, the length of the T2 DNA chain is about 80 μ. The length of the DNA in the bacterium *E. coli* is about 1000 μ, and the total length of the DNA chains in the chromosomes of a human cell is about 2 m, or the height of a man himself.

Replication (or repair) of DNA chains requires DNA-polymerase enzymes; transcription into RNA chains requires RNA polymerases. Presumably the enzymes force a complementary double-helix configuration between the parent and the growing daughter strands as each new base is added. The strands of DNA are thermally dissociated or "denatured" at temperatures between 55° and 80° C, chains with higher G-C content (more hydrogen bonds) "melting" at the higher temperatures. If the strands have not been broken chemically or mechanically, they can be "renatured" by slow cooling or annealing, reforming the double helix with full biological activity. sRNA has much base-pairing and is supposed to form a self-helix or "hairpin," but other single-stranded DNA's and RNA's are supposed to be simple "random-coil" polymers.

It is now known that in many organisms the DNA or RNA double helix bends around to form a large closed loop for at least part of the life cycle. The chains have special resistance to enzymatic attack, and special replication properties, in this loop form.

The translation of mRNA into protein occurs on ribosomal particles. The ribosome can be thought of as attached to the mRNA strand and moving along it, assembling successive amino acids into the designated polypeptide chain. The ribosomes are about 200 Å in diameter and four or five of them may be moving simultaneously along a 1500-Å strand of mRNA, forming a "polyribosome" of the kind seen in electron microscopes. Each ribosome has two parts, a small piece and a large one, "30s" and "50s" in the ultracentrifuge. Both contain RNA and protein, and both have attachment sites for different components of the process, so the ribosome is like a multifunctional enzyme or "assembly table" for the operation.

Chemically, the process requires: (*a*) attachment of each amino acid to specific sRNA's. This is done with an "activating enzyme." (*b*) attachment of the "activated" sRNA to the ribosome-messenger system when the messenger base-sequence calls for it; (*c*) linking of this sRNA amino acid onto the growing polypeptide chain at the carboxyl end, with the chain remaining attached to the sRNA through an ester-ribose linkage; (*d*) attachment of the next activated sRNA called for by the messenger; and (*e*) release of the previous sRNA, now "deactivated", from the peptide chain and from the ribosome as the new peptide bond is formed.

In this process, the amino acid sequence is determined by the messenger in the following way. There are 20 different amino acids in ordinary proteins, whose familiar abbreviations are Ala, Arg, Asp, Asn, Cys, Glu, Gln, Gly, His, Ilu, Leu, Lys, Met, Phe, Pro, Ser, Thr, Try, Tyr, and Val. Somehow, each can be attached by the activating enzymes only to its own types of sRNA. Each sRNA is then supposed to have on it a special sequence of three bases, called the "anticodon" for that amino acid, which is complementary to a messenger sequence of three bases called the "codon" for that amino acid. A complete messenger sequence of bases, if it makes "meaningful" codons, thus specifies a sequence of sRNA's and thereby the amino acid sequence of the assembled polypeptide.

The ribosomes and sRNA's and activating enzymes of one species can make proteins "correctly" from the mRNA of another, or from synthetic RNA chains. Thus, the code must be essentially universal on this planet, with the same codons designating the same amino acids in viruses, bacteria, yeasts, and the higher organisms. (Is this evolutionary accident, or the result of steric relations? The steric relations, if any, are not obvious.) That synthetic polyU chains —U–U–U–U–U–U— with *E. coli* ribosomal extracts would make polyphenylalanine—Phe–Phe–Phe—was the dramatic discovery of Nirenberg and Matthaei in 1961. It is now well established that each codon consists of three bases, probably adjacent, so this means the codon for Phe is UUU. The "dictionary" of codons for the other amino acids has been unexpectedly hard to work out, but it may be completed in the near future.

Our present picture of the control mechanism in the cell, which determines whether a particular mRNA and its protein will be synthesized or not, was proposed in 1961 by Jacob & Monod (10). They showed from genetic and mutational evidence that each enzyme must have a "regulator gene" that makes a "repressor substance." This substance can complex with certain cellular constituents called "corepressors" or "inducers." It then interacts, or fails to interact, with a particular "operator-gene" section on the DNA chain so as to "repress" or "derepress" (that is "induce") the manufacture of messenger-RNA from neighboring "structural genes" (Fig. 1).

An operator gene and the structural, or enzyme-making, genes that it controls make up together a single information unit called an "operon." They are all "turned on" or "turned off" together by the repressor mechanism, as though they formed a double-page of a book of information that could be either open or closed at that page. The enzymes in a single operon are frequently those involved in successive steps of the synthesis of the same product (e.g., histidine) or the degradation of the same metabolite (e.g., glucose). In these cases, the product (or the metabolite) itself may be the corepressor (or the inducer) so as to give a negative (or positive) feedback control of the turning off (or on) of the whole enzyme system.

New Results

DNA and RNA structure and copying.—The base-pairing that binds the DNA chains together is now being studied in small molecules. Single crystals of the G-C bases show the hydrogen bonding arrangement originally proposed by Watson and Crick. But it now appears that single crystals of the A-T and related bases do not have this arrangement, but have the A part "flipped over" into a "Hoogsteen structure" with bonding through the 5-ring rather than the 6-membered ring [Haschemeyer & Sobell (11); Mathews & Rich (12)]. This structure seems unfavorable for DNA because it would make the backbones irregular, but it might occur at special places [Crick (3); Miles (13)].

Circular loops of DNA have now been found in viruses and bacteria by electron microscopy [MacHattie & Thomas (14); Weil, Vinograd & Stoeckenius (15)] and in the double-stranded "replicative form" of the single-stranded ØX174 [Kleinschmidt, Burton & Sinsheimer (16); Chandler et al. (17)].

In the replication of DNA, Cairns (18) has shown by autoradiography that the two strands separate, forming a "Y," with simultaneous deposition of daughter strands along each of the old strands at the growing point (again as proposed by Watson and Crick). But the strands point oppositely, so this implies that the polymerase enzyme can add a base to the 3′-OH end of one daughter and to the 5′-PPP group of the other. Only the first of these reactions is known *in vitro*.

Cairns (19) also shows that the loop of bacterial DNA somehow remains unbroken during replication. The strands separate over some region, with a Y-point at each end. One of these points must be a "swivel" to permit the untwisting at the other, but no swivel-like narrowings of the DNA chain or "linkers" between sections are visible in nonreplicating loops (14).

The mechanisms for the regulation of DNA replication in bacteria have been analyzed in an important paper by Jacob, Brenner & Cuzin (20). They emphasize the probable attachment of the replicating system to the cell wall, and the possible effect of this on genetic transfer during bacterial mating.

Double-stranded RNA can be identified because it resists degradation by pancreatic RNase better than the single-stranded form, as was shown by Geiduschek, Moohr & Weiss (21). This form has now been found, presumably as the replicative form, in cells infected by a number of different viruses, including tobacco mosaic virus (TMV) [Shipp & Haselkorn (22)].

In 1963, several groups of workers showed that only one of the two complementary strands in DNA serves as a template for the transcription of mRNA messages *in vivo*, even though both strands acted as templates *in vitro*. Hayashi, Hayashi & Spiegelman (23) have now shown that this "asymmetric synthesis," with selection of one strand for transcription, is also obtained *in vitro* if the DNA is not fragmented but is kept in the circular double-stranded replicative form (of ØX174). They used their elegant technique of specific binding to a chromatographic column containing the complementary chain [reviewed by Spiegelman (24)]. Geiduschek, Tocchini-Valentini & Sarnat (25) have shown that phage-α will also give asymmetric RNA synthesis *in vitro* if its native conformation is preserved.

Autoradiography shows that RNA synthesis only occurs in the "puffed" sections of the chromosomes [Beerman & Clever (26)]. Electron-microscope autoradiography [Caro & van Tubergen (27)] of calf thymus cells likewise shows mRNA synthesis only in the more diffuse regions of the chromatin [Littau et al. (28)]. Chemical tests of the repressors of RNA synthesis in *E. coli* show a protein component [Gallant & Stapleton (29)], but there is a histone component in higher organisms [Allfrey, Faulkner & Mirsky (30)]. A hypothetical alternative to the Jacob-Monod control mechanism has been put forward by Stent (30a).

RNA-ribosome-protein relations.—Polyribosomes were first seen and identified functionally in 1963 by several groups of workers almost simultaneously [lists given by Burka & Marks (31) and by Eiserling et al. (32); general review by Rich (33)]. They have now been produced *in vitro* in an amino acid incorporating system prepared from *E. coli* extracts (32). Warner & Rich (34) have demonstrated that there is a single-growing polypeptide chain per ribosome. Meselson and co-workers (35) have shown that the nucleic acid parts of the ribosomes are not exchanged or replaced during protein synthesis.

The growing point of the polypeptide chain appears to be on the 50s subunit of the ribosome, along with the site that binds the sRNA [Gilbert (36); Cannon, Krug & Gilbert (37)]. But streptomycin experiments (to be discussed later) indicate that the point of contact between the sRNA and the mRNA is on the 30s subunit [Davies, Gilbert & Gorini (38)]. There may be two sites for binding sRNA's (for consecutive amino acids?) [Warner & Rich (39)]; and some say three [Wettstein & Noll (40); see also (3)].

In sRNA, it has been found that bromination and methylation affect the amino acid acceptor function and the amino acid transfer function differently, so they may involve different sites [Yu & Zamecnik (41); Weil et al. (42)]. Progress has been made in determining the base sequences in sRNA, partly by RNase enzymes that cleave the chain next to particular bases. Different sRNA's that accept different amino acids have now been shown to have different base sequences next to the terminal pCpCpA group at the acceptor end [Herbert, Smith & Wilson (43)], even though the sRNA structural model and the evidence assembled by Felsenfeld & Cantoni (44) puts the specific "anticodon" on the "hairpin loop" rather than at the terminal end (see Fig. 1). Perhaps the anticodon is mislocated; or perhaps the different sRNA base sequences are different for "accidental" reasons irrelevant to the anticodon sequence. [Added in proof: The first complete sRNA sequence has just been determined by Holley et al. (44a).]

Mutations and mechanisms.—The known mutational agents or mutagens are numerous. They include, among others, ultraviolet (UV) and particle radiation, the nitrogen mustards, peroxides, epoxides, hydroxylamine (HA), nitrous acid (NA), ethylmethanesulfonate (EMS), base-analogues like 2-aminopurine (AP) and fluorouracil (FU) and bromuracil (BU), and proflavine (PF) and other acridine dyes (1). The expected reactions of the chemical mutagens with the nucleic acid bases have been fairly well analyzed [Freese (45)]. For example, NA should deaminate the bases, simply converting C to U and A to G (through a hypoxanthine step). Since NA treatment of DNA produces a limited set of amino acid replacements in the proteins synthesized—for example, Pro only goes to Ser or Leu—this should then tell us which codons have more C or A and which have more U or G, and should give in fact a useful set of specific restrictions on the codon relations [Wittmann & Wittmann-Liebold (46); Yanofsky (47); and see Fraenkel-Conrat (48) for a general description]. But there have been internal contradictions, which raised the question whether some or all of the base changes were not caused by secondary reactions.

This question has been answered, and the base changes in mutations directly identified, in a beautiful series of experiments by Tessman, Poddar & Kumar (49) and Howard & Tessman (50). They eliminated many confusing effects by working *in vitro* with the single-strand DNA of the phages S13 and ØX174. They were able to check the identity of their mutations by producing and isolating back-mutations at the same site with other chemicals. In one case, they went back and forth serially for 12 mutation steps to make sure that no more than one site was involved.

They were able to classify their forward and back mutations into exclusive groups, which they could then identify with particular base changes from the reagents involved. The base changes they found were all "transitions," or replacements of pyrimidines by pyrimidines, and purines by purines, as Watson and Crick had originally proposed. But they found that NA and EMS mutate all four bases to varying degrees. Other mutagens were more selective, with BU tending to induce T→C and A→G, and with HA only mutating C→T. This work finally seems to give us reliable chemical tests for identifying base changes, and therefore the bases themselves, at particular mutation sites; and it may help clear up some of the codon contradictions, although the situation *in vivo* may be more complex [Tessman et al. (50a)].

A remarkable application of these methods is Howard & Tessman's proof (50), by reverse mutations, of the high specificity of UV-induced mutations. UV was shown to induce the change C→T in 11 out of 16 mutations; the remaining five were ambiguous, but were nontransitions, possibly deletions. Such a result was indicated earlier by the chemical evidence on UV effects [Dellweg & Wacker (51)], and by the two-strand UV results and photochemical inferences of Drake (52); but Howard & Tessman's work shows much more conclusively that the primary UV target is cytosine.

Proflavine or acridine dye mutations are different. These three-ring systems are about the same size as a base pair. Lerman (53) has given evidence from fluorescence polarization, changes of viscosity, and amino-group reactivity, that such dyes may intercalate between the base pairs of a DNA double helix (with some unwinding), although other authors favor external binding. In any case, they produce deletions and additions of bases (which intercalation during replication would seem to favor); and it was by successive mutations of this sort—shifting the codon sequence one, two, or three bases to the right or left, and producing functional protein again with a three-base shift—that Crick and co-workers (54) proved the triplet character of the code.

The nitrogen mustards may also produce mutants by base-deletion; and the dye acridine-orange on DNA, when irradiated with visible light, may produce photo-mutants of the base-deletion type, even without any other detectable chain-breaking, cross-linkage, or base-dimer formation [Freifelder (55)].

The nitrogen mustards and nitrous acid can also produce cross-links between DNA strands so that they resist thermal denaturation or strand separation. The cross-links are supposed to be covalent links between two guanine residues and with NA treatment, they are relatively frequent, with one cross-link per four deaminations [Becker, Zimmerman & Geiduschek (56)].

Irradiation of DNA by ultraviolet light or ionizing radiation in the fiber form at dry ice temperatures produces free radicals. Pershan et al. (57) have elegantly demonstrated by electron spin resonance methods, with an analysis of deuteration and fine-structure interactions, that these radicals are hydrogen-addition products of thymine, with the hydrogen added at the C_6 position, and the unpaired spin located at the C_5 position. In the Tessman results, cytosine was proved to be the primary UV target; perhaps this thymine radical formation is a parallel reaction, or perhaps it is connected with the deamination of an adjacent cytosine (C→U).

A major result in 1964 was the discovery of the enzymatic "repair" of thymine dimers in DNA. These intrastrand dimers apparently account for a large fraction of the gross ultraviolet damage. The thymines are covalently linked through their 5,6 bonds (perhaps following the production of the free radical) to form a cyclobutane ring which blocks DNA replication. However, it now appears that certain enzymes can excise the dimers from the strand (so that they appear in the medium), with the strand DNA then being reconstructed by a "repair enzyme" from the information on the complementary strand [Setlow & Carrier (58); Boyce & Howard-Flanders (59)]. Several other enzymatic processes, such as photoreactivation, are already known to assist recovery from UV damage [Harm (60)].

Schildkraut, Richardson & Kornberg (61) and Richardson, Inman & Kornberg (62) have made fundamental studies on the mechanisms of enzymatic synthesis and repair. They found that enzymatic replication of native DNA can lead to strangely branched and nondenaturable structures (oddly like the "transfer-twist" branches predicted earlier) [Platt (63)], presumably because of the growth of complementary strands around loose ends. DNA, with one end of each strand chewed off by an exonuclease, can be repaired to make a normal, unbranched double helix. It is not clear whether there is one enzyme, or two, for replication and repair (3), but repairing may be a major mechanism for preservation of DNA.

A quantum theoretical proposal about mutation mechanisms and their "application to mutation, tumors and aging" has been put forward by the distinguished theoretical chemist, Löwdin (64). He proposes that mutations are caused by keto-enol transformations of a base during replication (as Watson and Crick said)—but by the quantum-mechanical "tunneling" of protons "under" the activation barrier, rather than by the usual mechanism of thermal excitation over the barrier. Löwdin concludes, "It hence seems likely that the time-dependent proton tunneling may be the primary cause of the phenomenon of aging . . . the phenomenon may also be responsible for the occurrence of spontaneous tumors and cancer."

These are large claims. The idea of quantum-mechanical tunneling of a proton from one atom to the other in a hydrogen bond has a long history [Bernal & Fowler (65); Conway, Bockris & Linton (66); Eigen & de Maeyer (67); Haas & Hornig (68); Eigen (69)], but its relative importance, compared to thermal activation, is still uncertain. Eigen (69), discussing keto-enol rates and enzymatic hydrolysis, speaks of a "non-classical" mechanism, but he seems to mean cooperative charge movements rather than tunneling.

But this is not the real issue. Proton tunneling, in favorable cases, certainly exists [Haas & Hornig (68)], and might speed up the thermal keto-enol exchanges to some degree. The trouble is that it is not clear, in the Löwdin claims, why tunneling should change the equilibrium Boltzmann distribution between the keto and enol forms, which is what should determine how often a base error of this type is made. Even if tunneling did change the distribution, the question would still remain whether base mutations are any more important for tumors and aging than deletions, chromosomal aberrations, viral invasion, and the inflexibility of control mechanisms. In short, it is not clear, on several counts, what operational difference this special interpretation of keto-enol transformation rates will make to biology. However, this theory is now being used to interpret the reported effects of magnetic fields on tumors and aging in mice [Barnothy (70)]. Rein & Harris compute that tunneling may be important in excited states if not in ground states (70a).

Parallel of DNA sequence and protein sequence.—The proof of the colinearity of gene structure and protein structure, which has long been assumed, was one of the greatest triumphs of the year (3). Yanofsky and co-workers (71) showed that nine genetic mutants in the A protein of tryptophan synthetase of *E. coli*, a protein about 60 amino acids long, lay in the same sequence on the genetic map as the sequence of altered amino acids in the protein. And Brenner and co-workers (72) showed that in the head protein of phage T4, several hundred amino acid units long, the "amber" mutants (which terminate the polypeptide chain) produced longer fragments of the protein as the mutation sites moved further along the genetic map.

The larger relations between the (circular) genetic map and the life cycle of the T4 virus have been described by Edgar & Epstein (73).

The triplet code.—The determination of the sequence of three bases in the mRNA codon for each amino acid has tantalizing cryptographic aspects, which seemed to promise an early solution, but it has encountered many difficulties. They have come partly from the use of random polymer messenger chains; partly from the "degeneracy" of the code itself, with several codons for some amino acids; partly from "ambiguities," with the "wrong" amino acid coded at certain ion concentrations [Szer & Ochoa (74)]; and perhaps partly because of the ambiguities of the nitrous acid base changes mentioned earlier. Now several groups are using synthetic messenger chains of known sequence, the Doty group using short polynucleotides; the Khorana group using polymers of repeating sequences; and the Nirenberg group adding known terminal trinucleotides to RNA (3). No list of codons will be given here, since the situation is changing rapidly and more complete and conclusive lists will surely be available in a short time; but the most recent list and summary of the situation is that of Bernfield & Nirenberg (75).

The discovery of genetic recombination within a single codon (47) adds an interesting fillip to the problem and another useful constraint on codon relationships.

Antibiotic inhibition and coding errors.—Several of the well-known antibiotics block protein synthesis by different mechanisms. It is curious that the fungi from which they are derived have so many special methods of attack. Actinomycin suppresses all RNA synthesis (see 26) and interrupts the formation of ribosomes [Girard, Penman & Darnell (76)]. It is supposed to make specific hydrogen bonds to DNA, or to "block the smaller groove" between the strands so as to interfere with RNA transcription [Hamilton, Fuller & Reich (77); Reich (78)]. It does not bind to the hybrid DNA-RNA helix, and its bonding to DNA may require the "B" conformation of the helix [Haselkorn (79)].

Puromycin works by splitting the carboxyl bond between the growing polypeptide chain and the sRNA on the ribosome [Gilbert (80); Traut & Monro (81)], probably attaching itself to the end of the incomplete polypeptide [Nathans (82)]. Chloramphenicol may act at the same point by inhibiting the peptide-forming enzyme, and chlortetracycline (aureomycin) may act at a neighboring site (81). Cycloheximide acts by inhibiting the transfer of the amino acid from sRNA to the polypeptide chain [Ennis & Lubin (83)].

The remarkable new finding is that streptomycin acts to block protein synthesis by causing specific and extensive "misreadings" in the genetic code, making polyU-messenger,—U–U–U—, code for isoleucine, Ilu, instead of phenylalamine, Phe, for example [Davies, Gilbert & Gorini (38)]. Streptomycin is bound to the small 30s subunit of the ribosome [Davies (84); Cox, White & Flaks (85)], so this effect shows that "The translation mechanism involves not only the specific hydrogen bonds . . . but also the conformation of the site on the ribosome that holds the sRNA to the messenger" (38). Streptomycin presumably modifies this site, permitting a "wrong" sRNA to fit the messenger, but a mutation of the ribosome makes the fit independent of streptomycin, conferring streptomycin-resistance. A further mutation may even make it necessary to have streptomycin present (like a coenzyme) to obtain a correct fit; this could explain the occasional development of streptomycin dependence (38). The hydrogen-bonding fit between an m-RNA codon and an sRNA anticodon is therefore more complex and depends on the surroundings more than had been supposed.

Lederberg, Cavalli-Sforza & Lederberg (86) have noted further that the action of a suppressor mutation is affected by a mutation to streptomycin resistance, implying a similarity between suppressor genes and streptomycin drug action, and suggesting that some suppressors also may act by altering the ribosome. Beckwith (87, 88, 89) has studied the curious operator-negative, or O°, mutations in the lactose system, which seem to permit the making of mRNA but inhibit the process of translation into protein; he found them susceptible to external suppressors, suggesting that, like streptomycin, they operate at the level of reading the message. [See Attardi et al. (90).] Gorini has given a general review of the streptomycin problem (91).

Nonchromosomal inheritance.—Evidence is mounting that many subcellular structures or organelles outside the nucleus may be semi-autonomous bodies with their own hereditary systems. In 1963, Sager & Ishida (92),

Sager and Ramanis (93), and Chun, Vaughan & Rich (94) found DNA of a nonnuclear type in chloroplasts, and Luck & Reich (95) have now proved DNA is in mitochondria. Gibor & Granick (96) have reviewed the evidence that RNA and DNA are present in plastids and mitochondria, with multiple genes and with their own control mechanism; although such mechanisms must interact, of course, with the control mechanisms and products of the nuclear DNA. Sager (97, 98) has given more general reviews of the status of nonchromosomal genes and crossing experiments, emphasizing the inheritance of streptomycin sensitivity or dependence; this case, however, may be confused by the streptomycin-ribosome-coding interactions described earlier. [See also Jinks (98a), Ray & Hanawalt (98b), and McCalla (98c).]

A different kind of "cytoplasmic inheritance," long argued for by Weiss (99) [see also Salser (100)], has now been proved dramatically by Sonneborn (101, 102) who excised and rearranged structures in *Paramecium* and showed they were inherited, without any genetic changes! Cells made to have two or more cell mouths and gullets are not only viable, but can transmit the characteristic to their progeny at successive cell divisions and even through sexual reproduction. And cells with surface structures inverted [by the method of Beisson-Schecroun (103)] have perpetuated these changes for over a year, or through more than 700 cell generations.

These experiments help to answer a question that has always been unanswered by the genetic and protein synthesis studies, namely: What tells the enzyme products where to go? Why do they form a cell instead of a uniform soup? They are evidently organized by a process something like crystallization—"macrocrystallinity," Weiss has called it. Sonneborn prefers to name it "cytotaxis"; he says, ". . . during growth new surface units appear within an existing row of units and are oriented in the same way as the units already there. . . . This . . . is a second principle of cellular differentiation, one that is quite distinct from variable genic activity. The cell differences [here] . . . are not characterized by different kinds of substances or structures, but by different numbers or arrangements of structures. Their perpetuation shows that they can be decisive in cell differentiation. Self-assembly of genic products alone cannot account for this. The place and orientation of the assembly of genic products is also determined by pre-existing assemblies of molecules and structures."

Probably the "information content" represented by these pre-existing structures is very small by comparison with the genetic information, but it is of a different kind, and is evidently essential to cellular organization. Sonneborn's charming articles, showing how to manipulate and prove subcellular organizational inheritance, may be the start of a new era in our understanding of these relationships.

Protein replacements and evolution.—One of the most interesting chemical aspects of genetics has been the discovery of differences in the amino acid sequences of similar proteins made by different organisms. Zuckerkandl & Pauling (104–106) showed that the number of differences between two hemoglobin sequences was roughly proportional to the evolutionary divergence between the two organisms, as measured by the time since divergence occurred. The time scale is about 15 million years per amino acid replacement, in a chain of 150 amino acids. The complete elucidation of the structure of cytochrome *c* has now permitted the study of such "chemical paleogenetic" differences in the cytochromes from 12 species [Margoliash et al. (107–109); Smith et al. (110)]. The difference between the cytochromes *c* in man and chick is 12 amino acids; between man and tuna, 22; between man and yeast, 43. The cytochrome chain has 104 amino acids, representing 312 bases, so the man-yeast difference means divergence in about 15 per cent of these bases. A large symposium on this subject of Evolving Genes and Proteins was held in September 1964 and has been reported by Bryson & Vogel (111).

A single amino acid replacement in a protein may lead, of course, to malfunction or disease more often than to a useful change. Beadle's article (1) gives a fascinating summary of what we know about numerous "chemical diseases" of this type.

Protein structure.—Schachman (112) has given a thoughtful examination of the problems of tertiary structure, that is, of the folding and cross-linking of the secondary α-helices of the primary protein-chain sequence. He gives special attention to the controversy over so-called "hydrophobic bonding," in which one school (Klotz) predicts that the nonpolar amino acid residues should tend to be on the "exterior" of the protein to form crystalline hydrates with water (like the hydrates of many hydrocarbons); while the other school (Kauzmann) predicts a lower energy for the system if they are in a hydrophobic interior (like that of soap micelles) in contact with each other. The X-ray data on myoglobin and the hemoglobins [see Perutz (113) for a good general account] show the latter arrangement; but these proteins might be unusual. The 1963 Cold Spring Harbor symposium (5) had an especially large number of excellent papers from many laboratories on the tertiary structure of proteins; on "allosteric" interactions or change of three-dimensional conformation when proteins are bound to small molecules; and on complementation, or the joining together of incomplete proteins to give complete biochemical function.

Another volume carried three major articles on the structure and function of hemoglobins and myoglobin [Braunitzer et al. (114); Fanelli, Antonini & Caputo (115); Wyman (116)]. Fraenkel-Conrat (48) has also given a good popular account of the problems of genetics, sequence, and structure in the coat protein of tobacco mosaic virus.

The analytical developments of 1964 include the discovery by Shechter, Carver & Blout (117, 118) that optical rotatory dispersion data on proteins can be fitted more rationally by a two-term Drude equation than by the Moffitt equation. The two terms are approximately the differential contributions of helix and random-coil forms to the peptide $n-\pi$ and $\pi-\pi$ transitions near 225 mμ and 193 mμ, and the coefficients give two independent measures of the helix content. [But see also Yang (118a).]

An ingenious achievement of the year was the X-ray determination of the location of the azide ion, $N-N-N^-$, when attached to metmyoglobin. Stryer, Kendrew & Watson (119) used a Fourier difference method, which locates the light atoms directly, if they are added to or removed from a molecule whose X-ray analysis is already known, if the crystals are isomorphous. The azide ion is an inhibitor of heme enzymes, and it was proved (as expected) that it goes in to displace a water molecule from the heme iron; but it was interesting to find that there was a concomitant loss of a sulfate ion bound to the metmyoglobin 9 Å away. There may be many other cases where this method can be applied to determine directly the location of chemical substitutions in a large molecule.

But the year also brought disappointment in the finding that the high-resolution X-ray analysis of proteins other than myoglobin and hemoglobin may be much harder than expected. One symposium discussion ended with the conclusion, "Harker's maps on ribonuclease do not yield to a clear interpretation," and "Kraut now has a map at 4 Å resolution on chymotrypsinogen and it is essentially uninterpretable in terms of a chemically or biologically meaningful structure" [Ibers (120); Kraut, High & Sieker (121)]. The trouble is not only the lack of isomorphism of the heavy-metal derivatives necessary for the X-ray determinations, but also the low content of α-helices or other familiar substructures. Kraut and co-workers (121) say that greater reliance may have to be placed on model building and on computer simulation of models than had been supposed.

On the other hand, there begins to be a possibility of seeing direct images of molecules and their atoms, with a new type of electron microscope being developed at the Argonne National Laboratory by Crewe (122). It is a scanning microscope, designed with two new features: (*a*) "Strong-focusing lenses" for the electron beam that may permit a resolution approaching 1 Å. (*b*) Separate detection of the inelastically scattered electrons from C, N, and O atoms. The three pictures can be combined, say, in a color presentation, with the color designating either the average constitution or, at the hoped-for resolution, individual atoms. A supporting film of Be or Al would give inelastic electrons of a different energy and would not appear in the picture. This microscope is already in operation, and objective (*b*) has been accomplished at around 100 Å resolution. The possible effect of such a microscope on chemical and biochemical and subcellular analysis may be imagined.

Active sites in enzymes.—Neurath (123) has given a good general discussion of the relation between protein sequence and the structure and chemical function of the active sites of proteolytic enzymes like trypsin and chymotrypsin [Walsh et al. (124); Walsh & Neurath (125)]. In both these enzymes, the active sites seem to require the cooperative action of a sequence containing two histidines with another section containing a serine, and Bender and co-workers (126, 127) have proposed a quite specific model.

Some other proteins are even more complex. In carboxypeptidase-A, five functional groups, including a zinc atom, are involved in the active center [(123); Vallee (128)].

A general model for the correlation of structure and function was proposed in 1963 by Koshland (129). Burr & Koshland (130) have now shown how to get additional information on the binding to a substrate, by attaching "reporter groups" near to, but not on, the active site. In chymotrypsin, a nitrophenol dye located three residues away from this site gave significantly different spectra on binding to D and L isomers.

The problem of complementation of two incomplete or mutant proteins somewhat resembles the active-site problem with its cooperating sequences and groups. The action of multimeric enzymes which are inactive in the monomeric state also seems similar. This complementation problem was approached earlier by Kapuler & Bernstein (131), but Crick & Orgel (132) have now given a detailed alternative hypothesis, proposing "that complementation is usually due to the correction of the misfolding of one monomer (produced by the mutation) by some unaltered part of the other monomer," and they emphasize "that symmetry axes are likely to be involved." The cooperation of two antibody chains may also be related [Roholt, Onoue & Pressman (133); Metzger, Wofsy & Singer (134)]; Edelman & Gally (135) have proposed a detailed model.

The idea of "allosteric interactions" of proteins, or changes of their configuration and chemical behavior on binding to small molecules, coenzymes or substrates, dominates the thinking today. Koshland (136) long urged this kind of idea for enzymes with his hypothesis of "induced fit," but attention was focused on it more generally by the brilliant 1963 paper of Monod, Changeux & Jacob (137), who emphasized the necessity for such a mechanism in cellular control systems. The fine group of symposium papers on allosteric interactions (5) has already been mentioned.

The Central Problem

There are several reasons for believing that the central problem of biochemistry and, indeed, of chemistry in the next few years may be the relation of large-molecule sequence to spatial structure and function. Properties of large molecules go beyond the properties of their chemical subgroups [Platt (138)], but we do not understand the laws of this behavior. The problem is not limited to enzymes, but extends to structural and muscle proteins as well as to other biological polymers such as the nucleic acids, the polysaccharides, and so on. But it is the enzymes and heme proteins that give us in some ways the best material for study, with the most selective tests of purity, the best-determined sequences, the best location of chemical groups in the X-ray structures, the greatest knowledge of function and the greatest knowledge of alternative configurations, binding sites, and chemistry.

The nature of the problem is indicated in the lower half of Figure 1. When a functional protein of known sequence is assembled from the messenger-RNA information on the ribosomes, it evidently twists and folds into secondary and tertiary structures stabilized by hydrogen bonding, by covalent –S–S– cross-links and "hydrophobic bonding" and perhaps by other chemical forces, together holding the molecule in its biochemically effective spatial configuration. There may be, of course, a thermodynamic mixture of configurations, but one or more of them are presumably capable of binding to other molecules such as the appropriate coenzyme or a substrate molecule, either permanently or to catalyze a specific chemical reaction.

The problem for biochemistry is then the following: Given the amino acid sequence of the protein, what are the rules or laws (*a*) determining which parts will have α-helical or other secondary structure and determining how these parts will fold in space or connect up in a tertiary structure, say in aqueous solution; and what are the rules or laws; (*b*) determining what substrates or other molecules will be bound and what specific chemical reactions will be catalyzed? Evidently, we would also like to know how to use these laws in reverse. That is, given a biochemical function to be performed, such as the catalysis of a reaction step, say the addition of a methyl group to the 17 position on a keto steroid, are there reverse rules: (*b*′) determining the tertiary structure and folding of a possible catalytic enzyme for the process; and (*a*′) determining what amino acid sequence, or possible alternative sequences, such an enzyme would need?

For a time it was thought that the tertiary structure and folding might require special information put in at the moment of assembly, in addition to the sequence information. But the widespread reversibility of "denaturation" of secondary and tertiary structures (disrupted by heat or urea) has made it generally accepted that the secondary and tertiary configurations of the biochemically active forms of proteins are probably formed spontaneously, as the thermodynamically most stable arrangements for the given polypeptide sequence under these conditions [Epstein, Goldberger & Anfinsen (139)].

This removes the necessity of searching for an additional source of genetic assembly information, but it does not help much in giving us rules or laws of types (*a*) and (*b*), because the theoretical prediction of the thermodynamically most stable structure for any very long sequence would require exact evaluation of the stability of millions of alternative configurations. The only reasonable hope for working out such a problem is that it can be done in steps, by first finding rules for "subassembly" parts of the problem, which can then be put together, as Simon (140) emphasizes in discussing the "architecture of complexity." One simplifying possibility, for example, would be "sequential minimization," or the assumption that each amino acid added to the chain would generally take a configuration minimizing its energy with respect to the previously assembled part of the chain, with these configurations being disturbed afterwards only to fit a small number of additional structural principles, such as specific cross-linking.

Another way of breaking the problem into "subassemblies" has recently been proposed by Guzzo (141). He suggests thinking of the protein structure as being a set of α-helical subsections separated by specific "helix-interrupting groups," or "critical groups," that act somewhat like "hinges" between the helices. His proposed critical groups are His, Asp (not Asn), Glu, and Pro, as picked out by several aspects of their behavior. They may tend to disrupt α-helices [but see Blout (142)]; Pro disrupts them with its five-membered ring. And when a Pro has been removed in going from one functional protein to a viable mutant or variant protein, there always seem to be Asp or Glu groups at the same or adjacent positions, as though they could take over the role of Pro in helix interruption. These replacements are not just one-base changes; so it is not a matter of which mutations are chemically easy, but which ones are viable—on the grounds that general production of functional protein in a surviving species might be a sensitive test as to whether the tertiary structure is essentially unchanged.

Guzzo suggests that Glu, Asp, and His might be helix disrupters because they have short charged chains, with the polar groups perhaps close enough to the polypeptide backbone to change the water atmosphere and the stability of the hydrogen bonding of the α-helix. These groups are also found to occur in helical regions of the chains, but they might often be only "conditional" helix disrupters, with their disrupting function dependent on substrate or environment; that is, they may be necessary but not sufficient for helix interruption. This may seem to weaken the argument, unless it is remembered that "conditional" behavior of helix disrupters or helix formers is necessary in order to have allosteric changes, with the form dependent on the environment; consequently, "conditional" behavior is not to be shunned, but to be sought for, in analyzing this problem.

Whether this particular model is successful or not, the direction seems worth pursuing, and the attempt should be made to set up counter-models based on other subassembly partitions of the problem, so that experimental tests between different subassembly hypotheses may be possible.

Another, genetic, approach to the relation between groups some distance apart on the chain is suggested by the findings of Yanofsky, Horn & Thorpe (143). They found, in tryptophan synthetase protein, that the loss of function with certain amino acid mutants at two particular sites in one region could be reversed by mutants at two different sites in another region, with the "revertants" exactly 36 residues away in each case. They suggest that this compensation of one amino acid change by another one far down the chain (Ala→Val reverted by Leu→Arg, and Gly→Glu reverted by Tyr→Cys) might mean that these two regions are folded so as to be in contact in the tertiary structure.

In searching for these laws connecting protein sequence with spatial structure and with function, it is evident that there must be a coordinated attack from several directions, through biochemistry, through high-resolution X-ray analysis, and through model building. The possible role of models in this problem may not be sufficiently appreciated [Platt (144)]. Much of the biochemistry and molecular biology of the last 15 years has depended on the molecular-model studies of Pauling, Corey & Branscomb that led to the α-helix model for proteins, and on those of Watson & Crick that led to the DNA structure. It does not follow that models will continue to be equally valuable in going from these secondary-structure problems to the far more difficult problem of tertiary-structure configurations and their alternatives, but it is evidently important to see just how much can be done with this approach. The NIH-Biochemistry or Koltun models (144) expected to be available in 1965, will be lighter, cheaper, and more accurate than present atomic models and may make it possible for many more laboratories to get into such studies.

It seems possible that the discovery of sets of rules or general laws connecting polymer sequence with structure and function might play a role in biochemistry similar to that played by valence theory in organic chemistry a hundred years ago. It would permit for the first time a prediction of what substrates a protein or other polymer might react with and what the reactions would be; it would permit an analysis of what mutants or other alternative sequences or structures might do the same job; and eventually, it would permit the synthetic design of new sequences to carry out particular new reactions. So many reactions difficult by other methods—the fixation of nitrogen, for example—are done so easily and with such high yields at room temperatures by catalytic enzymes that it would be a great step forward if we could now begin to learn the rules for creating such catalysts ourselves.

In fact, one begins to see dimly that these rules might be part of some larger general laws of complementarity in the interactions of macromolecules. Many examples are now known which seem to suggest a general theorem: That every, or almost every, small molecule can be specifically "fitted," both for binding and for different catalyses, by some specific arrangements of side-groups on a long-chain polymer.

One kind of evidence for such a theorem is the frequent appearance, in biology, of specifically complementary pairs of molecules or groups, probably complementary in many cases by virtue of a steric fit between them. Examples are: DNA—DNA; DNA—RNA; sRNA anticodon—specific amino acid (if determined sterically); enzyme—coenzyme and substrate; embryological tissue-inducer—nuclear repressor substance; antibody—antigen or hapten; and mucopolysaccharide—mucopolysaccharide (in cell wall adhesion?). The chains in this list include nucleic acid chains, polypeptide chains, and polysaccharide chains, so that this kind of specific complementarity may be a general chain phenomenon, but one not generally recognized in ordinary organic chemistry because this has mainly been concerned with smaller molecules and with repeating "homopolymers" rather than with information-bearing sequences with high specificity.

A second kind of evidence for the idea that a complementary chain might be found that would fit every molecule, comes from the sequence of enzymes involved in the synthesis of a biological product such as histidine [Ames & Hartman (145)]. In going down a sequence of this sort, each intermediate is transformed into the next one by some small step, such as adding an amine group, making a keto group, cyclization, transferring a phosphate, and so on; yet at each step there is a new, identifiable, specific enzyme for that step. Obviously any molecule can be transformed into any other one by some sequence of such small steps; and this suggests the theorem that if we start with a protein or enzyme that "fits" some one molecule, there must be a related protein that will fit it after a one-step transformation, and so on for any number of steps, until we reach a protein that will fit any other specified molecule. This idea may tend to be supported further by the fact that mammals can create fairly specific antibodies against almost any foreign antigen or hapten molecule that is injected. (It is true that soil bacteria seem unable to digest our synthetic "nonbiodegradable" detergents, but it may be that their exposure so far has been too short for the evolution of detergentases to deal with such alien systems.)

There would be no need to invoke any special new laws to account for general complementarity. There is complementarity in ordinary chemistry, of acids and bases, of donors and acceptors, and, in a different sense, of polarity and nonpolarity, or of hydrophilic and hydrophobic groups. This is rather nonspecific, with each donor, for example, capable of complexing with dozens of acceptors. But if these various groups are assembled into a long molecule, the arrangement of the groups can become highly specific and highly discriminatory in binding to a second molecule. All that is necessary for the theorem is then to say that every kind of molecule or molecular combination differs from every other kind in its shape, polarity, and affinities, and that therefore there must be some long complementary chain that will wrap around or bind better to it than to any other molecule. This would make the DNA—DNA or DNA—RNA base-complementarity in the restricted geometry of the nucleic-acid double helix, a special case of a much more general and interesting phenomenon.

The attempt to establish such general laws relating to high-information chemical specificity and its steric and spatial requirements could become one of the most exciting branches of chemistry in the next few years. Many new techniques will be needed. Genetic systems offer an extremely powerful amplification of individual molecular events of this high-complexity, high-information type [Platt (146)]—sometimes with amplification factors of up to 10^{18} in 24 hr—and therefore might become the crucial tool for the chemist to use in such an undertaking.

LITERATURE CITED

1. Beadle, G. W., in *1964 Britannica Book of the Year*, 44–72 (Encyclopedia Britannica, Chicago, Ill., 1964)
2. Watson, J. D., in *Volume du cinquantenaire de la société de chimie biologique*, 1–21 (Maurice Declume, Lons-le-Saunier, France, 1964)
3. Crick, F. H. C., in *Proc. Plenary Sessions Intern. Congr. Biochem., 6th*, 109–28 (Federation Am. Soc. Exptl. Biol., Washington, D. C., 1964)
4. Asimov, I., *The Genetic Code* (New American Library, New York, 1963)
5. Synthesis and Structure of Macromolecules, *Cold Spring Harbor Symp. Quant. Biol. 1963*, **28** (1964)
6. Davidson, J. N., and Cohn, W. E., Eds., *Progress in Nucleic Acid Research* (Academic Press, New York, **I,** 1963; **II,** 1964)
7. Neurath, H., Ed., *The Proteins, Composition, Structure and Function* (Academic Press, New York, **I,** 1963; **II,** 1964)
8. Stent, G. S., *Molecular Biology of Bacterial Viruses* (Freeman, San Francisco, 1964)
9. Platt, J. R., in *Horizons in Biochemistry*, 167–87 (Kasha, M., and Pullman, B., Eds., Academic Press, New York, 1962)
10. Jacob, F., and Monod, J., *J. Mol. Biol.*, **3,** 318–45 (1961)
11. Haschemeyer, A. E. V., and Sobell, H. M., *Proc. Natl. Acad. Sci. U. S.*, **50,** 872–77 (1963); *Nature*, **202,** 969–71 (1964)
12. Mathews, F. S., and Rich, A., *J. Mol. Biol.* **8,** 89 (1964)
13. Miles, H. T., *Proc. Natl. Acad. Sci.*, **51,** 1104–16 (1964)
14. MacHattie, L. A., and Thomas, C. A., Jr., *Science*, **144,** 1142–45 (1964)
15. Weil, R., Vinograd, J., and Stoeckenius, W., *Proc. Natl. Acad. Sci. U.S.*, **50,** 730 (1963)
16. Kleinschmidt, A. K., Burton, A., and Sinsheimer, R. L., *Science*, **142,** 961, 962–63 (1963)
17. Chandler, B., Hayashi, M., Hayashi,

M. N., and Spiegelman, S., *Science*, **143**, 47–49 (1964)

18. Cairns, J., *J. Mol. Biol.*, **6**, 208–13 (1963)
19. Cairns, J., *Cold Spring Harbor. Symp. Quant. Biol.*, **28**, 43–46 (1963)
20. Jacob, F., Brenner, S., and Cuzin, F., *Cold Spring Harbor Symp. Quant. Biol.*, **28**, 329–48 (1963)
21. Geiduschek, E. P., Moohr, J. W., and Weiss, S. B., *Proc. Natl. Acad. Sci. U. S.*, **48**, 1078-86 (1962)
22. Shipp, W., and Haselkorn, R., *Proc. Natl. Acad. Sci. U.S.*, **52**, 401–8 (1964)
23. Hayashi, M., Hayashi, M. N., and Spiegelman, S., *Proc. Natl. Acad. Sci. U. S.*, **51**, 351–58 (1964)
24. Spiegelman, S., *Sci. Am.*, **210**, No. 5, 48–56 (May 1964)
25. Geiduschek, E. P., Tocchini-Valentini, G. P., and Sarnat, M. T., *Proc. Natl. Acad. Sci. U. S.*, **52**, 486–93 (1964)
26. Beermann, W., and Clever, U., *Sci. Am.*, **210**, No. 4, 50–58 (Apr. 1964)
27. Caro, L., and van Tubergen, R. P., *J. Cell Biol.*, **15**, 173 (1962)
28. Littau, V. C., Allfrey, V. G., Frenster, J. H., and Mirsky, A. E., *Proc. Natl. Acad. Sci. U. S.*, **52**, 93–100 (1964)
29. Gallant, J., and Stapleton, R., *J. Mol. Biol.*, **8**, 431–41, 442–51 (1964)
30. Allfrey, V. G., Faulkner, R., and Mirsky, A. E., *Proc. Natl. Acad. Sci. U.S.*, **51**, 786–94 (1964)

30a. Stent, G. S., *Science*, **144**, 816–20 (1964)

31. Burka, E. R., and Marks, P. A., *J. Mol. Biol.*, **9**, 439–51 (1964)
32. Eiserling, F., Levin, J. G., Byrne, R., Karlsson, U., Nirenberg, M. W., and Sjostrand, F. S., *J. Mol. Biol.*, **10**, 536–40 (1964)
33. Rich, A., *Sci. Am.*, **209**, No. 6, 44–53 (Dec. 1964)
34. Warner, J. R., and Rich, A., *J. Mol. Biol.*, **10**, 202–11 (1964)
35. Meselson, M., Nomura, M., Brenner, S., Davern, C., and Schlessinger, D., *J. Mol. Biol.*, **9**, 696–711 (1964)
36. Gilbert, W., *J. Mol. Biol.*, **6**, 389–403 (1963)
37. Cannon, M., Krug, R., and Gilbert, W., *J. Mol. Biol.*, **7**, 360–78 (1963)
38. Davies, J., Gilbert, W., and Gorini, L., *Proc. Natl. Acad. Sci. U.S.*, **51**, 883–90 (1964)
39. Warner, J. R., and Rich, A., *Proc. Natl. Acad. Sci. U.S.*, **51**, 1134–41 (1964)
40. Wettstein, F. O., and Noll, H., *J. Mol. Biol.*, **11**, 35–53 (1965)
41. Yu, C.-T., and Zamecnik, P. C., *Science*, **144**, 856–59 (1964)
42. Weil, J.-H., Befort, N., Rether, B., and Ebel, J.-P., *Biochem. Biophys. Res. Commun.*, **15**, 447–52 (1964)
43. Herbert, E., Smith, C. J., and Wilson, C. W., *J. Mol. Biol.*, **9**, 376–94 (1964)
44. Felsenfeld, G., and Cantoni, G. L., *Proc. Natl. Acad. Sci. U. S.*, **51**, 818–26 (1964)

44a. Holley, R. W., Apgar, J., Everett, G. A., Madison, J. T., Marquisee, M., Merrill, S. H., Penswick, J. R., and Zamir, A., *Science*, **147**, 1462–65 (1965)

45. Freese, E., in *Molecular Genetics*, Part I, 207 (Taylor, H., Ed., Academic Press, New York, 1963)
46. Wittmann, H. G., and Wittmann-Liebold, B., *Cold Spring Harbor Symp. Quant. Biol.*, **28**, 589–95 (1963)
47. Yanofsky, C., *Cold Spring Harbor Symp. Quant. Biol.*, **28**, 581–88 (1963)
48. Fraenkel-Conrat, H., *Sci. Am.*, **211**, No. 4, 46–54 (Oct. 1964)
49. Tessman, I., Poddar, R. K., and Kumar, S., *J. Mol. Biol.*, **9**, 352–63 (1964)
50. Howard, B. D., and Tessman, I., *J. Mol. Biol.*, **9**, 364–71, 372–75 (1964)

50a. Tessman, I., Ishiwa, H., and Kumar, S., *Science*, **148**, 507–8 (1965)

51. Dellweg, H., and Wacker, A., *Z. Naturforsch.*, **17**(b), 827 (1962)
52. Drake, J. W., *J. Mol. Biol.*, **6**, 268 (1963)
53. Lerman, L. S., *Proc. Natl. Acad. Sci. U.S.*, **49**, 94 (1963); *J. Cell. Comp. Physiol.*, **64**, Suppl. 1, 1–18 (1964); *J. Mol. Biol.*, **10**, 367–80 (1964)
54. Crick, F. H. C., Barnett, L., Brenner, S., and Watts-Tobin, R. J., *Nature*, **192**, 1227 (1961)
55. Freifelder, D. (Private communication)
56. Becker, E. F., Jr., Zimmerman, B. K., and Geiduschek, E. P., *J. Mol. Biol.*, **8**, 377–91 (1964)
57. Pershan, P. S., Shulman, R. G., Wyluda, B. J., and Eisinger, J., *Physics*, **1**, 163–82 (1964)
58. Setlow, R. B., and Carrier, W. L., *Proc. Natl. Acad. Sci. U. S.*, **51**, 226–30 (1964)
59. Boyce, R. P., and Howard-Flanders, P., *Proc. Natl. Acad. Sci. U.S.*, **51**, 293–300 (1964)
60. Harm, W., in *Repair from Genetic Radiation Damage*, 107–18 (F. H. Sobels, Ed., Macmillan, New York, 1963)

61. Schildkraut, C. L., Richardson, C. C., and Kornberg, A., *J. Mol. Biol.*, **9** 24–45 (1964)
62. Richardson, C. C., Inman, R. B., and Kornberg, A., *J. Mol. Biol.*, **9,** 46–69 (1964)
63. Platt, J. R., *Proc. Natl. Acad. Sci. U.S.*, **41,** 181–83 (1955)
64. Löwdin, P.-O., *Rev. Mod. Phys.*, **35,** 724–32 (1963)
65. Bernal, J. D., and Fowler R. H., *J. Chem. Phys.*, **1,** 515 (1933)
66. Conway, B. E., Bockris, J. O'M., and Linton, H., *J. Chem. Phys.*, **24,** 834 (1956)
67. Eigen, M., and de Maeyer, L., *Proc. Roy. Soc.* (*London*), *A*, **247,** 505 (1958)
68. Haas, C., and Hornig, D. F., *J. Chem. Phys.*, **32,** 1763 (1960)
69. Eigen, M., *Pure. Appl. Chem.*, **6,** 97 (1963)
70. Barnothy, M. F., *Biophys. J.*, **5,** Suppl. 1, 25 (1965)
70a. Rein, R., and Harris, F. H., *Science*, **146,** 649–50 (1964); *J. Chem. Phys.*, **42,** 2177–80 (1965)
71. Yanofsky, C., Carlton, B. C., Guest, J. R., Helinski, D. R., and Henning, U., *Proc. Natl. Acad. Sci. U. S.*, **51,** 266–72 (1964)
72. Sarabhai, A. S., Stretton, A. O. W., Brenner, S., and Bolle, A., *Nature*, **201,** 13 (1964)
73. Edgar, R. S., and Epstein, R. H., *Sci. Am.*, **212,** No. 2, 71–78 (Feb. 1965)
74. Szer, W., and Ochoa, S., *J. Mol. Biol.*, **8,** 823–34 (1964)
75. Bernfield, M. R., and Nirenberg, M. W., *Science*, **147,** 479–84 (1965)
76. Girard, M., Penman, S., and Darnell, J. E., *Proc. Natl. Acad. Sci. U. S.*, **51,** 205–11 (1964)
77. Hamilton, L. D., Fuller, W., and Reich, E., *Nature*, **198,** 538 (1963)
78. Reich, E., *Science*, **143,** 684 (1964)
79. Haselkorn, R., *Cold Spring Harbor Symp. Quant. Biol.*, **28,** 91–93 (1963)
80. Gilbert, W., *J. Mol. Biol.*, **6,** 389 (1963)
81. Traut, R. R., and Monro, R. E., *J. Mol. Biol.*, **10,** 63–71 (1964)
82. Nathans, D., *Proc. Natl. Acad. Sci. U. S.*, **51,** 585–92 (1964)
83. Ennis, H. L., and Lubin, M., *Science*, **146,** 1474–76 (1964)
84. Davies, J., *Proc. Natl. Acad. Sci. U. S.*, **51,** 659–64 (1964)
85. Cox, E. C., White, J. R., and Flaks, J. G., *Proc. Natl. Acad. Sci. U. S.*, **51,** 703–9 (1964)
86. Lederberg, E. M., Cavalli-Sforza, L., and Lederberg, J., *Proc. Natl. Acad. Sci. U. S.*, **51,** 678–82 (1964)
87. Beckwith, J. R., in *Proc. Intern. Congr. Genetics*, XI, *The Hague, 1963*
88. Beckwith, J. R., in *Structure and Function of the Genetic Material* (*Abhandl. Deut. Akad. Wiss. Berlin*, 1964)
89. Beckwith, J. R., *J. Mol. Biol.* **8,** 427–30 (1964); *Biochim. Biophys. Acta*, **76,** 162–70 (1963)
90. Attardi, G., Naono, S., Rouviere, J., Jacob, F., and Gros, F., *Cold Spring Harbor Symp. Quant. Biol.*, **28,** 363–72 (1963)
91. Gorini, L., *New Scientist*, **24,** 776–79 (1964)
92. Sager, R., and Ishida, M. R., *Proc. Natl. Acad. Sci. U. S.*, **50,** 725 (1963)
93. Sager, R., and Ramanis, Z., *Proc. Natl. Acad. Sci. U. S.*, **50,** 260 (1963)
94. Chun, E. H. L., Vaughan, N. H., Jr., and Rich, A., *J. Mol. Biol.*, **7,** 130–41 (1963)
95. Luck, D. J. L., and Reich, E., *Proc. Natl. Acad. Sci. U. S.*, **52,** 931–38 (1964)
96. Gibor, A., and Granick, S., *Science*, **145,** 890–97 (1964)
97. Sager, R., in *Function and Structure of Micro-Organisms* (Pollock, M. R., and Richmond, M. H., Eds., Cambridge Univ. Press, Cambridge, 1964)
98. Sager, R., *Sci. Am.*, **212,** No. 1, 70–79 (Jan. 1964)
98a. Jinks, J. L., *Extrachromosomal Inheritance* (Prentice-Hall, Englewood Cliffs, N. J., 1964)
98b. Ray, D. S., and Hanawalt, P. C., *J. Mol. Biol.*, **9,** 812–24 (1964)
98c. McCalla, D. R., *J. Protozool.*, **12,** 34–40 (1965); *Science*, **148,** 497–99 (1965)
99. Weiss, P., in *The Molecular Control of Cellular Activity* (Allen, J. M., Ed., McGraw, New York, 1962)
100. Salser, W., *Perspectives Biol. Med.*, **4,** 177–98 (1961)
101. Sonneborn, T. M., in *The Nature of Biological Diversity*, 165 (Allen, J. M., Ed., McGraw, New York, 1963)
102. Sonneborn, T. M., *Proc. Natl. Acad. Sci. U. S.*, **51,** 915–29 (1964)
103. Beisson-Schecroun, J., in *Proc. Intern. Congr. Cell Biol.*, *XI*, *The Hague, 1964*
103a. Beisson, J., and Sonneborn, T. M., *Proc. Natl. Acad. Sci. U.S.*, **53,** 275–82 (1965)
104. Zuckerkandl, E., and Pauling, L., in *Horizons in Biochemistry*, 189–225

(Kasha, M., and Pullman, B., Eds., Academic Press, New York, 1962)

105. Pauling, L., and Zuckerkandl, E., *Acta Chem. Scand.*, **17,** 59–79 (1963)
106. Zuckerkandl, E., and Pauling, L., in *Problems of Evolutionary and Industrial Biochemistry* (Dedicated to A. I. Oparin, Science Press, Acad. Sci. USSR, Moscow, 1964) [English version, *J. Theoret. Biol.*, **8,** 357–66 (1965)]
107. Margoliash, E., *Proc. Natl. Acad. Sci. U. S.*, **50,** 672 (1963)
108. Margoliash, E., Needleman, S. B., and Stewart, J. W., *Acta Chem. Scand.*, **17,** S250 (1963)
109. Margoliash, E., and Smith, E. L., *Abstr. Intern. Congr. Biochem., 6th, Washington, 1964,* 206–7
110. Smith, E. L., Matsubara, H., McDowall, M., and Rothfus, J. A., *Science*, **140,** 385 (1963)
111. Bryson, V., and Vogel, H. J., *Science*, **147,** 68–71 (1965)
112. Schachman, H., *Cold Spring Harbor Symp. Quant. Biol.*, **28,** 409–30 (1963)
113. Perutz, M. F., *Sci. Am.*, **211,** No. 5, 64–76 (Nov. 1964)
114. Braunitzer, G., Hilse, K., Rudloff, V., and Nilschmann, N., *Advan. Protein Chem.*, **19,** 1–73 (1964)
115. Fanelli, A. R., Antonini, E., and Caputo, A., *Advan. Protein Chem.*, **19,** 74–223 (1964)
116. Wyman, J., Jr., *Advan. Protein Chem.*, **19,** 224–86 (1964)
117. Schechter, E., and Blout, E. R., *Proc. Natl. Acad. Sci. U. S.*, **51,** 695–703, 794–801 (1964)
118. Schechter, E., Carver, J. P., and Blout, E. R., *Proc. Natl. Acad. Sci. U. S.*, **51,** 1029–36 (1964)
118a. Yang, J.-T., *Proc. Natl. Acad. Sci. U.S.*, **53,** 438–45 (1965)
119. Stryer, L., Kendrew, J. C., and Watson, H. C., *J. Mol. Biol.*, **8,** 96–104 (1964)
120. Ibers, J. A., *Phys. Today*, No. 1, 28–36 (Jan. 1965)
121. Kraut, J., High, D. F., and Sieker, L. C. *Proc. Natl. Acad. Sci. U. S.*, **51,** 839–45 (1964)
122. Crewe, A. V., *J. Microscopie*, **2,** 369–71 (1963)
123. Neurath, H., *Sci. Am.*, **211,** No. 6, 68–79 (Dec. 1964)
124. Walsh, K. A., Kauffman, D. L., Kumar, K. S. V. S., and Neurath, H., *Proc. Natl. Acad. Sci. U. S.*, **51,** 301–8 (1964)
125. Walsh, K. A., and Neurath, H., *Proc. Natl. Acad. Sci. U. S.*, **52,** 884–89 (1964)
126. Bender, M. L., Killheffer, J. V., Jr., and Kézdy, F. J., *J. Am. Chem. Soc.*, **86,** 5330–33 (1964)
127. Bender, M. L., and Brubacher, L. J., *J. Am. Chem. Soc.*, **86,** 5333–34 (1964)
128. Vallee, B. L., *Federation Proc.*, **23,** Part 1, 8–17 (1964)
129. Koshland, D. E., Jr., *Science*, **142,** 1533–41 (1963)
130. Burr, M., and Koshland, D. E., Jr., *Proc. Natl. Acad. Sci. U. S.*, **52,** 1017–24 (1964)
131. Kapuler, A. M., and Bernstein, H., *J. Mol. Biol.*, **6,** 443 (1963)
132. Crick, F. H. C., and Orgel, L. E., *J. Mol. Biol.*, **8,** 161–65 (1964)
133. Roholt, O., Onoue, K., and Pressman, D., *Proc. Natl. Acad. Sci. U. S.*, **51,** 173–78 (1964)
134. Metzger, H., Wofsy, L., and Singer, S. J., *Proc. Natl. Acad. Sci. U. S.*, **51,** 612–18 (1964)
135. Edelman, G. M., and Gally, J. A., *Proc. Natl. Acad. Sci. U. S.*, **51,** 846–53 (1964)
136. Koshland, D. E., Jr., *Cold Spring Harbor Symp. Quant. Bicl.*, **28,** 473–82 (1963)
137. Monod, J., Changeux, J., and Jacob, F., *J. Mol. Biol.*, **6,** 306–29 (1963)
138. Platt, J. R., *J. Theoret. Biol.*, **1,** 342–58 (1961)
139. Epstein, C. J., Goldberger, R. F., and Anfinsen, C. B., *Cold Spring Harbor Symp. Quant. Biol.*, **28,** 439–50 (1963)
140. Simon, H. A., *Proc. Am. Phil. Soc.*, **106,** 467–82 (1962)
141. Guzzo, A. V. *Biophys. J.* (In press)
142. Blout, E. R., in *Polyamino Acids, Polypeptides and Proteins*, 275–79 (Stahmann, M. A., Ed., Univ. Wisconsin Press, Madison, Wis., 1962)
143. Yanofsky, C., Horn, V., and Thorpe, D., *Science*, **146,** 1593–94 (1964)
144. Platt, J. R., *Science*, **131,** 1309–10 (1960)
145. Ames, B. N., and Hartman, P. E., *Cold Spring Harbor Symp. Quant. Biol.*, **28,** 349–56 (1963)
146. Platt, J. R., *Perspectives Biol. Med.*, **2,** 24–45 (1958)

AUTHOR INDEX

D

E

F

K

Y

SUBJECT INDEX

I

CUMULATIVE INDEXES

VOLUMES 12-16

INDEX OF CONTRIBUTING AUTHORS

INDEX OF CHAPTER TITLES

INDEX OF CHAPTER TITLES

VOLUMES 12-16